QUANTUM CHEMISTRY

third edition

Ira N. Levine
Chemistry Department
Brooklyn College
City University of New York
Brooklyn, New York

Allyn and Bacon, Inc.
Boston   London   Sydney   Toronto

This book is part of

The Allyn and Bacon Chemistry Series

and was developed under the co-consulting editorship of

Daryle H. Busch
Harrison Shull

Library of Congress Cataloging in Publication Data

Levine, Ira N., date
  Quantum chemistry.

  (The Allyn and Bacon chemistry series)
  Bibliography: p.
  Includes index.
  1. Quantum chemistry. I. Title. II. Series.
QD462.L48 1983   541.2'8   82-24291
ISBN 0-205-07793-5
ISBN 0-205-07952-0 (International)

Printed in the United States of America.

10 9 8 7 6 5 4 3 2 1   88 87 86 85 84 83

To my students: Abe Auerbach, Andrew Auerbach, Joseph Barbuto, David Baron, Sene Bauman, Howard Becker, Michael Beitchman, Anna Berne, Susan Bienenfeld, Mark Blackman, Toby Block, Allen Bloom, Diza Braksmayer, Paul Brumer, Lynn Caporale, Richard Carter, Shi-ching Chang, Ching-hong Chen, Joseph Cincotta, Robert Curran, Joseph D'Amore, Barry DuRon, Myron Elgart, Stephen Engel, Larry Filler, Seymour Fishman, Donald Franceschetti, Mark Freilich, Michael Freshwater, Tobi Eisenstein Fried, Joel Friedman, Kenneth Friedman, Aryeh Frimer, Mark Froimowitz, Paul Gallant, Mark Gold, Stephen Goldman, Neil Goodman, Roy Goodman, Isaac Gorbaty, Michael Gross, Sheila Handler, Warren Hirsch, Richard Hom, Kuo-zong Hong, Fu-juan Hsu, Jong-chin Hwan, Leonard Itzkowitz, Kirby Juengst, Abraham Karkowsky, Michael Kittay, Barry Kohn, David Kurnit, Athanasios Ladas, Alan Lambowitz, Yedidyah Langsam, Surin Laosooksathit, Stephen Lemont, Elliot Lerner, Joel Liebman, Steven Lipp, Dennis Lynch, Tom McDonough, Pietro Mangiaracina, Louis Maresca, Allen Marks, Ira Michaels, Paul Mogolesko, Irving Nadler, Stuart Nagourney, Harold Nelson, Eli Pines, Jerry Polesuk, Arlene Gallanter Pollin, James Pollin, Lahanda Punyasena, Robert Richman, Bruce Rosenberg, Robert Rundberg, Edward Sachs, David Schaeffer, Gary Schneier, Neil Schweid, Judith Rosenkranz Selwyn, Gunnar Senum, Steven Shaya, Allen Sheffron, Lawrence Shore, Alvin Silverstein, Barry Siskind, Jerome Solomon, Henry Sperling, Helen Sussman, David Trauber, King-hung Tse, Alan Waldman, Robert Washington, Janet Weaver, Ken Zaner.

CONTENTS

v

PREFACE

This book is intended for first-year graduate and advanced undergraduate introductory courses in quantum chemistry.

New material in the third edition includes the particle in a rectangular well (Section 2.4), vibration of molecules (Section 4.3), time-dependent perturbation theory and the interaction of radiation and matter (Sections 9.9 and 9.10), the Condon–Slater formulas (Section 11.8), population analysis (Section 15.5), the empirical force field method (Section 15.15), comparison of quantum-mechanical methods (Section 15.16), the generalized valence-bond method (Section 15.18), the MNDO, PCILO, PRDDO, Xα, pseudopotential, and DIM methods (Section 15.14), the STO-3G, 4-31G, and 6-31G* ab initio basis sets (Sections 15.3 and 15.16), Walsh diagrams (Section 15.8), van der Waals molecules (Section 13.6), the noncrossing rule (Section 13.6), Koopmans' theorem (Section 15.4), Hess–Schaad resonance energies (Section 15.11), isodesmic reactions (Section 15.16), and energy gradients (Section 15.16). The discussion of configuration interaction in Sections 11.3, 13.18, and 15.22 has been expanded substantially. A brief introduction to matrices is given in Section 15.21, but matrices are not used elsewhere in the book.

I have tried to make explanations clear and complete, without glossing over difficult or subtle points. Derivations are given with enough detail to make them easy to follow, and resort to the frustrating phrase "it can be shown that" is avoided wherever possible. The aim is to give students a solid understanding of the physical and mathematical aspects of quantum mechanics and molecular electronic structure. The book is designed to be useful to students in all branches of chemistry, not just future quantum chemists. However, the presentation is such that those who do go on in quantum chemistry will have a good foundation and will not be hampered by misconceptions.

One of the major obstacles faced by many chemistry students in learning quantum mechanics is their unfamiliarity with much of the required mathematics. In this text I have included detailed treatments of operators, differential equations, simultaneous linear equations, and other needed topics. Rather than putting all the mathematics in an introductory chapter or in a series of appendixes, I have integrated the mathematics with the physics and chemistry. Immediate application of the mathematics to solving some quantum-mechanical problem should provide better motivation for the student than would be provided by separate study of the mathematics. I have also kept in mind the limited physics background of many chemistry students by reviewing topics they are likely to be unclear on.

A large number of problems of varying degrees of difficulty are included, together with answers for many of them.

This book has benefited from the reviews and suggestions of Professors Leland Allen, N. Colin Baird, Uldis Blukis, James Bolton, Melvyn Feinberg, David Goldberg, Hans Jaffé, Harry King, Peter Kollman, Joel Liebman, Frank Meeks, Robert Metzger, William Palke, Kenneth Sando, and Harrison Shull. I wish to thank these people and several anonymous reviewers.

I would appreciate receiving any suggestions that readers may have for improving the book.

I. N. L.

1 *THE SCHROEDINGER EQUATION*

1.1 *QUANTUM CHEMISTRY*

In the late seventeenth century, Isaac Newton discovered classical mechanics, the laws of motion of macroscopic objects. In the early twentieth century, physicists found that classical mechanics does not correctly describe the behavior of very small particles such as the electrons and nuclei of atoms and molecules; the behavior of such particles is described by a set of laws called *quantum mechanics.*

Quantum chemistry applies quantum mechanics to problems in chemistry. The influence of quantum chemistry is felt in all branches of chemistry. Physical chemists use quantum mechanics to calculate (with the aid of statistical mechanics) thermodynamic properties (e.g., entropy, heat capacity) of gases; to interpret molecular spectra, thereby allowing experimental determination of molecular properties (e.g., bond lengths and bond angles, dipole moments, barriers to internal rotation, energy differences between conformational isomers); to calculate molecular properties theoretically; to calculate properties of transition states in chemical reactions, thereby allowing estimation of rate constants; to understand intermolecular forces; and to deal with bonding in solids.

Organic chemists use quantum mechanics to estimate the relative stabilities of molecules, to calculate properties of reaction intermediates, to investigate the mechanisms of chemical reactions, to predict aromaticity of compounds, and to analyze NMR spectra.

Analytical chemists use spectroscopic methods extensively. The frequencies and intensities of lines in a spectrum can be properly understood and interpreted only through use of quantum mechanics.

Inorganic chemists use ligand-field-theory, an approximate quantum-

1

mechanical method, to predict and explain the properties of transition-metal complex ions.

Although the large size of biologically important molecules makes quantum-mechanical calculations on them extremely difficult, biochemists are beginning to benefit from quantum-mechanical studies of conformations of biological molecules, enzyme–substrate binding, and solvation of biological molecules.

1.2   HISTORICAL BACKGROUND OF QUANTUM MECHANICS

The development of quantum mechanics began in 1900 in connection with the study of the light emitted by heated solids, so we begin by discussing the nature of light.

In 1801 Thomas Young gave convincing experimental evidence for the wave nature of light by showing that light exhibited diffraction and interference when passed through two adjacent pinholes. (See *Halliday and Resnick*, Chapter 45. References with the authors' names italicized are listed in the Bibliography.)

About 1860 James Clerk Maxwell developed four equations, known as Maxwell's equations, which unified the laws of electricity and magnetism. Furthermore, Maxwell's equations predicted that an accelerated electric charge would radiate energy in the form of electromagnetic waves consisting of oscillating electric and magnetic fields. The speed predicted by Maxwell's equations for these waves turned out to be the same as the experimentally measured speed of light. Maxwell concluded that light is an electromagnetic wave.

In 1888 Hertz detected radio waves produced by accelerated electric charges in a spark, as predicted by Maxwell's equations. This convinced physicists that light is indeed an electromagnetic wave.

All electromagnetic waves travel at speed $c = 2.998 \times 10^{10}$ cm/sec in vacuum. The frequency v and wavelength λ of a wave are related by

$$\lambda v = c$$

Various conventional labels are applied to electromagnetic waves depending on their frequency. In order of increasing frequency one has radio waves, microwaves, infrared radiation, visible light, ultraviolet radiation, X-rays, and gamma rays. We shall use the term *light* to denote any kind of electromagnetic radiation.

In the late 1800s, physicists measured the intensity of light at various frequencies emitted by a heated blackbody at a fixed temperature. (A *blackbody* absorbs all light falling on it. A good approximation to a blackbody is a cavity with a tiny hole.) When physicists used the electromagnetic-wave model of light and statistical mechanics to predict the intensity-versus-frequency curve for emitted blackbody radiation, they found a result in complete disagreement with the high-frequency portion of the experimental curves.

In 1900 Max Planck developed a theory that gave excellent agreement with the observed blackbody-radiation curves. In his theory Planck assumed that the atoms of the blackbody could emit light energy only in amounts given by hv, where v is the radiation's frequency and h is a proportionality constant (called *Planck's constant*). The value $h = 6.6 \times 10^{-27}$ erg $\cdot$ sec $= 6.6 \times 10^{-34}$ J $\cdot$ sec gave curves that agreed with the experimental blackbody curves. Planck's work marks the beginning of quantum mechanics.

Planck's hypothesis that only certain quantities of light energy could be emitted (i.e., that the emission was *quantized*) was in direct contradiction to all previous ideas of physics. The energy of a wave is related to its amplitude, and the amplitude varies continuously from zero on up. Moreover, according to Newtonian mechanics, the energy of a material body can vary continuously; hence physicists expected the energy of an atom to vary continuously. If one expects the energies of atoms and of electromagnetic waves to vary continuously, one also expects the electromagnetic radiation emitted by atoms to vary continuously. However, only with the hypothesis of quantized energy emission does one obtain the correct blackbody-radiation curves.

The second application of energy quantization was to the photoelectric effect. In the *photoelectric effect*, light shining on a metal causes emission of electrons. The energy of a wave is proportional to its intensity and is not related to its frequency, so the electromagnetic-wave picture of light leads one to expect that the kinetic energy of an emitted photoelectron would increase as the light intensity increases but would not change as the light frequency changes. Instead, one observes that the kinetic energy of an emitted electron is independent of the light's intensity but increases as the light's frequency increases.

In 1905 Einstein showed that these observations could be explained by viewing light as composed of particlelike entities (called *photons*), with each photon having an energy

$$E_{photon} = h\nu \tag{1.1}$$

When an electron in the metal absorbs a photon, part of the absorbed photon energy is used to overcome the forces holding the electron in the metal, and the remainder appears as kinetic energy of the electron after it has left the metal. Conservation of energy gives

$$h\nu = W + \tfrac{1}{2}mv^2$$

where W is the minimum energy needed by an electron to escape the metal (the metal's *work function*), and $\tfrac{1}{2}mv^2$ is the maximum kinetic energy of an emitted electron. An increase in the light's frequency ν increases the photon energy and hence increases the kinetic energy of the emitted electron. An increase in light intensity at fixed frequency increases the rate at which photons strike the metal and hence increases the rate of emission of electrons, but does not change the kinetic energy of each emitted electron.

The photoelectric effect shows that light can exhibit particlelike behavior in addition to the wavelike behavior it shows in diffraction experiments.

Now let us consider the structure of matter.

In the late nineteenth century, investigations of electric discharge tubes and natural radioactivity showed that atoms and molecules are composed of charged particles. Electrons have a negative charge. The proton has a positive charge equal in magnitude but opposite in sign to the electron charge and is 1836 times as heavy as the electron. The third constituent of atoms, the neutron (discovered in 1932), is neutral and slightly heavier than the proton.

Starting in 1909, Rutherford, Geiger, and Marsden carried out a series of experiments in which they passed a beam of alpha particles through a thin metal foil and observed the deflections of the particles by allowing them to fall on a fluorescent screen. Alpha particles are positively charged helium nuclei obtained from natural radioactive decay. Rutherford observed that most of the alpha particles passed through the foil essentially undeflected, but, surprisingly, a few underwent large deflections, some being deflected backwards. To get large deflections, one needs a very close approach between the charges, so that the Coulombic repulsive force is great. If the positive charge were spread throughout the atom (as J. J. Thomson had proposed in 1904), once the high-energy alpha particle penetrated the atom, the repulsive force would fall off, becoming zero at the center of the atom, according to classical electrostatics. Hence Rutherford concluded that such large deflections could occur only if the positive charge were concentrated in a tiny, heavy nucleus.

An atom contains a tiny (10^{-13} to 10^{-12} cm radius) heavy nucleus consisting of neutrons and Z protons, where Z is the atomic number. Outside the nucleus there are Z electrons. The charged particles interact according to Coulomb's law. (The nucleons are held together in the nucleus by strong, short-range nuclear forces, which will not concern us.) The radius of an atom is about one angstrom (1 Å $\equiv 10^{-8}$ cm $= 10^{-10}$ m), as shown, for example, by results from the kinetic theory of gases. Molecules have more than one nucleus.

The chemical properties of atoms and molecules are determined by their electronic structure, and so the question arises as to the nature of the motions and energies of the electrons. Since the nucleus is much more massive than the electron, we expect the motion of the nucleus to be slight compared to the electrons' motions.

In 1911 Rutherford proposed his planetary model of the atom in which the electrons revolved about the nucleus in various orbits, just as the planets revolve about the sun. However, there is a fundamental difficulty with this model. According to classical electromagnetic theory, an accelerated charged particle radiates energy in the form of electromagnetic (light) waves. An electron circling the nucleus at constant speed is being accelerated, since the direction of its velocity vector is continually changing. Hence the electrons in the Rutherford model should continually lose energy by radiation and therefore would spiral in toward the nucleus. Thus, according to classical (nineteenth century) physics, the Rutherford atom is unstable and would collapse.

A possible way out of this difficulty was proposed by Bohr in 1913, when he applied the concept of quantization of energy to the hydrogen atom. Bohr assumed that the energy of the electron in a hydrogen atom was quantized, with the electron constrained to move only on one of a number of allowed circles. When an electron makes a transition from one orbit (stationary state) to another, a quantum of light of frequency

$$\nu = \Delta E/h \qquad\qquad (1.2)$$

is absorbed or emitted, where ΔE is the energy difference between the two states (conservation of energy). With the assumption that an electron making a transition from a free (ionized) state to one of the bound orbits emits a photon whose frequency is an integral multiple of one-half the classical frequency of revolution of the electron in the bound orbit, Bohr used Newtonian mechanics to derive a formula for the hydrogen-atom energy levels. Using (1.2), he obtained agreement with the observed hydrogen spectrum. However, attempts to fit the helium spectrum using the Bohr theory failed. Moreover, the theory could not account for chemical bonds in molecules.

The basic difficulty in the Bohr model arises from the use of classical Newtonian mechanics to describe the electronic motions in atoms. The evidence of atomic spectra, which show discrete frequencies, indicates that only certain energies of motion are allowed; the electronic energy is "quantized." However, Newtonian mechanics allows a continuous range of energies. Quantization does occur in wave motion; for example, the fundamental and overtone frequencies of a violin string. Hence de Broglie suggested in 1923 that the motion of electrons might have a wave aspect; that an electron of mass m and speed v would have a wavelength λ associated with it, such that

$$\lambda = \frac{h}{mv} = \frac{h}{p} \tag{1.3}$$

where p is the linear momentum. De Broglie arrived at Eq. (1.3) by reasoning in analogy with photons. The energy of any particle (including a photon) can be expressed, according to Einstein's special theory of relativity, as $E = mc^2$, where c is the speed of light. Using (1.1), we get $mc^2 = hv = hc/\lambda$, so that $\lambda = h/mc = h/p$ for a photon traveling at speed c. Equation (1.3) is then the corresponding relation for an electron.

In 1927 Davisson and Germer experimentally confirmed de Broglie's hypothesis by reflecting electrons from metals and observing diffraction effects. In 1932 Stern observed the same effects with helium atoms and hydrogen molecules, thus verifying that the wave effects are not peculiar to electrons, but result from some general law of motion for microscopic particles. (The electron microscope is a practical application of the wave properties of electrons.)

Thus electrons behave in some respects like particles and in other respects like waves. We are faced with the apparently contradictory "wave-particle duality" of matter (and of light). How can an electron be both a particle, which is a localized entity, and a wave, which is nonlocalized? The answer is that an electron is neither a wave nor a particle, but something else. An accurate pictorial description of an electron's behavior is impossible using the wave or particle concept of classical physics. The concepts of classical physics have been developed from experience in the macroscopic world and do not provide a proper description of the microscopic world. Evolution has shaped the human brain to allow it to understand and deal effectively with macroscopic phenomena. The human nervous system was not developed to deal with phenomena at the atomic and molecular level, so it is not surprising if we cannot fully understand such phenomena.

Although both photons and electrons show an apparent "duality," they are not the same kinds of entities. Photons always travel at speed c and have zero rest mass; electrons always have $v < c$ and a nonzero rest mass. Photons must always be treated relativistically, but electrons whose speed is not too high can be treated nonrelativistically.

1.3   THE UNCERTAINTY PRINCIPLE

Let us consider what effect the wave-particle duality has on attempts to measure simultaneously the x coordinate and the x component of linear momentum of a microscopic particle. We start with a beam of particles with momentum p, traveling in the y direction, and we let the beam fall on a narrow slit. Behind this slit is a photographic plate. (See Fig. 1.1.)

Particles that pass through the slit of width w have an uncertainty w in their x coordinate at the time of going through the slit. Calling this spread in x values Δx, we have $\Delta x = w$.

Since microscopic particles have wave properties, they are diffracted by the slit producing (as would a light beam) a diffraction pattern on the plate. The height of the graph in Fig. 1.1 is a measure of the number of particles reaching a given point. The diffraction pattern indicates that when the particles were diffracted by the slit, their direction of motion was changed so that part of their momentum was transferred to the x direction. The x component of momentum is given by the projection of the momentum vector in the x direction. A particle deflected upward by an angle α has an x component of momentum $p \sin \alpha$. A particle deflected downward by an angle α has an x component of momentum $-p \sin \alpha$. Since most of the particles undergo deflections in the range $-\alpha$ to α, where α is the angle to the first minimum in the diffraction pattern, we shall take one-half the spread of momentum values in the central diffraction peak as a measure of the uncertainty Δp_x in the x component of momentum: $\Delta p_x = p \sin \alpha$.

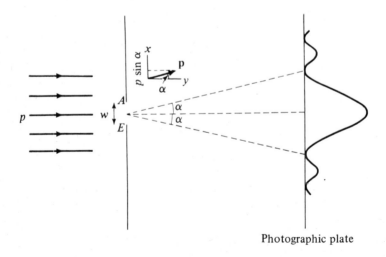

Photographic plate

Figure 1.1 *Diffraction of electrons by a slit*

Hence at the slit, where the measurement is made,

$$\Delta x\, \Delta p_x = pw \sin \alpha \qquad (1.4)$$

The angle α at which the first diffraction minimum occurs is readily calculated. The condition for the first minimum is that the difference in the distances traveled by particles passing through the slit at its upper edge and particles passing through the center of the slit be equal to $\frac{1}{2}\lambda$, where λ is the wavelength of the associated wave. Waves originating from the top of the slit are then exactly out of phase with waves originating from the center of the slit and they cancel each other. Waves originating from a point in the slit at a distance d below the slit midpoint cancel with waves originating at a distance d below the top of the slit. Drawing AC in Fig. 1.2 so that $AD = CD$, we have the difference in path length as BC. The distance from the slit to the screen is large compared with the slit width. Hence AD and BD are nearly parallel. This makes the angle ACB essentially a right angle, so that angle $BAC = \alpha$. The path difference BC is then $\frac{1}{2}w \sin \alpha$. Setting BC equal to $\frac{1}{2}\lambda$, we have $w \sin \alpha = \lambda$, and Eq. (1.4) becomes $\Delta x\, \Delta p_x = p\lambda$. The wavelength λ is given by the de Broglie relation $\lambda = h/p$, so that $\Delta x\, \Delta p_x = h$. Since the uncertainties have not been precisely defined, the equality sign is not really justified; instead we write

$$\Delta x\, \Delta p_x \approx h \qquad (1.5)$$

indicating that the product of the uncertainties in x and p_x is of the order of magnitude of Planck's constant. In Section 5.1 we will give a precise statistical definition of the uncertainties and a precise inequality to replace (1.5).

Although we have demonstrated (1.5) for only one experimental setup, its validity is general. No matter what attempts are made, the wave-particle duality of microscopic "particles" imposes a limit on our ability to measure simultaneously the position and momentum of such particles. The more precisely we determine the position, the less accurate is our determination of momentum. (In Fig. 1.1, $\sin \alpha = \lambda/w$, so that narrowing the slit increases the spread of the diffraction pattern.) This limitation is called the *uncertainty principle* and was discovered in 1927 by Heisenberg.

Because of the wave-particle duality, the act of measurement introduces an uncontrollable disturbance in the system being measured. We started with particles

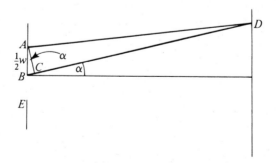

Figure 1.2 *Calculation of first diffraction minimum*

having a precise value of p_x (namely zero); by imposing the slit, we measured the x coordinate of the particles to an accuracy w, but this measurement introduced an uncertainty into the p_x values of the particles. The measurement changed the state of the system.

1.4   THE TIME-DEPENDENT SCHROEDINGER EQUATION

Classical mechanics is applicable only to macroscopic particles. For microscopic "particles" we require a new form of mechanics, which we will call *quantum mechanics* (or *wave mechanics*). We now consider some of the contrasts between classical and quantum mechanics. For simplicity we deal with a one-particle, one-dimensional system.

In classical mechanics the motion of the particle is governed by Newton's second law:

$$F = m \frac{d^2x}{dt^2} \tag{1.6}$$

where F is the force acting on the particle, m is its mass, and t is the time. Equation (1.6) contains the second derivative of the coordinate x with respect to time. To solve it we must carry out two integrations. This introduces two arbitrary constants c_1 and c_2 into the solution, so that

$$x = g(t, c_1, c_2) \tag{1.7}$$

where g is some function of time. We now ask: What information must we possess at a given time t_0 to be able to predict the future motion of the particle? If we know that at t_0 the particle is at point x_0, we have

$$x_0 = g(t_0, c_1, c_2) \tag{1.8}$$

Since we have two constants to determine, we need more information. Differentiating (1.7), we have

$$\frac{dx}{dt} = v = \frac{d}{dt} g(t, c_1, c_2)$$

where v is the particle's velocity. If we also know that at time t_0 the particle has velocity v_0, then we have the additional relation

$$v_0 = \frac{d}{dt} g(t, c_1, c_2) \Big|_{t=t_0} \tag{1.9}$$

We may then use (1.8) and (1.9) to solve for c_1 and c_2 in terms of x_0 and v_0. Knowing c_1 and c_2, we can use Eq. (1.7) to predict the exact future motion of the particle.

As an example of Eqs. (1.6) to (1.9), consider the vertical motion of a particle in the earth's gravitational field. Let the x axis point upward. The force on the particle is downward and is $F = -mg$, where g is the gravitational acceleration constant. Newton's second law (1.6) is $-mg = m \, d^2x/dt^2$, so $d^2x/dt^2 = -g$. A single integration gives $dx/dt = -gt + c_1$. The arbitrary constant c_1 can be found if we know that at time t_0 the particle had velocity v_0. Since $v = dx/dt$, we have $v_0 = -gt_0$

$+c_1$ and $c_1 = v_0 + gt_0$. Hence $dx/dt = -gt + gt_0 + v_0$. Integration of this equation gives $x = -\frac{1}{2}gt^2 + (gt_0 + v_0)t + c_2$. If we know that at time t_0 the particle had position x_0, then $x_0 = -\frac{1}{2}gt_0^2 + (gt_0 + v_0)t_0 + c_2$ and $c_2 = x_0 - \frac{1}{2}gt_0^2 - v_0 t_0$. The expression for x as a function of time becomes $x = -\frac{1}{2}gt^2 + (gt_0 + v_0)t + x_0 - \frac{1}{2}gt_0^2$ $- v_0 t_0$ or $x = x_0 - \frac{1}{2}g(t - t_0)^2 + v_0(t - t_0)$. Knowing x_0 and v_0 at time t_0, we can predict the future position of the particle.

The classical-mechanical potential energy V of a particle moving in one dimension is defined to satisfy

$$\partial V(x, t)/\partial x = -F(x, t)$$

For example, for a particle moving in the earth's gravitational field, $\partial V/\partial x = -F$ $= mg$ and $V = mgx + c$, where c is an arbitrary constant. We are free to set the zero level of potential energy wherever we please; choosing $c = 0$, we have $V = mgx$ as the potential-energy function.

The word *state* in classical mechanics means a specification of the position and velocity of each particle of the system at some instant of time, plus specification of the forces acting on the particles. According to Newton's second law, given the state of a system at any time, its future state and future motions are exactly determined, as shown by Eqs. (1.7)–(1.9). The impressive success of Newton's laws in explaining planetary motions led many philosophers to use Newton's laws as an argument for philosophical determinism. The mathematician and astronomer Laplace (1749–1827) assumed that the universe consisted of nothing but particles which obeyed Newton's laws. Therefore, given the state of the universe at some instant, the future motion of everything in the universe was completely determined. A super-being able to know the state of the universe at any instant could, in principle, calculate all future motions.

However, the Heisenberg uncertainty principle shows that we cannot determine simultaneously the exact position and velocity of a microscopic particle, so that the very knowledge required by classical mechanics for predicting the future motions of a system cannot be obtained. We must be content in quantum mechanics with something less than complete prediction of the exact future motion.

Our approach to quantum mechanics will be to *postulate* the basic principles and then use these postulates to deduce experimentally testable consequences such as the energy levels of atoms. To describe the *state* of a system in quantum mechanics, we postulate the existence of a function of the coordinates called the *wave function* (or *state function*) Ψ. Since the state will, in general, change with time, Ψ is also a function of time. For a one-particle, one-dimensional system, we have $\Psi = \Psi(x, t)$. The wave function contains all possible information about a system so that instead of speaking of "the state described by the wave function Ψ," we simply say, "the state Ψ." Newton's second law tells us how to find the future state of a classical-mechanical system from knowledge of the present state. To find the future state of a quantum-mechanical system from knowledge of the present state, we want an equation that tells us how the wave function changes with time. This equation is postulated to be

$$-\frac{\hbar}{i}\frac{\partial \Psi(x, t)}{\partial t} = -\frac{\hbar^2}{2m}\frac{\partial^2 \Psi(x, t)}{\partial x^2} + V(x, t)\Psi(x, t) \qquad (1.10)$$

where the constant $\hbar$ ("h-bar") is defined as

$$\hbar \equiv \frac{h}{2\pi} \qquad (1.11)$$

The concept of the wave function and the equation governing its change with time were discovered in 1926 by the Austrian physicist Schroedinger (1887–1961). In this equation, known as the *time-dependent Schroedinger equation* (or the *Schroedinger wave equation*), $i = \sqrt{-1}$, m is the mass of the particle, and $V(x, t)$ is the potential-energy function of the system.

The time-dependent Schroedinger equation contains the first derivative of the wave function with respect to time and allows us to calculate the future wave function (state) at any time, if we know the wave function at time t_0.

The wave function contains all the information we can possibly know about the system it describes. What information does Ψ give us about the result of a measurement of the x coordinate of the particle? We cannot expect Ψ to involve the definite specification of position that the state of a classical-mechanical system does. The correct answer to this question was provided by Born shortly after Schroedinger discovered the Schroedinger equation. Born postulated that

$$|\Psi(x, t)|^2 \, dx \qquad (1.12)$$

gives the *probability* at time t of finding the particle in the region of the x axis lying between x and $x + dx$. In (1.12) the bars denote the absolute value and dx is an infinitesimal length on the x axis. The function $|\Psi(x, t)|^2$ is the *probability density* for finding the particle at various places on the x axis. (A review of probability is given in Section 1.6.) For example, suppose that at some particular time t_0 the particle is in a state characterized by the wave function ae^{-bx^2}, where a and b are real constants. If we measure the particle's position at time t_0, we might get any value of x, because the probability density $a^2 e^{-2bx^2}$ is nonzero everywhere. Of course, values of x in the region around $x = 0$ are more likely to be found than other values, since $|\Psi|^2$ is a maximum at the origin.

To make a precise connection between $|\Psi|^2$ and experimental measurements, we would take a large number of identical noninteracting systems, each of which was in the same state Ψ. We would then make a measurement of position in each system. If we had n systems and made n measurements, and if dn_x denotes the number of measurements for which we found the particle between x and $x + dx$, then dn_x/n is the probability for finding the particle between x and $x + dx$. Thus

$$\frac{dn_x}{n} = |\Psi|^2 \, dx$$

and a graph of $(1/n) \, dn_x/dx$ versus x gives the probability density $|\Psi|^2$. It might be thought that we could find the probability-density function by taking one system which was in the state Ψ and repeatedly carrying out measurements of the particle's position. This procedure will not do because the process of measurement generally changes the state of a system; we saw an example of this in our discussion of the uncertainty principle (Section 1.3).

Quantum mechanics is basically *statistical* in nature. Knowing the state, we cannot predict the result of a position measurement with certainty; we can only predict the *probabilities* of various possible results. The Bohr theory of the hydrogen atom specified the precise path of the electron and is therefore not a correct quantum-mechanical picture.

Quantum mechanics does not say that an electron is distributed over a large region of space as a wave is distributed. Rather, it is the probability patterns (wave functions) used to describe the electron's motion that behave like waves and satisfy a wave equation.

The reader might ask how the wave function gives us information on other properties (e.g., momentum) besides the position. We postpone discussion on this point until later chapters.

The postulates of, say, thermodynamics (the first, second, and third laws of thermodynamics) are stated in terms of macroscopic experience and hence are fairly readily understood. The postulates of quantum mechanics are stated in terms of the microscopic world and appear quite abstract. You should not expect to have a full understanding of the postulates of quantum mechanics at first reading. As we treat various examples, understanding of the postulates will increase. (We have not yet introduced all the postulates of quantum mechanics. Further postulates will be introduced later.)

It may bother the reader that we wrote down the Schroedinger equation without any attempt to prove its plausibility. By using analogies between geometrical optics and classical mechanics on the one hand, and wave optics and quantum mechanics on the other hand, one can show the plausibility of the Schroedinger equation. Geometrical optics is an approximation to wave optics, valid when the wavelength of the light is much less than the size of the apparatus. (Recall its use in treating lenses and mirrors.) Likewise, classical mechanics is an approximation to wave mechanics, valid when the particle's wavelength is much less than the size of the apparatus. One can make a plausible guess as to how to get the proper equation for quantum mechanics from classical mechanics based on the known relation between the equations of geometrical and wave optics. Since most chemists are not particularly familiar with optics, these arguments have been omitted. In any case, such analogies can only make the Schroedinger equation seem *plausible*; they cannot be used to *derive* or *prove* this equation. The Schroedinger equation is a *postulate* of the theory, to be tested by agreement of its predictions with experiment. (Details of the reasoning that led Schroedinger to his equation are given in *Jammer*, Section 5.3.)

Quantum mechanics provides the law of motion for microscopic particles. Experimentally, macroscopic objects obey classical mechanics. Hence for quantum mechanics to be a valid theory, it should reduce to classical mechanics as we make the transition from microscopic to macroscopic particles. Quantum effects are associated with the de Broglie wavelength $\lambda = h/mv$. Since h is very small, the de Broglie wavelength of macroscopic objects is essentially zero. Thus in the limit $\lambda \to 0$, we expect the time-dependent Schroedinger equation to reduce to Newton's second law.[1] We can prove this to be so (see Problem 7.19).

---

[1] A similar situation holds in the relation between special relativity and classical mechanics. In the limit $v/c \to 0$, where c is the velocity of light, special relativity reduces to classical mechanics. The form of quantum mechanics that we will develop will be nonrelativistic. A complete integration of relativity with quantum mechanics has not been achieved.

Historically, quantum mechanics was first formulated in 1925 by Heisenberg, Born, and Jordan using matrices, several months before Schroedinger's 1926 formulation using differential equations. Schroedinger proved that the Heisenberg formulation (called *matrix mechanics*) is equivalent to the Schroedinger formulation (called *wave mechanics*).

1.5   *THE TIME-INDEPENDENT SCHROEDINGER EQUATION*

The time-dependent Schroedinger equation (1.10) is rather formidable looking. Fortunately, for many applications of quantum mechanics to chemistry, it is not necessary to deal with this equation; instead, the simpler time-independent Schroedinger equation is used. We now derive the time-independent from the time-dependent Schroedinger equation, for the one-particle, one-dimensional case.

We begin by restricting ourselves to the special case where the potential energy V is not a function of time but depends only on x. The time-dependent Schroedinger equation reads

$$-\frac{h}{i}\frac{\partial \Psi(x,t)}{\partial t} = -\frac{h^2}{2m}\frac{\partial^2 \Psi(x,t)}{\partial x^2} + V(x)\Psi(x,t) \tag{1.13}$$

We will look for those solutions of (1.13) that can be written as the product of a function of time and a function of x:

$$\Psi(x,t) = f(t)\psi(x) \tag{1.14}$$

Note that we use capital psi for the time-dependent wave function and lowercase psi for the factor which depends only on the coordinate x. States corresponding to wave functions of the form (1.14) possess certain properties (to be discussed shortly) which make them of great interest. Taking partial derivatives of (1.14), we have

$$\frac{\partial \Psi(x,t)}{\partial t} = \frac{df(t)}{dt}\psi(x), \qquad \frac{\partial^2 \Psi(x,t)}{\partial x^2} = f(t)\frac{d^2\psi(x)}{dx^2}$$

so that Eq. (1.13) becomes

$$-\frac{h}{i}\frac{df(t)}{dt}\psi(x) = -\frac{h^2}{2m}f(t)\frac{d^2\psi(x)}{dx^2} + V(x)\psi(x)f(t)$$

$$-\frac{h}{i}\frac{1}{f(t)}\frac{df(t)}{dt} = -\frac{h^2}{2m}\frac{1}{\psi(x)}\frac{d^2\psi(x)}{dx^2} + V(x) \tag{1.15}$$

where we divided by $f\psi$. In general, we expect the quantity to which each side of (1.15) is equal to be a certain function of x and t. However, the right side of (1.15) does not depend on t, so that the function to which each side of (1.15) is equal must be independent of t. The left side of (1.15) is independent of x, so this function must also be independent of x. Since the function is independent of both variables, x and t, it must be a constant. We call this constant E.

Equating the left side of (1.15) to E, we get

$$\frac{df(t)}{f(t)} = -\frac{iE}{\hbar}\,dt$$

Integrating both sides of this equation with respect to t, we have

$$\ln f(t) = -iEt/\hbar + C$$

where C is an arbitrary constant of integration. Hence

$$f(t) = e^C e^{-iEt/\hbar} = A e^{-iEt/\hbar}$$

where the arbitrary constant A has replaced e^C. Since A can be included as a factor in the function $\psi(x)$ which multiplies $f(t)$, there is no loss in generality in omitting it from $f(t)$. Thus we take

$$f(t) = e^{-iEt/\hbar}$$

Equating the right side of (1.15) to E, we have

$$-\frac{\hbar^2}{2m}\frac{d^2\psi(x)}{dx^2} + V(x)\psi(x) = E\psi(x) \tag{1.16}$$

$$\frac{d^2\psi(x)}{dx^2} + \frac{8\pi^2 m}{h^2}[E - V(x)]\psi(x) = 0$$

Equation (1.16) is the *time-independent Schroedinger equation* for a single particle of mass m moving in one dimension.[2]

What is the significance of the constant E? Since E occurs as $[E - V(x)]$, E has the same dimensions as V, so that E has the dimensions of energy. In fact, we postulate that E is the energy of the system. (This is a special case of a more general postulate to be discussed in a later chapter.) Thus for cases where the potential energy is a function of x only, there exist wave functions of the form

$$\Psi(x, t) = e^{-iEt/\hbar}\psi(x) \tag{1.17}$$

and these wave functions correspond to states of constant energy E. Much of our attention in the next few chapters will be devoted to finding the solutions of (1.16) for various systems.

The wave function in (1.17) is complex, but the quantity that is experimentally observable is the probability density $|\Psi(x, t)|^2$. The square of the absolute value of a complex quantity is given by the product of the quantity with its complex conjugate, the complex conjugate being formed by replacing i with $-i$ wherever it occurs.[3] Thus

$$|\Psi|^2 = \Psi^*\Psi$$

---

[2] Schroedinger actually developed the time-independent equation before the time-dependent equation. The relevant papers are E. Schroedinger, *Ann. Physik*, **79**, 361, 489 (1926); **80**, 437 (1926); **81**, 109 (1926).

[3] A review of complex numbers is provided in Section 1.7.

where the star denotes the complex conjugate. For the wave function (1.17), we have

$$|\Psi(x, t)|^2 = [e^{-iEt/\hbar}\psi(x)]^* \, e^{-iEt/\hbar}\psi(x)$$

$$= e^{iEt/\hbar}\psi^*(x)e^{-iEt/\hbar}\psi(x)$$

$$= e^0\psi^*(x)\psi(x) = \psi^*(x)\psi(x)$$

$$|\Psi(x, t)|^2 = |\psi(x)|^2 \tag{1.18}$$

[In deriving (1.18), we assumed that E is a real number so that $E = E^*$. This fact will be proved in Section 7.2.] Hence for states of the form (1.17), the probability density is given by $|\psi(x)|^2$ and does not change with time. Such states are called *stationary states*. Since the physically significant quantity is $|\Psi(x, t)|^2$, and since for stationary states $|\Psi(x, t)|^2 = |\psi(x)|^2$, the function $\psi(x)$ is often called the wave function, although the complete wave function of a stationary state is obtained by multiplying $\psi(x)$ by $e^{-iEt/\hbar}$. The term "stationary state" should not mislead the reader into thinking that a particle in a stationary state is at rest. What is stationary is the probability density $|\Psi|^2$, not the particle itself.

We will be concerned mostly with states of constant energy (stationary states) and hence will usually deal with the time-independent Schroedinger equation (1.16). For simplicity we will refer to this equation as "the Schroedinger equation." Note that the Schroedinger equation contains *two* unknowns, the allowed energies E and the allowed wave functions ψ. To solve for two unknowns, we need to impose additional conditions (called boundary conditions) on ψ besides requiring that it satisfy (1.16); the boundary conditions determine the allowed energies, since it turns out that only certain values of E allow ψ to meet the boundary conditions. This will become clearer when we discuss specific examples in later chapters.

1.6   PROBABILITY

Probability plays a fundamental role in quantum mechanics. In this section we review the mathematics of probability.

There has been much controversy about the proper definition of probability. One definition is the following: If an experiment has n equally probable outcomes, m of which are favorable to the occurrence of a certain event A, then the probability that A occurs is m/n. Note that this definition is circular, since it specifies equally *probable* outcomes when *probability* is what we are attempting to define. It is simply assumed that we can recognize equally probable outcomes. An alternative definition is based on actually performing the experiment many times. Suppose that we perform the experiment N times and that in M of these trials the event A occurs. The probability of A occurring is then defined as

$$\lim_{N \to \infty} \frac{M}{N}$$

Thus if we toss a coin repeatedly, the fraction of heads will approach 1/2 as we increase the number of tosses.

For example, suppose we pick a card at random from a deck and ask for the probability of drawing a heart. There are 52 cards and hence 52 equally probable outcomes. Since there are 13 hearts, there are 13 favorable outcomes. Hence $m/n = 13/52 = 1/4$. The probability for drawing a heart is 1/4. (Alternatively, we say the odds are 3 to 1 against drawing a heart.)

Sometimes we ask for the probability of two related events both occurring. For example, we may ask for the probability of drawing two hearts from a 52-card deck, assuming we do not replace the first card after it is drawn. There are 52 possible outcomes of the first draw, and for each of these possibilities there are 51 possible second draws. We have $52 \cdot 51$ possible outcomes. Since there are 13 hearts, there are $13 \cdot 12$ different ways to draw two hearts. The desired probability is $13 \cdot 12/52 \cdot 51 = 1/17$. This calculation illustrates the theorem: The probability that two events A and B both occur is the probability that A occurs, multiplied by the conditional probability that B then occurs, calculated with the assumption that A occurred. Thus if A is the probability of drawing a heart on the first draw, the probability of A is 13/52. The probability of drawing a heart on the second draw, given that the first draw yielded a heart, is 12/51 since there remain 12 hearts in the deck. The probability of drawing two hearts is then $(13/52)(12/51) = 1/17$, as found previously.

In quantum mechanics we must deal with probabilities involving a continuous variable; e.g., the x coordinate. It does not make much sense to talk about the probability of a particle being found *at* a particular point such as $x = 0.5000 \ldots$, since there are an infinite number of points on the x axis and for any finite number of measurements we make, the probability of getting *exactly* $0.5000 \ldots$ is vanishingly small. Instead we talk of the probability of finding the particle in a small interval of the x axis lying between x and $x + dx$, dx being an infinitesimal element of length. This probability will naturally be proportional to the length of the small interval, dx, and will vary for different regions of the x axis. Hence the probability that the particle will be found between x and $x + dx$ is equal to $g(x) \, dx$, where $g(x)$ is some function which tells how the probability varies over the x axis. The function $g(x)$ is called the *probability density*, since it is a probability per unit length. Since probabilities are real, nonnegative numbers, $g(x)$ must be a real function that is everywhere nonnegative. The wave function Ψ can take on negative and complex values and is not a probability density. Quantum mechanics postulates that the probability density is given by $|\Psi|^2$.

What is the probability that the particle lies in some finite region of space $a \leqslant x \leqslant b$? To find this probability we sum up the probabilities of finding the particle in all the infinitesimal regions lying between a and b. This is just the definition of the definite integral

$$\int_a^b |\Psi|^2 \, dx \qquad (1.19)$$

A probability of 1 represents certainty. Since it is certain that the particle is somewhere on the x axis, we have the requirement

$$\int_{-\infty}^{\infty} |\Psi|^2 \, dx = 1 \qquad (1.20)$$

When Ψ satisfies (1.20), it is said to be *normalized*.

1.7   COMPLEX NUMBERS

We have seen that the wave function can be complex, so we now review some properties of complex numbers.

If i is $\sqrt{-1}$, then we may write a complex number z as $z = x + iy$, where x and y are real numbers; x and y are called the real and imaginary parts of z: $x = \text{Re}(z)$, $y = \text{Im}(z)$. A convenient representation of z is as a point in the complex plane (Fig. 1.3), where the real part of z is plotted on the horizontal axis and the imaginary part on the vertical axis. This diagram immediately suggests defining two quantities which characterize the complex number z: the distance r of the point z from the origin is called the *absolute value* or modulus of z; the angle θ that the radius vector to the point z makes with the positive horizontal axis is called the *phase* or argument of z. We have

$$|z| = r = (x^2 + y^2)^{1/2}, \qquad \tan\theta = y/x$$

$$x = r\cos\theta, \qquad\qquad y = r\sin\theta$$

So we may write z as

$$z = r\cos\theta + ir\sin\theta = re^{i\theta} \tag{1.21}$$

since (Problem 4.10)

$$e^{i\theta} = \cos\theta + i\sin\theta \tag{1.22}$$

The *complex conjugate* of z, z^*, is defined as

$$z^* = x - iy = re^{-i\theta} \tag{1.23}$$

If z is a real number, its imaginary part is zero. Thus z is real if and only if $z = z^*$. Taking the complex conjugate twice, we get z back again, $(z^*)^* = z$. Forming the product of z and its complex conjugate, we have

$$zz^* = x^2 + y^2 = r^2 = |z|^2 \tag{1.24}$$

For the product and quotient of two complex numbers $z_1 = r_1 e^{i\theta_1}$ and

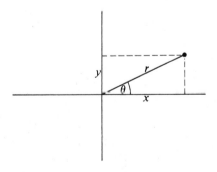

Figure 1.3 *Plot of a complex number* $z = x + iy$

$z_2 = r_2 e^{i\theta_2}$, we have

$$z_1 z_2 = r_1 r_2 e^{i(\theta_1 + \theta_2)}, \qquad \frac{z_1}{z_2} = \frac{r_1}{r_2} e^{i(\theta_1 - \theta_2)} \tag{1.25}$$

It is easy to prove, either directly from the definition of complex conjugate or from (1.25), that the complex conjugate of a product is the product of the complex conjugates:

$$(z_1 z_2)^* = z_1^* z_2^* \tag{1.26}$$

Likewise,

$$\left(\frac{z_1}{z_2}\right)^* = \frac{z_1^*}{z_2^*}, \qquad (z_1 + z_2)^* = z_1^* + z_2^*, \qquad (z_1 - z_2)^* = z_1^* - z_2^* \tag{1.27}$$

For the absolute values of products and quotients, it follows from (1.25) that

$$|z_1 z_2| = |z_1||z_2|, \qquad \left|\frac{z_1}{z_2}\right| = \frac{|z_1|}{|z_2|} \tag{1.28}$$

Therefore if ψ is a complex wave function, we have

$$|\psi^2| = |\psi|^2 = \psi^* \psi \tag{1.29}$$

We now obtain a formula for the nth roots of unity. We may take the phase of the number 1 to be 0 or 2π or 4π, etc.; hence

$$1 = e^{i2\pi k} \tag{1.30}$$

where k is any integer, zero, negative, or positive. Now consider the number ω, where

$$\omega = e^{i2\pi k/n} \tag{1.31}$$

n being a positive integer. Using (1.25) n times, we see that $\omega^n = 1$. Thus ω is an nth root of unity. There are n different complex nth roots of unity, and taking n successive values of the integer k gives us all of them:

$$\omega = e^{i2\pi k/n}, \qquad k = 0, 1, 2, \ldots, n-1 \tag{1.32}$$

Any other value of k besides those in (1.32) gives a number whose phase differs by an integral multiple of 2π from one of the numbers in (1.32) and hence is not a different root.

1.8   UNITS

There are currently two different systems of units in common use in science. In the cgs Gaussian system, the units of length, mass, and time are the centimeter (cm), gram (g), and second (sec). Force is measured in dynes (dyn) and energy in ergs. Coulomb's law for the magnitude of the force between two charges Q_1' and Q_2' separated by a distance r in vacuum is given by $F = Q_1' Q_2' / r^2$, where Q_1' and Q_2' are in statcoulombs (statC), also called electrostatic units of charge (esu).

In the International System (SI), the units of length, mass, and time are the meter (m), kilogram (kg), and second (s). Force is measured in newtons (N) and energy in joules (J). Coulomb's law is written as $F = Q_1 Q_2/4\pi\varepsilon_0 r^2$, where the charges Q_1 and Q_2 are in coulombs (C) and ε_0 is a constant (called the permittivity of vacuum) whose experimental value is $8.854 \times 10^{-12}\,C^2\,N^{-1}\,m^{-2}$. In the International System, charge is not expressible in terms of the mechanical units meters, kilograms, and seconds. Rather, a coulomb is defined as the amount of charge passing through a cross section of a wire in one second when that wire carries a current of one ampere (A); $1\,C = 1\,A\,s$, where the ampere is defined as the current needed to produce a certain force between two current-carrying wires. SI units are the officially recommended units of science, but the simple form of Coulomb's law in Gaussian units makes Gaussian units popular in quantum chemistry. It is best to be familiar with both systems of units.

In this book Coulomb's law is usually written as

$$F = Q_1' Q_2'/r^2 \tag{1.33}$$

One can think of this equation as being in Gaussian units with Q_1' and Q_2' in statcoulombs, r in centimeters, and F in dynes. Alternatively, one can also view Eq. (1.33) as being in SI units, with r in meters, F in newtons, and Q_1' and Q_2' as abbreviations for $Q_1/(4\pi\varepsilon_0)^{1/2}$ and $Q_2/(4\pi\varepsilon_0)^{1/2}$, where Q_1 and Q_2 are the charges in coulombs; we have

$$Q' = Q/(4\pi\varepsilon_0)^{1/2} \tag{1.34}$$

For the relation between Gaussian and SI quantities, see Table A.3 in the Appendix.

PROBLEMS

1.1 Calculate the force acting on an alpha particle passing a gold atomic nucleus at a distance of 0.0030 Å.

1.2 (a) Calculate the de Broglie wavelength of an electron moving at a speed 1/137th the speed of light. (b) Would you expect m in Eq. (1.3) to be the particle's rest mass or its relativistic mass?

1.3 Express each of the following units in terms of the fundamental units (cm, g, sec) of the Gaussian system: (a) dyne; (b) erg; (c) statcoulomb. Express each of the following units in terms of fundamental SI units (m, kg, s, A): (d) newton; (e) joule; (f) coulomb.

1.4 (a) The work function of Na is 2.28 eV, where $1\,eV = 1.602 \times 10^{-19}\,J$. (a) Calculate the maximum kinetic energy of photoelectrons emitted from Na exposed to 200 nm ultraviolet radiation. [1 nanometer (nm) $= 10^{-9}\,m$.] (b) Calculate the longest wavelength that will cause the photoelectric effect in Na.

1.5 What important probability-density function occurs in (a) the kinetic theory of gases? (b) the analysis of random errors of measurement?

1.6 Which of the following functions meet all the requirements of a probability-density function? (a) e^{ix}; (b) xe^{-x^2}; (c) e^{-x^2}.

1.7 When J. J. Thomson investigated electrons in cathode-ray tubes, he observed the behavior expected for particles obeying classical mechanics. (a) Electrons are accelerated through a potential difference of 1000 volts and passed through a collimating slit of width 0.100 cm. Calculate the diffraction angle α in Fig. 1.1. Use (6.121) and $\sin\alpha \approx \alpha$ for small α. (b) What slit width is needed to give $\alpha = 1.00°$ for 1000-volt electrons?

1.8 If the peak in the mass spectrum of C_2F_6 at mass number 138 is 100 units high, calculate the heights of the peaks at mass numbers 139 and 140. Isotopic abundances: ^{12}C—98.89%, ^{13}C—1.11%, ^{19}F—100%.

1.9 In bridge, each of the four players (A, B, C, D) receives 13 cards. Suppose A and C have 11 of the 13 spades between them. What is the probability that the remaining two spades are distributed so that B and D have one spade apiece?

1.10 The density of lead is 11 g/cm³. If lead atoms were cubes, what would be the length of each edge?

1.11 Verify that

$$\sin \theta = \frac{e^{i\theta} - e^{-i\theta}}{2i}, \qquad \cos \theta = \frac{e^{i\theta} + e^{-i\theta}}{2}$$

1.12 Find the cube roots of unity.

1.13 Find the absolute value and the phase of (a) i; (b) $ae^{i\pi/3}$; (c) $1 - 2i$.

1.14 In classical mechanics the kinetic energy of a particle is defined as $T \equiv \frac{1}{2}mv^2$. Use results from Section 1.4 to show that, for a particle moving vertically in the earth's gravitational field (with g assumed constant), $T + V = \frac{1}{2}mv_0^2 + mgx_0$, so $T + V$ is constant.

2 THE PARTICLE IN A BOX

2.1 DIFFERENTIAL EQUATIONS

Since the Schroedinger equation is a differential equation, we will review the mathematics of differential equations. In this section we consider only *ordinary* differential equations, which are those involving only one independent variable. Our differential equation will involve some sort of relation between an independent variable x, a dependent variable $y(x)$, and the first, second, ..., nth derivatives of y. We have

$$f(x, y, y', y'', \ldots, y^{(n)}) = 0 \qquad (2.1)$$

where f indicates some functional relation. An example is

$$y^{(iv)} + x(y')^2 + \sin x \cos y = 3e^x \qquad (2.2)$$

The *order* of a differential equation is the order of the highest derivative that occurs. Thus (2.1) is of nth order, while (2.2) is of fourth order.

A special kind of differential equation is the *linear* differential equation, which has the form

$$A_n(x)y^{(n)} + A_{n-1}(x)y^{(n-1)} + \cdots + A_0(x)y = g(x) \qquad (2.3)$$

where the A's are various functions of x. A differential equation that cannot be put in the form (2.3) is *nonlinear*. If $g(x) = 0$ in (2.3), the linear differential equation is said to be *homogeneous*; otherwise it is *inhomogeneous*. The one-dimensional Schroedinger equation (1.16) is a linear homogeneous differential equation of second order.

By dividing by the coefficient of y'', we can put any linear homogeneous second-order differential equation into the form

$$y'' + P(x)y' + Q(x)y = 0 \qquad (2.4)$$

Suppose we have two independent functions y_1 and y_2, each of which satisfies (2.4). By independent, we mean that y_2 is not simply a multiple of y_1. Then the general solution of (2.4) is

$$y = c_1 y_1 + c_2 y_2 \qquad (2.5)$$

where c_1 and c_2 are arbitrary constants. This is readily verified by substituting (2.5) into (2.4):

$$c_1 y_1'' + c_2 y_2'' + P(x)c_1 y_1' + P(x)c_2 y_2' + Q(x)c_1 y_1 + Q(x)c_2 y_2$$
$$= c_1 [y_1'' + P(x)y_1' + Q(x)y_1] + c_2 [y_2'' + P(x)y_2' + Q(x)y_2]$$
$$= c_1 \cdot 0 + c_2 \cdot 0 = 0$$

where the fact that y_1 and y_2 satisfy (2.4) has been used.

In general, the general solution of a differential equation of nth order has n arbitrary constants. To fix these constants, we may have *boundary conditions*, which are conditions that specify the value of y or various of its derivatives at a point or points. Thus if y represents the displacement of a vibrating string held fixed at two points, we know y must be zero at these points. We will discuss appropriate boundary conditions for the Schroedinger equation later.

An important case is the linear homogeneous second-order differential equation with *constant coefficients*:

$$y'' + py' + qy = 0 \qquad (2.6)$$

where p and q are constants. To solve (2.6) let us tentatively assume a solution of the form $y = e^{sx}$. (We are looking for a function whose derivatives when multiplied by constants will cancel out the original function. The exponential function repeats itself when differentiated, and is thus the correct choice.) Substitution in (2.6) gives

$$s^2 e^{sx} + pse^{sx} + qe^{sx} = 0$$
$$s^2 + ps + q = 0 \qquad (2.7)$$

Equation (2.7) is called the *auxiliary equation*. It is a quadratic equation with two roots s_1 and s_2 which, provided s_1 and s_2 are not equal, give two independent solutions to (2.6). Thus the general solution of (2.6) is

$$y = c_1 e^{s_1 x} + c_2 e^{s_2 x} \qquad (2.8)$$

2.2   PARTICLE IN A ONE-DIMENSIONAL BOX

Having obtained the solution of one kind of differential equation, let us look at a case where we can use this solution to solve the Schroedinger equation. We consider a particle in a one-dimensional box. By this we mean a particle subjected to a potential-energy function that is infinite everywhere along the x axis except for a line segment of length l, where the potential energy is zero. Such a system may seem physically unreal, but we shall later see that this model can be applied with some success to certain conjugated molecules; see Chapter 15 and Problem 2.10. We put the origin at the left end of the line segment (Fig. 2.1).

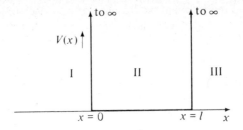

Figure 2.1 *Potential energy for the particle in a one-dimensional box*

We have three regions to consider. For regions I and III, the potential energy V equals infinity; the Schroedinger equation (1.16) for these regions is

$$\frac{d^2\psi}{dx^2} + \frac{2m}{\hbar^2}(E - \infty)\psi = 0$$

Neglecting E in comparison with ∞, we have

$$\frac{d^2\psi}{dx^2} = \infty\psi, \qquad \psi = \frac{1}{\infty}\frac{d^2\psi}{dx^2}$$

and we conclude that ψ is zero outside the box:

$$\psi_\text{I} = 0, \qquad \psi_\text{III} = 0 \tag{2.9}$$

For region II, x between zero and l, the potential energy V is zero, and the Schroedinger equation (1.16) becomes

$$\frac{d^2\psi_\text{II}}{dx^2} + \frac{2m}{\hbar^2}E\psi_\text{II} = 0 \tag{2.10}$$

where m is the mass of the particle and E is its total energy. We recognize (2.10) as a linear homogeneous second-order differential equation with constant coefficients. The auxiliary equation (2.7) gives

$$s^2 + 2mE\hbar^{-2} = 0$$

$$s = \pm(-2mE)^{1/2}\hbar^{-1} \tag{2.11}$$

The energy E is equal to the potential energy, which is zero, plus the kinetic energy, which is positive, so E is positive, and it is convenient to rewrite (2.11) as

$$s = \pm i(2mE)^{1/2}/\hbar \tag{2.12}$$

Using (2.8), we have

$$\psi_\text{II} = c_1 e^{i(2mE)^{1/2}x/\hbar} + c_2 e^{-i(2mE)^{1/2}x/\hbar} \tag{2.13}$$

Temporarily, let

$$\theta = (2mE)^{1/2}x/\hbar \tag{2.14}$$

$$\psi_\text{II} = c_1 e^{i\theta} + c_2 e^{-i\theta} \tag{2.15}$$

Using (1.22), we have

$$\psi_{II} = c_1 \cos\theta + ic_1 \sin\theta + c_2 \cos\theta - ic_2 \sin\theta$$

$$= (c_1 + c_2)\cos\theta + (ic_1 - ic_2)\sin\theta$$

$$= A \cos\theta + B \sin\theta$$

where A and B are new arbitrary constants. Hence

$$\psi_{II} = A \cos\left[\hbar^{-1}(2mE)^{1/2}x\right] + B \sin\left[\hbar^{-1}(2mE)^{1/2}x\right] \qquad \textbf{(2.16)}$$

Now we determine A and B by applying boundary conditions. It seems reasonable to postulate that the wave function will be continuous, i.e., that it will make no sudden jumps in value. If ψ is to be continuous at the point $x = 0$, then ψ_I and ψ_{II} must approach the same value at $x = 0$:

$$\lim_{x \to 0} \psi_I = \lim_{x \to 0} \psi_{II}$$

$$0 = \lim_{x \to 0} \left\{ A \cos\left[\hbar^{-1}(2mE)^{1/2}x\right] + B \sin\left[\hbar^{-1}(2mE)^{1/2}x\right] \right\}$$

$$0 = A$$

We have evaluated one arbitrary constant, and we now have

$$\psi_{II} = B \sin\left[(2\pi/h)(2mE)^{1/2}x\right] \qquad \textbf{(2.17)}$$

Applying the continuity condition at $x = l$, we get

$$B \sin\left[(2\pi/h)(2mE)^{1/2}l\right] = 0 \qquad \textbf{(2.18)}$$

B cannot be zero, because this would make the wave function zero everywhere—we would have an empty box. Therefore

$$\sin\left[(2\pi/h)(2mE)^{1/2}l\right] = 0$$

The zeros of the sine function occur at $0, \pm\pi, \pm2\pi, \pm3\pi, \ldots$. Hence

$$(2\pi/h)(2mE)^{1/2}l = \pm n\pi \qquad \textbf{(2.19)}$$

We must, however, reject the value zero for n, which from (2.19) corresponds to $E = 0$. For $E = 0$, the Schroedinger equation (2.10) reads $d^2\psi_{II}/dx^2 = 0$, so $d\psi_{II}/dx = c$ and $\psi_{II} = cx + d$, where c and d are constants. The boundary condition that $\psi_{II} = 0$ at $x = 0$ gives $d = 0$, and the condition that $\psi_{II} = 0$ at $x = l$ then gives $c = 0$. Thus $\psi_{II} = 0$ for $E = 0$, and $E = 0$ is not an allowed energy value. Solving (2.19) for E, we get

$$E = n^2 \frac{h^2}{8ml^2}, \qquad n = 1, 2, 3, \ldots \qquad \textbf{(2.20)}$$

Only the energy values (2.20) allow ψ to satisfy the boundary condition of continuity at $x = l$. Application of a boundary condition has forced us to the conclusion that the values of the energy are quantized; this is in striking contrast to the classical result that the particle in the box can have any nonnegative energy. Note also that there is a minimum value, greater than zero, for the energy of the particle.

Substitution of (2.19) into (2.17) gives for the wave function

$$\psi_{II} = B \sin\left(\frac{n\pi x}{l}\right), \qquad n = 1, 2, 3,\ldots \qquad (2.21)$$

[The use of the negative sign in front of $n\pi$ does not give us another independent solution. Since $\sin(-\theta) = -\sin\theta$ we would simply get a constant, -1, times the solution with the plus sign.]

The constant B in Eq. (2.21) is still arbitrary. To fix its value, we use the normalization requirement, Eq. (1.20):

$$\int_{-\infty}^{\infty} |\Psi|^2\, dx = \int_{-\infty}^{\infty} |\psi|^2\, dx = 1$$

$$\int_{-\infty}^{0} |\psi_I|^2\, dx + \int_{0}^{l} |\psi_{II}|^2\, dx + \int_{l}^{\infty} |\psi_{III}|^2\, dx = 1$$

$$|B|^2 \int_{0}^{l} \sin^2\left(\frac{n\pi x}{l}\right) dx = 1 = |B|^2 \frac{l}{2}$$

where the integral was evaluated by using

$$2\sin^2 t = 1 - \cos 2t \qquad (2.22)$$

We obtain

$$|B| = \left(\frac{2}{l}\right)^{1/2}$$

Note that we have determined only the absolute value of B. B could be $-\sqrt{2/l}$ as well as $+\sqrt{2/l}$. Moreover, B need not be a real number—we could use any complex number with absolute value $\sqrt{2/l}$; i.e., all we can say is that $B = (2/l)^{1/2} e^{i\alpha}$, where α is the phase of B. (See Section 1.7.) Choosing the phase to be zero, we finally write

$$\psi_{II} = \left(\frac{2}{l}\right)^{1/2} \sin\left(\frac{n\pi x}{l}\right), \qquad n = 1, 2, 3,\ldots \qquad (2.23)$$

Graphs of the wave functions and the probability densities are shown in Figs. 2.2 and 2.3.

Note that there are points (called *nodes*) where the wave function is zero. For each increase of one in the value of n (n is called a *quantum number*), we find one more node. The existence of nodes in ψ and $|\psi|^2$ may seem surprising. Thus for $n = 2$,

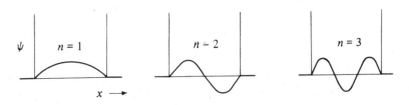

ψ $n = 1$ $n = 2$ $n = 3$

$x \longrightarrow$

Figure 2.2 *Graphs of ψ for the three lowest energy particle-in-a-box states*

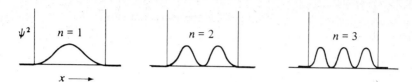

Figure 2.3 *Graphs of ψ^2 for the lowest particle-in-a-box states*

Fig. 2.3 says that there is zero probability of finding the particle in the center of the box at $x = l/2$. How can the particle get from one side of the box to the other without at any time being found in the center? This apparent paradox arises from attempting to understand the motion of microscopic particles using our everyday experience of the motions of macroscopic particles. However, as stated in Chapter 1, electrons and other microscopic "particles" cannot be fully and correctly described in terms of concepts of classical physics drawn from the macroscopic world.

Figure 2.3 shows that the probability of finding the particle at various places in the box is quite different from the classical result. Classically, a particle of fixed energy in a box bounces back and forth elastically between the two walls, moving at constant speed. Thus it is equally likely to be found at any point in the box. Quantum mechanically, we find a maximum in probability at the center of the box for the lowest energy level. As we go to higher energy levels with more nodes, the maxima and minima of probability come closer together, and the variations in probability along the length of the box ultimately become undetectable. For high quantum numbers we approach the classical result of uniform probability density. This result, that in the limit of large quantum numbers quantum mechanics goes over into classical mechanics, is known as the *Bohr correspondence principle*. Since Newtonian mechanics is valid for macroscopic bodies (moving at speeds much less than the speed of light), we expect nonrelativistic quantum mechanics to give the same answer as classical mechanics for macroscopic bodies. Because of the extremely small magnitude of Planck's constant, quantization of energy is unobservable for macroscopic bodies. Since the mass of the particle and the length of the box squared appear in the denominator of Eq. (2.20), a macroscopic object in a macroscopic box having a macroscopic energy of motion would have a huge value for n, and hence, according to the correspondence principle, would show classical behavior.

We have a whole set of wave functions, each one corresponding to a different value of the energy and characterized by the quantum number n, which may have integral values from one up. Let the subscript i denote a particular wave function with the value n_i for its quantum number:

$$\psi_i = \left(\frac{2}{l}\right)^{1/2} \sin\left(\frac{n_i \pi x}{l}\right), \qquad 0 < x < l$$

$$\psi_i = 0 \qquad\qquad\qquad \text{elsewhere}$$

Since the wave function has been normalized, we have

$$\int_{-\infty}^{\infty} \psi_i^* \psi_j \, dx = 1 \qquad \text{if } i = j \tag{2.24}$$

We now ask for the value of this integral when we use wave functions corresponding to *different* energy levels:

$$\int_{-\infty}^{\infty} \psi_i^* \psi_j \, dx = \int_0^l \left(\frac{2}{l}\right)^{1/2} \sin\left(\frac{n_i \pi x}{l}\right) \left(\frac{2}{l}\right)^{1/2} \sin\left(\frac{n_j \pi x}{l}\right) \, dx, \qquad n_i \neq n_j$$

Let $t = \pi x / l$:

$$\int_{-\infty}^{\infty} \psi_i^* \psi_j \, dx = \frac{2}{l} \int_0^{\pi} \sin n_i t \sin n_j t \, dt \cdot \frac{l}{\pi}$$

The integral may be evaluated with the identity

$$\sin n_i t \sin n_j t = \tfrac{1}{2} \cos (n_i - n_j)t - \tfrac{1}{2} \cos (n_i + n_j)t \qquad \textbf{(2.25)}$$

Hence

$$\int_{-\infty}^{\infty} \psi_i^* \psi_j \, dx = \frac{2}{\pi} \int_0^{\pi} \tfrac{1}{2} \cos (n_i - n_j)t \, dt - \frac{2}{\pi} \int_0^{\pi} \tfrac{1}{2} \cos (n_i + n_j)t \, dt = 0$$

since $\sin m\pi = 0$ for m an integer. We thus have

$$\int_{-\infty}^{\infty} \psi_i^* \psi_j \, dx = 0, \qquad i \neq j \qquad \textbf{(2.26)}$$

We say that ψ_i and ψ_j are *orthogonal* to each other for $i \neq j$. We can combine (2.24) and (2.26) by writing

$$\int_{-\infty}^{\infty} \psi_i^* \psi_j \, dx = \delta_{ij} \qquad \textbf{(2.27)}$$

The symbol δ_{ij} is called the Kronecker delta (after a mathematician); it equals one whenever the two indices i and j are equal, and equals zero when i and j are unequal:

$$\delta_{ij} = \begin{cases} 0 & \text{for } i \neq j \\ 1 & \text{for } i = j \end{cases} \qquad \textbf{(2.28)}$$

The property of the wave functions illustrated in (2.27) is called *orthonormality*. We proved orthonormality only for the particle-in-a-box wave functions; we shall later prove it more generally.

A more rigorous way to look at the particle in a box with infinite walls is to first treat the particle in a box with a finite jump in potential energy at the walls and then take the limit as the jump in potential energy becomes infinite. The results, when the limit is taken, will be the same as we have obtained (see Problem 2.12).

2.3   THE FREE PARTICLE IN ONE DIMENSION

By a free particle, we mean a particle subject to no forces whatever. For a free particle, the potential energy remains constant no matter what the value of x is. Since the choice of the zero level of energy is arbitrary, we may set $V(x) = 0$. The Schroedinger equation (1.16) becomes

$$\frac{d^2 \psi}{dx^2} + \frac{2m}{\hbar^2} E \psi = 0 \qquad \textbf{(2.29)}$$

Equation (2.29) is the same as Eq. (2.10) (except for the boundary conditions); therefore the general solution of (2.29) is (2.13):

$$\psi = c_1 e^{i(2mE)^{1/2}x/\hbar} + c_2 e^{-i(2mE)^{1/2}x/\hbar} \qquad (2.30)$$

What boundary condition might we impose? It seems reasonable to postulate (since $\psi^*\psi\,dx$ represents a probability) that ψ will remain finite as x goes to $\pm\infty$. If the energy E is less than zero, then this boundary condition will be violated, since for $E < 0$ we have

$$i(2mE)^{1/2} = i(-2m|E|)^{1/2} = i \cdot i \cdot (2m|E|)^{1/2} = -(2m|E|)^{1/2}$$

and therefore the first term in (2.30) will become infinite as x approaches minus infinity. Similarly, if E is negative, the second term in (2.30) becomes infinite as x approaches plus infinity. Thus the boundary condition requires

$$E \geqslant 0 \qquad (2.31)$$

for the free particle. The wave function is oscillatory and is a linear combination of a sine and a cosine term [Eq. (2.16)]. For the free particle, we do not get quantization of the energy; all nonnegative energies are allowed. Since we set $V = 0$, the energy E is in this case all kinetic energy. If we attempt to evaluate the arbitrary constants c_1 and c_2 by normalization, we will find that the integral $\int_{-\infty}^{\infty}\psi^*(x)\psi(x)\,dx$ is divergent. In other words, the free-particle wave function is not normalizable in the usual sense. This is to be expected on physical grounds, since there is no reason for the probability of finding the free particle to approach zero as x goes to $\pm\infty$.

The free-particle problem represents an unreal situation, since we could not actually have a particle that had no interaction with any other particle in the universe.

2.4   PARTICLE IN A RECTANGULAR WELL

We shall briefly discuss the problem of a particle in a one-dimensional box with walls of finite height (Fig. 2.4a). The potential-energy function is $V = V_0$ for

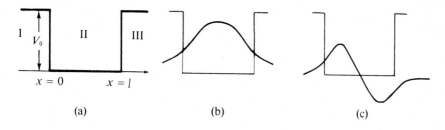

$$x = 0 \qquad x = l$$

(a)   (b)   (c)

Figure 2.4 (a) Potential energy for a particle in a one-dimensional rectangular well; (b) sketch of the ground-state wave function for this potential; (c) first excited-state wave function

$x < 0$, $V = 0$ for $0 \leqslant x \leqslant l$, and $V = V_0$ for $x > l$. There are two cases to consider, depending on whether the particle's energy E is less than or greater than V_0.

We first consider $E < V_0$. The Schroedinger equation (1.16) in regions I and III is $d^2\psi/dx^2 + (2m/\hbar^2)(E - V_0)\psi = 0$. This is a linear homogeneous differential equation with constant coefficients, and the auxiliary equation (2.7) is $s^2 + (2m/\hbar^2)(E - V_0) = 0$ with roots $s = \pm (2m/\hbar^2)^{1/2}(V_0 - E)^{1/2}$. Therefore

$$\psi_I = C \exp[(2m/\hbar^2)^{1/2}(V_0 - E)^{1/2}x] + D \exp[-(2m/\hbar^2)^{1/2}(V_0 - E)^{1/2}x]$$

$$\psi_{III} = F \exp[(2m/\hbar^2)^{1/2}(V_0 - E)^{1/2}x] + G \exp[-(2m/\hbar^2)^{1/2}(V_0 - E)^{1/2}x]$$

where C, D, F, and G are constants.

As in Section 2.3, we must prevent ψ_I from becoming infinite as $x \to -\infty$. Since we are assuming $E < V_0$, the quantity $(V_0 - E)^{1/2}$ is a real, positive number, and to keep ψ_I finite as $x \to -\infty$ we must have $D = 0$. Similarly, to keep ψ_{III} finite as $x \to +\infty$, we must have $F = 0$. Therefore

$$\psi_I = C \exp[(2m/\hbar^2)^{1/2}(V_0 - E)^{1/2}x], \qquad \psi_{III} = G \exp[-(2m/\hbar^2)^{1/2}(V_0 - E)^{1/2}x]$$

In region II, $V = 0$, the Schroedinger equation is (2.10), and its solution is (2.16):

$$\psi_{II} = A \cos[(2m/\hbar^2)^{1/2}E^{1/2}x] + B \sin[(2m/\hbar^2)^{1/2}E^{1/2}x]$$

To complete the problem, we must apply the boundary conditions. As with the particle in a box with infinite walls, we require the wave function to be continuous at $x = 0$ and at $x = l$; so $\psi_I(0) = \psi_{II}(0)$ and $\psi_{II}(l) = \psi_{III}(l)$. There are four arbitrary constants in the wave function, so more than these two boundary conditions are needed. As well as requiring ψ to be continuous, we shall require that its derivative $d\psi/dx$ be continuous everywhere. To justify this requirement, we note that if $d\psi/dx$ made a discontinuous change in value at a point, then its derivative (its instantaneous rate of change) $d^2\psi/dx^2$ would become infinite at that point. However, for the problem under consideration, the Schroedinger equation $d^2\psi/dx^2 = (2m/\hbar^2)(V - E)\psi$ does not contain anything infinite on the right side, so $d^2\psi/dx^2$ cannot become infinite. [For a more rigorous argument, see D. Branson, *Am. J. Phys.*, **47**, 1000 (1979).] Therefore $d\psi_I/dx = d\psi_{II}/dx$ at $x = 0$ and $d\psi_{II}/dx = d\psi_{III}/dx$ at $x = l$.

From $\psi_I(0) = \psi_{II}(0)$ we get $C = A$. From $\psi_I'(0) = \psi_{II}'(0)$ we get (Problem 2.11a) $B = (V_0 - E)^{1/2}A/E^{1/2}$. From $\psi_{II}(l) = \psi_{III}(l)$ we get a complicated equation that allows G to be found in terms of A. (The constant A is determined by normalization.)

Taking $\psi_{II}'(l) = \psi_{III}'(l)$, dividing it by $\psi_{II}(l) = \psi_{III}(l)$, and expressing B in terms of A, we obtain the following equation for the energy levels (Problem 2.11b):

$$\tan[(2m/\hbar^2)^{1/2}E^{1/2}l] = 2E^{1/2}(V_0 - E)^{1/2}/(2E - V_0) \qquad (2.32)$$

Defining the dimensionless constants ε and b as $\varepsilon \equiv E/V_0$ and $b \equiv (2mV_0/\hbar^2)^{1/2}l$, we write (2.32) as

$$\tan(b\varepsilon^{1/2}) = 2\varepsilon^{1/2}(1 - \varepsilon)^{1/2}/(2\varepsilon - 1) \qquad (2.33)$$

Only the particular values of E that satisfy (2.32) yield a wave function that is continuous and has a continuous derivative, so the energy levels are quantized for $E < V_0$. To find the allowed energy levels, one can plot the left and right sides of (2.33) on the same graph paper and find the intersection points of the two curves (see *Kauzmann*, p. 191). A detailed study (see *Merzbacher*, Section 6.8) shows that the number of allowed energy levels with $E < V_0$ is N, where N satisfies

$$N - 1 < (8mV_0)^{1/2} l/h \leqslant N$$

For example, if $V_0 = h^2/ml^2$, then $(8mV_0)^{1/2} l/h = 8^{1/2} = 2.83$ and $N = 3$.

Figure 2.4 shows ψ for the lowest two energy levels. The wave function is oscillatory inside the box and dies off exponentially outside the box. (It turns out that the number of nodes increases by one for each higher level.)

So far we have considered only states with $E < V_0$. For $E > V_0$, the quantity $(V_0 - E)^{1/2}$ is imaginary and instead of dying off to zero as x goes to $\pm \infty$, ψ_I and ψ_{III} oscillate (similar to the free-particle ψ). We no longer have any reason to set D in ψ_I and F in ψ_{III} equal to zero, and with these additional constants available to satisfy the boundary conditions on ψ and ψ', one finds that it is not necessary to restrict E to obtain properly behaved wave functions. Therefore all energies above V_0 are allowed.

States with $E < V_0$ are called bound states.

2.5 TUNNELING

For the particle in a rectangular well (Section 2.4), Fig. 2.4 and the equations for ψ_I and ψ_{III} show that for the bound states there is a nonzero probability of finding the particle in regions I and III, where its total energy E is less than its potential energy $V = V_0$. Classically, this behavior is not allowed. The classical equations $E = T + V$ and $T \geqslant 0$, where T is the kinetic energy, mean that E cannot be less than V in classical mechanics.

Consider a particle in a one-dimensional box with walls of finite height *and* finite thickness (Fig. 2.5). Classically, the particle cannot escape from the box unless its energy is greater than the potential-energy barrier V_0. However, a quantum-mechanical treatment (which is omitted) shows that there is a finite probability for a particle of total energy less than V_0 to be found outside the box.

The term *tunneling* denotes the penetration of a particle into a classically forbidden region (as in Fig. 2.4) or the passage of a particle through a potential-

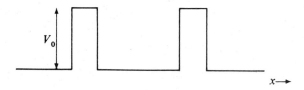

Figure 2.5 *Potential energy for a particle in a box of finite height and thickness*

energy barrier whose height exceeds the particle's energy. Since tunneling is a quantum effect, its probability of occurrence is greater the less classical is the behavior of the particle. Therefore tunneling is most prevalent with particles of small mass. (Note that the greater the mass m, the more rapidly the functions ψ_I and ψ_{III} of Section 2.4 die away to zero.) Electrons tunnel quite readily; hydrogen atoms tunnel more readily than heavier atoms.

The emission of alpha particles from a radioactive nucleus involves tunneling of the alpha particles through the potential-energy barrier produced by the short-range attractive nuclear forces and the Coulombic repulsive force between the daughter nucleus and the alpha particle. The NH_3 molecule is pyramidal. There is a potential-energy barrier to inversion of the molecule, with the potential-energy maximum occurring at the planar configuration. The hydrogen atoms can tunnel through this barrier, thereby inverting the molecule. In CH_3CH_3 there is a barrier to internal rotation, with a potential-energy maximum at the eclipsed position of the hydrogens. The hydrogens can tunnel through this barrier from one staggered position to the next. Tunneling of electrons is important in oxidation–reduction reactions and in electrode processes. Tunneling usually makes a significant contribution to the rate of chemical reactions that involve transfer of hydrogen atoms. For more on tunneling, see R. P. Bell, *The Tunnel Effect in Chemistry*, Chapman & Hall, London, 1980.

PROBLEMS

2.1 Solve $y''(x) + y'(x) - 2y(x) = 0$.

2.2 (a) For the case of equal roots of the auxiliary equation, $s_1 = s_2 = s$, we have found only one independent solution of the linear homogeneous second-order differential equation: e^{sx}. Verify that xe^{sx} is the second solution in this case. (b) Solve $y''(x) - 2y'(x) + y(x) = 0$.

2.3 Consider a macroscopic object of mass 1.0 g moving with speed 1.0 cm/sec in a one-dimensional box of length 1.0 cm; find the quantum number n.

2.4 Consider a particle with quantum number n moving in a one-dimensional box of length l. (a) Determine the probability of finding the particle in the left quarter of the box. (b) For what value of n is this probability a maximum? (c) What is the limit of this probability for $n \to \infty$? (d) What principle is illustrated in (c)?

2.5 An extremely crude picture of an electron in an atom or molecule treats it as a particle in a one-dimensional box whose length is of the order of the size of atoms and molecules. (a) For an electron in a box of length 1.0 Å, calculate the separation between the two lowest energy levels. (b) Calculate the wavelength of a photon corresponding to a transition between these two levels. (c) In what portion of the electromagnetic spectrum is this wavelength?

2.6 Write down the time-dependent wave function for a free particle with energy E.

2.7 For the particle in a one-dimensional box of length l, we could have put the coordinate origin at the center of the box. Find the wave functions and energy levels for this choice of origin.

2.8 Sketch rough graphs of ψ and of ψ^2 for the $n = 4$ and $n = 5$ particle-in-a-box states.

2.9 Integrals involving trigonometric functions can often be evaluated using the identities of Problem 1.11. Use the complex-exponential form of the sine function to verify Eq. (2.27) for the particle-in-a-box wave functions.

2.10 A crude treatment of the pi electrons of a conjugated molecule regards these electrons as moving in the particle-in-a-box potential of Fig. 2.1, where the box length is somewhat more than the length of the conjugated chain. The Pauli exclusion principle (Chapter 10) allows no more than two electrons to occupy each box level. (These two have opposite spins.) For butadiene $CH_2{=}CHCH{=}CH_2$, take the box length as 7.0 Å and use this model to estimate the wavelength of light absorbed when a pi electron is excited from the highest-occupied to the lowest-vacant box level. (The experimental value is 2170 Å.)

2.11 (a) For the particle in a rectangular well (Section 2.4), verify that $B = (V_0 - E)^{1/2} A / E^{1/2}$. (b) Verify Eq. (2.32).

2.12 For the particle in a rectangular well (Section 2.4), show that in the limit $V_0 \to \infty$ (a) Eq. (2.32) gives $E = n^2 h^2 / 8ml^2$ as in Eq. (2.20); (b) the wave function goes to Eqs. (2.9) and (2.21).

2.13 Sketch ψ for the next-lowest level in Fig. 2.4.

2.14 For an electron in a 15.0-eV-deep one-dimensional rectangular well of width 2.00 Å, calculate the number of bound states. Use (6.121).

2.15 For a particle in a one-dimensional rectangular well, (a) must there be at least one bound state? (b) is ψ'' continuous at $x = 0$?

2.16 Find the allowed bound energy levels for Problem 2.14 by using a programmable calculator or a computer to calculate the left side of (2.33) minus the right side (or to calculate each side separately) for values of ε going from 0 to 1 in small steps. [At $\varepsilon = 0$, Eq. (2.33) is satisfied, but this is not an allowed energy level since it turns out to lead to $\psi = 0$ everywhere.]

3 OPERATORS

3.1 OPERATORS

We now develop the theory of quantum mechanics in a more general fashion than previously. We begin by writing the one-particle, one-dimensional time-independent Schroedinger equation (1.16) in the form

$$\left[-\frac{\hbar^2}{2m}\frac{d^2}{dx^2} + V(x) \right]\psi(x) = E\psi(x) \tag{3.1}$$

The entity in brackets in (3.1) is an *operator*. Equation (3.1) suggests that we have an energy operator, which, operating on the wave function, gives us the wave function back again, but multiplied by the allowed values of the energy. We therefore discuss operators.

An *operator* is simply a rule whereby given some function we can find another, corresponding function. For example, let $\hat{D}$ be the operator that differentiates a function with respect to x. (We use a circumflex to indicate an operator.) Provided $f(x)$ is differentiable, we have as the result of operating on $f(x)$ with $\hat{D}$: $\hat{D}f(x) = f'(x)$. For example, $\hat{D}(x^2 + 3e^x) = 2x + 3e^x$. If $\hat{3}$ is the operator that multiplies a function by 3, then $\hat{3}(x^2 + 3e^x) = 3x^2 + 9e^x$. If SQRT is the operator that takes the square root of a function, then SQRT $(x^2 + 3e^x) = (x^2 + 3e^x)^{1/2}$.

We define the *sum* of two operators $\hat{A}$ and $\hat{B}$ by the equation

$$(\hat{A} + \hat{B})f(x) \equiv \hat{A}f(x) + \hat{B}f(x) \tag{3.2}$$

For example,

$$(\hat{D} + \hat{3})(x^2 + 3e^x) = (2x + 3e^x) + (3x^2 + 9e^x) = 2x + 3x^2 + 12e^x$$

32

We define the *product* of two operators $\hat{A}$ and $\hat{B}$ by the equation

$$\hat{A}\hat{B}f(x) \equiv \hat{A}[\hat{B}f(x)] \qquad (3.3)$$

In other words, we first operate on $f(x)$ with the operator on the right of the operator product, and then we take the resulting function and operate on it with the operator on the left of the operator product. For example,

$$3\hat{D}f(x) = 3[\hat{D}f(x)] = 3f'(x) = 3f'(x) \qquad (3.4)$$

For this example it would have made no difference in the final result whether we first applied one operator or the other. In general, we cannot assume that $\hat{A}\hat{B}$ and $\hat{B}\hat{A}$ have the same effect. Consider, for example, the operators d/dx and $\hat{x}$:

$$\hat{D}\hat{x}f(x) = \frac{d}{dx}[xf(x)] = f(x) + xf'(x) = (\hat{1} + \hat{x}\hat{D})f(x) \qquad (3.5)$$

$$\hat{x}\hat{D}f(x) = \hat{x}\left[\frac{d}{dx}f(x)\right] = xf'(x)$$

Thus $\hat{A}\hat{B}$ and $\hat{B}\hat{A}$ are different operators in this case.

We can develop an *operator algebra* as follows. If $\hat{A}$ and $\hat{B}$ are two operators such that $\hat{A}f = \hat{B}f$ for all functions f, we say that $\hat{A}$ and $\hat{B}$ are equal: $\hat{A} = \hat{B}$. For example, (3.5) shows that

$$\hat{D}\hat{x} = \hat{1} + \hat{x}\hat{D} \qquad (3.6)$$

The operator $\hat{1}$ (multiplication by one) is the *unit operator*. The operator $\hat{0}$ (multiplication by zero) is the *null operator*. We usually omit the circumflex over operators that are simply multiplication by a constant. We can transfer operators from one side of an operator equation to the other (Problem 3.28); thus (3.6) is equivalent to

$$\hat{D}\hat{x} - \hat{x}\hat{D} - 1 = 0$$

where we have omitted circumflexes over the null and unit operators.

Operators obey the associative law of multiplication:

$$\hat{A}(\hat{B}\hat{C}) = (\hat{A}\hat{B})\hat{C} \qquad (3.7)$$

The proof of (3.7) is outlined in Problem 3.2. As an example, let $\hat{A} = d/dx$, $\hat{B} = \hat{x}$, $\hat{C} = 3$. Using (3.6), we have

$$(\hat{A}\hat{B}) = \hat{D}\hat{x} = 1 + \hat{x}\hat{D}, \qquad [(\hat{A}\hat{B})\hat{C}]f = (1 + \hat{x}\hat{D})3f = 3f + 3xf'$$

$$(\hat{B}\hat{C}) = 3\hat{x}, \qquad [\hat{A}(\hat{B}\hat{C})]f = \hat{D}(3xf) = 3f + 3xf'$$

The major difference between operator algebra and ordinary algebra is that numbers obey the commutative law of multiplication, but operators do not necessarily do so; $ab = ba$ if a and b are numbers, but $\hat{A}\hat{B}$ and $\hat{B}\hat{A}$ are not necessarily equal operators. We define the *commutator* $[\hat{A}, \hat{B}]$ of the operators $\hat{A}$ and $\hat{B}$ as the operator $\hat{A}\hat{B} - \hat{B}\hat{A}$:

$$[\hat{A}, \hat{B}] \equiv \hat{A}\hat{B} - \hat{B}\hat{A} \qquad (3.8)$$

If $\hat{A}\hat{B} = \hat{B}\hat{A}$, we say that $\hat{A}$ and $\hat{B}$ *commute*. Since it makes no difference in what order we apply the operators $\hat{3}$ and d/dx, we have

$$\left[\hat{3}, \frac{d}{dx}\right] = \hat{3}\frac{d}{dx} - \frac{d}{dx}\hat{3} = 0$$

From Eq. (3.6) we have

$$\left[\frac{d}{dx}, \hat{x}\right] = \hat{D}\hat{x} - \hat{x}\hat{D} = 1 \qquad (3.9)$$

The *square* of an operator is defined as the product of the operator with itself: $\hat{A}^2 = \hat{A}\hat{A}$. Let us find the square of the differentiation operator:

$$\hat{D}^2 f(x) = \hat{D}(\hat{D}f) = \hat{D}f' = f''$$

$$\hat{D}^2 = \frac{d^2}{dx^2} \qquad (3.10)$$

As another example, the square of the operator that takes the complex conjugate of a function is equal to the unit operator, since taking the complex conjugate twice gives the original function. The nth power of an operator ($n = 1, 2, 3, \ldots$) is defined to mean applying the operator n times in succession.

It turns out that the operators occurring in quantum mechanics are linear. $\hat{A}$ is a *linear operator* if and only if it has the following two properties:

$$\hat{A}[f(x) + g(x)] = \hat{A}f(x) + \hat{A}g(x) \qquad (3.11)$$

$$\hat{A}[cf(x)] = c\hat{A}f(x) \qquad (3.12)$$

where f and g are arbitrary functions and c is an arbitrary constant. Examples of linear operators include $\hat{x}^2$, d/dx, and d^2/dx^2; the square-root operator SQRT is nonlinear.

Equation (2.3) defines the form of a linear differential equation. Using the differentiation operator $\hat{D}$, we can rewrite (2.3) as

$$[A_n(x)\hat{D}^n + A_{n-1}(x)\hat{D}^{n-1} + \cdots + A_0(x)]y(x) = g(x) \qquad (3.13)$$

The operator in brackets in (3.13) is linear.

Useful identities in linear-operator manipulations are

$$(\hat{A} + \hat{B})\hat{C} = \hat{A}\hat{C} + \hat{B}\hat{C} \qquad (3.14)$$

$$\hat{A}(\hat{B} + \hat{C}) = \hat{A}\hat{B} + \hat{A}\hat{C} \qquad (3.15)$$

The proof of the distributive law (3.14) follows from the definition of the product and sum of operators; we have

$$[(\hat{A} + \hat{B})\hat{C}]f(x) = (\hat{A} + \hat{B})(\hat{C}f) = \hat{A}(\hat{C}f) + \hat{B}(\hat{C}f)$$

$$= \hat{A}\hat{C}f + \hat{B}\hat{C}f = (\hat{A}\hat{C} + \hat{B}\hat{C})f$$

To gain facility in operator manipulations, we will work out the square of

$d/dx + \hat{x}$. We have

$$(\hat{D} + \hat{x})^2 f(x) = (\hat{D} + \hat{x})[(\hat{D} + \hat{x})f] = (\hat{D} + \hat{x})(f' + xf)$$

$$= f'' + f + xf' + xf' + x^2 f = (\hat{D}^2 + 2\hat{x}\hat{D} + \hat{x}^2 + 1)f(x)$$

$$(\hat{D} + \hat{x})^2 = \hat{D}^2 + 2\hat{x}\hat{D} + \hat{x}^2 + 1$$

Let us repeat this calculation, using only operator equations:

$$(\hat{D} + \hat{x})^2 = (\hat{D} + \hat{x})(\hat{D} + \hat{x}) = \hat{D}(\hat{D} + \hat{x}) + \hat{x}(\hat{D} + \hat{x})$$

$$= \hat{D}^2 + \hat{D}\hat{x} + \hat{x}\hat{D} + \hat{x}^2 = \hat{D}^2 + \hat{x}\hat{D} + 1 + \hat{x}\hat{D} + \hat{x}^2$$

$$= \hat{D}^2 + 2x\hat{D} + x^2 + 1$$

where (3.14), (3.15), and (3.6) have been used and the circumflex over the operator "multiplication by x" has been omitted.

Although we have considered only operators and functions in one dimension, x, we can readily extend all we have said to operators and functions in several dimensions.

3.2   EIGENFUNCTIONS AND EIGENVALUES

Suppose that the effect of operating on some function $f(x)$ with the operator $\hat{A}$ is simply to multiply $f(x)$ by a certain constant k. We then say that $f(x)$ is an *eigenfunction* of $\hat{A}$, with *eigenvalue* k. We have[1]

$$\hat{A}f(x) = kf(x) \qquad (3.16)$$

(*Eigen* is a German word meaning *characteristic*. "Eigenvalue" is a hybrid word; it has been suggested that "characteristicwert" would be just as suitable.) As an example of (3.16), e^{2x} is an eigenfunction of the operator d/dx with eigenvalue 2:

$$(d/dx)e^{2x} = 2e^{2x}$$

Let us find all the eigenfunctions and eigenvalues of d/dx. Equation (3.16) becomes

$$\frac{df(x)}{dx} = kf(x) \qquad (3.17)$$

$$df/f = k\,dx \qquad (3.18)$$

$$\ln f = kx + \text{constant}$$

$$f = e^{\text{constant}}e^{kx}$$

$$f = ce^{kx} \qquad (3.19)$$

The eigenfunctions of d/dx are given by (3.19). The eigenvalues are k, which can be

---

[1] As part of the definition, we shall require that $f(x)$ is not identically zero. By this we mean that although $f(x)$ may vanish at various points, it is not everywhere zero.

any number whatever and (3.17) will still be satisfied. The eigenfunctions contain an arbitrary multiplicative constant c. This is true for the eigenfunctions of any linear operator, as is easily shown: Using (3.12) and (3.16), we have

$$\hat{A}(cf) = c\hat{A}f = ckf = k(cf) \tag{3.20}$$

Thus if $f(x)$ is an eigenfunction of $\hat{A}$ with eigenvalue k, so is $cf(x)$. With each different value of k in (3.19), we get a different eigenfunction; eigenfunctions with the same value of k, but different values of c, are not independent of each other. One final point: Suppose we impose the boundary condition that the solutions of (3.17) remain finite as x goes to plus or minus infinity. Since k can be complex, we write it as $k = a + ib$, where a and b are real numbers. We then have $f(x) = ce^{ax}e^{ibx}$. The factor e^{ax} goes to infinity as x goes to infinity if a is positive; it goes to infinity as x goes to minus infinity if a is negative. Hence the boundary conditions require that $a = 0$, and we have pure imaginary eigenvalues, $k = ib$.

3.3   OPERATORS AND QUANTUM MECHANICS

We now consider the relationship between operators and quantum mechanics. Comparing Eq. (3.1) with (3.16), we see that the Schroedinger equation is an eigenvalue problem. The values of the energy E are the eigenvalues; the eigenfunctions are the wave functions ψ; the operator whose eigenfunctions and eigenvalues are desired is $-(\hbar^2/2m)\,d^2/dx^2 + V(x)$. This operator is called the *Hamiltonian operator* for the system.

Sir William Rowan Hamilton (1805–1865) devised an alternative form of Newton's equations of motion involving a function H, the Hamiltonian function for the system. For a system where the potential energy is a function of the coordinates only, the total energy remains constant with time, i.e., E is conserved. We shall restrict ourselves to such conservative systems. For conservative systems the classical-mechanical Hamiltonian function turns out to be simply the total energy expressed in terms of coordinates and conjugate momenta. For Cartesian coordinates x, y, z, the conjugate momenta are the components of linear momentum in the x, y, and z directions: p_x, p_y, and p_z.

Let us consider what the classical-mechanical Hamiltonian function is for a particle of mass m moving in one dimension, and subject to a potential energy $V(x)$. The Hamiltonian function is equal to the energy, which is composed of kinetic and potential energies. The familiar form of the kinetic energy, $\frac{1}{2}mv_x^2$, will not do, however, since we must express the Hamiltonian as a function of coordinates and momenta, not velocities. Since $v_x = p_x/m$, the form of the kinetic energy we want is $p_x^2/2m$. The Hamiltonian function is

$$H = \frac{p_x^2}{2m} + V(x) \tag{3.21}$$

The time-independent Schroedinger equation (3.1) indicates that corresponding to the Hamiltonian function (3.21) we have a quantum-mechanical operator

$$-\frac{\hbar^2}{2m}\frac{d^2}{dx^2} + V(x) \tag{3.22}$$

whose eigenvalues are the possible values of the system's energy. This correspondence between physical quantities in classical mechanics and operators in quantum mechanics is general. It is a fundamental postulate of quantum mechanics that to every physical property (e.g., the energy, the x coordinate, the momentum) there corresponds a quantum-mechanical operator. We further postulate that the operator corresponding to the property F is obtained by writing down the classical-mechanical expression for F as a function of Cartesian coordinates and corresponding momenta, and then making the following replacements. Each Cartesian coordinate q is replaced by the operator multiplication by that coordinate:

$$\hat{q} = q \cdot \tag{3.23}$$

Each Cartesian component of linear momentum p_q is replaced by the operator

$$\hat{p}_q = \frac{\hbar}{i}\frac{\partial}{\partial q} = -i\hbar\frac{\partial}{\partial q} \tag{3.24}$$

In (3.24), $i = \sqrt{-1}$ and $\partial/\partial q$ is the operator for the partial derivative with respect to the coordinate q. Note that $1/i = i/i^2 = i/(-1) = -i$.

Consider some examples. The operator corresponding to the x coordinate is multiplication by x:

$$\hat{x} = x \cdot \tag{3.25}$$

Also

$$\hat{y} = y \cdot \tag{3.26}$$

$$\hat{z} = z \cdot \tag{3.27}$$

The operators for the components of linear momentum are

$$\hat{p}_x = \frac{\hbar}{i}\frac{\partial}{\partial x}, \qquad \hat{p}_y = \frac{\hbar}{i}\frac{\partial}{\partial y}, \qquad \hat{p}_z = \frac{\hbar}{i}\frac{\partial}{\partial z} \tag{3.28}$$

For the operator corresponding to p_x^2, we have

$$\hat{p}_x^2 = \left(\frac{\hbar}{i}\frac{\partial}{\partial x}\right)^2 = \frac{\hbar}{i}\frac{\partial}{\partial x}\frac{\hbar}{i}\frac{\partial}{\partial x} = -\hbar^2\frac{\partial^2}{\partial x^2} \tag{3.29}$$

with similar expressions for $\hat{p}_y^2$ and $\hat{p}_z^2$.

Now consider the potential-energy and kinetic-energy operators in one dimension. Suppose we had a system with the potential-energy function $V(x) = ax^2$, where a is a constant. Replacing x with $x \cdot$, we see that the potential-energy operator is simply multiplication by ax^2:

$$\hat{V}(x) = ax^2 \cdot \tag{3.30}$$

In general, we have for any potential-energy function

$$\hat{V}(x) = V(x) \cdot \tag{3.31}$$

The classical-mechanical expression for the kinetic energy T is

$$T = p_x^2/2m \qquad (3.32)$$

Replacing p_x by the corresponding operator, we have

$$\hat{T} = -\frac{\hbar^2}{2m}\frac{\partial^2}{\partial x^2} = -\frac{\hbar^2}{2m}\frac{d^2}{dx^2} \qquad (3.33)$$

where (3.29) has been used, and the partial derivative becomes an ordinary derivative in one dimension. The classical-mechanical Hamiltonian (3.21) is

$$H = T + V \qquad (3.34)$$

The corresponding quantum-mechanical Hamiltonian (or energy) operator is

$$\hat{H} = \hat{T} + \hat{V} = -\frac{\hbar^2}{2m}\frac{d^2}{dx^2} + V(x) \qquad (3.35)$$

which agrees with (3.22). Note that all these operators are linear.

How are the quantum-mechanical operators related to the corresponding properties of a system? Each such operator has its own set of eigenfunctions and eigenvalues. If φ_i is an eigenfunction of $\hat{F}$ with eigenvalue a_i, we have [Eq. (3.16)]

$$\hat{F}\varphi_i = a_i\varphi_i \qquad (3.36)$$

(The subscript indicates that we have a whole set of possible eigenfunctions and eigenvalues: $i = 1, 2, 3, \ldots$.) The operator $\hat{F}$ is usually a differential operator, and (3.36) is a differential equation whose solutions give the eigenfunctions and eigenvalues. It is postulated that the eigenvalues a_i are the only values of the property F that the system can have; i.e., *a measurement of property F will necessarily yield one of the values a_i.* For example, the only values possible for the energy of a system are the eigenvalues of the energy (Hamiltonian) operator $\hat{H}$. Using ψ for the eigenfunctions of $\hat{H}$, we have for the eigenvalue equation (3.36)

$$\hat{H}\psi_i = E_i\psi_i \qquad (3.37)$$

Using the Hamiltonian (3.35), we obtain for a one-dimensional, one-particle system

$$\left[-\frac{\hbar^2}{2m}\frac{d^2}{dx^2} + V(x)\right]\psi_i = E_i\psi_i \qquad (3.38)$$

which is the time-independent Schroedinger equation, (3.1). Thus our postulates about operators are consistent with our previous work. We will later provide further justification of the choice (3.24) for the momentum operator, by showing that in the limiting transition to classical mechanics this choice yields $p_x = m(dx/dt)$, as it should. (See Problem 7.19.)

In Chapter 1 we postulated that the state of a quantum-mechanical system is specified by a state function $\Psi(x, t)$, which contains all the information we can know about the system. How does Ψ give us information about the property F? We postulate that *if Ψ is an eigenfunction of $\hat{F}$ with eigenvalue a_i, then a measurement of F is certain to yield the value a_i.* Consider, for example, the energy. The eigenfunctions of the energy operator are the solutions $\psi(x)$ of the time-independent Schroedinger equation (3.38). Suppose the system is in a stationary state with state function [Eq. (1.17)]

$$\Psi(x, t) = e^{-iEt/\hbar}\psi(x) \tag{3.39}$$

Is $\Psi(x, t)$ an eigenfunction of the energy operator $\hat{H}$? We have

$$\hat{H}\Psi(x, t) = \hat{H}e^{-iEt/\hbar}\psi(x) \tag{3.40}$$

$\hat{H}$ contains no derivatives with respect to time, and therefore does not affect the exponential factor in (3.40). We have

$$\hat{H}\Psi(x, t) = e^{-iEt/\hbar}\hat{H}\psi(x) = Ee^{-iEt/\hbar}\psi(x) = E\Psi(x, t)$$

$$\hat{H}\Psi = E\Psi \tag{3.41}$$

where (3.37) was used. Hence, for a stationary state, $\Psi(x, t)$ is an eigenfunction of $\hat{H}$, and we are certain to obtain the value E when we measure the energy.

As an example of another property, consider momentum. The eigenfunctions φ of $\hat{p}_x$ are found by solving

$$\hat{p}_x\varphi = k\varphi$$

$$\frac{\hbar}{i}\frac{d\varphi}{dx} = k\varphi \tag{3.42}$$

We find (Problem 3.25)

$$\varphi = Ae^{ikx/\hbar} \tag{3.43}$$

where A is an arbitrary constant. To keep φ finite for large $|x|$, the eigenvalues k must be real. Thus the eigenvalues of $\hat{p}_x$ are all the real numbers

$$-\infty < k < \infty \tag{3.44}$$

which is reasonable. It might seem surprising that the operator for the physical property momentum involves the imaginary number i. Actually, the presence of i in $\hat{p}_x$ ensures that the eigenvalues k are real; recall that the eigenvalues of d/dx are imaginary (Section 3.2).

Now consider the momentum of a particle in a box. The state function for a particle in a stationary state in a one-dimensional box is [Eqs. (3.39), (2.20), and (2.23)]

$$\Psi(x, t) = e^{-iEt/\hbar}\left(\frac{2}{l}\right)^{1/2}\sin\left(\frac{n\pi x}{l}\right) \tag{3.45}$$

$$E = n^2h^2/8ml^2 \tag{3.46}$$

Do we have a definite value of p_x; i.e., is $\Psi(x, t)$ an eigenfunction of $\hat{p}_x$? Looking at the eigenfunctions of $\hat{p}_x$, we see that there is no numerical value of the real constant k that will make the exponential function in (3.43) become a sine function, as in (3.45). Hence Ψ is not an eigenfunction of $\hat{p}_x$. We can verify this directly; we have

$$\hat{p}_x\Psi = \frac{\hbar}{i}\frac{\partial}{\partial x}e^{-iEt/\hbar}\left(\frac{2}{l}\right)^{1/2}\sin\left(\frac{n\pi x}{l}\right) = \frac{n\pi\hbar}{il}e^{-iEt/\hbar}\left(\frac{2}{l}\right)^{1/2}\cos\left(\frac{n\pi x}{l}\right)$$

Since $\hat{p}_x\Psi \neq \text{constant} \cdot \Psi$, the state function Ψ is not an eigenfunction of $\hat{p}_x$. How about p_x^2? We have [Eq. (3.29)]

$$\hat{p}_x^2 \Psi = -\hbar^2 \frac{\partial^2}{\partial x^2} e^{-iEt/\hbar} \left(\frac{2}{l}\right)^{1/2} \sin\left(\frac{n\pi x}{l}\right) = \frac{n^2\pi^2\hbar^2}{l^2} e^{-iEt/\hbar} \left(\frac{2}{l}\right)^{1/2} \sin\left(\frac{n\pi x}{l}\right)$$

$$\hat{p}_x^2 \Psi = \frac{n^2 h^2}{4l^2} \Psi \tag{3.47}$$

Hence a measurement of p_x^2 will always give the result $n^2h^2/4l^2$ when the particle is in the stationary state with quantum number n. This should come as no surprise: The potential energy in the box is zero, so that the Hamiltonian is

$$\hat{H} = \hat{T} + \hat{V} = \hat{T} = \hat{p}_x^2/2m \tag{3.48}$$

We then have [Eq. (3.41)]

$$\hat{H}\Psi = E\Psi = \frac{\hat{p}_x^2}{2m}\Psi$$

$$\hat{p}_x^2\Psi = 2mE\Psi = 2m\frac{n^2h^2}{8ml^2}\Psi = \frac{n^2h^2}{4l^2}\Psi \tag{3.49}$$

in agreement with (3.47). The only possible value for p_x^2 is

$$p_x^2 = n^2h^2/4l^2 \tag{3.50}$$

Equation (3.50) suggests that a measurement of p_x would necessarily yield one of the two values $\pm\frac{1}{2}nh/l$, corresponding to the particle moving to the right or to the left in the box. This plausible suggestion is not accurate. An analysis using the methods of Chapter 7 shows that there is a high probability that the measured value will be close to one of the two values $\pm\frac{1}{2}nh/l$, but that any value consistent with (3.44) can result from a measurement of p_x for the particle in a box; see Problem 7.30.

Comparing the free-particle wave function (2.30) with the eigenfunctions (3.43) of $\hat{p}_x$, we note the following physical interpretation: The first term in (2.30) corresponds to positive momentum and represents motion in the $+x$ direction, while the second term in (2.30) corresponds to negative momentum and represents motion in the $-x$ direction.

We postulated that a measurement of the property F can only give a result that is one of the eigenvalues of the operator $\hat{F}$. If the state function Ψ happens to be an eigenfunction of $\hat{F}$ with eigenvalue a, we are certain to get a in a measurement of F. However, suppose that Ψ is not one of the eigenfunctions of $\hat{F}$; what then? We still assert that *we will get one of the eigenvalues of $\hat{F}$ when we measure F, but we cannot say for certain which eigenvalue will be obtained.* We will later see (Chapter 7) that the *probabilities* for obtaining each of the possible eigenvalues of $\hat{F}$ can be predicted.

3.4   THE THREE-DIMENSIONAL MANY-PARTICLE SCHROEDINGER EQUATION

Up to now we have restricted ourselves to one-dimensional, one-particle systems. The operator formalism developed in the last section enables us to extend our work to three-dimensional many-particle systems. The time-dependent Schroedinger equation for the time development of the state function is postulated to be [cf. Eq. (1.10)]

$$i\hbar \frac{\partial \Psi}{\partial t} = \hat{H}\Psi \tag{3.51}$$

while the time-independent Schroedinger equation for the energy eigenfunctions and eigenvalues is

$$\hat{H}\psi = E\psi \tag{3.52}$$

For a one-particle, three-dimensional system, the classical-mechanical Hamiltonian is

$$H = T + V = \frac{1}{2m}(p_x^2 + p_y^2 + p_z^2) + V(x, y, z) \tag{3.53}$$

Introducing the quantum-mechanical operators [Eq. (3.29)], we have for the Hamiltonian operator

$$\hat{H} = -\frac{\hbar^2}{2m}\left(\frac{\partial^2}{\partial x^2} + \frac{\partial^2}{\partial y^2} + \frac{\partial^2}{\partial z^2}\right) + V(x, y, z) \tag{3.54}$$

The operator in parentheses in (3.54) is called the *Laplacian operator* ∇^2 (read as "del squared"):

$$\nabla^2 \equiv \frac{\partial^2}{\partial x^2} + \frac{\partial^2}{\partial y^2} + \frac{\partial^2}{\partial z^2} \tag{3.55}$$

The one-particle, three-dimensional time-independent Schroedinger equation is then

$$-\frac{\hbar^2}{2m}\nabla^2\psi + V\psi = E\psi \tag{3.56}$$

Now consider a three-dimensional system with n particles. Let particle i have mass m_i and coordinates (x_i, y_i, z_i), where $i = 1, 2, 3, \ldots, n$. The kinetic energy is the sum of the kinetic energies of the individual particles:

$$T = \frac{1}{2m_1}(p_{x_1}^2 + p_{y_1}^2 + p_{z_1}^2) + \frac{1}{2m_2}(p_{x_2}^2 + p_{y_2}^2 + p_{z_2}^2) + \cdots + \frac{1}{2m_n}(p_{x_n}^2 + p_{y_n}^2 + p_{z_n}^2) \tag{3.57}$$

where p_{x_i} is the x-component of the linear momentum of particle i, etc. The kinetic-energy operator is

$$\hat{T} = -\frac{\hbar^2}{2m_1}\left(\frac{\partial^2}{\partial x_1^2} + \frac{\partial^2}{\partial y_1^2} + \frac{\partial^2}{\partial z_1^2}\right) - \cdots - \frac{\hbar^2}{2m_n}\left(\frac{\partial^2}{\partial x_n^2} + \frac{\partial^2}{\partial y_n^2} + \frac{\partial^2}{\partial z_n^2}\right) \tag{3.58}$$

$$\hat{T} = -\sum_{i=1}^{n}\frac{\hbar^2}{2m_i}\nabla_i^2 \tag{3.59}$$

$$\nabla_i^2 \equiv \frac{\partial^2}{\partial x_i^2} + \frac{\partial^2}{\partial y_i^2} + \frac{\partial^2}{\partial z_i^2} \tag{3.60}$$

We will restrict ourselves to cases where the potential energy depends only on the $3n$ coordinates:

$$V = V(x_1, y_1, z_1, \ldots, x_n, y_n, z_n) \tag{3.61}$$

The Hamiltonian operator for an n-particle three-dimensional system is then

$$\hat{H} = -\sum_{i=1}^{n} \frac{\hbar^2}{2m_i} \nabla_i^2 + V(x_1, \ldots, z_n) \tag{3.62}$$

and the time-independent Schroedinger equation is

$$\left[-\sum_{i=1}^{n} \frac{\hbar^2}{2m_i} \nabla_i^2 + V(x_1, \ldots, z_n) \right] \psi = E\psi \tag{3.63}$$

where the time-independent wave function is a function of the $3n$ coordinates of the n particles:

$$\psi = \psi(x_1, y_1, z_1, \ldots, x_n, y_n, z_n) \tag{3.64}$$

The Schroedinger equation (3.63) is a linear partial differential equation.

As an example, consider a system of two particles interacting so that the potential energy is inversely proportional to the distance between them, with c being the proportionality constant. The Schroedinger equation (3.63) becomes

$$\left[-\frac{\hbar^2}{2m_1} \left(\frac{\partial^2}{\partial x_1^2} + \frac{\partial^2}{\partial y_1^2} + \frac{\partial^2}{\partial z_1^2} \right) - \frac{\hbar^2}{2m_2} \left(\frac{\partial^2}{\partial x_2^2} + \frac{\partial^2}{\partial y_2^2} + \frac{\partial^2}{\partial z_2^2} \right) \right.$$

$$\left. + \frac{c}{[(x_1 - x_2)^2 + (y_1 - y_2)^2 + (z_1 - z_2)^2]^{1/2}} \right] \psi = E\psi \tag{3.65}$$

$$\psi = \psi(x_1, y_1, z_1, x_2, y_2, z_2) \tag{3.66}$$

Although (3.65) looks formidable, we will solve it in Chapter 6.

For a one-particle, one-dimensional system, the Born postulate [Eq. (1.12)] states that $|\Psi(x, t)|^2\, dx$ is the probability of observing the particle between x and $x + dx$ at time t. We extend this postulate as follows. For a three-dimensional, one-particle system, the quantity

$$|\Psi(x, y, z, t)|^2\, dx\, dy\, dz \tag{3.67}$$

is the probability of finding the particle in the infinitesimal region of space with the x coordinate lying between x and $x + dx$, the y coordinate lying between y and $y + dy$, and the z coordinate lying between z and $z + dz$. Since the total probability of finding the particle is 1, the normalization condition is

$$\int_{-\infty}^{\infty} \int_{-\infty}^{\infty} \int_{-\infty}^{\infty} |\Psi|^2\, dx\, dy\, dz = 1 \tag{3.68}$$

For a three-dimensional n-particle system, we postulate that

$$|\Psi(x_1, y_1, z_1, x_2, y_2, z_2, \ldots, x_n, y_n, z_n, t)|^2\, dx_1\, dy_1\, dz_1\, dx_2\, dy_2\, dz_2 \cdots dx_n\, dy_n\, dz_n \tag{3.69}$$

is the probability at time t of simultaneously finding particle 1 in the infinitesimal rectangular box at (x_1, y_1, z_1) with edges dx_1, dy_1, dz_1, particle 2 in the infinitesimal box at (x_2, y_2, z_2) with edges $dx_2, dy_2, dz_2, \ldots$, particle n in the infinitesimal box at (x_n, y_n, z_n) with edges dx_n, dy_n, dz_n. The total probability of finding all the particles is 1, and the normalization condition is

$$\int_{-\infty}^{\infty} \int_{-\infty}^{\infty} \int_{-\infty}^{\infty} \cdots \int_{-\infty}^{\infty} \int_{-\infty}^{\infty} \int_{-\infty}^{\infty} |\Psi|^2 \, dx_1 \, dy_1 \, dz_1 \cdots dx_n \, dy_n \, dz_n = 1 \quad (3.70)$$

It is customary in quantum mechanics to denote integration over the full range of all the coordinates of a system by $\int dq$ or $\int d\tau$. A shorthand way of writing (3.68) or (3.70) is

$$\int |\Psi|^2 \, d\tau = 1 \quad (3.71)$$

Although (3.71) may look like an indefinite integral, it is understood to be a definite integral; the integration variables and their ranges are understood from the context in which the integral appears.

Since $|\Psi|^2 = |\psi|^2$ for a stationary state, we have for a stationary state

$$\int |\psi|^2 \, d\tau = 1 \quad (3.72)$$

3.5   THE PARTICLE IN A THREE-DIMENSIONAL BOX

For the present, we confine ourselves to one-particle problems. In this section we consider the three-dimensional case of the problem solved in Section 2.2, the particle in a box.

There are many possible shapes for a three-dimensional box. The box we consider is a rectangular parallelepiped with edges of length a, b, and c. We choose our coordinate system so that one corner of the box lies at the origin and the box lies in the first octant of space. Within the box, the potential energy is zero; outside the box, it is infinite:

$$V(x, y, z) = 0 \quad \text{in the region} \quad \begin{cases} 0 < x < a \\ 0 < y < b \\ 0 < z < c \end{cases} \quad (3.73)$$

$$V = \infty \quad \text{elsewhere}$$

We conclude at once that the wave function is zero outside the box. Within the box, we have zero for the potential-energy operator, and the Schroedinger equation is [Eq. (3.56)]

$$-\frac{\hbar^2}{2m} \left(\frac{\partial^2 \psi}{\partial x^2} + \frac{\partial^2 \psi}{\partial y^2} + \frac{\partial^2 \psi}{\partial z^2} \right) = E\psi \quad (3.74)$$

To solve (3.74) we start by assuming that the solution can be written as the product of a function of x alone times a function of y alone times a function of z alone:

$$\psi(x, y, z) = f(x) g(y) h(z) \quad (3.75)$$

It might seem that this assumption throws away solutions which are not of the form (3.75). However, it can be shown mathematically that, if we can find solutions of the form (3.75) that satisfy the boundary conditions, then there are no other solutions of the Schroedinger equation that will satisfy the boundary conditions. The method we are using to solve (3.74) is called, for obvious reasons, *separation of variables*.

From (3.75) we find

$$\frac{\partial^2 \psi}{\partial x^2} = f''(x)g(y)h(z), \qquad \frac{\partial^2 \psi}{\partial y^2} = f(x)g''(y)h(z), \qquad \frac{\partial^2 \psi}{\partial z^2} = f(x)g(y)h''(z)$$

(3.76)

We then substitute (3.75) and (3.76) into (3.74) to obtain

$$f''gh + fg''h + fgh'' + 2mh^{-2}Efgh = 0 \tag{3.77}$$

Dividing (3.77) by fgh, we have

$$\frac{f''}{f} + \frac{g''}{g} + \frac{h''}{h} + \frac{2m}{\hbar^2}E = 0 \tag{3.78}$$

$$\frac{f''(x)}{f(x)} = -\frac{g''(y)}{g(y)} - \frac{h''(z)}{h(z)} - \frac{2mE}{\hbar^2} \tag{3.79}$$

Let $k_x \equiv f''/f$. Since k_x equals $f''(x)/f(x)$, it must be independent of y and z. Equation (3.79) shows that k_x equals $-g''(y)/g(y) - h''(z)/h(z) - 2mE/\hbar^2$, and hence k_x must be independent of x. Being independent of x, y, and z, the quantity k_x must be a constant, and we have

$$f''(x)/f(x) = k_x \tag{3.80}$$

Since x, y, and z occur symmetrically in (3.78), we can use the same argument to conclude that

$$g''(y)/g(y) = k_y, \qquad h''(z)/h(z) = k_z \tag{3.81}$$

where k_y and k_z are two more constants. What do we know about the k's? If we substitute (3.80) and (3.81) into (3.78), we find

$$k_x + k_y + k_z = -2mE\hbar^{-2} \tag{3.82}$$

Looking at (3.82), we decide to change the form of our constants by defining three new constants, as follows:

$$k_x = -2mE_x\hbar^{-2}, \qquad k_y = -2mE_y\hbar^{-2}, \qquad k_z = -2mE_z\hbar^{-2} \tag{3.83}$$

where E_x, E_y, and E_z are constants having the dimensions of energy. Equation (3.82) becomes

$$E_x + E_y + E_z = E \tag{3.84}$$

Using (3.83) in (3.80)–(3.81), we have

$$\frac{d^2f(x)}{dx^2} + \frac{2m}{\hbar^2}E_xf(x) = 0 \tag{3.85}$$

$$\frac{d^2g(y)}{dy^2} + \frac{2m}{\hbar^2} E_y g(y) = 0 \tag{3.86}$$

$$\frac{d^2h(z)}{dz^2} + \frac{2m}{\hbar^2} E_z h(z) = 0 \tag{3.87}$$

We have converted the partial differential equation in three variables into three ordinary differential equations. What are the boundary conditions on (3.85)? Since the wave function vanishes outside the box, continuity of ψ requires that it vanish on the walls of the box. In particular, ψ must be zero on the wall of the box lying in the yz plane, where $x = 0$, and it must be zero on the parallel wall of the box, where $x = a$. Therefore

$$f(0) = 0, \qquad f(a) = 0 \tag{3.88}$$

Now compare Eq. (3.85) with Eq. (2.10) in Section 2.2, dealing with the particle in a one-dimensional box. The equations are the same in form, with E_x in (3.85) corresponding to E in (2.10). Are the boundary conditions the same? Yes, except that we have $x = a$ instead of $x = l$ as the second point where the independent variable vanishes. Thus we can use the work in Section 2.2 to write down as the solution [see Eqs. (2.23) and (2.20)]

$$f(x) = \left(\frac{2}{a}\right)^{1/2} \sin\left(\frac{n_x \pi x}{a}\right) \tag{3.89}$$

$$E_x = \frac{n_x^2 h^2}{8ma^2}, \qquad n_x = 1, 2, 3, \ldots \tag{3.90}$$

The same reasoning applied to the y and z equations gives

$$g(y) = \left(\frac{2}{b}\right)^{1/2} \sin\left(\frac{n_y \pi y}{b}\right), \qquad h(z) = \left(\frac{2}{c}\right)^{1/2} \sin\left(\frac{n_z \pi z}{c}\right) \tag{3.91}$$

$$E_y = \frac{n_y^2 h^2}{8mb^2}, \qquad n_y = 1, 2, 3, \ldots \tag{3.92}$$

$$E_z = \frac{n_z^2 h^2}{8mc^2}, \qquad n_z = 1, 2, 3, \ldots \tag{3.93}$$

From (3.84) we have for the energy

$$E = \frac{h^2}{8m}\left(\frac{n_x^2}{a^2} + \frac{n_y^2}{b^2} + \frac{n_z^2}{c^2}\right) \tag{3.94}$$

From (3.75) we have for the wave function in the box

$$\psi(x, y, z) = \left(\frac{8}{abc}\right)^{1/2} \sin\left(\frac{n_x \pi x}{a}\right) \sin\left(\frac{n_y \pi y}{b}\right) \sin\left(\frac{n_z \pi z}{c}\right) \tag{3.95}$$

There are three quantum numbers, n_x, n_y, n_z; we can attribute this to the three-dimensional nature of the problem. The three quantum numbers vary independently of one another.

Since the x, y, and z factors in the wave function are each independently normalized, the wave function is normalized:

$$\int_{-\infty}^{\infty}\int_{-\infty}^{\infty}\int_{-\infty}^{\infty} |\psi|^2 \, dx \, dy \, dz = \int_0^a |f(x)|^2 \, dx \int_0^b |g(y)|^2 \, dy \int_0^c |h(z)|^2 \, dz = 1$$

Suppose that $a = b = c$; we then have a cube. The energy levels are then

$$E = (h^2/8ma^2)(n_x^2 + n_y^2 + n_z^2) \tag{3.96}$$

Let us tabulate some of the allowed energies of a particle confined to a cube with infinitely strong walls:

$n_x n_y n_z$	111	211	121	112	122	212	221	113	131	311	222	
$E(8ma^2/h^2)$	3	6	6	6	9	9	9	11	11	11	12	(3.97)

Observe that states with different quantum numbers may have the same energy. Thus the states with the quantum numbers $(2, 1, 1)$, $(1, 2, 1)$, and $(1, 1, 2)$ all have the same energy. However, referring to Eq. (3.95), we see that these three sets of quantum numbers give three different, independent wave functions and therefore do represent different states of the system. When two or more independent wave functions correspond to states with the same energy eigenvalue, we have *degeneracy*. The *degree* of degeneracy of an energy level is the number of states that have that energy. Thus the second-lowest energy level of the particle in a cube is threefold degenerate. We got the degeneracy when we made the edges of the box equal; degeneracy is generally related to the *symmetry* of the system. Usually one does not find degeneracy in one-dimensional problems.

[In the statistical-mechanical evaluation of the molecular partition function of an ideal gas, the translational energy levels of each gas molecule are taken to be the levels of a particle in a three-dimensional rectangular box; see *Levine, Physical Chemistry*, Sections 22.6 and 22.7.]

3.6   DEGENERACY

Let us prove an important theorem about the wave functions of an n-fold degenerate energy level. We have n independent wave functions $\psi_1, \psi_2, \ldots, \psi_n$. Let W be the energy of the degenerate level:

$$\hat{H}\psi_1 = W\psi_1, \hat{H}\psi_2 = W\psi_2, \ldots, \hat{H}\psi_n = W\psi_n \tag{3.98}$$

We wish to prove that any linear combination

$$\varphi \equiv c_1\psi_1 + c_2\psi_2 + \cdots + c_n\psi_n \tag{3.99}$$

of the n wave functions of the *degenerate* level is an eigenfunction of the Hamiltonian with eigenvalue W; we must show $\hat{H}\varphi = W\varphi$ or

$$\hat{H}[c_1\psi_1 + c_2\psi_2 + \cdots + c_n\psi_n] = W[c_1\psi_1 + c_2\psi_2 + \cdots + c_n\psi_n] \tag{3.100}$$

Since the Hamiltonian is a linear operator, we can apply Eq. (3.11) $n-1$ times to the left side of (3.100), obtaining

$$\hat{H}[c_1\psi_1 + c_2\psi_2 + \cdots + c_n\psi_n] = \hat{H}c_1\psi_1 + \hat{H}c_2\psi_2 + \cdots + \hat{H}c_n\psi_n \quad \textbf{(3.101)}$$

Using Eqs. (3.12) and (3.98), we have

$$\hat{H}[c_1\psi_1 + c_2\psi_2 + \cdots + c_n\psi_n] = c_1\hat{H}\psi_1 + c_2\hat{H}\psi_2 + \cdots + c_n\hat{H}\psi_n$$

$$= c_1 W\psi_1 + c_2 W\psi_2 + \cdots + c_n W\psi_n$$

$$\hat{H}[c_1\psi_1 + c_2\psi_2 + \cdots + c_n\psi_n] = W[c_1\psi_1 + c_2\psi_2 + \cdots + c_n\psi_n] \quad \textbf{(3.102)}$$

which completes the proof.

Since any linear combination of the wave functions corresponding to a degenerate energy level is an eigenfunction of $\hat{H}$ with the same eigenvalue, we can construct an infinite number of different wave functions for any degenerate energy level. Actually, we are only interested in eigenfunctions that are linearly independent. The n functions $f_1, \ldots, f_n$ are said to be *linearly independent* if the equation $c_1f_1 + \cdots + c_nf_n = 0$ can only be satisfied with all the constants $c_1, \ldots, c_n$ equal to zero. This means that no member of the set of functions can be expressed as a linear combination of the remaining members. For example, the functions $f_1 = 3x, f_2 = 5x^2 - x, f_3 = x^2$ are not linearly independent, since $f_2 = 5f_3 - \frac{1}{3}f_1$. The functions $g_1 = 1, g_2 = x, g_3 = x^2$ are linearly independent, since none of them can be written as a linear combination of the other two. The *degree of degeneracy* of an energy level is equal to the number of linearly independent wave functions corresponding to that value of the energy.

3.7 AVERAGE VALUES

Consider the physical property F. In Section 3.3 we pointed out that when the state function Ψ is not an eigenfunction of the operator $\hat{F}$, a measurement of F will give one of a number of possible values. We now consider the average value of the property F for a system whose state is Ψ.

To determine the average value of F experimentally, we do the following. (Recall the discussion in Section 1.4.) We take a large number of separate systems, each in the same state Ψ, and we make a measurement of F in each system. The *average value* of F is defined as the arithmetic mean of the observed values. Let $f_1, f_2, \ldots, f_N$ be the observed values of F. The symbol for the average value is $\langle F \rangle$ or $\bar{F}$, and we have

$$\langle F \rangle = \bar{F} = \frac{\sum_{i=1}^{N} f_i}{N} \quad \textbf{(3.103)}$$

where N (the number of systems) is very large.

Instead of summing over the observed values of F, we can sum over all possible values of F, multiplying each possible value by the number of times it is observed, to obtain the equivalent expression

$$\langle F \rangle = \frac{\sum_f n_f f}{N} \quad \textbf{(3.104)}$$

where n_f is the number of times the value f is observed. An example will make this clear. Suppose a class of 9 students takes a quiz that has 5 questions, and that the students receive these grades: 0, 20, 20, 60, 60, 80, 80, 80, 100. Calculating the average grade according to (3.103), we have

$$\frac{1}{N} \sum_{i=1}^{N} f_i = \frac{0 + 20 + 20 + 60 + 60 + 80 + 80 + 80 + 100}{9} = 56 \qquad \textbf{(3.105)}$$

To calculate the average grade according to (3.104), we sum over the possible grades: 0, 20, 40, 60, 80, 100. We have

$$\frac{1}{N} \sum_{f} n_f f = \frac{1(0) + 2(20) + 0(40) + 2(60) + 3(80) + 1(100)}{9} = 56 \qquad \textbf{(3.106)}$$

We can write (3.104) as

$$\langle F \rangle = \sum_{f} \left(\frac{n_f}{N} \right) f \qquad \textbf{(3.107)}$$

Since N is very large, n_f / N is the probability of observing the value f (see Section 1.6); denoting this probability by P_f, we have

$$\langle F \rangle = \sum_{f} P_f f \qquad \textbf{(3.108)}$$

Now consider the average value of the x coordinate for a one-particle one-dimensional system in the state $\Psi(x, t)$. x takes on a continuous range of values, and the probability of observing the particle between x and $x + dx$ is $|\Psi|^2\, dx$. The summation over the infinitesimal probabilities is equivalent to an integration, so that (3.108) becomes

$$\langle x \rangle = \int_{-\infty}^{\infty} x |\Psi(x, t)|^2\, dx \qquad \textbf{(3.109)}$$

For the one-particle three-dimensional case, the probability of finding the particle in the volume element at point (x, y, z) with edges dx, dy, dz is

$$|\Psi(x, y, z, t)|^2\, dx\, dy\, dz \qquad \textbf{(3.110)}$$

If we want the probability that the particle is between x and $x + dx$, we must integrate (3.110) over all possible values of y and z, since the particle can have any values for its y and z coordinates while its x coordinate lies between x and $x + dx$. Hence in the three-dimensional case, (3.109) becomes

$$\langle x \rangle = \int_{-\infty}^{\infty} \left[\int_{-\infty}^{\infty} \int_{-\infty}^{\infty} |\Psi(x, y, z, t)|^2\, dy\, dz \right] x\, dx$$

$$\langle x \rangle = \int_{-\infty}^{\infty} \int_{-\infty}^{\infty} \int_{-\infty}^{\infty} |\Psi(x, y, z, t)|^2\, x\, dx\, dy\, dz \qquad \textbf{(3.111)}$$

Now consider the average value of some physical property $F(x, y, z)$ that is a function of the particle's coordinates. An example is the potential energy $V(x, y, z)$.

Reasoning along the same lines that gave us (3.111), we have for the average value of F

$$\langle F(x, y, z) \rangle = \int_{-\infty}^{\infty} \int_{-\infty}^{\infty} \int_{-\infty}^{\infty} |\Psi(x, y, z, t)|^2 F(x, y, z)\, dx\, dy\, dz \quad \textbf{(3.112)}$$

$$\langle F(x, y, z) \rangle = \int_{-\infty}^{\infty} \int_{-\infty}^{\infty} \int_{-\infty}^{\infty} \Psi^* F \Psi\, dx\, dy\, dz \qquad \textbf{(3.113)}$$

The form (3.113) might seem like a bit of whimsy, as it is no different from (3.112). In a moment we will see its significance.

In general, the property F depends on both coordinates *and* momenta:

$$F = F(x, y, z, p_x, p_y, p_z) \qquad \textbf{(3.114)}$$

for the one-particle three-dimensional case. How do we find the average value of F? We *postulate* that

$$\langle F \rangle = \int_{-\infty}^{\infty} \int_{-\infty}^{\infty} \int_{-\infty}^{\infty} \Psi^* F\left(x, y, z, \frac{\hbar}{i}\frac{\partial}{\partial x}, \frac{\hbar}{i}\frac{\partial}{\partial y}, \frac{\hbar}{i}\frac{\partial}{\partial z}\right) \Psi\, dx\, dy\, dz \quad \textbf{(3.115)}$$

$$\langle F \rangle = \int_{-\infty}^{\infty} \int_{-\infty}^{\infty} \int_{-\infty}^{\infty} \Psi^* \hat{F} \Psi\, dx\, dy\, dz \qquad \textbf{(3.116)}$$

where $\hat{F}$ is the quantum-mechanical operator for the property F. [Later we will provide some justification for this postulate, by using (3.116) to show that the time-dependent Schroedinger equation reduces to Newton's second law in the transition from quantum to classical mechanics; see Problem 7.19.] For the n-particle case, we postulate that

$$\langle F \rangle = \int \Psi^* \hat{F} \Psi\, d\tau \qquad \textbf{(3.117)}$$

where $\int d\tau$ indicates integration over the full range of the $3n$ coordinates. The state function in (3.117) must be normalized, since we took $\Psi^*\Psi$ as the probability density. It is important to have the operator properly sandwiched between Ψ^* and Ψ; the quantities $\hat{F}\Psi^*\Psi$ and $\Psi^*\Psi\hat{F}$ are not the same as $\Psi^*\hat{F}\Psi$, unless $\hat{F}$ is a function of coordinates only.

For a stationary state, we have

$$\Psi^* \hat{F} \Psi = e^{iEt/\hbar} \psi^* \hat{F} e^{-iEt/\hbar} \psi = e^0 \psi^* \hat{F} \psi = \psi^* \hat{F} \psi \qquad \textbf{(3.118)}$$

since $\hat{F}$ contains no time derivatives and does not affect the time factor. Hence for a stationary state,

$$\langle F \rangle = \int \psi^* \hat{F} \psi\, d\tau \qquad \textbf{(3.119)}$$

Consider the special case where Ψ is an eigenfunction of $\hat{F}$: $\hat{F}\Psi = k\Psi$. Equation (3.117) becomes

$$\langle F \rangle = \int \Psi^* \hat{F} \Psi\, d\tau = \int \Psi^* k \Psi\, d\tau = k \int \Psi^* \Psi\, d\tau = k \qquad \textbf{(3.120)}$$

since Ψ is normalized. The result (3.120) is reasonable, since k is the only possible value we can find for F when we make a measurement.

From Eq. (3.117) it readily follows that the average value of a sum is the sum of the average values:

$$\langle F+G \rangle = \langle F \rangle + \langle G \rangle \qquad (3.121)$$

where F and G are any two properties. However, it is not necessarily true that the average value of a product is the product of the average values:

$$\langle FG \rangle \neq \langle F \rangle \langle G \rangle \qquad (3.122)$$

The term *expectation value* is often used instead of average value. The expectation value is not necessarily one of the possible values we might observe; e.g., American families average 2.1 children.

Consider some average values for the particle in a three-dimensional box. For the average value of the x coordinate, Eqs. (3.119) and (3.75) give

$$\langle x \rangle = \int \int \int f^*g^*h^* xfgh \, dx \, dy \, dz \qquad (3.123)$$

$$\langle x \rangle = \int_0^a x|f(x)|^2 \, dx \int_0^b |g(y)|^2 \, dy \int_0^c |h(z)|^2 \, dz = \int_0^a x|f(x)|^2 \, dx$$

since $g(y)$ and $h(z)$ are separately normalized. For the lowest energy state, Eq. (3.89) gives

$$\langle x \rangle = \frac{2}{a} \int_0^a x \sin^2 \left(\frac{\pi x}{a} \right) dx = \frac{a}{2} \qquad (3.124)$$

where we used (2.22) to evaluate the integral. A glance at Fig. 2.3 shows that (3.124) is reasonable. For the average value of p_x, we have

$$\langle p_x \rangle = \int \int \int f^*g^*h^* \frac{\hbar}{i} \frac{\partial}{\partial x} [f(x)g(y)h(z)] \, dx \, dy \, dz$$

$$\langle p_x \rangle = \frac{\hbar}{i} \int_0^a f^*(x)f'(x) \, dx \int_0^b |g(y)|^2 \, dy \int_0^c |h(z)|^2 \, dz$$

$$\langle p_x \rangle = \frac{\hbar}{i} \int_0^a f(x)f'(x) \, dx = \frac{\hbar}{2i} f^2(x) \Big|_0^a = 0 \qquad (3.125)$$

where (3.88) was used. The result is reasonable, since the particle is just as likely to be headed in the $+x$ direction as in the $-x$ direction.

3.8   REQUIREMENTS FOR AN ACCEPTABLE WAVE FUNCTION

In solving the particle in a box, we required ψ to be continuous. We now discuss other requirements the wave function must satisfy.

Since $\psi^*\psi \, dq$ is a probability, we want to be able to normalize the wave function by choosing a suitable *normalization constant* as a multiplier of the wave function. However, we can do this only if the integral over all space $\int \psi^*\psi \, dq$ exists.

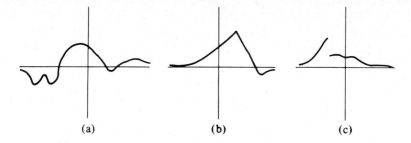

Figure 3.1 Function (a) is continuous, and its first derivative is continuous. Function (b) is continuous, but its first derivative has a discontinuity. Function (c) is discontinuous.

If this integral exists we say that ψ is *quadratically integrable*; thus we generally
demand that ψ be quadratically integrable. The important exception is a particle that
is not bound. Thus the wave functions for the unbound states of the hydrogen atom
and for a free particle are not quadratically integrable.

Since $\psi^*\psi$ is the probability density, it must be single-valued; it would be
embarrassing if our theory gave two different values for the probability of finding a
particle at a certain point. If we demand that ψ be single-valued, then surely $\psi^*\psi$ will
be single-valued. It is possible to have ψ multivalued [e.g., $\psi(q) = -1, +1, i$] and
still have $\psi^*\psi$ single-valued. We will, however, demand single-valuedness for ψ.

In addition to demanding that ψ be continuous, we generally also require that
all the partial derivatives $\partial\psi/\partial x$, $\partial\psi/\partial y$, etc., be continuous. (See Fig. 3.1.) Referring
back to Section 2.2, however, we note that for the particle in a box there is a
discontinuity in the derivative of the wave function at the walls of the box; ψ and
$d\psi/dx$ are zero everywhere outside the box, but from Eq. (2.23) we see that $d\psi/dx$
does not become zero at the walls. The discontinuity in ψ' is due to the infinite jump
in potential energy at the walls of the box. For a box with walls of finite height, ψ' is
continuous at the walls (Section 2.4).

In line with the requirement of quadratic integrability, it is sometimes stated
that we require the wave function to be finite everywhere, including infinity. However, this is usually a much stronger requirement than quadratic integrability, and, in
fact, it turns out that some of the relativistic wave functions for the hydrogen atom
are infinite at the origin, but are quadratically integrable; occasionally one
encounters nonrelativistic wave functions that are infinite at the origin. Thus the
fundamental requirement is quadratic integrability, rather than finiteness.

We require that the eigenfunctions of any operator representing a physical
quantity meet the above requirements. A function meeting these requirements is said
to be *well behaved*.

3.9   SUMMARY OF POSTULATES

We have presented the postulates of quantum mechanics in an unsystematic
way, so it might be helpful to list the postulates introduced so far.

(a) The state of a system is described by a function Ψ (the state function or wave function) of the coordinates and the time; Ψ is single-valued, continuous, and (except for unbound states) is quadratically integrable.

(b) To each physical property G of a system, there corresponds an operator $\hat{G}$. This operator is found by taking the classical-mechanical expression for the property in terms of Cartesian coordinates and momenta and replacing each coordinate x by $x \cdot$ and each momentum component p_x by $(\hbar/i)\partial/\partial x$.

(c) The only possible values that can result from measurements of the property G are the eigenvalues g_i of the equation $\hat{G}\varphi_i = g_i\varphi_i$.

(d) The average value of the property G is given by $\langle G \rangle = \int \Psi^*\hat{G}\Psi \, d\tau$, where Ψ is the system's state function.

(e) The state function of an undisturbed system changes with time according to $-(\hbar/i)(\partial\Psi/\partial t) = \hat{H}\Psi$, where $\hat{H}$ is the Hamiltonian operator (the energy operator) of the system.

(f) For a three-dimensional n-particle system, the quantity (3.69) is the probability of finding the system's particles in the infinitesimal regions of space listed after (3.69).

A revised set of postulates will be given in Chapter 7.

PROBLEMS

3.1 Let $\hat{D} = d/dx$. Verify that

$$(\hat{D} + x)(\hat{D} - x) = \hat{D}^2 - x^2 - 1$$

3.2 By repeated application of the definition of the product of two operators show that

$$[(\hat{A}\hat{B})\hat{C}]f = \hat{A}[\hat{B}(\hat{C}f)]$$
$$[\hat{A}(\hat{B}\hat{C})]f = \hat{A}[\hat{B}(\hat{C}f)]$$

thus proving associativity, Eq. (3.7).

3.3 (a) Show that $(\hat{A} + \hat{B})^2 = (\hat{B} + \hat{A})^2$, for any two operators (linear or nonlinear). (b) Under what conditions is

$$(\hat{A} + \hat{B})^2 = \hat{A}^2 + 2\hat{A}\hat{B} + \hat{B}^2$$

3.4 What do you suppose we mean by the zeroth power of an operator?

3.5 Classify these operators as linear or nonlinear: (a) $3x^2 d^2/dx^2$; (b) $(\ )^2$; (c) $\int dx$; (d) exp; (e) $\sum_{x=1}^{n}$.

3.6 (a) Give an example of an operator that satisfies Eq. (3.11) but does not satisfy (3.12). (b) Give an example of an operator that satisfies (3.12) but not (3.11).

3.7 Prove that the product of two linear operators is a linear operator.

3.8 Verify the commutator identity: $[\hat{A}, \hat{B}] = -[\hat{B}, \hat{A}]$.

3.9 Evaluate

$$\left[\frac{d^2}{dx^2}, ax^2 + bx + c\right]$$

3.10 (a) If $\hat{A}$ is linear, show that

$$\hat{A}(bf + cg) = b\hat{A}f + c\hat{A}g \qquad (3.126)$$

where b and c are arbitrary constants and f and g are arbitrary functions. (b) If (3.126) is true, show that $\hat{A}$ is linear.

3.11 Evaluate the commutators: (a) $[\hat{x}, \hat{p}_x]$; (b) $[\hat{x}, \hat{p}_x^2]$; (c) $[\hat{x}, \hat{p}_y]$; (d) $[\hat{x}, \hat{V}(x, y, z)]$; (e) $[\hat{x}, \hat{H}]$, where the Hamiltonian operator is given by Eq. (3.54); (f) $[\hat{x}\hat{y}\hat{z}, \hat{p}_x^2]$.

3.12 The *Laplace transform operator* $\hat{L}$ is defined by

$$\hat{L}f(x) = \int_0^\infty e^{-px}f(x)\,dx$$

(a) Is $\hat{L}$ linear? (b) Evaluate $\hat{L}(1)$. (c) Evaluate $\hat{L}e^{ax}$, assuming that $p > a$.

3.13 We define the *translation operator* $\hat{T}_h$ by

$$\hat{T}_hf(x) = f(x + h) \qquad (3.127)$$

(a) Is $\hat{T}_h$ a linear operator? (b) Evaluate $(\hat{T}_1^2 - 3\hat{T}_1 + 2)x^2$.

3.14 We define the operator $e^{\hat{A}}$ by the equation

$$e^{\hat{A}} = \hat{1} + \hat{A} + \frac{\hat{A}^2}{2!} + \frac{\hat{A}^3}{3!} + \cdots = \sum_{k=0}^\infty \frac{\hat{A}^k}{k!} \qquad (3.128)$$

Show that $e^{\hat{D}} = \hat{T}_1$ where $\hat{D} = d/dx$, and $\hat{T}_1$ is defined by (3.127).

3.15 Which of the following functions are eigenfunctions of d^2/dx^2? (a) e^x; (b) x^2; (c) $\sin x$; (d) $3\cos x$; (e) $\sin x + \cos x$. Give the eigenvalue for each eigenfunction.

3.16 For the particle confined to a box with dimensions a, b, and c, find the following values for the state with quantum numbers n_x, n_y, n_z. (a) $\langle x \rangle$; (b) $\langle y \rangle$, $\langle z \rangle$. Use symmetry considerations and the answer to part (a). (c) $\langle p_x \rangle$; (d) $\langle x^2 \rangle$. Is $\langle x^2 \rangle = \langle x \rangle^2$? Is $\langle xy \rangle = \langle x \rangle \langle y \rangle$?

3.17 For the particle in a cubic box, what is the degree of degeneracy of the energy levels with the following values of $8ma^2E/h^2$? (a) 12; (b) 14; (c) 27.

3.18 Find the eigenfunctions of

$$-\frac{\hbar^2}{2m}\frac{d^2}{dx^2}$$

If the eigenfunctions are to remain finite for $x \to \pm \infty$, what are the allowed eigenvalues?

3.19 If ψ is a normalized wave function, what are its cgs units for (a) the one-particle one-dimensional case; (b) the one-particle three-dimensional case; (c) the n-particle three-dimensional case?

3.20 Which of the following are sets of linearly independent functions? (a) x, x^2, x^6; (b) 8, x, x^2, $3x^2 - 1$; (c) $\sin x$, $\cos x$; (d) $\sin z$, $\cos z$, $\tan z$; (e) $\sin x$, $\cos x$, e^{ix}; (f) $\sin^2 x$, $\cos^2 x$, 1; (g) $\sin^2 x$, $\cos^2 y$, 1.

3.21 The wave function of a particle in a three-dimensional box is an eigenfunction of which of these operators? (a) $\hat{p}_x$; (b) $\hat{p}_x^2$; (c) $\hat{p}_z^2$; (d) $\hat{x}$.

3.22 Which of the following functions, when multiplied by a normalization constant, would be acceptable one-dimensional wave functions for a bound particle? (a) e^{-x}; (b) e^{-x^2}; (c) xe^{-x^2}; (d) ie^{-x^2}; (e) $f(x) = e^{-x^2}$ for $x < 0$, $f(x) = 2e^{-x^2}$ for $x \geqslant 0$.

3.23 Solve the one-particle three-dimensional time-independent Schroedinger equation for the free particle.

3.24 If ψ is an unnormalized wave function and N is a constant such that $N\psi$ is normalized, express $|N|$ in terms of ψ.

3.25 Fill in the details leading to (3.43) and (3.44) as the eigenfunctions and eigenvalues of $\hat{p}_x$.

3.26 Give the quantum-mechanical operators for the following physical quantities. (a) p_y^3; (b) $xp_y - yp_x$; (c) $(xp_y - yp_x)^2$.

3.27 For an n-particle three-dimensional system in the state Ψ, what is the probability of finding particle 1 with x coordinate between 0 and 2?

3.28 Prove that if $\hat{A} + \hat{B} = \hat{C}$, then $\hat{A} = \hat{C} - \hat{B}$.

3.29 Prove that Eq. (3.15) holds for linear operators.

3.30 The terms *state* and *energy level* are not synonymous in quantum mechanics. For the particle in a cubic box, consider the energy range $E < 15h^2/8ma^2$. (a) How many states lie in this range? (b) How many energy levels lie in this range?

3.31 (a) Write a computer program that will print all sets of positive integers n_x, n_y, n_z for which $n_x^2 + n_y^2 + n_z^2 \leq 60$, will print the sets in order of increasing $n_x^2 + n_y^2 + n_z^2$, and will print the $n_x^2 + n_y^2 + n_z^2$ values. (b) What is the degeneracy of the particle-in-a-cubic-box level with $n_x^2 + n_y^2 + n_z^2 = 54$?

4

THE HARMONIC OSCILLATOR

4.1 *POWER-SERIES SOLUTION OF DIFFERENTIAL EQUATIONS*

So far we have considered only cases where the potential energy $V(x)$ is a constant. This makes the Schroedinger equation a second-order linear homogeneous differential equation with *constant* coefficients, which we know how to solve. However, we want to deal with the wave equation for cases in which V varies with x. A useful approach in this case is to attempt a power-series solution of the differential equation.

To illustrate the method, consider the differential equation

$$y''(x) + c^2 y(x) = 0 \qquad (4.1)$$

where c^2 is a real, positive number. Of course, this differential equation has *constant* coefficients, but we can solve it with the power-series method if we want. Let us first obtain the solution by using the auxiliary equation, which is $s^2 + c^2 = 0$. We find $s = \pm ic$. Recalling the work in Section 2.2 [Eqs. (2.10) and (4.1) are the same], we get trigonometric solutions when the roots of the auxiliary equation are pure imaginary:

$$y = A \cos(cx) + B \sin(cx) \qquad (4.2)$$

where A and B are the constants of integration. A different form of (4.2) is sometimes useful, namely

$$y = D \sin(cx + e) \qquad (4.3)$$

where D and e are arbitrary constants. Using the formula for the sine of the sum of two angles, we can show that (4.3) is equivalent to (4.2).

Now let us solve (4.1) using the power-series method. We start by assuming

55

that the solution can be expanded in a Taylor series (see Problem 4.8) about $x = 0$; i.e., we assume that

$$y(x) = \sum_{n=0}^{\infty} a_n x^n = a_0 + a_1 x + a_2 x^2 + a_3 x^3 + \cdots \tag{4.4}$$

where the a's are coefficients to be determined so as to satisfy (4.1). Differentiating (4.4), we have

$$y'(x) = a_1 + 2a_2 x + 3a_3 x^2 + \cdots = \sum_{n=1}^{\infty} n a_n x^{n-1} \tag{4.5}$$

where we have assumed that term-by-term differentiation is valid for the series. (This is not always true for infinite series.) For y'' we have

$$y''(x) = 2a_2 + 3(2)a_3 x + \cdots = \sum_{n=2}^{\infty} n(n-1)a_n x^{n-2} \tag{4.6}$$

Substituting (4.4) and (4.6) into (4.1), we get

$$\sum_{n=2}^{\infty} n(n-1)a_n x^{n-2} + \sum_{n=0}^{\infty} c^2 a_n x^n = 0 \tag{4.7}$$

We want to combine the two sums in (4.7). Provided certain conditions are met, we can add two infinite series term by term to get their sum:

$$\sum_{j=0}^{\infty} b_j x^j + \sum_{j=0}^{\infty} c_j x^j = \sum_{j=0}^{\infty} (b_j + c_j) x^j \tag{4.8}$$

To apply (4.8) to the two sums in (4.7), we want the summation limits in each sum to be the same and the powers of x to be the same. We therefore make a change of summation index in the first sum in (4.7), setting $n = k + 2$:

$$\sum_{n=2}^{\infty} n(n-1)a_n x^{n-2} = \sum_{k=0}^{\infty} (k+2)(k+1)a_{k+2} x^k \tag{4.9}$$

$$\sum_{n=2}^{\infty} n(n-1)a_n x^{n-2} = \sum_{n=0}^{\infty} (n+2)(n+1)a_{n+2} x^n \tag{4.10}$$

Sometimes students are puzzled as to how (4.10) follows from (4.9). The point is that the summation index is a *dummy variable*; it makes no difference what letter we use to denote this variable. An example will make this clear. Consider the two sums

$$\sum_{i=1}^{3} c_i x^i \quad \text{and} \quad \sum_{j=1}^{3} c_j x^j \tag{4.11}$$

Because only the dummy variables in the two sums differ, the sums are equal to each other. This is easy to see if we write them out:

$$\sum_{i=1}^{3} c_i x^i = c_1 x + c_2 x^2 + c_3 x^3 \quad \text{and} \quad \sum_{j=1}^{3} c_j x^j = c_1 x + c_2 x^2 + c_3 x^3$$

In going from (4.9) to (4.10), we simply changed the symbol denoting the summation index from k to n.

The integration variable in a definite integral is also a dummy variable, since the value of a definite integral is unaffected by what letter we use for this variable. Thus

$$\int_a^b f(x)\,dx = \int_a^b f(t)\,dt \tag{4.12}$$

Using (4.10) in (4.7), we find, after applying (4.8),

$$\sum_{n=0}^{\infty} [(n+2)(n+1)a_{n+2} + c^2 a_n]x^n = 0 \tag{4.13}$$

If (4.13) is to be true for all values of x, then the coefficient of each power of x must vanish. To see this, consider the equation

$$\sum_{j=0}^{\infty} b_j x^j = 0 \tag{4.14}$$

Putting $x = 0$ in (4.14) shows that $b_0 = 0$. Taking the first derivative of (4.14) with respect to x and then putting $x = 0$ shows that $b_1 = 0$. Taking the nth derivative and putting $x = 0$ gives $b_n = 0$. Thus from (4.13) we have

$$(n+2)(n+1)a_{n+2} + c^2 a_n = 0 \tag{4.15}$$

$$a_{n+2} = -\frac{c^2}{(n+1)(n+2)}a_n \tag{4.16}$$

An equation like (4.16) is called a *recursion relation*. Using (4.16), if we know the value of a_0, we can find $a_2, a_4, a_6, \dots$. If we know a_1, we can find $a_3, a_5, a_7, \dots$. Since there is no restriction on the values of a_0 and a_1, they are arbitrary constants, which we denote by A and Bc:

$$a_0 = A, \qquad a_1 = Bc \tag{4.17}$$

Using (4.16) we find for the coefficients

$$a_0 = A, \qquad a_2 = -\frac{c^2 A}{1 \cdot 2}, \qquad a_4 = \frac{c^4 A}{4 \cdot 3 \cdot 2 \cdot 1}, \qquad a_6 = -\frac{c^6 A}{6!}, \dots$$

$$a_{2k} = (-1)^k \frac{c^{2k} A}{(2k)!}, \qquad k = 0, 1, 2, 3, \dots \tag{4.18}$$

$$a_1 = Bc, \qquad a_3 = -\frac{c^3 B}{2 \cdot 3}, \qquad a_5 = \frac{c^5 B}{5 \cdot 4 \cdot 3 \cdot 2}, \qquad a_7 = -\frac{c^7 B}{7!}, \dots$$

$$a_{2k+1} = (-1)^k \frac{c^{2k+1} B}{(2k+1)!}, \qquad k = 0, 1, 2, \dots \tag{4.19}$$

We have

$$y = \sum_{n=0}^{\infty} a_n x^n = \sum_{n=0,2,4,\dots}^{\infty} a_n x^n + \sum_{n=1,3,5,\dots}^{\infty} a_n x^n$$

$$y = A \sum_{k=0}^{\infty} (-1)^k \frac{c^{2k} x^{2k}}{(2k)!} + B \sum_{k=0}^{\infty} (-1)^k \frac{c^{2k+1} x^{2k+1}}{(2k+1)!} \tag{4.20}$$

The two series in (4.20) are the Taylor series for $\cos (cx)$ and $\sin (cx)$ (Problem 4.9); hence, in agreement with (4.2), we have

$$y = A \cos (cx) + B \sin (cx) \tag{4.21}$$

4.2   THE ONE-DIMENSIONAL HARMONIC OSCILLATOR

In this section we will increase our quantum-mechanical repertoire by solving the Schroedinger equation for the one-dimensional harmonic oscillator. This system is important as a model for molecular vibrations.

Before looking at the wave mechanics of the harmonic oscillator, we review the classical treatment. We have a single particle of mass m attracted toward the origin by a force proportional to its displacement from the origin:

$$F_x = -kx \tag{4.22}$$

F_x is the x component of the force on the particle. This is also the total force in this one-dimensional problem. Newton's second law gives

$$-kx = m \frac{d^2x}{dt^2} \tag{4.23}$$

where t is the time. Equation (4.23) is the same as Eq. (4.1) with $c^2 = k/m$; hence the solution is [Eq. (4.3)]

$$x = A \sin (2\pi vt + b) \tag{4.24}$$

where A and b are the integration constants and the vibration frequency v is

$$v = \frac{1}{2\pi} \left(\frac{k}{m} \right)^{1/2} \tag{4.25}$$

k is the *force constant*. Now consider the energy. The potential energy V is related to the components of force in the three-dimensional case by

$$F_x = -\frac{\partial V}{\partial x}, \qquad F_y = -\frac{\partial V}{\partial y}, \qquad F_z = -\frac{\partial V}{\partial z} \tag{4.26}$$

Equation (4.26) is the definition of potential energy. Since we have a one-dimensional problem, we have

$$F_x = -\frac{dV}{dx} = -kx \tag{4.27}$$

Integrating (4.27), we have

$$V = \tfrac{1}{2} kx^2 + C \tag{4.28}$$

where C is the constant of integration. The potential energy always has an arbitrary additive constant. We choose $C = 0$; hence

$$V = \tfrac{1}{2} kx^2 = 2\pi^2 v^2 m x^2 \tag{4.29}$$

The graph of $V(x)$ is a parabola—Fig. 4.2. The kinetic energy T is

$$T = \tfrac{1}{2}m\left(\frac{dx}{dt}\right)^2 \tag{4.30}$$

and can be evaluated by differentiating (4.24) with respect to t. Adding T and V, we find for the total energy

$$E = T + V = \tfrac{1}{2}kA^2 = 2\pi^2 v^2 m A^2 \tag{4.31}$$

where the identity $\sin^2 \theta + \cos^2 \theta = 1$ was used.

Now for the quantum-mechanical treatment. The Hamiltonian is

$$\hat{H} = -\frac{\hbar^2}{2m}\frac{d^2}{dx^2} + 2\pi^2 v^2 m x^2 = -\frac{\hbar^2}{2m}\left(\frac{d^2}{dx^2} - \alpha^2 x^2\right) \tag{4.32}$$

where to save time in writing, we defined α as

$$\alpha \equiv 2\pi v m/\hbar \tag{4.33}$$

The Schroedinger equation $\hat{H}\psi = E\psi$ reads, after multiplication by $2m/\hbar^2$,

$$\frac{d^2\psi}{dx^2} + (2mE\hbar^{-2} - \alpha^2 x^2)\psi = 0 \tag{4.34}$$

We might now attempt a power-series solution of (4.34) using the methods of the last section. If we do now try a power series for ψ of the form (4.4), we will find that it leads to a three-term recursion relation, which is much more difficult to deal with than a two-term recursion relation like Eq. (4.15). We therefore attempt to modify the form of (4.34) so as to get a two-term recursion relation when we try a series solution. A substitution that will achieve this purpose is (see Problem 4.17)

$$\psi = e^{-\alpha x^2/2} f(x) \tag{4.35}$$

This equation is simply the definition of a new function $f(x)$, which replaces $\psi(x)$ as the unknown function to be solved for. Differentiating (4.35) twice, we have

$$\psi'' = e^{-\alpha x^2/2}\left(f'' - 2\alpha x f' - \alpha f + \alpha^2 x^2 f\right) \tag{4.36}$$

Substituting (4.35) and (4.36) into (4.34), we find

$$f''(x) - 2\alpha x f'(x) + (2mE\hbar^{-2} - \alpha)f(x) = 0 \tag{4.37}$$

Now we try a series solution for $f(x)$:

$$f(x) = \sum_{n=0}^{\infty} c_n x^n \tag{4.38}$$

Assuming the validity of term-by-term differentiation of (4.38), we get

$$f'(x) = \sum_{n=1}^{\infty} n c_n x^{n-1} = \sum_{n=0}^{\infty} n c_n x^{n-1} \tag{4.39}$$

[The first term in the second sum in (4.39) is zero.] Also

$$f''(x) = \sum_{n=2}^{\infty} n(n-1)c_n x^{n-2} = \sum_{j=0}^{\infty} (j+2)(j+1)c_{j+2}x^j$$

$$= \sum_{n=0}^{\infty} (n+2)(n+1)c_{n+2}x^n \tag{4.40}$$

where we made the substitution $j = n - 2$ and then changed the summation index from j to n. [Compare Eqs. (4.9) and (4.10).] Substitution into (4.37) gives

$$\sum_{n=0}^{\infty} (n+2)(n+1)c_{n+2}x^n - 2\alpha \sum_{n=0}^{\infty} nc_n x^n + (2mEh^{-2}-\alpha)\sum_{n=0}^{\infty} c_n x^n = 0$$

$$\sum_{n=0}^{\infty} [(n+2)(n+1)c_{n+2} - 2\alpha nc_n + (2mEh^{-2}-\alpha)c_n]x^n = 0 \tag{4.41}$$

Setting the coefficient of x^n equal to zero [for the same reason as in Eq. (4.13)], we have

$$c_{n+2} = \frac{(\alpha + 2\alpha n - 2mEh^{-2})}{(n+1)(n+2)} c_n \tag{4.42}$$

which is the desired two-term recursion relation. Equation (4.42) has the same form as (4.16), in that knowing c_n we can calculate c_{n+2}; we thus have two arbitrary constants: c_0 and c_1. If we set c_1 equal to zero, then we will have as a solution a power series containing only even powers of x, multiplied by the exponential factor:

$$\psi = e^{-\alpha x^2/2}f(x) = e^{-\alpha x^2/2}\sum_{n=0,2,4,\ldots}^{\infty} c_n x^n = e^{-\alpha x^2/2}\sum_{l=0}^{\infty} c_{2l}x^{2l} \tag{4.43}$$

If we set c_0 equal to zero, we get another independent solution:

$$\psi = e^{-\alpha x^2/2}\sum_{n=1,3,\ldots}^{\infty} c_n x^n = e^{-\alpha x^2/2}\sum_{l=0}^{\infty} c_{2l+1}x^{2l+1} \tag{4.44}$$

The general solution of the Schroedinger equation is [recall Eq. (2.5)]

$$\psi = Ae^{-\alpha x^2/2}\sum_{l=0}^{\infty} c_{2l+1}x^{2l+1} + Be^{-\alpha x^2/2}\sum_{l=0}^{\infty} c_{2l}x^{2l} \tag{4.45}$$

where A and B are arbitrary constants.

We now must see if the boundary conditions on the wave function lead to any restrictions on the solution. To see how the two infinite series behave for large x, we examine the ratio of successive coefficients in each series. The ratio of the coefficient of x^{2l+2} to that of x^{2l} in the second series is [set $n = 2l$ in Eq. (4.42)]

$$\frac{c_{2l+2}}{c_{2l}} = \frac{\alpha + 4\alpha l - 2mEh^{-2}}{(2l+1)(2l+2)} \tag{4.46}$$

Assuming that for large values of x the later terms in the series are the dominant ones, we look at the ratio (4.46) for large values of l:

$$\frac{c_{2l+2}}{c_{2l}} \sim \frac{4\alpha l}{(2l)(2l)} = \frac{\alpha}{l} \qquad \text{for } l \text{ large} \tag{4.47}$$

Setting $n = 2l + 1$ in (4.42), we find that for large l, the ratio of successive coefficients in the first series is also α/l. Now consider the power-series expansion for the function $e^{\alpha x^2}$. Using (Problem 4.10)

$$e^z = \sum_{n=0}^{\infty} \frac{z^n}{n!} = 1 + z + \frac{z^2}{2!} + \cdots \tag{4.48}$$

we get

$$e^{\alpha x^2} = 1 + \alpha x^2 + \cdots + \frac{\alpha^l x^{2l}}{l!} + \frac{\alpha^{l+1} x^{2l+2}}{(l+1)!} + \cdots \tag{4.49}$$

The ratio of the coefficients of x^{2l+2} and x^{2l} in this series is

$$\frac{\alpha^{l+1}}{(l+1)!} \div \frac{\alpha^l}{l!} = \frac{\alpha}{l+1} \sim \frac{\alpha}{l} \qquad \text{for large } l \tag{4.50}$$

Thus the ratio of successive coefficients in each of the infinite series in the solution (4.45) is the same as in the series for $e^{\alpha x^2}$ for large l. We conclude that for large x, each series goes as $e^{\alpha x^2}$. [This is not a rigorous proof. A proper mathematical derivation is given in H. A. Buchdahl, *Am. J. Phys.*, **42**, 47 (1974); see also M. Bowen and J. Coster, *Am. J. Phys.*, **48**, 307 (1980).]

If each series goes as $e^{\alpha x^2}$, then (4.45) shows that ψ will behave as $e^{\alpha x^2/2}$ for large x. The wave function will become infinite as x goes to infinity and will not be quadratically integrable. If we could somehow break off the series after a finite number of terms, then the factor $e^{-\alpha x^2/2}$ would ensure that ψ went to zero as x became infinite. (Using l'Hospital's rule, it is easy to show that

$$\lim_{x \to \infty} x^p e^{-\alpha x^2/2} = 0$$

where p is any finite power.) In order to have one of the series break off after a finite number of terms, the coefficient of c_n in the recursion relation (4.42) must become zero for some value of n, say for $n = v$. This makes $c_{v+2}, c_{v+4}, \ldots$ all equal to zero, and one of the series in (4.45) will have a finite number of terms. In the recursion relation (4.42) there is one quantity whose value is not yet fixed, but can be adjusted to make the coefficient of c_v vanish; this quantity is the energy E. Setting the coefficient of c_v equal to zero in (4.42), we get

$$\alpha + 2\alpha v - 2mEh^{-2} = 0$$

$$2mEh^{-2} = (2v+1)2\pi vmh^{-1}$$

$$E = (v + \tfrac{1}{2})hv, \qquad v = 0, 1, 2, \ldots \tag{4.51}$$

and the recursion relation (4.42) becomes

$$c_{n+2} = \frac{2\alpha(n-v)}{(n+1)(n+2)} c_n \tag{4.52}$$

By quantizing the energy according to (4.51), we have made one of the series break off after a finite number of terms. To get rid of the other infinite series in (4.45), we must set the arbitrary constant that multiplies it equal to zero. This leaves us with a wave function that is $e^{-\alpha x^2/2}$ times a finite power series containing only even or only odd powers of x, depending on whether v is even or odd, respectively.

The quantum number v must be a nonnegative integer, and we have a series of equally spaced energy levels. As in the particle in a box, *it is the boundary conditions that force us to quantize the energy.* A graph of a solution to (4.34) for any value of E that differs from the values in (4.51) would show ψ becoming infinite as x goes to infinity.

The state of lowest energy is called the *ground state* (or *normal state*). The harmonic-oscillator ground-state energy is nonzero; this energy, $\frac{1}{2}h\nu$, is called the *zero-point energy.* (This would be the vibrational energy of diatomic molecules at absolute zero.) The zero-point energy can be understood from the uncertainty principle. If the lowest state had an energy of zero, both its potential and kinetic energies (which are nonnegative) would have to be zero. Zero kinetic energy would mean that the momentum was exactly zero, so that Δp_x would be zero. Zero potential energy would mean that the particle was always located at the origin, so that Δx would be zero. But we cannot have both Δx and Δp_x equal to zero. Hence the necessity for a nonzero ground-state energy. Similar considerations apply for the particle in a box.

Before considering the wave functions, we define the concept of even and odd functions. If $f(x)$ satisfies

$$f(-x) = f(x) \qquad (4.53)$$

then f is an *even function* of x. Thus x^2 and e^{-bx^2} are both even functions of x since $(-x)^2 = x^2$ and $e^{-b(-x)^2} = e^{-bx^2}$. The graph of an even function is symmetric about the y axis (e.g., see Fig. 4.1a); hence

$$\int_{-a}^{+a} f(x)\, dx = 2 \int_0^a f(x)\, dx \qquad \text{for } f(x) \text{ even} \qquad (4.54)$$

If $g(x)$ satisfies

$$g(-x) = -g(x) \qquad (4.55)$$

then g is an *odd function* of x. Examples are $x, 1/x, xe^{x^2}$. Setting $x = 0$ in (4.55), we see that an odd function must be zero at $x = 0$, provided $g(0)$ is defined and single-valued. The graph of an odd function has the general appearance of Fig. 4.1b. Because positive contributions on one side of the y axis are cancelled by corresponding negative contributions on the other side, we have

$$\int_{-a}^{+a} g(x)\, dx = 0 \qquad \text{for } g(x) \text{ odd} \qquad (4.56)$$

It is easy to show that the product of two even functions or of two odd functions is an even function, while the product of an even and an odd function is an odd function.

Coming back to the harmonic-oscillator eigenfunctions, we state that any eigenfunction is either an even or an odd function. The exponential factor is an even function of x. If v is even, then the polynomial factor contains only even powers of x, which makes ψ an even function. If v is odd, then the polynomial factor contains only odd powers of x, and ψ, being the product of an even and an odd function, is odd.

Let us find the wave functions for the lowest levels. For the ground state, $v = 0$, there are no odd powers of x in the polynomial factor, and the recursion relation (4.52) shows $c_2 = c_4 = \cdots = 0$. Hence, using the value of v as a subscript on ψ, we have

$$\psi_0 = c_0 e^{-\alpha x^2/2} \tag{4.57}$$

We fix c_0 by normalization:

$$1 = \int_{-\infty}^{\infty} |c_0|^2 e^{-\alpha x^2} \, dx = 2|c_0|^2 \int_{0}^{\infty} e^{-\alpha x^2} \, dx$$

where Eq. (4.54) has been used. Using the integral (A.5) in the Appendix, we find

$$|c_0| = (\alpha/\pi)^{1/4} \tag{4.58}$$

$$\psi_0 = \left(\frac{\alpha}{\pi}\right)^{1/4} e^{-\alpha x^2/2} \tag{4.59}$$

if we choose the phase of the normalization constant to be zero. The wave function (4.59) is a Gaussian function (Fig. 4.1a).

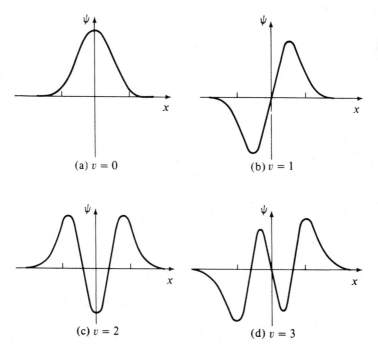

(a) $v = 0$

(b) $v = 1$

(c) $v = 2$

(d) $v = 3$

Figure 4.1 *Harmonic-oscillator wave functions. The same scale is used for all graphs; the points marked on the abscissas are for* $\alpha^{1/2}x = \pm 2$.

For $v = 1$ there are no even powers of x in the polynomial factor, and c_3, c_5, etc. all vanish; hence

$$\psi_1 = c_1 x e^{-\alpha x^2/2} \tag{4.60}$$

After normalization [the required integral is Eq. (A.6)] we have

$$\psi_1 = \left(\frac{4\alpha^3}{\pi}\right)^{1/4} x e^{-\alpha x^2/2} \tag{4.61}$$

This function is graphed in Fig. 4.1b.

For $v = 2$ there are no odd powers of x, and the even coefficients after c_2 all vanish; hence

$$\psi_2 = (c_0 + c_2 x^2) e^{-\alpha x^2/2} \tag{4.62}$$

From (4.52) we have for $v = 2$

$$c_2 = \frac{2\alpha(-2)c_0}{1 \cdot 2} = -2\alpha c_0 \tag{4.63}$$

Hence

$$\psi_2 = c_0(1 - 2\alpha x^2) e^{-\alpha x^2/2} \tag{4.64}$$

Evaluating c_0 by normalization, we find

$$\psi_2 = \left(\frac{\alpha}{4\pi}\right)^{1/4} (2\alpha x^2 - 1) e^{-\alpha x^2/2} \tag{4.65}$$

The number of nodes in the wave function equals the quantum number v. It can be proven (see *Pauling and Wilson*, p. 61) that in a one-dimensional problem the number of nodes interior to the boundary points is zero for the ground-state ψ and increases by one for each successive excited state. The boundary points for the harmonic oscillator are $\pm \infty$.

The polynomial factors in the harmonic-oscillator wave functions are well known in mathematics and are called *Hermite polynomials*, after a French mathematician. (See Problem 4.14.)

A state whose wave function is an even function is said to have *even parity*; if the wave function is odd, then the state is of *odd parity*. It can be shown (see Section 7.5) that if the potential energy $V(x)$ is an even function, then any wave function must be either an even function or an odd function.

According to the quantum-mechanical solution, there is some probability of finding the particle at any point on the x axis (except at the nodes). Classically, the particle is confined to the region where the potential energy does not exceed the total energy; this is the region from $-a$ to $+a$ in Fig. 4.2. It might seem that by saying the particle can be found outside the classically allowed region, we are allowing it to have negative kinetic energy. Actually there is no paradox in the quantum-mechanical view. To verify that the particle is in the classically forbidden region, we must measure its position. This measurement changes the state of the system (Sections 1.3 and 1.4); the interaction of the oscillator with the measuring apparatus transfers sufficient energy to the oscillator for it to be in the classically forbidden region. An accurate measurement of x introduces a large uncertainty in the momentum and

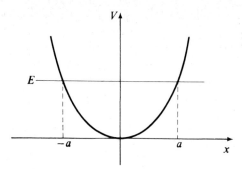

Figure 4.2 *The classically allowed ($|x| \leqslant a$) and forbidden ($|x| > a$) regions for the harmonic oscillator*

hence in the kinetic energy. (Penetration of classically forbidden regions was previously discussed in Sections 2.4 and 2.5.)

Note from Fig. 4.1 that ψ oscillates in the classically allowed region [which is $-(2v+1)^{1/2} \leqslant \alpha^{1/2}x \leqslant (2v+1)^{1/2}$; Problem 4.21] and decreases exponentially to zero in the classically forbidden region. We previously saw this behavior for the particle in a rectangular well (Section 2.4).

Figure 4.1 shows that as we go to higher-energy states of the harmonic oscillator, ψ and $|\psi|^2$ tend to have maxima further and further from the origin; since $V = \frac{1}{2}kx^2$ increases as we go further from the origin, the average potential energy $\langle V \rangle = \int_{-\infty}^{\infty} |\psi|^2 V \, dx$ increases as the quantum number increases. The average kinetic energy is $\langle T \rangle = -(\hbar^2/2m)\int_{-\infty}^{\infty} \psi^*\psi'' dx$. Integration by parts gives (Problem 7.3b) $\langle T \rangle = (\hbar^2/2m)\int_{-\infty}^{\infty} |d\psi/dx|^2 dx$. The higher number of nodes in states with higher quantum number produces a faster rate of change of ψ, so $\langle T \rangle$ increases as the quantum number increases.

4.3   VIBRATION OF MOLECULES

We shall see in Section 13.1 that to an excellent approximation one can treat separately the motions of the electrons and the motions of the nuclei of a molecule. (This is due to the much heavier mass of the nuclei.) One first imagines the nuclei to be held stationary and solves a Schroedinger equation for the electronic energy U. (U also includes the energy of nuclear repulsion.) For a diatomic (two-atom) molecule, the electronic energy U depends on the distance R between the nuclei, $U = U(R)$, and the U versus R curve has the typical appearance of Fig. 13.1.

After finding $U(R)$, one then solves a Schroedinger equation for nuclear motion, using $U(R)$ as the potential energy for nuclear motion. For a diatomic molecule, the nuclear Schroedinger equation is a two-particle equation. We shall see in Section 6.2 that when the potential energy of a two-particle system depends only on the distance between the particles, the energy of the system is the sum of (a) the kinetic energy of translational motion of the entire system through space and

(b) the energy of internal motion of the particles relative to each other. The classical expression for the two-particle internal energy turns out to be the sum of the potential energy of interaction between the particles and the kinetic energy of a hypothetical particle whose mass is $m_1 m_2/(m_1 + m_2)$ (where m_1 and m_2 are the masses of the two particles) and whose coordinates are the coordinates of one particle relative to the other. The quantity $m_1 m_2/(m_1 + m_2)$ is called the *reduced mass* μ.

The internal motion of a diatomic molecule consists of *vibration*, corresponding to a change in the distance R between the two nuclei, and *rotation*, corresponding to a change in the spatial orientation of the line joining the nuclei. It turns out that to a good approximation one can usually treat the vibrational and rotational motions separately. The rotational energy levels are considered in Section 6.3. Here we consider the vibrational levels.

The Schroedinger equation for the vibration of a diatomic molecule has a kinetic-energy operator for the hypothetical particle of mass $\mu = m_1 m_2/(m_1 + m_2)$ and a potential-energy term given by $U(R)$. If we place the origin to coincide with the minimum point of the U curve in Fig. 13.1 and take the zero of potential energy at the energy of this minimum point, then the lower portion of the $U(R)$ curve will nearly coincide with the potential-energy curve of a harmonic oscillator with the appropriate force constant k (see Fig. 4.3 and Problem 4.26). The minimum in the $U(R)$ curve occurs at the equilibrium distance R_e between the nuclei.

The harmonic-oscillator force constant k in Eq. (4.28) is obtained as

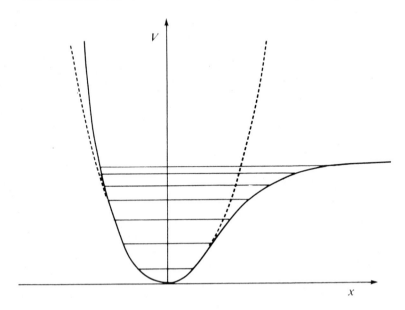

Figure 4.3 *Potential energy for vibration of a diatomic molecule (solid curve) and for a harmonic oscillator (dashed curve). Also shown are the bound-state vibrational energy levels for the diatomic molecule.*

$k = d^2V/dx^2$, and the harmonic-oscillator curve essentially coincides with the $U(R)$ curve at $R = R_e$, so the molecular force constant is $k = d^2U/dR^2|_{R = R_e}$ (see also Problem 4.26). [Differences in nuclear mass have virtually no effect on the electronic-energy curve $U(R)$, so different isotopic species of the same molecule have essentially the same force constant k.]

We expect, therefore, that a reasonable approximation to the vibrational energy levels E_{vib} of a diatomic molecule would be the harmonic-oscillator vibrational energy levels; Eqs. (4.51) and (4.25) give

$$E_{vib} \approx (v + \tfrac{1}{2})h\nu_e, \qquad v = 0, 1, 2, \ldots \qquad (4.66)$$

$$\nu_e = \frac{1}{2\pi}\left(\frac{k}{\mu}\right)^{1/2}, \qquad \mu = \frac{m_1 m_2}{m_1 + m_2}, \qquad k = \frac{d^2U}{dR^2}\bigg|_{R = R_e} \qquad (4.67)$$

ν_e is called the equilibrium vibrational frequency. This approximation should be best for the lower vibrational levels. As v increases, the nuclei spend more and more time in regions far from their equilibrium separation. For such regions the potential energy deviates substantially from that of a harmonic oscillator and the harmonic-oscillator approximation is poor. Instead of being equally spaced, one finds that the vibrational levels of a diatomic molecule come closer and closer together as v increases (Fig. 4.3). Eventually the vibrational energy is sufficiently large to dissociate the diatomic molecule into atoms that are not bound to each other. A more accurate expression for the molecular vibrational energy that allows for the anharmonicity of the vibration is

$$E_{vib} = (v + \tfrac{1}{2})h\nu_e - (v + \tfrac{1}{2})^2 h\nu_e x_e \qquad (4.68)$$

where the anharmonicity constant $\nu_e x_e$ is positive in nearly all cases.

Using the time-dependent Schroedinger equation, one finds (see *Levine, Molecular Spectroscopy*, Section 4.4) that the most probable vibrational transitions when a diatomic molecule is exposed to electromagnetic radiation are those where v changes by ± 1; furthermore, the vibration must cause a change in the molecule's dipole moment in order for absorption or emission of electromagnetic radiation to occur. Hence homonuclear diatomics (H_2, N_2, Cl_2, etc.) cannot undergo transitions between vibrational levels by absorption of radiation. Equation (1.2), the approximate equation (4.66), and the *selection rule* $\Delta v = 1$ for absorption of radiation show that a diatomic molecule with vibrational frequency ν_e will most strongly absorb light of frequency ν_{light} given approximately by

$$\nu_{light} = (E_2 - E_1)/h \approx [(v_2 + \tfrac{1}{2})h\nu_e - (v_1 + \tfrac{1}{2})h\nu_e]/h = (v_2 - v_1)\nu_e = \nu_e$$

The values of k and μ for diatomic molecules are such that ν_{light} falls in the infrared region of the spectrum. Transitions with $\Delta v = 2, 3, \ldots$ also occur, but these (called *overtones*) are much weaker than the $\Delta v = 1$ absorption.

Use of the more accurate equation (4.68) gives (Problem 4.23)

$$\nu_{light} = \nu_e - 2\nu_e x_e (v_1 + 1) \qquad (4.69)$$

where v_1 is the quantum number of the lower level and $\Delta v = 1$.

The relative population of two molecular energy levels is given by the

Boltzmann distribution law (see any physical chemistry text) as

$$\frac{N_i}{N_j} = \frac{g_i}{g_j} e^{-(E_i - E_j)/kT}$$ (4.70)

where energy levels i and j have energies E_i and E_j, degeneracies g_i and g_j, and are populated by N_i and N_j molecules and where k is Boltzmann's constant and T the absolute temperature.

The magnitude of $v = (1/2\pi)(k/\mu)^{1/2}$ is such that for light diatomics (e.g., H_2, HCl, CO) only the $v = 0$ vibrational level is significantly populated at room temperature. For heavy diatomics (e.g., I_2) there is significant room-temperature population of one or more excited vibrational levels.

The vibrational absorption spectrum of a polar diatomic molecule consists of a $v = 0 \rightarrow 1$ band, much weaker overtone bands ($v = 0 \rightarrow 2, 0 \rightarrow 3, \dots$), and, if there is significant population of $v > 0$ levels, *hot bands* such as $v = 1 \rightarrow 2, 2 \rightarrow 3$. Each band corresponding to a particular vibrational transition consists of several closely spaced lines; each such line corresponds to a different change in rotational state simultaneous with the change in vibrational state; each line is the result of a vibration–rotation transition.

The SI unit for spectroscopic frequencies is the *hertz* (Hz), defined by $1 \text{ Hz} \equiv 1 \text{ sec}^{-1}$. Multiples such as the megahertz (MHz) equal to 10^6 Hz and the gigahertz (GHz) equal to 10^9 Hz are often used. Infrared absorption lines are usually specified by giving their wave number σ, defined as $\sigma = 1/\lambda$, where λ is the wavelength.

In the harmonic-oscillator approximation, the quantum-mechanical energy levels of a polyatomic molecule turn out to be $E_{\text{vib}} = \Sigma_i (v_i + \frac{1}{2}) h v_i$, where the v_i's are the frequencies of the normal modes of vibration of the molecule and v_i is the vibrational quantum number of the ith normal mode; each v_i takes on the values $0, 1, 2, \dots$ independently of the values of the other vibrational quantum numbers. A linear molecule with n atoms has $3n - 5$ normal modes; a nonlinear molecule has $3n - 6$ normal modes. (See *Levine, Molecular Spectroscopy*, Chapter 6 for details.)

PROBLEMS

4.1 For the ground state of the one-dimensional harmonic oscillator, find the average value of the kinetic energy and of the potential energy; verify that $\langle T \rangle = \langle V \rangle$ in this case.

4.2 Which of the following are even functions? are odd functions? (a) $\sin x$; (b) $\cos x$; (c) $\tan x$; (d) e^x; (e) 13; (f) $x \cosh x$; (g) $2 - 2x$.

4.3 Prove the statements made in this chapter about products of even and odd functions.

4.4 Draw rough graphs of ψ and of ψ^2 for the $v = 5$ state of the one-dimensional harmonic oscillator without finding the explicit formula for ψ.

4.5 The three-dimensional harmonic oscillator has the potential-energy function

$$V = \tfrac{1}{2}k_x x^2 + \tfrac{1}{2}k_y y^2 + \tfrac{1}{2}k_z z^2$$

where the k's are three force constants. Find the energy eigenvalues by solving the Schroedinger equation.

4.6 Given that $f(x)$ is an even function that is differentiable at the origin, find $f'(0)$.

4.7 Verify the normalization factors for the $v = 1$ and $v = 2$ harmonic-oscillator wave functions.

4.8 Provided certain conditions are met, we can expand the function $f(x)$ in an infinite power series about the point $x = a$:

$$f(x) = \sum_{n=0}^{\infty} c_n (x-a)^n \tag{4.71}$$

Differentiate (4.71) m times, and then set $x = a$ to show that $c_n = f^{(n)}(a)/n!$, thus giving the familiar *Taylor series*:

$$f(x) = \sum_{n=0}^{\infty} \frac{f^{(n)}(a)}{n!} (x-a)^n \tag{4.72}$$

4.9 (a) Use (4.72) to derive the first few terms in the Taylor-series expansion about $x = 0$ for the function $\sin x$, and infer the general formula. (b) Differentiate the Taylor series in (a) to obtain the Taylor series for $\cos x$.

4.10 (a) Obtain the Taylor-series expansion about $x = 0$ for e^x. (b) Use the Taylor series (about $x = 0$) of $\sin x$, $\cos x$, and e^x to verify Eq. (1.22).

4.11 Point out the similarities and differences between the particle-in-a-box and the harmonic-oscillator wave functions and energies.

4.12 For the $v = 1$ harmonic-oscillator state, find the most likely position(s) of the particle.

4.13 (a) Obtain the recursion relation for the coefficients c_n in the power-series solution of $(1 - x^2)y''(x) - 2xy'(x) + 3y(x) = 0$. (b) Express c_4 in terms of c_0 and c_5 in terms of c_1.

4.14 The *Hermite polynomials* are defined by

$$H_n(z) = (-1)^n e^{z^2} \frac{d^n e^{-z^2}}{dz^n}$$

(a) Verify that

$$H_0 = 1, \qquad H_1 = 2z, \qquad H_2 = 4z^2 - 2, \qquad H_3 = 8z^3 - 12z$$

(b) The Hermite polynomials obey the relation (*Pauling and Wilson*, pp. 77–79)

$$zH_n(z) = nH_{n-1}(z) + \tfrac{1}{2}H_{n+1}(z)$$

Verify this identity for $n = 0, 1$, and 2. (c) The normalized harmonic-oscillator wave functions can be written as (*Pauling and Wilson*, pp. 79–80)

$$\psi_v(x) = (2^v v!)^{-1/2} \left(\frac{\alpha}{\pi}\right)^{1/4} e^{-\alpha x^2/2} H_v(\alpha^{1/2}x) \tag{4.73}$$

Verify (4.73) for the three lowest states.

4.15 What single-valued function is both odd and even?

4.16 Find $\langle x \rangle$ for the harmonic-oscillator state with quantum number v.

4.17 When a second-order linear homogeneous differential equation is written in the form (2.4), any point at which $P(x)$ or $Q(x)$ becomes infinite is called a *singular point* (or *singularity*). In solving a differential equation by the power-series method, one can often find the proper substitution to give a two-term recursion relation by examining the differential equation near its singularities. For the harmonic-oscillator Schroedinger equation (4.34), the singularities are at $x = \pm \infty$. (To check whether $x = \infty$ is a singular point, one substitutes $z = 1/x$ and examines the coefficients at $z = 0$.) Verify that $\exp(-\alpha x^2/2)$ is an approximate solution of (4.34) for very large $|x|$.

4.18 Find the eigenvalues and eigenfunctions of $\hat{H}$ for a one-dimensional system with $V(x) = \infty$ for $x < 0$, $V(x) = \frac{1}{2}kx^2$ for $x \geqslant 0$.

4.19 (a) A certain system in a certain stationary state has $\psi = Ne^{-ax^4}$. (N is the normalization constant.) Find the system's potential-energy function $V(x)$ and its energy E. [*Hint:* The zero level of energy is arbitrary, so choose $V(0) = 0$]. (b) Sketch $V(x)$. (c) Is this the ground-state ψ? Explain.

4.20 Show that adding a constant C to the potential energy leaves the stationary-state wave functions unchanged and simply adds C to the energy eigenvalues.

4.21 For the harmonic-oscillator state with quantum number v, what range of the x coordinate is allowed classically?

4.22 (a) The infrared absorption spectrum of $^1H^{35}Cl$ has its strongest band at 8.65×10^{13} Hz. Calculate the force constant of the bond in this molecule. (b) Find the approximate zero-point vibrational energy of $^1H^{35}Cl$. (c) Predict the frequency of the strongest infrared band of $^1H^{37}Cl$.

4.23 (a) Verify (4.69). (b) Find the corresponding equation for the $v = 0 \rightarrow v_2$ transition.

4.24 The $v = 0 \rightarrow 1$ and $v = 0 \rightarrow 2$ bands of $^1H^{35}Cl$ occur at 2885.98 cm^{-1} and 5667.98 cm^{-1}. (a) Calculate v_e/c and $v_e x_e/c$ for this molecule. (b) Predict the wave number of the $v = 0 \rightarrow 3$ band of $^1H^{35}Cl$.

4.25 (a) The $v = 0 \rightarrow 1$ band of LiH occurs at 1359 cm^{-1}. Calculate the ratio of the $v = 1$ to $v = 0$ populations at $25°C$ and at $200°C$. (b) Do the same as in (a) for ICl, whose strongest infrared band occurs at 381 cm^{-1}.

4.26 Show that if one expands $U(R)$ in Fig. 4.3 about $R = R_e$ and neglects terms containing $(R - R_e)^3$ and higher powers (these terms are small for R near R_e), then one obtains a harmonic-oscillator potential with $k = d^2U/dR^2|_{R = R_e}$.

4.27 The one-dimensional double-well potential has $V = \infty$ for $x < -a$, $V = 0$ for $-a \leqslant x \leqslant -b$, $V = V_0$ for $-b < x < b$, $V = 0$ for $b \leqslant x \leqslant a$, and $V = \infty$ for $x > a$ (where a, b, and V_0 are positive constants). (a) Sketch V. (b) Use general properties of wave functions discussed in Chapters 2 and 4 to sketch ψ for the lowest four bound states for a finite value of V_0. (c) Do the same as in (b) for $V_0 = \infty$.

5 ANGULAR MOMENTUM

5.1 SIMULTANEOUS MEASUREMENT OF SEVERAL PROPERTIES

In this chapter we will discuss angular momentum, and in the next chapter we will show that for the stationary states of the hydrogen atom the magnitude of the electron's angular momentum is constant. As a preliminary, we consider what criterion we can use to decide which properties of a system can be simultaneously assigned definite values.

In Section 3.3 we postulated that if the state function Ψ is an eigenfunction of the operator $\hat{A}$ with eigenvalue s, then a measurement of the physical property A is certain to yield the result s. If Ψ is simultaneously an eigenfunction of the two operators $\hat{A}$ and $\hat{B}$,

$$\hat{A}\Psi = s\Psi, \qquad \hat{B}\Psi = t\Psi \tag{5.1}$$

we can simultaneously assign definite values to the physical quantities A and B. When will it be possible for Ψ to be simultaneously an eigenfunction of two different operators? In Chapter 7 we will prove the following two theorems. First, a necessary condition for the existence of a complete set of simultaneous eigenfunctions of two operators is that the operators commute with each other. (The word *complete* is used here in a certain technical sense, which we won't worry about until Chapter 7.) Conversely, if $\hat{A}$ and $\hat{B}$ are two commuting operators that correspond to physical quantities, then there exists a complete set of functions that are eigenfunctions of both $\hat{A}$ and $\hat{B}$. Thus if $[\hat{A}, \hat{B}] = 0$, then Ψ can be an eigenfunction of both $\hat{A}$ and $\hat{B}$.

To facilitate the calculation of commutators, we list some identities; these are easily proved by writing out the commutators in detail.

$$[\hat{A}, \hat{B}] = -[\hat{B}, \hat{A}] \tag{5.2}$$

71

$$[\hat{A}, \hat{A}^n] = 0, \qquad n = 1, 2, 3, \ldots \tag{5.3}$$

$$[k\hat{A}, \hat{B}] = [\hat{A}, k\hat{B}] = k[\hat{A}, \hat{B}] \tag{5.4}$$

$$[\hat{A}, \hat{B} + \hat{C}] = [\hat{A}, \hat{B}] + [\hat{A}, \hat{C}] \tag{5.5}$$

$$[\hat{A} + \hat{B}, \hat{C}] = [\hat{A}, \hat{C}] + [\hat{B}, \hat{C}] \tag{5.6}$$

$$[\hat{A}, \hat{B}\hat{C}] = [\hat{A}, \hat{B}]\hat{C} + \hat{B}[\hat{A}, \hat{C}] \tag{5.7}$$

$$[\hat{A}\hat{B}, \hat{C}] = [\hat{A}, \hat{C}]\hat{B} + \hat{A}[\hat{B}, \hat{C}] \tag{5.8}$$

where k is a constant and the operators are assumed to be linear.

Let us use these identities to work out some of the commutators in Problem 3.11. Our starting point is Eq. (3.9):

$$\left[\frac{\partial}{\partial x}, x\right] = 1 \tag{5.9}$$

We have

$$[\hat{x}, \hat{p}_x] = \left[x, \frac{\hbar}{i}\frac{\partial}{\partial x}\right] = \frac{\hbar}{i}\left[x, \frac{\partial}{\partial x}\right] = -\frac{\hbar}{i}\left[\frac{\partial}{\partial x}, x\right] = -\frac{\hbar}{i}$$

$$[\hat{x}, \hat{p}_x] = i\hbar \tag{5.10}$$

$$[\hat{x}, \hat{p}_x^2] = [\hat{x}, \hat{p}_x]\hat{p}_x + \hat{p}_x[\hat{x}, \hat{p}_x] = i\hbar \cdot \frac{\hbar}{i}\frac{\partial}{\partial x} + \frac{\hbar}{i}\frac{\partial}{\partial x} \cdot i\hbar$$

$$[\hat{x}, \hat{p}_x^2] = 2\hbar^2 \frac{\partial}{\partial x} \tag{5.11}$$

For a one-particle three-dimensional system, we have

$$[\hat{x}, \hat{H}] = [\hat{x}, \hat{T} + \hat{V}] = [\hat{x}, \hat{T}] + [\hat{x}, \hat{V}(x, y, z)] = [\hat{x}, \hat{T}]$$

$$= \left[\hat{x}, \frac{1}{2m}(\hat{p}_x^2 + \hat{p}_y^2 + \hat{p}_z^2)\right]$$

$$= \frac{1}{2m}[\hat{x}, \hat{p}_x^2] + \frac{1}{2m}[\hat{x}, \hat{p}_y^2] + \frac{1}{2m}[\hat{x}, \hat{p}_z^2]$$

$$= \frac{1}{2m} \cdot 2\hbar^2 \frac{\partial}{\partial x} + 0 + 0$$

$$[\hat{x}, \hat{H}] = \frac{\hbar^2}{m}\frac{\partial}{\partial x} = \frac{i\hbar}{m}\hat{p}_x \tag{5.12}$$

We leave it to you to show that

$$[\hat{p}_x, \hat{H}] = \frac{\hbar}{i}\frac{\partial V(x, y, z)}{\partial x} \tag{5.13}$$

The above commutators have important physical consequences. Since $[\hat{x}, \hat{p}_x] \neq 0$, we cannot expect the state function to be simultaneously an eigenfunction of $\hat{x}$ and of $\hat{p}_x$; hence we cannot simultaneously assign definite values to x and p_x, in agreement with the uncertainty principle. Since $\hat{x}$ and $\hat{H}$ do not commute, we cannot expect to assign definite values to the energy and the x coordinate at the same time. A stationary state (which has a definite energy) shows a spread of possible values for x, the probabilities for observing various values of x being given by the Born postulate.

For a state function Ψ that is not an eigenfunction of $\hat{A}$, we get various possible outcomes when we measure A in identical systems. We want some measure of the spread or dispersion in the set of observed values A_i. If $\langle A \rangle$ is the average of these values, then the deviation of each measurement from the average is $A_i - \langle A \rangle$. If we averaged all the deviations, we would get zero, since positive and negative deviations would cancel. Hence to make all deviations positive, we square them. The average of the squares of the deviations is called the *variance* of A, symbolized in statistics by σ_A^2 and in quantum mechanics by $(\Delta A)^2$:

$$(\Delta A)^2 \equiv \sigma_A^2 \equiv \langle (A - \langle A \rangle)^2 \rangle = \int \Psi^* (\hat{A} - \langle A \rangle)^2 \Psi \, d\tau \qquad \textbf{(5.14)}$$

where (3.117) was used. The positive square root of the variance is called the *standard deviation*, σ_A or ΔA. The standard deviation is the most commonly used measure of spread, and we shall take it as the measure of the "uncertainty" in A. The definition (5.14) is equivalent to (Problem 5.1)

$$(\Delta A)^2 = \langle A^2 \rangle - \langle A \rangle^2 \qquad \textbf{(5.15)}$$

For the product of the standard deviations of two properties, it can be shown (see Problem 7.35) that

$$\Delta A \, \Delta B \geqslant \frac{1}{2} \left| \int \Psi^* [\hat{A}, \hat{B}] \Psi \, d\tau \right| \qquad \textbf{(5.16)}$$

If $\hat{A}$ and $\hat{B}$ commute, then the integral in (5.16) is zero, and we have the possibility of having ΔA and ΔB both zero, in agreement with the previous discussion. As an example of (5.16), we find, using (5.10),

$$\Delta x \, \Delta p_x \geqslant \frac{1}{2} \left| \int \Psi^* i\hbar \Psi \, d\tau \right| = \tfrac{1}{2}\hbar |i| \left| \int \Psi^* \Psi \, d\tau \right| \qquad \textbf{(5.17)}$$

$$\Delta x \, \Delta p_x \geqslant \tfrac{1}{2}\hbar \qquad \textbf{(5.18)}$$

Equation (5.18) is the quantitative statement of the Heisenberg uncertainty principle. Consider, for example, the ground state of the particle in a three-dimensional box. Using the results of Problem 3.16 and Eqs. (3.124), (3.125), (3.120), and (3.47), we have

$$\langle x \rangle = \frac{a}{2} \qquad \langle x^2 \rangle = a^2 \left(\frac{1}{3} - \frac{1}{2\pi^2} \right) \qquad \textbf{(5.19)}$$

$$\langle p_x \rangle = 0 \qquad \langle p_x^2 \rangle = h^2/4a^2 \qquad \textbf{(5.20)}$$

From (5.15) we find

$$\Delta x = \frac{a}{2\pi}\left(\frac{\pi^2 - 6}{3}\right)^{1/2} \qquad \Delta p_x = \frac{h}{2a} \tag{5.21}$$

$$\Delta x \, \Delta p_x = \frac{h}{2}\left(\frac{\pi^2 - 6}{3}\right)^{1/2} = 0.57h > \frac{h}{2}$$

There is also an uncertainty relation involving energy and time:

$$\Delta E \, \Delta t \geqslant \tfrac{1}{2}h \tag{5.22}$$

Some texts state that (5.22) is derived from (5.16) by taking $i h \, \partial/\partial t$ as the energy operator and multiplication by t as the time operator. However, the energy operator is the Hamiltonian $\hat{H}$, and not $i h \, \partial/\partial t$. Moreover, time is not an observable but is a parameter in quantum mechanics; hence there is no quantum-mechanical time operator. (The noun *observable* in quantum mechanics means a physically measurable property of a system.) Equation (5.22) must be derived by a special treatment, which we omit. (See *Blokhintsev*, Sections 99 and 113.) The derivation of (5.22) shows that Δt is to be interpreted as the lifetime of the state whose energy is uncertain by ΔE. It is often stated that Δt in (5.22) is the duration of the energy measurement; however, Aharonov and Bohm have shown that "energy can be measured reproducibly in an arbitrarily short time." [See Y. Aharonov and D. Bohm, *Phys. Rev.*, **122**, 1649 (1961); **134**, B1417 (1964).]

Now consider the possibility of simultaneously assigning definite values to *three* physical quantities: A, B, and C. Suppose

$$[\hat{A}, \hat{B}] = 0 \tag{5.23}$$

$$[\hat{A}, \hat{C}] = 0 \tag{5.24}$$

Is this sufficient to ensure that there exist simultaneous eigenfunctions of all three operators? Equation (5.23) ensures that we can construct a common set of eigenfunctions for $\hat{A}$ and $\hat{B}$; Eq. (5.24) ensures that we can construct a common set of eigenfunctions for $\hat{A}$ and $\hat{C}$. If these two sets of eigenfunctions are the same, then we will have a common set of eigenfunctions for all three operators. Hence we ask: Is the set of eigenfunctions of the linear operator $\hat{A}$ uniquely determined (apart from arbitrary multiplicative constants)? The answer is, in general, no. If there is more than one independent eigenfunction corresponding to each eigenvalue of $\hat{A}$ (i.e., degeneracy), then any linear combination of the eigenfunctions of the degenerate eigenvalue is an eigenfunction of $\hat{A}$ (Section 3.6). It might well be that the proper linear combinations needed to give eigenfunctions of $\hat{B}$ would differ from the linear combinations that give eigenfunctions of $\hat{C}$. It turns out that if we are to have a common complete set of eigenfunctions of all three operators, we require that

$$[\hat{B}, \hat{C}] = 0 \tag{5.25}$$

in addition to (5.23) and (5.24). To have a complete set of functions that are simultaneous eigenfunctions of several operators, we must have each operator commute with every other operator.

5.2   *VECTORS*

In the next section we will solve the eigenvalue problem for angular momentum, which is a vector property. We therefore first review vectors.

Physical properties (e.g., mass, length, energy) that are completely specified by their magnitude are called *scalars*. Physical properties (e.g., force, velocity, momentum) that require specification of both magnitude and direction are called *vectors*. A vector is represented by a directed line segment whose length and direction give the magnitude and direction of the property. We use boldface type for vectors.

We define the sum of two vectors **A** and **B** by the following procedure: Slide the first vector so that its tail touches the head of the second vector, keeping the direction of the first vector fixed. Then draw a new vector from the tail of the second vector to the head of the first vector. See Fig. 5.1. The product of a vector and a scalar, $c\mathbf{A}$, is defined as a vector of length $|c|$ times the length of **A** with the same direction as **A** if c is positive, or the opposite direction to **A** if c is negative.

To obtain an algebraic (as well as geometric) way of representing vectors, we set up Cartesian coordinates in space. We draw a vector of unit length directed along the positive x axis and call it **i**. (No connection with $i = \sqrt{-1}$.) Unit vectors in the positive y and z directions are called **j** and **k** (Fig. 5.2). To represent any vector **A** in

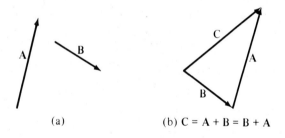

(a)   (b) C = A + B = B + A

Figure 5.1 *Addition of vectors*

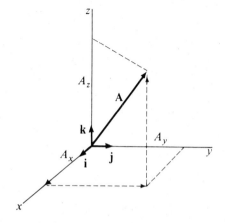

Figure 5.2 *Unit vectors and components of* **A**

terms of the three unit vectors, we first slide $\mathbf{A}$ so that its tail is at the origin, preserving its direction during this process. We then find the projections of $\mathbf{A}$ on the x, y, and z axes: A_x, A_y, and A_z. From the definition of vector addition it follows that (Fig. 5.2)

$$\mathbf{A} = A_x\mathbf{i} + A_y\mathbf{j} + A_z\mathbf{k} \tag{5.26}$$

To specify $\mathbf{A}$ it is sufficient to specify its three components: (A_x, A_y, A_z). We can therefore define a vector in three-dimensional space as an ordered set of three numbers. The advantage of this definition is that it can be extended to more than three dimensions. Thus a vector in five-dimensional "space" is an ordered set of five numbers. Such abstract vector spaces (which can be of infinite dimension) are used in advanced formulations of quantum mechanics.

Two vectors $\mathbf{A}$ and $\mathbf{B}$ are equal if and only if all the corresponding components are equal: $A_x = B_x$, $A_y = B_y$, $A_z = B_z$. Hence a vector equation is equivalent to three scalar equations.

To add two vectors analytically, we add corresponding components:

$$\mathbf{A} + \mathbf{B} = A_x\mathbf{i} + A_y\mathbf{j} + A_z\mathbf{k} + B_x\mathbf{i} + B_y\mathbf{j} + B_z\mathbf{k}$$

$$\mathbf{A} + \mathbf{B} = (A_x + B_x)\mathbf{i} + (A_y + B_y)\mathbf{j} + (A_z + B_z)\mathbf{k} \tag{5.27}$$

Also, if c is a scalar, then

$$c\mathbf{A} = cA_x\mathbf{i} + cA_y\mathbf{j} + cA_z\mathbf{k} \tag{5.28}$$

The magnitude A of a vector $\mathbf{A}$ is its length, and is therefore a scalar. Often the notation $|\mathbf{A}|$ is used for the magnitude of $\mathbf{A}$.

The *dot product* (or *scalar product*) $\mathbf{A} \cdot \mathbf{B}$ of two vectors is defined by

$$\mathbf{A} \cdot \mathbf{B} = |\mathbf{A}|\,|\mathbf{B}| \cos\theta = \mathbf{B} \cdot \mathbf{A} \tag{5.29}$$

where θ is the angle between the vectors. The dot product, being the product of three scalars, is a scalar. Note that $|\mathbf{A}|\cos\theta$ is the projection of $\mathbf{A}$ on $\mathbf{B}$. From the definition of vector addition, it follows that the projection of the vector $(\mathbf{A} + \mathbf{B})$ on some vector $\mathbf{C}$ is the sum of the projections of $\mathbf{A}$ and of $\mathbf{B}$ on $\mathbf{C}$. Hence

$$(\mathbf{A} + \mathbf{B}) \cdot \mathbf{C} = \mathbf{A} \cdot \mathbf{C} + \mathbf{B} \cdot \mathbf{C} \tag{5.30}$$

Since the three unit vectors $\mathbf{i}$, $\mathbf{j}$, and $\mathbf{k}$ are each of unit length and are mutually perpendicular, we have

$$\mathbf{i} \cdot \mathbf{i} = \mathbf{j} \cdot \mathbf{j} = \mathbf{k} \cdot \mathbf{k} = \cos(0) = 1$$

$$\mathbf{i} \cdot \mathbf{j} = \mathbf{j} \cdot \mathbf{k} = \mathbf{k} \cdot \mathbf{i} = \cos\left(\frac{\pi}{2}\right) = 0 \tag{5.31}$$

We can use (5.31) and the distributive law (5.30) to get an important formula for the dot product of two vectors:

$$\mathbf{A} \cdot \mathbf{B} = (A_x\mathbf{i} + A_y\mathbf{j} + A_z\mathbf{k}) \cdot (B_x\mathbf{i} + B_y\mathbf{j} + B_z\mathbf{k}) \tag{5.32}$$

$$\mathbf{A} \cdot \mathbf{B} = A_xB_x + A_yB_y + A_zB_z \tag{5.33}$$

where six of the nine terms in the dot product are zero.

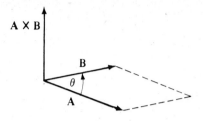

Figure 5.3 *Cross product of two vectors*

Consider the dot product of a vector with itself. From (5.29) we have

$$\mathbf{A} \cdot \mathbf{A} = |\mathbf{A}|^2 \tag{5.34}$$

Using (5.33), we therefore have

$$|\mathbf{A}| = (A_x^2 + A_y^2 + A_z^2)^{1/2} \tag{5.35}$$

For three-dimensional vectors, there is another type of product. The *vector product* or *cross product* $\mathbf{A} \times \mathbf{B}$ is a vector whose magnitude is

$$|\mathbf{A} \times \mathbf{B}| = |\mathbf{A}| \, |\mathbf{B}| \sin \theta \tag{5.36}$$

whose line segment is perpendicular to the plane defined by $\mathbf{A}$ and $\mathbf{B}$, and whose direction is such that $\mathbf{A}$, $\mathbf{B}$, and $\mathbf{A} \times \mathbf{B}$ form a right-handed system (just as the x, y, and z axes form a right-handed system). (See Fig. 5.3.) From the definition it follows that

$$\mathbf{B} \times \mathbf{A} = -\mathbf{A} \times \mathbf{B} \tag{5.37}$$

Also, it can be shown that (*Taylor and Mann*, pp. 281–282)

$$\mathbf{A} \times (\mathbf{B} + \mathbf{C}) = \mathbf{A} \times \mathbf{B} + \mathbf{A} \times \mathbf{C} \tag{5.38}$$

For the three unit vectors, we have

$$\mathbf{i} \times \mathbf{i} = \mathbf{j} \times \mathbf{j} = \mathbf{k} \times \mathbf{k} = \sin (0) = 0 \tag{5.39}$$

$$\mathbf{i} \times \mathbf{j} = \mathbf{k}, \qquad \mathbf{j} \times \mathbf{i} = -\mathbf{k}$$

$$\mathbf{j} \times \mathbf{k} = \mathbf{i}, \qquad \mathbf{k} \times \mathbf{j} = -\mathbf{i} \tag{5.40}$$

$$\mathbf{k} \times \mathbf{i} = \mathbf{j}, \qquad \mathbf{i} \times \mathbf{k} = -\mathbf{j}$$

Using these equations and the distributive property (5.38), we find

$$\mathbf{A} \times \mathbf{B} = (A_x \mathbf{i} + A_y \mathbf{j} + A_z \mathbf{k}) \times (B_x \mathbf{i} + B_y \mathbf{j} + B_z \mathbf{k}) \tag{5.41}$$

$$\mathbf{A} \times \mathbf{B} = (A_y B_z - A_z B_y)\mathbf{i} + (A_z B_x - A_x B_z)\mathbf{j} + (A_x B_y - A_y B_x)\mathbf{k}$$

As a mnemonic device, we can write (see Section 8.3)

$$\mathbf{A} \times \mathbf{B} = \begin{vmatrix} \mathbf{i} & \mathbf{j} & \mathbf{k} \\ A_x & A_y & A_z \\ B_x & B_y & B_z \end{vmatrix} = \mathbf{i} \begin{vmatrix} A_y & A_z \\ B_y & B_z \end{vmatrix} - \mathbf{j} \begin{vmatrix} A_x & A_z \\ B_x & B_z \end{vmatrix} + \mathbf{k} \begin{vmatrix} A_x & A_y \\ B_x & B_y \end{vmatrix} \tag{5.42}$$

We define the vector operator *del* as

$$\nabla \equiv \mathbf{i}\frac{\partial}{\partial x} + \mathbf{j}\frac{\partial}{\partial y} + \mathbf{k}\frac{\partial}{\partial z} \qquad (5.43)$$

From Eq. (3.28) the operator for the linear-momentum vector is

$$\hat{\mathbf{p}} = -i\hbar\nabla \qquad (5.44)$$

We define the *gradient* of a function $g(x, y, z)$ as the result of operating on the function with del:

$$\mathbf{grad}\, g(x, y, z) \equiv \nabla g(x, y, z) \equiv \mathbf{i}\frac{\partial g}{\partial x} + \mathbf{j}\frac{\partial g}{\partial y} + \mathbf{k}\frac{\partial g}{\partial z} \qquad (5.45)$$

The gradient of a scalar function is a vector function. Physically, the vector $\nabla g(x, y, z)$ represents the spatial rate of change of the function g; the x component of ∇g is the rate of change of g with respect to x, etc. It can be shown that the vector ∇g points in the direction in which the rate of change of g is greatest. From Eq. (4.26), we have

$$\mathbf{F} = -\nabla V(x, y, z) = -\mathbf{i}\frac{\partial V}{\partial x} - \mathbf{j}\frac{\partial V}{\partial y} - \mathbf{k}\frac{\partial V}{\partial z} \qquad (5.46)$$

For example, the force field of a point charge is radially directed, and the rate of change of potential energy of another charge in this field is greatest in the radial direction.

The *divergence* of a vector function is the scalar function found by taking the dot product of the vector operator del and the vector function:

$$\text{div}\,\mathbf{A} \equiv \nabla \cdot \mathbf{A} = \left(\mathbf{i}\frac{\partial}{\partial x} + \mathbf{j}\frac{\partial}{\partial y} + \mathbf{k}\frac{\partial}{\partial z}\right) \cdot (A_x\mathbf{i} + A_y\mathbf{j} + A_z\mathbf{k})$$

$$\text{div}\,\mathbf{A} \equiv \frac{\partial A_x}{\partial x} + \frac{\partial A_y}{\partial y} + \frac{\partial A_z}{\partial z} \qquad (5.47)$$

For the divergence of the gradient of a function, we have

$$\text{div}\,[\mathbf{grad}\, g(x, y, z)] = \left(\mathbf{i}\frac{\partial}{\partial x} + \mathbf{j}\frac{\partial}{\partial y} + \mathbf{k}\frac{\partial}{\partial z}\right) \cdot \left(\mathbf{i}\frac{\partial g}{\partial x} + \mathbf{j}\frac{\partial g}{\partial y} + \mathbf{k}\frac{\partial g}{\partial z}\right)$$

$$\nabla \cdot \nabla g = \frac{\partial^2 g}{\partial x^2} + \frac{\partial^2 g}{\partial y^2} + \frac{\partial^2 g}{\partial z^2} = \nabla^2 g \qquad (5.48)$$

which gives the Laplacian [Eq. (3.55)].

Finally, suppose that the components of a vector are each functions of some parameter t:

$$A_x = A_x(t), \qquad A_y = A_y(t), \qquad A_z = A_z(t) \qquad (5.49)$$

We define the derivative of the vector with respect to t as

$$\frac{d\mathbf{A}}{dt} = \mathbf{i}\frac{dA_x}{dt} + \mathbf{j}\frac{dA_y}{dt} + \mathbf{k}\frac{dA_z}{dt} \qquad (5.50)$$

5.3 ANGULAR MOMENTUM OF A ONE-PARTICLE SYSTEM

In Section 3.3 we found the eigenfunctions and eigenvalues for the linear-momentum operator $\hat{p}_x$. In this section we consider the same problem for angular momentum. Angular momentum is extremely important in the quantum mechanics of atomic structure. We begin by reviewing the classical mechanics of angular momentum.

Consider a moving particle of mass m. We set up a Cartesian coordinate system that is fixed in space. Let $\mathbf{r}$ be the vector from the origin to the instantaneous position of the particle. We have

$$\mathbf{r} = \mathbf{i}x + \mathbf{j}y + \mathbf{k}z \tag{5.51}$$

where x, y, and z are the particle's coordinates at a given instant. These coordinates are functions of time, and defining the velocity vector $\mathbf{v}$ as the time derivative of the position vector, we have [Eq. (5.50)]

$$\mathbf{v} = \frac{d\mathbf{r}}{dt} = \mathbf{i}\frac{dx}{dt} + \mathbf{j}\frac{dy}{dt} + \mathbf{k}\frac{dz}{dt} \tag{5.52}$$

$$v_x = \frac{dx}{dt}, \qquad v_y = \frac{dy}{dt}, \qquad v_z = \frac{dz}{dt} \tag{5.53}$$

We define the *linear momentum* vector $\mathbf{p}$ by

$$\mathbf{p} = m\mathbf{v} \tag{5.54}$$

$$p_x = mv_x, \qquad p_y = mv_y, \qquad p_z = mv_z \tag{5.55}$$

The particle's *angular momentum* $\mathbf{L}$ is defined in classical mechanics as

$$\mathbf{L} = \mathbf{r} \times \mathbf{p} \tag{5.56}$$

$$\mathbf{L} = \begin{vmatrix} \mathbf{i} & \mathbf{j} & \mathbf{k} \\ x & y & z \\ p_x & p_y & p_z \end{vmatrix} \tag{5.57}$$

$$L_x = yp_z - zp_y, \qquad L_y = zp_x - xp_z, \qquad L_z = xp_y - yp_x \tag{5.58}$$

The angular-momentum vector $\mathbf{L}$ is perpendicular to the plane defined by the particle's position vector $\mathbf{r}$ and its velocity $\mathbf{v}$ (recall Fig. 5.3).

The torque τ acting on a particle is defined as the cross product of $\mathbf{r}$ and the force acting on the particle:

$$\tau = \mathbf{r} \times \mathbf{F} \tag{5.59}$$

It is readily shown that (*Halliday and Resnick*, Section 12-3)

$$\tau = \frac{d\mathbf{L}}{dt} \tag{5.60}$$

When there is no torque acting on the particle, the rate of change of its angular momentum is zero; i.e., its angular momentum is constant (or conserved). For a planet orbiting the sun, the gravitational force is radially directed; since the cross

product of two parallel vectors is zero, there is no torque on the planet and its
angular momentum is conserved.

Now for the quantum-mechanical treatment. In quantum mechanics there
are two kinds of angular momenta: *orbital* angular momentum results from the
motion of a particle through space, and is the analog of the classical-mechanical
quantity **L**; *spin* angular momentum (Chapter 10) is an intrinsic property of many
microscopic particles, and has no classical-mechanical analog. We are now
considering only orbital angular momentum. We get the quantum-mechanical
operators for the components of orbital angular momentum of a particle by
replacing the quantities in (5.58) by their corresponding operators
[Eqs. (3.25)–(3.28)]. We find

$$\hat{L}_x = -i\hbar \left(y \frac{\partial}{\partial z} - z \frac{\partial}{\partial y} \right) \tag{5.61}$$

$$\hat{L}_y = -i\hbar \left(z \frac{\partial}{\partial x} - x \frac{\partial}{\partial z} \right) \tag{5.62}$$

$$\hat{L}_z = -i\hbar \left(x \frac{\partial}{\partial y} - y \frac{\partial}{\partial x} \right) \tag{5.63}$$

(Since $\hat{y}\hat{p}_z = \hat{p}_z\hat{y}$, etc., we do not run into any problems of noncommutativity in
constructing these operators.) Using

$$\hat{L}^2 = |\hat{\mathbf{L}}|^2 = \hat{\mathbf{L}} \cdot \hat{\mathbf{L}} = \hat{L}_x^2 + \hat{L}_y^2 + \hat{L}_z^2 \tag{5.64}$$

we can construct the operator for the square of the angular-momentum magnitude
from the operators in (5.61)–(5.63).

Since the commutation relations determine which observables can be
assigned definite values, we investigate these relations for angular momentum.
Operating on some function $f(x, y, z)$ with $\hat{L}_y$, we have

$$\hat{L}_y f = -i\hbar \left(z \frac{\partial f}{\partial x} - x \frac{\partial f}{\partial z} \right) \tag{5.65}$$

Operating on this last equation with $\hat{L}_x$, we have

$$\hat{L}_x \hat{L}_y f = -\hbar^2 \left(y \frac{\partial f}{\partial x} + yz \frac{\partial^2 f}{\partial z \partial x} - yx \frac{\partial^2 f}{\partial z^2} - z^2 \frac{\partial^2 f}{\partial y \partial x} + zx \frac{\partial^2 f}{\partial y \partial z} \right) \tag{5.66}$$

Similarly,

$$\hat{L}_x f = -i\hbar \left(y \frac{\partial f}{\partial z} - z \frac{\partial f}{\partial y} \right) \tag{5.67}$$

$$\hat{L}_y \hat{L}_x f = -\hbar^2 \left(zy \frac{\partial^2 f}{\partial x \partial z} - z^2 \frac{\partial^2 f}{\partial x \partial y} - xy \frac{\partial^2 f}{\partial z^2} + x \frac{\partial f}{\partial y} + xz \frac{\partial^2 f}{\partial z \partial y} \right) \tag{5.68}$$

Subtracting (5.68) from (5.66), we have

$$\hat{L}_x \hat{L}_y f - \hat{L}_y \hat{L}_x f = -\hbar^2 \left(y \frac{\partial f}{\partial x} - x \frac{\partial f}{\partial y} \right) \tag{5.69}$$

$$[\hat{L}_x, \hat{L}_y] = i\hbar \hat{L}_z \tag{5.70}$$

where we have used relations such as

$$\frac{\partial^2 f}{\partial z \, \partial x} = \frac{\partial^2 f}{\partial x \, \partial z} \tag{5.71}$$

which are true for well-behaved functions. We could use the same procedure to find $[\hat{L}_y, \hat{L}_z]$ and $[\hat{L}_z, \hat{L}_x]$, but we can save time by noting a certain kind of symmetry in (5.61)–(5.63). By a *cyclic permutation* of x, y, and z we mean replacing x by y, replacing y by z, and replacing z by x. If we carry out a cyclic permutation in $\hat{L}_x$, we get $\hat{L}_y$; a cyclic permutation in $\hat{L}_y$ gives $\hat{L}_z$; and $\hat{L}_z$ is transformed into $\hat{L}_x$ by a cyclic permutation. Hence by carrying out two successive cyclic permutations on (5.70), we have

$$[\hat{L}_y, \hat{L}_z] = i\hbar \hat{L}_x, \qquad [\hat{L}_z, \hat{L}_x] = i\hbar \hat{L}_y \tag{5.72}$$

Now we evaluate the commutators of $\hat{L}^2$ with each of its components, using commutator identities of Section 5.1.

$$\begin{aligned} [\hat{L}^2, \hat{L}_x] &= [\hat{L}_x^2 + \hat{L}_y^2 + \hat{L}_z^2, \hat{L}_x] \\ &= [\hat{L}_x^2, \hat{L}_x] + [\hat{L}_y^2, \hat{L}_x] + [\hat{L}_z^2, \hat{L}_x] \\ &= [\hat{L}_y^2, \hat{L}_x] + [\hat{L}_z^2, \hat{L}_x] \\ &= [\hat{L}_y, \hat{L}_x]\hat{L}_y + \hat{L}_y[\hat{L}_y, \hat{L}_x] + [\hat{L}_z, \hat{L}_x]\hat{L}_z + \hat{L}_z[\hat{L}_z, \hat{L}_x] \\ &= -i\hbar \hat{L}_z \hat{L}_y - i\hbar \hat{L}_y \hat{L}_z + i\hbar \hat{L}_y \hat{L}_z + i\hbar \hat{L}_z \hat{L}_y \end{aligned}$$

$$[\hat{L}^2, \hat{L}_x] = 0 \tag{5.73}$$

Since a cyclic permutation of x, y, and z leaves $\hat{L}^2 = \hat{L}_x^2 + \hat{L}_y^2 + \hat{L}_z^2$ unchanged, if we carry out two such permutations on (5.73), we get

$$[\hat{L}^2, \hat{L}_y] = 0, \qquad [\hat{L}^2, \hat{L}_z] = 0 \tag{5.74}$$

To which of the quantities L^2, L_x, L_y, L_z can we assign definite values simultaneously? Because $\hat{L}^2$ commutes with each of its components, we can specify an exact value for L^2 and any *one* component. However, because no two components of $\hat{L}$ commute with each other, we cannot specify more than one component simultaneously. (There is one exception to this statement which will be discussed shortly.) It is traditional to take L_z as the component of angular momentum that will be specified along with L^2. Note that in specifying $L^2 = |\mathbf{L}|^2$, we are not specifying the vector $\mathbf{L}$, only its magnitude. A complete specification of $\mathbf{L}$ requires simultaneous specification of each of its three components, which we usually cannot do. In classical mechanics when angular momentum is conserved, each of its three components has a definite value. In quantum mechanics when angular momentum is conserved, only its magnitude and one of its components are specifiable.

We could now attempt to find the eigenvalues and common eigenfunctions of $\hat{L}^2$ and $\hat{L}_z$ by using the forms for these operators in Cartesian coordinates. However, we would find that the partial differential equations obtained would not be separable. For this reason we carry out a transformation to spherical polar coordinates (Fig. 5.4). The coordinate r is the distance from the origin to the point (x, y, z). The angle θ is the angle the vector $\mathbf{r}$ makes with the positive z axis. The angle

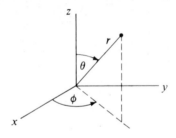

Figure 5.4 *Spherical polar coordinates*

that the projection of **r** in the xy plane makes with the positive x axis is φ. (Mathematics texts often interchange θ and φ.) A little trigonometry gives

$$x = r \sin \theta \cos \varphi, \qquad y = r \sin \theta \sin \varphi, \qquad z = r \cos \theta \qquad \textbf{(5.75)}$$

$$r^2 = x^2 + y^2 + z^2 \qquad\qquad\qquad\qquad\qquad\qquad\qquad\qquad \textbf{(5.76)}$$

$$\cos \theta = \frac{z}{(x^2 + y^2 + z^2)^{1/2}} \qquad\qquad\qquad\qquad\qquad\qquad \textbf{(5.77)}$$

$$\tan \varphi = y/x \qquad\qquad\qquad\qquad\qquad\qquad\qquad\qquad \textbf{(5.78)}$$

To transform the angular-momentum operators to spherical polar coordinates, we must transform $\partial/\partial x$, $\partial/\partial y$, and $\partial/\partial z$ into these coordinates. [This transformation may be skimmed if desired. Begin reading again after Eq. (5.93).]

To perform this transformation, we use the *chain rule*. Suppose we have a function of r, θ, and φ: $f(r, \theta, \varphi)$. If we carry out a change of independent variables by substituting

$$r = r(x, y, z), \qquad \theta = \theta(x, y, z), \qquad \varphi = \varphi(x, y, z) \qquad \textbf{(5.79)}$$

into f, we transform it into a function of x, y, and z:

$$f[r(x, y, z), \theta(x, y, z), \varphi(x, y, z)] = g(x, y, z) \qquad \textbf{(5.80)}$$

For example, suppose that

$$f(r, \theta, \varphi) = 3r \cos \theta + 2 \tan^2 \varphi$$

Using (5.76), (5.77), and (5.78), we have

$$g(x, y, z) = 3z + 2y^2 x^{-2}$$

The chain rule tells us how the partial derivatives of $g(x, y, z)$ are related to those of $f(r, \theta, \varphi)$. In fact,

$$\left(\frac{\partial g}{\partial x} \right)_{y,z} = \left(\frac{\partial f}{\partial r} \right)_{\theta,\varphi} \left(\frac{\partial r}{\partial x} \right)_{y,z} + \left(\frac{\partial f}{\partial \theta} \right)_{r,\varphi} \left(\frac{\partial \theta}{\partial x} \right)_{y,z} + \left(\frac{\partial f}{\partial \varphi} \right)_{r,\theta} \left(\frac{\partial \varphi}{\partial x} \right)_{y,z} \qquad \textbf{(5.81)}$$

$$\left(\frac{\partial g}{\partial y} \right)_{x,z} = \left(\frac{\partial f}{\partial r} \right)_{\theta,\varphi} \left(\frac{\partial r}{\partial y} \right)_{x,z} + \left(\frac{\partial f}{\partial \theta} \right)_{r,\varphi} \left(\frac{\partial \theta}{\partial y} \right)_{x,z} + \left(\frac{\partial f}{\partial \varphi} \right)_{r,\theta} \left(\frac{\partial \varphi}{\partial y} \right)_{x,z} \qquad \textbf{(5.82)}$$

$$\left(\frac{\partial g}{\partial z} \right)_{x,y} = \left(\frac{\partial f}{\partial r} \right)_{\theta,\varphi} \left(\frac{\partial r}{\partial z} \right)_{x,y} + \left(\frac{\partial f}{\partial \theta} \right)_{r,\varphi} \left(\frac{\partial \theta}{\partial z} \right)_{x,y} + \left(\frac{\partial f}{\partial \varphi} \right)_{r,\theta} \left(\frac{\partial \varphi}{\partial z} \right)_{x,y} \qquad \textbf{(5.83)}$$

To convert these equations to operator equations, we delete f and g. However, it would not do to write, for example,

$$\frac{\partial}{\partial r}\left(\frac{\partial r}{\partial x}\right)_{y,z}$$

for the first term on the right side of the operator equation corresponding to (5.81), because this would imply that $\partial/\partial r$ was to operate on

$$\left[\left(\frac{\partial r}{\partial x}\right)_{y,z} f\right]$$

whereas, according to (5.81), it should operate only on f. For the operator equation corresponding to (5.81), we thus write

$$\frac{\partial}{\partial x} = \left(\frac{\partial r}{\partial x}\right)_{y,z}\frac{\partial}{\partial r} + \left(\frac{\partial \theta}{\partial x}\right)_{y,z}\frac{\partial}{\partial \theta} + \left(\frac{\partial \varphi}{\partial x}\right)_{y,z}\frac{\partial}{\partial \varphi} \tag{5.84}$$

with similar equations for $\partial/\partial y$ and $\partial/\partial z$. The task now is to evaluate the partial derivatives such as $(\partial r/\partial x)_{y,z}$. Taking the partial derivative of (5.76) with respect to x, at constant y and z, we have

$$2r\left(\frac{\partial r}{\partial x}\right)_{y,z} = 2x = 2r \sin\theta \cos\varphi$$

$$\left(\frac{\partial r}{\partial x}\right)_{y,z} = \sin\theta \cos\varphi \tag{5.85}$$

Differentiating (5.76) with respect to y and with respect to z, we find

$$\left(\frac{\partial r}{\partial y}\right)_{x,z} = \sin\theta \sin\varphi, \qquad \left(\frac{\partial r}{\partial z}\right)_{x,y} = \cos\theta \tag{5.86}$$

From (5.77), we find

$$-\sin\theta\left(\frac{\partial\theta}{\partial x}\right)_{y,z} = -\frac{xz}{r^3} \tag{5.87}$$

$$\left(\frac{\partial\theta}{\partial x}\right)_{y,z} = \frac{\cos\theta \cos\varphi}{r} \tag{5.88}$$

Also

$$\left(\frac{\partial\theta}{\partial y}\right)_{x,z} = \frac{\cos\theta \sin\varphi}{r}, \qquad \left(\frac{\partial\theta}{\partial z}\right)_{x,y} = -\frac{\sin\theta}{r} \tag{5.89}$$

From (5.78) we have

$$\left(\frac{\partial\varphi}{\partial x}\right)_{y,z} = -\frac{\sin\varphi}{r\sin\theta}, \qquad \left(\frac{\partial\varphi}{\partial y}\right)_{x,z} = \frac{\cos\varphi}{r\sin\theta}, \qquad \left(\frac{\partial\varphi}{\partial z}\right)_{x,y} = 0 \tag{5.90}$$

Substituting (5.85), (5.88), and (5.90) into (5.84), we find

$$\frac{\partial}{\partial x} = \sin\theta \cos\varphi\,\frac{\partial}{\partial r} + \frac{\cos\theta \cos\varphi}{r}\frac{\partial}{\partial \theta} - \frac{\sin\varphi}{r\sin\theta}\frac{\partial}{\partial \varphi} \tag{5.91}$$

Similarly,

$$\frac{\partial}{\partial y} = \sin\theta\sin\varphi\,\frac{\partial}{\partial r} + \frac{\cos\theta\sin\varphi}{r}\frac{\partial}{\partial\theta} + \frac{\cos\varphi}{r\sin\theta}\frac{\partial}{\partial\varphi} \tag{5.92}$$

$$\frac{\partial}{\partial z} = \cos\theta\,\frac{\partial}{\partial r} - \frac{\sin\theta}{r}\frac{\partial}{\partial\theta} \tag{5.93}$$

At long last, we are ready to express the angular-momentum components in spherical polar coordinates. Substituting (5.75), (5.92), and (5.93) into (5.61), we have

$$\hat{L}_x = -i\hbar\left[r\sin\theta\sin\varphi\left(\cos\theta\,\frac{\partial}{\partial r} - \frac{\sin\theta}{r}\frac{\partial}{\partial\theta}\right)\right.$$

$$\left. - r\cos\theta\left(\sin\theta\sin\varphi\,\frac{\partial}{\partial r} + \frac{\cos\theta\sin\varphi}{r}\frac{\partial}{\partial\theta} + \frac{\cos\varphi}{r\sin\theta}\frac{\partial}{\partial\varphi}\right)\right]$$

$$\hat{L}_x = i\hbar\left(\sin\varphi\,\frac{\partial}{\partial\theta} + \cot\theta\cos\varphi\,\frac{\partial}{\partial\varphi}\right) \tag{5.94}$$

Also

$$\hat{L}_y = -i\hbar\left(\cos\varphi\,\frac{\partial}{\partial\theta} - \cot\theta\sin\varphi\,\frac{\partial}{\partial\varphi}\right) \tag{5.95}$$

$$\hat{L}_z = -i\hbar\,\frac{\partial}{\partial\varphi} \tag{5.96}$$

By squaring each of $\hat{L}_x$, $\hat{L}_y$, and $\hat{L}_z$, and then adding their squares, we can construct $\hat{L}^2$ [Eq. (5.64)]. The result is (Problem 5.10)

$$\hat{L}^2 = -\hbar^2\left(\frac{\partial^2}{\partial\theta^2} + \cot\theta\,\frac{\partial}{\partial\theta} + \frac{1}{\sin^2\theta}\frac{\partial^2}{\partial\varphi^2}\right) \tag{5.97}$$

Although the angular-momentum operators depend on all three Cartesian coordinates, x, y, and z, they involve only the two spherical polar coordinates θ and φ.

We now find the common eigenfunctions of $\hat{L}^2$ and $\hat{L}_z$, which we denote by Y. Since these operators involve θ and φ only, Y will be a function of these two coordinates: $Y = Y(\theta, \varphi)$. (Of course, since the operators are linear, we can multiply Y by an arbitrary function of r and still have an eigenfunction of $\hat{L}^2$ and $\hat{L}_z$.) We must solve

$$\hat{L}_z Y(\theta, \varphi) = bY(\theta, \varphi) \tag{5.98}$$

$$\hat{L}^2 Y(\theta, \varphi) = cY(\theta, \varphi) \tag{5.99}$$

where b and c are the eigenvalues of $\hat{L}_z$ and $\hat{L}^2$.

Using the $\hat{L}_z$ operator, we have

$$-i\hbar\,\frac{\partial}{\partial\varphi}Y(\theta, \varphi) = bY(\theta, \varphi) \tag{5.100}$$

Since the operator in (5.100) does not involve θ, we try a separation of variables, writing

$$Y(\theta, \varphi) = S(\theta)T(\varphi) \tag{5.101}$$

Equation (5.100) becomes

$$-i\hbar \frac{\partial}{\partial \varphi}[S(\theta)T(\varphi)] = bS(\theta)T(\varphi)$$

$$-i\hbar S(\theta)\frac{dT(\varphi)}{d\varphi} = bS(\theta)T(\varphi)$$

$$\frac{dT(\varphi)}{T(\varphi)} = \frac{ib}{\hbar}d\varphi$$

$$T(\varphi) = Ae^{ib\varphi/\hbar} \qquad \textbf{(5.102)}$$

where A is an arbitrary constant. Is T suitable as an eigenfunction? The answer is no, since it is not, in general, a single-valued function. If we add 2π to φ, we will still be at the same point in space, and hence we want no change in T when this is done. For T to be single-valued, we have the restriction

$$T(\varphi + 2\pi) = T(\varphi) \qquad \textbf{(5.103)}$$

$$Ae^{ib\varphi/\hbar}e^{ib2\pi/\hbar} = Ae^{ib\varphi/\hbar}$$

$$e^{ib2\pi/\hbar} = 1 \qquad \textbf{(5.104)}$$

To satisfy $e^{i\alpha} = \cos\alpha + i\sin\alpha = 1$, we must have $\alpha = 2\pi m$, where

$$m = 0, \pm 1, \pm 2, \pm \cdots$$

Therefore (5.104) gives

$$2\pi b/\hbar = 2\pi m$$

$$b = m\hbar, \qquad m = \cdots -2, -1, 0, 1, 2, \cdots \qquad \textbf{(5.105)}$$

and (5.102) becomes

$$T(\varphi) = Ae^{im\varphi}, \qquad m = 0, \pm 1, \pm 2, \ldots \qquad \textbf{(5.106)}$$

The eigenvalues for the z component of angular momentum are quantized.

We fix A by normalizing T. First let us consider normalizing some function F of r, θ, and φ. The ranges of the independent variables are (see Fig. 5.4)

$$0 \leqslant r \leqslant \infty, \qquad 0 \leqslant \theta \leqslant \pi, \qquad 0 \leqslant \varphi \leqslant 2\pi$$

The infinitesimal volume element in spherical polar coordinates corresponding to $dx\,dy\,dz$ in Cartesian coordinates is (*Taylor and Mann*, p. 417)

$$r^2 \sin\theta\, dr\, d\theta\, d\varphi \qquad \textbf{(5.107)}$$

Hence we have

$$\int_0^\infty \left[\int_0^\pi \left[\int_0^{2\pi} |F^2(r, \theta, \varphi)|\, d\varphi \right] \sin\theta\, d\theta \right] r^2\, dr = 1 \qquad \textbf{(5.108)}$$

If it happens that F has the form

$$F(r, \theta, \varphi) = R(r)S(\theta)T(\varphi) \qquad \textbf{(5.109)}$$

then (5.108) becomes

$$\int_0^\infty |R^2(r)| r^2\, dr \int_0^\pi |S^2(\theta)| \sin\theta\, d\theta \int_0^{2\pi} |T^2(\varphi)|\, d\varphi = 1 \qquad \text{(5.110)}$$

and it is convenient to normalize each factor of F separately:

$$\int_0^\infty |R^2| r^2\, dr = 1, \qquad \int_0^\pi |S^2| \sin\theta\, d\theta = 1, \qquad \int_0^{2\pi} |T^2|\, d\varphi = 1 \quad \text{(5.111)}$$

(We did the same sort of thing for the wave function of the particle in a three-dimensional box.) Therefore

$$\int_0^{2\pi} (Ae^{im\varphi})^* \, Ae^{im\varphi}\, d\varphi = 1 = |A|^2 \int_0^{2\pi} d\varphi \qquad \text{(5.112)}$$

$$|A| = (2\pi)^{-1/2} \qquad \text{(5.113)}$$

$$T(\varphi) = \frac{1}{\sqrt{2\pi}}\, e^{im\varphi}, \qquad m = 0, \pm 1, \pm 2, \ldots \qquad \text{(5.114)}$$

We now solve (5.99) for the eigenvalues of $\hat{L}^2$. Using (5.97), (5.101), and (5.114), we have

$$-\hbar^2 \left(\frac{\partial^2}{\partial\theta^2} + \cot\theta\, \frac{\partial}{\partial\theta} + \frac{1}{\sin^2\theta}\, \frac{\partial^2}{\partial\varphi^2} \right) \left(S(\theta)\, \frac{1}{\sqrt{2\pi}}\, e^{im\varphi} \right) = cS(\theta)\, \frac{1}{\sqrt{2\pi}}\, e^{im\varphi}$$

$$\frac{d^2 S}{d\theta^2} + \cot\theta\, \frac{dS}{d\theta} - \frac{m^2}{\sin^2\theta}\, S = -\frac{c}{\hbar^2}\, S \qquad \text{(5.115)}$$

To solve (5.115), we carry out some rather unexciting manipulations, which may be skimmed on a first reading. First, for convenience, we change the independent variable by making the substitution

$$w = \cos\theta \qquad \text{(5.116)}$$

This transforms S into some new function of w:

$$S(\theta) = G(w) \qquad \text{(5.117)}$$

The chain rule gives

$$\frac{dS}{d\theta} = \frac{dG}{dw}\, \frac{dw}{d\theta} = -\sin\theta\, \frac{dG}{dw} = -(1-w^2)^{1/2}\, \frac{dG}{dw} \qquad \text{(5.118)}$$

To calculate $d^2 S/d\theta^2$, we use some operator algebra:

$$\frac{d}{d\theta} = -(1-w^2)^{1/2}\, \frac{d}{dw}$$

$$\frac{d^2}{d\theta^2} = (1-w^2)^{1/2}\, \frac{d}{dw}\, (1-w^2)^{1/2}\, \frac{d}{dw}$$

$$\frac{d^2}{d\theta^2} = (1-w^2)\, \frac{d^2}{dw^2} + (1-w^2)^{1/2}\, (\tfrac{1}{2})\, (1-w^2)^{-1/2}\, (-2w)\, \frac{d}{dw}$$

$$\frac{d^2 S}{d\theta^2} = (1-w^2)\, \frac{d^2 G}{dw^2} - w\, \frac{dG}{dw} \qquad \text{(5.119)}$$

Using (5.119), (5.118), and

$$\cot \theta = \frac{\cos \theta}{\sin \theta} = \frac{w}{(1 - w^2)^{1/2}}$$

we find that (5.115) becomes

$$(1 - w^2)\frac{d^2 G}{dw^2} - 2w\frac{dG}{dw} + \left[\frac{c}{\hbar^2} - \frac{m^2}{(1 - w^2)}\right]G(w) = 0 \qquad \textbf{(5.120)}$$

The range of w is $-1 \leqslant w \leqslant 1$.

To get a two-term recursion relation when we try a power-series solution, we make the following change of dependent variable:

$$G(w) = (1 - w^2)^{|m|/2} H(w) \qquad \textbf{(5.121)}$$

Differentiating (5.121), we evaluate G' and G'', and (5.120) becomes, after we divide by $(1 - w^2)^{|m|/2}$,

$$(1 - w^2)H'' - 2(|m| + 1)wH' + [c\hbar^{-2} - |m|(|m| + 1)]H = 0 \qquad \textbf{(5.122)}$$

We now try a power-series for H:

$$H(w) = \sum_{j=0}^{\infty} a_j w^j \qquad \textbf{(5.123)}$$

Differentiating [compare Eqs. (4.38)–(4 40)], we have

$$H'(w) = \sum_{j=0}^{\infty} j a_j w^{j-1}$$

$$H''(w) = \sum_{j=0}^{\infty} j(j-1)a_j w^{j-2} = \sum_{j=0}^{\infty} (j+2)(j+1)a_{j+2}w^j$$

Substitution of these power series into (5.122) yields (after combining sums)

$$\sum_{j=0}^{\infty} \left[(j+2)(j+1)a_{j+2} + \left(-j^2 - j - 2|m|j + \frac{c}{\hbar^2} - |m|^2 - |m| \right)a_j \right]w^j = 0$$

Setting the coefficient of w^j equal to zero, we have the recursion relation

$$a_{j+2} = \frac{[(j + |m|)(j + |m| + 1) - c/\hbar^2]}{(j+1)(j+2)} a_j \qquad \textbf{(5.124)}$$

Just as in the harmonic-oscillator case, the general solution of (5.122) is an arbitrary linear combination of a series of even powers (whose coefficients are determined by a_0) and a series of odd powers (whose coefficients are determined by a_1). It can be shown[1] that the infinite series defined by the recursion relation (5.124) does not give well-behaved eigenfunctions. Hence as in the harmonic-oscillator case, we must cause one of the series to break off, its last term being $a_k w^k$. We eliminate the other series by setting a_0 or a_1 equal to zero, depending on whether k is odd or even.

---

[1] Many texts point out that the infinite series diverges at $w = \pm 1$. However, this is not sufficient cause to reject the infinite series, since the eigenfunctions might be quadratically integrable, even though infinite at two points. For a careful discussion, see M. Whippen, *Am. J. Phys.*, **34**, 656 (1966).

Setting the coefficient of a_k in (5.124) equal to zero, we have

$$c = \hbar^2 (k + |m|) (k + |m| + 1), \qquad k = 0, 1, 2, \ldots \qquad (5.125)$$

Since $|m|$ takes on the values $0, 1, 2, \ldots$, the quantity $k + |m|$ takes on the values $0, 1, 2, \ldots$. We therefore define the quantum number l as

$$l \equiv k + |m| \qquad (5.126)$$

and the allowed eigenvalues for the square of the magnitude of angular momentum are

$$c = l(l+1)\hbar^2, \qquad l = 0, 1, 2, \ldots \qquad (5.127)$$

From (5.126) it follows that $|m| \leqslant l$. The possible values for m are thus

$$m = -l, \ -l+1, \ -l+2, \ \ldots, \ -1, 0, 1, \ \ldots, l-2, l-1, l \qquad (5.128)$$

Taking the positive square root of (5.127), we have for the magnitude of the angular momentum

$$[l(l+1)]^{1/2}\hbar \qquad (5.129)$$

Since $l \geqslant |m|$, Eq. (5.129) shows that the magnitude of the total angular momentum is greater than the magnitude of the z component $|m|\hbar$ (except for $l = 0$). If it were possible to have the total angular momentum equal to its z component, this would mean that the x and y components were zero, and we would have specified all three components of **L**. However, since the components of angular momentum do not commute with each other, we cannot do this. The one exception is when l is zero. In this case $|\mathbf{L}|^2 = L_x^2 + L_y^2 + L_z^2$ has zero for its eigenvalue, and it must be true that all three components L_x, L_y, and L_z have zero eigenvalues. From Eq. (5.16) the uncertainties in angular-momentum components satisfy

$$\Delta L_x \, \Delta L_y \geqslant \frac{1}{2} \left| \int \Psi^* [\hat{L}_x, \hat{L}_y] \Psi \, dq \right| = \frac{\hbar}{2} \left| \int \Psi^* \hat{L}_z \Psi \, dq \right| \qquad (5.130)$$

and two similar equations obtained by cyclic permutation. When the eigenvalues of $\hat{L}_z$, $\hat{L}_x$, and $\hat{L}_y$ are zero, $\hat{L}_x \Psi = 0$, $\hat{L}_y \Psi = 0$, $\hat{L}_z \Psi = 0$, the right-hand sides of (5.130) and the two similar equations are zero, and having $\Delta L_x = \Delta L_y = \Delta L_z = 0$ is permitted. But what about the statement in Section 5.1 that to have simultaneous eigenfunctions of two operators, the operators must commute? The answer is that this theorem refers to the possibility of having the whole set of eigenfunctions of one operator be eigenfunctions of the other operator. Thus even though $\hat{L}_x$ and $\hat{L}_z$ do not commute, it is possible to have *some* of the eigenfunctions of $\hat{L}_z$ (those with $l = 0 = m$) be eigenfunctions of $\hat{L}_x$. However, it is impossible to have *all* of the $\hat{L}_z$ eigenfunctions also be eigenfunctions of $\hat{L}_x$.

Let us find some of the angular-momentum eigenfunctions. From Eqs. (5.116), (5.117), (5.121), (5.123), (5.125), and (5.126), the theta factor in the eigenfunctions is

$$S_{l,m}(\theta) = \sin^{|m|}(\theta) \sum_{\substack{j = 1, 3, \ldots \\ \text{or } j = 0, 2, \ldots}}^{l - |m|} a_j \cos^j \theta \qquad (5.131)$$

where the sum is over even or odd values of j, depending on whether $l - |m|$ is even or odd.

For $l = 0$, we have $m = 0$, and the theta factor is

$$S_{0,0}(\theta) = a_0 \tag{5.132}$$

Normalizing [Eq. (5.111)], we have

$$\int_0^\pi |a_0^2| \sin\theta \, d\theta = 1 = 2|a_0^2|$$

$$|a_0| = 2^{-1/2} \tag{5.133}$$

The eigenfunction is [Eqs. (5.101) and (5.114)]

$$Y_l^m(\theta, \varphi) = S_{l,m}(\theta)T(\varphi) = \frac{1}{\sqrt{2\pi}} S_{l,m}(\theta)e^{im\varphi} \tag{5.134}$$

$$Y_0^0(\theta, \varphi) = \frac{1}{\sqrt{4\pi}} \tag{5.135}$$

[Obviously, (5.135) is an eigenfunction of the operators $\hat{L}^2$, $\hat{L}_x$, $\hat{L}_y$, and $\hat{L}_z$, Eqs. (5.94)–(5.97).] For $l = 0$, there is no angular dependence in the eigenfunction— we say that the eigenfunctions are *spherically symmetric* for $l = 0$.

For $l = 1$, the possible values for m are $-1, 0, 1$. For $|m| = 1$,

$$S_{1,\pm 1}(\theta) = a_0 \sin\theta \tag{5.136}$$

Note that a_0 in (5.136) is not necessarily the same as a_0 in (5.132). Normalization gives

$$1 = |a_0^2| \int_0^\pi \sin^2\theta \sin\theta \, d\theta = |a_0^2| \int_{-1}^1 (1 - w^2) \, dw$$

$$|a_0| = \sqrt{3}/2 \tag{5.137}$$

where the substitution $w = \cos\theta$ was made. For $m = 0$, we find

$$S_{1,0}(\theta) = (3/2)^{1/2} \cos\theta \tag{5.138}$$

To find further functions we may need the recursion relation (5.124), which, using (5.127), becomes

$$a_{j+2} = \frac{[(j + |m|)(j + |m| + 1) - l(l + 1)]}{(j + 1)(j + 2)} a_j \tag{5.139}$$

The functions $S_{l,m}(\theta)$ are well known in mathematics, and are *associated Legendre functions* multiplied by a normalization constant. The associated Legendre functions $P_l^{|m|}(w)$ are defined by

$$P_l^{|m|}(w) = \frac{1}{2^l l!}(1 - w^2)^{|m|/2}\frac{d^{l+|m|}}{dw^{l+|m|}}(w^2 - 1)^l, \qquad l = 0, 1, 2, \ldots \tag{5.140}$$

These functions are related to the *Legendre polynomials* $P_l(w)$, which are defined by

$$P_l(w) = \frac{1}{2^l l!} \frac{d^l}{dw^l} (w^2 - 1)^l, \qquad l = 0, 1, 2, \ldots \qquad (5.141)$$

From the definitions, we have

$$P_l^{|m|}(w) = (1 - w^2)^{|m|/2} \frac{d^{|m|}}{dw^{|m|}} P_l(w), \qquad P_l^0(w) = P_l(w) \qquad (5.142)$$

It can be shown that (*Pauling and Wilson*, p. 129)

$$S_{l,m}(\theta) = \left[\frac{(2l+1)}{2} \frac{(l-|m|)!}{(l+|m|)!} \right]^{1/2} P_l^{|m|}(\cos \theta) \qquad (5.143)$$

Equations (5.143) and (5.140) give the explicit formula for the normalized theta factor in the angular-momentum eigenfunctions. The first few Legendre polynomials are

$$\begin{aligned} P_0(w) &= 1 & P_2(w) &= \tfrac{1}{2}(3w^2 - 1) \\ P_1(w) &= w & P_3(w) &= \tfrac{1}{2}(5w^3 - 3w) \end{aligned} \qquad (5.144)$$

Some associated Legendre functions are

$$\begin{aligned} P_0^0(w) &= 1 & P_2^0(w) &= \tfrac{1}{2}(3w^2 - 1) \\ P_1^0(w) &= w & P_2^1(w) &= 3w(1 - w^2)^{1/2} \\ P_1^1(w) &= (1 - w^2)^{1/2} & P_2^2(w) &= 3 - 3w^2 \end{aligned} \qquad (5.145)$$

Using (5.143), we construct Table 5.1, which gives the theta factor in the angular-momentum eigenfunctions.

The eigenfunctions of $\hat{L}^2$ and $\hat{L}_z$ are called *spherical harmonics* (or surface

Table 5.1 $S_{l,m}(\theta)$

$l = 0$:	$S_{0,0}$	$= \tfrac{1}{2}\sqrt{2}$
$l = 1$:	$S_{1,0}$	$= \tfrac{1}{2}\sqrt{6}\cos\theta$
	$S_{1,\pm 1}$	$= \tfrac{1}{2}\sqrt{3}\sin\theta$
$l = 2$:	$S_{2,0}$	$= \tfrac{1}{4}\sqrt{10}\,(3\cos^2\theta - 1)$
	$S_{2,\pm 1}$	$= \tfrac{1}{2}\sqrt{15}\sin\theta\cos\theta$
	$S_{2,\pm 2}$	$= \tfrac{1}{4}\sqrt{15}\sin^2\theta$
$l = 3$:	$S_{3,0}$	$= \tfrac{3}{4}\sqrt{14}\,(\tfrac{5}{3}\cos^3\theta - \cos\theta)$
	$S_{3,\pm 1}$	$= \tfrac{1}{8}\sqrt{42}\sin\theta\,(5\cos^2\theta - 1)$
	$S_{3,\pm 2}$	$= \tfrac{1}{4}\sqrt{105}\sin^2\theta\cos\theta$
	$S_{3,\pm 3}$	$= \tfrac{1}{8}\sqrt{70}\sin^3\theta$

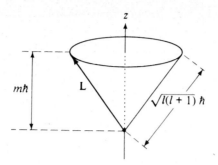

Figure 5.5 *Orientation of* **L**

harmonics) and are given by the formula

$$Y_l^m(\theta, \varphi) = \left[\frac{2l+1}{4\pi} \frac{(l-|m|)!}{(l+|m|)!} \right]^{1/2} P_l^{|m|}(\cos\theta)e^{im\varphi} \tag{5.146}$$

The phase of the normalization constant of the spherical harmonics is arbitrary (see Section 1.7); many physics texts use a different phase convention than in (5.146). Thus Y_l^m differs from text to text by a minus sign.

Summarizing our results, we see that the one-particle orbital angular-momentum eigenfunctions and eigenvalues are

$$\hat{L}^2 Y_l^m(\theta, \varphi) = l(l+1)\hbar^2 Y_l^m(\theta, \varphi), \qquad l = 0, 1, 2, \ldots \tag{5.147}$$

$$\hat{L}_z Y_l^m(\theta, \varphi) = m\hbar Y_l^m(\theta, \varphi), \qquad m = -l, -l+1, \ldots, l-1, l \tag{5.148}$$

where the eigenfunctions are given by (5.146). (Often the symbol m_l is used instead of m for the L_z quantum number.)

Since we cannot specify L_x and L_y, the vector **L** can lie anywhere on the surface of a cone whose axis is the z axis, whose altitude is $m\hbar$, and whose slant height is $\sqrt{l(l+1)}\hbar$ (Fig. 5.5). The possible orientations of **L** with respect to the z axis for the case $l = 1$ are shown in Fig. 5.6. For each eigenvalue of $\hat{L}^2$ there are $2l+1$

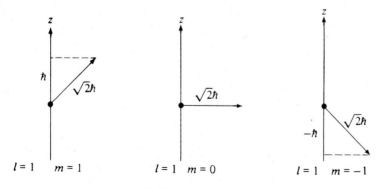

Figure 5.6 *Orientations of* **L** *for* $l = 1$

different eigenfunctions Y_l^m, corresponding to the $2l + 1$ values of m. We say that the $\hat{L}^2$ eigenvalues are $(2l + 1)$-fold degenerate. (The term *degeneracy* is applicable to the eigenvalues of any operator, not just the Hamiltonian.)

Of course, there is nothing special about the z axis; all directions of space are equivalent. If we had chosen to specify L^2 and L_x (rather than L_z), we would have gotten the same eigenvalues for L_x as we found for L_z. However, it is easier to solve the $\hat{L}_z$ eigenvalue equation because $\hat{L}_z$ has a simple form in spherical polar coordinates (which involve the angle of rotation φ about the z axis).

5.4   THE LADDER-OPERATOR METHOD FOR ANGULAR MOMENTUM

We obtained the eigenvalues of $\hat{L}^2$ and $\hat{L}_z$ by expressing these orbital angular-momentum operators as differential operators and solving the resulting differential equations. We now demonstrate that it is possible to solve for these eigenvalues using only the operator commutation relations. The work in this section is applicable to any operators that satisfy the angular-momentum commutation relations. In particular, it applies to spin angular momentum (Chapter 10), as well as orbital angular momentum. The operator manipulations that follow are abstract and may be difficult to fully grasp at first reading. However, the process of finding the eigenvalues of operators without using their explicit forms has an elegance that will reward the persistent student.

We used the letter L for orbital angular momentum. Here we will use the letter M to indicate that we are dealing with any kind of angular momentum. We have three linear operators $\hat{M}_x$, $\hat{M}_y$, and $\hat{M}_z$, and all we know about them is that they obey the commutation relations:

$$[\hat{M}_x, \hat{M}_y] = i\hbar\hat{M}_z, \qquad [\hat{M}_y, \hat{M}_z] = i\hbar\hat{M}_x, \qquad [\hat{M}_z, \hat{M}_x] = i\hbar\hat{M}_y \quad \text{(5.149)}$$

We define the operator $\hat{M}^2$ as

$$\hat{M}^2 = \hat{M}_x^2 + \hat{M}_y^2 + \hat{M}_z^2 \tag{5.150}$$

Our problem is to find the eigenvalues of $\hat{M}^2$ and $\hat{M}_z$.

We begin by evaluating the commutators of $\hat{M}^2$ with its components, using Eqs. (5.149) and (5.150). The work is identical with that used to derive Eqs. (5.73)–(5.74), and we have

$$[\hat{M}^2, \hat{M}_x] = [\hat{M}^2, \hat{M}_y] = [\hat{M}^2, \hat{M}_z] = 0 \tag{5.151}$$

Hence we can have simultaneous eigenfunctions of $\hat{M}^2$ and $\hat{M}_z$.

Next we define two new operators— the *raising operator* $\hat{M}_+$ and the *lowering operator* $\hat{M}_-$:

$$\hat{M}_+ \equiv \hat{M}_x + i\hat{M}_y \tag{5.152}$$

$$\hat{M}_- \equiv \hat{M}_x - i\hat{M}_y \tag{5.153}$$

These are examples of *ladder operators*. The reason for the terminology will become clear shortly. Let us investigate their properties. We have

$$\hat{M}_+\hat{M}_- = (\hat{M}_x + i\hat{M}_y)(\hat{M}_x - i\hat{M}_y) = \hat{M}_x(\hat{M}_x - i\hat{M}_y) + i\hat{M}_y(\hat{M}_x - i\hat{M}_y)$$

$$= \hat{M}_x^2 - i\hat{M}_x\hat{M}_y + i\hat{M}_y\hat{M}_x + \hat{M}_y^2 = \hat{M}^2 - \hat{M}_z^2 + i[\hat{M}_y, \hat{M}_x]$$

$$\hat{M}_+\hat{M}_- = \hat{M}^2 - \hat{M}_z^2 + \hbar\hat{M}_z \qquad (5.154)$$

Similarly, we find

$$\hat{M}_-\hat{M}_+ = \hat{M}^2 - \hat{M}_z^2 - \hbar\hat{M}_z \qquad (5.155)$$

For the commutators of these operators with $\hat{M}_z$, we have

$$[\hat{M}_+, \hat{M}_z] = [\hat{M}_x + i\hat{M}_y, \hat{M}_z] = [\hat{M}_x, \hat{M}_z] + i[\hat{M}_y, \hat{M}_z]$$

$$= -i\hbar\hat{M}_y - \hbar\hat{M}_x$$

$$[\hat{M}_+, \hat{M}_z] = -\hbar\hat{M}_+ \qquad (5.156)$$

$$\hat{M}_+\hat{M}_z = \hat{M}_z\hat{M}_+ - \hbar\hat{M}_+ \qquad (5.157)$$

Similarly, we find

$$\hat{M}_-\hat{M}_z = \hat{M}_z\hat{M}_- + \hbar\hat{M}_- \qquad (5.158)$$

Using Y for the common eigenfunctions of $\hat{M}^2$ and $\hat{M}_z$, we have

$$\hat{M}^2 Y = cY \qquad (5.159)$$

$$\hat{M}_z Y = bY \qquad (5.160)$$

where c and b are the eigenvalues. Operating on Eq. (5.160) with the raising operator, we get

$$\hat{M}_+\hat{M}_z Y = \hat{M}_+ bY \qquad (5.161)$$

Using Eq. (5.157) and the fact that $\hat{M}_+$ is linear, we have

$$(\hat{M}_z\hat{M}_+ - \hbar\hat{M}_+)Y = b\hat{M}_+ Y$$

$$\hat{M}_z(\hat{M}_+ Y) = (b + \hbar)(\hat{M}_+ Y) \qquad (5.162)$$

This last equation says that the function $\hat{M}_+ Y$ is an eigenfunction of $\hat{M}_z$ with eigenvalue $b + \hbar$. In other words, operating on the eigenfunction Y with the raising operator converts Y into another eigenfunction of $\hat{M}_z$ with eigenvalue $\hbar$ higher than the eigenvalue of Y. If we now apply the raising operator to (5.162) and use (5.157) again, we find similarly

$$\hat{M}_z(\hat{M}_+^2 Y) = (b + 2\hbar)(\hat{M}_+^2 Y)$$

Repeated application of the raising operator gives

$$\hat{M}_z(\hat{M}_+^k Y) = (b + k\hbar)(\hat{M}_+^k Y), \qquad k = 0, 1, 2, \ldots \qquad (5.163)$$

If we operate on (5.160) with the lowering operator and apply (5.158), we find in the same manner

$$\hat{M}_z(\hat{M}_- Y) = (b - \hbar)(\hat{M}_- Y) \qquad (5.164)$$

$$\hat{M}_z(\hat{M}_-^k Y) = (b - k\hbar)(\hat{M}_-^k Y) \qquad (5.165)$$

Thus by using the raising and lowering operators on the eigenfunction with eigenvalue b, we generate a ladder of eigenvalues, the difference from step to step being $\hbar$:

$$\vdots$$
$$\underline{}\,b + 2\hbar$$
$$\underline{}\,b + \hbar$$
$$\underline{}\,b$$
$$\underline{}\,b - \hbar$$
$$\underline{}\,b - 2\hbar$$
$$\vdots$$

The functions $\hat{M}_{\pm}^{k} Y$ are eigenfunctions of $\hat{M}_z$ with eigenvalues $b \pm k\hbar$. We now show that these functions are also eigenfunctions of $\hat{M}^2$, all with the *same* eigenvalue c:

$$\hat{M}_z \hat{M}_{\pm}^{k} Y = (b \pm k\hbar)\hat{M}_{\pm}^{k} Y \tag{5.166}$$

$$\hat{M}^2 \hat{M}_{\pm}^{k} Y = c\hat{M}_{\pm}^{k} Y, \qquad k = 0, 1, 2, \ldots \tag{5.167}$$

To prove (5.167) we first show that $\hat{M}^2$ commutes with $\hat{M}_+$ and $\hat{M}_-$:

$$[\hat{M}^2, \hat{M}_{\pm}] = [\hat{M}^2, \hat{M}_x \pm i\hat{M}_y] = [\hat{M}^2, \hat{M}_x] \pm i[\hat{M}^2, \hat{M}_y] = 0 \pm 0 = 0$$

We also have

$$[\hat{M}^2, \hat{M}_{\pm}^2] = [\hat{M}^2, \hat{M}_{\pm}]\hat{M}_{\pm} + \hat{M}_{\pm}[\hat{M}^2, \hat{M}_{\pm}] = 0 + 0 = 0 \tag{5.168}$$

and it follows by induction that

$$[\hat{M}^2, \hat{M}_{\pm}^{k}] = 0 \qquad \text{or} \qquad \hat{M}^2 \hat{M}_{\pm}^{k} = \hat{M}_{\pm}^{k} \hat{M}^2, \qquad k = 0, 1, 2, \ldots \tag{5.169}$$

If we operate on (5.159) with $\hat{M}_{\pm}^{k}$ and use (5.169), we get

$$\hat{M}_{\pm}^{k} \hat{M}^2 Y = \hat{M}_{\pm}^{k} c Y$$

$$\hat{M}^2 (\hat{M}_{\pm}^{k} Y) = c(\hat{M}_{\pm}^{k} Y) \tag{5.170}$$

which is what we wanted to prove.

Next we show that the set of eigenvalues of $\hat{M}_z$ generated using the ladder operators must be bounded. For the particular eigenfunction Y with eigenvalue b, we have

$$\hat{M}_z Y = bY \tag{5.171}$$

and for the set of eigenfunctions and eigenvalues generated by the ladder operators, we have

$$\hat{M}_z Y_k = b_k Y_k \tag{5.172}$$

where[2]

$$Y_k = \hat{M}_{\pm}^{k} Y \tag{5.173}$$

$$b_k = b \pm k\hbar \tag{5.174}$$

---

[2] Application of $\hat{M}_+$ (or $\hat{M}_-$) destroys the normalization of Y, so that Y_k is not normalized. For the normalization constant, see Problem 10.13.

Operating on (5.172) with $\hat{M}_z$, we have

$$\hat{M}_z^2 Y_k = b_k \hat{M}_z Y_k$$

$$\hat{M}_z^2 Y_k = b_k^2 Y_k \tag{5.175}$$

Now subtract (5.175) from (5.167), and use (5.173) and (5.150):

$$\hat{M}^2 Y_k - \hat{M}_z^2 Y_k = c Y_k - b_k^2 Y_k$$

$$(\hat{M}_x^2 + \hat{M}_y^2) Y_k = (c - b_k^2) Y_k \tag{5.176}$$

The operator $\hat{M}_x^2 + \hat{M}_y^2$ corresponds to a nonnegative physical quantity and hence has nonnegative eigenvalues. Therefore (5.176) implies that

$$c - b_k^2 \geqslant 0$$

$$\sqrt{c} \geqslant |b_k|$$

$$\sqrt{c} \geqslant b_k \geqslant -\sqrt{c}, \qquad k = 0, \pm 1, \pm 2, \ldots \tag{5.177}$$

Since c remains constant as k varies, (5.177) shows that the set of eigenvalues b_k is bounded above and below. Let b_{max} and b_{min} denote the maximum and minimum values of b_k. Y_{max} and Y_{min} will be the corresponding eigenfunctions:

$$\hat{M}_z Y_{max} = b_{max} Y_{max} \tag{5.178}$$

$$\hat{M}_z Y_{min} = b_{min} Y_{min} \tag{5.179}$$

Now operate on (5.178) with the raising operator, and use (5.157):

$$\hat{M}_+ \hat{M}_z Y_{max} = b_{max} \hat{M}_+ Y_{max}$$

$$\hat{M}_z (\hat{M}_+ Y_{max}) = (b_{max} + \hbar)(\hat{M}_+ Y_{max}) \tag{5.180}$$

This last equation seems to contradict the statement that b_{max} is the largest eigenvalue of $\hat{M}_z$, since it says that $\hat{M}_+ Y_{max}$ is an eigenfunction of $\hat{M}_z$ with eigenvalue $b_{max} + \hbar$. The only way out of this contradiction is to have the function $\hat{M}_+ Y_{max}$ vanish. (We always reject zero as an eigenfunction on physical grounds.) Thus

$$\hat{M}_+ Y_{max} = 0 \tag{5.181}$$

Operating on (5.181) with the lowering operator, and using (5.155) and (5.178), we have

$$\hat{M}_- \hat{M}_+ Y_{max} = 0$$

$$(\hat{M}^2 - \hat{M}_z^2 - \hbar \hat{M}_z) Y_{max} = 0$$

$$(c - b_{max}^2 - \hbar b_{max}) Y_{max} = 0$$

$$c - b_{max}^2 - \hbar b_{max} = 0$$

$$c = b_{max}^2 + \hbar b_{max} \tag{5.182}$$

A similar argument shows that

$$\hat{M}_- Y_{min} = 0 \tag{5.183}$$

and by applying the raising operator to this equation and using (5.154), we find

$$c = b_{min}^2 - \hbar b_{min} \tag{5.184}$$

Subtracting this last equation from (5.182), we have

$$b_{max}^2 + \hbar b_{max} + (\hbar b_{min} - b_{min}^2) = 0 \tag{5.185}$$

This is a quadratic equation in the unknown b_{max}, and using the usual formula (it still works in quantum mechanics) we find

$$b_{max} = -b_{min}, \qquad b_{max} = b_{min} - \hbar \tag{5.186}$$

The second root is rejected, since it says that b_{max} is less than b_{min}. So

$$b_{min} = -b_{max} \tag{5.187}$$

Moreover, from (5.174) we know that b_{max} and b_{min} differ by an integral multiple of $\hbar$:

$$b_{max} - b_{min} = n\hbar, \qquad n = 0, 1, 2, \ldots \tag{5.188}$$

Substituting (5.187) in (5.188), we have

$$b_{max} = \tfrac{1}{2} n\hbar \tag{5.189}$$

$$b_{max} = j\hbar, \qquad j = 0, \tfrac{1}{2}, 1, \tfrac{3}{2}, 2, \ldots \tag{5.190}$$

$$b_{min} = -j\hbar$$

$$b = -j\hbar, (-j+1)\hbar, (-j+2)\hbar, \ldots, (j-2)\hbar, (j-1)\hbar, j\hbar \tag{5.191}$$

and from (5.182) we find

$$c = j(j+1)\hbar^2, \qquad j = 0, \tfrac{1}{2}, 1, \tfrac{3}{2}, \ldots \tag{5.192}$$

We have found the eigenvalues of $\hat{M}^2$ and $\hat{M}_z$ using just the commutation relations. However, comparing (5.191)–(5.192) with Eqs. (5.147) and (5.148) shows that in addition to integral values for the angular-momentum quantum number ($l = 0$, 1, 2,...) we now also have the possibility for half-integral values ($j = 0, \tfrac{1}{2}, 1, \tfrac{3}{2}, \ldots$). This perhaps suggests that there might be another kind of angular momentum besides orbital angular momentum. In Chapter 10 we will see that spin angular momentum can have half-integral, as well as integral, quantum numbers. The particular form of the orbital angular-momentum differential operators rules out half-integral values.

The ladder-operator method can be used to solve other eigenvalue problems; see Problem 5.17.

PROBLEMS

5.1 Derive Eq. (5.15) for the standard deviation of an observable.

5.2 For the ground state of the one-dimensional harmonic oscillator, compute the standard deviations Δx and Δp_x, and check that the uncertainty principle is obeyed. Use the results of Problem 4.1 to save time.

5.3 Show that the standard deviation ΔA is zero when Ψ is an eigenfunction of $\hat{A}$.

5.4 Verify the commutator identities (5.2)–(5.8).

5.5 Let **A** have the components $(3, -2, 6)$; let **B** have the components $(-1, 4, 4)$. Find $|\mathbf{A}|$, $|\mathbf{B}|$, $\mathbf{A} + \mathbf{B}$, $\mathbf{A} - \mathbf{B}$, $\mathbf{A} \cdot \mathbf{B}$, $\mathbf{A} \times \mathbf{B}$. Find the angle between **A** and **B**.

5.6 Use the vector dot product to find the obtuse angle between two diagonals of a cube. What is the chemical significance of this angle?

5.7 Show that

$$\nabla^2 [f(x, y, z)g(x, y, z)] = g\nabla^2 f + 2\nabla f \cdot \nabla g + f\nabla^2 g$$

5.8 Let $f = 2x^2 - 5xyz + z^2 - 1$. Find $\mathbf{grad}\, f$. Find $\nabla^2 f$.

5.9 Find $\nabla \cdot \mathbf{r}$, where $\mathbf{r}$ is given by (5.51).

5.10 Derive Eq. (5.97) for $\hat{L}^2$ from Eqs. (5.94)–(5.96).

5.11 Derive the formula for $S_{2,0}$ (Table 5.1) in two ways: (a) by using (5.143); (b) by using the recursion relation and normalization.

5.12 Calculate the possible angles between **L** and the z axis for $l = 2$.

5.13 Let w be the variable defined as the number of heads that show when two coins are tossed simultaneously. Find $\langle w \rangle$ and σ_w. [*Hint:* Use (3.108) and (5.15).]

5.14 Show that the spherical harmonics are eigenfunctions of the operator $(\hat{L}_x^2 + \hat{L}_y^2)$. (The proof is short.) What are the eigenvalues?

5.15 Show that the three commutation relations (5.70) and (5.72) are equivalent to the single relation

$$\hat{\mathbf{L}} \times \hat{\mathbf{L}} = i\hbar \hat{\mathbf{L}}$$

5.16 Apply the lowering operator three times in succession to $Y_1^1 (\theta, \varphi)$ and verify that we obtain functions that are proportional to Y_1^0, Y_1^{-1}, and zero.

5.17 The one-dimensional harmonic-oscillator Hamiltonian is

$$\hat{H} = \frac{\hat{p}_x^2}{2m} + 2\pi^2 v^2 m\hat{x}^2$$

The raising and lowering operators for this problem are defined as

$$\hat{A}_+ \equiv \frac{1}{(2m)^{1/2}} [\hat{p}_x + 2\pi i v m\hat{x}], \qquad \hat{A}_- \equiv \frac{1}{(2m)^{1/2}} [\hat{p}_x - 2\pi i v m\hat{x}]$$

Show that

$$\hat{A}_+ \hat{A}_- = \hat{H} - \tfrac{1}{2}hv, \qquad \hat{A}_- \hat{A}_+ = \hat{H} + \tfrac{1}{2}hv$$

$$[\hat{A}_+, \hat{A}_-] = -hv$$

$$[\hat{H}, \hat{A}_+] = hv\hat{A}_+, \qquad [\hat{H}, \hat{A}_-] = -hv\hat{A}_-$$

Show that $\hat{A}_+$ and $\hat{A}_-$ are indeed ladder operators and that the eigenvalues are spaced at intervals of hv. Since both the kinetic energy and the potential energy are nonnegative, we expect the energy eigenvalues to be nonnegative. Hence there must be a state of minimum energy. Operate on the wave function for this state first with $\hat{A}_-$ and then with $\hat{A}_+$ and show that the lowest energy eigenvalue is $\tfrac{1}{2}hv$. Finally, conclude that

$$E = (n + \tfrac{1}{2})hv, \qquad n = 0, 1, 2, \ldots$$

5.18 Consider the following objection to the conclusions of Section 5.4. It might be argued that although we proved that the quantities

$$-j\hbar, (-j+1)\hbar, \ldots, j\hbar$$

are eigenvalues of $\hat{M}_z$, there might be other eigenvalues spaced between the eigenvalues we found. How can this objection be answered?

5.19 The *curl* of a vector function $\mathbf{A}$ is defined by $\mathbf{curl}\,\mathbf{A} \equiv \mathbf{\nabla} \times \mathbf{A}$. Prove that $\mathbf{curl}\,\mathbf{grad}\,g(x, y, z) = 0$ for all well-behaved functions g.

5.20 Consider the following incorrect derivation. Differentiation of x in (5.75) gives $\partial x / \partial r = \sin\theta \cos\varphi$. Then, since $\partial r / \partial x = 1/(\partial x / \partial r)$, we have $(\partial r / \partial x)_{y,z} = 1/(\sin\theta \cos\varphi)$ (?). But this result disagrees with (5.85). Find the error in this reasoning.

6 THE HYDROGEN ATOM

6.1 THE CENTRAL-FORCE PROBLEM

Before tackling the hydrogen atom, we shall consider the more general problem of a single particle moving under a central force. The results of this section will be applicable to any central-force problem, e.g., the hydrogen atom (Section 6.4), the isotropic three-dimensional harmonic oscillator (Problem 6.23).

A *central force* is derived from a potential-energy function that is spherically symmetric, i.e., that is a function only of the distance of the particle from the origin: $V = V(r)$. The relation between force and potential energy is given by Eq. (5.46). The partial derivatives in (5.46) can be evaluated by the chain rule [Eqs. (5.81)–(5.83)]. Since, in this case, V is a function of r only, we have

$$\left(\frac{\partial V}{\partial \theta}\right)_{r,\varphi} = \left(\frac{\partial V}{\partial \varphi}\right)_{r,\theta} = 0 \tag{6.1}$$

Hence

$$\left(\frac{\partial V}{\partial x}\right)_{y,z} = \frac{dV}{dr}\left(\frac{\partial r}{\partial x}\right)_{y,z} = \frac{x}{r}\frac{dV}{dr} \tag{6.2}$$

$$\left(\frac{\partial V}{\partial y}\right)_{x,z} = \frac{y}{r}\frac{dV}{dr}, \quad \left(\frac{\partial V}{\partial z}\right)_{x,y} = \frac{z}{r}\frac{dV}{dr} \tag{6.3}$$

where Eqs. (5.85) and (5.86) have been used. Equation (5.46) becomes

$$\mathbf{F} = -\frac{1}{r}\frac{dV}{dr}(x\mathbf{i} + y\mathbf{j} + z\mathbf{k}) = -\frac{dV(r)}{dr}\frac{\mathbf{r}}{r} \tag{6.4}$$

A central force has only a radial component.

Now we consider the quantum mechanics of a single particle subject to a central force. The Hamiltonian operator is

$$\hat{H} = \hat{T} + \hat{V} = -(\hbar^2/2m)\nabla^2 + V(r) \tag{6.5}$$

where ∇^2 is given by Eq. (3.55). Since V is spherically symmetric, we will work in spherical polar coordinates. Hence we want to transform the Laplacian operator to these coordinates. We already have the forms of the operators $\partial/\partial x, \partial/\partial y$, and $\partial/\partial z$ in these coordinates [Eqs. (5.91)–(5.93)], and by squaring each of these operators and then adding their squares, we get the Laplacian. This calculation is left as an exercise. The result is (Problem 6.12)

$$\nabla^2 = \frac{\partial^2}{\partial r^2} + \frac{2}{r}\frac{\partial}{\partial r} + \frac{1}{r^2}\frac{\partial^2}{\partial \theta^2} + \frac{1}{r^2}\cot\theta\,\frac{\partial}{\partial\theta} + \frac{1}{r^2\sin^2\theta}\frac{\partial^2}{\partial\varphi^2} \tag{6.6}$$

Looking back to (5.97), which gives the operator for the square of the magnitude of the orbital angular momentum of a single particle, $\hat{L}^2$, we see that

$$\nabla^2 = \frac{\partial^2}{\partial r^2} + \frac{2}{r}\frac{\partial}{\partial r} - \frac{1}{r^2\hbar^2}\hat{L}^2 \tag{6.7}$$

The Hamiltonian (6.5) becomes

$$\hat{H} = -\frac{\hbar^2}{2m}\left(\frac{\partial^2}{\partial r^2} + \frac{2}{r}\frac{\partial}{\partial r}\right) + \frac{1}{2mr^2}\hat{L}^2 + V(r) \tag{6.8}$$

In classical mechanics a particle subject to a central force has its angular momentum conserved (Section 5.3). In quantum mechanics we might ask whether we can have states with definite values for both the energy and the angular momentum. To have the set of eigenfunctions of $\hat{H}$ also be eigenfunctions of $\hat{L}^2$, the commutator $[\hat{H}, \hat{L}^2]$ must vanish. We have

$$[\hat{H}, \hat{L}^2] = [\hat{T}, \hat{L}^2] + [\hat{V}, \hat{L}^2] \tag{6.9}$$

$$[\hat{T}, \hat{L}^2] = \left[-\frac{\hbar^2}{2m}\left(\frac{\partial^2}{\partial r^2} + \frac{2}{r}\frac{\partial}{\partial r}\right) + \frac{1}{2mr^2}\hat{L}^2, \hat{L}^2\right] \tag{6.10}$$

$$[\hat{T}, \hat{L}^2] = -\frac{\hbar^2}{2m}\left[\frac{\partial^2}{\partial r^2} + \frac{2}{r}\frac{\partial}{\partial r}, \hat{L}^2\right] + \frac{1}{2m}\left[\frac{1}{r^2}\hat{L}^2, \hat{L}^2\right] \tag{6.11}$$

Recall that $\hat{L}^2$ involves only θ and φ and not r [Eq. (5.97)]. Hence it commutes with any operator that involves only r. [To reach this conclusion we must use relations like (5.71) with x and z replaced by r and θ.] Thus the first commutator in (6.11) is zero. Moreover, since any operator commutes with itself, the second commutator in (6.11) is zero. Hence

$$[\hat{T}, \hat{L}^2] = 0 \tag{6.12}$$

Now, since $\hat{L}^2$ does not involve r, and V involves only r, we have

$$[\hat{V}(r), \hat{L}^2] = 0$$

Therefore

$$[\hat{H}, \hat{L}^2] = 0 \tag{6.13}$$

The Hamiltonian commutes with $\hat{L}^2$ when the potential-energy function is independent of θ and φ.

Consider now the $\hat{L}_z$ operator, Eq. (5.96). Since $\hat{L}_z$ does not involve r, and since it commutes with $\hat{L}^2$ [Eq. (5.74)], we see that $\hat{L}_z$ commutes with the Hamiltonian (6.8)

$$[\hat{H}, \hat{L}_z] = 0 \tag{6.14}$$

We can, therefore, have a set of simultaneous eigenfunctions of $\hat{H}$, $\hat{L}^2$, and $\hat{L}_z$ for the central-force problem. Using ψ to denote these common eigenfunctions, we have

$$\hat{H}\psi = E\psi \tag{6.15}$$

$$\hat{L}^2\psi = l(l+1)\hbar^2\psi, \qquad l = 0, 1, 2, \ldots \tag{6.16}$$

$$\hat{L}_z\psi = m\hbar\psi, \qquad m = -l, -l+1, \ldots, l \tag{6.17}$$

where Eqs. (5.147) and (5.148) have been used.

Using (6.8) and (6.16), we have for the Schroedinger equation (6.15)

$$\left[-\frac{\hbar^2}{2m}\left(\frac{\partial^2}{\partial r^2} + \frac{2}{r}\frac{\partial}{\partial r}\right) + \frac{1}{2mr^2}\hat{L}^2 + V(r) \right]\psi = E\psi \tag{6.18}$$

$$\left[-\frac{\hbar^2}{2m}\left(\frac{\partial^2}{\partial r^2} + \frac{2}{r}\frac{\partial}{\partial r}\right) + \frac{l(l+1)\hbar^2}{2mr^2} + V(r) \right]\psi = E\psi \tag{6.19}$$

The eigenfunctions of $\hat{L}^2$ are the spherical harmonics $Y_l^m(\theta, \varphi)$, and since $\hat{L}^2$ does not involve r, we can multiply Y_l^m by an arbitrary function of r and still have eigenfunctions of $\hat{L}^2$. Therefore

$$\psi = R(r)Y_l^m(\theta, \varphi) \tag{6.20}$$

Using (6.20) in (6.19), we then divide both sides by Y_l^m to obtain an ordinary differential equation for the unknown function $R(r)$:

$$-\frac{\hbar^2}{2m}\left(R'' + \frac{2}{r}R'\right) + \frac{l(l+1)\hbar^2}{2mr^2}R + V(r)R = ER(r) \tag{6.21}$$

We have shown that for *any* one-particle problem with a spherically symmetric potential-energy function $V(r)$, the wave function ψ is a product of a radial factor and a spherical harmonic. The radial factor $R(r)$ satisfies Eq. (6.21). By using some specific form for $V(r)$ in (6.21), we can solve it for some particular problem.

6.2   REDUCTION OF THE TWO-PARTICLE PROBLEM TO A ONE-PARTICLE PROBLEM

The hydrogen atom contains two particles, the proton and the electron. So far we have considered only one-particle problems. We now show how we can frequently reduce a two-particle problem to a one-particle problem.

Consider the classical-mechanical treatment of two particles of masses m_1

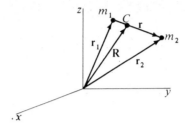

Figure 6.1 *A two-particle system*

and m_2. We specify their positions by the radius vectors $\mathbf{r}_1$ and $\mathbf{r}_2$ drawn from the origin of a Cartesian coordinate system (Fig. 6.1). Particles one and two have coordinates (x_1, y_1, z_1) and (x_2, y_2, z_2). We draw the vector $\mathbf{r} = \mathbf{r}_2 - \mathbf{r}_1$ from particle one to two and denote the components of $\mathbf{r}$ by x, y, and z:

$$x = x_2 - x_1, \qquad y = y_2 - y_1, \qquad z = z_2 - z_1 \qquad (6.22)$$

The coordinates x, y, and z are called the *relative* (or *internal*) coordinates. We now draw the vector $\mathbf{R}$ from the origin to the center of mass, point C, and denote the coordinates of C by X, Y, and Z:

$$\mathbf{R} = \mathbf{i}X + \mathbf{j}Y + \mathbf{k}Z \qquad (6.23)$$

The definition of the center of mass of this two-particle system gives

$$X = \frac{m_1 x_1 + m_2 x_2}{m_1 + m_2}, \qquad Y = \frac{m_1 y_1 + m_2 y_2}{m_1 + m_2}, \qquad Z = \frac{m_1 z_1 + m_2 z_2}{m_1 + m_2} \qquad (6.24)$$

These three equations are equivalent to the vector equation

$$\mathbf{R} = \frac{m_1 \mathbf{r}_1 + m_2 \mathbf{r}_2}{m_1 + m_2} \qquad (6.25)$$

We also have

$$\mathbf{r} = \mathbf{r}_2 - \mathbf{r}_1 \qquad (6.26)$$

We regard (6.25) and (6.26) as simultaneous linear equations in the two unknowns $\mathbf{r}_1$ and $\mathbf{r}_2$, and solve for them to get

$$\mathbf{r}_1 = \mathbf{R} - \frac{m_2}{m_1 + m_2}\mathbf{r}, \qquad \mathbf{r}_2 = \mathbf{R} + \frac{m_1}{m_1 + m_2}\mathbf{r} \qquad (6.27)$$

Equations (6.25) and (6.26) represent a transformation of coordinates from $x_1, y_1, z_1, x_2, y_2, z_2$ to X, Y, Z, x, y, z. Consider what happens to the Hamiltonian under this transformation. Let an overhead dot indicate differentiation with respect to time; the velocity of particle one is [Eq. (5.52)]

$$\mathbf{v}_1 = \frac{d\mathbf{r}_1}{dt} = \dot{\mathbf{r}}_1 \qquad (6.28)$$

The kinetic energy is the sum of the kinetic energies of the two particles:

$$T = \tfrac{1}{2}m_1 |\dot{\mathbf{r}}_1|^2 + \tfrac{1}{2}m_2 |\dot{\mathbf{r}}_2|^2 \qquad (6.29)$$

Introducing the time derivatives of Eqs. (6.27) into (6.29), we have

$$T = \tfrac{1}{2}m_1 \left(\dot{\mathbf{R}} - \frac{m_2}{m_1 + m_2}\dot{\mathbf{r}} \right) \cdot \left(\dot{\mathbf{R}} - \frac{m_2}{m_1 + m_2}\dot{\mathbf{r}} \right)$$
$$+ \tfrac{1}{2}m_2 \left(\dot{\mathbf{R}} + \frac{m_1}{m_1 + m_2}\dot{\mathbf{r}} \right) \cdot \left(\dot{\mathbf{R}} + \frac{m_1}{m_1 + m_2}\dot{\mathbf{r}} \right)$$

where (5.34) has been used. Using the distributive law for the dot products, we find, after simplifying,

$$T = \tfrac{1}{2}(m_1 + m_2)|\dot{\mathbf{R}}|^2 + \frac{1}{2}\frac{m_1 m_2}{(m_1 + m_2)}|\dot{\mathbf{r}}|^2 \tag{6.30}$$

Let M be the total mass of the system:

$$M \equiv m_1 + m_2 \tag{6.31}$$

We define the *reduced mass* μ of the two-particle system as

$$\mu \equiv \frac{m_1 m_2}{m_1 + m_2} \tag{6.32}$$

Then

$$T = \tfrac{1}{2}M|\dot{\mathbf{R}}|^2 + \tfrac{1}{2}\mu|\dot{\mathbf{r}}|^2 \tag{6.33}$$

The first term in (6.33) is the kinetic energy due to translational motion of the whole system of mass M. The second term is kinetic energy of internal motion of the two particles. This internal motion is of two types. The distance r between the two particles can change (vibration), and the direction of the $\mathbf{r}$ vector can change (rotation). Note that

$$|\dot{\mathbf{r}}| = \left| \frac{d}{dt}\mathbf{r} \right| \neq \frac{d}{dt}|\mathbf{r}| \tag{6.34}$$

Corresponding to the original coordinates $x_1, y_1, z_1, x_2, y_2, z_2$, we had six linear momenta:

$$p_{x_1} = m_1 \dot{x}_1, \ldots, p_{z_2} = m_2 \dot{z}_2 \tag{6.35}$$

Comparing Eqs. (6.29) and (6.33), we define the six linear momenta for the new coordinates X, Y, Z, x, y, z as

$$p_X \equiv M\dot{X}, \qquad p_Y \equiv M\dot{Y}, \qquad p_Z \equiv M\dot{Z} \tag{6.36}$$

$$p_x \equiv \mu\dot{x}, \qquad p_y \equiv \mu\dot{y}, \qquad p_z \equiv \mu\dot{z} \tag{6.37}$$

We define two new momentum vectors as

$$\mathbf{p}_M \equiv \mathbf{i}M\dot{X} + \mathbf{j}M\dot{Y} + \mathbf{k}M\dot{Z} \tag{6.38}$$

$$\mathbf{p}_\mu \equiv \mathbf{i}\mu\dot{x} + \mathbf{j}\mu\dot{y} + \mathbf{k}\mu\dot{z} \tag{6.39}$$

Introducing these momenta into (6.33), we have

$$T = \frac{|\mathbf{p}_M|^2}{2M} + \frac{|\mathbf{p}_\mu|^2}{2\mu} \tag{6.40}$$

Now consider the potential energy. We make the restriction that V is a function *only* of the relative coordinates x, y, and z of the two particles:

$$V = V(x, y, z) \tag{6.41}$$

An example of (6.41) is two charged particles interacting according to Coulomb's law [see Eq. (3.65)]. With this restriction on V, the Hamiltonian function is

$$H = \frac{p_M^2}{2M} + \left[\frac{p_\mu^2}{2\mu} + V(x, y, z) \right] \tag{6.42}$$

Now suppose we had a system composed of a particle of mass M subject to no forces and a particle of mass μ subject to the potential-energy function $V(x, y, z)$, and further suppose that there was no interaction between these particles. If (X, Y, Z) are the coordinates of the particle of mass M, and (x, y, z) are the coordinates of the particle of mass μ, what is the Hamiltonian of this hypothetical system? Clearly it is identical with (6.42). Since our hypothetical system has no interaction between the two particles, we could treat the motion of each separately. The particle of mass M, being subject to no forces, would simply undergo translational motion at constant velocity, and its contribution to the total energy of the system would be some constant. We could then treat separately the motion of the hypothetical particle of mass μ, using the potential-energy function in brackets in (6.42).

Thus we have reduced our original two-body problem to two separate one-body problems: (1) The translational motion of the system as a whole, which simply adds a constant to the total energy. (2) The internal motion, which may be dealt with by considering a hypothetical particle of mass μ subject to the potential-energy function $V(x, y, z)$.

Up to now in this section, we have been working with classical mechanics. We now want to show that this same separation holds in quantum mechanics. To this end, we will prove an important theorem. Let us suppose that our system is composed of two particles that do not interact with each other. We use q_1 to symbolize the coordinates (x_1, y_1, z_1) of particle one, and q_2 to symbolize the coordinates (x_2, y_2, z_2) of particle two. Let $\hat{H}_1$ and $\hat{H}_2$ be the Hamiltonian operators for particles one and two, respectively. Since there is no interaction between the particles, the Hamiltonian for the whole system is

$$\hat{H} = \hat{H}_1 + \hat{H}_2$$

and the Schroedinger equation reads

$$(\hat{H}_1 + \hat{H}_2)\psi(q_1, q_2) = E\psi(q_1, q_2) \tag{6.43}$$

We attempt a solution of (6.43) by separation of variables, setting

$$\psi(q_1, q_2) = G_1(q_1)G_2(q_2) \tag{6.44}$$

We have

$$\hat{H}_1 G_1(q_1)G_2(q_2) + \hat{H}_2 G_1(q_1)G_2(q_2) = EG_1(q_1)G_2(q_2) \tag{6.45}$$

Now $\hat{H}_1$ involves only the coordinate and momentum operators of particle one.

Therefore

$$\hat{H}_1[G_1(q_1)G_2(q_2)] = G_2(q_2)\hat{H}_1 G_1(q_1) \tag{6.46}$$

since, as far as $\hat{H}_1$ is concerned, G_2 is a constant. Using (6.46) and a similar equation for $\hat{H}_2$, we find that Eq. (6.45) becomes

$$G_2(q_2)\hat{H}_1 G_1(q_1) + G_1(q_1)\hat{H}_2 G_2(q_2) = EG_1(q_1)G_2(q_2)$$

$$\frac{\hat{H}_1 G_1(q_1)}{G_1(q_1)} + \frac{\hat{H}_2 G_2(q_2)}{G_2(q_2)} = E \tag{6.47}$$

Now by the same arguments used in connection with Eq. (3.78), we conclude that each term on the left in (6.47) must be a constant. Using E_1 and E_2 to denote these constants, we have

$$\frac{\hat{H}_1 G_1(q_1)}{G_1(q_1)} = E_1, \qquad \frac{\hat{H}_2 G_2(q_2)}{G_2(q_2)} = E_2$$

$$E = E_1 + E_2 \tag{6.48}$$

In other words, when our system is composed of two noninteracting particles, we can reduce the two-particle problem to two separate one-particle problems by solving

$$\hat{H}_1 G_1(q_1) = E_1 G_1(q_1), \qquad \hat{H}_2 G_2(q_2) = E_2 G_2(q_2) \tag{6.49}$$

The total energy is the sum of the individual energies of each particle, according to (6.48).

This result is easily generalized to any number of noninteracting particles. For n such particles, we have

$$\hat{H} = \hat{H}_1 + \hat{H}_2 + \cdots + \hat{H}_n$$

$$\psi(q_1, q_2, \ldots, q_n) = G_1(q_1)G_2(q_2)\cdots G_n(q_n)$$

$$E = E_1 + E_2 + \cdots + E_n \tag{6.50}$$

$$\hat{H}_i G_i = E_i G_i, \qquad i = 1, 2, \ldots, n$$

Our results are also applicable to a single particle whose Hamiltonian is the sum of separate terms for each coordinate:

$$\hat{H} = \hat{H}_x(\hat{x}, \hat{p}_x) + \hat{H}_y(\hat{y}, \hat{p}_y) + \hat{H}_z(\hat{z}, \hat{p}_z)$$

In this case, we conclude that the wave functions and energies are

$$\psi(x, y, z) = F(x)G(y)K(z), \qquad E = E_x + E_y + E_z$$

$$\hat{H}_x F(x) = E_x F(x), \qquad \hat{H}_y G(y) = E_y G(y), \qquad \hat{H}_z K(z) = E_z K(z)$$

Examples include the particle in a three-dimensional box (Section 3.5), the three-dimensional free particle (Problem 3.23), and the three-dimensional harmonic oscillator (Problem 4.5).

Now consider again the two-particle quantum-mechanical problem of masses m_1 and m_2 interacting so that the potential energy is a function of their relative coordinates only [Eqs. (6.41) and (6.22)]. By making the coordinate transformation of Eqs. (6.25) and (6.26), we can write the Hamiltonian as the sum of the Hamiltonians of two independent hypothetical particles [Eq. (6.42)]. Therefore the total energy of the system will be the sum of the energies of these two particles. One particle will have mass M and will move as a free particle ($V = 0$) with some nonnegative energy. The second hypothetical particle will have mass μ, and its Hamiltonian operator $\hat{H}_\mu$ is formed from the two terms in brackets in (6.42):

$$\hat{H}_\mu = \frac{\hat{p}_\mu^2}{2\mu} + \hat{V}(x, y, z) \tag{6.51}$$

6.3   THE TWO-PARTICLE RIGID ROTOR

Although we are now equipped to solve the Schroedinger equation for the hydrogen atom, we will first solve a simpler problem: that of the two-particle rigid rotor. By this we mean a two-particle system with the particles held at a fixed distance from each other by a rigid massless rod of length d. For this problem the vector $\mathbf{r}$ in Fig. 6.1 has the constant magnitude $|\mathbf{r}| = d$. Therefore (see Section 6.2) the kinetic energy of internal motion is wholly rotational energy. The energy of the rotor is wholly kinetic, and we have

$$V = 0 \tag{6.52}$$

Equation (6.52) is a special case of Eq. (6.41), and we may therefore use the results of the last section to separate off the translational motion of the system as a whole. We will concern ourselves only with the rotational energy. The Hamiltonian for the rotation is given by (6.51) as

$$\hat{H} = \frac{\hat{p}_\mu^2}{2\mu} = -\frac{\hbar^2}{2\mu} \nabla^2, \qquad \mu = \frac{m_1 m_2}{m_1 + m_2} \tag{6.53}$$

where m_1 and m_2 are the masses of the two particles. The coordinates of the fictitious particle are the relative coordinates of m_1 and m_2, as given by Eq. (6.22).

Instead of the relative Cartesian coordinates, x, y, z, it will prove more fruitful to use the relative spherical polar coordinates, r, θ, φ. The r coordinate is equal to the magnitude of the $\mathbf{r}$ vector in Fig. 6.1, and since m_1 and m_2 are constrained to remain a fixed distance apart, we have $r = d$. Thus the problem is equivalent to a particle of mass μ constrained to move on the surface of a sphere of radius d. Because the radial coordinate is constant, the wave function will be a function of θ and φ only. Hence the first two terms of the Laplacian operator in (6.8) will give zero when operating on the wave function and may be omitted. (Looking at things in a slightly different way, we note that the operators in (6.8) that involve r derivatives correspond to the kinetic energy of radial motion, and since there is no radial motion, the r derivatives are omitted from the Hamiltonian.)

Since $V = 0$ is a special case of $V = V(r)$, the results of Section 6.1 tell us that the eigenfunctions are given by (6.20), with the r factor omitted:

$$\psi = Y_l^m(\theta, \varphi) \tag{6.54}$$

The Hamiltonian operator is [Eqs. (6.8) and (5.97)]

$$\hat{H} = -\frac{\hbar^2}{2\mu}\left[\frac{1}{d^2\sin\theta}\frac{\partial}{\partial\theta}\left(\sin\theta\,\frac{\partial}{\partial\theta}\right)+\frac{1}{d^2\sin^2\theta}\frac{\partial^2}{\partial\varphi^2}\right] \tag{6.55}$$

$$\hat{H} = \frac{1}{2\mu d^2}\,\hat{L}^2 \tag{6.56}$$

Using (6.16), we have

$$\hat{H}\psi = E\psi$$

$$\frac{1}{2\mu d^2}\,\hat{L}^2\,Y_J^m(\theta,\varphi) = EY_J^m(\theta,\varphi) \tag{6.57}$$

$$\frac{1}{2\mu d^2}\,J(J+1)\hbar^2\,Y_J^m(\theta,\varphi) = EY_J^m(\theta,\varphi)$$

$$E = \frac{J(J+1)\hbar^2}{2\mu d^2}, \qquad J = 0,1,2,\ldots \tag{6.58}$$

where we use J rather than l for the rotational quantum number.

The *moment of inertia I* of a system of n particles about some particular axis in space is defined as

$$I \equiv \sum_{i=1}^{n} m_i\rho_i^2 \tag{6.59}$$

where m_i is the mass of the ith particle and ρ_i is the perpendicular distance from this particle to the axis. The value of I depends on the choice of axis. For the two-particle rigid rotor, we choose our axis to be a line that passes through the center of mass and is perpendicular to the line joining m_1 and m_2 (Fig. 6.2). If we place the rotor so that the center of mass, point C, lies at the origin of a Cartesian coordinate system and the line joining m_1 and m_2 lies on the x axis, then C will have the coordinates $(0, 0, 0)$, m_1 will have the coordinates $(-\rho_1, 0, 0)$, and m_2 will have the coordinates $(\rho_2, 0, 0)$. Using these coordinates in (6.24), we find

$$m_1\rho_1 = m_2\rho_2 \tag{6.60}$$

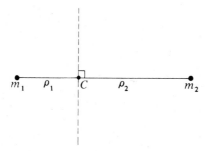

Figure 6.2 *Axis (dashed line) for calculating the moment of inertia of a two-particle rigid rotor*

The moment of inertia of the rotor about the axis we have chosen is

$$I = m_1 \rho_1^2 + m_2 \rho_2^2 \tag{6.61}$$

Using (6.60), we transform Eq. (6.61) to (see Problem 6.13)

$$I = \frac{m_1 m_2}{m_1 + m_2} (\rho_1 + \rho_2)^2 = \mu d^2 \tag{6.62}$$

where μ is the reduced mass of the system and d is the distance between m_1 and m_2. The allowed energy levels of the rotor are

$$E = \frac{J(J+1)\hbar^2}{2I}, \qquad J = 0, 1, 2, \ldots \tag{6.63}$$

The lowest level is $E = 0$, so we have no zero-point rotational energy. Having zero rotational energy and therefore zero angular momentum for the rotor does not violate the uncertainty principle—recall the discussion following Eq. (5.129). Note that E increases as $J^2 + J$, so the spacing between adjacent rotational levels increases as J increases.

Are the rigid-rotor energy levels degenerate? The energy depends only on J, but the wave function depends on J and m. For each value of J, we can have $2J + 1$ values of m, ranging from $-J$ to J; hence the levels are $(2J + 1)$-fold degenerate. Physically, this corresponds to the different possible orientations of the angular-momentum vector of the rotor about a space-fixed axis.

The angles θ and φ which occur in the wave function are relative coordinates of the two point masses. If we set up a Cartesian coordinate system with the origin at the center of mass of the rotor, θ and φ will be as shown in Fig. 6.3. This coordinate system undergoes the same translational motion as the center of mass of the rotor but does not rotate in space.

The rotational levels of a diatomic molecule can be well approximated by the two-particle rigid-rotor energies (6.63). It is found (*Levine, Molecular Spectroscopy,*

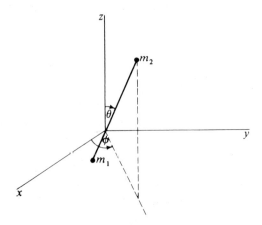

Figure 6.3 *Coordinate system for the two-particle rigid rotor*

Section 4.4) that the allowed pure-rotational spectroscopic transitions involve a change of ± 1 in J; in addition, a molecule must have a nonzero dipole moment in order to show a pure-rotational spectrum. In a pure-rotational transition only the rotational quantum number changes. [Vibration–rotation transitions (Section 4.3) involve simultaneous changes in vibrational and rotational quantum numbers.] The spacing between adjacent low-lying rotational levels is significantly less than that between adjacent vibrational levels, and the pure-rotational spectrum falls in the microwave (or the far-infrared) region. The frequencies of the pure-rotational spectral lines of a diatomic molecule are then (approximately)

$$\nu = \frac{E(J+1)-E(J)}{h} = \frac{[(J+1)(J+2)-J(J+1)]h}{8\pi^2 I} = 2(J+1)B$$

where the *rotational constant B* is defined as $h/8\pi^2 I$.

The spacings between the diatomic rotational levels (6.63) for low and moderate values of J are generally less than or of the same order of magnitude as kT at room temperature, so the Boltzmann distribution law (4.70) shows that many rotational levels are significantly populated at room temperature. Absorption of radiation by diatomic molecules having $J = 0$ (the $J = 0 \rightarrow 1$ transition) gives a line at the frequency $2B$; absorption by molecules having $J = 1$ (the $J = 1 \rightarrow 2$ transition) gives a line at $4B$; absorption by $J = 2$ molecules gives a line at $6B$; etc. Measurement of the rotational absorption frequencies allows B to be found; from B we get the molecule's moment of inertia I, and from I we get the bond distance d.

As noted in Section 4.3, isotopic species such as $^1H^{35}Cl$ and $^1H^{37}Cl$ have virtually the same electronic energy curve $U(R)$ and so have virtually the same equilibrium bond distance. However, the different isotopic masses produce different moments of inertia and hence different rotational absorption frequencies.

(For the rotational energies of polyatomic molecules, see *Levine, Molecular Spectroscopy*, Chapter 5.)

6.4   *THE HYDROGEN ATOM*

The hydrogen atom consists of a proton and an electron. If e represents the charge on the proton ($e = +1.6 \times 10^{-19}$ C), then the electron's charge is $-e$. (From time to time, scientists have speculated that the proton and electron charges might not be exactly equal in magnitude. In 1959 Lyttleton and Bondi pointed out that if a difference of 10^{-36} coulomb existed between the magnitudes of these charges, then the observed rate of expansion of the universe could be accounted for by electrostatic repulsions between the charged galaxies. Subsequent experimental work[1] showed that if such a difference exists, it is much less than the amount required by the Lyttleton-Bondi hypothesis.) We will assume the electron and proton to be two point masses whose interaction is given by Coulomb's law. In our

---

[1] J. G. King, *Phys. Rev. Letters*, **5**, 562 (1960); H. F. Dylla and J. G. King, *Phys. Rev. A*, **7**, 1224 (1973).

discussions of atoms and molecules, we will be considering isolated systems, ignoring interatomic or intermolecular interactions. The isolated hydrogen atom is a two-particle system, and we will use the results of Section 6.2 to treat it.

Instead of treating just the hydrogen atom, we consider a slightly more general problem: the *hydrogenlike atom*. By this we mean a system consisting of one electron and a nucleus of charge Ze. For $Z = 1$, we have the hydrogen atom; for $Z = 2$, the He^+ ion; for $Z = 3$, the Li^{2+} ion, etc. The hydrogenlike atom is the single most important system in quantum chemistry. An exact solution of the Schroedinger equation for atoms with more than one electron cannot be obtained because of the interelectronic repulsions. If, as a first approximation, we ignore these repulsions, then the electrons can be treated independently. (See Section 6.2.) The atomic wave function will be approximated by a product of one-electron functions, which will be hydrogenlike wave functions. We call such one-electron functions *atomic orbitals*; we will use atomic orbitals to construct approximate molecular wave functions, as well as to discuss many-electron atoms. Let (x, y, z) be the coordinates of the electron relative to the nucleus, and let $\mathbf{r} = \mathbf{i}x + \mathbf{j}y + \mathbf{k}z$. The Coulomb's law force on the electron in the hydrogenlike atom is

$$\mathbf{F} = -\frac{Ze'^2}{r^2}\frac{\mathbf{r}}{r} \qquad (6.64)$$

The minus sign indicates an attractive force. The quantity e' can be viewed either as the proton charge in statcoulombs or as $e' \equiv e/(4\pi\varepsilon_0)^{1/2}$, where e is the proton charge in coulombs. (See Section 1.8.)

[The possibility of small deviations from Coulomb's law has been considered. Experiments have shown that if the Coulomb's law force is written as being proportional to r^{-2+s}, then $|s| < 10^{-15}$. It can be shown that a deviation from Coulomb's law would imply a nonzero rest mass for the photon; see A. S. Goldhaber and M. M. Nieto, *Rev. Mod. Phys.*, **43**, 277 (1971). There is no evidence for a nonzero photon rest mass, and data indicate that any such mass must be less than 10^{-48} g; L. Davis et al., *Phys. Rev. Lett.*, **35**, 1402 (1975).]

The force in (6.64) is central, and comparison with Eq. (6.4) gives $dV(r)/dr = Ze'^2/r^2$. Integration gives

$$V = Ze'^2 \int \frac{dr}{r^2} = -\frac{Ze'^2}{r} \qquad (6.65)$$

where the integration constant has been taken as 0 to make $V = 0$ at infinite separation between the charges. For any two charges Q_1 and Q_2 separated by distance r_{12}, Eq. (6.65) becomes

$$V = Q_1' Q_2'/r_{12} \qquad (6.66)$$

Since the potential energy of this two-particle system depends only on the relative coordinates of the particles, we can apply the results of Section 6.2 to reduce the problem to two one-particle problems. The translational motion of the atom as a whole simply adds some constant to the total energy, and we will not concern ourselves with it. To deal with the internal motion of the system, we introduce our fictitious particle of mass μ, where

$$\mu = \frac{m_e m_N}{m_e + m_N} \tag{6.67}$$

where m_e and m_N are the electronic and nuclear masses. The particle of reduced mass μ moves subject to the potential-energy function (6.65), and its coordinates (r, θ, φ) are the spherical polar coordinates of one particle relative to the other (Fig. 6.4).

The Hamiltonian for the internal motion is [Eq. (6.51)]

$$\hat{H} = -\frac{\hbar^2}{2\mu}\nabla^2 - \frac{Ze'^2}{r} \tag{6.68}$$

Since V is a function of the r coordinate only, we have a one-particle central-force problem, and we may apply the results of Section 6.1. Using Eqs. (6.20) and (6.21), we have for the wave function

$$\psi(r, \theta, \varphi) = R(r)Y_l^m(\theta, \varphi), \qquad l = 0, 1, 2, \ldots, \qquad |m| \leq l \tag{6.69}$$

where Y_l^m is a spherical harmonic, and the radial function $R(r)$ satisfies

$$-\frac{\hbar^2}{2\mu}\left(R'' + \frac{2}{r}R'\right) + \frac{l(l+1)\hbar^2}{2\mu r^2}R - \frac{Ze'^2}{r}R = ER(r) \tag{6.70}$$

To save time in writing, we define the constant a as

$$a \equiv \hbar^2/\mu e'^2 \tag{6.71}$$

and (6.70) becomes

$$R'' + \frac{2}{r}R' + \left[\frac{2E}{ae'^2} + \frac{2Z}{ar} - \frac{l(l+1)}{r^2}\right]R = 0 \tag{6.72}$$

We could now try a power-series solution of (6.72), but we would get a three-term rather than a two-term recursion relation. We therefore seek a substitution that will lead to a two-term recursion relation. It turns out that the proper substitution

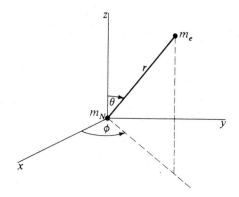

Figure 6.4 *Relative spherical polar coordinates*

can be found by examining the behavior of the solution for large values of r. For large r, (6.72) becomes

$$R'' + \frac{2E}{ae'^2} R = 0, \qquad r \text{ large} \tag{6.73}$$

which may be solved using the auxiliary equation. The solutions are

$$\exp\left[\pm(-2E/ae'^2)^{1/2}r\right] \tag{6.74}$$

Suppose that E is positive. The quantity under the square-root sign in (6.74) is negative, and the factor multiplying r is imaginary:

$$R(r) \sim e^{\pm i\sqrt{2\mu E}r/\hbar}, \qquad E \geqslant 0 \tag{6.75}$$

where (6.71) has been used. The symbol $\sim$ in (6.75) indicates that we are giving the behavior of $R(r)$ for large values of r; this is called the *asymptotic* behavior of the function. Note the resemblance of (6.75) to Eq. (2.30), the free-particle wave function. Equation (6.75) does not give the complete radial factor in the wave function for positive energies. Further study[2] shows that the radial function for $E \geqslant 0$ remains finite for all values of r, no matter what the value of E. Thus just as for the free particle, all nonnegative energies of the hydrogen atom are allowed. Physically, these eigenfunctions correspond to states in which the electron is not bound to the nucleus. (A classical-mechanical analogy is a comet moving in a hyperbolic orbit about the sun. The comet is not bound and makes but one visit to the solar system.) Since we get continuous rather than discrete allowed values for $E \geqslant 0$, the positive-energy eigenfunctions are called *continuum* eigenfunctions. The angular part of a continuum wave function is, of course, a spherical harmonic. Like the free-particle wave functions, the continuum eigenfunctions are not normalizable in the usual sense.

We now consider the *bound* states of the hydrogen atom, with $E < 0$. In this case, the quantity under the square-root sign in (6.74) is positive. Since we want the wave functions to remain finite as r goes to infinity, we prefer the minus sign in (6.74), and in order to get a two-term recursion relation, we make the substitution

$$R(r) = e^{-cr}K(r) \tag{6.76}$$

$$c \equiv \left(-\frac{2E}{ae'^2}\right)^{1/2} \tag{6.77}$$

where e in (6.76) stands for the base of natural logarithms, and not the proton charge. Use of the substitution (6.76) will guarantee nothing about the behavior of the wave function for large r. The differential equation we obtain from this substitution will still have two linearly independent solutions. We can make any substitution we please in a differential equation; in fact, we could make the substitution $R(r) = e^{+cr}J(r)$ and still wind up with the correct eigenfunctions and eigenvalues. The relation between J and K would naturally be $J(r) = e^{-2cr}K(r)$.

---

[2] *Bethe and Salpeter*, pp. 21–24.

Proceeding with (6.76), we evaluate R' and R'', substitute into (6.72), multiply through by $r^2 e^{cr}$, and use (6.77) to obtain the following differential equation for $K(r)$:

$$r^2 K'' + (2r - 2cr^2)K' + [(2Za^{-1} - 2c)r - l(l+1)]K = 0 \qquad (6.78)$$

We could now substitute a power series of the form

$$K = \sum_{k=0}^{\infty} c_k r^k \qquad (6.79)$$

for K; if we did we would find that, in general, the first few coefficients in (6.79) are zero. If c_s is the first nonzero coefficient, (6.79) can be written as

$$K = \sum_{k=s}^{\infty} c_k r^k, \qquad c_s \neq 0 \qquad (6.80)$$

Letting $j \equiv k - s$, and then defining b_j as $b_j \equiv c_{j+s}$, we have

$$K = \sum_{j=0}^{\infty} c_{j+s} r^{j+s} = r^s \sum_{j=0}^{\infty} b_j r^j, \qquad b_0 \neq 0 \qquad (6.81)$$

(Although the various substitutions we are making might seem arbitrary, they are standard procedure in solving differential equations by power series.) In (6.81), s is an integer whose value is to be determined by substitution into the differential equation. We thus set

$$K(r) = r^s M(r) \qquad (6.82)$$

$$M(r) = \sum_{j=0}^{\infty} b_j r^j, \qquad b_0 \neq 0 \qquad (6.83)$$

Evaluating K' and K'' from (6.82) and substituting into (6.78), we get

$$r^2 M'' + [(2s+2)r - 2cr^2]M' + [s^2 + s + (2Za^{-1} - 2c - 2cs)r - l(l+1)]M = 0 \qquad (6.84)$$

To determine s, let us look at (6.84) for $r = 0$. From (6.83) we have

$$M(0) = b_0, \qquad M'(0) = b_1, \qquad M''(0) = 2b_2 \qquad (6.85)$$

Using (6.85) in (6.84), we find for $r = 0$

$$b_0[s^2 + s - l^2 - l] = 0 \qquad (6.86)$$

Since b_0 is not zero, the terms in brackets must vanish:

$$s^2 + s - l^2 - l = 0 \qquad (6.87)$$

This is a quadratic equation in the unknown s, with the roots

$$s = l, \qquad s = -l - 1 \qquad (6.88)$$

These roots correspond to the two linearly independent solutions of the differential equation. Let us examine them from the standpoint of proper behavior of the wave function. From Eqs. (6.76), (6.82), and (6.83), we have

$$R(r) = e^{-cr} r^s \sum_{j=0}^{\infty} b_j r^j \tag{6.89}$$

Since $e^{-cr} = 1 - cr + \cdots$, the function $R(r)$ behaves for small r as $b_0 r^s$. For the root $s = l$, $R(r)$ behaves properly at the origin. However, for $s = -l-1$, $R(r)$ is proportional to

$$\frac{1}{r^{l+1}} \tag{6.90}$$

for small r. Since $l = 0, 1, 2, \ldots$, the root $s = -l-1$ makes the radial factor in the wave function infinite at the origin. Many texts take this as sufficient reason for rejecting this root. However, this is not a good argument, since for the *relativistic* hydrogen atom, the $l = 0$ eigenfunctions are infinite at $r = 0$. Let us therefore look at (6.90) from the standpoint of quadratic integrability, since we certainly require the bound-state eigenfunctions to be normalizable.

The normalization integral [Eq. (5.111)] for the radial functions that behave like (6.90) looks like

$$\int_0^{\infty} |R|^2 r^2 \, dr \approx \int_0^{\infty} \frac{1}{r^{2l}} \, dr \tag{6.91}$$

for small r. The behavior of the integral at the lower limit of integration is

$$\frac{1}{r^{2l-1}} \bigg|_{r=0} \tag{6.92}$$

For $l = 1, 2, 3, \ldots$, (6.92) is infinite, and the normalization integral is infinite. Hence we must reject the root $s = -l-1$ for $l \geqslant 1$. However, for $l = 0$, (6.92) is finite, and there is no trouble with quadratic integrability. Thus there is a solution to the radial equation that behaves as r^{-1} for small r and is quadratically integrable.

Further study of this solution shows that it corresponds to an energy value that the experimental hydrogen-atom spectrum shows does not exist. Thus the r^{-1} solution must be rejected, but there is some dispute over the reason for doing so. We will state the two reasons most often given, omitting the detailed arguments. One view[3] is that the $1/r$ solution satisfies the Schroedinger equation everywhere in space *except* at the origin and hence must be rejected. A second view[4] is that the $1/r$ solution must be rejected because the Hamiltonian operator is not Hermitian with respect to it. (In Chapter 7 we will define Hermitian operators and show that we require quantum-mechanical operators to be Hermitian.)

Taking the first root in (6.88), we have for the radial factor (6.89)

$$R(r) = e^{-cr} r^l M(r) \tag{6.93}$$

With $s = l$, Eq. (6.84) becomes

$$rM'' + [2l + 2 - 2cr]M' + [2Za^{-1} - 2c - 2cl]M = 0 \tag{6.94}$$

---

[3] *Dirac*, p. 156.
[4] *Merzbacher*, Section 10.5; B. H. Armstrong and E. A. Power, *Am. J. Phys.*, **31**, 262 (1963).

From (6.83) we have

$$M(r) = \sum_{j=0}^{\infty} b_j r^j \tag{6.95}$$

$$M' = \sum_{j=0}^{\infty} jb_j r^{j-1} = \sum_{j=1}^{\infty} jb_j r^{j-1} = \sum_{k=0}^{\infty} (k+1)b_{k+1} r^k = \sum_{j=0}^{\infty} (j+1)b_{j+1} r^j$$

$$M'' = \sum_{j=0}^{\infty} j(j-1)b_j r^{j-2} = \sum_{j=1}^{\infty} j(j-1)b_j r^{j-2} = \sum_{k=0}^{\infty} (k+1)k b_{k+1} r^{k-1}$$

$$= \sum_{j=0}^{\infty} (j+1)j b_{j+1} r^{j-1} \tag{6.96}$$

Substituting these expressions in (6.94), we have after combining sums

$$\sum_{j=0}^{\infty} \left[j(j+1)b_{j+1} + 2(l+1)(j+1)b_{j+1} + \left(\frac{2Z}{a} - 2c - 2cl - 2cj \right) b_j \right] r^j = 0 \tag{6.97}$$

Setting the coefficient of r^j equal to zero, we get the recursion relation

$$b_{j+1} = \frac{(2c + 2cl + 2cj - 2Za^{-1})}{j(j+1) + 2(l+1)(j+1)} b_j \tag{6.98}$$

We now must examine the behavior of the infinite series (6.95) for large r. Since for large r the behavior of the series is determined by the terms with large j, we examine the ratio b_{j+1}/b_j for large j:

$$\frac{b_{j+1}}{b_j} \sim \frac{2cj}{j^2} = \frac{2c}{j} \qquad \text{for } j \text{ large} \tag{6.99}$$

Now consider the power series for e^{2cr}:

$$e^{2cr} = 1 + 2cr + \cdots + \frac{(2c)^j r^j}{j!} + \frac{(2c)^{j+1} r^{j+1}}{(j+1)!} + \cdots \tag{6.100}$$

The ratio of successive powers of r in (6.100) is

$$\frac{(2c)^{j+1}}{(j+1)!} \cdot \frac{j!}{(2c)^j} = \frac{2c}{(j+1)} \sim \frac{2c}{j} \qquad \text{for } j \text{ large} \tag{6.101}$$

which is the same as (6.99) for large j. This indicates that for large r, the infinite series (6.95) behaves like e^{2cr}. For large r the radial function (6.93) behaves like

$$R(r) \sim e^{-cr} r^l e^{2cr} = r^l e^{cr} \tag{6.102}$$

Therefore $R(r)$ will become infinite as r goes to infinity and will not be quadratically integrable. The only way to avoid this "infinity catastrophe" (as in the harmonic-oscillator case) is to have the series terminate after a finite number of terms, in which case the e^{-cr} factor will ensure that the wave function goes to zero as r goes to infinity. Let the last term in the series be $b_k r^k$. Then, to have $b_{k+1}, b_{k+2}, \cdots$ all vanish, the fraction multiplying b_j in the recursion relation (6.98) must vanish when $j = k$; we have

$$2c(k+l+1) = 2Za^{-1}, \qquad k = 0, 1, 2, \ldots \qquad \textbf{(6.103)}$$

k and l are integers, and we now define a new integer n by

$$n \equiv k+l+1, \qquad n = 1, 2, 3, \ldots \qquad \textbf{(6.104)}$$

From (6.104) the quantum number l must satisfy

$$l \leqslant n-1 \qquad \textbf{(6.105)}$$

Hence l ranges from 0 to $n-1$. Equation (6.103) reads

$$cn = Za^{-1} \qquad \textbf{(6.106)}$$

Using (6.77) and (6.71) in (6.106), we find

$$E = -\frac{Z^2}{n^2}\left(\frac{e'^2}{2a}\right) = -\frac{Z^2 \mu e'^4}{2n^2 \hbar^2} \qquad \textbf{(6.107)}$$

These are the bound-state energy levels of the hydrogenlike atom, and they are discrete. Figure 6.5 shows the potential-energy curve [Eq. (6.65)] and some of the allowed energy levels for the hydrogen atom ($Z = 1$). The crosshatching indicates that all positive energies are allowed.

It turns out that all changes in n are allowed in light absorption and emission; the reciprocal wavelengths of hydrogen-atom spectral lines are then

$$\frac{1}{\lambda} = \frac{v}{c} = \frac{E_2 - E_1}{hc} = \frac{e'^2}{2ahc}\left(\frac{1}{n_1^2} - \frac{1}{n_2^2}\right) \equiv R_H\left(\frac{1}{n_1^2} - \frac{1}{n_2^2}\right)$$

where $R_H = 109{,}677.6 \text{ cm}^{-1}$ is the *Rydberg constant* for hydrogen.

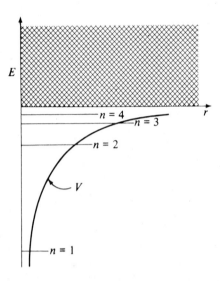

Figure 6.5 *Energy levels for the hydrogen atom*

Are the hydrogen-atom energy levels degenerate? For the bound states, the energy (6.107) depends only on n. However, the wave function (6.69) depends on all three quantum numbers n, l, and m, whose allowed values are [Eqs. (6.104), (6.105), (5.147), and (5.148)]

$$n = 1, 2, 3, \ldots \tag{6.108}$$

$$l = 0, 1, 2, \ldots, n-1 \tag{6.109}$$

$$m = -l, -l+1, \ldots, 0, \ldots, l-1, l \tag{6.110}$$

States with different values of l or m, but the same value of n, have the same energy; we have degeneracy, except for $n = 1$, where l and m must both be 0. For a given value of n, we can have n different values of l. For each of these values of l, we can have $2l + 1$ values of m. The degree of degeneracy of a given bound-state level is then

$$\sum_{l=0}^{n-1} (2l + 1) = \sum_{l=0}^{n-1} (2l) + \sum_{l=0}^{n-1} 1 \tag{6.111}$$

The second sum on the right side of (6.111) has n terms, each term having the value 1. Therefore it is equal to n. For the first sum, we have

$$\sum_{l=0}^{n-1} (2l) = 2 \sum_{l=0}^{n-1} l = 2 \sum_{l=1}^{n-1} l \tag{6.112}$$

The sum of the integers from 1 to k is (Problem 6.25) $\frac{1}{2}k(k + 1)$; thus the sum of the integers from 1 to $n-1$ is $\frac{1}{2}(n-1)n$. We have

$$\sum_{l=0}^{n-1} (2l + 1) = 2 \cdot \tfrac{1}{2}(n-1)n + n = n^2 \tag{6.113}$$

and the degeneracy of the discrete levels for the hydrogenlike atom is n^2 (spin considerations being omitted). For the continuum levels, it turns out that for a given energy, there is no restriction on the maximum value of l; hence these levels are infinity-fold degenerate.

The radial equation for the hydrogen atom can also be solved by the use of ladder operators (also known as *factorization*); see Z. W. Salsburg, *Am. J. Phys.*, **33**, 36 (1965).

6.5   THE BOUND-STATE HYDROGEN-ATOM WAVE FUNCTIONS

Using (6.106), we have for the recursion relation (6.98)

$$b_{j+1} = \frac{2Z}{na} \frac{(j+l+1-n)}{(j+1)(j+2l+2)} b_j \tag{6.114}$$

The discussion preceding Eq. (6.104) shows that the highest power of r in the polynomial $M(r)$ [Eq. (6.95)] is $k = n-l-1$; hence using (6.106), the radial factor (6.93) is

$$R_{nl}(r) = r^l e^{-Zr/na} \sum_{j=0}^{n-l-1} b_j r^j \tag{6.115}$$

where a is given by (6.71). The complete wave function is [Eq. (6.69)]

$$\psi_{nlm} = R_{nl}(r)Y_l^m(\theta, \varphi) = R_{nl}(r)S_{lm}(\theta)\frac{1}{\sqrt{2\pi}}e^{im\varphi} \qquad (6.116)$$

where the first few theta functions are given in Table 5.1.

How many nodes does $R(r)$ have? The radial function is zero at $r = \infty$, at $r = 0$ for $l \neq 0$, and at values of r that make $M(r)$ vanish. $M(r)$ is a polynomial of degree $(n - l - 1)$, and it can be shown that the roots of $M(r) = 0$ are all real and positive. Thus aside from the origin and infinity, there are $(n - l - 1)$ nodes in $R(r)$. The nodes of the spherical harmonics are discussed in Problem 6.21.

For the ground state of the hydrogenlike atom we have $n = 1$, $l = 0$, $m = 0$. The radial factor is

$$R_{10}(r) = b_0 e^{-Zr/a} \qquad (6.117)$$

The constant b_0 is determined by normalization [Eq. (5.111)]:

$$|b_0|^2 \int_0^\infty e^{-2Zr/a}r^2\, dr = 1 \qquad (6.118)$$

Using the Appendix integral (A.4), we find

$$R_{10}(r) = 2\left(\frac{Z}{a}\right)^{3/2}e^{-Zr/a} \qquad (6.119)$$

Multiplying by Y_0^0, we have as the ground-state wave function

$$\psi_{100} = \frac{1}{\pi^{1/2}}\left(\frac{Z}{a}\right)^{3/2}e^{-Zr/a} \qquad (6.120)$$

The energy of the ground state is given by (6.107) with $n = 1$. A convenient unit for electronic energies is the *electron volt* (eV), defined as the kinetic energy acquired by an electron accelerated through a potential difference of one volt (V). Potential difference is defined as energy per unit charge; since $e = 1.6022 \times 10^{-19}$ C and 1 volt coulomb = 1 joule = 10^7 ergs, we have

$$1\text{ eV} = 1.6022 \times 10^{-19}\text{ J} = 1.6022 \times 10^{-12}\text{ erg} \qquad (6.121)$$

Substituting the values of the physical constants into (6.107), we find for the hydrogen-atom $(Z = 1)$ ground-state energy (Problem 6.10)

$$E = -13.598 \text{ electron volts} \approx -13.60 \text{ eV} \qquad (6.122)$$

a number worth remembering. This is the minimum energy needed to ionize a hydrogen atom in its ground state.

The energies and wave functions of the hydrogen atom involve the reduced mass μ_H, given by (6.67) with $m_N = m_p$, the proton mass. Since $m_p = 1836.1 m_e$, we have

$$\mu_H = \frac{1836.1}{1837.1}m_e = 0.99946 m_e \qquad (6.123)$$

The reduced mass is quite close to the electron mass. In fact, many texts do not bother to use the reduced mass but simply use the electron mass in the Schroedinger equation for the hydrogen atom. Physically, this corresponds to assuming that the proton mass is infinite in comparison to the electron mass and that all the internal motion is motion of the electron—taking the limit of (6.67) as m_p goes to infinity, we find

$$\lim_{m_p \to \infty} \frac{m_p m_e}{m_p + m_e} = \frac{m_p m_e}{m_p} = m_e \tag{6.124}$$

The error introduced by using the electron mass for the reduced mass is about one part in two thousand for the hydrogen atom. For heavier atoms the error introduced by assuming an infinitely heavy nucleus is even less than this. Also, for many-electron atoms, the form of the correction for nuclear motion is quite complicated. For these reasons we will in future assume an infinitely heavy nucleus and simply use the electron mass in writing the Schroedinger equation for atoms.

If we replace the reduced mass of the hydrogen atom by the electron mass, the quantity a defined by (6.71) becomes

$$a_0 \equiv \frac{\hbar^2}{m_e e'^2} = 0.5292 \text{ Å} \tag{6.125}$$

where the subscript zero indicates use of the electron mass instead of the reduced mass. For historical reasons a_0 is called the *Bohr radius*; it was the radius of the circle in which the electron moved in the ground state of the hydrogen atom, according to the Bohr theory. Of course, since the ground-state wave function (6.120) is nonzero for all finite values of r, there is some probability of finding the electron at any distance from the nucleus. The electron is certainly not confined to a circle. (In a 1959 freshman chemistry text, there is the amusing statement that the electron is confined to the surface of a sphere.) Use of the electron mass instead of the reduced mass to calculate the ground-state energy of the hydrogen atom gives -13.606 eV, which is in error by 7 parts in 13,000.

Let us have a look at an interesting feature of the ground-state wave function, Eq. (6.120). For points on the x axis, where $y = 0$ and $z = 0$, the coordinate r in (5.76) equals $(x^2)^{1/2} = |x|$, and

$$\psi_{100}(x, 0, 0) = \frac{1}{\pi^{1/2}} \left(\frac{Z}{a} \right)^{3/2} e^{-Z|x|/a} \tag{6.126}$$

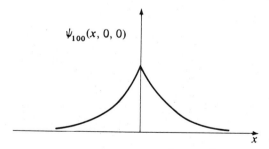

Figure 6.6 *Cusp in the hydrogen atom ground-state wave function*

Figure 6.6 shows how (6.126) varies along the x axis. Although the wave function is continuous at the origin, the slope of the tangent to the curve is positive at the left of the origin but negative at its right. Thus $\partial\psi/\partial x$ is discontinuous at the origin; we say that the wave function has a *cusp* at the origin. The cusp is present because the potential energy (6.65) becomes infinite at the origin. (Recall the discontinuous slope of the particle-in-a-box wave functions at the walls of the box.)

We denoted the hydrogen-atom bound-state wave functions by three subscripts which give the values of n, l, and m. We now introduce a slightly different notation, in which the value of l is indicated by a letter:

letter	s	p	d	f	g	h	i	k	$\cdots$
l	0	1	2	3	4	5	6	7	$\cdots$

$$\text{(6.127)}$$

The letters s, p, d, f are of spectroscopic origin, standing for sharp, principal, diffuse, fundamental. After these we go alphabetically, except that j is omitted. Preceding the code letter for l, we write the value of n. Thus the ground-state wave function ψ_{100} is called ψ_{1s}.

For $n = 2$, we have the following states: $\psi_{200}, \psi_{21-1}, \psi_{210}, \psi_{211}$. We denote ψ_{200} as ψ_{2s}. To distinguish the three $2p$ functions, we use a subscript giving the value of m: $\psi_{2p_{-1}}, \psi_{2p_0}, \psi_{2p_{+1}}$. The radial factor in the wave function depends on n and l, but not on m, as can be seen from (6.115). Each of the three $2p$ wave functions thus has the same radial factor. The $2s$ and $2p$ radial factors may be found in the usual manner from (6.115) and (6.114), followed by normalization. The results are given in Table 6.1. Note that the exponential factor in these radial functions is not the same as in the R_{1s} function, Eq. (6.119). The complete wave function is found by multiplying the radial factor by the appropriate spherical harmonic. Using (6.116), Table 6.1, and Table 5.1, we have

$$\psi_{2s} = \frac{1}{\pi^{1/2}}\left(\frac{Z}{2a}\right)^{3/2}\left(1 - \frac{Zr}{2a}\right)e^{-Zr/2a} \qquad \text{(6.128)}$$

$$\psi_{2p_{-1}} = \frac{1}{8\pi^{1/2}}\left(\frac{Z}{a}\right)^{5/2} re^{-Zr/2a}\sin\theta\, e^{-i\varphi} \qquad \text{(6.129)}$$

$$\psi_{2p_0} = \frac{1}{\pi^{1/2}}\left(\frac{Z}{2a}\right)^{5/2} re^{-Zr/2a}\cos\theta \qquad \text{(6.130)}$$

$$\psi_{2p_1} = \frac{1}{8\pi^{1/2}}\left(\frac{Z}{a}\right)^{5/2} re^{-Zr/2a}\sin\theta\, e^{i\varphi} \qquad \text{(6.131)}$$

Table 6.1 lists some of the normalized radial factors in the hydrogenlike wave functions. Figure 6.7 gives graphs of some of the radial functions. The r^l factor makes the radial functions zero at $r = 0$, except for s states.

The probability of finding the electron in the region of space where its r coordinate is between r and $r + dr$, its θ coordinate is between θ and $\theta + d\theta$, and its φ coordinate is between φ and $\varphi + d\varphi$ is [Eq. (5.107)]

$$|\psi|^2\, d\tau = [R_{nl}(r)]^2 |Y_l^m(\theta, \varphi)|^2 r^2 \sin\theta\, dr\, d\theta\, d\varphi \qquad \text{(6.132)}$$

Table 6.1 *Radial factors in the hydrogenlike-atom wave functions*

$$R_{1s} = 2\left(\frac{Z}{a}\right)^{3/2} e^{-Zr/a}$$

$$R_{2s} = \frac{1}{\sqrt{2}}\left(\frac{Z}{a}\right)^{3/2}\left(1 - \frac{Zr}{2a}\right)e^{-Zr/2a}$$

$$R_{2p} = \frac{1}{2\sqrt{6}}\left(\frac{Z}{a}\right)^{5/2} re^{-Zr/2a}$$

$$R_{3s} = \frac{2}{3\sqrt{3}}\left(\frac{Z}{a}\right)^{3/2}\left(1 - \frac{2Zr}{3a} + \frac{2Z^2 r^2}{27a^2}\right)e^{-Zr/3a}$$

$$R_{3p} = \frac{8}{27\sqrt{6}}\left(\frac{Z}{a}\right)^{3/2}\left(\frac{Zr}{a} - \frac{Z^2 r^2}{6a^2}\right)e^{-Zr/3a}$$

$$R_{3d} = \frac{4}{81\sqrt{30}}\left(\frac{Z}{a}\right)^{7/2} r^2 e^{-Zr/3a}$$

We now ask: What is the probability of the electron having its r coordinate between r and $r + dr$ with no restriction on the values of θ and φ? We are asking for the probability of finding the electron in a thin spherical shell centered at the origin, of inner radius r and outer radius $r + dr$. We must thus add up the probabilities (6.132) for all possible values of θ and φ, keeping r fixed. This amounts to integrating (6.132) over θ and φ. Hence the probability of finding the electron between r and $r + dr$ is

$$[R_{nl}(r)]^2 r^2 \, dr \int_0^{2\pi} \int_0^{\pi} |Y_l^m(\theta, \varphi)|^2 \sin\theta \, d\theta \, d\varphi = [R_{nl}(r)]^2 r^2 \, dr \quad \textbf{(6.133)}$$

since the spherical harmonics are normalized:

$$\int_0^{2\pi} \int_0^{\pi} |Y_l^m(\theta, \varphi)|^2 \sin\theta \, d\theta \, d\varphi = 1 \quad \textbf{(6.134)}$$

as can be seen from (5.112), (5.111), and (5.101). The function $R^2(r)r^2$, which determines the probability of finding the electron at a distance r from the nucleus, is called the *radial distribution function*; see Fig. 6.8. Although $R_{1s}(r)$ is not zero at the origin, the 1s radial distribution function is zero at $r = 0$ because of the r^2 factor; the volume of the thin spherical shell becomes zero as r goes to zero. The maximum in the radial distribution function for the 1s state of hydrogen is at $r = a$.

The factor $e^{im\varphi}$ makes the spherical harmonics imaginary, except when $m = 0$. Instead of working with imaginary wave functions such as (6.129) and (6.131), chemists often use real wave functions formed by taking certain linear combinations of the complex functions. The justification for this procedure is given by the theorem of Section 3.6: Any linear combination of eigenfunctions of a degenerate energy level

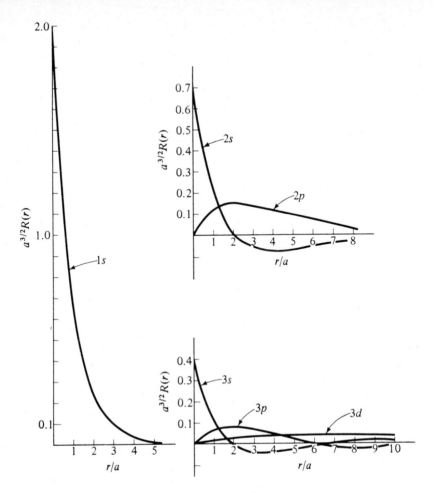

Figure 6.7 *Graphs of the radial factor* $R_{nl}(r)$ *in the hydrogen-atom* ($Z = 1$) *wave functions. The same scale is used in all the graphs. (In some well-known texts, these functions are not properly drawn to scale.)*

is an eigenfunction of the Hamiltonian with the same eigenvalue. Since the energy of the hydrogen atom does not depend on m, the $2p_1$ and $2p_{-1}$ states belong to a degenerate energy level; any linear combination of them is an eigenfunction of the Hamiltonian with the same energy eigenvalue.

There are essentially two different ways to combine these two functions to obtain a real function. One way is

$$\psi_{2p_x} = \frac{1}{\sqrt{2}} (\psi_{2p_{-1}} + \psi_{2p_1}) = \frac{1}{4\sqrt{2\pi}} \left(\frac{Z}{a}\right)^{5/2} re^{-Zr/2a} \sin\theta \cos\varphi \quad (6.135)$$

where we have used (6.129), (6.131), and $e^{\pm i\varphi} = \cos\varphi \pm i \sin\varphi$. The $1/\sqrt{2}$ factor

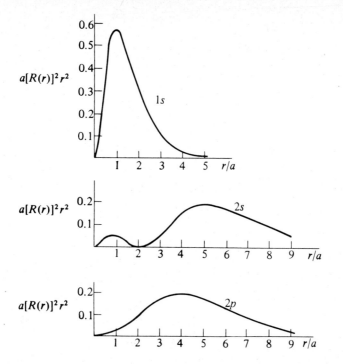

Figure 6.8 *Plots of the radial distribution function* $[R_{nl}(r)]^2 r^2$ *for the hydrogen atom*

normalizes ψ_{2p_x}:

$$\int |\psi_{2p_x}|^2 \, d\tau = \frac{1}{2} \left(\int |\psi_{2p_{-1}}|^2 \, d\tau + \int |\psi_{2p_1}|^2 \, d\tau + \int \psi^*_{2p_{-1}} \psi_{2p_1} \, d\tau + \int \psi^*_{2p_1} \psi_{2p_{-1}} \, d\tau \right)$$

$$= \tfrac{1}{2}(1 + 1 + 0 + 0) = 1 \qquad (6.136)$$

Here we have used the fact that ψ_{2p_1} and $\psi_{2p_{-1}}$ are normalized and are orthogonal to each other, since

$$\int_0^{2\pi} (e^{-i\varphi})^* e^{i\varphi} \, d\varphi = \int_0^{2\pi} e^{2i\varphi} \, d\varphi = 0 \qquad (6.137)$$

The designation ψ_{2p_x} for (6.135) becomes clearer if we note that (5.75) gives

$$\psi_{2p_x} = \frac{1}{4\sqrt{2\pi}} \left(\frac{Z}{a} \right)^{5/2} x e^{-Zr/2a} \qquad (6.138)$$

The second way of combining the functions is

$$\psi_{2p_y} \equiv \frac{1}{i\sqrt{2}} (\psi_{2p_1} - \psi_{2p_{-1}}) = \frac{1}{4\sqrt{2\pi}} \left(\frac{Z}{a} \right)^{5/2} r \sin\theta \sin\varphi \, e^{-Zr/2a} \qquad (6.139)$$

$$\psi_{2p_y} = \frac{1}{4\sqrt{2\pi}} \left(\frac{Z}{a} \right)^{5/2} y e^{-Zr/2a} \qquad (6.140)$$

The function ψ_{2p_0} is real and is often denoted by

$$\psi_{2p_0} = \psi_{2p_z} = \frac{1}{\sqrt{\pi}} \left(\frac{Z}{2a}\right)^{5/2} z e^{-Zr/2a} \qquad (6.141)$$

Table 6.2 *Real hydrogenlike wave functions*

$$\psi_{1s} = \frac{1}{\pi^{1/2}} \left(\frac{Z}{a}\right)^{3/2} e^{-Zr/a}$$

$$\psi_{2s} = \frac{1}{4(2\pi)^{1/2}} \left(\frac{Z}{a}\right)^{3/2} \left(2 - \frac{Zr}{a}\right) e^{-Zr/2a}$$

$$\psi_{2p_z} = \frac{1}{4(2\pi)^{1/2}} \left(\frac{Z}{a}\right)^{5/2} r e^{-Zr/2a} \cos\theta$$

$$\psi_{2p_x} = \frac{1}{4(2\pi)^{1/2}} \left(\frac{Z}{a}\right)^{5/2} r e^{-Zr/2a} \sin\theta \cos\varphi$$

$$\psi_{2p_y} = \frac{1}{4(2\pi)^{1/2}} \left(\frac{Z}{a}\right)^{5/2} r e^{-Zr/2a} \sin\theta \sin\varphi$$

$$\psi_{3s} = \frac{1}{81(3\pi)^{1/2}} \left(\frac{Z}{a}\right)^{3/2} \left(27 - 18\frac{Zr}{a} + 2\frac{Z^2 r^2}{a^2}\right) e^{-Zr/3a}$$

$$\psi_{3p_z} = \frac{2^{1/2}}{81\pi^{1/2}} \left(\frac{Z}{a}\right)^{5/2} \left(6 - \frac{Zr}{a}\right) r e^{-Zr/3a} \cos\theta$$

$$\psi_{3p_x} = \frac{2^{1/2}}{81\pi^{1/2}} \left(\frac{Z}{a}\right)^{5/2} \left(6 - \frac{Zr}{a}\right) r e^{-Zr/3a} \sin\theta \cos\varphi$$

$$\psi_{3p_y} = \frac{2^{1/2}}{81\pi^{1/2}} \left(\frac{Z}{a}\right)^{5/2} \left(6 - \frac{Zr}{a}\right) r e^{-Zr/3a} \sin\theta \sin\varphi$$

$$\psi_{3d_{z^2}} = \frac{1}{81(6\pi)^{1/2}} \left(\frac{Z}{a}\right)^{7/2} r^2 e^{-Zr/3a} (3\cos^2\theta - 1)$$

$$\psi_{3d_{xz}} = \frac{2^{1/2}}{81\pi^{1/2}} \left(\frac{Z}{a}\right)^{7/2} r^2 e^{-Zr/3a} \sin\theta \cos\theta \cos\varphi$$

$$\psi_{3d_{yz}} = \frac{2^{1/2}}{81\pi^{1/2}} \left(\frac{Z}{a}\right)^{7/2} r^2 e^{-Zr/3a} \sin\theta \cos\theta \sin\varphi$$

$$\psi_{3d_{x^2-y^2}} = \frac{1}{81(2\pi)^{1/2}} \left(\frac{Z}{a}\right)^{7/2} r^2 e^{-Zr/3a} \sin^2\theta \cos 2\varphi$$

$$\psi_{3d_{xy}} = \frac{1}{81(2\pi)^{1/2}} \left(\frac{Z}{a}\right)^{7/2} r^2 e^{-Zr/3a} \sin^2\theta \sin 2\varphi$$

where capital Z stands for the number of protons in the nucleus and small z is the z coordinate of the electron. We leave it to you to show that ψ_{2p_x}, ψ_{2p_y}, and ψ_{2p_z} are mutually orthogonal. Note that ψ_{2p_z} is zero in the xy plane, positive above this plane, and negative below it.

The functions $\psi_{2p_{-1}}$ and ψ_{2p_1} are eigenfunctions of $\hat{L}^2$ with the *same* eigenvalue: $2\hbar^2$. The reasoning of Section 3.6 shows that the linear combinations (6.135) and (6.139) are also eigenfunctions of $\hat{L}^2$ with eigenvalue $2\hbar^2$. However, $\psi_{2p_{-1}}$ and ψ_{2p_1} are eigenfunctions of $\hat{L}_z$ with *different* eigenvalues: $-\hbar$ and $+\hbar$. Therefore ψ_{2p_x} and ψ_{2p_y} are not eigenfunctions of $\hat{L}_z$.

We can extend this procedure to construct real wave functions for higher states. Since m ranges from $-l$ to $+l$, for each imaginary function containing the factor $e^{-i|m|\varphi}$, there is a function with the same value of n and l but having the factor $e^{+i|m|\varphi}$. Addition and subtraction of these functions gives two real functions, one with the factor $\cos(|m|\varphi)$, the other with the factor $\sin(|m|\varphi)$. Table 6.2 lists these real wave functions for the hydrogenlike atom. The subscripts on these functions come from similar considerations as for the $2p_x$, $2p_y$, and $2p_z$ functions. For example,

$$\psi_{3d_{x^2-y^2}} \equiv 2^{-1/2}(\psi_{3d_{+2}} + \psi_{3d_{-2}}) \tag{6.142}$$

$$= \frac{1}{81\sqrt{2\pi}} \left(\frac{Z}{a}\right)^{7/2} e^{-Zr/3a} r^2 \sin^2\theta (\cos^2\varphi - \sin^2\varphi)$$

$$= \frac{1}{81\sqrt{2\pi}} \left(\frac{Z}{a}\right)^{7/2} e^{-Zr/3a}(x^2 - y^2) \tag{6.143}$$

where we have used Table 6.1 for the radial factor, Table 5.1 for the theta factor, and $e^{im\varphi}/\sqrt{2\pi}$ for the phi factor.

The real hydrogenlike functions are derived from the complex functions by replacing $e^{im\varphi}/(2\pi)^{1/2}$ with $\pi^{-1/2}\sin|m|\varphi$ or $\pi^{-1/2}\cos|m|\varphi$ for $m \neq 0$; for $m = 0$ the φ factor is $1/(2\pi)^{1/2}$ for both real and complex functions.

6.6 HYDROGENLIKE ORBITALS

We shall refer to the hydrogenlike wave functions as hydrogenlike *orbitals*. These functions have been derived for a one-electron atom, and we cannot expect to use them to get a truly accurate representation of the wave function of a many-electron atom. We will consider the use of the orbital concept to approximate many-electron atomic wave functions in Chapter 11. For the present we restrict ourselves to one-electron atoms.

There are two fundamentally different ways of depicting orbitals: Method I is to draw graphs of the functions; Method II is to draw contour surfaces of constant probability density. There is considerable confusion between the two methods; many texts draw graphs of the angular part of the wave function and then say that these are orbital contour surfaces, which is wrong.

First consider drawing graphs (Method I). To graph the variation of ψ as a function of the three independent variables r, θ, and φ, we need four dimensions; the three-dimensional nature of our world prevents us from drawing such a graph. Instead, we draw graphs of the factors in ψ. Graphing $R(r)$ versus r, we get the curves of Fig. 6.7. Such graphs contain no information on the angular variation of ψ.

Now consider graphs of $S(\theta)$. We have (Table 5.1)

$$S_{0,0} = 1/\sqrt{2}, \qquad S_{1,0} = \tfrac{1}{2}\sqrt{6}\cos\theta$$

We can graph these functions using two-dimensional Cartesian coordinates, plotting S on the vertical axis and θ on the horizontal axis. $S_{0,0}$ gives a horizontal straight line, and $S_{1,0}$ gives a cosine curve. More commonly, S is graphed using plane polar coordinates. The variable θ is the angle with the positive z axis, and $S(\theta)$ is the distance from the origin to the point on the graph. For $S_{0,0}$ we naturally get a circle; for $S_{1,0}$ we obtain two tangent circles (Fig. 6.9). The negative sign on the lower circle of the graph of $S_{1,0}$ indicates that for $\tfrac{1}{2}\pi < \theta \leqslant \pi$, $S_{1,0}$ is negative. Strictly speaking, in graphing $\cos\theta$ we only get the upper circle, which is traced out twice; to get two tangent circles we must graph $|\cos\theta|$.

Instead of graphing the angular factors separately, we can draw a single graph that plots $|S(\theta)T(\varphi)|$ as a function of θ and φ; we will use spherical polar coordinates, and the distance from the origin to a point on the graph will be $|S(\theta)T(\varphi)|$. For an s state, ST is independent of the angles, and we get a sphere of radius $1/(4\pi)^{1/2}$ for our graph. For a p_z state, $ST = \tfrac{1}{2}\sqrt{3/\pi}\cos\theta$, and the graph of $|ST|$ consists of two spheres with centers on the z axis and tangent at the origin (Fig. 6.10). No doubt Fig. 6.10 is familiar. Many texts say this gives the shape of a p_z orbital, which is wrong; Fig. 6.10 is simply a *graph* of the *angular factor* in a p_z wave function. Graphs of the p_x and p_y angular factors give tangent spheres lying on the x and y axes, respectively. If we graph $S^2 T^2$ in spherical polar coordinates, we get surfaces with the familiar figure-eight cross sections; again, these are graphs and not orbital shapes.

Now consider Method II, drawing contour surfaces of constant probability density. No doubt you are familiar with geographical contour maps, which show lines of constant altitude. We will draw surfaces in space, on each of which the value of $|\psi|^2$, the probability density, is constant. Naturally, if $|\psi|^2$ is constant on a given

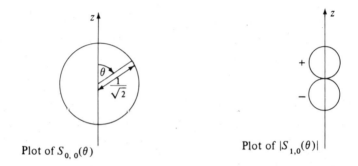

Plot of $S_{0,0}(\theta)$ Plot of $|S_{1,0}(\theta)|$

Figure 6.9 *Polar graphs of the θ factors in the hydrogen-atom wave functions*

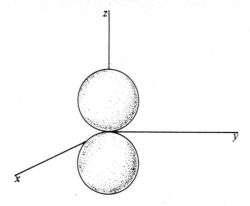

Figure 6.10 *Graph of* $|Y_1^0(\theta, \varphi)|$, *the angular factor for a* p_z *wave function*

surface, $|\psi|$ is also constant on that surface; the contour surfaces for $|\psi|^2$ and for $|\psi|$ are identical.

For an s orbital, the wave function depends only on r, so that a contour surface is a surface of constant r, i.e., a sphere centered at the origin. To pin down the size of an orbital, we take a contour surface within which the total probability of finding the electron is, say, 90 percent; thus we want

$$\int_V |\psi|^2 \, d\tau = 0.90 \qquad\qquad (6.144)$$

where V is the volume enclosed by the orbital contour surface.

Let us obtain the cross section of the $2p_y$ hydrogenlike orbital in the yz plane. In this plane, $\varphi = \pi/2$ (Fig. 6.4), and $\sin \varphi = 1$; hence Table 6.2 gives for this orbital in the yz plane

$$|\psi_{2p_y}| = k^{5/2}\pi^{-1/2}re^{-kr}|\sin\theta| \qquad\qquad (6.145)$$

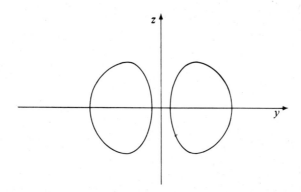

Figure 6.11 *Contour of a* $2p_y$ *orbital*

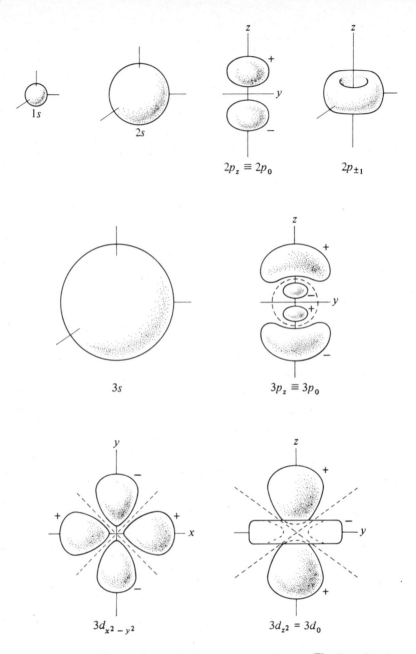

Figure 6.12 *Shapes of some hydrogen-atom orbitals. The 2s orbital has a spherical node, which is not visible; the 3s orbital has two such nodes. The 3p$_z$ orbital has a spherical node (indicated by the dashed line) and a nodal plane (the xy plane). The 3d$_{z^2}$ orbital has two nodal cones; the 3d$_{x^2-y^2}$ orbital has two nodal planes. Note that the view shown is not the same for the various orbitals. The relative*

where $k = Z/2a$. To find the orbital cross section, we use plane polar coordinates to plot (6.145) for a fixed value of ψ; r is the distance from the origin, and θ is the angle with the z axis. The result for a typical contour (Problem 6.18) is shown in Fig. 6.11. Since

$$ye^{-kr} = y \exp[-k(x^2 + y^2 + z^2)^{1/2}]$$

we see that ψ_{2p_y} is a function of y and $(x^2 + z^2)$; hence, on a circle centered on the y axis and parallel to the xz plane, ψ_{2p_y} is constant. Thus a three-dimensional contour surface may be developed by rotating the cross section in Fig. 6.11 about the y axis, giving a pair of distorted ellipsoids. The shape of a real $2p$ orbital is two separated, distorted ellipsoids, and not two tangent spheres. Although the two tangent spheres are only a crude approximation to the true orbital shape, they will probably continue to be used by many chemists.

Now consider the shape of the complex orbitals $\psi_{2p_{\pm 1}}$. We have

$$\psi_{2p_{\pm 1}} = k^{5/2} \pi^{-1/2} re^{-kr} \sin\theta e^{\pm i\varphi} \tag{6.146}$$

$$|\psi_{2p_{\pm 1}}| = k^{5/2}\pi^{-1/2}e^{-kr}r|\sin\theta| \tag{6.147}$$

and the two orbitals have the same shape. Since the right sides of (6.147) and (6.145) are identical, we conclude that Fig. 6.11 also gives the cross section of the $2p_{\pm 1}$ orbitals in the yz plane. Since [Eq. (5.75)]

$$e^{-kr}r|\sin\theta| = \exp[-k(x^2 + y^2 + z^2)^{1/2}](x^2 + y^2)^{1/2}$$

we see that $|\psi_{2p_{\pm 1}}|$ is a function of z and $(x^2 + y^2)$; so we get the three-dimensional orbital shape by rotating Fig. 6.11 about the z axis. This gives a doughnut-shaped surface. Various hydrogenlike orbital surfaces are shown in Fig. 6.12.

Schroedinger's original interpretation of $|\psi|^2$ was that the electron was "smeared out" into a charge cloud. If we consider an electron passing from one medium to another, we find that $|\psi|^2$ is nonzero in both mediums. According to the charge-cloud interpretation, this would mean that part of the electron was reflected and part transmitted. However, experimentally one never detects a fraction of an electron; electrons behave as indivisible entities. This difficulty is removed by the Born interpretation, according to which the values of $|\psi|^2$ in the two mediums give the *probabilities* for reflection and transmission. The orbital shapes we have drawn give the regions of space in which the total probability of finding the electron is 90 percent.

Figure 6.13 represents the probability density in the yz plane for various orbitals; the number of dots in a given region is proportional to the value of $|\psi|^2$ in that region. Rotation of these diagrams about the vertical (z) axis gives the three-dimensional probability density. The $2s$ orbital has a constant for its angular factor

---

signs of the wave function are indicated. The three other real 3d orbitals of Table 6.2 have the same shape as the $3d_{x^2-y^2}$ orbital, but different spatial orientations. [For a set of orthonormal d orbitals all five of which have the same shape, see R. E. Powell, J. Chem. Educ., 45, 45 (1968).]

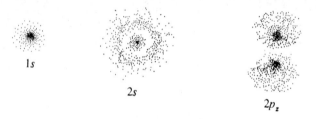

Figure 6.13 *Probability densities for some hydrogen-atom states.*
[For accurate stereo plots, see D. T. Cromer, J. Chem. Educ., **45***,*
626 (1968).]

and hence has no angular nodes; for this orbital we have $n - l - 1 = 1$, indicating one
radial node. The sphere on which $\psi_{2s} = 0$ is evident in Fig. 6.13.

6.7   THE ZEEMAN EFFECT

In 1896 Zeeman observed that application of an external magnetic field
caused a splitting of atomic spectral lines. We will consider this *Zeeman effect* for the
hydrogen atom; we begin by reviewing magnetism.

Magnetic fields arise from moving electric charges. A charge Q with velocity $\mathbf{v}$
gives rise to a magnetic field $\mathbf{B}$ at point P in space, such that

$$\mathbf{B} = \frac{\mu_0}{4\pi} \frac{Q\mathbf{v} \times \mathbf{r}}{r^3} \tag{6.148}$$

where $\mathbf{r}$ is the vector from Q to point P and where μ_0 (the *permeability of vacuum*) is
defined as $4\pi \times 10^{-7}\,\mathrm{N\,C^{-2}\,s^2}$. [Equation (6.148) is valid only for a nonaccelerated
charge moving with a speed much less than the speed of light.] The vector $\mathbf{B}$ is called
the *magnetic induction* or *magnetic flux density*. (It was formerly believed that the
vector $\mathbf{H}$ was the fundamental magnetic field vector, so $\mathbf{H}$ was called the *magnetic
field strength*. It is now known that $\mathbf{B}$ is the fundamental magnetic vector.) Equation
(6.148) is in SI units with Q in coulombs and $\mathbf{B}$ in teslas (T), where $1\,\mathrm{T}$
$= 1\,\mathrm{N\,C^{-1}\,m^{-1}\,s}$. Only SI units will be used in this section.

Two electric charges $+Q$ and $-Q$ separated by a small distance b constitute
an electric dipole. The *electric dipole moment* is defined as a vector from $-Q$ to $+Q$
with magnitude Qb. For a small planar loop of electric current, it turns out that the
magnetic field generated by the moving charges of the current is given by the same
mathematical expression as that giving the electric field due to an electric dipole,
except that the electric dipole moment is replaced by the *magnetic dipole moment* $\boldsymbol{\mu}$; $\boldsymbol{\mu}$
is a vector of magnitude IA, where I is the current flowing in a loop of area A; the
direction of $\boldsymbol{\mu}$ is perpendicular to the plane of the current loop. [An electric charge Q
is an electric monopole; one might wonder whether magnetic monopoles exist. The
possible detection of a magnetic monopole was reported in 1982. See *Physics Today*,
June 1982, p. 17; *Newsweek*, May 10, 1982, p. 46; *Time*, May 10, 1982, p. 96.]

Consider the magnetic (dipole) moment associated with a charge Q moving in a circle of radius r with speed v. The current is the charge flow per unit time. The circumference of the circle is $2\pi r$, and the time for one revolution is $2\pi r/v$. Hence

$$I = Qv/2\pi r \qquad (6.149)$$

The magnitude of $\boldsymbol{\mu}$ is

$$|\boldsymbol{\mu}| = IA = (Qv/2\pi r)\pi r^2 = Qvr/2 = Qrp/2m \qquad (6.150)$$

where m is the mass of the charged particle and p is its linear momentum. Since the radius vector $\mathbf{r}$ is perpendicular to $\mathbf{p}$, we have

$$\boldsymbol{\mu}_L = \frac{Q\mathbf{r}\times\mathbf{p}}{2m} = \frac{Q}{2m}\mathbf{L} \qquad (6.151)$$

where the definition of orbital angular momentum $\mathbf{L}$ was used and where the subscript on $\boldsymbol{\mu}$ indicates that it arises from the orbital motion of the particle. Although we derived (6.151) for the special case of circular motion, its validity is general. For an electron, $Q = -e$, and the magnetic moment due to its orbital motion is

$$\boldsymbol{\mu}_L = -\frac{e}{2m}\mathbf{L} \qquad (6.152)$$

The magnitude of $\mathbf{L}$ is given by (5.129), and the magnitude of the orbital magnetic moment of an electron with orbital-angular-momentum quantum number l is

$$|\boldsymbol{\mu}_L| = \frac{e\hbar}{2m}[l(l+1)]^{1/2} = \beta_e[l(l+1)]^{1/2} \qquad (6.153)$$

The constant $e\hbar/2m$ is called the *Bohr magneton* β_e:

$$\beta_e = e\hbar/2m = 9.274 \times 10^{-24} \text{ J/T} \qquad (6.154)$$

Now consider applying an external magnetic field to the hydrogen atom. The energy of interaction between a magnetic dipole $\boldsymbol{\mu}$ and an external magnetic field $\mathbf{B}$ is [*Halliday and Resnick*, Eq. (33–12)]

$$E_B = -\boldsymbol{\mu} \cdot \mathbf{B} \qquad (6.155)$$

Using Eq. (6.152), we have

$$E_B = \frac{e}{2m}\mathbf{L} \cdot \mathbf{B} \qquad (6.156)$$

We take the z axis along the direction of the applied field: $\mathbf{B} = B\mathbf{k}$, where $\mathbf{k}$ is a unit vector in the z direction. We have

$$E_B = \frac{e}{2m}B(L_x\mathbf{i} + L_y\mathbf{j} + L_z\mathbf{k}) \cdot \mathbf{k} = \frac{e}{2m}BL_z = \frac{\beta_e}{\hbar}BL_z$$

where L_z is the z component of orbital angular momentum. We now replace L_z by the operator $\hat{L}_z$ to give the following additional term in the Hamiltonian, resulting from the external magnetic field

$$\hat{H}_B = \beta_e B\hbar^{-1}\hat{L}_z \qquad (6.157)$$

The Schroedinger equation for the hydrogen atom in a magnetic field is

$$(\hat{H} + \hat{H}_B)\psi = E\psi \tag{6.158}$$

where $\hat{H}$ is the hydrogen-atom Hamiltonian in the absence of an external field. We readily verify that the solutions of Eq. (6.158) are the complex hydrogenlike wave functions (6.69):

$$(\hat{H} + \hat{H}_B)R(r)Y_l^m(\theta, \varphi) = \hat{H}R\,Y_l^m + \beta_e \hbar^{-1} B\hat{L}_z R\,Y_l^m$$

$$= \left(-\frac{Z^2\,e'^2}{n^2\,2a} + \beta_e Bm \right)R\,Y_l^m \tag{6.159}$$

where Eqs. (6.107) and (5.148) were used. Thus there is an additional term $\beta_e\,Bm$ in the energy, and the external magnetic field removes the m degeneracy. For obvious reasons, m is often called the *magnetic quantum number*. Actually the observed energy shifts do *not* match the predictions of Eq. (6.159), because of the existence of electron spin magnetic moment (Chapter 10 and Section 11.7).

In Chapter 5 we found that in quantum mechanics **L** lies on the surface of a cone. A *classical*-mechanical treatment (*Halliday and Resnick*, Sections 13-2 and 37-7) of the motion of **L** in an applied magnetic field shows that the field exerts a torque on $\boldsymbol{\mu}_L$, causing **L** to revolve about the direction of **B** at a constant frequency given by $|\boldsymbol{\mu}_L|B/2\pi|\mathbf{L}|$, while maintaining a constant angle with **B**. This gyroscopic motion is called *precession*. In quantum mechanics a complete specification of **L** is impossible; however, one finds that $\langle \mathbf{L} \rangle$ precesses about the field direction (*Dicke and Wittke*, Section 12-3; *Blokhintsev*, Section 62).

PROBLEMS

6.1 The lowest observed microwave absorption frequency of $^{12}C^{16}O$ is 115271 MHz. (a) Compute the bond distance in $^{12}C^{16}O$. (b) Predict the next two lowest microwave absorption frequencies of $^{12}C^{16}O$. (c) Predict the lowest microwave absorption frequency of $^{13}C^{16}O$. (d) For $^{12}C^{16}O$ at 25°C, calculate the ratio of the $J = 1$ population to the $J = 0$ population. Repeat for the $J = 2$ to $J = 0$ ratio. (Don't forget degeneracy.)

6.2 (a) For the hydrogen-atom ground state, find the probability of finding the electron farther than twice the Bohr radius from the nucleus. (b) For the hydrogen-atom ground state, find the probability of finding the electron in the classically forbidden region. (c) Find the radius of the sphere defining the $1s$ hydrogen orbital, using the 90 percent probability definition.

6.3 What is the value of the angular-momentum quantum number l for a t orbital?

6.4 For the ground state of the hydrogenlike atom, find: (a) the average value of r; (b) the most probable value of r. (c) Find $\langle r \rangle$ for a $2p$ state.

6.5 Where is the probability density a maximum for the hydrogen-atom ground state?

6.6 A stationary-state wave function is an eigenfunction of the Hamiltonian operator $\hat{H} = \hat{T} + \hat{V}$. Students sometimes erroneously believe that ψ is an eigenfunction of $\hat{T}$ and of $\hat{V}$. For the ground state of the hydrogen atom, verify directly that ψ is not an eigenfunction of $\hat{T}$ or of $\hat{V}$, but is an eigenfunction of $(\hat{T} + \hat{V})$. (Can you think of a problem we solved where ψ *is* an eigenfunction of $\hat{T}$ and of $\hat{V}$?)

6.7 Show that $\langle T \rangle + \langle V \rangle = E$ for a stationary state.

6.8 Compute $\langle V \rangle$, $\langle T \rangle$, and $\langle T \rangle / \langle V \rangle$ for the hydrogen-atom ground state. (Use the result of Problem 6.7 to shorten the work.)

6.9 Calculate in electron volts the ground-state energy of positronium—an "atom" which consists of a positron and an electron. (The positron has charge $+e$ and mass equal to the electronic mass.)

6.10 (a) Verify Eq. (6.122) by substituting Gaussian values of the physical constants into (6.107). (b) Do the same as in (a) using SI values.

6.11 Assign each of the following observed vacuum wavelengths to a transition between two hydrogen-atom levels:

6564.7 Å,   4862.7 Å,   4341.7 Å,   4102.9 Å   (Balmer series).

Predict the wavelengths of the next two lines in this series and the wavelength of the series limit. (Balmer was a Swiss mathematician who, in 1885, came up with an empirical formula that fitted lines of the hydrogen spectrum.)

6.12 Verify Eq. (6.6) for the Laplacian in spherical polar coordinates.

6.13 Verify Eq. (6.62). [Begin by multiplying and dividing the right side of (6.61) by $m_1 m_2 / (m_1 + m_2)$. Then use (6.60).]

6.14 Calculate the ratio of the electrical and gravitational forces between a proton and an electron. Is neglect of the gravitational force justified?

6.15 Derive the 2s and 2p radial hydrogenlike functions.

6.16 For the hydrogen-atom ground state, use the value of $\langle T \rangle$ from Problem 6.8 to calculate the root-mean-square speed of the electron; find the numerical value of $\langle v^2 \rangle^{1/2}/c$, where c is the speed of light.

6.17 For the particle in a box with infinitely high walls and for the harmonic oscillator, there are no continuum eigenfunctions, whereas for the hydrogen atom we do have continuum functions. Explain this in terms of the nature of the potential-energy function for each problem.

6.18 Show that the maximum value for ψ_{2p_y} [Eq. (6.145)] is $k^{3/2}\pi^{-1/2}e^{-1}$. Use Eq. (6.145) to plot the $2p_y$ contour for which $\psi = 0.316\,\psi_{\text{max}}$.

6.19 For which hydrogen-atom states is ψ nonzero at the nucleus?

6.20 If we were to ignore the interelectronic repulsion in helium, what would be its ground-state energy and wave function? (See the latter part of Section 6.2.) Compute the percent error in the energy; the experimental He ground-state energy is -79.0 eV. (Assume an infinitely heavy nucleus.)

6.21 For the *real* hydrogenlike functions: (a) What is the shape of the $n - l - 1$ nodal surfaces for which the radial factor is zero? (b) The nodal surfaces for which the φ factor vanishes are of the form $\varphi = $ constant; thus they are planes perpendicular to the xy plane. How many such planes are there? (Values of φ that differ by π are considered to be part of the same plane.) (c) It can be shown that there are $l - m$ surfaces on which the θ factor vanishes. What is the shape of these surfaces? (d) How many nodal surfaces are there for the real hydrogenlike wave functions?

6.22 (a) The *Laguerre polynomials* are defined as

$$L_q(z) \equiv e^z \frac{d^q}{dz^q}(e^{-z}z^q)$$

Verify that

$$L_0 = 1, \; L_1 = -z+1, \; L_2 = z^2 - 4z + 2, \; L_3 = -z^3 + 9z^2 - 18z + 6$$

(b) The *associated Laguerre polynomials* are usually defined as

$$L_q^s(z) = \frac{d^s}{dz^s} L_q(z)$$

Verify that

$$L_q^0(z) = L_q(z), \; L_1^1(z) = -1, \; L_2^1(z) = 2z - 4, \; L_2^2(z) = 2$$
$$L_3^1(z) = -3z^2 + 18z - 18, \; L_3^2(z) = -6z + 18, \; L_3^3(z) = -6$$

(c) The radial factor in the hydrogenlike wave functions can be shown to be[5]

$$R_{nl}(r) = -\left[\frac{4Z^3}{n^4 a^3} \frac{(n-l-1)!}{[(n+l)!]^3} \right]^{1/2} \left(\frac{2Zr}{na} \right)^l e^{-Zr/na} L_{n+l}^{2l+1} \left(\frac{2Zr}{na} \right)$$

Verify this equation for the $1s$, $2s$, and $2p$ states.

6.23 If the three force constants in Problem 4.5 each have the same value, we have a three-dimensional isotropic harmonic oscillator. (a) State why the wave functions for this case can be written in the form

$$F(r)G(\theta, \varphi) \qquad\qquad (6.160)$$

(b) What is the function G? (c) Write down a differential equation for $F(r)$. (d) Use the results of Problem 4.5 to show that the ground-state wave function has the form (6.160), and verify that the ground-state $F(r)$ satisfies the differential equation in (c). (e) Give the expression for the energy eigenvalues. (f) Give the degeneracies of the lowest three energy levels.

6.24 Sketch rough contours of constant $|\psi|$ for each of the following states of a particle in a two-dimensional square box: $n_x n_y = 11; 12; 21; 22$. What are you reminded of?

6.25 To find the degeneracy of the hydrogen-atom energy levels, we used

$$\sum_{k=1}^{n} k = \frac{n(n+1)}{2} \qquad\qquad (6.161)$$

Prove this result by adding corresponding terms of the two series: $1, 2, 3, \ldots, n$ and $n, n-1, n-2, \ldots, 1$. [Legend has it that Gauss used (6.161) at age six to add up a sum given him by his arithmetic teacher.]

6.26 The hydrogenlike wave functions $2p_1$, $2p_0$, and $2p_{-1}$ can be characterized as those $2p$ functions that are eigenfunctions of $\hat{L}_z$. What operators can we use to characterize the functions $2p_x$, $2p_y$, and $2p_z$, and what are the corresponding eigenvalues?

6.27 Given that $\hat{A}f = af$ and $\hat{A}g = bg$, where f and g are functions and a and b are constants, under what condition(s) is the linear combination $c_1 f + c_2 g$ an eigenfunction of the linear operator $\hat{A}$?

6.28 State which of the three operators $\hat{H}$, $\hat{L}^2$, and $\hat{L}_z$ each of the following functions is an eigenfunction of: (a) $2p_z$; (b) $2p_x$; (c) $2p_1$.

6.29 Each hydrogen-atom line of Problem 6.11 shows a very weak nearby satellite line. Two of the satellites occur at the vacuum wavelengths 6562.9 Å and 4861.4 Å. (a) Explain their origin. (The person who first answered this question got a Nobel Prize.) (b) Calculate the other two satellite wavelengths.

---

[5] *Pauling and Wilson*, p. 132.

6.30 On the x axis the ground-state hydrogen-atom wave function is given by (6.126). Calculate the limit that the derivative of (6.126) approaches (a) as $x \to 0$ from the left; (b) as $x \to 0$ from the right.

6.31 Name a quantum-mechanical system for which the spacing between adjacent bound-state energy levels (a) remains constant as E increases; (b) increases as E increases; (c) decreases as E increases.

7 THEOREMS OF QUANTUM MECHANICS

7.1 INTRODUCTION

The Schroedinger equation for the one-electron atom (Chapter 6) is exactly solvable; however, because of the interelectronic repulsion terms in the Hamiltonian, the Schroedinger equation for many-electron atoms and molecules is not separable in any coordinate system and cannot be solved exactly. Hence we must seek approximate methods of solution. The two main approximation methods, the variation method and perturbation theory, will be developed in Chapters 8 and 9; to derive these methods, we must develop further the theory of quantum mechanics, which is what we will do in this chapter. Although much of the work will be of an abstract nature, and perhaps not so easy to master, the theorems involved are basic to an understanding of quantum mechanics. We will conclude this chapter with a summary of the postulates of quantum mechanics.

Before getting started, we introduce some notations for the integrals with which we will be dealing. The definite integral over all space of an operator sandwiched between two functions frequently occurs, and various abbreviations are used:

$$\int \varphi_m^* \hat{A} \varphi_n \, d\tau \equiv \langle \varphi_m | \hat{A} | \varphi_n \rangle \equiv (\varphi_m | \hat{A} | \varphi_n) \equiv \langle m | \hat{A} | n \rangle \tag{7.1}$$

where φ_m and φ_n are two functions. If it is clear what functions are meant, we can use just the indices, as indicated in (7.1). The above notation, introduced by Dirac, is called *bracket notation*. Another notation is

$$\int \varphi_m^* \hat{A} \varphi_n \, d\tau \equiv A_{mn} \tag{7.2}$$

136

The notations A_{mn} and $\langle m|\hat{A}|n\rangle$ imply that we use the complex conjugate of the function whose letter appears first. An integral such as $\int \varphi_m^* \hat{A} \varphi_n \, d\tau$ is called a *matrix element* of the operator $\hat{A}$. Matrices are rectangular arrays of numbers and obey certain rules of combination (see Section 15.21).

For the integral over all space between two functions, we write

$$\int \varphi_m^* \varphi_n \, d\tau \equiv \langle \varphi_m | \varphi_n \rangle \equiv \langle m|n \rangle \tag{7.3}$$

Another notation for the integral (7.3) is (φ_m, φ_n). Since

$$\left[\int \varphi_m^* \varphi_n \, d\tau \right]^* = \int \varphi_n^* \varphi_m \, d\tau \tag{7.4}$$

we have the identity

$$\langle m|n \rangle^* = \langle n|m \rangle \tag{7.5}$$

In particular,

$$\langle m|m \rangle^* = \langle m|m \rangle \tag{7.6}$$

7.2   HERMITIAN OPERATORS

We stated in Section 3.1 that the operators representing physical quantities are linear. There is another requirement such operators must meet, which we now discuss.

Let $\hat{A}$ be the linear operator representing the physical property A. For the average value of A, Eq. (3.117) gives

$$\langle A \rangle = \int \Psi^* \hat{A} \Psi \, d\tau \tag{7.7}$$

where Ψ is the state function of the system. The average value of a physical quantity must be a real number; therefore we demand that

$$\langle A \rangle = \langle A \rangle^* \tag{7.8}$$

$$\int \Psi^* \hat{A} \Psi \, d\tau = \int \Psi (\hat{A} \Psi)^* \, d\tau \tag{7.9}$$

Equation (7.9) must hold for any function Ψ that can represent a possible state of the system; i.e., it must hold for all well-behaved functions Ψ. A linear operator that satisfies (7.9) for all well-behaved functions is called a *Hermitian operator*.

Many texts define a Hermitian operator as an operator that satisfies

$$\int f^* \hat{A} g \, d\tau = \int g (\hat{A} f)^* \, d\tau \tag{7.10}$$

for all well-behaved functions f and g. [Note especially that on the left side of (7.10) $\hat{A}$ operates on g, but on the right side $\hat{A}$ operates on f.] For the special case $f = g$, (7.10) reduces to (7.9). Equation (7.10) is apparently a more stringent requirement than (7.9), but we will prove that (7.10) is a consequence of (7.9), so that the two definitions of a Hermitian operator are equivalent.

We begin the proof by setting $\Psi = f + cg$ in (7.9), where c is an arbitrary parameter; this gives

$$\int (f+cg)^* \hat{A}(f+cg)\, d\tau = \int (f+cg)[\hat{A}(f+cg)]^*\, d\tau \qquad \text{(7.11)}$$

$$\int (f^* + c^*g^*)\hat{A}f\, d\tau + \int (f^* + c^*g^*)\hat{A}cg\, d\tau$$

$$= \int (f+cg)(\hat{A}f)^*\, d\tau + \int (f+cg)(\hat{A}cg)^*\, d\tau$$

$$\int f^*\hat{A}f\, d\tau + c^* \int g^*\hat{A}f\, d\tau + c \int f^*\hat{A}g\, d\tau + cc^* \int g^*\hat{A}g\, d\tau$$

$$= \int f(\hat{A}f)^*\, d\tau + c \int g(\hat{A}f)^*\, d\tau + c^* \int f(\hat{A}g)^*\, d\tau + c^*c \int g(\hat{A}g)^*\, d\tau$$

By virtue of (7.9), the first terms on each side of this last equation are equal to each other; likewise the last terms on each side are equal. Hence

$$c^* \int g^*\hat{A}f\, d\tau + c \int f^*\hat{A}g\, d\tau = c \int g(\hat{A}f)^*\, d\tau + c^* \int f(\hat{A}g)^*\, d\tau \qquad \text{(7.12)}$$

Setting $c = 1$ in (7.12), we have

$$\int g^*\hat{A}f\, d\tau + \int f^*\hat{A}g\, d\tau = \int g(\hat{A}f)^*\, d\tau + \int f(\hat{A}g)^*\, d\tau \qquad \text{(7.13)}$$

Setting $c = i$ in (7.12), we have, after dividing by i,

$$-\int g^*\hat{A}f\, d\tau + \int f^*\hat{A}g\, d\tau = \int g(\hat{A}f)^*\, d\tau - \int f(\hat{A}g)^*\, d\tau \qquad \text{(7.14)}$$

We now add (7.13) and (7.14) to get (7.10). This completes the proof. Therefore a Hermitian operator $\hat{A}$ possesses the property that

$$\int \varphi_i^* \hat{A}\varphi_j\, d\tau = \int \varphi_j (\hat{A}\varphi_i)^*\, d\tau \qquad \text{(7.15)}$$

where φ_i and φ_j are arbitrary well-behaved functions. Using the bracket and matrix-element notations, we write

$$\langle \varphi_i | \hat{A} | \varphi_j \rangle = \langle \varphi_j | \hat{A} | \varphi_i \rangle^* \qquad \text{(7.16)}$$

$$\langle i | \hat{A} | j \rangle = \langle j | \hat{A} | i \rangle^* \qquad \text{(7.17)}$$

$$A_{ij} = (A_{ji})^* \qquad \text{(7.18)}$$

Let us show that some of the operators we have been using are indeed Hermitian. For simplicity we will work in one dimension. To prove that an operator is Hermitian, it suffices to show that it satisfies (7.9) for all well-behaved functions; however, we will make things a bit tougher on ourselves by proving that (7.15) is satisfied.

First consider the potential-energy operator. The right side of (7.15) is

$$\int_{-\infty}^{\infty} \varphi_j(x)[V(x)\varphi_i(x)]^* \, dx \tag{7.19}$$

We have $V^* = V$, since the potential energy is a real function. Moreover, the order of the factors in (7.19) is immaterial. Hence

$$\int \varphi_j[V\varphi_i]^* \, dx = \int \varphi_j V^* \varphi_i^* \, dx = \int \varphi_i^* V \varphi_j \, dx \tag{7.20}$$

which proves V is Hermitian.

The operator for the x component of linear momentum is given by Eq. (3.28). For this operator, the left side of (7.15) is

$$-i\hbar \int_{-\infty}^{\infty} \varphi_i^*(x) \frac{d\varphi_j(x)}{dx} \, dx \tag{7.21}$$

Now we use the formula for integration by parts:

$$\int_a^b f(x) \frac{dg(x)}{dx} \, dx = f(x)g(x) \Big|_a^b - \int_a^b g(x) \frac{df(x)}{dx} \, dx \tag{7.22}$$

$$f(x) = -i\hbar \varphi_i^*(x), \qquad g(x) = \varphi_j(x)$$

$$-i\hbar \int_{-\infty}^{\infty} \varphi_i^* \frac{d\varphi_j}{dx} \, dx = -i\hbar \varphi_i^* \varphi_j \Big|_{-\infty}^{\infty} + i\hbar \int_{-\infty}^{\infty} \varphi_j(x) \frac{d\varphi_i^*(x)}{dx} \, dx \tag{7.23}$$

Now φ_i and φ_j are well-behaved functions; hence they vanish at $x = \pm \infty$. Therefore, (7.23) becomes

$$\int_{-\infty}^{\infty} \varphi_i^* \left(-i\hbar \frac{d\varphi_j}{dx} \right) dx = \int_{-\infty}^{\infty} \varphi_j \left(-i\hbar \frac{d\varphi_i}{dx} \right)^* dx \tag{7.24}$$

which is the same as (7.15) and proves that $\hat{p}_x$ is Hermitian. The proof that the kinetic-energy operator is Hermitian is left to the reader. The sum of two Hermitian operators can be shown to be Hermitian; hence the Hamiltonian operator $\hat{H} = \hat{T} + \hat{V}$ is Hermitian.

We now prove some theorems about the eigenvalues and eigenfunctions of Hermitian operators.

Since the eigenvalues of the operator $\hat{A}$ corresponding to the physical quantity A are the possible results of a measurement of A, these eigenvalues should all be real numbers. We now prove this, using the Hermitian property of $\hat{A}$. We start by setting $\varphi_j = \varphi_i$ in (7.15):

$$\int \varphi_i^* \hat{A} \varphi_i \, d\tau = \int \varphi_i (\hat{A} \varphi_i)^* \, d\tau \tag{7.25}$$

Now let us suppose that φ_i is an eigenfunction of $\hat{A}$ with eigenvalue b:

$$\hat{A}\varphi_i = b\varphi_i \tag{7.26}$$

Equation (7.25) becomes

$$b\int \varphi_i^* \varphi_i \, d\tau = \int \varphi_i (b\varphi_i)^* \, d\tau = b^* \int \varphi_i \varphi_i^* \, d\tau \tag{7.27}$$

$$(b - b^*)\int |\varphi_i|^2 \, d\tau = 0 \tag{7.28}$$

Since the integrand $|\varphi_i|^2$ is never negative, the only way the integral in (7.28) could be zero would be if φ_i were zero for all values of the coordinates. However, we always reject zero as an eigenfunction on physical grounds; hence the integral in (7.28) cannot be zero. Therefore $(b - b^*) = 0$, or $b = b^*$, which completes the proof.

To help develop a familiarity with bracket notation, we will repeat the proof that eigenvalues of Hermitian operators are real, using bracket notation. We begin by setting $i = j$ in (7.17):

$$\langle i|\hat{A}|i\rangle = \langle i|\hat{A}|i\rangle^* \tag{7.29}$$

Using (7.26), we have

$$\langle i|b|i\rangle = \langle i|b|i\rangle^* \tag{7.30}$$

$$b\langle i|i\rangle = b^*\langle i|i\rangle^* = b^*\langle i|i\rangle$$

$$(b - b^*)\langle i|i\rangle = 0$$

$$b = b^* \tag{7.31}$$

where (7.6) was used.

We showed that the particle-in-a-box wave functions are mutually orthogonal. We now prove the general theorem that *the eigenfunctions of a Hermitian operator are, or can be chosen to be, mutually orthogonal.* Given

$$\hat{B}F = sF, \quad \hat{B}G = tG \tag{7.32}$$

where F and G are two independent eigenfunctions of $\hat{B}$, we want to prove

$$\int F^*G \, d\tau \equiv \langle F|G\rangle = 0 \tag{7.33}$$

We begin with Eq. (7.17), which expresses the Hermitian nature of $\hat{B}$:

$$\langle F|\hat{B}|G\rangle = \langle G|\hat{B}|F\rangle^* \tag{7.34}$$

Using (7.32), we have

$$\langle F|t|G\rangle = \langle G|s|F\rangle^*$$

$$t\langle F|G\rangle = s^*\langle G|F\rangle^* \tag{7.35}$$

Since eigenvalues of Hermitian operators are real, we have $s^* = s$. Using (7.5), we have

$$t\langle F|G\rangle = s\langle F|G\rangle$$

$$(t - s)\langle F|G\rangle = 0 \tag{7.36}$$

If $s \neq t$, then

$$\langle F | G \rangle = 0 \tag{7.37}$$

We have proved that two eigenfunctions of a Hermitian operator that correspond to *different* eigenvalues are orthogonal. The question now is this: Can we have two independent eigenfunctions that have the *same* eigenvalue? The answer is yes. In the case of *degeneracy*, we have the same eigenvalue for more than one independent eigenfunction. Therefore we can only be certain that two independent eigenfunctions of a Hermitian operator are orthogonal to each other if they do not correspond to a degenerate eigenvalue. We now show that in the case of degeneracy, we may *construct* eigenfunctions that will be orthogonal to one another. We shall use the theorem proved in Section 3.6, that any linear combination of eigenfunctions corresponding to a degenerate eigenvalue is an eigenfunction with the same eigenvalue. Let us therefore suppose that F and G are independent eigenfunctions that have the same eigenvalue:

$$\hat{B}F = sF, \qquad \hat{B}G = sG \tag{7.38}$$

We take linear combinations of F and G to form two new eigenfunctions φ_1 and φ_2 that will be orthogonal to each other. We choose φ_1 equal to F, and for φ_2 we write

$$\varphi_2 = G + cF, \qquad \varphi_1 = F \tag{7.39}$$

The constant c will be chosen to ensure orthogonality. We want

$$\int \varphi_1^* \varphi_2 \, d\tau = 0 \tag{7.40}$$

$$\int F^*(G + cF) \, d\tau = \int F^* G \, d\tau + c \int F^* F \, d\tau = 0 \tag{7.41}$$

Hence choosing

$$c = -\int F^* G \, d\tau \Big/ \int F^* F \, d\tau \tag{7.42}$$

we have two orthogonal eigenfunctions φ_1 and φ_2 corresponding to the degenerate eigenvalue. This procedure (called Schmidt orthogonalization) can be extended to the case of n-fold degeneracy, to give n linearly independent orthogonal eigenfunctions corresponding to the degenerate eigenvalue.

   Thus, although there is no guarantee that the eigenfunctions of a degenerate eigenvalue are orthogonal, we can always *choose* them to be orthogonal, if we desire, by using the Schmidt (or some other) orthogonalization method. In fact, unless stated otherwise, we will always assume that we have chosen the eigenfunctions to be orthogonal:

$$\int \varphi_i^* \varphi_j \, d\tau = 0, \qquad i \neq j \tag{7.43}$$

where φ_i and φ_j are independent eigenfunctions of a Hermitian operator.

   We can usually multiply an eigenfunction by a suitable constant to normalize it, and we shall assume, unless stated otherwise, that all eigenfunctions are normalized:

$$\int \varphi_i^* \varphi_i \, d\tau = 1 \tag{7.44}$$

The exception is where the eigenvalues form a continuum, rather than a discrete set of values; in this case, the eigenfunctions are not quadratically integrable. Examples are the linear-momentum eigenfunctions, the free-particle energy eigenfunctions, and the hydrogen-atom continuum energy eigenfunctions.

Using the Kronecker delta, we can combine (7.43) and (7.44) into one equation:

$$\int \varphi_i^* \varphi_j \, d\tau = \langle i | j \rangle = \delta_{ij} \tag{7.45}$$

where φ_i and φ_j are eigenfunctions of some Hermitian operator.

As an example, consider the spherical harmonics. We shall prove that

$$\int_0^{2\pi} \int_0^{\pi} [Y_l^m(\theta, \varphi)]^* \, Y_{l'}^{m'}(\theta, \varphi) \sin\theta \, d\theta \, d\varphi = \delta_{l,l'} \, \delta_{m,m'} \tag{7.46}$$

where the $\sin\theta$ factor comes from the volume element in spherical polar coordinates, (5.107). The spherical harmonics are eigenfunctions of the Hermitian operator $\hat{L}^2$ [Eq. (5.147)]. Since eigenfunctions of a Hermitian operator belonging to different eigenvalues are orthogonal, we conclude that the integral in (7.46) is zero unless $l = l'$. Similarly Eq. (5.148) allows us to conclude that this integral is zero unless $m = m'$. Also, the multiplicative constant in (5.146) has been chosen so that the spherical harmonics are normalized [Eq. (6.134)]. Therefore (7.46) is valid.

7.3　EXPANSION IN TERMS OF EIGENFUNCTIONS

In the previous section, we proved the orthogonality of the eigenfunctions of a Hermitian operator. We now discuss another important property of these functions; this property allows us to expand an arbitrary well-behaved function in terms of these eigenfunctions.

We have often used the Taylor-series expansion [Eq. (4.72)] of a function as a linear combination of the nonnegative integral powers of $(x - a)$. Can we expand a function as a linear combination of some other set of functions besides 1, $(x - a)$, $(x - a)^2, \ldots$? The answer is yes, as was first shown by Fourier in 1807. A Fourier series is an expansion of a function as a linear combination of an infinite number of sine and cosine functions. We will not go into detail about Fourier series but will simply look at one example. Let us consider expanding a function in terms of the particle-in-a-box wave functions, which, according to (2.23), are

$$\psi_n = \left(\frac{2}{l}\right)^{1/2} \sin\left(\frac{n\pi x}{l}\right), \qquad n = 1, 2, 3, \ldots \tag{7.47}$$

for x between 0 and l. What are our chances for representing an arbitrary function $f(x)$, in the interval $0 \leqslant x \leqslant l$, by a series of the form

$$f(x) = \sum_{n=1}^{\infty} a_n \psi_n = \left(\frac{2}{l}\right)^{1/2} \sum_{n=1}^{\infty} a_n \sin\left(\frac{n\pi x}{l}\right), \qquad 0 \leqslant x \leqslant l \qquad (7.48)$$

Substituting $x = 0$ and $x = l$ in (7.48), we have the restriction that $f(0) = 0 = f(l)$. In other words, $f(x)$ must satisfy the same boundary conditions as the ψ_n. We will also assume that $f(x)$ is finite, single-valued, and continuous, but not necessarily differentiable. With these assumptions it can be shown that the expansion (7.48) is valid. We will not prove (7.48) but will simply illustrate its use to represent a function.

Before we can apply (7.48) to a specific $f(x)$, we must derive an expression for the expansion coefficients a_n. We start by multiplying (7.48) by ψ_m^*:

$$\psi_m^* f(x) = \sum_{n=1}^{\infty} a_n \psi_m^* \psi_n = \left(\frac{2}{l}\right) \sum_{n=1}^{\infty} a_n \sin\left(\frac{n\pi x}{l}\right) \sin\left(\frac{m\pi x}{l}\right) \qquad (7.49)$$

Now we integrate (7.49) from 0 to l. Assuming the validity of interchanging the integration and the infinite summation, we have

$$\int_0^l \psi_m^* f(x)\, dx = \sum_{n=1}^{\infty} a_n \int_0^l \psi_m^* \psi_n\, dx \qquad (7.50)$$

$$= \sum_{n=1}^{\infty} a_n \left(\frac{2}{l}\right) \int_0^l \sin\left(\frac{n\pi x}{l}\right) \sin\left(\frac{m\pi x}{l}\right) dx \qquad (7.51)$$

Recall that we proved the orthonormality of the particle-in-a-box wave functions [Eq. (2.27)]. Therefore (7.50) becomes

$$\int_0^l \psi_m^* f(x)\, dx = \sum_{n=1}^{\infty} a_n \delta_{mn} \qquad (7.52)$$

The type of sum in (7.52) occurs often. Writing it in detail, we have

$$\sum_{n=1}^{\infty} a_n \delta_{mn} = a_1 \delta_{m,1} + a_2 \delta_{m,2} + \cdots + a_m \delta_{m,m} + a_{m+1} \delta_{m,m+1} + \cdots$$

$$= 0 + 0 + \cdots + a_m + 0 + \cdots$$

$$\sum_{n=1}^{\infty} a_n \delta_{mn} = a_m \qquad (7.53)$$

Thus since δ_{mn} is zero except when the summation index n is equal to m, all terms but one vanish, and (7.52) becomes

$$a_m = \int_0^l \psi_m^* f(x)\, dx \qquad (7.54)$$

which is the desired expression. Using (7.54) in (7.48), we have

$$f(x) = \sum_{n=1}^{\infty} \left[\int_0^l \psi_n^* f(x)\, dx\right] \psi_n(x) \qquad (7.55)$$

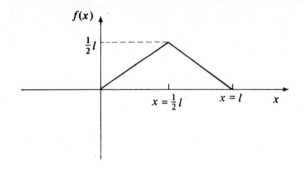

Figure 7.1 *Function to be expanded in terms of particle-in-a-box functions*

We now use (7.48) to represent a specific function, the function of Fig. 7.1, which is defined by

$$f(x) = x \qquad \text{for} \quad 0 \leqslant x \leqslant l/2$$
$$f(x) = l - x \quad \text{for} \quad \tfrac{1}{2}l \leqslant x \leqslant l \tag{7.56}$$

To evaluate the expansion coefficients, we use (7.54), (7.47), and (7.56):

$$a_n = \left(\frac{2}{l}\right)^{1/2} \int_0^l \sin\left(\frac{n\pi x}{l}\right) f(x)\, dx$$

$$= \left(\frac{2}{l}\right)^{1/2} \int_0^{l/2} x \sin\left(\frac{n\pi x}{l}\right) dx + \left(\frac{2}{l}\right)^{1/2} \int_{l/2}^l (l - x) \sin\left(\frac{n\pi x}{l}\right) dx$$

Using the Appendix integral (A.1), we find

$$a_n = \frac{(2l)^{3/2}}{n^2 \pi^2} \sin\left(\frac{n\pi}{2}\right) \tag{7.57}$$

Using (7.57) in the expansion (7.48), we have [note that $\sin(n\pi/2)$ vanishes for n even and equals $+1$ or -1 for n odd]

$$f(x) = \frac{4l}{\pi^2}\left[\sin\left(\frac{\pi x}{l}\right) - \frac{1}{3^2}\sin\left(\frac{3\pi x}{l}\right) + \frac{1}{5^2}\sin\left(\frac{5\pi x}{l}\right) - \cdots\right] \tag{7.58}$$

$$f(x) = \frac{4l}{\pi^2}\sum_{n=1}^{\infty} (-1)^{n+1} \sin\left[(2n-1)\frac{\pi x}{l}\right]\frac{1}{(2n-1)^2}$$

where $f(x)$ is given by (7.56). Let us check the accuracy of (7.58) at $x = \tfrac{1}{2}l$. We have

$$f\left(\frac{l}{2}\right) = \frac{4l}{\pi^2}\left(1 + \frac{1}{3^2} + \frac{1}{5^2} + \frac{1}{7^2} + \cdots\right) \tag{7.59}$$

Let us tabulate the right side of (7.59) as a function of the number of terms we take in the infinite series:

Number of terms	1	2	3	4	5	6	20
Right side of (7.59)	$0.405l$	$0.450l$	$0.467l$	$0.475l$	$0.480l$	$0.483l$	$0.495l$

If we take an infinite number of terms, the series should sum to $\frac{1}{2}l$, which is the value of $f(\frac{1}{2}l)$. Assuming the validity of the series, we have the interesting result that the infinite sum in parentheses in (7.59) equals $\pi^2/8$.

We have seen an example of the expansion of a function in terms of a set of functions—the particle-in-a-box wave functions. There are many different sets of functions that can be used to expand an arbitrary function. Consider a set of functions $\varphi_1, \varphi_2, \ldots, \varphi_i, \ldots$. We say that the φ_i form a *complete set* if it is possible to expand any well-behaved function f that obeys the same boundary conditions as the φ_i, as a linear combination of the φ_i, according to

$$f = \sum_i a_i \varphi_i \qquad (7.60)$$

where the a_i are constants. By virtue of theorems of Fourier analysis (which we have not proved), it can be shown that the particle-in-a-box energy eigenfunctions form a complete set. Now let φ_i be the set of eigenfunctions of *any* linear Hermitian operator that represents a physical quantity. We *postulate* that these φ_i form a complete set. (Completeness of the eigenfunctions can usually be proved in the one-dimensional case, but it must be postulated for multidimensional systems.) Thus any well-behaved function that satisfies the same boundary conditions as the ω_i can be expanded according to (7.60). Equation (7.48) is an example of (7.60).

Recall that the harmonic-oscillator wave functions are given by a Hermite polynomial times an exponential factor [Eq. (4.73)]. Therefore, by virtue of the expansion postulate, we can expand any well-behaved function $f(x)$, for which $f(\pm\infty) = 0$, as

$$f(x) = \left(\frac{\alpha}{\pi}\right)^{1/4} \sum_{n=0}^{\infty} \frac{a_n}{(2^n n!)^{1/2}} H_n(\alpha^{1/2}x) e^{-\alpha x^2/2} \qquad \text{for } -\infty \leqslant x \leqslant \infty \quad (7.61)$$

How about using the hydrogen-atom bound-state wave functions to expand an arbitrary function $f(r, \theta, \varphi)$? The answer is that these functions do *not* form a complete set, and we cannot expand f using them. To have a complete set, we must use *all* the eigenfunctions of a particular Hermitian operator. In addition to the bound-state eigenfunctions of the hydrogen-atom Hamiltonian, we have the continuum eigenfunctions, corresponding to ionized states. If we include the continuum eigenfunctions along with the bound-state eigenfunctions, then we have a complete set. (For the particle-in-a-box and the harmonic oscillator, there are no continuum functions.) Equation (7.60) implies an integration over the continuum eigenfunctions, if there are any. Thus if $\psi_{nlm}(r, \theta, \varphi)$ is a bound-state wave function of the hydrogen atom and $\psi_{Elm}(r, \theta, \varphi)$ is a continuum eigenfunction, then (7.60) becomes

$$f(r, \theta, \varphi) = \sum_{n=1}^{\infty} \sum_{l=0}^{n-1} \sum_{m=-l}^{l} a_{nlm} \psi_{nlm}(r, \theta, \varphi) + \sum_{l=0}^{\infty} \sum_{m=-l}^{l} \int_0^{\infty} a_{lm}(E) \psi_{Elm}(r, \theta, \varphi) \, dE$$

As another example, consider the eigenfunctions of $\hat{p}_x$ [Eq. (3.43)]:

$$\varphi_k = e^{ikx/\hbar}, \qquad -\infty < k < \infty \tag{7.62}$$

Here the eigenvalues are all continuous, and the eigenfunction expansion (7.60) of an arbitrary function f becomes

$$f(x) = \int_{-\infty}^{\infty} a(k) e^{ikx/\hbar} dk \tag{7.63}$$

The reader with a good mathematical background may recognize the integral in (7.63) as very nearly the Fourier transform of $a(k)$.

Let us evaluate the expansion coefficients in (7.60). The work is identical to that involved in deriving (7.54). We multiply (7.60) by φ_j^* and integrate over all space to get

$$\int \varphi_j^* f \, d\tau = \sum_i a_i \int \varphi_j^* \varphi_i \, d\tau = \sum_i a_i \delta_{ij} \tag{7.64}$$

$$a_j = \int \varphi_j^* f \, d\tau \tag{7.65}$$

where the orthonormality of the eigenfunctions, Eq. (7.45), was used. The expansion (7.60) becomes

$$f = \sum_i \left[\int \varphi_i^* f \, d\tau \right] \varphi_i = \sum_i \langle \varphi_i | f \rangle \varphi_i \tag{7.66}$$

A useful theorem is as follows. Let φ_i be the complete set of eigenfunctions of the Hermitian operator $\hat{A}$. Let f be a function that is an eigenfunction of $\hat{A}$: $\hat{A}f = kf$. Then if f is expressed as a linear combination of the φ_i's according to (7.60), the only nonzero expansion coefficients a_i are those for which φ_i has the eigenvalue k. (There may be several φ_i's with eigenvalue k because of degeneracy.) Thus in the expansion of f, we include only those eigenfunctions that have the same eigenvalue as f. The proof of this theorem follows at once from (7.65); if f and φ_j in (7.65) correspond to different eigenvalues of the Hermitian operator $\hat{A}$, they will be orthogonal [Eq. (7.37)], and a_j will vanish.

We shall occasionally use a notation (called *ket* notation) in which the function f is denoted by the symbol $|f\rangle$. There doesn't seem to be any point to this notation, but in advanced formulations of quantum mechanics, it takes on a special significance. In ket notation Eq. (7.66) reads

$$|f\rangle = \sum_i |\varphi_i\rangle \langle \varphi_i | f \rangle = \sum_i |i\rangle \langle i | f \rangle \tag{7.67}$$

Ket notation is conveniently used to specify eigenfunctions by listing their eigenvalues. For example, the hydrogen-atom wave function with quantum numbers n, l, m is denoted by $\psi_{nlm} = |nlm\rangle$.

The contents of Sections 7.2 and 7.3 can be summarized by the statement that *the eigenfunctions of a Hermitian operator form a complete, orthonormal set, and the eigenvalues are real.*

7.4   *EIGENFUNCTIONS OF COMMUTING OPERATORS*

In Section 5.1 we made some statements about simultaneous eigenfunctions of two different operators. We now prove these statements.

First we show that if there exists a common *complete set* of eigenfunctions for two linear operators, then the operators commute. Let $\hat{A}$ and $\hat{B}$ denote two linear operators that possess a common complete set of eigenfunctions φ_i. We have

$$\hat{A}\varphi_i = s_i\varphi_i, \qquad \hat{B}\varphi_i = t_i\varphi_i \tag{7.68}$$

where s_i and t_i are the eigenvalues. We must prove that

$$[\hat{A}, \hat{B}] = \hat{0} \tag{7.69}$$

Equation (7.69) is an operator equation; for two operators to be equal, the results of operating with either of them on an arbitrary function f must be the same; we must therefore show that

$$(\hat{A}\hat{B} - \hat{B}\hat{A})f = \hat{0}f = 0 \tag{7.70}$$

where f is an arbitrary function. We begin the proof by expanding f (assuming that it obeys the proper boundary conditions) in terms of the complete set of eigenfunctions φ_i:

$$f = \sum_i c_i\varphi_i \tag{7.71}$$

Operating on each side of (7.71) with $\hat{A}\hat{B} - \hat{B}\hat{A}$, we have

$$(\hat{A}\hat{B} - \hat{B}\hat{A})f = (\hat{A}\hat{B} - \hat{B}\hat{A})\sum_i c_i\varphi_i \tag{7.72}$$

Since $\hat{A}\hat{B}$ and $\hat{B}\hat{A}$ are linear operators (Problem 3.7), we have

$$(\hat{A}\hat{B} - \hat{B}\hat{A})f = \sum_i c_i(\hat{A}\hat{B} - \hat{B}\hat{A})\varphi_i = \sum_i c_i[\hat{A}(\hat{B}\varphi_i) - \hat{B}(\hat{A}\varphi_i)]$$

where we have used the definitions of the sum and the product of operators. Use of (7.68) yields

$$(\hat{A}\hat{B} - \hat{B}\hat{A})f = \sum_i c_i[\hat{A}t_i\varphi_i - \hat{B}s_i\varphi_i] = \sum_i c_i(s_it_i\varphi_i - t_is_i\varphi_i) = 0$$

which completes the proof that $\hat{A}$ and $\hat{B}$ commute if they have a common complete set of eigenfunctions.

It is sometimes erroneously stated that if there exists a common eigenfunction of $\hat{A}$ and $\hat{B}$, then they commute. We have already seen an example that shows this statement to be false; in Section 5.3 we saw that the spherical harmonic Y_0^0 is an eigenfunction of both $\hat{L}_z$ and $\hat{L}_x$ even though these two operators do not commute. It is instructive to examine the so-called proof that is often given of this erroneous statement. Let φ be the common eigenfunction:

$$\hat{A}\varphi = s\varphi, \qquad \hat{B}\varphi = t\varphi$$

We have
$$\hat{A}\hat{B}\varphi = \hat{A}t\varphi = st\varphi$$
$$\hat{B}\hat{A}\varphi = \hat{B}s\varphi = ts\varphi = st\varphi$$
$$\hat{A}\hat{B}\varphi = \hat{B}\hat{A}\varphi \qquad (7.73)$$

The "proof" is completed by cancelling φ from each side of (7.73) to get

$$\hat{A}\hat{B} = \hat{B}\hat{A} \,(?) \qquad (7.74)$$

It is in going from (7.73) to (7.74) that the error occurs. Just because the two operators $\hat{A}\hat{B}$ and $\hat{B}\hat{A}$ give the same result when acting on the single function φ, there is no reason to conclude that $\hat{A}\hat{B} = \hat{B}\hat{A}$. (For example, d/dx and d^2/dx^2 give the same result when operating on e^x, but d/dx is certainly not equal to d^2/dx^2.) The two operators must give the same result when acting on *every* (well-behaved) function before one can conclude that they are equal. Thus even though $\hat{A}$ and $\hat{B}$ do not commute, there might exist one or more common eigenfunctions of $\hat{A}$ and $\hat{B}$. However, we cannot have a common *complete* set of eigenfunctions of two noncommuting operators, as we proved earlier in this section.

We have shown that *if there exists a common complete set of eigenfunctions of $\hat{A}$ and $\hat{B}$, then they commute.* We now prove the converse theorem: *If $\hat{A}$ and $\hat{B}$ commute, we can select a common complete set of eigenfunctions for them.* Let φ_i and t_i be the eigenfunctions and eigenvalues of $\hat{B}$:

$$\hat{B}\varphi_i = t_i\varphi_i \qquad (7.75)$$

Operating on both sides of this equation with $\hat{A}$, we have

$$\hat{A}\hat{B}\varphi_i = \hat{A}t_i\varphi_i \qquad (7.76)$$

Since $\hat{A}$ and $\hat{B}$ commute and $\hat{A}$ is linear, we have

$$\hat{B}(\hat{A}\varphi_i) = t_i(\hat{A}\varphi_i) \qquad (7.77)$$

This equation states that the function $\hat{A}\varphi_i$ is also an eigenfunction of the operator $\hat{B}$ with the same eigenvalue t_i as the eigenfunction φ_i. Let us for the moment assume that the eigenvalues of $\hat{B}$ are nondegenerate—that for any given eigenvalue t_i there corresponds one and only one linearly independent eigenfunction. If this is so, then the two eigenfunctions φ_i and $\hat{A}\varphi_i$, which correspond to the same eigenvalue t_i, must be linearly dependent, i.e., one must simply be a multiple of the other:

$$\hat{A}\varphi_i = c_i\varphi_i \qquad (7.78)$$

where c_i is some constant. This equation states that the functions φ_i are eigenfunctions of $\hat{A}$, which is what we wanted to prove. In Section 7.3 we postulated that the eigenfunctions of any operator that represents a physical quantity form a complete set; hence the φ_i form a complete set.

We have just proved the desired theorem for the nondegenerate case, but what about the degenerate case? Let the level with eigenvalue t_i be n-fold degenerate. We know from Eq. (7.77) that $\hat{A}\varphi_i$ is an eigenfunction of $\hat{B}$ with eigenvalue t_i. Hence it must be some linear combination of the n linearly independent eigenfunctions corresponding to the eigenvalue t_i. From this statement we cannot conclude that φ_i is an eigenfunction of $\hat{A}$. However, by taking suitable linear combinations of the n linearly independent eigenfunctions, we can construct a new set of n linearly independent eigenfunctions of $\hat{B}$ that will also be eigenfunctions of $\hat{A}$. For the proof that this can be done, see *Merzbacher*, Section 8.5.

Thus when $\hat{A}$ and $\hat{B}$ commute, it is always possible to *select* a common set of eigenfunctions for them. As an example, consider the hydrogen atom, where the operators $\hat{L}_z$ and $\hat{H}$ were shown to commute. If we desired, we could take the phi factor in the eigenfunctions of $\hat{H}$ as $\sin m\varphi$ and $\cos m\varphi$. In this case, we will not have eigenfunctions of $\hat{L}_z$, except for $m = 0$. However, the linear combinations

$$R(r) S(\theta) [\cos m\varphi + i \sin m\varphi] = RSe^{im\varphi}, \qquad m = -l, \ldots, +l$$

give us eigenfunctions of $\hat{L}_z$ that are still eigenfunctions of $\hat{H}$ by virtue of the theorem in Section 3.6.

Extension of the above proofs to the case of more than two operators shows that for a set of Hermitian operators $\hat{A}, \hat{B}, \hat{C}, \ldots$ there exists a common complete set of eigenfunctions if and only if every pair of operators commutes.

Let us prove a useful theorem about matrix elements. Let $\hat{A}$ be a Hermitian operator with eigenfunctions φ_i:

$$\hat{A}\varphi_i = s_i \varphi_i \qquad (7.79)$$

Let $\hat{B}$ be an operator that commutes with $\hat{A}$. We assert that

$$B_{ij} = \int \varphi_i^* \hat{B}\varphi_j \, d\tau = 0 \qquad \text{for } s_i \neq s_j \qquad (7.80)$$

We have

$$\int \varphi_i^* \hat{B}\varphi_j \, d\tau = \frac{1}{s_j} \int \varphi_i^* \hat{B}s_j\varphi_j \, d\tau$$

$$B_{ij} = \frac{1}{s_j} \int \varphi_i^* \hat{B}\hat{A}\varphi_j \, d\tau = \frac{1}{s_j} \int \varphi_i^* \hat{A}\hat{B}\varphi_j \, d\tau$$

$$s_j B_{ij} = \langle \varphi_i | \hat{A} | \hat{B}\varphi_j \rangle$$

Using the Hermitian property of $\hat{A}$ [Eq. (7.16)], we have

$$s_j B_{ij} = \langle \hat{B}\varphi_j | \hat{A} | \varphi_i \rangle^*$$

Using (7.79) and the fact that the eigenvalues of $\hat{A}$ are real, we have

$$s_j B_{ij} = s_i^* \langle \hat{B}\varphi_j | \varphi_i \rangle^*$$

$$s_j B_{ij} = s_i \langle \varphi_i | \hat{B}\varphi_j \rangle = s_i B_{ij}$$

$$B_{ij}(s_j - s_i) = 0 \qquad (7.81)$$

We conclude that (7.80) holds.

A proof of the uncertainty principle is outlined in Problem 7.35.

7.5 PARITY

There are certain quantum-mechanical operators that have no classical analog. One of these is the parity operator. Recall that the harmonic-oscillator wave functions are either even or odd; we will show how this property is related to the parity operator.

The parity operator $\hat{\Pi}$ is defined in terms of its effect on an arbitrary function f:

$$\hat{\Pi} f(x, y, z) = f(-x, -y, -z) \tag{7.82}$$

The parity operator replaces each Cartesian coordinate with its negative. For example, $\hat{\Pi}(x^2 - ze^{ay}) = x^2 + ze^{-ay}$.

As with any quantum-mechanical operator, we are interested in the eigenvalues c_i and the eigenfunctions g_i of the parity operator:

$$\hat{\Pi} g_i = c_i g_i \tag{7.83}$$

The key to the problem is to calculate the square of $\hat{\Pi}$:

$$\hat{\Pi}^2 f(x, y, z) = \hat{\Pi}[\hat{\Pi} f(x, y, z)]$$

$$\hat{\Pi}^2 f(x, y, z) = \hat{\Pi}[f(-x, -y, -z)]$$

$$\hat{\Pi}^2 f(x, y, z) = f(x, y, z)$$

Since f is arbitrary, we conclude that $\hat{\Pi}^2$ equals the unit operator:

$$\hat{\Pi}^2 = \hat{1} \tag{7.84}$$

We now operate on (7.83) with $\hat{\Pi}$, to get

$$\hat{\Pi}\hat{\Pi} g_i = \hat{\Pi} c_i g_i$$

Since $\hat{\Pi}$ is linear (Problem 7.15), we have

$$\hat{\Pi}^2 g_i = c_i \hat{\Pi} g_i$$

$$\hat{\Pi}^2 g_i = c_i^2 g_i \tag{7.85}$$

where the eigenvalue equation (7.83) has been used. Since $\hat{\Pi}^2$ is the unit operator, the left side of (7.85) is simply g_i, and we have

$$g_i = c_i^2 g_i \tag{7.86}$$

The function g_i cannot be zero everywhere (zero is always rejected as an eigenfunction on physical grounds); hence we can divide (7.86) by g_i to obtain $c_i^2 = 1$, and

$$c_i = \pm 1 \tag{7.87}$$

The eigenvalues of $\hat{\Pi}$ are $+1$ and -1. (Note that this derivation is applicable to any operator whose square is the unit operator.)

What are the eigenfunctions g_i? The eigenvalue equation (7.83) reads

$$\hat{\Pi} g_i(x, y, z) = \pm g_i(x, y, z)$$

$$g_i(-x, -y, -z) = \pm g_i(x, y, z) \tag{7.88}$$

If the eigenvalue is $+1$, then

$$g_i(-x, -y, -z) = g_i(x, y, z) \tag{7.89}$$

so that g_i is an even function. If the eigenvalue is -1, then g_i is odd:

$$g_i(-x, -y, -z) = -g_i(x, y, z) \qquad (7.90)$$

Hence the eigenfunctions are all possible well-behaved even and odd functions.

When the parity operator commutes with the Hamiltonian operator $\hat{H}$, we can select a common set of eigenfunctions for these operators, as proved in Section 7.4. The eigenfunctions of $\hat{H}$ are the stationary-state wave functions ψ_i. Hence when

$$[\hat{\Pi}, \hat{H}] = 0 \qquad (7.91)$$

the wave functions ψ_i can be chosen to be eigenfunctions of $\hat{\Pi}$. We just proved that the eigenfunctions of $\hat{\Pi}$ are either even or odd; hence when (7.91) holds, each wave function can be chosen to be either even or odd. Let us find out when the parity and Hamiltonian operators commute.

We have, for a one-particle system,

$$[\hat{H}, \hat{\Pi}] = [\hat{T}, \hat{\Pi}] + [\hat{V}, \hat{\Pi}] = -\frac{\hbar^2}{2m}\left[\frac{\partial^2}{\partial x^2}, \hat{\Pi}\right] - \frac{\hbar^2}{2m}\left[\frac{\partial^2}{\partial y^2}, \hat{\Pi}\right]$$
$$-\frac{\hbar^2}{2m}\left[\frac{\partial^2}{\partial z^2}, \hat{\Pi}\right] + [\hat{V}, \hat{\Pi}] \qquad (7.92)$$

Since

$$\hat{\Pi}\left[\frac{\partial^2}{\partial x^2}\varphi(x, y, z)\right] = \frac{\partial}{\partial(-x)}\frac{\partial}{\partial(-x)}\varphi(-x, -y, -z) = \frac{\partial^2}{\partial x^2}\varphi(-x, -y, -z)$$
$$= \frac{\partial^2}{\partial x^2}\hat{\Pi}\varphi(x, y, z)$$

where φ is any function, we conclude that

$$\left[\frac{\partial^2}{\partial x^2}, \hat{\Pi}\right] = 0$$

Similar equations hold for the y and z coordinates, and (7.92) becomes

$$[\hat{H}, \hat{\Pi}] = [\hat{V}, \hat{\Pi}]$$

Now

$$\hat{\Pi}[V(x, y, z)\varphi(x, y, z)] = V(-x, -y, -z)\varphi(-x, -y, -z) \qquad (7.93)$$

If the potential energy is an even function, i.e., if $V(-x, -y, -z) = V(x, y, z)$, then (7.93) becomes

$$\hat{\Pi}[V(x, y, z)\varphi(x, y, z)] = V(x, y, z)\varphi(-x, -y, -z)$$
$$= V(x, y, z)\hat{\Pi}\varphi(x, y, z)$$

so that $[\hat{V}, \hat{\Pi}] = 0$. Hence when the potential energy is an even function, the parity operator commutes with the Hamiltonian:

$$[\hat{H}, \hat{\Pi}] = 0 \qquad \text{if } V \text{ is even} \qquad (7.94)$$

These results are easily extended to the n-particle case. For an n-particle system, the parity operator is defined by

$$\hat{\Pi} f(x_1, y_1, z_1, \ldots, x_n, y_n, z_n) = f(-x_1, -y_1, -z_1, \ldots, -x_n, -y_n, -z_n)$$

It is easy to see that (7.91) holds when

$$V(x_1, y_1, z_1, \ldots, x_n, y_n, z_n) = V(-x_1, -y_1, -z_1, \ldots, -x_n, -y_n, -z_n)$$

If V satisfies this equation, we say that it is an even function of the $3n$ coordinates. In summary, when the potential energy V is an even function, we can choose the wave functions so that each ψ_i is either even or odd. A function that is either even or odd is said to have a definite parity.

  If there is no degeneracy in the energy levels (as is usually true in one-dimensional problems), then there is only one independent wave function ψ_i corresponding to each energy E_i, and there is no element of choice (apart from an arbitrary multiplicative constant) in the wave functions. Thus for the nondegenerate case, the stationary-state wave functions must be of definite parity when V is an even function. For example, the one-dimensional harmonic oscillator has $V = \frac{1}{2}kx^2$, which is an even function, and we found the wave functions to be of definite parity.

  For the degenerate case, we have an element of choice in the wave functions, since an arbitrary linear combination of the functions corresponding to the degenerate level is an eigenfunction of $\hat{H}$. For a degenerate energy level, by taking appropriate linear combinations we can choose wave functions that are of definite parity, but there is no necessity that they be of definite parity.

  Parity aids in evaluating integrals. In Chapter 4 we showed that

$$\int_{-\infty}^{\infty} f(x)\, dx = 0 \tag{7.95}$$

when $f(x)$ is an odd function. Let us extend this result to the $3n$-dimensional case. An odd function of $3n$ variables satisfies

$$g(-x_1, -y_1, -z_1, \ldots, -x_n, -y_n, -z_n) = -g(x_1, y_1, z_1, \ldots, x_n, y_n, z_n)$$

If g is an odd function of the $3n$ variables, then

$$\int_{-\infty}^{\infty} \cdots \int_{-\infty}^{\infty} g(x_1, \ldots, z_n)\, dx_1 \cdots dz_n = 0 \tag{7.96}$$

where the integration is over the $3n$ coordinates. This equation holds because the contribution to the integral from the value of g at $(x_1, y_1, z_1, \ldots, x_n, y_n, z_n)$ is cancelled by the contribution from $(-x_1, -y_1, -z_1, \ldots, -x_n, -y_n, -z_n)$.

  A more general case is where the integrand is an odd function of some (but not necessarily all) of the variables. Let f be a function such that

$$f(-q_1, -q_2, \ldots, -q_k, q_{k+1}, q_{k+2}, \ldots, q_m) =$$
$$-f(q_1, q_2, \ldots, q_k, q_{k+1}, q_{k+2}, \ldots, q_m) \tag{7.97}$$

where $1 \leqslant k \leqslant m$. We assert that if f obeys (7.97), then

$$\int_{-\infty}^{\infty} \cdots \int_{-\infty}^{\infty} f(q_1, \ldots, q_m)\, dq_1 \cdots dq_m = 0 \tag{7.98}$$

The proof of (7.98) is simple. The integral in (7.98) can be written as

$$\int_{-\infty}^{\infty} \cdots \int_{-\infty}^{\infty} \left[\int_{-\infty}^{\infty} \cdots \int_{-\infty}^{\infty} f(q_1, \ldots, q_k, q_{k+1}, \ldots, q_m)\, dq_1 \cdots dq_k \right] dq_{k+1} \cdots dq_m$$

$$\text{(7.99)}$$

As far as the multiple integral in brackets is concerned, q_{k+1} through q_m are constants. By virtue of (7.97), the contributions from

$$f(-q_1, \ldots, -q_k, q_{k+1}, \ldots, q_m) \qquad \text{and} \qquad f(q_1, \ldots, q_k, q_{k+1}, \ldots, q_m)$$

cancel, so that the bracketed integral in (7.99) vanishes and (7.98) follows.

7.6   MEASUREMENT AND THE SUPERPOSITION OF STATES

Quantum mechanics can be regarded as a scheme for calculating the probabilities of the various possible outcomes of a measurement. For example, if we know the state function $\Psi(x, t)$, then the probability that a measurement at time t of the particle's position yields a value between x and $x + dx$ is given by $|\Psi(x, t)|^2\, dx$. We now consider measurement of the general property G; our aim is to find out how to use Ψ to calculate the probabilities for each possible result of a measurement of G. The results of this section lie at the very heart of quantum mechanics.

We will deal with an n-particle system and will use q to symbolize the $3n$ coordinates. We have postulated that the eigenvalues g_i of the operator $\hat{G}$ are the only possible results of a measurement of the property G. Using φ_i for the eigenfunctions of $\hat{G}$, we have

$$\hat{G}\varphi_i(q) = g_i \varphi_i(q) \qquad \text{(7.100)}$$

We postulated in Section 7.3 that the eigenfunctions of any linear Hermitian operator that represents a physical observable form a complete set. Since the φ_i form a complete set, we can expand any well-behaved function in terms of them; in particular, we can expand the state function Ψ in terms of the φ_i:

$$\Psi(q, t) = \sum_i c_i(t)\, \varphi_i(q) \qquad \text{(7.101)}$$

To allow for the change of Ψ with time, the expansion coefficients c_i vary with time.

Now $|\Psi|^2$ is a probability density, so that we require

$$\int \Psi^* \Psi\, d\tau = 1 \qquad \text{(7.102)}$$

Substituting (7.101) into the normalization condition, we get

$$1 = \int \sum_i c_i^* \varphi_i^* \sum_i c_i \varphi_i\, d\tau = \int \sum_i c_i^* \varphi_i^* \sum_j c_j \varphi_j\, d\tau$$

$$1 = \int \sum_i \sum_j c_i^* c_j \varphi_i^* \varphi_j\, d\tau \qquad \text{(7.103)}$$

[It is important to use different symbols for the two dummy summation indices in (7.103). For example, consider the following product of two sums:

$$\sum_{i=1}^{2} a_i \sum_{i=1}^{2} b_i = (a_1 + a_2)(b_1 + b_2) = a_1 b_1 + a_1 b_2 + a_2 b_1 + a_2 b_2$$

If we carelessly write

$$\sum_{i=1}^{2} a_i \sum_{i=1}^{2} b_i \overset{\text{(wrong)}}{=} \sum_{i=1}^{2}\sum_{i=1}^{2} a_i b_i = \sum_{i=1}^{2} (a_1 b_1 + a_2 b_2) = 2(a_1 b_1 + a_2 b_2)$$

we get the wrong answer. The correct way to write the product is

$$\sum_{i=1}^{2} a_i \sum_{i=1}^{2} b_i = \sum_{i=1}^{2} a_i \sum_{j=1}^{2} b_j = \sum_{i=1}^{2}\sum_{j=1}^{2} a_i b_j$$

$$= \sum_{i=1}^{2} (a_i b_1 + a_i b_2) = a_1 b_1 + a_1 b_2 + a_2 b_1 + a_2 b_2$$

which gives the right answer.] Assuming the validity of interchanging the infinite summation and integration in (7.103), we have

$$\sum_i \sum_j c_i^* c_j \int \varphi_i^* \varphi_j \, d\tau = 1$$

Since $\hat{G}$ is Hermitian, the eigenfunctions φ_i are orthonormal [Eq. (7.45)]; hence

$$\sum_i \sum_j c_i^* c_j \delta_{ij} = 1$$

$$\sum_i |c_i|^2 = 1 \tag{7.104}$$

We will point out the significance of (7.104) shortly.

Recall the postulate (Section 3.7) that if Ψ is the normalized state function of a system, then the average value of the property G is

$$\langle G \rangle = \int \Psi^*(q, t)\, \hat{G} \Psi(q, t)\, d\tau$$

Using the expansion (7.101) in the average-value expression, we have

$$\langle G \rangle = \int \sum_i c_i^* \varphi_i^* \, \hat{G} \sum_j c_j \varphi_j \, d\tau = \sum_i \sum_j c_i^* c_j \int \varphi_i^* \, \hat{G} \varphi_j \, d\tau \tag{7.105}$$

Applying (7.100), we get

$$\langle G \rangle = \sum_i \sum_j c_i^* c_j g_j \int \varphi_i^* \varphi_j \, d\tau = \sum_i \sum_j c_i^* c_j g_j \delta_{ij}$$

$$\langle G \rangle = \sum_i |c_i|^2 g_i \tag{7.106}$$

How do we interpret (7.106)? We postulated (Section 3.3) that the eigenvalues of an operator are the only possible numbers we can get when we measure the property that the operator represents. In any measurement of G, we will get one of the values g_i. Now recall Eq. (3.108):

$$\langle G \rangle = \sum_i P_{g_i} g_i \qquad (7.107)$$

where P_{g_i} is the probability of getting g_i in a measurement of G. Comparing (7.106) and (7.107), we conclude that $|c_i|^2$ *is the probability*[1] *of getting the value g_i in a measurement of the property G*; the quantity c_i is the coefficient of the eigenfunction φ_i in the expansion (7.101). The $|c_i|^2$ sum to 1, as probabilities should [Eq. (7.104)].

When can we predict the result of a measurement of G with certainty? We can do this if all the coefficients c_i are zero, except one: $c_i = 0$, $i \neq k$; $c_k \neq 0$. For this case, Eq. (7.104) gives $|c_k|^2 = 1$, so that we are certain to find the result g_k. In this case, the state function (7.101) reduces to

$$\Psi = \varphi_k$$

We may thus view the expansion (7.101) as expressing the general state Ψ as a *superposition* of the eigenstates φ_i of the operator $\hat{G}$; each eigenstate φ_i corresponds to the value g_i for the property G. The degree to which any function φ_i occurs in the expansion of Ψ, as measured by $|c_i|^2$, determines the probability of getting the value g_i in a measurement of G.

How do we calculate the expansion coefficients c_i, so that we can get the probabilities $|c_i|^2$? We multiply (7.101) by φ_j^* and integrate over all space to get

$$\int \varphi_j^* \Psi \, d\tau = \sum_i c_i \int \varphi_j^* \varphi_i \, d\tau = \sum_i c_i \delta_{ij}$$

$$c_j = \int \varphi_j^* \Psi \, d\tau = \langle \varphi_j | \Psi \rangle \qquad (7.108)$$

The probability of finding the value g_j in a measurement of G is

$$|c_j|^2 = \left| \int \varphi_j^* \Psi \, d\tau \right|^2 = |\langle \varphi_j | \Psi \rangle|^2 \qquad (7.109)$$

The quantity $\langle \varphi_j | \Psi \rangle$ is called a *probability amplitude*.

Thus if we know the state of the system, as determined by the state function Ψ, we can use (7.109) to predict the probabilities for the various possible outcomes of a measurement of any property G. (Determination of the eigenfunctions φ_j and eigenvalues g_j is a mathematical problem.)

(To determine experimentally the probability of finding g_j on measurement, we take a large number of identical, noninteracting systems, each in the same state Ψ, and carry out a measurement of G on each system. If n_j of the measurements yield g_j,

---

[1] However, if g_i is a degenerate eigenvalue, we must add together the probabilities for all the states with eigenvalue g_i.

then

$$\frac{n_j}{n} = |\langle \varphi_j | \Psi \rangle|^2 \qquad (7.110)$$

where n is the total number of measurements made.)

We illustrate our conclusions with an example. Let the property G be the linear momentum p_x. The eigenfunctions and eigenvalues in (7.100) are [see Eqs. (3.43) and (3.44)]

$$\varphi_i = e^{ik_i x/\hbar}, \qquad g_i = k_i \qquad (7.111)$$

where k_i is any real number. Now consider a free particle moving in one dimension and having energy E; its state function is, according to (2.30) and (3.39),

$$\Psi = a_1 e^{-iEt/\hbar} e^{i(2mE)^{1/2}x/\hbar} + a_2 e^{-iEt/\hbar} e^{-i(2mE)^{1/2}x/\hbar} \qquad (7.112)$$

$$\Psi = c_1 e^{ik_1 x/\hbar} + c_2 e^{ik_2 x/\hbar} \qquad (7.113)$$

where

$$c_1 \equiv a_1 e^{-iEt/\hbar}, \qquad c_2 \equiv a_2 e^{-iEt/\hbar}$$

$$k_1 \equiv (2mE)^{1/2}, \qquad k_2 \equiv -(2mE)^{1/2}$$

Comparing this equation with (7.101), we see that the free-particle state function is a superposition of two of the eigenfunctions of $\hat{p}_x$; the probability that a measurement of p_x yields the value $k_1 = (2mE)^{1/2}$ is, according to the italicized statement following (7.107),

$$|c_1|^2 = |a_1 e^{-iEt/\hbar}|^2 = |a_1|^2$$

while the probability that a measurement of p_x yields $-(2mE)^{1/2}$ is

$$|c_2|^2 = |a_2|^2$$

The probability of getting any other value in a measurement of p_x is zero, since eigenfunctions corresponding to eigenvalues other than k_1 and k_2 do not occur in the expansion (7.113).

If G is a property that takes on a continuous range of eigenvalues (e.g., position), the summation in the expansion (7.101) is replaced by an integration over the values of g:

$$\Psi = \int c_g \varphi_g(q) \, dg \qquad (7.114)$$

and the quantity

$$|\langle \varphi_g(q) | \Psi \rangle|^2$$

is interpreted as a probability density; i.e., the probability of finding a value of G between g and $g + dg$ for a system in the state Ψ is

$$|\langle \varphi_g(q) | \Psi(q, t) \rangle|^2 \, dg \qquad (7.115)$$

We now consider the superposition of energy eigenstates. When the Hamiltonian is independent of time, we have shown that there are solutions of the time-dependent Schroedinger equation of the form

$$\Psi_n(q, t) = e^{-iE_n t/\hbar} \psi_n(q) \qquad (7.116)$$

where the $\psi_n(q)$ are eigenfunctions of $\hat{H}$:

$$\hat{H}\psi_n(q) = E_n\psi_n(q) \tag{7.117}$$

Let the function Ψ be a superposition of the stationary states Ψ_n:

$$\Psi = \sum_n c_n\Psi_n = \sum_n c_n e^{-iE_nt/\hbar}\psi_n \tag{7.118}$$

The linear combination (7.118) is, of course, not an eigenfunction of $\hat{H}$ (unless all the stationary states in the sum belong to the same degenerate energy level), but it *is* a solution of the time-dependent Schroedinger equation. The time-dependent Schroedinger equation (3.51) is

$$\left[\frac{\hbar}{i}\frac{\partial}{\partial t} + \hat{H}\right]\Psi = 0 \tag{7.119}$$

Since the operator in brackets in (7.119) is linear, it follows that a linear combination [such as (7.118)] of solutions of (7.119) is a solution of (7.119). Also, a linear combination of well-behaved functions is a well-behaved function. Hence the superposition (7.118) of energy eigenstates is a possible (nonstationary) state of the system. Because of the completeness of the eigenfunctions of $\hat{H}$, any state function can be written in the form (7.118). The state function (7.118) represents a state that does not have a definite energy; rather, when we make an energy measurement, the probability of getting E_n is, according to the results of this section,

$$|c_n e^{-iE_nt/\hbar}|^2 = |c_n|^2$$

What determines whether a system is in a stationary state such as (7.116) or a nonstationary state such as (7.118)? The answer is that the past history of the system determines its present state. For example, if we take a system that is in a stationary state and expose it to radiation, the time-dependent Schroedinger equation shows that the radiation causes the state to change to a nonstationary state; see Section 9.9.

7.7 POSITION EIGENFUNCTIONS

We have derived the eigenfunctions of the linear-momentum and angular-momentum operators. We now ask: What are the eigenfunctions of the position operator?

We have

$$\hat{x} = x \cdot$$

Denoting the position eigenfunctions by $\varphi_a(x)$, we write

$$x\varphi_a(x) = a\varphi_a(x) \tag{7.120}$$

where a symbolizes the possible eigenvalues. It follows that

$$(x-a)\varphi_a(x) = 0 \tag{7.121}$$

We conclude from (7.121) that

$$\varphi_a(x) = 0 \quad \text{for } x \neq a \tag{7.122}$$

Moreover, since an eigenfunction that is zero everywhere is unacceptable, we have

$$\varphi_a(x) \neq 0 \quad \text{for } x = a \tag{7.123}$$

These conclusions make sense. If the state function is an eigenfunction of $\hat{x}$ with eigenvalue a, $\Psi = \varphi_a(x)$, we know (Section 7.6) that a measurement of x is certain to give the value a; this can only be true if the probability density $|\Psi|^2$ is zero for $x \neq a$, in agreement with (7.122).

Before considering further properties of $\varphi_a(x)$, we define the *Heaviside step function* $H(x)$ by (see Fig. 7.2)

$$H(x) = 1 \quad \text{for } x > 0$$
$$H(x) = 0 \quad \text{for } x < 0 \tag{7.124}$$
$$H(x) = \tfrac{1}{2} \quad \text{for } x = 0$$

We next define the *Dirac delta function* $\delta(x)$ as the derivative of the Heaviside step function:

$$\delta(x) \equiv \frac{dH(x)}{dx} \tag{7.125}$$

From (7.124) and (7.125), we have at once (see also Fig. 7.2)

$$\delta(x) = 0 \quad \text{for } x \neq 0 \tag{7.126}$$

Since $H(x)$ makes a sudden jump at $x = 0$, its derivative is infinite at the origin:

$$\delta(x) = \infty \quad \text{for } x = 0 \tag{7.127}$$

We can generalize these equations slightly by setting $x = t - a$, and then changing the symbol t to x; Eqs. (7.124)–(7.127) become

$$H(x - a) = 1, \qquad x > a \tag{7.128}$$
$$H(x - a) = 0, \qquad x < a \tag{7.129}$$
$$H(x - a) = \tfrac{1}{2}, \qquad x = a \tag{7.130}$$

$$\delta(x - a) = \frac{d}{dx}H(x - a) \tag{7.131}$$

$$\delta(x - a) = 0, \quad x \neq a \tag{7.132}$$
$$\delta(x - a) = \infty, \quad x = a \tag{7.133}$$

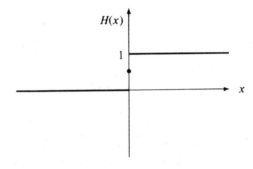

Figure 7.2 *The Heaviside step function*

Now consider the following integral:

$$\int_{-\infty}^{\infty} f(x)\delta(x-a)\,dx$$

We evaluate it using integration by parts:

$$\int u\,dv = uv - \int v\,du$$

$$u = f(x), \quad dv = \delta(x-a)\,dx$$

Using (7.131), we have

$$du = f'(x)\,dx, \quad v = H(x-a)$$

$$\int_{-\infty}^{\infty} f(x)\delta(x-a)\,dx = f(x)H(x-a)\Big|_{-\infty}^{\infty} - \int_{-\infty}^{\infty} H(x-a)f'(x)\,dx$$

$$\int_{-\infty}^{\infty} f(x)\delta(x-a)\,dx = f(\infty) - \int_{-\infty}^{\infty} H(x-a)f'(x)\,dx \qquad (7.134)$$

where (7.128) and (7.129) have been used. Since $H(x-a)$ vanishes for $x < a$, Eq. (7.134) becomes

$$\int_{-\infty}^{\infty} f(x)\delta(x-a)\,dx = f(\infty) - \int_{a}^{\infty} H(x-a)f'(x)\,dx$$

$$= f(\infty) - \int_{a}^{\infty} f'(x)\,dx = f(\infty) - f(x)\Big|_{a}^{\infty}$$

$$\int_{-\infty}^{\infty} f(x)\delta(x-a)\,dx = f(a) \qquad (7.135)$$

Comparing (7.135) with the equation

$$\sum_{j} c_{j}\delta_{ij} = c_{i} \qquad (7.136)$$

we see that the Dirac delta function plays the same role in an integral that the Kronecker delta plays in a sum. The special case of (7.135) with $f(x) = 1$ is

$$\int_{-\infty}^{\infty} \delta(x-a)\,dx = 1$$

The properties (7.132) and (7.133) of the Dirac delta function are in accord with the properties (7.122) and (7.123) for the position eigenfunctions. Let us therefore tentatively set

$$\varphi_{a}(x) = \delta(x-a) \qquad (7.137)$$

To verify (7.137) we now show it to be in accord with the Born postulate that $|\Psi(a,t)|^2\,da$ is the probability of observing a value of x between a and $a+da$. According to (7.115) this probability is given by

$$|\langle\varphi_{a}(x)|\Psi(x,t)\rangle|^{2}\,da = \left|\int_{-\infty}^{\infty} \varphi_{a}^{*}(x)\Psi(x,t)\,dx\right|^{2} da \qquad (7.138)$$

Using (7.137) and then (7.135), we have for (7.138)

$$\left|\int_{-\infty}^{\infty} \delta(x-a)\Psi(x,t)\,dx\right|^{2} da = |\Psi(a,t)|^{2}\,da \qquad (7.139)$$

Q.E.D.

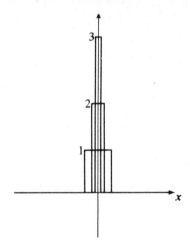

Figure 7.3 *Functions that approximate* $\delta(x)$ *with successively increasing accuracy. The area under each curve is* 1.

Since the quantity a in $\delta(x-a)$ can have any real value, the eigenvalues of $\hat{x}$ form a continuum: $-\infty < a < \infty$. As usual for continuum eigenfunctions, $\delta(x-a)$ is not quadratically integrable; we have, using (7.135),

$$\int_{-\infty}^{\infty} |\delta(x-a)|^2 \, dx = \int_{-\infty}^{\infty} \delta(x-a)\delta(x-a) \, dx = \delta(a-a) = \delta(0) = \infty$$

Summarizing, the eigenfunctions and eigenvalues of position are

$$\hat{x}\delta(x-a) = a\,\delta(x-a) \qquad\qquad \textbf{(7.140)}$$

where a is any real number.

(The delta function is rather badly behaved, and consequently the various manipulations we performed are lacking in rigor and would make a mathematician shudder. However, one can put use of the delta function on a rigorous basis by considering it to be the limiting case of a function that becomes successively more peaked at the origin; see Fig. 7.3.)

7.8   THE POSTULATES OF QUANTUM MECHANICS

We have introduced the postulates of quantum mechanics as we have gone along. In this section we summarize them. There is no unique set of postulates; we will simply use a convenient set.

Postulate 1. The state of a system is described by a function Ψ of the coordinates and the time. This function, called the state function (or wave function), contains all the information that can be determined about the system. We further postulate that Ψ is single-valued, continuous, and quadratically integrable. (For continuum states the quadratic integrability requirement is omitted.)

The designation "wave function" for Ψ is perhaps not the best choice. A physical wave moving in three-dimensional space is a function of the three spatial coordinates and the time; however, for an n-particle system, the function Ψ is a function of $3n$ spatial coordinates and the time. Hence for a many-particle system, we cannot interpret Ψ as any sort of physical wave. The state function is best thought of as a function from which we can calculate various properties of the system; the nature of the information that Ψ contains is the subject of Postulate 5 and its consequences.

> ***Postulate 2.*** To every physical observable there corresponds a linear Hermitian operator. To find this operator, write down the classical-mechanical expression for the observable in terms of Cartesian coordinates[2] and corresponding linear-momentum components, and then replace each coordinate x by the operator $x\cdot$ and each momentum component p_x by the operator $-i\hbar\partial/\partial x$.

We saw in Section 7.2 that the restriction to Hermitian operators arises from the requirement that average values of physical quantities be real numbers. The requirement of linearity is closely connected to the superposition of states discussed in Section 7.6. In our derivation of (7.106) for the average value of a property G for a state that was expanded as a superposition of the eigenfunctions of $\hat{G}$, the linearity of $\hat{G}$ played a key role. In (7.105) we had to use the linearity condition:

$$\hat{G}(c_1\varphi_1 + c_2\varphi_2 + \cdots) = c_1\hat{G}\varphi_1 + c_2\hat{G}\varphi_2 + \cdots$$

When the classical quantity contains a product of a Cartesian coordinate and its conjugate momentum, we run into the problem of noncommutativity in constructing the correct quantum-mechanical operator. Several different rules have been proposed to handle this case. See J. R. Shewell, *Am. J. Phys.*, **27**, 16 (1959); E. H. Kerner and W. G. Sutcliffe, *J. Math. Phys.*, **11**, 391 (1970).

> ***Postulate 3.*** The only possible values that can result from measurements of the physical observable G are the eigenvalues g_i of the equation
>
> $$\hat{G}\varphi_i = g_i\varphi_i$$
>
> where $\hat{G}$ is the operator corresponding to the property G. The eigenfunctions φ_i are required to be well-behaved.

Our main concern is with the energy levels of atoms and molecules. These are given by the eigenvalues of the energy operator, the Hamiltonian $\hat{H}$. The eigenvalue equation for $\hat{H}$, $\hat{H}\psi = E\psi$, is the time-independent Schroedinger equation.

---

[2]   The process of finding the quantum-mechanical operators in coordinates other than Cartesian involves complications. See K. Simon, *Am. J. Phys.*, **33**, 60 (1965); G. R. Gruber, *Found. Phys.*, **1**, 227 (1971).

However, finding the possible values of any property involves solving an eigenvalue equation.

Postulate 4. If $\hat{G}$ is any linear Hermitian operator that represents a physical observable, then the eigenfunctions φ_i of the eigenvalue equation

$$\hat{G}\varphi_i = g_i\varphi_i$$

form a complete set.

This postulate is more a mathematical than a physical postulate. Since there is no mathematical proof (except in various special cases) of the completeness of the eigenfunctions of a linear Hermitian operator, we must assume completeness. Postulate 4 allows us to expand the wave function for any state as a superposition of the eigenfunctions of any quantum-mechanical operator:

$$\Psi = \sum_i c_i \varphi_i = \sum_i |\varphi_i\rangle \langle \varphi_i|\Psi\rangle \qquad (7.141)$$

Postulate 5. If $\Psi(q, t)$ is the normalized state function of a system at time t, then the average value of a physical observable G at time t is

$$\langle G \rangle = \int \Psi^* \hat{G} \Psi \, d\tau \qquad (7.142)$$

The definition of the quantum-mechanical average value is given in Section 3.7 and should not be confused with the time average used in classical mechanics.

From Postulates 4 and 5, we showed in Section 7.6 that the probability of observing the value g_i in a measurement of G at time t is given by

$$P_{g_i} = \left| \int \varphi_i^* \Psi \, d\tau \right|^2 = |\langle \varphi_i|\Psi\rangle|^2 \qquad (7.143)$$

where $\hat{G}\varphi_i = g_i\varphi_i$. If the state function Ψ happens to be one of the eigenfunctions of $\hat{G}, \Psi = \varphi_k$, then (7.143) becomes

$$P_{g_i} = \left| \int \varphi_i^* \varphi_k \, d\tau \right|^2 = |\delta_{ik}|^2 = \delta_{ik} \qquad (7.144)$$

where we have used the orthonormality of the eigenfunctions of a Hermitian operator. Equation (7.144) says that we are certain to observe the value g_k when $\Psi = \varphi_k$.

Postulate 6. The time development of the state of an undisturbed system is given by the Schroedinger time-dependent equation

$$-\frac{\hbar}{i} \frac{\partial \Psi}{\partial t} = \hat{H}\Psi \qquad (7.145)$$

where $\hat{H}$ is the Hamiltonian (i.e., energy) operator of the system.

The time-dependent Schroedinger equation is a first-order differential equation in the time, so that, just as in classical mechanics, the present state of an undisturbed system determines the future state. However, unlike knowledge of the state in classical mechanics, knowledge of the state in quantum mechanics involves only a knowledge of the *probabilities* for various possible outcomes of a measurement. Thus, suppose we have several identical noninteracting systems, each having the same state function $\Psi(t_0)$ at time t_0. If we leave each system to itself, then the state function for each system will change in accord with (7.145); since each system has the same Hamiltonian, each system will have the same state function $\Psi(t_1)$ at any future time t_1. However, suppose that at time t_2 we make a measurement of property G in each system; although each system has the same state function $\Psi(t_2)$ at the instant the measurement begins, we will not get the same result for each system. Rather, we will get a spread of possible values g_i, where g_i are the eigenvalues of $\hat{G}$. The relative number of times we get each g_i can be calculated from the quantities $|c_i|^2$, where

$$\Psi(t_2) = \sum_i c_i \varphi_i$$

with φ_i being the eigenfunctions of $\hat{G}$.

In quantum mechanics the state function of a system changes in two ways.[3] First, there is the continuous, causal change with time given by the time-dependent Schroedinger equation. Second, there is the sudden, discontinuous, probabilistic change that occurs when a measurement is made on the system; this kind of change cannot be predicted with certainty, since the result of a measurement cannot be predicted with certainty. The sudden change in Ψ caused by a measurement is called the *reduction of the wave function. A measurement of property G that yields the result g_k changes the state function to φ_k, the eigenfunction of $\hat{G}$ with eigenvalue g_k.* Let us look at an example. Suppose that at time t we carry out a measurement of the particle's position. Let $\Psi(x, t_-)$ be the state function of the particle the instant before the measurement is made (Fig. 7.4a). We further suppose that the result of the measurement is that the particle is found to be in the small region of space

$$a < x < a + da \qquad\qquad \textbf{(7.146)}$$

We ask: What is the state function $\Psi(x, t_+)$ the instant after the measurement? To answer this question, suppose we were to make a second measurement of position at time t_+. Since t_+ differs from the time t of the first measurement by an infinitesimal amount, we must still find that the particle is confined to the region (7.146). (If the particle moved a finite distance in an infinitesimal amount of time, it would have infinite velocity, which is unacceptable.) Since $|\Psi(x, t_+)|^2$ is the probability density for finding various values of x, we conclude that $\Psi(x, t_+)$ must be zero outside the region (7.146) and must look something like Fig. 7.4b. Thus the position measurement at time t has reduced Ψ from a function that is spread out over all space to one that is localized in the region (7.146). The change from $\Psi(x, t_-)$ to $\Psi(x, t_+)$ is a probabilistic change. (A science-fiction story involving the reduction of Ψ is "A

---

[3] For a discussion see E. P. Wigner, *Am. J. Phys.*, **31**, 6 (1963).

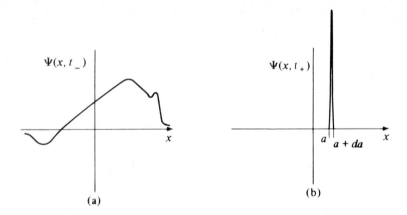

$\Psi(x, t_-)$

x

(a)

$\Psi(x, t_+)$

a $a + da$ x

(b)

Figure 7.4 *Reduction of the wave function caused by a measurement of position*

Jury of Five" in *Element 79*, F. Hoyle, Signet, New American Library, New York, 1968.)

The measurement process is one of the most controversial areas in quantum mechanics. Just how and at what stage in the measurement process reduction occurs is unclear. Some physicists take the reduction of Ψ as an additional quantum-mechanical postulate, while others claim it is a theorem derivable from the other postulates. A few physicists reject the idea of reduction (see M. Jammer, *The Philosophy of Quantum Mechanics*, Wiley, New York, 1974, Section 11.4). Ballentine advocates Einstein's statistical-ensemble interpretation of quantum mechanics, in which the wave function does not describe the state of a single system (as in the orthodox interpretation) but gives a statistical description of a collection of a large number of systems each prepared in the same way (an ensemble); in this interpretation, the need for reduction of the wave function does not occur. [See L. E. Ballentine, *Am. J. Phys.*, **40**, 1763 (1972); *Rev. Mod. Phys.*, **42**, 358 (1970).]

"For the majority of physicists the problem of finding a consistent and plausible quantum theory of measurement is still unsolved.... The immense diversity of opinion . . . concerning quantum measurements . . . [is] a reflection of the fundamental disagreement as to the interpretation of quantum mechanics as a whole." (M. Jammer, *The Philosophy of Quantum Mechanics*, pp. 519, 521.)

The probabilistic nature of quantum mechanics has disturbed many eminent physicists including Einstein, de Broglie, and Schroedinger. These physicists and others have suggested that quantum mechanics may not furnish a complete description of physical reality; rather, the probabilistic laws of quantum mechanics might be simply a reflection of deterministic laws that operate at a subquantum-mechanical level and that involve "hidden variables." An analogy given by the physicist Bohm is the Brownian motion of a dust particle in air. The particle undergoes random fluctuations of position, and its motion is not completely determined by its position and velocity. Of course, Brownian motion is a result of collisions with the gas molecules, and is determined by variables existing on the level of molecular motion. Analogously, the motions of electrons might be determined by hidden variables existing on a subquantum-mechanical level. The orthodox interpretation (often called the Copenhagen interpretation) of quantum mechanics, which was developed by Heisenberg and Bohr, denies the existence of hidden variables and asserts that the laws of quantum mechanics provide a complete description of physical reality. (Hidden-variables theories are discussed in F. J. Belinfante, *A Survey of Hidden-Variables Theories*, Pergamon, Elmsford, N.Y., 1973.)

In 1964 J. S. Bell proved that in certain experiments any possible local hidden-variable theory must make predictions different from those that quantum mechanics makes. (In a local theory, two systems very far from each other act independently of each other.) The results of such experiments are in agreement with quantum-mechanical predictions, thus providing very strong evidence against all deterministic local hidden-variable theories. [These experiments are summarized in J. F. Clauser and A. Shimony, *Rep. Prog. Phys.*, **41**, 1881 (1978); F. M. Pipkin, *Adv. At. Mol. Phys.*, **14**, 281 (1978); B. Hiley, *New Scientist*, **85**, 746 (1980); A. Aspect, *Phys. Rev. Lett.*, **47**, 460 (1981). There seems to be a fundamental randomness in nature, not explainable in any deterministic way.

Further analysis by Bell and others shows that the results of these experiments and the predictions of quantum mechanics are incompatible with a view of the world in which both realism and locality hold. Realism (also called objectivity) is the doctrine that external reality exists and has definite properties independent of whether or not this reality is observed by us. Locality excludes instantaneous action-at-a-distance and asserts that any influence from one system to another must travel at a speed that does not exceed the speed of light. Clauser and Shimony stated that quantum mechanics leads to the "philosophically startling" conclusion that we must either "totally abandon the realistic philosophy of most working scientists, or dramatically revise our concept of space-time" to permit "some kind of action-at-a-distance." (J. F. Clauser and A. Shimony, *Rep. Prog. Phys.*, **41**, 1881; see also B. d'Espagnat, *Scientific American*, Nov. 1979, p. 158.)

Although the experimental predictions of quantum mechanics are not arguable, its conceptual interpretation is still the subject of heated debate, with several different viewpoints being advocated. An excellent bibliography with commentary on this subject is B. S. DeWitt and R. N. Graham, *Am. J. Phys.*, **39**, 724 (1971). Some other references are M. Jammer, *The Philosophy of Quantum Mechanics*, Wiley, New York, 1974; B. d'Espagnat, *Conceptual Foundations of Quantum Mechanics*, 2nd ed., Benjamin, New York, 1976; L. E. Ballentine et al., *Physics Today*, April 1971, p. 36; H. P. Stapp, *Am. J. Phys.*, **40**, 1098 (1972); **42**, 83 (1974).

The relation between quantum mechanics and the mind has been the subject of much speculation. Wigner argued that the reduction of the wave function occurs when the result of a measurement enters the consciousness of an observer and thus "the being with consciousness must have a different role in quantum mechanics than the inanimate measuring device." He believes it likely that conscious beings obey different laws of nature than inanimate objects and proposes that scientists look for unusual effects of consciousness acting on matter. [E. P. Wigner, "Remarks on the

Mind-Body Question," in *The Scientist Speculates*, I. J. Good, ed., Capricorn, New York, 1965, p. 284; *Proc. Amer. Phil. Soc.*, **113**, 95 (1969); *Found. Phys.*, **1**, 35 (1970).]

The biophysicist and biochemist Harold Morowitz noted that " . . . biologists . . . have been moving relentlessly toward the hardcore materialism that characterized nineteenth-century physics. At the same time, physicists, faced with compelling experimental evidence, have been moving away from strictly mechanical models of the universe to a view that sees the mind as playing an integral role in all physical events." (H. J. Morowitz, *Psychology Today*, Aug. 1980, p. 12.)

Schmidt constructed an electronic random-number generator based on the radioactive decay of ^{90}Sr (a probabilistic quantum-mechanical process) and presented evidence for human precognition of and psychokinetic influence on the numbers generated by this device. [H. Schmidt, *New Scientist*, **50**, 757 (1971); *J. Parapsychol.*, **33**, 99 (1969); **34**, 175 (1970); *Physics Today*, Oct. 1971, p. 13; *Bull. Am. Phys. Soc.*, **24**, 38 (1979).] Several investigators have replicated Schmidt's results, but several others have not found any evidence for psychic phenomena using similar devices. A failure of locality thereby permitting action-at-a-distance has been suggested as an explanation for supposed psychic phenomena.

Some physicists see parallels between ideas of Eastern mystical thought and certain concepts of quantum mechanics, relativity, and elementary-particle physics. (See G. Zukav, *The Dancing Wu Li Masters: An Overview of the New Physics*, Morrow, New York, 1979.) Physicists in California have formed two groups to explore the relations between the ideas of modern physics, the nature of consciousness, and Eastern mystical thought (*Newsweek*, July 23, 1979, p. 85). The majority of physicists are skeptical of the value of such philosophizing.

We now resume consideration of the time-dependent Schroedinger equation (7.145).

If the Hamiltonian is independent of time, we have the possibility of states of definite energy E. For such states the state function must satisfy

$$\hat{H}\Psi = E\Psi \tag{7.147}$$

and the time-dependent Schroedinger equation becomes

$$-\frac{\hbar}{i}\frac{\partial \Psi}{\partial t} = E\Psi$$

which integrates to $\Psi = Be^{-iEt/\hbar}$, where B, the integration "constant," is independent of time. The function Ψ depends on the coordinates and the time, so that B is some function of the coordinates, which we designate as $\psi(q)$. We have

$$\Psi(q, t) = e^{-iEt/\hbar}\psi(q) \tag{7.148}$$

for a state of constant energy. The function $\psi(q)$ satisfies the time-independent Schroedinger equation

$$\hat{H}\psi(q) = E\psi(q) \tag{7.149}$$

which follows from (7.147) and (7.148). The factor $\exp(-iEt/\hbar)$ simply indicates a change in the phase of the wave function $\Psi(q, t)$ with time and has no direct physical significance; hence we generally refer to $\psi(q)$ as *"the wave function."* The Hamiltonian operator plays a unique role in quantum mechanics in that it occurs in the fundamental dynamical equation, the time-dependent Schroedinger equation; the eigenstates of $\hat{H}$ (known as stationary states) have the special property that the probability density $|\Psi|^2$ is independent of time.

You might be wondering about the absence from our list of the Born postulate that $|\Psi(x, t)|^2 \, dx$ is the probability at time t for finding the particle between x and $x + dx$; this postulate is actually a consequence of Postulate 5, as we now demonstrate. From the definition of average value in Section 3.7, we derived Eq. (3.108):

$$\langle G \rangle = \sum_g P_g g$$

where P_g is the probability of observing the value g in a measurement of the property G. This equation is for a variable that takes on discrete values; the corresponding equation for a continuous variable is

$$\langle G \rangle = \int P(g) g \, dg \qquad (7.150)$$

where $P(g)$ is the probability density for observing various values of G. For the average value of the x coordinate, Eq. (7.150) reads

$$\langle x \rangle = \int P(x) x \, dx$$

According to Postulate 5, we have

$$\langle x \rangle = \int \Psi^* \hat{x} \Psi \, dx = \int |\Psi|^2 x \, dx$$

Comparing these last two equations, we conclude that $|\Psi|^2$ is the probability density for the x coordinate of a one-dimensional system.

In Chapter 10 we will introduce two further postulates, dealing with spin and the Pauli principle.

PROBLEMS

7.1 True or false: (a) The state function is always equal to a function of time multiplied by a function of the coordinates. (b) In both classical and quantum mechanics, knowledge of the present state of an isolated system allows its future state to be calculated. (c) The state function is always an eigenfunction of the Hamiltonian. (d) Any linear combination of eigenfunctions of the Hamiltonian is an eigenfunction of the Hamiltonian. (e) If the state function is not an eigenfunction of the operator $\hat{A}$, then a measurement of the property A might give a value that is not one of the eigenvalues of $\hat{A}$. (f) The probability density is independent of time for a stationary state. (g) If two operators do not commute, then they cannot possess any common eigenfunctions. (h) If two operators commute, then every eigenfunction of one must be an eigenfunction of the other.

7.2 Let $\hat{A}$ and $\hat{B}$ be Hermitian operators and let c be a real constant. Show that $c\hat{A}$ and $\hat{A} + \hat{B}$ are Hermitian.

7.3 Show that (a) d^2/dx^2 is Hermitian; (b) $\langle T_x \rangle = (\hbar^2/2m)\int |\partial\Psi/\partial x|^2 \, d\tau$.

7.4 Which of the following operators are Hermitian: d/dx, $i(d/dx)$, $4d^2/dx^2$, $i(d^2/dx^2)$?

7.5 Let $\hat{A}$ be a Hermitian operator. Show that $\langle A^2 \rangle = \int |\hat{A}\psi|^2 \, d\tau$ and therefore $\langle A^2 \rangle \geqslant 0$. [This result can be used to derive Eq. (5.177) more rigorously than in the text. Thus since $\hat{M}^2 - \hat{M}_z^2 = \hat{M}_x^2 + \hat{M}_y^2$, we have $\langle M^2 \rangle - \langle M_z^2 \rangle = \langle M_x^2 \rangle + \langle M_y^2 \rangle$. Now $\langle M^2 \rangle = c$, $\langle M_z^2 \rangle = b_k^2$, and by the theorem of this exercise $\langle M_x^2 \rangle \geqslant 0$, $\langle M_y^2 \rangle \geqslant 0$. Hence $c - b_k^2 \geqslant 0$.]

7.6 Verify that $\hat{L}_z$ is Hermitian using: (a) spherical polar coordinates; (b) Cartesian coordinates.

7.7 Extend the Schmidt orthogonalization procedure to the case of threefold degeneracy.

7.8 Evaluate the coefficients in the sine-series expansion (7.48) for the parabolic function $f(x) = x(l - x)$. What is the result at $x = \frac{1}{2}l$? Use the first five terms of the result to approximate π^3.

7.9 What operator is shown to be Hermitian by $\langle m|n \rangle = \langle n|m \rangle^*$?

7.10 For the hydrogenlike atom, we have $V = -Ze'^2(x^2 + y^2 + z^2)^{-1/2}$ so that the potential energy is an even function of the coordinates. (a) What is the parity of ψ_{2s}? (b) What is the parity of ψ_{2p_x}? (c) Consider $\psi_{2s} + \psi_{2p_x}$. Is it an eigenfunction of $\hat{H}$? Does it have definite parity?

7.11 Let $\hat{\Pi}$ be the parity operator and m be a positive integer. What is $\hat{\Pi}^m$ if m is even? if m is odd?

7.12 Let $\hat{\Pi}$ be the parity operator and let $\psi_i(x)$ be a normalized harmonic-oscillator wave function. We define the matrix elements Π_{ij} as

$$\Pi_{ij} = \int_{-\infty}^{\infty} \psi_i^* \hat{\Pi} \psi_j \, dx$$

Show that $\Pi_{ij} = 0$ for $i \neq j$, and $\Pi_{ii} = \pm 1$.

7.13 Let $\hat{T}$ be a linear operator such that $\hat{T}^n = \hat{I}$, where n is a positive integer, and no lower power of $\hat{T}$ equals $\hat{I}$. Find the eigenvalues of $\hat{T}$.

7.14 On the basis of parity considerations, determine which of the following integrals must be zero:

$$\text{(a)} \int \psi_{2s}^* x \psi_{2p_x} \, d\tau \qquad \text{(b)} \int \psi_{2s}^* x^2 \psi_{2p_x} \, d\tau \qquad \text{(c)} \int \psi_{2p_y}^* x \psi_{2p_x} \, d\tau$$

where the wave functions are the hydrogenlike orbitals.

7.15 (a) Show that the parity operator is linear. (b) Show that the parity operator is Hermitian; a proof in one dimension is sufficient.

7.16 (a) In Cartesian coordinates the parity operator $\hat{\Pi}$ corresponds to the transformation of variables $x \rightarrow -x$, $y \rightarrow -y$, $z \rightarrow -z$. Show that in spherical polar coordinates the parity operator corresponds to the transformation $r \rightarrow r$, $\theta \rightarrow \pi - \theta$, $\varphi \rightarrow \varphi + \pi$. (b) Show that $\hat{\Pi} e^{im\varphi} = (-1)^m e^{im\varphi}$. (c) Use Eq. (5.131) to show that

$$\hat{\Pi} S_{l,\,m}(\theta) = (-1)^{l-|m|} S_{l,\,m}(\theta) = (-1)^{l-m} S_{l,\,m}(\theta)$$

(d) Combine the results of (b) and (c) to conclude that the spherical harmonic $Y_l^m(\theta, \varphi)$ is an even function if l is even and is an odd function if l is odd.

7.17 Which of the following operators meet the requirements for a quantum-mechanical operator that is to represent a physical quantity: (a) $(\quad)^{1/2}$; (b) d/dx; (c) d^2/dx^2; (d) $i(d/dx)$?

7.18 Consider an operator $\hat{A}$ that contains the time as a parameter. We are interested in how the average value of the property A changes with time, and we have

$$\frac{d\langle A\rangle}{dt} = \frac{d}{dt}\int \Psi^* \hat{A}\Psi\, d\tau$$

The definite integral on the right side of this equation is a function of the parameter t, and it is generally a valid mathematical operation to calculate its derivative with respect to t by differentiating the integrand with respect to t:

$$\frac{d\langle A\rangle}{dt} = \int \frac{\partial}{\partial t}[\Psi^* \hat{A}\Psi]\, d\tau = \int \frac{\partial \Psi^*}{\partial t}\hat{A}\Psi\, d\tau + \int \Psi^* \frac{\partial \hat{A}}{\partial t}\Psi\, d\tau + \int \Psi^* \hat{A}\frac{\partial \Psi}{\partial t}\, d\tau$$

Use the time-dependent Schroedinger equation, its complex conjugate, and the Hermitian property of $\hat{H}$ to show that

$$\frac{d\langle A\rangle}{dt} = \int \Psi^* \frac{\partial \hat{A}}{\partial t}\Psi\, d\tau + \frac{i}{\hbar}\int \Psi^*(\hat{H}\hat{A} - \hat{A}\hat{H})\Psi\, d\tau$$

$$\frac{d\langle A\rangle}{dt} = \left\langle\frac{\partial \hat{A}}{\partial t}\right\rangle + \frac{i}{\hbar}\langle[\hat{H}, \hat{A}]\rangle \qquad (7.151)$$

7.19 Use (7.151) and (5.12) to show that

$$\frac{d}{dt}\langle x\rangle = \frac{\langle p_x\rangle}{m} = \frac{1}{m}\int \Psi^* \frac{\hbar}{i}\frac{\partial \Psi}{\partial x}\, d\tau \qquad (7.152)$$

From (7.152) it follows that

$$\frac{d^2\langle x\rangle}{dt^2} = \frac{1}{m}\frac{d}{dt}\langle p_x\rangle$$

Use (7.151), (5.13), and (4.26) to show that

$$\langle F_x\rangle = m\frac{d^2\langle x\rangle}{dt^2} \qquad (7.153)$$

If we consider a classical-mechanical particle, its wave function will be large only in a very small region corresponding to its position, and we may then drop the averages in (7.153) to obtain Newton's second law. Thus classical mechanics is a special case of quantum mechanics. Equation (7.153) is known as *Ehrenfest's theorem*, after the physicist who derived it in 1927.

7.20 In accord with the probability interpretation of $\Psi^*\Psi$, we have normalized the wave function: $\int \Psi^*\Psi\, d\tau = 1$. Assuming that particles are neither created nor destroyed, we expect this normalization to hold for all time, i.e.,

$$\frac{d}{dt}\int \Psi^*\Psi\, d\tau = 0 \qquad (7.154)$$

Prove (7.154) as a special case of (7.151).

7.21 For a hydrogen atom in a p state, the possible outcomes of a measurement of L_z are $-\hbar$, 0, and $\hbar$. For each of the following wave functions, give the probabilities for each of these three results: (a) ψ_{2p_z}; (b) ψ_{2p_x}; (c) ψ_{2p_y}; (d) $\psi_{2p_{+1}}$.

7.22 Verify the following Kronecker-delta identities:

$$\delta_{ij} = \delta_{ji} \qquad [\delta_{ij}]^2 = \delta_{ij}$$

$$\delta_{ij}\delta_{ik} = \delta_{ij}\delta_{jk} \qquad \delta_{i+n,j} = \delta_{i,j-n}$$

7.23 Give the value of each of the following integrals:

$$\text{(a)} \int_{-\infty}^{\infty} \delta(x)\, dx \qquad\qquad \text{(b)} \int_{-\infty}^{-1} \delta(x)\, dx$$

$$\text{(c)} \int_{-1}^{1} \delta(x)\, dx \qquad\qquad \text{(d)} \int_{1}^{2} f(x)\delta(x-3)\, dx$$

7.24 What is the value of $\displaystyle\int_{0}^{\infty} f(x)\delta(x)\, dx$?

7.25 (a) If $\hat{A}$ and $\hat{B}$ are Hermitian operators, prove that their product $\hat{A}\hat{B}$ is Hermitian if and only if $\hat{A}$ and $\hat{B}$ commute. (b) If $\hat{A}$ and $\hat{B}$ are Hermitian, prove that $\frac{1}{2}(\hat{A}\hat{B} + \hat{B}\hat{A})$ is Hermitian. (c) Is $\hat{x}\hat{p}_x$ Hermitian? (d) Is $\frac{1}{2}(\hat{x}\hat{p}_x + \hat{p}_x\hat{x})$ Hermitian?

7.26 Since the parity operator is Hermitian (Problem 7.15), two eigenfunctions of $\hat{\Pi}$ that correspond to different eigenvalues must be orthogonal. Show directly that this is so.

7.27 Consider the integral $\langle v_2|x|v_1 \rangle$, where the functions are one-dimensional harmonic-oscillator wave functions with quantum numbers v_2 and v_1. Under what conditions do parity considerations allow us to conclude that this integral must be zero? Might the integral be zero in other cases as well? (This integral is important in discussing radiative transitions.)

7.28 Consider a particle in a nonstationary state in a one-dimensional box (Fig. 2.1) of length l. Suppose that at time t_0 its state function is the parabolic function

$$\Psi(t_0) = Nx(l-x), \qquad 0 \leqslant x \leqslant l$$

where N is the normalization constant. If at time t_0 we were to make a measurement of the particle's energy, what would be the possible outcomes of the measurement, and what would be the probability for each such outcome?

7.29 Use the results of Problem 7.28 to evaluate

$$\sum_{m=0}^{\infty} \frac{1}{(2m+1)^6}$$

7.30 (a) Show that for a particle in a one-dimensional box (Fig. 2.1) of length l, the probability of observing a value of p_x between p and $p+dp$ is

$$\frac{4|N|^2 s^2}{l(s^2 - b^2)^2}[1 - (-1)^n \cos bl]\, dp \qquad\qquad (7.155)$$

where $s \equiv n\pi l^{-1}$ and $b \equiv p\hbar^{-1}$. The constant N is to be chosen so that the integral from minus infinity to infinity of (7.155) is unity. (b) Evaluate (7.155) for $p = \pm nh/2l$. [At these values of p_x, the denominator of (7.155) is zero and the probability reaches a large but finite value.]

7.31 A measurement yields $2^{1/2}\hbar$ for the magnitude of a particle's orbital angular momentum. If L_x is now measured, what are the possible outcomes?

7.32 The functions in Fig. 7.3 approximate the Dirac delta function. Draw graphs of the corresponding functions that approximate the Heaviside step function with successively increasing accuracy.

7.33 The energy of most stars results from the fusion of hydrogen nuclei to helium nuclei.[4] The interior of the sun (a typical star) is at $15 \times 10^6\,°C$. At this temperature, virtually

---

[4] After a certain fraction of hydrogen has been used up, a star will fuse helium nuclei to form heavier nuclei, ultimately forming ^{56}Fe, the most stable nuclide. Still heavier nuclei are formed by neutron capture by ^{56}Fe followed by beta decay. It is believed that all elements heavier than helium were formed mainly by stars; except for hydrogen, most of the atomic nuclei in our bodies originated in the interiors of stars.

no nuclei have sufficient kinetic energy to overcome the electrostatic repulsion between nuclei and approach each other closely enough to undergo a fusion reaction. Therefore when Eddington proposed in 1920 that nuclear fusion is the source of stellar energy, his idea was rejected. Explain why fusion does occur in stars, despite the above-mentioned apparent difficulty.

7.34 Quantum mechanics postulates that the present state of an undisturbed system determines its future state. Consider the special case of a system with a time-independent Hamiltonian $\hat{H}$. Suppose it is known that at time t_0 the state function is $\Psi(x, t_0)$. Use Eqs. (7.145) and (7.101) to show that

$$\Psi(x, t) = \sum_j \langle \psi_j(x) | \Psi(x, t_0) \rangle \exp[-iE_j(t - t_0)/\hbar] \psi_j(x)$$

where $\hat{H}\psi_j = E_j\psi_j$.

7.35 Prove the uncertainty principle (5.16) for any two Hermitian operators $\hat{A}$ and $\hat{B}$, as follows. (a) Define functions f and g as $f = (\hat{A} - \langle A \rangle)\Psi$ and $g = i(\hat{B} - \langle B \rangle)\Psi$. Prove that $\langle f | f \rangle = \langle f | f \rangle^* = (\Delta A)^2$ and $\langle g | g \rangle = (\Delta B)^2$. (b) Now let $I \equiv \langle f + sg | f + sg \rangle$, where s is defined as an arbitrary *real* parameter. The integrand $|f + sg|^2$ of I is everywhere nonnegative and I must therefore be positive, unless it happens that $f = -sg$. We have two possible cases: either (1) $f = -sg$, or (2) $f \neq -sg$. Show directly that for case (1), we have $4\langle f | f \rangle \langle g | g \rangle = [\langle f | g \rangle + \langle g | f \rangle]^2$. For case (2) we have $I = as^2 + bs + c > 0$, where a, b, and c are certain integrals. This inequality means that the equation $as^2 + bs + c = 0$ can have no real roots for s, which (using the quadratic formula) means that $4ac > b^2$. Show that this inequality leads to $4\langle f | f \rangle \langle g | g \rangle > [\langle f | g \rangle + \langle g | f \rangle]^2$. (c) Combining cases (1) and (2), we have shown that $4\langle f | f \rangle \langle g | g \rangle \geq [\langle g | f \rangle^* + \langle f | g \rangle^*]^2$. Show that this leads to the uncertainty principle (5.16).

8 THE VARIATION METHOD

8.1 THE VARIATION PRINCIPLE

We now begin the study of the approximation methods needed to deal with the time-independent Schroedinger equation for systems of several interacting particles. This chapter deals with the *variation method*, which allows us to obtain an approximation to the ground-state energy of a system without solving the Schroedinger equation.

The variation method is based on the following theorem: Given a system with Hamiltonian operator $\hat{H}$, if φ is *any normalized well-behaved function that satisfies the boundary conditions of the problem*, it is true that

$$\int \varphi^* \hat{H} \varphi \, d\tau \geq E_0 \tag{8.1}$$

where E_0 is the true value of the lowest energy eigenvalue of $\hat{H}$. The significance of this theorem is that it allows us to calculate an upper bound for the ground-state energy.

We begin the proof of (8.1) by defining the integral I as

$$I = \int \varphi^* (\hat{H} - E_0) \varphi \, d\tau \tag{8.2}$$

$$I = \int \varphi^* \hat{H} \varphi \, d\tau - E_0 \int \varphi^* \varphi \, d\tau = \int \varphi^* \hat{H} \varphi \, d\tau - E_0 \tag{8.3}$$

since φ is normalized. If we prove that $I \geq 0$, we will have proved (8.1). Now let ψ_i

and E_i be the true eigenfunctions and eigenvalues of $\hat{H}$:

$$\hat{H}\psi_i = E_i\psi_i \tag{8.4}$$

Since the eigenfunctions ψ_i form a complete set, we can expand φ in terms of the ψ_i:

$$\varphi = \sum_k a_k \psi_k \tag{8.5}$$

(Note that this requires that φ satisfy the same boundary conditions as the ψ_i.) We substitute this expansion into (8.2) to get

$$I = \int \sum_k a_k^* \psi_k^* (\hat{H} - E_0) \sum_j a_j \psi_j \, d\tau = \int \sum_k a_k^* \psi_k^* \sum_j (\hat{H} - E_0) a_j \psi_j \, d\tau \tag{8.6}$$

Applying (8.4), and assuming the validity of interchanging the integration and the infinite summations, we get

$$I = \int \sum_k a_k^* \psi_k^* \sum_j a_j (E_j - E_0) \psi_j \, d\tau = \sum_k \sum_j a_k^* a_j (E_j - E_0) \int \psi_k^* \psi_j \, d\tau \tag{8.7}$$

Use of the orthonormality of the eigenfunctions gives

$$I = \sum_k \sum_j a_k^* a_j (E_j - E_0) \delta_{kj} \tag{8.8}$$

We perform the sum over j, and, as usual, the Kronecker delta factor makes all terms zero except the one with $j = k$, giving

$$I = \sum_k a_k^* a_k (E_k - E_0) = \sum_k |a_k|^2 (E_k - E_0) \tag{8.9}$$

By hypothesis, E_0 is the lowest eigenvalue; hence $E_k - E_0 \geq 0$. Moreover $|a_k|^2 \geq 0$. Thus all the terms in the sum on the right of (8.9) are nonnegative, and we have the desired result that $I \geq 0$.

Suppose we have a function φ that is *not* normalized. To apply the variation theorem, we multiply φ by a normalization constant N, so that $N\varphi$ is normalized. Replacing φ by $N\varphi$ in (8.1), we have

$$|N|^2 \int \varphi^* \hat{H} \varphi \, d\tau \geq E_0 \tag{8.10}$$

The constant N is determined by

$$\int (N\varphi)^* N\varphi \, d\tau = |N|^2 \int \varphi^* \varphi \, d\tau = 1 \tag{8.11}$$

$$|N|^2 = 1 \Big/ \int \varphi^* \varphi \, d\tau \tag{8.12}$$

Hence

$$\frac{\int \varphi^* \hat{H} \varphi \, d\tau}{\int \varphi^* \varphi \, d\tau} \geq E_0 \tag{8.13}$$

where φ is any well-behaved function (not necessarily normalized) that satisfies the boundary conditions of the problem.

The function φ is called a *trial variation function*, and the integral in (8.1) is called the *variational integral*. To arrive at a good approximation to the ground-state energy E_0, we try many trial variation functions and look for the one that gives the lowest value of the variational integral. From (8.1) the lower the value of the variational integral, the better the approximation we have to E_0. (One way to disprove quantum mechanics would be to find a trial variation function that made the variational integral less than E_0 for some system where E_0 was known either experimentally, or from the exact solution of the Schroedinger equation.)

Let ψ_0 be the true ground-state wave function:

$$\hat{H}\psi_0 = E_0\psi_0 \tag{8.14}$$

If we happened to be lucky enough to hit upon a trial variation function that was equal to ψ_0, then, using (8.14) in (8.1), we see that the variational integral will be equal to E_0. Thus the ground-state wave function gives the minimum value of the variational integral. We therefore expect that the lower the value of the variational integral, the closer the trial variational function will approach the true ground-state wave function. However, it turns out that the variational integral approaches E_0 a lot faster than the trial variation function approaches ψ_0, and it is possible to get a rather good approximation to E_0 using a rather poor φ.

In practice, what is usually done is to put several parameters into the trial function φ, and then vary these parameters so as to minimize the variational integral. Successful use of the variation method depends on the ability to make a shrewd choice (or guess) for the trial function.

Let us look at some examples of the variation method. Although the real utility of the method is for problems to which we do not know the true solutions, we will here consider problems that are exactly solvable so that we can judge the accuracy of our results.

We will use the variation method to get an upper bound to the ground-state energy of the particle in a one-dimensional box of length l (Section 2.2). The wave function is zero outside the box, and the boundary conditions require that $\psi = 0$ at $x = 0$ and at $x = l$. The trial variation function must satisfy these boundary conditions. A simple function satisfying these conditions is the parabolic function

$$\varphi = x(l-x) \tag{8.15}$$

for $0 \leqslant x \leqslant l$, and $\varphi = 0$ outside the box. Since we have not normalized φ, we use (8.13). Inside the box the Hamiltonian is $-(\hbar^2/2m)d^2/dx^2$. For the numerator and denominator of (8.13), we have

$$\int \varphi^*\hat{H}\varphi \, d\tau = -\frac{\hbar^2}{2m}\int_0^l (lx - x^2)\frac{d^2}{dx^2}(lx - x^2)\,dx = -\frac{\hbar^2}{m}\int_0^l (x^2 - lx)\,dx$$

$$\int \varphi^*\hat{H}\varphi \, d\tau = \frac{\hbar^2 l^3}{6m}$$

$$\int \varphi^*\varphi \, d\tau = \int_0^l x^2(l-x)^2\,dx = \frac{l^5}{30}$$

Substituting in (8.13), we get

$$\frac{5h^2}{4\pi^2 l^2 m} \geq E_0 \tag{8.16}$$

From Eq. (2.20), E_0 is $h^2/8ml^2$ and the percent error is

$$\frac{(5/4\pi^2) - (1/8)}{1/8} \times 100\% = 1.3\%$$

The preceding example did not have a parameter in the trial function. We now give an example involving the one-dimensional harmonic oscillator to illustrate this method. We expect the wave function to go to zero as x goes to $\pm\infty$. The function e^{-x} has the proper behavior at $+\infty$ but becomes infinite at $-\infty$. The function e^{-x^2} has the proper behavior at $\pm\infty$. However, it is not satisfactory from a dimensional standpoint, since the power to which we raise e must be dimensionless. This can be seen from the Taylor-series expansion (4.48). All the terms on the right of (4.48) must have the same dimensions. Therefore z must have the same dimensions as 1, i.e., z is dimensionless. The quantity α, defined in Eq. (4.33), has the dimensions $1/(\text{length})^2$; hence αx^2 is dimensionless. This suggests a trial function of the form $e^{-\alpha x^2}$. Note that α is a constant, not a variable parameter. We now introduce some flexibility into the trial function by use of a parameter. Perhaps we might multiply the exponential function by terms involving powers of x. We might try $(1 + bx)e^{-\alpha x^2}$, where b is a parameter. This function is, however, neither even nor odd, whereas we know that when $V(x)$ is an even function, the wave functions must be either even or odd. We therefore try

$$\varphi = (1 + c\alpha x^2)e^{-\alpha x^2} \tag{8.17}$$

where c is a parameter. The factor α multiplying cx^2 was thrown in to simplify things by making c dimensionless. The Hamiltonian is given by Eq. (4.32). Using (4.54), we have

$$\int \varphi^* \varphi \, dx = 2 \int_0^\infty (1 + 2c\alpha x^2 + c^2\alpha^2 x^4)e^{-2\alpha x^2} \, dx \tag{8.18}$$

Using the Appendix integrals (A.5) and (A.6), we find

$$\int \varphi^* \varphi \, dx = \left(\frac{\pi}{2\alpha}\right)^{1/2}\left(1 + \frac{c}{2} + \frac{3}{16}c^2\right) \tag{8.19}$$

$$\int \varphi^* \hat{H} \varphi \, dx = \frac{h^2}{2m} \int_{-\infty}^\infty (1 + c\alpha x^2)e^{-\alpha x^2}$$

$$\times \left(-\frac{d^2}{dx^2} + \alpha^2 x^2\right)(e^{-\alpha x^2} + x^2 c\alpha e^{-\alpha x^2}) \, dx \tag{8.20}$$

$$\int \varphi^* \hat{H} \varphi \, dx = \frac{h^2}{m}\left(\frac{\pi\alpha}{2}\right)^{1/2}\left(\frac{43}{128}c^2 - \frac{1}{16}c + \frac{5}{8}\right) \tag{8.21}$$

$$\frac{\int \varphi^* \hat{H} \varphi \, dx}{\int \varphi^* \varphi \, dx} = hv\frac{(43c^2 - 8c + 80)}{(24c^2 + 64c + 128)} \tag{8.22}$$

Now we vary c to minimize (8.22). A necessary condition for minimizing (8.22) is that

$$\frac{\partial}{\partial c}\left[\frac{\int \varphi^* \hat{H} \varphi \, dx}{\int \varphi^* \varphi \, dx}\right] = 0 \qquad (8.23)$$

Differentiation of (8.22) with respect to c gives

$$\frac{(24c^2 + 64c + 128)(86c - 8) - (43c^2 - 8c + 80)(48c + 64)}{(24c^2 + 64c + 128)^2} = 0$$

$$23c^2 + 56c - 48 = 0 \qquad (8.24)$$

$$c = -3.107, \ +0.6718 \qquad (8.25)$$

Substituting these roots in (8.22), we find that 0.6718 gives the smaller value, and using it we have

$$E_0 \leqslant 0.517 \, hv \qquad (8.26)$$

The (true) value of E_0 is $\frac{1}{2}hv$, and our error is 3.4 percent.

Instead of using a polynomial containing a parameter to multiply $e^{-\alpha x^2}$, as in (8.17), we might put the parameter in the exponent, taking

$$\varphi = e^{-b\alpha x^2} \qquad (8.27)$$

If we used (8.27) as a trial function, we would find (Problem 8.1) that minimization of the variational integral leads to $b = \frac{1}{2}$ and an energy

$$E_0 \leqslant \frac{1}{2}hv \qquad (8.28)$$

Recall [Eq. (4.59)] that with $b = \frac{1}{2}$, (8.27) is the true ground-state wave function.

8.2   EXTENSION OF THE VARIATION METHOD

The variation method as presented in the last section has two major limitations. First, it provides information about only the *ground*-state energy and wave function. Second, it provides only an upper bound to the ground-state energy. We now discuss some extensions of the variation method. (See also Section 8.5.)

Consider how we might extend the variation method to obtain an estimate for the energy of the first excited state. We number the stationary states of the system $0, 1, 2, \ldots$ in order of increasing energy, so that

$$E_0 \leqslant E_1 \leqslant E_2 \leqslant \cdots \qquad (8.29)$$

We define I_1 as [cf. Eq. (8.2)]

$$I_1 = \int \varphi^* (\hat{H} - E_1) \varphi \, d\tau \qquad (8.30)$$

where φ is a normalized function satisfying the boundary conditions of the problem. We expand φ as

$$\varphi = \sum_k a_k \psi_k \qquad (8.31)$$

where the ψ_k are the true wave functions. Carrying out the same steps as we did following Eq. (8.5), we are led to an equation that is the same as (8.9), except that E_0 is replaced by E_1, so that

$$I_1 = \sum_k |a_k|^2 (E_k - E_1) = |a_0|^2 (E_0 - E_1) + |a_2|^2 (E_2 - E_1) + |a_3|^2 (E_3 - E_1) + \cdots$$

$$(8.32)$$

All the terms in (8.32) are nonnegative *except* the first term. From Eq. (7.65) the coefficients in the expansion (8.31) are given by

$$a_k = \int \psi_k^* \varphi \, d\tau \qquad (8.33)$$

If we restrict ourselves to variational functions φ that are *orthogonal* to the true ground-state wave function ψ_0, Eq. (8.33) gives

$$a_0 = \int \psi_0^* \varphi \, d\tau = 0 \qquad (8.34)$$

For normalized trial functions orthogonal to ψ_0, the first term in the sum (8.32) vanishes, and we conclude that $I_1 \geqslant 0$, or

$$\int \varphi^* \hat{H} \varphi \, d\tau \geqslant E_1 \qquad \text{if} \quad \int \psi_0^* \varphi \, d\tau = 0 \qquad (8.35)$$

Thus (8.35) provides a method for obtaining an upper bound for the energy of the first excited state, E_1. However, the restriction (8.34) makes the method troublesome to apply.

For certain problems it is possible to rigorously guarantee that (8.34) holds even though we do not know the true ground-state wave function. One example is a one-dimensional problem for which V is an even function of x. In this case the ground-state wave function is always an even function, while the first excited-state wave function is odd. (All the wave functions must be of definite parity. The ground-state wave function is nodeless, and, since an odd function vanishes at the origin, the ground-state wave function must be even. The first excited-state wave function has one node and must be odd.) Therefore for odd trial functions, it must be true that

$$\int_{-\infty}^{\infty} \psi_0^* \varphi \, dx = 0 \qquad (8.36)$$

(The function ψ_0^* times the odd function φ gives an odd integrand, and the integral of an odd function from $-\infty$ to ∞ is zero.)

Another example is a particle moving in a central field (Section 6.1). The form of the potential energy might be such that we could not solve for the radial factor $R(r)$ in the eigenfunction. However, we know that the angular factor is a spherical harmonic [Eq. (6.20)] and that spherical harmonics with different values of l are orthogonal. Thus we can get an upper bound to the energy of the lowest state with any given angular momentum l, by using the factor $Y_l^m(\theta, \varphi)$ in the trial function.

This result depends on the extension of (8.35) to higher excited states, namely

$$\frac{\int \varphi^* \hat{H} \varphi \, d\tau}{\int \varphi^* \varphi \, d\tau} \geq E_{k+1} \qquad \text{if} \quad \int \psi_0^* \varphi \, d\tau = \int \psi_1^* \varphi \, d\tau = \cdots = \int \psi_k^* \varphi \, d\tau = 0 \qquad (8.37)$$

Several procedures have been devised that yield *lower* bounds for energy levels, but these procedures are not so easy to apply and have not been used very often. (For a survey see the article by F. Weinhold in *Advances in Quantum Chemistry*, Vol. 6, P.-O. Löwdin, ed., Academic Press, New York, 1972, p. 299.)

8.3   DETERMINANTS

We will shortly consider a kind of variational function that gives rise to an equation involving a determinant. Therefore, we now review the properties of determinants.

A determinant is a square array of n^2 quantities (called elements); the value of the determinant is calculated from its elements in a manner to be given shortly. The number n is the *order* of the determinant. Using a_{ij} to represent a typical element, we write the nth-order determinant as

$$\det(a_{ij}) = \begin{vmatrix} a_{11} & a_{12} & a_{13} & \cdots & a_{1n} \\ a_{21} & a_{22} & a_{23} & \cdots & a_{2n} \\ a_{31} & a_{32} & a_{33} & \cdots & a_{3n} \\ \vdots & \vdots & \vdots & \vdots & \vdots \\ a_{n1} & a_{n2} & a_{n3} & \cdots & a_{nn} \end{vmatrix} \qquad (8.38)$$

The vertical lines in (8.38) have nothing to do with absolute value. Before considering how the value of the nth-order determinant is defined, we consider determinants of first, second, and third orders.

A first-order determinant has one element, and its value is simply the value of that element. Thus

$$|a_{11}| = a_{11} \qquad (8.39)$$

where the vertical lines indicate a determinant and not absolute value.

A second-order determinant has 4 elements, and its value is defined by

$$\begin{vmatrix} a_{11} & a_{12} \\ a_{21} & a_{22} \end{vmatrix} = a_{11}a_{22} - a_{12}a_{21} \qquad (8.40)$$

The value of a third-order determinant is defined by

$$\begin{vmatrix} a_{11} & a_{12} & a_{13} \\ a_{21} & a_{22} & a_{23} \\ a_{31} & a_{32} & a_{33} \end{vmatrix} = a_{11} \begin{vmatrix} a_{22} & a_{23} \\ a_{32} & a_{33} \end{vmatrix} - a_{12} \begin{vmatrix} a_{21} & a_{23} \\ a_{31} & a_{33} \end{vmatrix} + a_{13} \begin{vmatrix} a_{21} & a_{22} \\ a_{31} & a_{32} \end{vmatrix} \qquad (8.41)$$

$$= a_{11}a_{22}a_{33} - a_{11}a_{32}a_{23} - a_{12}a_{21}a_{33} + a_{12}a_{31}a_{23}$$
$$+ a_{13}a_{21}a_{32} - a_{13}a_{31}a_{22} \qquad (8.42)$$

A third-order determinant is evaluated by writing down the elements of the top row with alternating plus and minus signs and then multiplying each element by a certain second-order determinant; the second-order determinant that multiplies a given element is found by crossing out the row and column (of the third-order determinant) in which that element appears. The $(n-1)$-order determinant obtained by striking out the ith row and the jth column (of the nth-order determinant) is called the *minor* of the element a_{ij}. We define the *cofactor* of a_{ij} as the minor of a_{ij} times the factor $(-1)^{i+j}$. Thus (8.41) states that the third-order determinant is evaluated by multiplying each element of the top row by its cofactor and adding up the three products. [Note that (8.40) conforms to this evaluation by means of cofactors, since the cofactor of a_{11} in (8.40) is a_{22}, and the cofactor of a_{12} is $-a_{21}$.] A numerical example is

$$\begin{vmatrix} 5 & 10 & 2 \\ 0.1 & 3 & 1 \\ 0 & 4 & 4 \end{vmatrix} = 5 \begin{vmatrix} 3 & 1 \\ 4 & 4 \end{vmatrix} - 10 \begin{vmatrix} 0.1 & 1 \\ 0 & 4 \end{vmatrix} + 2 \begin{vmatrix} 0.1 & 3 \\ 0 & 4 \end{vmatrix}$$

$$= 5(8) - 10(0.4) + 2(0.4) = 36.8$$

We denote the minor of a_{ij} by M_{ij} and the cofactor of a_{ij} by C_{ij}. We have

$$C_{ij} = (-1)^{i+j} M_{ij} \tag{8.43}$$

The expansion (8.41) of the third-order determinant can be written as

$$\det(a_{ij}) = \begin{vmatrix} a_{11} & a_{12} & a_{13} \\ a_{21} & a_{22} & a_{23} \\ a_{31} & a_{32} & a_{33} \end{vmatrix} = a_{11}C_{11} + a_{12}C_{12} + a_{13}C_{13} \tag{8.44}$$

Let us note that we may expand the third-order determinant by using the elements of any row and the corresponding cofactors. For example, using the second row to expand the third-order determinant, we have

$$\det(a_{ij}) = a_{21}C_{21} + a_{22}C_{22} + a_{23}C_{23} \tag{8.45}$$

$$\det(a_{ij}) = -a_{21} \begin{vmatrix} a_{12} & a_{13} \\ a_{32} & a_{33} \end{vmatrix} + a_{22} \begin{vmatrix} a_{11} & a_{13} \\ a_{31} & a_{33} \end{vmatrix} - a_{23} \begin{vmatrix} a_{11} & a_{12} \\ a_{31} & a_{32} \end{vmatrix} \tag{8.46}$$

and expansion of the second-order determinants shows that (8.46) is equal to (8.42). We may also use the elements of any column and the corresponding cofactors to expand the determinant, as can be readily verified. Thus for the third-order determinant, we can write

$$\det(a_{ij}) = a_{k1}C_{k1} + a_{k2}C_{k2} + a_{k3}C_{k3} = \sum_{l=1}^{3} a_{kl}C_{kl}, \qquad k = 1 \text{ or } 2 \text{ or } 3$$

$$\det(a_{ij}) = a_{1k}C_{1k} + a_{2k}C_{2k} + a_{3k}C_{3k} = \sum_{l=1}^{3} a_{lk}C_{lk}, \qquad k = 1 \text{ or } 2 \text{ or } 3$$

The first expansion uses one of the rows, while the second uses one of the columns. We now define determinants of higher order by an analogous row (or column) expansion. Thus the fourth-order determinant is defined by

$$\det(a_{ij}) = \sum_{l=1}^{4} a_{kl} C_{kl}, \qquad k = 1 \text{ or } 2 \text{ or } 3 \text{ or } 4 \qquad (8.47)$$

$$\det(a_{ij}) = \sum_{l=1}^{4} a_{lk} C_{lk}, \qquad k = 1 \text{ or } 2 \text{ or } 3 \text{ or } 4 \qquad (8.48)$$

Extending the definition to the nth-order determinant, we have

$$\det(a_{ij}) = \sum_{l=1}^{n} a_{kl} C_{kl} = \sum_{l=1}^{n} a_{lk} C_{lk}, \qquad k = 1 \text{ or } 2 \text{ or } \dots \text{ or } n \qquad (8.49)$$

We now state some theorems.[1]

   I. If every element of a row (or column) of a determinant is zero, the value of the determinant is zero.

  II. Interchanging any two rows (or columns) multiplies the value of a determinant by -1.

 III. If any two rows (or columns) of a determinant are identical, the determinant has the value zero.

 IV. Multiplication of each element of any one row (or any one column) by some constant k multiplies the value of the determinant by k.

  V. Addition to each element of one row of the same constant multiple of the corresponding element of another row leaves the value of the determinant unchanged. (This theorem also applies to the addition of a multiple of one column to another column.)

 VI. The interchange of all corresponding rows and columns leaves the value of the determinant unchanged.

These theorems are useful in evaluating higher-order determinants. An example is the evaluation of

$$B = \begin{vmatrix} 1 & 2 & 3 & 4 \\ 4 & 1 & 2 & 3 \\ 3 & 4 & 1 & 2 \\ 2 & 3 & 4 & 1 \end{vmatrix} \qquad (8.50)$$

Addition of -2 times the elements of row one to the corresponding elements of row four changes row four to $0, -1, -2, -7$. Then, addition of -3 times row one to row three and -4 times row one to row two gives

$$B = \begin{vmatrix} 1 & 2 & 3 & 4 \\ 0 & -7 & -10 & -13 \\ 0 & -2 & -8 & -10 \\ 0 & -1 & -2 & -7 \end{vmatrix} = 1 \begin{vmatrix} -7 & -10 & -13 \\ -2 & -8 & -10 \\ -1 & -2 & -7 \end{vmatrix}$$

_____

[1] For proofs see *Sokolnikoff and Redheffer*, pp. 702–707.

where we expanded B in terms of elements of the first column. Subtracting twice row three from row two and seven times row three from row one, we have

$$B = \begin{vmatrix} 0 & 4 & 36 \\ 0 & -4 & 4 \\ -1 & -2 & -7 \end{vmatrix} = (-1) \begin{vmatrix} 4 & 36 \\ -4 & 4 \end{vmatrix} = -(16+144) = -160 \quad \textbf{(8.51)}$$

The determinant (8.50) is a special kind, called a *circulant*. A *circulant* (or *cyclic determinant*) has only n independent elements—they appear in the first row, and succeeding rows are formed by successive cyclic permutations of these elements. The nth-order circulant $C(a_1, a_2, \ldots, a_n)$ is

$$C(a_1, a_2, a_3, \ldots, a_n) = \begin{vmatrix} a_1 & a_2 & a_3 & \cdots & a_n \\ a_n & a_1 & a_2 & \cdots & a_{n-1} \\ a_{n-1} & a_n & a_1 & \cdots & a_{n-2} \\ \cdot & \cdot & \cdot & \cdots & \cdot \\ a_2 & a_3 & a_4 & \cdots & a_1 \end{vmatrix} \quad \textbf{(8.52)}$$

It can be shown[2] that

$$C(a_1, a_2, a_3, \ldots, a_n) = \prod_{k=1}^{n} (a_1 + \omega_k a_2 + \omega_k^2 a_3 + \cdots + \omega_k^{n-1} a_n) \quad \textbf{(8.53)}$$

where $\omega_1, \omega_2, \ldots, \omega_n$ are the n different nth roots of unity [Eq. (1.32)]:

$$\omega_k = e^{2\pi i k/n}, \qquad k = 1, 2, \ldots, n, \qquad i = \sqrt{-1} \quad \textbf{(8.54)}$$

The definition of the product notation used in (8.53) is

$$\prod_{k=1}^{n} b_k = b_1 b_2 b_3 \cdots b_n$$

As an illustration of (8.53), let us use it to evaluate the determinant (8.50). This is a fourth-order determinant, and the four fourth roots of unity are $1, -1, i, -i$. We have

$$C(1, 2, 3, 4) = (1 + 1 \cdot 2 + 1 \cdot 3 + 1 \cdot 4)(1 - 1 \cdot 2 + 1 \cdot 3 - 1 \cdot 4)$$
$$\times (1 + i \cdot 2 - 1 \cdot 3 - i \cdot 4)(1 - i \cdot 2 - 1 \cdot 3 + i \cdot 4)$$
$$= -160$$

which checks with (8.51). Circulants occur in the quantum chemistry of molecules such as benzene.

The diagonal of a determinant that runs from the top left to the lower right is

---

[2] T. Muir, *A Treatise on the Theory of Determinants*, Dover, New York, 1960, pp. 442–445.

the *principal diagonal*. For a determinant all of whose elements are zero except those on the principal diagonal, we have

$$
\begin{vmatrix}
a_{11} & 0 & 0 & \cdots & 0 \\
0 & a_{22} & 0 & \cdots & 0 \\
0 & 0 & a_{33} & \cdots & 0 \\
\cdot & \cdot & \cdot & \cdots & \cdot \\
0 & 0 & 0 & \cdots & a_{nn}
\end{vmatrix}
= a_{11}
\begin{vmatrix}
a_{22} & 0 & \cdots & 0 \\
0 & a_{33} & \cdots & 0 \\
\cdot & \cdot & \cdots & \cdot \\
0 & 0 & \cdots & a_{nn}
\end{vmatrix}
$$

$$
= a_{11}a_{22}
\begin{vmatrix}
a_{33} & 0 & \cdots & 0 \\
0 & a_{44} & \cdots & 0 \\
\cdot & \cdot & \cdots & \cdot \\
0 & 0 & \cdots & a_{nn}
\end{vmatrix}
$$

$$
= \cdots = a_{11}a_{22}a_{33}\cdots a_{nn} \qquad (8.55)
$$

and this determinant equals the product of the diagonal elements.

A determinant in which all elements are zero except the elements of the principal diagonal and the elements immediately above and below this diagonal is called a *continuant*. (The name arises from a relation between this type of determinant and continued fractions.) There is a special kind of continuant, whose value we will need, in which the elements immediately above the principal diagonal are all equal, those on the principal diagonal are all equal, and those immediately below the principal diagonal are all equal. It can be shown[3] that the *n*th-order continuant of this form has the value

$$
\begin{vmatrix}
a & b & 0 & 0 & \cdot & \cdot & \cdot & \cdot & 0 \\
c & a & b & 0 & \cdot & \cdot & \cdot & \cdot & 0 \\
0 & c & a & b & \cdot & \cdot & \cdot & \cdot & 0 \\
\cdot & \cdot & \cdot & \cdot & \cdot & \cdot & \cdot & \cdot & \cdot \\
0 & \cdot & \cdot & \cdot & \cdot & 0 & c & a & b \\
0 & \cdot & \cdot & \cdot & \cdot & 0 & 0 & c & a
\end{vmatrix}_n
= \prod_{j=1}^{n}\left[a - 2(bc)^{1/2}\cos\left(\frac{j\pi}{n+1}\right)\right] \qquad (8.56)
$$

The subscript on the determinant denotes its order.

We say that a determinant is in *diagonal form* when the only nonzero elements are those on the principal diagonal. We showed that a determinant in diagonal form is equal to the product of the diagonal elements. We now consider a determinant whose only nonzero elements occur in square blocks centered about the principal diagonal. Such a determinant is in *block-diagonal form*. We assert that if we regard each square block as a determinant, then a block-diagonal determinant is equal to

---

[3] T. Muir, *The Theory of Determinants in the Historical Order of Development*, Dover, New York, 1960, vol. 4, p. 401.

the product of all the blocks. For example,

$$
\begin{vmatrix}
a & b & 0 & 0 & 0 & 0 \\
c & d & 0 & 0 & 0 & 0 \\
0 & 0 & e & 0 & 0 & 0 \\
0 & 0 & 0 & f & g & h \\
0 & 0 & 0 & i & j & k \\
0 & 0 & 0 & l & m & n
\end{vmatrix}
=
\begin{vmatrix} a & b \\ c & d \end{vmatrix}
(e)
\begin{vmatrix} f & g & h \\ i & j & k \\ l & m & n \end{vmatrix}
\tag{8.57}
$$

(The dashed lines outline the blocks.) The proof of (8.57) is simple. Expanding in terms of elements of the top row, we have

$$
\begin{vmatrix}
a & b & 0 & \cdots & 0 \\
c & d & 0 & \cdots & 0 \\
0 & 0 & e & \cdots & \cdot \\
\vdots & \vdots & \vdots & \ddots & \vdots \\
0 & 0 & & \cdots & \cdot
\end{vmatrix}
= a
\begin{vmatrix}
d & 0 & \cdots & 0 \\
0 & e & \cdots & \cdot \\
\vdots & \vdots & \ddots & \vdots \\
0 & & \cdots & \cdot
\end{vmatrix}
- b
\begin{vmatrix}
c & 0 & \cdots & 0 \\
0 & e & \cdots & \cdot \\
\vdots & \vdots & \ddots & \vdots \\
0 & & \cdots & \cdot
\end{vmatrix}
$$

and expanding the two determinants on the right side of this equation in terms of elements of the top rows, we get

$$
ad
\begin{vmatrix}
e & \cdot & \cdot & \cdot \\
\cdot & & & \cdot \\
\cdot & & & \cdot \\
\cdot & & & \cdot
\end{vmatrix}
- bc
\begin{vmatrix}
e & \cdot & \cdot & \cdot \\
\cdot & & & \cdot \\
\cdot & & & \cdot \\
\cdot & & & \cdot
\end{vmatrix}
=
\begin{vmatrix} a & b \\ c & d \end{vmatrix}
\begin{vmatrix}
e & \cdot & \cdot & \cdot \\
\cdot & & & \cdot \\
\cdot & & & \cdot \\
\cdot & & & \cdot
\end{vmatrix}
\tag{8.58}
$$

A similar expansion of the second determinant on the right side of (8.58) completes the proof.

8.4   *SIMULTANEOUS LINEAR EQUATIONS*

Determinants find application in the solution of systems of linear equations. Consider a system of n linear equations in n unknowns:

$$
\begin{aligned}
a_{11}x_1 + a_{12}x_2 + \cdots + a_{1n}x_n &= b_1 \\
a_{21}x_1 + a_{22}x_2 + \cdots + a_{2n}x_n &= b_2 \\
\cdot \quad \cdot \quad \cdot \quad \cdots \quad \cdot \quad \cdot & \\
a_{n1}x_1 + a_{n2}x_2 + \cdots + a_{nn}x_n &= b_n
\end{aligned}
\tag{8.59}
$$

where the a's and b's are known constants and $x_1, x_2, \ldots, x_n$ are the unknowns. If at least one of the b's in (8.59) is not zero, we have a system of *in-*

homogeneous linear equations. Such a system can be solved by Cramer's rule.[4] Let det (a_{ij}) be the determinant of the coefficients of the unknowns in (8.59). Cramer's rule states that x_k ($k = 1, 2, \ldots, n$) is equal to the determinant obtained by replacing the kth column of det (a_{ij}) with the elements $b_1, b_2, \ldots, b_n$, divided by det (a_{ij}):

$$x_k = \frac{\begin{vmatrix} a_{11} & a_{12} & \cdots & a_{1,k-1} & b_1 & a_{1,k+1} & \cdots & a_{1n} \\ a_{21} & a_{22} & \cdots & a_{2,k-1} & b_2 & a_{2,k+1} & \cdots & a_{2n} \\ \cdot & \cdot & \cdots & \cdot & \cdot & \cdot & \cdots & \cdot \\ a_{n1} & a_{n2} & \cdots & a_{n,k-1} & b_k & a_{n,k+1} & \cdots & a_{nn} \end{vmatrix}}{\begin{vmatrix} a_{11} & a_{12} & \cdots & a_{1n} \\ a_{21} & a_{22} & \cdots & a_{2n} \\ \cdot & \cdot & \cdots & \cdot \\ a_{n1} & a_{n2} & \cdots & a_{nn} \end{vmatrix}}, \quad k = 1, 2, \ldots, n \tag{8.60}$$

For example, application of Cramer's rule to the system

$$2x + 10y = 9$$

$$1.5x - 6y = 0$$

gives

$$x = \frac{\begin{vmatrix} 9 & 10 \\ 0 & -6 \end{vmatrix}}{\begin{vmatrix} 2 & 10 \\ 1.5 & -6 \end{vmatrix}} = 2, \quad y = \frac{\begin{vmatrix} 2 & 9 \\ 1.5 & 0 \end{vmatrix}}{\begin{vmatrix} 2 & 10 \\ 1.5 & -6 \end{vmatrix}} = 0.5$$

Cramer's rule is of considerable theoretical significance, but for actual solution of a given set of simultaneous equations, successive elimination of unknowns is much more efficient. One systematic procedure (Gauss reduction) is this: Divide each equation in (8.59) by the coefficient of x_1 in that equation; then subtract the first equation from every other equation; this eliminates x_1 from all equations but the first. Now divide each of equations 2, $\ldots$, n by the coefficient of x_2 in that equation and then subtract equation 2 from each of equations 3, $\ldots$, n, thereby eliminating x_2 from all equations but the first and second. Continue in this manner. Ultimately equation n will contain only x_n, equation $n-1$ only x_{n-1} and x_n, etc. The value of x_n from equation n is substituted into equation $n-1$ to give x_{n-1}, etc.

Now we consider the case where all the b's in (8.59) are zero. We have a system of linear *homogeneous* equations:

$$a_{11}x_1 + a_{12}x_2 + \cdots + a_{1n}x_n = 0$$

$$a_{21}x_1 + a_{22}x_2 + \cdots + a_{2n}x_n = 0$$

$$\cdot \quad \cdot \quad \cdot \quad \cdots \quad \cdot \quad \cdot \tag{8.61}$$

$$a_{n1}x_1 + a_{n2}x_2 + \cdots + a_{nn}x_n = 0$$

---

[4] For a proof see *Sokolnikoff and Redheffer*, p. 708.

One obvious solution of (8.61) is $x_1 = x_2 = \cdots = x_n = 0$, which is called the *trivial solution*. If the determinant of the coefficients in (8.61) is not equal to zero, $\det (a_{ij}) \neq 0$, then we can use Cramer's rule (8.60) to solve for the unknowns, and we find $x_k = 0, k = 1, 2, \ldots, n$, since the determinant in the numerator of (8.60) has a column all of whose elements are zero. Thus when $\det (a_{ij}) \neq 0$, the only solution is the trivial solution (which is of no interest). For there to be a nontrivial solution of a system of n linear homogeneous equations in n unknowns, the determinant of the coefficients must be zero. It can also be shown[5] that this condition is sufficient to ensure the existence of a nontrivial solution. We therefore have the extremely important theorem: *A system of n linear homogeneous equations in n unknowns has a nontrivial solution if and only if the determinant of the coefficients is zero.*

Suppose that $\det (a_{ij}) = 0$, so that (8.61) has a nontrivial solution. How do we go about finding it? With $\det (a_{ij}) = 0$, Cramer's rule (8.60) gives $x_k = 0/0, k = 1, \ldots, n$, which is indeterminate. Thus Cramer's rule is of no immediate help. We also observe that if $x_1 = d_1, x_2 = d_2, \ldots, x_n = d_n$ is a solution of (8.61), then so is $x_1 = cd_1, x_2 = cd_2, \ldots, x_n = cd_n$, where c is an arbitrary constant. This is easily seen, since

$$a_{11}cd_1 + a_{12}cd_2 + \cdots + a_{1n}cd_n = c(a_{11}d_1 + a_{12}d_2 + \cdots + a_{1n}d_n) = c \cdot 0 = 0$$

etc. Therefore the solution to the linear homogeneous system of equations will contain an arbitrary constant, and we cannot determine a unique value for each unknown. To solve (8.61), we therefore assign an arbitrary value to any one of the unknowns, say x_n; we set $x_n = c$, where c is an arbitrary constant. Having assigned a value to x_n, we transfer the last term in each of the equations of (8.61) to the right-hand side to get

$$
\begin{aligned}
a_{11}x_1 &+ a_{12}x_2 + \cdots + a_{1,\,n-1}x_{n-1} &= -a_{1,\,n}c \\
a_{21}x_1 &+ a_{22}x_2 + \cdots + a_{2,\,n-1}x_{n-1} &= -a_{2,\,n}c \\
&\phantom{+ a_{22}x_2} \cdots \\
a_{n-1,\,1}x_1 &+ a_{n-1,\,2}x_2 + \cdots + a_{n-1,\,n-1}x_{n-1} &= -a_{n-1,\,n}c \\
a_{n1}x_1 &+ a_{n2}x_2 + \cdots + a_{n,\,n-1}x_{n-1} &= -a_{nn}c
\end{aligned}
\tag{8.62}
$$

We now have n equations in $n - 1$ unknowns, which is one more equation than we need. We therefore discard any one of the equations of (8.62), say the last one. This gives a system of $n - 1$ linear *inhomogeneous* equations in $n - 1$ unknowns. We can then apply Cramer's rule (8.60) to solve for $x_1, x_2, \ldots, x_{n-1}$. Since the constants on the right side of the equations in (8.62) all contain the factor c, Theorem IV in Section 8.3 shows that all the unknowns contain this arbitrary constant as a factor. The form of the solution is therefore

$$x_1 = ce_1, x_2 = ce_2, \ldots, x_{n-1} = ce_{n-1}, x_n = c \tag{8.63}$$

where $e_1, \ldots, e_{n-1}$ are numbers and c is an arbitrary constant.

---

[5] T. L. Wade, *The Algebra of Vectors and Matrices*, Addison-Wesley, Reading, Mass., 1951, p. 146.

There is one possible hitch to the procedure just outlined. It might happen that the determinant of the coefficients of the inhomogeneous system of $n-1$ equations in $n-1$ unknowns [(8.62) with the last equation omitted] is zero. In this case, Cramer's rule would have a zero in the denominator and would be of no use. We could attempt to get around this difficulty by initially assigning the arbitrary value to another of the unknowns rather than to x_n. We could also try discarding some other equation of (8.62), rather than the last one. What we are looking for, really, is a nonvanishing determinant of order $n-1$, formed from the determinant of the coefficients of the system (8.61) by striking out a row and a column. If such a determinant exists, then by the procedure given, with the right choice of the equation to be discarded and the right choice of the unknown to be assigned an arbitrary value, we can solve the system and will obtain solutions of the form (8.63). If no such determinant exists, we must assign arbitrary values to two of the unknowns and attempt to proceed from there. Thus it might happen that the solution to (8.61) would contain two (or even more) arbitrary constants.

8.5   LINEAR VARIATION FUNCTIONS

A special kind of variation function widely used in the study of molecules is the linear variation function. A *linear variation function* is a linear combination of n linearly independent functions $f_1, f_2, \ldots, f_n$:

$$\varphi = c_1 f_1 + c_2 f_2 + \cdots + c_n f_n = \sum_{j=1}^{n} c_j f_j \tag{8.64}$$

where φ is the trial variation function and the coefficients c_j are parameters to be determined by minimizing the variational integral. The functions f_j must satisfy the boundary conditions of the problem. We will restrict ourselves to *real* φ so that the c_j's and f_j's are all real.

We now apply the variation theorem (8.13). For the real linear variation function, we have

$$\int \varphi^* \varphi \, d\tau = \int \sum_{j=1}^{n} c_j f_j \sum_{k=1}^{n} c_k f_k \, d\tau = \sum_{j=1}^{n} \sum_{k=1}^{n} c_j c_k \int f_j f_k \, d\tau \tag{8.65}$$

We define the *overlap integral* S_{jk} as

$$S_{jk} \equiv \int f_j^* f_k \, d\tau \tag{8.66}$$

We have

$$\int \varphi^* \varphi \, d\tau = \sum_{j=1}^{n} \sum_{k=1}^{n} c_j c_k S_{jk} \tag{8.67}$$

Note that S_{jk} is not necessarily equal to δ_{jk}, since there is no reason to suppose that the functions f_j are mutually orthogonal. They are not necessarily the eigenfunctions of any operator. For the numerator of (8.13), we have

$$\int \varphi^* \hat{H} \varphi \, d\tau = \int \sum_{j=1}^{n} c_j f_j \, \hat{H} \sum_{k=1}^{n} c_k f_k \, d\tau = \sum_{j=1}^{n} \sum_{k=1}^{n} c_j c_k \int f_j \hat{H} f_k \, d\tau \quad \textbf{(8.68)}$$

and using the abbreviation

$$H_{jk} \equiv \int f_j^* \hat{H} f_k \, d\tau \qquad\qquad \textbf{(8.69)}$$

we write

$$\int \varphi^* \hat{H} \varphi \, d\tau = \sum_{j=1}^{n} \sum_{k=1}^{n} c_j c_k H_{jk} \qquad\qquad \textbf{(8.70)}$$

The variational integral W is

$$W \equiv \frac{\int \varphi^* \hat{H} \varphi \, d\tau}{\int \varphi^* \varphi \, d\tau} = \frac{\sum_{j=1}^{n} \sum_{k=1}^{n} c_j c_k H_{jk}}{\sum_{j=1}^{n} \sum_{k=1}^{n} c_j c_k S_{jk}} \qquad\qquad \textbf{(8.71)}$$

$$W \sum_{j=1}^{n} \sum_{k=1}^{n} c_j c_k S_{jk} = \sum_{j=1}^{n} \sum_{k=1}^{n} c_j c_k H_{jk} \qquad\qquad \textbf{(8.72)}$$

We now minimize W, so as to approach as closely as we can to E_0 ($W \geqslant E_0$). The variational integral W is a function of the n independent variables $c_1, c_2, \ldots, c_n$:

$$W = W(c_1, c_2, \ldots, c_n) \qquad\qquad \textbf{(8.73)}$$

A necessary condition for a minimum in W is that its partial derivatives with respect to each of the variables must be zero:

$$\frac{\partial W}{\partial c_i} = 0, \qquad i = 1, 2, \ldots, n \qquad\qquad \textbf{(8.74)}$$

We now differentiate (8.72) partially with respect to each c_i to obtain n equations:

$$\frac{\partial W}{\partial c_i} \sum_{j=1}^{n} \sum_{k=1}^{n} c_j c_k S_{jk} + W \frac{\partial}{\partial c_i} \sum_{j=1}^{n} \sum_{k=1}^{n} c_j c_k S_{jk} = \frac{\partial}{\partial c_i} \sum_{j=1}^{n} \sum_{k=1}^{n} c_j c_k H_{jk},$$
$$i = 1, 2, \ldots, n \quad \textbf{(8.75)}$$

Now

$$\frac{\partial}{\partial c_i} \sum_{j=1}^{n} \sum_{k=1}^{n} c_j c_k S_{jk} = \sum_{j=1}^{n} \sum_{k=1}^{n} \left[\frac{\partial}{\partial c_i}(c_j c_k) \right] S_{jk}$$
$$= \sum_{j=1}^{n} \sum_{k=1}^{n} \left(c_k \frac{\partial c_j}{\partial c_i} + c_j \frac{\partial c_k}{\partial c_i} \right) S_{jk} \qquad \textbf{(8.76)}$$

The c_j's are independent variables, and therefore

$$\frac{\partial c_j}{\partial c_i} = 0 \quad \text{if } j \neq i, \qquad \frac{\partial c_j}{\partial c_i} = 1 \quad \text{if } j = i \qquad \textbf{(8.77)}$$

$$\frac{\partial c_j}{\partial c_i} = \delta_{ij} \qquad\qquad \textbf{(8.78)}$$

We then have

$$\frac{\partial}{\partial c_i} \sum_{j=1}^{n} \sum_{k=1}^{n} c_j c_k S_{jk} = \sum_{k=1}^{n} \sum_{j=1}^{n} c_k \delta_{ij} S_{jk} + \sum_{j=1}^{n} \sum_{k=1}^{n} c_j \delta_{ik} S_{jk}$$

$$= \sum_{k=1}^{n} c_k S_{ik} + \sum_{j=1}^{n} c_j S_{ji} \qquad (8.79)$$

where we have evaluated one of the sums in each double summation using Eq. (7.53). Now according to Eq. (7.5), we have

$$S_{ji} = S_{ij}^* = S_{ij} \qquad (8.80)$$

where the last equality follows because we are dealing with real functions. Hence

$$\frac{\partial}{\partial c_i} \sum_{j=1}^{n} \sum_{k=1}^{n} c_j c_k S_{jk} = \sum_{k=1}^{n} c_k S_{ik} + \sum_{j=1}^{n} c_j S_{ij} = \sum_{k=1}^{n} c_k S_{ik} + \sum_{k=1}^{n} c_k S_{ik}$$

$$\frac{\partial}{\partial c_i} \sum_{j=1}^{n} \sum_{k=1}^{n} c_j c_k S_{jk} = 2 \sum_{k=1}^{n} c_k S_{ik} \qquad (8.81)$$

By replacing S_{jk} by H_{jk} in each of these manipulations, we get

$$\frac{\partial}{\partial c_i} \sum_{j=1}^{n} \sum_{k=1}^{n} c_j c_k H_{jk} = 2 \sum_{k=1}^{n} c_k H_{ik} \qquad (8.82)$$

This result depends on the fact that

$$H_{ji} = H_{ij}^* = H_{ij} \qquad (8.83)$$

which is true because $\hat{H}$ is a Hermitian operator and because we are dealing with real functions and a real Hamiltonian.

Substituting Eqs. (8.74), (8.81), and (8.82) in (8.75), we have

$$2W \sum_{k=1}^{n} c_k S_{ik} = 2 \sum_{k=1}^{n} c_k H_{ik}, \qquad i = 1, 2, \dots, n \qquad (8.84)$$

$$\sum_{k=1}^{n} [(H_{ik} - S_{ik} W)c_k] = 0, \qquad i = 1, 2, \dots, n \qquad (8.85)$$

We have n simultaneous linear homogeneous equations in the n unknowns $c_1, c_2, \dots, c_n$. Thus for $n = 2$,

$$(H_{11} - S_{11} W)c_1 + (H_{12} - S_{12} W)c_2 = 0$$
$$(H_{21} - S_{21} W)c_1 + (H_{22} - S_{22} W)c_2 = 0 \qquad (8.86)$$

From the theorem of the last section, for there to be a solution of the system of linear homogeneous equations besides $0 = c_1 = c_2 = \cdots = c_n$ (which would make the variation function φ zero), the determinant of the coefficients must vanish. For $n = 2$, we have

$$\begin{vmatrix} H_{11} - S_{11} W & H_{12} - S_{12} W \\ H_{21} - S_{21} W & H_{22} - S_{22} W \end{vmatrix} = 0 \qquad (8.87)$$

and for the general case det $(H_{ij} - S_{ij}W) = 0$, or

$$\begin{vmatrix} H_{11} - S_{11}W & H_{12} - S_{12}W & \cdots & H_{1n} - S_{1n}W \\ H_{21} - S_{21}W & H_{22} - S_{22}W & \cdots & H_{2n} - S_{2n}W \\ \vdots & \vdots & \vdots & \vdots \\ H_{n1} - S_{n1}W & H_{n2} - S_{n2}W & \cdots & H_{nn} - S_{nn}W \end{vmatrix} = 0 \qquad (8.88)$$

Expansion of the determinant in (8.88) gives an algebraic equation of degree n in the unknown W, with n roots (which may be shown to be real). Arranging these roots in order of increasing value, we denote them as

$$W_0 \leqslant W_1 \leqslant W_2 \leqslant \cdots \leqslant W_{n-1} \qquad (8.89)$$

If we number the states of the system in order of increasing energy, we have

$$E_0 \leqslant E_1 \leqslant E_2 \leqslant \cdots \leqslant E_{n-1} \leqslant E_n \leqslant E_{n+1} \leqslant \cdots$$

where the E's denote the true energies of various states. From the variation theorem, we know that

$$E_0 \leqslant W_0 \qquad (8.90)$$

Moreover, it can be proved that[6]

$$E_1 \leqslant W_1, E_2 \leqslant W_2, \ldots, E_{n-1} \leqslant W_{n-1} \qquad (8.91)$$

In other words, the linear variation method provides upper bounds to the energies of the first n states of the system; we use the roots $W_0, W_1, \ldots, W_{n-1}$ as approximations to the energies of the n lowest states. If we want approximations to the energies of more states, we increase the number of functions f_k in the trial function φ. It can also be shown that increasing the number of functions f_k will increase (or cause no change in) the accuracy of the previously obtained energies. A major difficulty in this method is in solving Eq. (8.88) for W. For example, we might want to include as many as 100 terms in the variation function. Solving a 100th-degree equation by hand is out of the question. The advent of the electronic computer after World War II has been a godsend for the quantum chemist.

If we want an approximation to the wave function of, say, the ground state, we take the value of W_0 we have found and substitute it in the set of equations (8.85), which we then solve for the coefficients $c_1, c_2, \ldots, c_n$. As stated in the previous section, all we can determine is the ratio of coefficients. We solve for $c_2, \ldots, c_n$ in terms of c_1, and then determine c_1 by normalization. Use of higher roots of (8.88) in (8.85) gives approximations to excited-state wave functions. We can prove (Problem 8.15) that the approximate wave functions obtained form an orthogonal set.

Solution of (8.88) (which is called the *secular equation*) is facilitated by having as many of the determinant's elements equal to zero as possible. If the functions f_j are orthogonal, the off-diagonal S_{ij}'s vanish. One can frequently make some of the off-

---

[6] J. K. L. MacDonald, *Phys. Rev.*, **43**, 830 (1933); R. H. Young, *Int. J. Quantum Chem.*, **6**, 596 (1972).

diagonal H_{ij}'s vanish by choosing the functions f_j as eigenfunctions of some operator $\hat{A}$ that commutes with $\hat{H}$; if f_i and f_j correspond to different eigenvalues of $\hat{A}$, then H_{ij} vanishes (Section 7.4). We shall use this method in Chapter 15 to simplify molecular variational calculations. (See also Section 9.6.)

Equations (8.85) and (8.88) are also valid when the restriction that the variation function be real is removed (Problem 8.12).

PROBLEMS

8.1 For the ground state of the harmonic oscillator, use the trial variation function (8.27), and show that minimization of the variational integral gives $b = \frac{1}{2}$ and that Eq. (8.28) holds.

8.2 Consider an integral whose integrand is a function of a parameter p. The integral itself is a function of this parameter, and we have

$$\int_a^b f(x, p)\, dx = G(p)$$

Provided the integrand is reasonably well-behaved, we may differentiate under the integral sign to evaluate $G'(p)$, and we have (*Taylor and Mann*, p. 582)

$$G'(p) = \int_a^b \frac{\partial f(x, p)}{\partial p}\, dx$$

where a and b are constants. Use this theorem to derive Eq. (A.6) from (A.5).

8.3 Apply the function of Eq. (7.56) as a trial variation function for the ground state of the particle in a box. Note that $f''(x)$ is infinite at $x = \frac{1}{2} l$ because of the discontinuity in $f'(x)$ at this point. Therefore in evaluating $\int f^* \hat{H} f\, dx$ we run into difficulty in evaluating the integral of ff''. One way around this problem is to first show that

$$\int_0^l ff''\, dx = -\int_0^l (f')^2\, dx = -\int_0^{l/2} (f')^2\, dx - \int_{l/2}^l (f')^2\, dx \qquad (8.92)$$

for any function obeying the boundary conditions. Then, using the expression on the right of (8.92), we can calculate the variational integral. Prove (8.92), and then calculate the percent error in the ground-state energy, using this triangular function. Note that there is no parameter in this trial function. [If you are ambitious, try this alternative procedure: Note that $f'(x)$ involves the Heaviside step function (Section 7.7), and therefore $f''(x)$ involves the Dirac delta function. Use the properties of the delta function to evaluate $\int_0^l ff''\, dx$, and find the percent error using this triangular function.]

Functions with discontinuities in their first derivatives have been used as trial variation functions in atomic and molecular quantum mechanics.[7]

8.4 If we use the normalized trial variation function

$$\varphi = \left(\frac{3}{l^3}\right)^{1/2} x, \qquad 0 \leqslant x \leqslant l$$

for the particle in a one-dimensional box, we find that the variational integral has the value zero, which is *less* than the true ground-state energy. What is wrong?

---

[7] H. M. James and A. S. Coolidge, *Phys. Rev.*, **49**, 688 (1936); M. J. S. Dewar and C. Wulfman, *J. Chem. Phys.*, **29**, 158 (1958); L. C. Snyder and R. G. Parr, *J. Chem. Phys.*, **34**, 1661 (1961); C. A. Coulson and C. S. Sharma, *Proc. Roy. Soc.*, **272A**, 1 (1963).

8.5 (a) For the ground state of the hydrogen atom, use the Gaussian trial function

$$\varphi = e^{-cr^2/a_0^2} \tag{8.93}$$

Find the optimum value of c and the percent error in the energy. (Gaussian variational functions are often used in molecular quantum mechanics; see Section 15.3.) (b) Multiply the function (8.93) by the spherical harmonic Y_2^0, and then minimize the variational integral. This yields an upper bound to the energy of which hydrogen-atom state?

8.6 Prove that the value of a determinant all of whose elements below the principal diagonal are zero is equal to the product of the diagonal elements.

8.7 Evaluate

$$\begin{vmatrix} 2 & 5 & 1 & 3 \\ 8 & 0 & 4 & -1 \\ 6 & 6 & 6 & 1 \\ 5 & -2 & -2 & 2 \end{vmatrix}$$

8.8 (a) Consider some permutation of the integers $1, 2, 3, \ldots, n$. The permutation is an *even permutation* if an even number of interchanges of pairs of integers restores the permutation to the natural order $1, 2, 3, \ldots, n$. An *odd permutation* requires an odd number of interchanges of pairs to reach the natural order. For example, the permutation 3124 is even, since two interchanges restore it to the natural order: $3124 \rightarrow 1324 \rightarrow 1234$. Write down and classify (even or odd) all permutations of 123. (b) Verify that the definition (8.42) of the third-order determinant is equivalent to

$$\begin{vmatrix} a_{11} & a_{12} & a_{13} \\ a_{21} & a_{22} & a_{23} \\ a_{31} & a_{32} & a_{33} \end{vmatrix} = \sum (\pm 1) a_{1i} a_{2j} a_{3k}$$

where ijk is one of the permutations of the integers 123, the sum is over the 3! different permutations of these integers, and the sign of each term is plus or minus, depending on whether the permutation is even or odd. (c) How would we define the nth-order determinant using this type of definition?

8.9 Solve each set of simultaneous equations:

$$\begin{array}{ll} & x + 2y + 3z = 0 \qquad\qquad x + 2y + 3z = 0 \\ \text{(a)} \ 3x + y + 2z = 0 \quad \text{(b)} \quad x - y + z = 0 \\ & 2x + 3y + z = 0 \qquad\qquad 7x - y + 11z = 0 \end{array}$$

8.10 How many terms are there in the expansion of an nth-order determinant?

8.11 Solve the second-order determinantal equation (8.87) for the special case where $H_{11} = H_{22}$ and $S_{11} = S_{22}$. Then solve for c_1/c_2 for each of the two roots.

8.12 Let the variation function of (8.64) be complex. Then $c_j = a_j + ib_j$, where a_j and b_j are real numbers. There are $2n$ parameters to be varied, namely the a_j's and b_j's. (a) Use the chain rule to show that the minimization conditions $\partial W / \partial a_i = 0, \partial W / \partial b_i = 0$ are equivalent to the conditions $\partial W / \partial c_i = 0, \partial W / \partial c_i^* = 0$. (b) Show that minimization of W leads to Eq. (8.85) and its complex conjugate, which may be discarded. Hence Eqs. (8.85) and (8.88) are valid for complex variation functions.

8.13 Apply the linear trial variation function

$$\varphi = c_1 x^2 (l - x) + c_2 x (l - x)^2, \qquad 0 \leqslant x \leqslant l$$

to the particle in a one-dimensional box. Calculate the percent errors for the $n = 1$ and 2 states. Sketch $x^2(l - x)$, $x(l - x)^2$, and the two approximate wave functions you find.

8.14 In 1971 a paper was published that applied the normalized variation function $N \exp(-br^2/a_0^2 - cr/a_0)$ to the hydrogen atom and stated that minimization of the variational integral with respect to the parameters b and c yielded an energy 0.7 percent above the true ground-state energy (for infinite nuclear mass). Without doing any calculations, state why this result must be in error.

8.15 We wish to prove that the approximate wave functions obtained in the linear variation method are orthogonal. Let the approximate function φ_α have the value W_α for the variational integral and the coefficients $c_j^{(\alpha)}$ in (8.64). [We add α to distinguish the n different φ's.] We rewrite (8.85) as

$$\sum_k [(\langle f_i | \hat{H} | f_k \rangle - \langle f_i | f_k \rangle W_\alpha) c_k^{(\alpha)}] = 0, \qquad i = 1, \ldots, n \qquad \textbf{(8.94)}$$

(a) Show that $\langle f_i | \hat{H} - W_\alpha | \varphi_\alpha \rangle = 0$ by showing this integral to equal the left side of (8.94). (b) Use the result of (a) to show $\langle \varphi_\beta | \hat{H} - W_\alpha | \varphi_\alpha \rangle = 0$ and $\langle \varphi_\alpha | \hat{H} - W_\beta | \varphi_\beta \rangle^* = 0$ for all α and β. (c) Equate the two integrals in (b), and use the Hermitian property of $\hat{H}$ to show that $\langle \varphi_\beta | \varphi_\alpha \rangle (W_\beta - W_\alpha) = 0$. We conclude that for $W_\alpha \neq W_\beta$, φ_α and φ_β are orthogonal. (For $W_\alpha = W_\beta$, we can form orthogonal linear combinations of φ_α and φ_β that will have the same value for the variational integral.)

8.16 Write a computer program that will perform Gauss reduction on a system of up to 10 linear simultaneous inhomogeneous equations in 10 unknowns. Test it on a couple of simple examples.

9 *PERTURBATION THEORY*

9.1 *INTRODUCTION*

We now discuss the second major quantum-mechanical approximation method, perturbation theory.

Suppose we have a system with a time-independent Hamiltonian $\hat{H}$ and that we are unable to solve the Schroedinger equation

$$\hat{H}\psi_n = E_n\psi_n \tag{9.1}$$

for the eigenfunctions and eigenvalues of the bound stationary states. Suppose also that the Hamiltonian $\hat{H}$ is only slightly different from the Hamiltonian $\hat{H}^0$ of a system whose Schroedinger equation

$$\hat{H}^0\psi_n^{(0)} = E_n^{(0)}\psi_n^{(0)} \tag{9.2}$$

we are able to solve. An example is the one-dimensional anharmonic oscillator with Hamiltonian

$$\hat{H} = -\frac{\hbar^2}{2m}\frac{d^2}{dx^2} + \tfrac{1}{2}kx^2 + cx^3 + dx^4 \tag{9.3}$$

The Hamiltonian (9.3) is closely related to the Hamiltonian of the harmonic oscillator, which is

$$\hat{H}^0 = -\frac{\hbar^2}{2m}\frac{d^2}{dx^2} + \tfrac{1}{2}kx^2 \tag{9.4}$$

If the constants c and d in (9.3) are small, it is natural to suppose that the eigenfunctions and eigenvalues of the anharmonic oscillator are closely related to the eigenfunctions and eigenvalues of the harmonic oscillator.

We will call the system with Hamiltonian $\hat{H}^0$ the *unperturbed* system; the system with Hamiltonian $\hat{H}$ is the *perturbed* system. The difference between the two Hamiltonians is the *perturbation, $\hat{H}'$*:

$$\hat{H}' \equiv \hat{H} - \hat{H}^0$$

$$\hat{H} = \hat{H}^0 + \hat{H}' \tag{9.5}$$

(Of course the prime does not refer to differentiation.) For the anharmonic oscillator with Hamiltonian (9.3), the perturbation on the related harmonic oscillator is

$$\hat{H}' = cx^3 + dx^4 \tag{9.6}$$

Our task is to relate the unknown eigenvalues and eigenfunctions of the perturbed system to the known eigenvalues and eigenfunctions of the unperturbed system. As an aid in doing so, we will imagine that the perturbation is applied in small steps, giving a continuous change from the unperturbed to the perturbed system. Mathematically, this corresponds to introducing a parameter λ into the Hamiltonian, so that

$$\hat{H} = \hat{H}^0 + \lambda\hat{H}' \tag{9.7}$$

When λ is zero, we have the unperturbed system. As λ increases toward one, the perturbation grows larger, and at $\lambda = 1$ the perturbation is fully "turned on." We have introduced λ as a convenience in relating the perturbed and unperturbed eigenfunctions, and ultimately we will set $\lambda = 1$, thereby eliminating it.

Sections 9.1–9.8 deal with time-independent Hamiltonians and stationary states. Section 9.9 deals with time-dependent perturbations.

9.2   NONDEGENERATE PERTURBATION THEORY

The perturbation treatments of degenerate and nondegenerate energy levels differ. In this section we examine the effect of a perturbation on a nondegenerate level. If some of the energy levels of the unperturbed system are degenerate while others are nondegenerate, the treatment in this section will be applicable to the nondegenerate levels only.

Let $\psi_n^{(0)}$ be the wave function of some particular unperturbed nondegenerate level with energy $E_n^{(0)}$. Let ψ_n be the perturbed wave function into which $\psi_n^{(0)}$ is converted when the perturbation is applied. From (9.1) and (9.7), the Schroedinger equation for the perturbed state is

$$(\hat{H}^0 + \lambda\hat{H}')\psi_n = E_n\psi_n \tag{9.8}$$

Since the Hamiltonian in (9.8) depends on the parameter λ, both the eigenfunction ψ_n and the eigenvalue E_n depend on λ:

$$\psi_n = \psi_n(\lambda, q) \tag{9.9}$$

$$E_n = E_n(\lambda) \tag{9.10}$$

where q indicates the system's spatial coordinates. We now expand ψ_n and E_n as

Taylor series in powers of λ:

$$\psi_n = \psi_n|_{\lambda=0} + \frac{\partial \psi_n}{\partial \lambda}\bigg|_{\lambda=0} \lambda + \frac{\partial^2 \psi_n}{\partial \lambda^2}\bigg|_{\lambda=0} \frac{\lambda^2}{2!} + \cdots \qquad (9.11)$$

$$E_n = E_n|_{\lambda=0} + \frac{dE_n}{d\lambda}\bigg|_{\lambda=0} \lambda + \frac{d^2 E_n}{d\lambda^2}\bigg|_{\lambda=0} \frac{\lambda^2}{2!} + \cdots \qquad (9.12)$$

By hypothesis, when λ goes to zero, ψ_n and E_n go to $\psi_n^{(0)}$ and $E_n^{(0)}$:

$$\psi_n|_{\lambda=0} = \psi_n^{(0)} \qquad (9.13)$$

$$E_n|_{\lambda=0} = E_n^{(0)} \qquad (9.14)$$

We introduce the following abbreviations:

$$\psi_n^{(k)} = \frac{1}{k!} \frac{\partial^k \psi_n}{\partial \lambda^k}\bigg|_{\lambda=0}, \qquad k = 1, 2, \ldots \qquad (9.15)$$

$$E_n^{(k)} = \frac{1}{k!} \frac{d^k E_n}{d\lambda^k}\bigg|_{\lambda=0}, \qquad k = 1, 2, \ldots \qquad (9.16)$$

Equations (9.11) and (9.12) become

$$\psi_n = \psi_n^{(0)} + \lambda \psi_n^{(1)} + \lambda^2 \psi_n^{(2)} + \cdots + \lambda^k \psi_n^{(k)} + \cdots \qquad (9.17)$$

$$E_n = E_n^{(0)} + \lambda E_n^{(1)} + \lambda^2 E_n^{(2)} + \cdots + \lambda^k E_n^{(k)} + \cdots \qquad (9.18)$$

We call $\psi_n^{(k)}$ and $E_n^{(k)}$ the kth-order corrections to the wave function and energy. We will assume that the series (9.17) and (9.18) converge for $\lambda = 1$, and we hope that for a small perturbation, taking just the first few terms of the series will provide a good approximation to the true energy and wave function.

Substituting (9.17) and (9.18) into (9.8), we have

$$(\hat{H}^0 + \lambda \hat{H}') (\psi_n^{(0)} + \lambda \psi_n^{(1)} + \lambda^2 \psi_n^{(2)} + \cdots)$$
$$= (E_n^{(0)} + \lambda E_n^{(1)} + \lambda^2 E_n^{(2)} + \cdots) (\psi_n^{(0)} + \lambda \psi_n^{(1)} + \lambda^2 \psi_n^{(2)} + \cdots) \qquad (9.19)$$

Collecting like powers of λ, we have

$$\hat{H}^0 \psi_n^{(0)} + \lambda(\hat{H}' \psi_n^{(0)} + \hat{H}^0 \psi_n^{(1)}) + \lambda^2 (\hat{H}^0 \psi_n^{(2)} + \hat{H}' \psi_n^{(1)}) + \cdots$$
$$= E_n^{(0)} \psi_n^{(0)} + \lambda(E_n^{(1)} \psi_n^{(0)} + E_n^{(0)} \psi_n^{(1)}) + \lambda^2 (E_n^{(2)} \psi_n^{(0)} + E_n^{(1)} \psi_n^{(1)} + E_n^{(0)} \psi_n^{(2)}) + \cdots \qquad (9.20)$$

Now (assuming suitable convergence) for the two series on each side of (9.20) to be equal to each other for all values of λ, the coefficients of like powers of λ in the two series must be equal. [For the proof see the argument following Eq. (4.13).]

Equating the coefficients of the λ^0 terms, we have

$$\hat{H}^0 \psi_n^{(0)} = E_n^{(0)} \psi_n^{(0)} \qquad (9.21)$$

which is the Schroedinger equation for the unperturbed problem, Eq. (9.2), and gives us no new information.

Equating the coefficients of the λ^1 terms, we have

$$\hat{H}'\psi_n^{(0)} + \hat{H}^0\psi_n^{(1)} = E_n^{(1)}\psi_n^{(0)} + E_n^{(0)}\psi_n^{(1)} \tag{9.22}$$

$$(\hat{H}^0 - E_n^{(0)})\psi_n^{(1)} = (E_n^{(1)} - \hat{H}')\psi_n^{(0)} \tag{9.23}$$

This equation contains $\psi_n^{(1)}$, which is something we do not know. Since $\hat{H}^0$ is Hermitian, the set of eigenfunctions of the unperturbed system is a complete set of known functions; hence we expand $\psi_n^{(1)}$ using the unperturbed wave functions:

$$\psi_n^{(1)} = \sum_j a_j \psi_j^{(0)} \tag{9.24}$$

Equation (9.23) becomes

$$\sum_j a_j(\hat{H}^0\psi_j^{(0)} - E_n^{(0)}\psi_j^{(0)}) = (E_n^{(1)} - \hat{H}')\psi_n^{(0)} \tag{9.25}$$

$$\sum_j a_j(E_j^{(0)} - E_n^{(0)})\psi_j^{(0)} = (E_n^{(1)} - \hat{H}')\psi_n^{(0)} \tag{9.26}$$

where we used (9.2). We now multiply (9.26) by $\psi_m^{(0)*}$ and then integrate over all space, obtaining

$$\int \psi_m^{(0)*} \sum_j a_j(E_j^{(0)} - E_n^{(0)})\psi_j^{(0)} \, d\tau = \int \psi_m^{(0)*}(E_n^{(1)} - \hat{H}')\psi_n^{(0)} \, d\tau$$

$$\sum_j a_j(E_j^{(0)} - E_n^{(0)}) \int \psi_m^{(0)*}\psi_j^{(0)} \, d\tau = E_n^{(1)} \int \psi_m^{(0)*}\psi_n^{(0)} \, d\tau - \int \psi_m^{(0)*}\hat{H}'\psi_n^{(0)} \, d\tau$$

$$\sum_j a_j(E_j^{(0)} - E_n^{(0)})\delta_{mj} = E_n^{(1)}\delta_{mn} - \int \psi_m^{(0)*}\hat{H}'\psi_n^{(0)} \, d\tau \tag{9.27}$$

since the unperturbed eigenfunctions are orthonormal:

$$\langle \psi_m^{(0)} | \psi_j^{(0)} \rangle = \delta_{mj} \tag{9.28}$$

Performing the sum, only the term with $j = m$ survives, and we have

$$a_m(E_m^{(0)} - E_n^{(0)}) = E_n^{(1)}\delta_{mn} - \int \psi_m^{(0)*}\hat{H}'\psi_n^{(0)} \, d\tau \tag{9.29}$$

We have two cases to consider: either $m = n$ or $m \neq n$. For $m = n$, the left side of (9.29) vanishes, and we have

$$E_n^{(1)} = \int \psi_n^{(0)*}\hat{H}'\psi_n^{(0)} \, d\tau = \langle \psi_n^{(0)} | \hat{H}' | \psi_n^{(0)} \rangle = H'_{nn} \tag{9.30}$$

Thus we have the not-unreasonable result that the first-order correction to the energy is found by averaging the perturbation $\hat{H}'$ over the appropriate unperturbed wave function. Equation (9.30) is extremely important.

Setting $\lambda = 1$ in (9.18), we have

$$E_n \approx E_n^{(0)} + E_n^{(1)} = E_n^{(0)} + \int \psi_n^{(0)*}\hat{H}'\psi_n^{(0)} \, d\tau \tag{9.31}$$

Let us find the first-order correction to the wave function. For $m \neq n$, Eq. (9.29) becomes

$$a_m(E_m^{(0)} - E_n^{(0)}) = -\int \psi_m^{(0)*} \hat{H}' \psi_n^{(0)} \, d\tau, \qquad m \neq n \qquad (9.32)$$

By hypothesis, the level $E_n^{(0)}$ is nondegenerate, so that $E_m^{(0)} \neq E_n^{(0)}$ for $m \neq n$, and we may divide by $(E_m^{(0)} - E_n^{(0)})$ to obtain

$$a_m = \frac{\int \psi_m^{(0)*} \hat{H}' \psi_n^{(0)} \, d\tau}{E_n^{(0)} - E_m^{(0)}} = \frac{H'_{mn}}{E_n^{(0)} - E_m^{(0)}}, \qquad m \neq n \qquad (9.33)$$

The first-order correction to the wave function is, according to (9.24),

$$\psi_n^{(1)} = \sum_m a_m \psi_m^{(0)} \qquad (9.34)$$

where we changed the dummy summation index. The $\psi_m^{(0)}$ are the known unperturbed wave functions. The coefficients a_m are given by (9.33), except for a_n, the coefficient of $\psi_n^{(0)}$. We might determine a_n by normalizing the perturbed wave function $\psi_n^{(0)} + \psi_n^{(1)}$, but this will not fully specify a_n since we can always multiply a wave function by an arbitrary phase factor $e^{i\alpha}$. A detailed investigation (which we omit) shows that the choice of a_n will not affect the expressions for any of the energy corrections. The conventional procedure is to set a_n (and similar constants which arise in calculating higher-order corrections to the wave function) equal to zero. If desired, the perturbed wave function may be normalized at the end of the calculation. With $a_n = 0$, Eqs. (9.33) and (9.34) give the first-order correction to the wave function as

$$\psi_n^{(1)} = \sum_{m \neq n} \frac{\int \psi_m^{(0)*} \hat{H}' \psi_n^{(0)} \, d\tau}{E_n^{(0)} - E_m^{(0)}} \psi_m^{(0)} \qquad (9.35)$$

Setting $\lambda = 1$ in (9.17), and using just the first-order correction, we have as our approximation to the perturbed wave function

$$\psi_n \approx \psi_n^{(0)} + \sum_{m \neq n} \frac{H'_{mn}}{E_n^{(0)} - E_m^{(0)}} \psi_m^{(0)} \qquad (9.36)$$

The symbol $\sum_{m \neq n}$ in (9.36) means that we sum over all the unperturbed states except state n.

We now consider the second-order correction to the energy. Equating coefficients of the λ^2 terms in (9.20), we have

$$(\hat{H}^0 - E_n^{(0)})\psi_n^{(2)} = E_n^{(2)} \psi_n^{(0)} + (E_n^{(1)} - \hat{H}')\psi_n^{(1)} \qquad (9.37)$$

We expand the second-order correction to the wave function as

$$\psi_n^{(2)} = \sum_j b_j \psi_j^{(0)} \qquad (9.38)$$

and (9.37) becomes [after (9.2) is used]

$$\sum_j b_j(E_j^{(0)} - E_n^{(0)})\psi_j^{(0)} = E_n^{(2)} \psi_n^{(0)} + (E_n^{(1)} - \hat{H}')\psi_n^{(1)} \qquad (9.39)$$

Multiplying by $\psi_m^{(0)*}$ and integrating, we get

$$\sum_j b_j(E_j^{(0)} - E_n^{(0)})\delta_{mj} = E_n^{(2)}\delta_{mn} + E_n^{(1)}\int \psi_m^{(0)*}\psi_n^{(1)}\,d\tau - \int \psi_m^{(0)*}\hat{H}'\psi_n^{(1)}\,d\tau$$

where (9.28) has been used. Performing the sum, we find

$$b_m(E_m^{(0)} - E_n^{(0)}) = E_n^{(2)}\delta_{mn} + E_n^{(1)}\int \psi_m^{(0)*}\psi_n^{(1)}\,d\tau - \int \psi_m^{(0)*}\hat{H}'\psi_n^{(1)}\,d\tau \quad \textbf{(9.40)}$$

To determine $\psi_n^{(2)}$ we would consider the case $m \neq n$. We omit doing so, and only consider $m = n$. With $m = n$, Eq. (9.40) becomes

$$0 = E_n^{(2)} + E_n^{(1)}\int \psi_n^{(0)*}\psi_n^{(1)}\,d\tau - \int \psi_n^{(0)*}\hat{H}'\psi_n^{(1)}\,d\tau \quad \textbf{(9.41)}$$

Note from this last equation that to determine the *second*-order correction to the energy, we only have to know the *first*-order correction to the wave function. In fact, it can be shown that knowledge of $\psi_n^{(1)}$ suffices to determine $E_n^{(3)}$ also; in general, it can be shown that if we know the corrections to the wave function through the kth order, then we can compute the corrections to the energy through order $2k + 1$ (see *Bates*, Vol. I, p. 184).

Resuming consideration of (9.41), we have two integrals to do. From (9.35) we have

$$\psi_n^{(1)} = \sum_{k \neq n} \frac{H'_{kn}}{E_n^{(0)} - E_k^{(0)}}\psi_k^{(0)} \quad \textbf{(9.42)}$$

so that the first integral in (9.41) becomes

$$\int \psi_n^{(0)*}\psi_n^{(1)}\,d\tau = \sum_{k \neq n}\int \psi_n^{(0)*}\frac{H'_{kn}}{E_n^{(0)} - E_k^{(0)}}\psi_k^{(0)}\,d\tau \quad \textbf{(9.43)}$$

The quantities H'_{kn} are constants, since they are the values of certain definite integrals:

$$H'_{kn} \equiv \int \psi_k^{(0)*}\hat{H}'\psi_n^{(0)}\,d\tau$$

We have

$$\int \psi_n^{(0)*}\psi_n^{(1)}\,d\tau = \sum_{k \neq n}\frac{H'_{kn}}{E_n^{(0)} - E_k^{(0)}}\int \psi_n^{(0)*}\psi_k^{(0)}\,d\tau = \sum_{k \neq n}\frac{H'_{kn}}{E_n^{(0)} - E_k^{(0)}}\delta_{nk}$$

Because the sum does not include the term with $k = n$, the Kronecker delta causes all the terms to vanish, and we have

$$\int \psi_n^{(0)*}\psi_n^{(1)}\,d\tau = 0 \quad \textbf{(9.44)}$$

Equation (9.44) is a consequence of our choosing $a_n = 0$.

Equation (9.41) becomes

$$E_n^{(2)} = \int \psi_n^{(0)*}\hat{H}'\psi_n^{(1)}\,d\tau \quad \textbf{(9.45)}$$

Using (9.42) in (9.45), we have

$$E_n^{(2)} = \sum_{k \neq n} \frac{H'_{kn}}{E_n^{(0)} - E_k^{(0)}} \int \psi_n^{(0)*} \hat{H}' \psi_k^{(0)} \, d\tau = \sum_{k \neq n} \frac{H'_{kn} H'_{nk}}{E_n^{(0)} - E_k^{(0)}}$$

Since $\hat{H}'$ is Hermitian, we have $H'_{nk} = (H'_{kn})^*$; and

$$E_n^{(2)} = \sum_{k \neq n} \frac{|H'_{kn}|^2}{E_n^{(0)} - E_k^{(0)}} = \sum_{k \neq n} \frac{|\langle \psi_k^{(0)} | \hat{H}' | \psi_n^{(0)} \rangle|^2}{E_n^{(0)} - E_k^{(0)}} \qquad \textbf{(9.46)}$$

It might be thought that the result (9.46) depended on the choice of $a_n = 0$, since with $a_n \neq 0$, Eq. (9.44) is no longer true. However, if instead of (9.42) we had written

$$\psi_n^{(1)} = a_n \psi_n^{(0)} + \sum_{k \neq n} \frac{H'_{kn}}{E_n^{(0)} - E_k^{(0)}} \psi_k^{(0)} \qquad \textbf{(9.47)}$$

where a_n is left completely unspecified, and had used (9.47) in (9.41), we would still have come up with Eq. (9.46) for $E_n^{(2)}$. (See Problem 9.4.)

Using (9.46), our approximation to the energy of the perturbed state becomes [Eq. (9.18) with $\lambda = 1$]

$$E_n \approx E_n^{(0)} + H'_{nn} + \sum_{k \neq n} \frac{|H'_{nk}|^2}{E_n^{(0)} - E_k^{(0)}} \qquad \textbf{(9.48)}$$

For formulas for higher-order corrections to the energy, see *Bates*, Vol. I, pp. 181–185. (The form of perturbation theory developed in this section is known as Rayleigh–Schroedinger perturbation theory; other approaches exist.)

Let us discuss our results, Eqs. (9.48) and (9.36). From (9.36) we see that the effect of the perturbation on the wave function $\psi_n^{(0)}$ is to "mix in" contributions from other states $\psi_m^{(0)}$, $m \neq n$. Because of the factor $1/(E_n^{(0)} - E_m^{(0)})$, the most important contributions (aside from $\psi_n^{(0)}$) to the perturbed wave function come from states nearest in energy to state n.

To evaluate the first-order correction to the energy, we only have to evaluate the single integral H'_{nn}, whereas to evaluate the second-order energy correction, we must evaluate the matrix elements of $\hat{H}'$ between the nth state and all other states k, and then perform the infinite sum in (9.46). In many cases it is impossible to evaluate exactly the second-order energy correction. Third-order and higher-order energy corrections are even more difficult to deal with.

The sums in (9.36) and (9.48) are sums over different states rather than sums over different energy values. If some of the energy levels (other than the nth) are degenerate, we must include a term in the sums for each linearly independent wave function corresponding to the degenerate levels.

The reason we have a sum over states in (9.36) and (9.48) is that we require a complete set of functions for the expansion (9.38), and therefore we must include all linearly independent wave functions in the sum. If the unperturbed problem has continuum wave functions (e.g., the hydrogen atom), we must also include an integration over the continuum functions, if we are to have a complete set. If $\psi_E^{(0)}$

denotes an unperturbed continuum wave function of energy $E^{(0)}$, then (9.42) and (9.46) become

$$\psi_n^{(1)} = \sum_{k \neq n} \frac{H'_{kn}}{E_n^{(0)} - E_k^{(0)}} \psi_k^{(0)} + \int \frac{H'_{E,n}}{E_n^{(0)} - E^{(0)}} \psi_E^{(0)} \, dE^{(0)} \qquad \textbf{(9.42a)}$$

$$E_n^{(2)} = \sum_{k \neq n} \frac{|H'_{kn}|^2}{E_n^{(0)} - E_k^{(0)}} + \int \frac{|H'_{E,n}|^2}{E_n^{(0)} - E^{(0)}} \, dE^{(0)} \qquad \textbf{(9.46a)}$$

$$H'_{E,n} = \int \psi_E^{(0)*} \hat{H}' \psi_n^{(0)} \, d\tau$$

(Sometimes the symbol S is used to indicate summation over the discrete states and integration over the continuum states.) The integrals in (9.42a) and (9.46a) are over the range of continuum-state energies, e.g., from zero to infinity for the hydrogen atom. The existence of continuum states in the unperturbed problem makes evaluation of the second-order energy correction even more difficult.

Note that the unperturbed wave functions are normalized [Eq. (9.28)]. Hence in using (9.30) and (9.46), we must use *normalized* unperturbed wave functions.

9.3   PERTURBATION TREATMENT OF THE GROUND STATE OF THE HELIUM ATOM

The helium atom has two electrons and a nucleus of charge $+2e$. We will consider the nucleus to be at rest (Section 6.5) and will place the origin of the coordinate system at the nucleus. The coordinates of electrons 1 and 2 are (x_1, y_1, z_1) and (x_2, y_2, z_2); see Fig. 9.1.

If we take the nuclear charge to be $+Ze$ instead of $+2e$, we can treat helium-like ions such as H^-, Li^+, Be^{2+}. The Hamiltonian operator is

$$\hat{H} = -\frac{\hbar^2}{2m} \nabla_1^2 - \frac{\hbar^2}{2m} \nabla_2^2 - \frac{Ze'^2}{r_1} - \frac{Ze'^2}{r_2} + \frac{e'^2}{r_{12}} \qquad \textbf{(9.49)}$$

where m is the mass of the electron, r_1 and r_2 are the distances of electrons 1 and 2 from the nucleus, and r_{12} is the distance from electron 1 to 2. The first two terms are

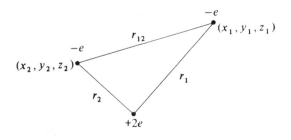

Figure 9.1 The helium atom

the operators for the electrons' kinetic energy; the third and fourth terms are the potential energies of attraction between the electrons and the nucleus; the final term is the potential energy of interelectronic repulsion [Eq. (6.66)]. (Note that the potential energy of a system of interacting particles cannot be written as the sum of potential energies of the individual particles; the potential energy is a property of the system as a whole.)

The Schroedinger equation involves six independent variables, three coordinates for each electron. In spherical polar coordinates,

$$\psi = \psi(r_1, \theta_1, \varphi_1, r_2, \theta_2, \varphi_2) \tag{9.50}$$

The operator ∇_1^2 is given by Eq. (6.6) with r_1, θ_1, φ_1 replacing r, θ, φ. The variable r_{12} in Cartesian coordinates is $r_{12} = [(x_1 - x_2)^2 + (y_1 - y_2)^2 + (z_1 - z_2)^2]^{1/2}$, and by using the relations between Cartesian and spherical polar coordinates, we can express r_{12} in terms of the coordinates of (9.50).

Because of the $1/r_{12}$ term, the Schroedinger equation for helium cannot be separated in any coordinate system, and we must use approximation methods. To use the perturbation method, we must separate the Hamiltonian (9.49) into two parts, $\hat{H}^0$ and $\hat{H}'$, where $\hat{H}^0$ is the Hamiltonian of an exactly solvable problem. If we choose

$$\hat{H}^0 = -\frac{\hbar^2}{2m}\nabla_1^2 - \frac{Ze'^2}{r_1} - \frac{\hbar^2}{2m}\nabla_2^2 - \frac{Ze'^2}{r_2} \tag{9.51}$$

$$\hat{H}' = \frac{e'^2}{r_{12}} \tag{9.52}$$

our unperturbed Hamiltonian is the sum of two hydrogenlike Hamiltonians, one for each electron:

$$\hat{H}^0 = \hat{H}_1^0 + \hat{H}_2^0 \tag{9.53}$$

$$\hat{H}_1^0 \equiv -\frac{\hbar^2}{2m}\nabla_1^2 - \frac{Ze'^2}{r_1}, \qquad \hat{H}_2^0 \equiv -\frac{\hbar^2}{2m}\nabla_2^2 - \frac{Ze'^2}{r_2} \tag{9.54}$$

The unperturbed system is a helium atom in which the two electrons exert no forces on each other. Although there is no physical reality to such a system, this does not prevent us from applying the mathematics of perturbation theory to this system to obtain an approximation to the actual helium atom.

Since the unperturbed Hamiltonian (9.53) is the sum of the Hamiltonians for two independent particles, we can use the results of Eqs. (6.43)–(6.49) to conclude that the unperturbed wave functions have the form

$$\psi^{(0)}(r_1, \theta_1, \varphi_1, r_2, \theta_2, \varphi_2) = F_1(r_1, \theta_1, \varphi_1)F_2(r_2, \theta_2, \varphi_2) \tag{9.55}$$

and the unperturbed energies are

$$E^{(0)} = E_1 + E_2 \tag{9.56}$$

$$\hat{H}_1^0 F_1 = E_1 F_1, \qquad \hat{H}_2^0 F_2 = E_2 F_2 \tag{9.57}$$

Since the Hamiltonians $\hat{H}_1^0$ and $\hat{H}_2^0$ are hydrogenlike Hamiltonians, the solutions of

(9.57) are the hydrogenlike eigenfunctions and eigenvalues. From Eq. (6.107) we have

$$E_1 = -\frac{Z^2}{n_1^2}\frac{e'^2}{2a_0}, \qquad E_2 = -\frac{Z^2}{n_2^2}\frac{e'^2}{2a_0} \tag{9.58}$$

$$E^{(0)} = -Z^2\left(\frac{1}{n_1^2} + \frac{1}{n_2^2}\right)\frac{e'^2}{2a_0}, \qquad \begin{array}{l} n_1 = 1, 2, 3, \ldots \\ n_2 = 1, 2, 3, \ldots \end{array} \tag{9.59}$$

where a_0 is the Bohr radius. Equation (9.59) gives the zeroth-order energies of states with both electrons bound to the nucleus; we also have continuum states.

The lowest level has $n_1 = 1$, $n_2 = 1$, and its zeroth-order wave function is [Eq. (6.120)]

$$\psi_{1s^2}^{(0)} = \frac{1}{\pi^{1/2}}\left(\frac{Z}{a_0}\right)^{3/2}e^{-Zr_1/a_0}\cdot\frac{1}{\pi^{1/2}}\left(\frac{Z}{a_0}\right)^{3/2}e^{-Zr_2/a_0} \tag{9.60}$$

where the subscript indicates that both electrons are in hydrogenlike $1s$ orbitals. (Note that the procedure of assigning electrons to orbitals and writing the atomic wave function as the product of one-electron orbital functions is an *approximation*.) The energy of this unperturbed ground state is

$$E_{1s^2}^{(0)} = -Z^2(2)\frac{e'^2}{2a_0} \tag{9.61}$$

The quantity $-\frac{1}{2}e'^2/a_0$ is the ground-state energy of the hydrogen atom (taking the nucleus to be infinitely heavy) and equals -13.606 eV (Section 6.5). For helium $Z = 2$, and

$$E_{1s^2}^{(0)} = -108.8 \text{ eV} \tag{9.62}$$

How does this zeroth-order energy compare with the true ground-state energy of helium? The experimental first ionization potential of He is 24.6 volts. The second ionization potential of He is easily calculated theoretically, since it is the ionization potential of the hydrogenlike ion He$^+$ and is equal to $2^2(13.606$ volts$)$ $= 54.4$ volts. If we choose the zero of energy as the completely ionized atom [this choice is implicit in (9.49)], then the ground-state energy of the helium atom is $-(24.6 + 54.4)$ electron volts $= -79.0$ eV. Thus the zeroth-order energy is in error by 38 percent. We should have expected such a large error, since the perturbation term e'^2/r_{12} is not particularly small.

The next step is to evaluate the first-order perturbation correction to the energy. The unperturbed ground state is nondegenerate, and we have

$$E^{(1)} = \int \psi^{(0)*}\hat{H}'\psi^{(0)}\,d\tau$$

$$E^{(1)} = \frac{Z^6 e'^2}{\pi^2 a_0^6}\int_0^{2\pi}\int_0^{2\pi}\int_0^{\pi}\int_0^{\pi}\int_0^{\infty}\int_0^{\infty}e^{-2Zr_1/a_0}e^{-2Zr_2/a_0}\frac{1}{r_{12}}r_1^2\sin\theta_1$$

$$\times r_2^2\sin\theta_2\,dr_1\,dr_2\,d\theta_1\,d\theta_2\,d\varphi_1\,d\varphi_2 \tag{9.63}$$

The volume element for this two-electron problem contains the coordinates of both electrons, $d\tau = d\tau_1\, d\tau_2$. [The evaluation of the integral may be skimmed, if desired; begin reading again at Eq. (9.67).]

We must now express $1/r_{12}$ in terms of the spherical polar coordinates of the two electrons. If we try using

$$1/r_{12} = [(x_1 - x_2)^2 + (y_1 - y_2)^2 + (z_1 - z_2)^2]^{-1/2}$$

together with the relations between Cartesian and spherical polar coordinates, we will find that we will be unable to do the integration. Instead, we use an expansion of $1/r_{12}$ in terms of spherical harmonics. We will not derive this expansion[1] but will simply write it down and use it:

$$\frac{1}{r_{12}} = \begin{cases} \dfrac{1}{r_1}\displaystyle\sum_{l=0}^{\infty}\sum_{m=-l}^{l}\frac{4\pi}{2l+1}\left(\frac{r_2}{r_1}\right)^l [Y_l^m(\theta_1,\varphi_1)]^* Y_l^m(\theta_2,\varphi_2) & \text{if } r_1 \geqslant r_2 \\[3mm] \dfrac{1}{r_2}\displaystyle\sum_{l=0}^{\infty}\sum_{m=-l}^{l}\frac{4\pi}{2l+1}\left(\frac{r_1}{r_2}\right)^l [Y_l^m(\theta_1,\varphi_1)]^* Y_l^m(\theta_2,\varphi_2) & \text{if } r_2 \geqslant r_1 \end{cases} \qquad (9.64)$$

Although the expansion looks formidable, it makes evaluation of the integral relatively simple. Using (9.64) and $Y_0^0 = 1/(4\pi)^{1/2}$, we have

$$E^{(1)} = \frac{16Z^6 e'^2}{a_0^6}\sum_{l=0}^{\infty}\sum_{m=-l}^{l}\frac{1}{2l+1}\int_0^{\infty}\int_0^{\infty} e^{-2Zr_1/a_0}e^{-2Zr_2/a_0}\frac{r_<^l}{r_>^{l+1}}r_1^2 r_2^2\, dr_1\, dr_2$$

$$\times \int_0^{2\pi}\int_0^{\pi} [Y_l^m(\theta_1,\varphi_1)]^* Y_0^0(\theta_1,\varphi_1)\sin\theta_1\, d\theta_1\, d\varphi_1$$

$$\times \int_0^{2\pi}\int_0^{\pi} [Y_0^0(\theta_2,\varphi_2)]^* Y_l^m(\theta_2,\varphi_2)\sin\theta_2\, d\theta_2\, d\varphi_2 \qquad (9.65)$$

where $r_<$ means the smaller of r_1 and r_2 and where $r_>$ means the larger of r_1 and r_2. Recalling the orthonormality of the spherical harmonics [Eq. (7.46)], we have

$$E^{(1)} = \frac{16Z^6 e'^2}{a_0^6}\sum_{l=0}^{\infty}\sum_{m=-l}^{l}\frac{1}{2l+1}\int_0^{\infty}\int_0^{\infty}(\cdots)\, dr_1\, dr_2\, \delta_{l,0}\,\delta_{m,0}\,\delta_{l,0}\,\delta_{m,0}$$

The Kronecker deltas cause all terms to vanish except the single term with $m = 0 = l$, so that

$$E^{(1)} = \frac{16Z^6 e'^2}{a_0^6}\int_0^{\infty}\int_0^{\infty} e^{-2Zr_1/a_0}e^{-2Zr_2/a_0}\frac{1}{r_>}r_1^2 r_2^2\, dr_1\, dr_2$$

If we integrate first over r_1, then in the range $0 \leqslant r_1 \leqslant r_2$, we have $r_> = r_2$; in the range $r_2 \leqslant r_1 \leqslant \infty$, we have $r_> = r_1$. Thus

$$E^{(1)} = \frac{16Z^6 e'^2}{a_0^6}\int_0^{\infty} e^{-2Zr_2/a_0}r_2^2\left[\int_0^{r_2} e^{-2Zr_1/a_0}\frac{r_1^2}{r_2}\, dr_1 + \int_{r_2}^{\infty} e^{-2Zr_1/a_0}\frac{r_1^2}{r_1}\, dr_1\right] dr_2$$

$$E^{(1)} = \frac{16Z^6 e'^2}{a_0^6}\int_0^{\infty} e^{-2Zr_2/a_0}r_2\left(\int_0^{r_2} e^{-2Zr_1/a_0}r_1^2\, dr_1\right) dr_2$$

$$+ \frac{16Z^6 e'^2}{a_0^6}\int_0^{\infty} e^{-2Zr_2/a_0}r_2^2\left(\int_{r_2}^{\infty} e^{-2Zr_1/a_0}r_1\, dr_1\right) dr_2 \qquad (9.66)$$

[1] *Eyring, Walter, and Kimball*, p. 369.

Using the indefinite integrals (A.2) and (A.3) in the Appendix, we can do the r_1 integrals to obtain r_2 integrals, which are evaluated using (A.4). The result is

$$E^{(1)} = \frac{5Z}{8}\left(\frac{e'^2}{a_0}\right) \qquad (9.67)$$

Recalling that $\frac{1}{2}e'^2/a_0 = 13.606$ eV and putting $Z = 2$, we find for the first-order perturbation energy correction for the helium ground state

$$E^{(1)} = \tfrac{10}{4}(13.606 \text{ eV}) = 34.0 \text{ eV} \qquad (9.68)$$

Our approximation to the energy is now

$$E^{(0)} + E^{(1)} = -108.8 \text{ eV} + 34.0 \text{ eV} = -74.8 \text{ eV} \qquad (9.69)$$

which, compared with the experimental value -79.0 eV, is in error by 5.3 percent.

To evaluate the first-order correction to the wave function and higher-order corrections to the energy requires evaluating the matrix elements of $1/r_{12}$ between the ground unperturbed state and all excited states (including the continuum) and performing the appropriate summations and integrations. No one has yet figured out how to evaluate directly all of the contributions to $E^{(2)}$. Note that the effect of $\psi^{(1)}$ is to mix into the wave function contributions from other configurations besides $1s^2$; we call this *configuration interaction*. The largest contribution to the true ground-state wave function of He will naturally come from the $1s^2$ configuration, which is the unperturbed (zeroth-order) wave function.

It is possible to obtain very accurate estimates of $E^{(2)}$ and higher-order energies by an indirect method that is a combination of the variation and perturbation methods. We will outline this method, which was developed and applied to helium by Hylleraas. Consider the integral

$$\int [\varphi^{(1)*}(\hat{H}^0 - E^{(0)})\varphi^{(1)} + \varphi^{(1)*}(\hat{H}' - E^{(1)})\psi^{(0)} + \psi^{(0)*}(\hat{H}' - E^{(1)})\varphi^{(1)}]\, d\tau \quad (9.70)$$

where $\varphi^{(1)}$ is any well-behaved function satisfying the appropriate boundary conditions and where the other symbols have their usual meanings. Suppose we are lucky enough to guess a function $\varphi^{(1)}$ that is the true first-order correction to the wave function; setting $\varphi^{(1)} = \psi^{(1)}$ and using Eq. (9.23), we see that (9.70) becomes

$$\int \psi^{(0)*}[\hat{H}' - E^{(1)}]\psi^{(1)}\, d\tau \qquad (9.71)$$

which [Eq. (9.41)] is $E^{(2)}$, the second-order energy correction. Moreover, it is possible to prove (see *Hameka*, p. 223) that for any function $\varphi^{(1)}$ satisfying the boundary conditions, the integral (9.70) is equal to or greater than the second-order correction to the ground-state energy. We can therefore carry out a variational procedure to obtain an upper bound to $E^{(2)}$ for the ground state. We take $\varphi^{(1)}$ to be a trial function containing several parameters, and we vary the parameters so as to minimize the integral (9.70). This integral is then an approximation to $E^{(2)}$, and $\varphi^{(1)}$ is an approximation to $\psi^{(1)}$. We can then use $\varphi^{(1)}$ to obtain an approximation to $E^{(3)}$ also. Similar variational integrals may be used to find approximations to higher-order corrections to the ground-state wave function and energy.

Hylleraas's work on helium using this *variation–perturbation* method has been extended by Scherr and Knight,[2] who used 100-term trial functions to obtain extremely accurate approximations to the wave-function corrections through sixth order and thus to the energy corrections through thirteenth order! The second-order correction $E^{(2)}$ turns out to be -4.3 eV, and $E^{(3)}$ is $+0.1$ eV. Through third order, we have for the ground-state energy

$$E \approx -108.8\,\text{eV} + 34.0\,\text{eV} - 4.3\,\text{eV} + 0.1\,\text{eV} = -79.0\,\text{eV} \qquad (9.72)$$

which agrees (within 0.1 eV) with the experimental value -79.0 eV. Including corrections through thirteenth order, Scherr and Knight obtained a ground-state energy of helium of $-2.90372433(e'^2/a_0)$, which is nearly as good as the value $-2.90372438(e'^2/a_0)$ obtained from the purely variational calculations described in the next section.

9.4   VARIATION TREATMENTS OF THE GROUND STATE OF HELIUM

Although it might seem that the variation and perturbation methods represent quite different approaches, they are actually related in several ways. As one example, suppose we are interested in the ground-state energy of a system with Hamiltonian $\hat{H}$. The perturbation approach would first split $\hat{H}$ into two parts, $\hat{H} = \hat{H}^0 + \hat{H}'$, and then solve the unperturbed problem

$$\hat{H}^0 \psi_g^{(0)} = E_g^{(0)} \psi_g^{(0)} \qquad (9.73)$$

where the subscript g indicates the ground state. Including only the first-order energy correction, the perturbation-theory approximation to the true ground-state energy is

$$E_g \approx E_g^{(0)} + \int \psi_g^{(0)*} \hat{H}' \psi_g^{(0)} \, d\tau \qquad (9.74)$$

The variation method would try various normalized functions φ to minimize the variational integral

$$\int \varphi^* \hat{H} \varphi \, d\tau \geq E_g \qquad (9.75)$$

Suppose we used a variation function that was the zeroth-order ground-state wave function of the perturbation treatment: $\varphi = \psi_g^{(0)}$. The variational integral would then be

$$\int \varphi^* \hat{H} \varphi \, d\tau = \int \psi_g^{(0)*} (\hat{H}^0 + \hat{H}') \psi_g^{(0)} \, d\tau = \int \psi_g^{(0)*} (E_g^{(0)} + \hat{H}') \psi_g^{(0)} \, d\tau$$

$$\int \varphi^* \hat{H} \varphi \, d\tau = E_g^{(0)} + \int \psi_g^{(0)*} \hat{H}' \psi_g^{(0)} \, d\tau \qquad (9.76)$$

---

[2] C. W. Scherr and R. E. Knight, *Rev. Mod. Phys.*, **35**, 436 (1963). For calculations of the energy corrections through twenty-first order, see J. Midtdal, *Phys. Rev.*, **138**, A1010 (1965).

since $\psi_g^{(0)}$ is normalized. Comparison with the perturbation-theory expression (9.74) shows that the variation method gives the same result if we use $\psi_g^{(0)}$ as the trial function.

With this discussion as background, we consider variation functions for the helium ground state. If we used $\psi_g^{(0)}$ [Eq. (9.60)] for the trial function, we would get the first-order perturbation result -74.8 eV. To improve on this result, we use a trial function of the form of (9.60) but having a variational parameter. We try the normalized function

$$\varphi = \frac{1}{\pi}\left(\frac{\zeta}{a_0}\right)^3 e^{-\zeta r_1/a_0}\, e^{-\zeta r_2/a_0} \qquad (9.77)$$

which is obtained from (9.60) by replacing the true atomic number Z by a variational parameter ζ (zeta). ζ has a simple physical interpretation. Since one electron tends to screen the other from the nucleus, each electron is subject to an effective nuclear charge somewhat less than the full nuclear charge Z. If one electron fully shielded the other from the nucleus, we would have an effective nuclear charge of $Z-1$; since both electrons are in the same orbital, they will be only partly effective in shielding each other. We thus expect ζ to lie between $Z-1$ and Z.

We now evaluate the variational integral. To expedite things, we rewrite the Hamiltonian (9.49) as

$$\hat{H} = \left[-\frac{\hbar^2}{2m}\nabla_1^2 - \frac{\zeta e'^2}{r_1} - \frac{\hbar^2}{2m}\nabla_2^2 - \frac{\zeta e'^2}{r_2}\right] + (\zeta - Z)\frac{e'^2}{r_1} + (\zeta - Z)\frac{e'^2}{r_2} + \frac{e'^2}{r_{12}} \qquad (9.78)$$

where we have added and subtracted the terms involving zeta. The terms in brackets in (9.78) are the sum of two hydrogenlike Hamiltonians for nuclear charge ζ; moreover, the trial function (9.77) is the product of two hydrogenlike $1s$ functions for nuclear charge ζ. Therefore, when these terms operate on φ, we will have an eigenvalue equation, the eigenvalue being the sum of two hydrogenlike $1s$ energies for nuclear charge ζ:

$$\left[-\frac{\hbar^2}{2m}\nabla_1^2 - \frac{\zeta e'^2}{r_1} - \frac{\hbar^2}{2m}\nabla_2^2 - \frac{\zeta e'^2}{r_2}\right]\varphi = -\zeta^2(2)\frac{e'^2}{2a_0}\varphi \qquad (9.79)$$

Using (9.78) and (9.79), we have

$$\int \varphi^*\hat{H}\varphi\, d\tau = -\zeta^2\frac{e'^2}{a_0}\int \varphi^*\varphi\, d\tau + (\zeta - Z)e'^2\int\frac{\varphi^*\varphi}{r_1}\, d\tau$$

$$+ (\zeta - Z)e'^2\int\frac{\varphi^*\varphi}{r_2}\, d\tau + e'^2\int\frac{\varphi^*\varphi}{r_{12}}\, d\tau \qquad (9.80)$$

Let f_1 be a normalized $1s$ hydrogenlike orbital for nuclear charge ζ, occupied by electron 1; let f_2 be the same function for electron 2:

$$f_1 = \frac{1}{\pi^{1/2}}\left(\frac{\zeta}{a_0}\right)^{3/2} e^{-\zeta r_1/a_0}, \qquad f_2 = \frac{1}{\pi^{1/2}}\left(\frac{\zeta}{a_0}\right)^{3/2} e^{-\zeta r_2/a_0} \qquad (9.81)$$

Noting that $\varphi = f_1 f_2$, we now evaluate the integrals in (9.80):

$$\int \varphi^* \varphi \, d\tau = \int \int f_1^* f_2^* f_1 f_2 \, d\tau_1 \, d\tau_2 = \int f_1^* f_1 \, d\tau_1 \int f_2^* f_2 \, d\tau_2 = 1$$

$$\int \frac{\varphi^* \varphi}{r_1} \, d\tau = \int \frac{f_1^* f_1}{r_1} \, d\tau_1 \int f_2^* f_2 \, d\tau_2 = \int \frac{f_1^* f_1}{r_1} \, d\tau_1$$

$$\int \frac{\varphi^* \varphi}{r_1} \, d\tau = \frac{1}{\pi} \frac{\zeta^3}{a_0^3} \int_0^\infty e^{-2\zeta r_1 / a_0} \frac{r_1^2}{r_1} \, dr_1 \int_0^\pi \sin \theta_1 \, d\theta_1 \int_0^{2\pi} d\varphi_1 = \frac{\zeta}{a_0}$$

where the Appendix integral (A.4) was used. Also

$$\int \frac{\varphi^* \varphi}{r_2} \, d\tau = \int \frac{f_2^* f_2}{r_2} \, d\tau_2 = \int \frac{f_1^* f_1}{r_1} \, d\tau_1 = \frac{\zeta}{a_0} \tag{9.82}$$

since it doesn't matter whether we use the label 1 or 2 on the dummy variables in the definite integral. Finally we must evaluate

$$e'^2 \int \frac{\varphi^* \varphi}{r_{12}} \, d\tau \tag{9.83}$$

We recognize (9.83) as the same integral (9.63) that occurred in the perturbation treatment, except that Z is replaced by ζ; hence from (9.67)

$$e'^2 \int \frac{\varphi^* \varphi}{r_{12}} \, d\tau = \frac{5\zeta e'^2}{8 a_0} \tag{9.84}$$

The variational integral (9.80) thus has the value

$$\int \varphi^* \hat{H} \varphi \, d\tau = (\zeta^2 - 2Z\zeta + \tfrac{5}{8}\zeta) \frac{e'^2}{a_0} \tag{9.85}$$

As a check, if we set $\zeta = Z$ in (9.85), we get the first-order perturbation-theory result, (9.61) plus (9.67).

We now vary ζ to minimize the variational integral:

$$\frac{\partial}{\partial \zeta} \int \varphi^* \hat{H} \varphi \, d\tau = (2\zeta - 2Z + \tfrac{5}{8}) \frac{e'^2}{a_0} = 0$$

$$\zeta = Z - \tfrac{5}{16} \tag{9.86}$$

As anticipated, the effective nuclear charge lies between Z and $Z - 1$. Using (9.86) and (9.85), we get

$$\int \varphi^* \hat{H} \varphi \, d\tau = (-Z^2 + \tfrac{5}{8}Z - \tfrac{25}{256}) \frac{e'^2}{a_0} = -(Z - \tfrac{5}{16})^2 \frac{e'^2}{a_0} \tag{9.87}$$

Putting $Z = 2$, we get as our approximation to the ground-state energy $-(27/16)^2 e'^2 / a_0 = -(729/256)27.21 \text{ eV} = -77.5 \text{ eV}$, as compared to the true value -79.0 eV. Use of ζ instead of Z has reduced the error from 5.3 percent to 1.9 percent. In accord with the variation theorem, the true ground-state energy is less than the variational integral.

How can we improve our variational result? We might try a function that had the general form of (9.77), i.e., a product of two functions, one for each electron:

$$\varphi = g(1)g(2) \tag{9.88}$$

However, we could try a variety of functions g in (9.88), instead of the single exponential used in (9.77). A systematic procedure for finding the function g that gives the lowest value of the variational integral will be discussed in Section 11.1. This procedure shows that for the best possible choice of g in (9.88), the variational integral equals -77.9 eV, which is still in error by 1.4 percent. We might ask why (9.88) does not cause the variational integral to converge to the true ground-state energy, no matter what form we try for g. The answer is that when we write the trial function as the product of separate functions for each electron, we are making an approximation. Because of the e'^2/r_{12} term in the Hamiltonian, the Schroedinger equation for helium is not separable, and the true ground-state wave function cannot be written as the product of separate functions for each electron. To reach the true ground-state energy, we must go beyond a function of the form (9.88).

The Bohr model gave the correct energies for the hydrogen atom but failed when applied to helium. Hence in the early days of quantum mechanics, it was important to show that the new theory could give an accurate treatment of helium. The pioneering work on the helium ground state was done by Hylleraas in the years 1928–1930. He used variational functions that contained the interelectronic distance r_{12} explicitly. This provides an effective way of taking into account the effects of one electron on the motion of the other. One function Hylleraas used is

$$\varphi = N[e^{-\zeta r_1/a_0}e^{-\zeta r_2/a_0}(1 + br_{12})] \tag{9.89}$$

where N is the normalization constant and ζ and b are the variational parameters. Since

$$r_{12} = [(x_2 - x_1)^2 + (y_2 - y_1)^2 + (z_2 - z_1)^2]^{1/2} \tag{9.90}$$

the function (9.89) goes beyond the simple product form (9.88). Minimization of the variational integral with respect to the parameters gives $\zeta = 1.849$, $b = 0.364$, and a ground-state energy of -78.7 eV, in error by 0.3 eV. The $1 + br_{12}$ term makes the wave function larger for large values of r_{12}; this is as it should be, because the repulsion between the electrons makes it energetically more favorable for the electrons to avoid each other. A more complicated function used by Hylleraas is

$$\varphi = e^{-\zeta r_1/a_0}e^{-\zeta r_2/a_0}\sum_{i,j,k}c_{ijk}(r_1 + r_2)^i(r_1 - r_2)^j r_{12}^k \tag{9.91}$$

where the summation is over nonnegative integral values of i, j, and k. Using a function containing six terms in the sum, Hylleraas obtained an energy only 0.01 eV above the true ground-state energy.

Hylleraas's work has been extended by others. Pekeris[3] used a function of the form (9.91) containing 1078(!!) terms. The calculations involved a determinant of order 1078 and were carried out on an electronic computer. Pekeris found a ground-

---

[3] C. L. Pekeris, *Phys. Rev.*, **115**, 1216 (1959).

state energy of $-2.903724375(e'^2/a_0)$. With relativistic corrections and corrections for nuclear motion, this gave an ionization energy of $198{,}310.69\,\mathrm{cm}^{-1}$ compared to the experimental value $198{,}310.82 \pm 0.15\,\mathrm{cm}^{-1}$. Schwartz[4] used half-integral as well as integral values for i in (9.91); with a 189-term function, he bettered Pekeris's result by obtaining $-2.903724376(e'^2/a_0)$. Schwartz estimates his result to be within $0.000000001(e'^2/a_0)$ of the true nonrelativistic ground-state energy.

A variational calculation on the Li ground state using a 60-term function and including r_{12}, r_{13}, and r_{23} explicitly in the trial function[5] gave an energy of $-7.47802(e'^2/a_0)$ as compared to the experimental value $-7.47807(e'^2/a_0)$. Variational calculations with functions involving r_{ij} become quite difficult for many-electron atoms because of the large number of terms and difficult integrals that must be dealt with.

Approximation of the helium ground-state wave function as a product of hydrogenlike $1s$ orbitals with effective nuclear charge ζ [Eq. (9.77)] provides a simple physical picture that is in accord with the usual chemical concepts; however, this function does not give a very accurate value for the energy. When we use a function like (9.91) with a hundred or so terms, we get an extremely accurate wave function and energy, but we lose the simple physical interpretation. This is a general occurrence in quantum chemistry; "the more accurate the calculations become the more the concepts tend to vanish into thin air."[6]

9.5   PERTURBATION THEORY FOR A DEGENERATE ENERGY LEVEL

We now consider the perturbation treatment of an energy level that is n-fold degenerate. We have n linearly independent unperturbed wave functions corresponding to the degenerate level. We shall use the labels $1, 2, \ldots, n$ for the states of the degenerate level; we do not mean to imply that these are necessarily the lowest lying states.

The unperturbed Schroedinger equation is

$$\hat{H}^0 \psi_j^{(0)} = E_j^{(0)} \psi_j^{(0)} \tag{9.92}$$

and we have

$$E_1^{(0)} = E_2^{(0)} = \cdots = E_n^{(0)} \tag{9.93}$$

For the perturbed problem, we have

$$\hat{H}\psi_j = E_j \psi_j \tag{9.94}$$

$$\hat{H} = \hat{H}^0 + \lambda \hat{H}' \tag{9.95}$$

As λ goes to zero, the eigenvalues of (9.94) go to the eigenvalues of (9.92). Figure 9.2 illustrates this for a hypothetical system with six states and a threefold-degenerate unperturbed level. Note that the effect of the perturbation is generally to split the

---

[4] C. Schwartz, *Phys. Rev.*, **128**, 1146 (1962). See also K. Frankowski and C. L. Pekeris, *Phys. Rev.*, **146**, 46 (1966).

[5] S. Larsson, *Phys. Rev.*, **169**, 49 (1968).

[6] R. S. Mulliken, *J. Chem. Phys.*, **43**, S2 (1965).

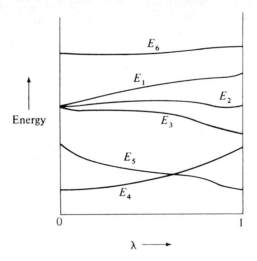

Figure 9.2 *Effect of a perturbation on energy levels*

degenerate energies. In some cases, the perturbation may have no effect on the degeneracy or may only partly remove the degeneracy. Figure 9.2 is a *correlation diagram*; it shows the correspondence between states of the perturbed and unperturbed problems. Later on, we will consider correlation diagrams relating the states of molecules to the states of the atoms from which they are formed. (The word *correlation* is also used to refer to the effects of one electron on the motion of another; do not confuse the two different usages.)

We express the correspondence between the perturbed and unperturbed energies by writing

$$\lim_{\lambda \to 0} E_j = E_j^{(0)} \qquad (9.96)$$

As λ goes to zero, the eigenfunctions satisfying (9.94) approach eigenfunctions satisfying (9.92). Does this mean that $\lim_{\lambda \to 0} \psi_j = \psi_j^{(0)}$? In general, the answer is no. If $E_j^{(0)}$ is nondegenerate, there is a unique normalized eigenfunction $\psi_j^{(0)}$ of $\hat{H}^0$ with eigenvalue $E_j^{(0)}$, and we can be sure that $\lim_{\lambda \to 0} \psi_j = \psi_j^{(0)}$. However, if $E_j^{(0)}$ is the eigenvalue of the n-fold degenerate level, then (Section 3.6) any function of the form

$$c_1 \psi_1^{(0)} + c_2 \psi_2^{(0)} + \cdots + c_n \psi_n^{(0)} \qquad (9.97)$$

will be a solution of (9.92) with eigenvalue $E_j^{(0)}$. The set of linearly independent normalized functions $\psi_1^{(0)}, \psi_2^{(0)}, \ldots, \psi_n^{(0)}$, which we use as eigenfunctions corresponding to the states of the degenerate level, is not unique; using (9.97), we can construct an infinite number of sets of n linearly independent normalized eigenfunctions for the degenerate level. As far as the unperturbed problem is concerned, one such set is as good as another. For example, for the three degenerate $2p$ states of the hydrogen atom, we can use the $2p_1$, $2p_0$, and $2p_{-1}$ functions, the $2p_x$, $2p_y$, and $2p_z$ functions, or some other set of three linearly independent functions constructed as linear combinations of the members of one of the preceding sets. For the perturbed eigenfunctions that correspond to the n-fold degenerate unperturbed level, all we can say is that

$$\lim_{\lambda \to 0} \psi_j = \sum_{i=1}^{n} c_i \psi_i^{(0)}, \qquad 1 \leqslant j \leqslant n \tag{9.98}$$

Our first task is thus to determine the *correct* zeroth-order wave functions for the perturbation $\hat{H}'$. Calling these correct zeroth-order functions $\varphi_j^{(0)}$, we have

$$\varphi_j^{(0)} = \sum_{i=1}^{n} c_i \psi_i^{(0)}, \qquad 1 \leqslant j \leqslant n \tag{9.99}$$

For each particular $\varphi_j^{(0)}$, we have a different set of coefficients c_i in (9.99); we could indicate this by using a pair of subscripts on the coefficients, calling them c_{ij}, but to keep the notation uncluttered, we have not done so. The right set of zeroth-order functions will depend on what the perturbation $\hat{H}'$ is.

The treatment of the n-fold degenerate level proceeds like the nondegenerate treatment of Section 9.2, except that instead of $\psi_j^{(0)}$ we use $\varphi_j^{(0)}$. Instead of Eqs. (9.17) and (9.18), we have

$$\psi_j = \varphi_j^{(0)} + \lambda \psi_j^{(1)} + \lambda^2 \psi_j^{(2)} + \cdots, \qquad j = 1, 2, \ldots, n \tag{9.100}$$

$$E_j = E_n^{(0)} + \lambda E_j^{(1)} + \lambda^2 E_j^{(2)} + \cdots, \qquad j = 1, 2, \ldots, n \tag{9.101}$$

where Eq. (9.93) has been used. Substituting these expansions into the Schroedinger equation (9.94), we obtain

$$(\hat{H}^0 + \lambda \hat{H}')(\varphi_j^{(0)} + \lambda \psi_j^{(1)} + \lambda^2 \psi_j^{(2)} + \cdots)$$
$$= (E_n^{(0)} + \lambda E_j^{(1)} + \lambda^2 E_j^{(2)} + \cdots)(\varphi_j^{(0)} + \lambda \psi_j^{(1)} + \lambda^2 \psi_j^{(2)} + \cdots) \tag{9.102}$$

Equating the coefficients of λ^0 in (9.102), we have

$$\hat{H}^0 \varphi_j^{(0)} = E_n^{(0)} \varphi_j^{(0)}$$

$$\hat{H}^0 \sum_{i=1}^{n} c_i \psi_i^{(0)} = E_n^{(0)} \sum_{i=1}^{n} c_i \psi_i^{(0)} \tag{9.103}$$

In view of (9.92) and (9.93), Eq. (9.103) is satisfied and gives no new information. Equating the coefficients of λ^1, we have

$$\hat{H}' \varphi_j^{(0)} + \hat{H}^0 \psi_j^{(1)} = E_n^{(0)} \psi_j^{(1)} + E_j^{(1)} \varphi_j^{(0)}$$

$$(E_n^{(0)} - \hat{H}^0)\psi_j^{(1)} = \sum_{i=1}^{n} c_i (\hat{H}' - E_j^{(1)})\psi_i^{(0)} \tag{9.104}$$

Expanding and substituting into (9.104), we have

$$\psi_j^{(1)} = \sum_{k=1}^{\infty} a_k \psi_k^{(0)} \tag{9.105}$$

$$\sum_{k=1}^{\infty} a_k (E_n^{(0)} - \hat{H}^0)\psi_k^{(0)} = \sum_{i=1}^{n} c_i (\hat{H}' - E_j^{(1)})\psi_i^{(0)} \tag{9.106}$$

The sum on the left side is over all the unperturbed states, while the sum on the right side is over the n unperturbed states that correspond to the n-fold degenerate energy level we are treating. Applying (9.92), we have

$$\sum_{k=1}^{\infty} a_k (E_n^{(0)} - E_k^{(0)}) \psi_k^{(0)} = \sum_{i=1}^{n} c_i (\hat{H}' - E_j^{(1)}) \psi_i^{(0)} \qquad (9.107)$$

As we did with Eq. (9.26), we multiply by $\psi_m^{(0)*}$ and integrate:

$$\sum_{k=1}^{\infty} a_k (E_n^{(0)} - E_k^{(0)}) \int \psi_m^{(0)*} \psi_k^{(0)} \, d\tau = \sum_{i=1}^{n} c_i \left(\int \psi_m^{(0)*} \hat{H}' \psi_i^{(0)} \, d\tau - E_j^{(1)} \int \psi_m^{(0)*} \psi_i^{(0)} \, d\tau \right) \qquad (9.108)$$

Consider (9.108) when m is one of the states corresponding to the n-fold degenerate level under consideration, i.e., $1 \leqslant m \leqslant n$. The sum on the left side can be split into two sums, as follows:

$$\sum_{k=1}^{n} a_k (E_n^{(0)} - E_k^{(0)}) \int \psi_m^{(0)*} \psi_k^{(0)} \, d\tau + \sum_{k=n+1}^{\infty} a_k (E_n^{(0)} - E_k^{(0)}) \int \psi_m^{(0)*} \psi_k^{(0)} \, d\tau$$

For $1 \leqslant k \leqslant n$, we have $E_k^{(0)} = E_n^{(0)}$, so that each term of the first sum is zero. In the second sum, we have $k > n$, while by hypothesis $1 \leqslant m \leqslant n$, so that $\psi_k^{(0)}$ and $\psi_m^{(0)}$ are eigenfunctions of a Hermitian operator (the unperturbed Hamiltonian) corresponding to different eigenvalues. Hence for all terms in the second sum, $\psi_k^{(0)}$ and $\psi_m^{(0)}$ are orthogonal, and every term in this sum vanishes. We have shown that when m is one of the states of the n-fold degenerate level, the sum on the left side of (9.108) is zero; hence

$$\sum_{i=1}^{n} \left[\int \psi_m^{(0)*} \hat{H}' \psi_i^{(0)} \, d\tau - E_j^{(1)} \int \psi_m^{(0)*} \psi_i^{(0)} \, d\tau \right] c_i = 0, \qquad m = 1, 2, \ldots, n \qquad (9.109)$$

This is a set of linear homogeneous equations in the n unknowns $c_1, c_2, \ldots, c_n$. For the set of equations to have a nontrivial solution, the determinant of the coefficients must vanish (Section 8.4):

$$\det \left[\int \psi_m^{(0)*} \hat{H}' \psi_i^{(0)} \, d\tau - E_j^{(1)} \int \psi_m^{(0)*} \psi_i^{(0)} \, d\tau \right] = 0 \qquad (9.110)$$

$$\det (H'_{mi} - S_{mi} E_j^{(1)}) = 0 \qquad (9.111)$$

$$H'_{mi} \equiv \int \psi_m^{(0)*} \hat{H}' \psi_i^{(0)} \, d\tau, \qquad S_{mi} \equiv \int \psi_m^{(0)*} \psi_i^{(0)} \, d\tau \qquad (9.112)$$

The determinant is of order n, the degeneracy of the unperturbed level we are considering.

Consider the overlap integrals S_{mi}, $1 \leqslant m \leqslant n$, $1 \leqslant i \leqslant n$. We can always choose the eigenfunctions corresponding to a degenerate eigenvalue to be orthogonal (Section 7.2):

$$S_{mi} = \delta_{mi} \qquad (9.113)$$

and (9.111) becomes

$$\det(H'_{mi} - \delta_{mi} E_j^{(1)}) = 0 \qquad (9.114)$$

[As a check, note that for $n = 1$, Eq. (9.114) reduces to (9.30).]

Equation (9.114) is known as the *secular equation*. Writing out the *secular determinant* in detail, we have

$$\begin{vmatrix} H'_{11} - E_j^{(1)} & H'_{12} & H'_{13} & \cdots & H'_{1n} \\ H'_{21} & H'_{22} - E_j^{(1)} & H'_{23} & \cdots & H'_{2n} \\ H'_{31} & H'_{32} & H'_{33} - E_j^{(1)} & \cdots & H'_{3n} \\ \cdot & \cdot & \cdot & \cdots & \cdot \\ H'_{n1} & H'_{n2} & H'_{n3} & \cdots & H'_{nn} - E_j^{(1)} \end{vmatrix} = 0 \quad (9.115)$$

The secular equation is an algebraic equation of nth degree in $E_j^{(1)}$. It has n roots, $E_1^{(1)}, E_2^{(1)}, \ldots, E_n^{(1)}$, which are the first-order corrections to the energy of the n-fold degenerate unperturbed level. If the roots are all different, then the first-order perturbation correction has split the n-fold degenerate unperturbed level into n different perturbed levels of energies (correct through first order):

$$E_n^{(0)} + E_1^{(1)}, \quad E_n^{(0)} + E_2^{(1)}, \quad \ldots, \quad E_n^{(0)} + E_n^{(1)} \qquad (9.116)$$

It may happen that two or more roots of the secular equation are equal, so that the degeneracy is not completely removed in first order. In the rest of this section, we assume that all the roots of (9.115) are different.

Having determined the n first-order energy corrections, we go back to the set of equations (9.109) to determine the unknowns c_i, which are the coefficients determining the correct zeroth-order wave functions. Writing (9.109) out, we have

$$(H'_{11} - E_j^{(1)})c_1 + H'_{12}c_2 + \cdots + H'_{1n}c_n = 0$$

$$H'_{21}c_1 + (H'_{22} - E_j^{(1)})c_2 + \cdots + H'_{2n}c_n = 0$$

$$\cdots\cdots\cdots\cdots\cdots\cdots\cdots\cdots\cdots\cdots\cdots\cdots\cdots\cdots\cdots\cdots\cdots \qquad (9.117)$$

$$H'_{n1}c_1 + H'_{n2}c_2 + \cdots + (H'_{nn} - E_j^{(1)})c_n = 0$$

where we have assumed orthonormality, Eq. (9.113).

To determine the zeroth-order wave function

$$\varphi_j^{(0)} = c_1 \psi_1^{(0)} + c_2 \psi_2^{(0)} + \cdots + c_n \psi_n^{(0)}$$

corresponding to the root $E_j^{(1)}$, we solve (9.117) for $c_2, c_3, \ldots, c_n$ in terms of c_1 and then determine c_1 by the normalization requirement:

$$\int \varphi_j^{(0)*} \varphi_j^{(0)} \, d\tau = 1 = \sum_{k=1}^{n} \sum_{l=1}^{n} c_k^* c_l \int \psi_k^{(0)*} \psi_l^{(0)} \, d\tau = \sum_{k=1}^{n} \sum_{l=1}^{n} c_k^* c_l \delta_{kl}$$

$$\sum_{k=1}^{n} |c_k|^2 = 1 \qquad (9.118)$$

For each root $E_j^{(1)}$, $j = 1, 2, \ldots, n$, we have a different set of coefficients $c_1, c_2, \ldots, c_n$ giving a different correct zeroth-order wave function.

Using procedures similar to those for the nondegenerate case, we can now find the first-order corrections to the correct zeroth-order wave functions and the second-order energy corrections. We omit the details[7] and simply give the results. The coefficients in (9.105) determining the first-order wave-function correction are

$$a_k = \frac{\langle \psi_k^{(0)}|\hat{H}'|\varphi_j^{(0)}\rangle}{E_n^{(0)} - E_k^{(0)}}, \qquad k > n, \qquad 1 \leqslant j \leqslant n \qquad \textbf{(9.119)}$$

Equation (9.119) is the same as the corresponding equation (9.33) for the nondegenerate case, except that we must use the correct zeroth-order wave function $\varphi_j^{(0)}$ in the integral. Equation (9.119) gives only the correction resulting from mixing in of wave functions from states other than those belonging to the degenerate unperturbed level; we omit the formula for a_k with $k \leqslant n$. The second-order energy correction is

$$E_j^{(2)} = \sum_{k > n} \frac{|\langle \psi_k^{(0)}|\hat{H}'|\varphi_j^{(0)}\rangle|^2}{E_n^{(0)} - E_k^{(0)}}, \qquad 1 \leqslant j \leqslant n \qquad \textbf{(9.120)}$$

which is the same as (9.46) in the nondegenerate case, except that we must use the correct zeroth-order wave function in the integrals. In Section 9.6 we will show that Eq. (9.30) for the first-order energy correction in the nondegenerate case holds also for the degenerate case, if we use the correct zeroth-order wave function:

$$E_j^{(1)} = \langle \varphi_j^{(0)}|\hat{H}'|\varphi_j^{(0)}\rangle, \qquad 1 \leqslant j \leqslant n \qquad \textbf{(9.121)}$$

Thus the equations for the degenerate case are essentially the same as for the nondegenerate case, provided we use the correct zeroth-order wave functions.

As an example, consider the effect of a perturbation $\hat{H}'$ on the lowest degenerate energy level of a particle in a cubic box. We have three states corresponding to this level: $\psi_{211}^{(0)}$, $\psi_{121}^{(0)}$, and $\psi_{112}^{(0)}$. These unperturbed wave functions are orthonormal, and the secular equation (9.114) is

$$\begin{vmatrix} \langle 211|\hat{H}'|211\rangle - E_j^{(1)} & \langle 211|\hat{H}'|121\rangle & \langle 211|\hat{H}'|112\rangle \\ \langle 121|\hat{H}'|211\rangle & \langle 121|\hat{H}'|121\rangle - E_j^{(1)} & \langle 121|\hat{H}'|112\rangle \\ \langle 112|\hat{H}'|211\rangle & \langle 112|\hat{H}'|121\rangle & \langle 112|\hat{H}'|112\rangle - E_j^{(1)} \end{vmatrix} = 0$$

Solving this equation, we find the first-order energy corrections:

$$E_1^{(1)}, \qquad E_2^{(1)}, \qquad E_3^{(1)} \qquad \textbf{(9.122)}$$

The triply degenerate unperturbed level is split into three levels of energies (through first order):

$$(6h^2/8ma^2) + E_1^{(1)}, \qquad (6h^2/8ma^2) + E_2^{(1)}, \qquad (6h^2/8ma^2) + E_3^{(1)}$$

Using each of the roots (9.122), we get a different set of simultaneous equations (9.117). Solving each set, we find three sets of coefficients, which determine the three correct zeroth-order wave functions.

---

[7] *Bates*, Vol. I, pp. 197–198; *Hameka*, pp. 208–209; Problem 9.7.

The secular determinant (9.111) has the same general form as the determinant (8.88) occurring in the linear variation method. Hence the determinant in (8.88) is often called a secular determinant.

9.6 *SIMPLIFICATION OF THE SECULAR EQUATION*

For an unperturbed level with a high degree of degeneracy, the secular equation (9.115) becomes troublesome to solve. Things are simplified if some of the off-diagonal elements of the secular determinant are zero. In the most favorable case, all off-diagonal elements are zero, so that

$$
\begin{vmatrix}
H'_{11} - E_j^{(1)} & 0 & 0 & \cdots & 0 \\
0 & H'_{22} - E_j^{(1)} & 0 & \cdots & 0 \\
0 & 0 & H'_{33} - E_j^{(1)} & \cdots & 0 \\
\cdot & \cdot & \cdot & \cdots & \cdot \\
0 & 0 & 0 & \cdots & H'_{nn} - E_j^{(1)}
\end{vmatrix} = 0 \quad \textbf{(9.123)}
$$

$$
(H'_{11} - E_j^{(1)})(H'_{22} - E_j^{(1)})(H'_{33} - E_j^{(1)}) \ldots (H'_{nn} - E_j^{(1)}) = 0
$$
$$
E_1^{(1)} = H'_{11}, \quad E_2^{(1)} = H'_{22}, \quad E_3^{(1)} = H'_{33}, \quad \ldots, \quad E_n^{(1)} = H'_{nn} \quad \textbf{(9.124)}
$$

Now we want to determine the correct zeroth-order wave functions. We will assume that the roots (9.124) are all different. For the root $E_j^{(1)} = H'_{11}$, the system of equations (9.117) is

$$
0 = 0
$$
$$
(H'_{22} - H'_{11})c_2 = 0
$$
$$
\cdots\cdots\cdots\cdots\cdots\cdots
$$
$$
(H'_{nn} - H'_{11})c_n = 0
$$

$$\textbf{(9.125)}$$

Since we are assuming unequal roots, the factors of $c_2, c_3, \ldots, c_n$ in these equations are all nonzero; hence

$$
c_2 = 0, \quad c_3 = 0, \quad \ldots, \quad c_n = 0 \quad \textbf{(9.126)}
$$

The normalization condition (9.118) gives $c_1 = 1$. The correct zeroth-order wave function corresponding to the first-order perturbation energy correction H'_{11} is then [Eq. (9.99)]

$$
\varphi_1^{(0)} = \psi_1^{(0)} \quad \textbf{(9.127)}
$$

For the root H'_{22}, the same reasoning that led to (9.127) gives for the correct zeroth-order wave function

$$
\varphi_2^{(0)} = \psi_2^{(0)} \quad \textbf{(9.128)}
$$

Using each of the remaining roots, we find similarly

$$
\varphi_3^{(0)} = \psi_3^{(0)}, \quad \ldots, \quad \varphi_n^{(0)} = \psi_n^{(0)} \quad \textbf{(9.129)}
$$

In other words, when the secular determinant is in diagonal form, the initially assumed functions $\psi_1^{(0)}, \psi_2^{(0)}, \ldots, \psi_n^{(0)}$ are the correct zeroth-order wave functions for the particular perturbation $\hat{H}'$.

The converse is also true; if the initially assumed functions are the correct zeroth-order functions, then the secular determinant is in diagonal form. This is seen as follows: From $\varphi_1^{(0)} = \psi_1^{(0)}$, we know that the coefficients in the expansion

$$\varphi_1^{(0)} = \sum_{i=1}^{n} c_i \psi_i^{(0)}$$

are $c_1 = 1, c_2 = c_3 = \cdots = c_n = 0$, so that for $j = 1$, the set of simultaneous equations (9.117) becomes

$$H'_{11} - E_1^{(1)} = 0$$
$$H'_{21} = 0$$
$$\ldots\ldots\ldots \tag{9.130}$$
$$H'_{n1} = 0$$

Applying the same reasoning to the remaining functions $\varphi_j^{(0)}$, we conclude that $H'_{im} = 0$ for $i \neq m$. Hence use of the correct zeroth-order functions makes the secular determinant diagonal. Note also that the first-order corrections to the energy can be found by averaging the perturbation over the correct zeroth-order wave functions:

$$E_j^{(1)} = \int \varphi_j^{(0)*} \hat{H}' \varphi_j^{(0)} \, d\tau \tag{9.131}$$

a result mentioned in Eq. (9.121).

Often, instead of being in diagonal form, the secular determinant is in block-diagonal form. For example, we might have

$$\begin{vmatrix} H'_{11} - E_j^{(1)} & H'_{12} & 0 & 0 \\ H'_{21} & H'_{22} - E_j^{(1)} & 0 & 0 \\ 0 & 0 & H'_{33} - E_j^{(1)} & H'_{34} \\ 0 & 0 & H'_{43} & H'_{44} - E_j^{(1)} \end{vmatrix} = 0 \tag{9.132}$$

The secular equation (9.132) becomes (see Section 8.3)

$$\begin{vmatrix} H'_{11} - E_j^{(1)} & H'_{12} \\ H'_{21} & H'_{22} - E_j^{(1)} \end{vmatrix} \cdot \begin{vmatrix} H'_{33} - E_j^{(1)} & H'_{34} \\ H'_{43} & H'_{44} - E_j^{(1)} \end{vmatrix} = 0 \tag{9.133}$$

and its roots are found from

$$\begin{vmatrix} H'_{11} - E_j^{(1)} & H'_{12} \\ H'_{21} & H'_{22} - E_j^{(1)} \end{vmatrix} = 0 \tag{9.134}$$

$$\begin{vmatrix} H'_{33} - E_j^{(1)} & H'_{34} \\ H'_{43} & H'_{44} - E_j^{(1)} \end{vmatrix} = 0 \tag{9.135}$$

Let the roots of (9.134) be $E_1^{(1)}$ and $E_2^{(1)}$ and the roots of (9.135) be $E_3^{(1)}$ and $E_4^{(1)}$. For $E_1^{(1)}$ the set of equations (9.117) becomes

$$\left.\begin{array}{l} (H'_{11} - E_1^{(1)})c_1 + H'_{12}c_2 = 0 \\ H'_{21}c_1 + (H'_{22} - E_1^{(1)})c_2 = 0 \end{array}\right\} \qquad (9.136)$$

$$\left.\begin{array}{l} (H'_{33} - E_1^{(1)})c_3 + H'_{34}c_4 = 0 \\ H'_{43}c_3 + (H'_{44} - E_1^{(1)})c_4 = 0 \end{array}\right\} \qquad (9.137)$$

Because $E_1^{(1)}$ is a root of (9.134), the set of equations (9.136) has the determinant of its coefficients equal to zero and therefore possesses a nontrivial solution. However (assuming there are no equal roots of the secular equation), $E_1^{(1)}$ is not a root of (9.135), so that the determinant of the coefficients of the set of equations (9.137) is nonzero. Therefore (9.137) has only the trivial solution $c_3 = c_4 = 0$. The correct zeroth-order wave function $\varphi_1^{(0)}$ is then

$$\varphi_1^{(0)} = c_1 \psi_1^{(0)} + c_2 \psi_2^{(0)}$$

Similar reasoning shows that $\varphi_2^{(0)}$ is a linear combination of $\psi_1^{(0)}$ and $\psi_2^{(0)}$, while $\varphi_3^{(0)}$ and $\varphi_4^{(0)}$ are each linear combinations of $\psi_3^{(0)}$ and $\psi_4^{(0)}$. When the secular equation is in block-diagonal form, it factors into two or more smaller secular equations, and the set of simultaneous equations (9.117) breaks up into two or more smaller sets of simultaneous equations.

Conversely, if we have, say, a fourfold-degenerate unperturbed level, and we happen to know that $\varphi_1^{(0)}$ and $\varphi_2^{(0)}$ are each linear combinations of $\psi_1^{(0)}$ and $\psi_2^{(0)}$ only, while $\varphi_3^{(0)}$ and $\varphi_4^{(0)}$ are each linear combinations of $\psi_3^{(0)}$ and $\psi_4^{(0)}$ only, we are led to two 2×2 secular determinants rather than a 4×4 secular determinant. This is true because instead of (9.99) we have the two sets of equations

$$\varphi_j^{(0)} = \sum_{i=1}^{2} c_i \psi_i^{(0)}, \qquad j = 1, 2$$

$$\varphi_j^{(0)} = \sum_{i=3}^{4} c_i \psi_i^{(0)}, \qquad j = 3, 4$$

How can we choose the right zeroth-order wave functions in advance and thereby simplify the secular equation? Suppose there is an operator $\hat{A}$ that commutes with both $\hat{H}^0$ and $\hat{H}'$. Then we can choose the unperturbed functions to be eigenfunctions of $\hat{A}$. Because $\hat{A}$ commutes with $\hat{H}'$, this choice of unperturbed functions will make the integrals H'_{ij} vanish if $\psi_i^{(0)}$ and $\psi_j^{(0)}$ belong to different eigenvalues of $\hat{A}$ [see Eq. (7.80)]. Thus if the eigenvalues of $\hat{A}$ for $\psi_1^{(0)}, \psi_2^{(0)}, \dots, \psi_n^{(0)}$ are all different, the secular determinant will be in diagonal form, and we will have the right zeroth-order wave functions. If some of the eigenvalues of $\hat{A}$ are the same, we will get block-diagonal rather than diagonal form. In general, the correct zeroth-order functions will be linear combinations of those unperturbed functions that have the same eigenvalue of $\hat{A}$. (This is to be expected since $\hat{A}$ commutes with $\hat{H} = \hat{H}^0 + \hat{H}'$, so the perturbed eigenfunctions of $\hat{H}$ can be chosen to be eigenfunctions of $\hat{A}$.) For an application of these ideas, see Problem 9.13, dealing with the Stark effect in the hydrogen atom.

9.7   PERTURBATION TREATMENT OF THE FIRST EXCITED STATES OF HELIUM

In Section 9.3 we considered the ground state of the helium atom according to perturbation theory. In this section we treat the lowest excited states. The unperturbed energy is given by Eq. (9.59). The lowest excited unperturbed states have $n_1 = 1$, $n_2 = 2$ or $n_1 = 2$, $n_2 = 1$ with an energy

$$E^{(0)} = -\frac{5Z^2}{8}\left(\frac{e'^2}{a_0}\right) \tag{9.138}$$

Recall that the $n = 2$ level of a hydrogenlike atom is fourfold degenerate, the 2s and three 2p states all having the same energy. The first excited unperturbed energy level is eightfold degenerate; the eight unperturbed wave functions are [Eq. (9.55)]

$$\begin{array}{ll}
\psi_1^{(0)} = 1s(1)2s(2) & \psi_5^{(0)} = 1s(1)2p_y(2) \\
\psi_2^{(0)} = 2s(1)1s(2) & \psi_6^{(0)} = 2p_y(1)1s(2) \\
\psi_3^{(0)} = 1s(1)2p_x(2) & \psi_7^{(0)} = 1s(1)2p_z(2) \\
\psi_4^{(0)} = 2p_x(1)1s(2) & \psi_8^{(0)} = 2p_z(1)1s(2)
\end{array} \tag{9.139}$$

where $1s(1)2s(2)$ signifies the product of a hydrogenlike 1s function for electron one and a hydrogenlike 2s function for electron two. The explicit form of $\psi_8^{(0)}$, for example, is (Table 6.2)

$$\psi_8^{(0)} = \frac{1}{4(2\pi)^{1/2}}\left(\frac{Z}{a_0}\right)^{5/2} r_1 e^{-Zr_1/2a_0}\cos\theta_1 \cdot \frac{1}{\pi^{1/2}}\left(\frac{Z}{a_0}\right)^{3/2} e^{-Zr_2/a_0}$$

We have chosen to use the real 2p hydrogenlike orbitals, rather than the complex ones.

Since we have a degenerate unperturbed level, we must deal with a secular equation. For us to use (9.115), the functions $\psi_1^{(0)}, \psi_2^{(0)}, \ldots, \psi_8^{(0)}$ must be orthonormal. This condition is met. For example,

$$\int \psi_1^{(0)*}\psi_1^{(0)}\, d\tau = \int\int 1s(1)^*2s(2)^*1s(1)\,2s(2)\, d\tau_1\, d\tau_2$$

$$= \int |1s(1)|^2\, d\tau_1 \int |2s(2)|^2\, d\tau_2 = 1\cdot 1 = 1$$

$$\int \psi_3^{(0)*}\psi_5^{(0)}\, d\tau = \int |1s(1)|^2\, d\tau_1 \int 2p_x(2)^*2p_y(2)\, d\tau_2 = 1\cdot 0 = 0$$

where the orthonormality of the hydrogenlike orbitals has been used.

The secular determinant contains $8^2 = 64$ elements. The operator $\hat{H}'$ is Hermitian, so that $H'_{ij} = (H'_{ji})^*$. Also, since $\hat{H}'$ and $\psi_1^{(0)}, \ldots, \psi_8^{(0)}$ are all real, we have $(H'_{ji})^* = H'_{ji}$, so that $H'_{ij} = H'_{ji}$. The secular determinant is symmetric about the principal diagonal. This cuts the labor of evaluating integrals almost in half.

By using parity considerations, we can show that most of the integrals H'_{ij} are zero. First consider H'_{13}:

$$H'_{13} = \int_{-\infty}^{\infty}\int_{-\infty}^{\infty}\int_{-\infty}^{\infty}\int_{-\infty}^{\infty}\int_{-\infty}^{\infty}\int_{-\infty}^{\infty} 1s(1)\,2s(2)\frac{e'^2}{r_{12}}$$
$$\times 1s(1)2p_x(2)\,dx_1\,dy_1\,dz_1\,dx_2\,dy_2\,dz_2$$

An s hydrogenlike function depends only on $r = (x^2 + y^2 + z^2)^{1/2}$ and is therefore an even function. The $2p_x(2)$ function is an odd function of x_2 [Eq. (6.138)]. r_{12} is given by (9.90), and if we invert all six coordinates, r_{12} is unchanged:

$$r_{12} \to [(-x_1 + x_2)^2 + (-y_1 + y_2)^2 + (-z_1 + z_2)^2]^{1/2} = r_{12}$$

Hence on inverting all six coordinates, the integrand of H'_{13} goes into minus itself. We conclude [Eq. (7.96)] that $H'_{13} = 0$. The same reasoning yields $H'_{14} = H'_{15} = H'_{16} = H'_{17} = H'_{18} = 0$ and $H'_{23} = H'_{24} = H'_{25} = H'_{26} = H'_{27} = H'_{28} = 0$. Now consider H'_{35}:

$$H'_{35} = \int_{-\infty}^{\infty} \cdots \int_{-\infty}^{\infty} 1s(1)2p_x(2)\frac{e'^2}{r_{12}}1s(1)2p_y(2)\,dx_1 \cdots dz_2 \qquad \textbf{(9.140)}$$

Consider the effect of inverting the coordinates x_1 and x_2: $x_1 \to -x_1, x_2 \to -x_2$. This transformation will leave r_{12} unchanged. The functions $1s(1)$ and $2p_y(2)$ will be unaffected. However, $2p_x(2)$ will go over to minus itself, so that the net effect will be to change the integrand of H'_{35} into minus itself. Therefore [Eq. (7.98)] $H'_{35} = 0$. Likewise $H'_{36} = H'_{37} = H'_{38} = 0$ and $H'_{45} = H'_{46} = H'_{47} = H'_{48} = 0$. By considering the transformation $y_1 \to -y_1, y_2 \to -y_2$, we see that $H'_{57} = H'_{58} = H'_{67} = H'_{68} = 0$. The secular equation is therefore

$$\begin{vmatrix} b_{11} & H'_{12} & 0 & 0 & 0 & 0 & 0 & 0 \\ H'_{12} & b_{22} & 0 & 0 & 0 & 0 & 0 & 0 \\ 0 & 0 & b_{33} & H'_{34} & 0 & 0 & 0 & 0 \\ 0 & 0 & H'_{34} & b_{44} & 0 & 0 & 0 & 0 \\ 0 & 0 & 0 & 0 & b_{55} & H'_{56} & 0 & 0 \\ 0 & 0 & 0 & 0 & H'_{56} & b_{66} & 0 & 0 \\ 0 & 0 & 0 & 0 & 0 & 0 & b_{77} & H'_{78} \\ 0 & 0 & 0 & 0 & 0 & 0 & H'_{78} & b_{88} \end{vmatrix} = 0$$

$$b_{ii} \equiv H'_{ii} - E^{(1)}, \qquad i = 1, 2, \ldots, 8$$

The secular determinant is in block-diagonal form and factors into four determinants, each of second order. We conclude that the correct zeroth-order functions have the form

$$\varphi_1^{(0)} = c_1\psi_1^{(0)} + c_2\psi_2^{(0)}, \qquad \varphi_2^{(0)} = \bar{c}_1\psi_1^{(0)} + \bar{c}_2\psi_2^{(0)}$$
$$\varphi_3^{(0)} = c_3\psi_3^{(0)} + c_4\psi_4^{(0)}, \qquad \varphi_4^{(0)} = \bar{c}_3\psi_3^{(0)} + \bar{c}_4\psi_4^{(0)}$$
$$\varphi_5^{(0)} = c_5\psi_5^{(0)} + c_6\psi_6^{(0)}, \qquad \varphi_6^{(0)} = \bar{c}_5\psi_5^{(0)} + \bar{c}_6\psi_6^{(0)} \qquad \textbf{(9.141)}$$
$$\varphi_7^{(0)} = c_7\psi_7^{(0)} + c_8\psi_8^{(0)}, \qquad \varphi_8^{(0)} = \bar{c}_7\psi_7^{(0)} + \bar{c}_8\psi_8^{(0)}$$

where the unbarred coefficients correspond to one root of each second-order determinant and the barred coefficients correspond to the second root.

The first determinant is

$$\begin{vmatrix} H'_{11} - E^{(1)} & H'_{12} \\ H'_{12} & H'_{22} - E^{(1)} \end{vmatrix} = 0 \qquad (9.142)$$

We have

$$H'_{11} = \int_{-\infty}^{\infty} \cdots \int_{-\infty}^{\infty} 1s(1)2s(2) \frac{e'^2}{r_{12}} 1s(1)\, 2s(2)\, dx_1 \cdots dz_2$$

$$H'_{11} = \int\int [1s(1)]^2 [2s(2)]^2 \frac{e'^2}{r_{12}}\, d\tau_1\, d\tau_2$$

$$H'_{22} = \int\int [1s(2)]^2 [2s(1)]^2 \frac{e'^2}{r_{12}}\, d\tau_1\, d\tau_2$$

The integration variables are dummy variables and may be given any symbols whatever. Let us relabel the integration variables in H'_{22} as follows: We interchange x_1 and x_2, interchange y_1 and y_2, and interchange z_1 and z_2. This relabeling leaves r_{12} [Eq. (9.90)] unchanged, so that

$$H'_{22} = \int\int [1s(1)]^2 [2s(2)]^2 \frac{e'^2}{r_{12}}\, d\tau_2\, d\tau_1 = H'_{11} \qquad (9.143)$$

(The same argument shows $H'_{33} = H'_{44}$, $H'_{55} = H'_{66}$, $H'_{77} = H'_{88}$.)

We denote H'_{11} by the symbol J_{1s2s}:

$$H'_{11} = J_{1s2s} = \int\int [1s(1)]^2 [2s(2)]^2 \frac{e'^2}{r_{12}}\, d\tau_1\, d\tau_2 \qquad (9.144)$$

This is an example of a *Coulomb integral*, the name arising from the fact that J_{1s2s} is equal to the electrostatic energy of repulsion between an electron with probability density function $[1s]^2$ and an electron with probability density function $[2s]^2$. The integral H'_{12} is denoted by K_{1s2s}:

$$H'_{12} = K_{1s2s} = \int\int 1s(1)2s(2) \frac{e'^2}{r_{12}} 2s(1)1s(2)\, d\tau_1\, d\tau_2 \qquad (9.145)$$

This is an *exchange integral*—the functions on the left and right of e'^2/r_{12} differ from each other by an exchange of electrons one and two. The general definitions of the Coulomb integral J_{ij} and the exchange integral K_{ij} are

$$J_{ij} \equiv \langle f_i(1)f_j(2)|e'^2/r_{12}|f_i(1)f_j(2)\rangle, \qquad K_{ij} \equiv \langle f_i(1)f_j(2)|e'^2/r_{12}|f_j(1)f_i(2)\rangle \qquad (9.146)$$

where the integrals go over the full range of the spatial coordinates of electrons 1 and 2 and f_i and f_j are spatial orbitals.

Substituting (9.143)–(9.145) into (9.142) and expanding the determinant, we get

$$[J_{1s2s} - E^{(1)}]^2 = [K_{1s2s}]^2$$

$$E_1^{(1)} = J_{1s2s} + K_{1s2s}, \qquad E_2^{(1)} = J_{1s2s} - K_{1s2s} \qquad \text{(9.147)}$$

We now find the coefficients of the correct zeroth-order wave functions corresponding to these two roots of the secular equation. Using $E_1^{(1)}$, we have for (9.117)

$$-K_{1s2s}c_1 + K_{1s2s}c_2 = 0$$

$$K_{1s2s}c_1 - K_{1s2s}c_2 = 0$$

so that $c_1 = c_2$. Normalizing, we have

$$\int |\varphi_1^{(0)}|^2 \, d\tau = \int [c_1\psi_1^{(0)} + c_1\psi_2^{(0)}]^2 \, d\tau = c_1^2 + c_1^2 = 1$$

$$c_1 = 1/\sqrt{2}$$

where the orthonormality of $\psi_1^{(0)}$ and $\psi_2^{(0)}$ has been used. The zeroth-order wave function corresponding to $E_1^{(1)}$ is then

$$\varphi_1^{(0)} = 2^{-1/2}(\psi_1^{(0)} + \psi_2^{(0)}) = 2^{-1/2}[1s(1)2s(2) + 2s(1)1s(2)] \qquad \text{(9.148)}$$

In like manner, we find as the function corresponding to $E_2^{(1)}$

$$\varphi_2^{(0)} = 2^{-1/2}(\psi_1^{(0)} - \psi_2^{(0)}) = 2^{-1/2}[1s(1)2s(2) - 2s(1)1s(2)] \qquad \text{(9.149)}$$

We have three more second-order determinants to deal with:

$$\begin{vmatrix} H'_{33} - E^{(1)} & H'_{34} \\ H'_{34} & H'_{33} - E^{(1)} \end{vmatrix} = 0 \qquad \text{(9.150)}$$

$$\begin{vmatrix} H'_{55} - E^{(1)} & H'_{56} \\ H'_{56} & H'_{55} - E^{(1)} \end{vmatrix} = 0 \qquad \text{(9.151)}$$

$$\begin{vmatrix} H'_{77} - E^{(1)} & H'_{78} \\ H'_{78} & H'_{77} - E^{(1)} \end{vmatrix} = 0 \qquad \text{(9.152)}$$

Consider H'_{33} and H'_{55}:

$$H'_{33} = \int_{-\infty}^{\infty} \cdots \int_{-\infty}^{\infty} 1s(1)2p_x(2) \frac{e'^2}{r_{12}} 1s(1)2p_x(2) \, dx_1 \cdots dz_2$$

$$H'_{55} = \int_{-\infty}^{\infty} \cdots \int_{-\infty}^{\infty} 1s(1)2p_y(2) \frac{e'^2}{r_{12}} 1s(1)2p_y(2) \, dx_1 \cdots dz_2$$

These two integrals are equal—the only difference between them involves replacement of $2p_x(2)$ by $2p_y(2)$—and these two orbitals differ only in their orientation in space. More formally, if we relabel the dummy integration variables in H'_{33} according to the scheme $x_2 \to y_2, y_2 \to x_2, x_1 \to y_1, y_1 \to x_1$, then r_{12} is unaffected and H'_{33} is transformed to H'_{55}. Similar reasoning shows H'_{77} is equal to these two integrals. Introducing the symbol J_{1s2p} for these Coulomb integrals, we have

$$H'_{33} = H'_{55} = H'_{77} = J_{1s2p} = \int\int 1s(1)2p_z(2)\frac{e'^2}{r_{12}}1s(1)2p_z(2)\,d\tau_1\,d\tau_2$$

Also, the exchange integrals involving the $2p$ orbitals are equal:

$$H'_{34} = H'_{56} = H'_{78} = K_{1s2p} = \int\int 1s(1)2p_z(2)\frac{e'^2}{r_{12}}2p_z(1)1s(2)\,d\tau_1\,d\tau_2$$

The three determinants (9.150)–(9.152) are thus identical and have the form

$$\begin{vmatrix} J_{1s2p}-E^{(1)} & K_{1s2p} \\ K_{1s2p} & J_{1s2p}-E^{(1)} \end{vmatrix} = 0 \qquad (9.153)$$

The determinant is similar to (9.146), and by analogy with (9.147)–(9.149), we have

$$E^{(1)}_3 = E^{(1)}_5 = E^{(1)}_7 = J_{1s2p}+K_{1s2p} \qquad (9.154)$$

$$E^{(1)}_4 = E^{(1)}_6 = E^{(1)}_8 = J_{1s2p}-K_{1s2p} \qquad (9.155)$$

$$
\begin{aligned}
\varphi^{(0)}_3 &= 2^{-1/2}[1s(1)\,2p_x(2)+1s(2)\,2p_x(1)] \\
\varphi^{(0)}_4 &= 2^{-1/2}[1s(1)\,2p_x(2)-1s(2)\,2p_x(1)] \\
\varphi^{(0)}_5 &= 2^{-1/2}[1s(1)\,2p_y(2)+1s(2)\,2p_y(1)] \\
\varphi^{(0)}_6 &= 2^{-1/2}[1s(1)\,2p_y(2)-1s(2)\,2p_y(1)] \\
\varphi^{(0)}_7 &= 2^{-1/2}[1s(1)\,2p_z(2)+1s(2)\,2p_z(1)] \\
\varphi^{(0)}_8 &= 2^{-1/2}[1s(1)\,2p_z(2)-1s(2)\,2p_z(1)]
\end{aligned}
\qquad (9.156)
$$

The electrostatic repulsion e'^2/r_{12} between the electrons has partly removed the degeneracy; the hypothetical eightfold-degenerate unperturbed level has been split into two nondegenerate levels associated with the configuration $1s2s$ and two triply degenerate levels associated with the configuration $1s2p$. (It might be thought that higher-order energy corrections would further resolve the degeneracy; actually application of an external magnetic field is required to completely remove the degeneracy.)

We must now evaluate the integrals in (9.147) and (9.154). As might be expected, we employ the expansion (9.64). [Note that (9.63) is a Coulomb integral, J_{1s1s}.] We find (Problem 9.10)

$$
J_{1s2s} = \frac{17}{81}\frac{Ze'^2}{a_0}, \qquad J_{1s2p} = \frac{59}{243}\frac{Ze'^2}{a_0}
$$

$$
K_{1s2s} = \frac{16}{729}\frac{Ze'^2}{a_0}, \qquad K_{1s2p} = \frac{112}{6561}\frac{Ze'^2}{a_0}
\qquad (9.157)
$$

Figure 9.3 is a correlation diagram for the first group of helium excited states. For $Z = 2$, we have

$$E^{(0)} = -\frac{20}{8}\left(\frac{e'^2}{a_0}\right) = -68.0 \text{ eV} \qquad (9.158)$$

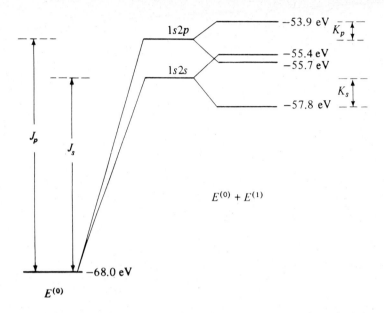

Figure 9.3 *Correlation diagram for the first excited levels of the helium atom*

$$J_{1s2s} = 11.4 \text{ eV}, \qquad K_{1s2s} = 1.2 \text{ eV} \qquad \textbf{(9.159)}$$

$$J_{1s2p} = 13.2 \text{ eV}, \qquad K_{1s2p} = 0.9 \text{ eV} \qquad \textbf{(9.160)}$$

The results of the first-order energy corrections seem to indicate that the lower of the two levels of the $1s2p$ configuration lies below the higher of the two levels of the $1s2s$ configuration. Experimental study of the helium spectrum reveals that this is not so. The error is due, of course, to neglect of the higher-order perturbation energy corrections.

Knight and Scherr have calculated the second-order and third-order energy corrections, $E^{(2)}$ and $E^{(3)}$, for these four excited levels.[8] In Fig. 9.4 we give their results, rounded off to the nearest 0.1 eV. They used the variational perturbation method outlined in Section 9.3. Comparison with experiment shows that the Knight and Scherr values differ from the true values by 0.01 to 0.09 eV. Figure 9.4 shows that Fig. 9.3 is quite inaccurate. Since the perturbation e'^2/r_{12} is not really very small, we cannot expect great accuracy from a perturbation treatment that includes only the $E^{(1)}$ correction to the energy.

Note that the first-order correction to the wave function, $\psi^{(1)}$, will include contributions from other configurations (configuration interaction). When we say that a level belongs to the configuration $1s2s$, we are indicating the configuration that makes the largest contribution to the true wave function.

---

[8] R. E. Knight and C. W. Scherr, *Rev. Mod. Phys.*, **35**, 431 (1963); for energy corrections through seventeenth order, see F. C. Sanders and C. W. Scherr, *Phys. Rev.*, **181**, 84 (1969).

Figure 9.4 $E^{(0)} + E^{(1)} + E^{(2)} + E^{(3)}$ *for the first excited levels of helium*

The perturbation treatment of the first excited states of helium illustrates several important points, which we now discuss. We started with the eight degenerate zeroth-order functions (9.139). These functions have three kinds of degeneracy. There is the degeneracy between hydrogenlike functions with the same n, but different l; the $2s$ and the $2p$ functions have the same energy. There is the degeneracy between hydrogenlike functions with the same n and l, but different m; the $2p_1, 2p_0,$ and $2p_{-1}$ functions have the same energy. (For convenience we used the real functions $2p_{x,y,z}$, but we could have started with the functions $2p_{1,0,-1}$.) Finally, there is the degeneracy between functions that differ only in the interchange of the two electrons between the orbitals; the functions $\psi_1^{(0)} = 1s(1)2s(2)$ and $\psi_2^{(0)} = 1s(2)2s(1)$ have the same energy. We refer to this last kind of degeneracy as *exchange degeneracy*. When the interelectronic repulsion e'^2/r_{12} was introduced as a perturbation, the exchange degeneracy and the degeneracy associated with the quantum number l were removed. The degeneracy associated with m remained, however; each $1s2p$ helium level is triply degenerate, and we could just as well have used the $2p_1, 2p_0,$ and $2p_{-1}$ orbitals instead of the real orbitals in constructing the correct zeroth-order wave functions. Let us consider the reasons for the removal of the l and the exchange degeneracy.

The interelectronic repulsion in helium causes the $2s$ orbital energy to be less than the $2p$ energy. We can give a simple physical explanation for this fact. Figures 6.7 and 6.8 show that a $2s$ electron has a greater probability than a $2p$ electron of being closer to the nucleus than the $1s$ electron(s). A $2s$ electron will not be as effectively shielded from the nucleus by the $1s$ electron(s), and will therefore have a lower energy than a $2p$ electron. [According to Eq. (6.107), the greater the nuclear charge, the lower the energy.] Mathematically, the difference between the $1s2s$ and the $1s2p$ energies results from the Coulomb integral J_{1s2s} being smaller than J_{1s2p}. These Coulomb integrals represent the electrostatic repulsion between the appropriate charge distributions. When the $2s$ electron penetrates the charge distribution of the $1s$ electron, it feels a repulsion from only the unpenetrated part of the $1s$ charge distribution (recall that this point came up in the discussion of Rutherford scattering in Section 1.1); hence the $1s$–$2s$ electrostatic repulsion is less than the $1s$–$2p$ repulsion, and the $1s2s$ levels lie below the $1s2p$ levels. The interelectronic repulsion in many-electron atoms lifts the l degeneracy, and the orbital energies for the same value of n increase with increasing l.

Now consider the removal of the exchange degeneracy. The functions (9.139) with which we began the perturbation treatment have each electron assigned to a definite orbital. For example, the function $\psi_1^{(0)} = 1s(1)\,2s(2)$ has electron 1 in the $1s$ orbital and electron 2 in the $2s$ orbital; for $\psi_2^{(0)}$ the opposite is true. The secular determinant was not diagonal, so that the initial functions were not the correct zeroth-order wave functions. The correct zeroth-order functions do not assign each electron to a definite orbital. Thus the first two correct zeroth-order functions are

$$\varphi_1^{(0)} = 2^{-1/2}[1s(1)\,2s(2) + 1s(2)\,2s(1)] \tag{9.161}$$

$$\varphi_2^{(0)} = 2^{-1/2}[1s(1)\,2s(2) - 1s(2)\,2s(1)] \tag{9.162}$$

so that we cannot say which orbital electron 1 is in, for either $\varphi_1^{(0)}$ or $\varphi_2^{(0)}$. This feature of the wave functions of systems containing more than one electron is a result of the indistinguishability of identical particles in quantum mechanics and will be discussed further in the next chapter. The functions (9.161) and (9.162) have different energies, so that the exchange degeneracy is removed when we use the correct zeroth-order functions. Note that the function (9.162) becomes zero when the two electrons are at the same point in space, since if $r_1 = r_2$, then $1s(r_1)\,2s(r_2) = 1s(r_2)\,2s(r_1)$. This means that the probability of having the two electrons close together is smaller for the state (9.162) than for (9.161). Thus $\langle e'^2/r_{12} \rangle$, the average interelectronic repulsion, is less for (9.162), and hence it lies lower in energy.

Sometimes one speaks of an "exchange phenomenon." If we *know* that at a given instant of time electron 1 is in the $1s$ orbital and electron 2 is in the $2s$ orbital, then it is possible to show[9] that after a time interval $h/4K_{1s2s}$ has elapsed, the two electrons have exchanged states so that electron 1 is in the $2s$ orbital and electron 2 is in the $1s$ orbital. After another such interval, the electrons are back in their original orbitals, etc. The exchange "phenomenon" should *not* be thought of as a real physical process, because the assumption that it is possible to *know* which electron is in which orbital at a given instant of time is a violation of the principle of indistinguishability of identical particles.

The perturbation treatment of the helium excited states was given by Heisenberg in 1926, and his name is closely associated with the subject of exchange. Other excited states of helium can be treated by perturbation theory along the lines of the treatment of the $1s2s$ and $1s2p$ configurations.

9.8 COMPARISON OF THE VARIATION AND PERTURBATION METHODS

The perturbation method is applicable to all the states of an atom or molecule, whereas the variation method is restricted to the ground state (which is the state of greatest chemical interest) and possibly the first few excited states. However, it is usually extremely difficult to evaluate the infinite sums over discrete states and integrals over continuum states that are necessary to evaluate the second-order and higher-order perturbation energy corrections. (See Problem 9.16 for an approximate method.)

---

[9] *Bethe and Salpeter*, p. 132.

In the perturbation method, one can calculate the energy much more accurately (to order $2k + 1$) than the wave function (to order k). The same situation holds in the variation method, where one can get a rather good energy with a rather inaccurate wave function; if one attempts to calculate properties other than the energy, the results will generally not be as reliable as the calculated energy.

Although most calculations of molecular wave functions have been done using the variation method, there has been a revival of interest in applying the perturbation method to such problems.[10]

9.9   TIME-DEPENDENT PERTURBATION THEORY

In spectroscopy we start with a system in some stationary state, expose it to electromagnetic radiation (light), and then observe whether the system has made a transition to another stationary state. The radiation produces a time-dependent potential-energy term in the Hamiltonian, so we must use the time-dependent Schroedinger equation. The most convenient approach here is an approximate one called time-dependent perturbation theory.

Let the system (atom or molecule) have the time-independent Hamiltonian $\hat{H}^0$ in the absence of the radiation (or other time-dependent perturbation) and let $\hat{H}'(t)$ be the perturbation. The time-independent Schroedinger equation for the unperturbed problem is

$$\hat{H}^0 \psi_k^0 = E_k^0 \psi_k^0 \tag{9.163}$$

where E_k^0 and ψ_k^0 are the stationary-state energies and wave functions. The time-dependent Schroedinger equation (7.145) in the presence of the radiation is

$$-\frac{\hbar}{i} \frac{\partial \Psi}{\partial t} = (\hat{H}^0 + \hat{H}') \Psi \tag{9.164}$$

where the state function Ψ depends on the spatial and spin coordinates (symbolized by q) and on the time: $\Psi = \Psi(q, t)$. (See Chapter 10 for a discussion of spin coordinates.)

First suppose that $\hat{H}'$ is absent. The unperturbed time-dependent Schroedinger equation is

$$-(\hbar/i)\partial \Psi^0 / \partial t = \hat{H}^0 \Psi^0 \tag{9.165}$$

The system's possible stationary state functions are given by (7.148) as $\Psi_k^0 = \exp(-iE_k^0 t/\hbar)\psi_k^0$, where the ψ_k^0 functions are the eigenfunctions of $\hat{H}^0$ [Eq. (9.163)]. Each Ψ_k^0 is a solution of (9.165). Moreover, the linear combination

$$\Psi^0 = \sum_k c_k \Psi_k^0 = \sum_k c_k \exp(-iE_k^0 t/\hbar)\psi_k^0 \tag{9.166}$$

[10] For a discussion of developments in perturbation theory, see J. O. Hirschfelder, W. Byers Brown, and S. T. Epstein, *Advances in Quantum Chemistry*, Academic Press, New York, 1964, vol. 1, p. 255.

(with the c_k's being arbitrary time-independent constants) is a solution of the time-dependent Schroedinger equation (9.165). To prove this assertion, we substitute (9.166) into the left side of (9.165) and use (9.163) to give

$$-(\hbar/i)\partial\Psi^0/\partial t = \sum_k c_k E_k^0 \exp(-iE_k^0 t/\hbar)\psi_k^0 = \sum_k c_k \exp(-iE_k^0 t/\hbar)\hat{H}^0\psi_k^0$$

$$= \hat{H}^0 \sum_k c_k \exp(-iE_k^0 t/\hbar)\psi_k^0 = \hat{H}^0\Psi^0$$

Q.E.D. The functions Ψ_k^0 form a complete set (since they are the eigenfunctions of the Hermitian operator $\hat{H}^0$), so any solution of (9.165) can be expressed as (9.166). Hence (9.166) is the general solution of the time-dependent Schroedinger equation (9.165), where $\hat{H}^0$ is independent of time.

Now suppose that $\hat{H}'(t)$ is present. The function (9.166) is no longer a solution of the time-dependent Schroedinger equation. However, because the unperturbed functions Ψ_k^0 form a complete set, the true state function Ψ can at any instant of time be expanded as a linear combination of the Ψ_k^0 functions according to $\Psi = \sum_k b_k \Psi_k^0$. Because $\hat{H}$ is time-dependent, Ψ will change with time and the expansion coefficients b_k will change with time. Therefore

$$\Psi = \sum_k b_k(t) \exp(-iE_k^0 t/\hbar)\psi_k^0 \tag{9.167}$$

In the limit $\hat{H}'(t) \to 0$, the expansion (9.167) reduces to (9.166).

Substitution of (9.167) into the time-dependent Schroedinger equation (9.164) and use of (9.163) gives

$$-\frac{\hbar}{i}\sum_k \frac{db_k}{dt}\exp(-iE_k^0 t/\hbar)\psi_k^0 + \sum_k E_k^0 b_k \exp(-iE_k^0 t/\hbar)\psi_k^0 =$$

$$\sum_k b_k \exp(-iE_k^0 t/\hbar)E_k^0\psi_k^0 + \sum_k b_k \exp(-iE_k^0 t/\hbar)\hat{H}'\psi_k^0$$

$$-\frac{\hbar}{i}\sum_k \frac{db_k}{dt}\exp(-iE_k^0 t/\hbar)\psi_k^0 = \sum_k b_k \exp(-iE_k^0 t/\hbar)\hat{H}'\psi_k^0$$

We now multiply by ψ_m^{0*} and integrate over the spatial and spin coordinates. Using the orthonormality of the unperturbed wave functions, $\langle \psi_m^0 | \psi_k^0 \rangle = \delta_{mk}$, we get

$$-\frac{\hbar}{i}\sum_k \frac{db_k}{dt}\exp(-iE_k^0 t/\hbar)\delta_{mk} = \sum_k b_k \exp(-iE_k^0 t/\hbar)\langle \psi_m^0 | \hat{H}' | \psi_k^0 \rangle$$

Because of the δ_{mk} factor, all terms but one in the sum on the left are zero, and the left side equals $-(\hbar/i)(db_m/dt)\exp(-iE_m^0 t/\hbar)$. We get

$$\frac{db_m}{dt} = -\frac{i}{\hbar}\sum_k b_k \exp[i(E_m^0 - E_k^0)t/\hbar]\langle \psi_m^0 | \hat{H}' | \psi_k^0 \rangle \tag{9.168}$$

Let us suppose that the perturbation $\hat{H}'(t)$ was applied at time $t = 0$ and that before the perturbation was applied, the system was in stationary state n with energy

E_n^0. Therefore at $t = 0$ we have $\Psi = \exp(-iE_n^0 t/\hbar)\psi_n^0$, so that the $t = 0$ values of the expansion coefficients in (9.167) are $b_n(0) = 1$ and $b_k(0) = 0$ for $k \neq n$; thus

$$b_k(0) = \delta_{kn} \tag{9.169}$$

To facilitate the solution of (9.168) we shall make an approximation. We shall assume that the perturbation $\hat{H}'$ is small and acts for only a short time. Under these conditions, the change in the expansion coefficients b_k from their initial values at the time the perturbation is applied will be small. To a good approximation, we can replace the expansion coefficients on the right of (9.168) by their initial values (9.169). This gives

$$\frac{db_m}{dt} \approx -\frac{i}{\hbar} \exp[i(E_m^0 - E_n^0)t/\hbar] \langle \psi_m^0 | \hat{H}' | \psi_n^0 \rangle$$

Usually the perturbation $\hat{H}'$ acts for a limited period of time, say, from $t = 0$ to $t = t'$. Integrating from $t = 0$ to t' and using (9.169), we get

$$b_m(t') \approx \delta_{mn} - \frac{i}{\hbar} \int_0^{t'} \exp[i(E_m^0 - E_n^0)t/\hbar] \langle \psi_m^0 | \hat{H}' | \psi_n^0 \rangle \, dt \tag{9.170}$$

Use of the approximate result (9.170) for the expansion coefficients in (9.167) gives the desired approximation to the state function at time t' for the case that the time-dependent perturbation $\hat{H}'$ is applied at $t = 0$ to a system in a stationary state n. [As with time-independent perturbation theory, one can go to higher-order approximations (see *Fong*, pp. 234–244), but we shall not do so.]

For times after t', the perturbation has ceased to act, and $\hat{H}' = 0$. Equation (9.168) gives $db_m/dt = 0$ for $t > t'$, so $b_m = b_m(t')$ for $t \geq t'$. Therefore for times after exposure to the perturbation, the state function Ψ is [Eq. (9.167)]

$$\Psi = \sum_m b_m(t') \exp(-iE_m^0 t/\hbar)\psi_m^0 \tag{9.171}$$

Ψ is a superposition of the eigenfunctions ψ_m^0 of the energy operator $\hat{H}^0$, the expansion coefficients being $b_m \exp(-iE_m^0 t/\hbar)$. [Compare Eqs. (9.171) and (7.101).] The work of Section 7.6 tells us that a measurement of the system's energy at a time after t' will give one of the eigenvalues E_m^0 of the energy operator $\hat{H}^0$, and the probability of getting E_m^0 equals the square of the absolute value of the expansion coefficient that multiplies ψ_m^0, i.e., it equals $|b_m(t') \exp(-iE_m^0 t/\hbar)|^2 = |b_m(t')|^2$.

The time-dependent perturbation changes the state function from $\exp(-iE_n^0 t/\hbar)\psi_n^0$ to the superposition (9.171). Measurement of the energy then changes Ψ to one of the energy eigenfunctions $\exp(-iE_m^0 t/\hbar)\psi_m^0$ (reduction of the wave function—Section 7.8). The net result is a transition from stationary state n to stationary state m, the probability of such a transition being given by $|b_m(t')|^2$.

9.10   INTERACTION OF RADIATION AND MATTER

We now consider the interaction of an atom (or molecule) with electromagnetic radiation. A proper quantum-mechanical approach would treat both the atom and the radiation quantum-mechanically, but we shall simplify things by using the classical picture of the light as an electromagnetic wave of oscillating electric and magnetic fields.

A detailed investigation, which we omit (see *Levine, Molecular Spectroscopy,* Section 3.2), shows that usually the interaction between the radiation's magnetic field and the atom's charges is much weaker than the interaction between the radiation's electric field and the charges, so we shall consider only the latter interaction. (In NMR spectroscopy the important interaction is between the magnetic dipole moments of the nuclei and the radiation's magnetic field. We will not consider this case.)

Let the electric field $\mathscr{E}$ of the electromagnetic wave point in the x direction only. (This is plane-polarized radiation.) The electric field is defined as the force per unit charge, so the force on charge Q_i is $F = Q_i \mathscr{E}_x = -dV/dx$, where (4.26) was used. Integration gives the potential energy of interaction between the radiation's electric field and the charge as $V = -Q_i \mathscr{E}_x x$, where the arbitrary integration constant was taken as zero. For a system of several charges, $V = -\sum_i Q_i x_i \mathscr{E}_x$. This is the time-dependent perturbation $\hat{H}'(t)$. The space and time dependence of the electric field of an electromagnetic wave traveling in the z direction with wavelength λ and frequency v is given by (*Halliday and Resnick,* Section 41-8) $\mathscr{E}_x = \mathscr{E}_0 \sin(2\pi v t - 2\pi z/\lambda)$, where $\mathscr{E}_0$ is the maximum value of $\mathscr{E}_x$ (the amplitude). Therefore

$$\hat{H}'(t) = -\mathscr{E}_0 \sum_i Q_i x_i \sin(2\pi v t - 2\pi z_i/\lambda)$$

where the sum goes over all the electrons and nuclei of the atom or molecule.

Defining ω and ω_{mn} as

$$\omega \equiv 2\pi v, \qquad \omega_{mn} \equiv (E_m^0 - E_n^0)/\hbar \qquad (9.172)$$

and substituting $\hat{H}'(t)$ into (9.170), we get the coefficients in the expansion (9.167) of the state function Ψ as

$$b_m \approx \delta_{mn} + \frac{i\mathscr{E}_0}{\hbar} \int_0^{t'} \exp(i\omega_{mn}t) \left\langle \psi_m^0 \left| \sum_i Q_i x_i \sin(\omega t - 2\pi z_i/\lambda) \right| \psi_n^0 \right\rangle dt$$

Consider the integral $\langle \psi_m^0 | \sum_i \cdots | \psi_n^0 \rangle$ in this equation. This integral is over all space, but significant contributions to its magnitude come only from regions where ψ_m^0 and ψ_n^0 are of significant magnitude. In regions well outside the atom or molecule, ψ_m^0 and ψ_n^0 are vanishingly small, and such regions can be ignored. Let the coordinate origin be chosen within the atom or molecule. Since regions well outside the atom can be ignored, the coordinate z_i can be considered to have a maximum magnitude of the order of several angstroms. For ultraviolet light the wavelength λ is of the order of 10^3 Å; for visible, infrared, microwave, and radio-frequency radiation, λ is even larger. Hence $2\pi z_i/\lambda$ is very small and can be neglected, and this leaves $\sum_i Q_i x_i \sin \omega t$ in the integral.

Use of the identity (Problem 1.11) $\sin \omega t = (e^{i\omega t} - e^{-i\omega t})/2i$ gives

$$b_m(t') \approx \delta_{mn} + \frac{\mathscr{E}_0}{2\hbar} \left\langle \psi_m^0 \left| \sum_i Q_i x_i \right| \psi_n^0 \right\rangle \int_0^{t'} [e^{i(\omega_{mn} + \omega)t} - e^{i(\omega_{mn} - \omega)t}] \, dt$$

Using $\int_0^{t'} e^{at}\,dt = a^{-1}(e^{at'} - 1)$, we get

$$b_m(t') \approx \delta_{mn} + \frac{\mathscr{E}_0}{2\hbar i}\left\langle \psi_m^0 \left| \sum_i Q_i x_i \right| \psi_n^0 \right\rangle \left[\frac{e^{i(\omega_{mn}+\omega)t'} - 1}{\omega_{mn}+\omega} - \frac{e^{i(\omega_{mn}-\omega)t'} - 1}{\omega_{mn}-\omega} \right] \quad (9.173)$$

For $m \neq n$, the δ_{mn} term equals zero.

As noted at the end of Section 9.9, $|b_m(t')|^2$ gives the probability of a transition to state m from state n. There are two cases where this probability becomes of significant magnitude. If $\omega_{mn} = \omega$, the denominator of the second fraction in brackets is zero and this fraction's absolute value is large (but not infinite—see Problem 9.18). If $\omega_{mn} = -\omega$, the first fraction has a zero denominator and a large absolute value.

For $\omega_{mn} = \omega$, Eq. (9.172) gives $E_m^0 - E_n^0 = h\nu$. Exposure of the atom to radiation of frequency ν has produced a transition from stationary state n to stationary state m, where (since ν is positive) $E_m^0 > E_n^0$. We might suppose that the energy for this transition came from the absorption by the system of a photon of energy $h\nu$. This supposition is confirmed by a fully quantum-mechanical treatment (called *quantum field theory*) in which the radiation is treated quantum-mechanically rather than classically. We have *absorption* of radiation with a consequent increase in the system's energy.

For $\omega_{mn} = -\omega$, we get $E_n^0 - E_m^0 = h\nu$. Exposure to radiation of frequency ν has induced a transition from stationary state n to stationary state m, where (since ν is positive) $E_n^0 > E_m^0$. The system has gone to a lower energy level, and a quantum-field-theory treatment shows that a photon of energy $h\nu$ is emitted in this process. This is *stimulated emission* of radiation. Stimulated emission is the basis of the laser.

A defect of our treatment is that it does not predict *spontaneous emission*, the emission of a photon by a system not exposed to radiation, the system falling to a lower energy level in the process. Quantum field theory does predict spontaneous emission.

Note from (9.173) that the probability of absorption is proportional to $|\langle \psi_m^0 | \sum_i Q_i x_i | \psi_n^0 \rangle|^2$. The quantity $\sum_i Q_i x_i$ is the x component of the system's dipole-moment operator $\hat{\mathbf{d}}$ (see Section 13.12 for details), which is [Eqs. (13.147) and (13.148)] $\hat{\mathbf{d}} = \mathbf{i}\sum_i Q_i x_i + \mathbf{j}\sum_i Q_i y_i + \mathbf{k}\sum_i Q_i z_i = \mathbf{i}\hat{d}_x + \mathbf{j}\hat{d}_y + \mathbf{k}\hat{d}_z$, where $\mathbf{i}$, $\mathbf{j}$, $\mathbf{k}$ are unit vectors along the axes and $\hat{d}_x$, $\hat{d}_y$, $\hat{d}_z$ are the components of $\hat{\mathbf{d}}$. We assumed polarized radiation with an electric field in the x direction only. If the radiation has electric-field components in the y and z directions also, then the probability of absorption will be proportional to

$$|\langle \psi_m^0 | \hat{d}_x | \psi_n^0 \rangle|^2 + |\langle \psi_m^0 | \hat{d}_y | \psi_n^0 \rangle|^2 + |\langle \psi_m^0 | \hat{d}_z | \psi_n^0 \rangle|^2 = |\langle \psi_m^0 | \hat{\mathbf{d}} | \psi_n^0 \rangle|^2$$

where Eq. (5.35) was used. The integral $\langle \psi_m^0 | \hat{\mathbf{d}} | \psi_n^0 \rangle = \mathbf{d}_{mn}$ is the *transition (dipole) moment*.

When $\mathbf{d}_{mn} = 0$, the transition between states m and n with absorption or emission of radiation is said to be *forbidden*. *Allowed* transitions have $\mathbf{d}_{mn} \neq 0$. [Because of approximations made in the derivation of (9.173), forbidden transitions may have some small probability of occurring.]

Consider, for example, the particle in a one-dimensional box (Section 2.2).

The transition dipole moment is $\langle \psi_m^0 | Qx | \psi_n^0 \rangle$, where Q is the particle's charge and x is its coordinate and where $\psi_m^0 = (2/l)^{1/2} \sin (m\pi x/l)$ and $\psi_n^0 = (2/l)^{1/2} \sin (n\pi x/l)$. Evaluation of this integral (Problem 9.19) shows it is nonzero only when $m - n = \pm 1, \pm 3, \pm 5, \ldots$ and is zero when $m - n = 0, \pm 2, \ldots$. The *selection rule* for a particle in a one-dimensional box is that the quantum number must change by an odd integer when radiation is absorbed or emitted.

Evaluation of the transition moment for the harmonic oscillator and for the two-particle rigid rotor gives the selection rules stated in Sections 4.3 and 6.3.

The quantity $|b_m|^2$ in (9.173) is sharply peaked at $\omega = \omega_{mn}$ and $\omega = -\omega_{mn}$, but there is a nonzero probability that a transition will occur when ω is not precisely equal to $|\omega_{mn}|$, i.e., when $h\nu$ is not precisely equal to $|E_m^0 - E_n^0|$. This fact is related to the energy–time uncertainty relation (5.22). States with a finite lifetime have an uncertainty in their energy.

Radiation is not the only example of a time-dependent perturbation that produces transitions between states. When an atom or molecule comes close to another atom or molecule, it suffers a time-dependent perturbation that can change its state. Selection rules derived for radiative transitions need not apply to collision processes, since $\hat{H}'(t)$ differs.

PROBLEMS

9.1 For the anharmonic oscillator with Hamiltonian (9.3), calculate $E_0^{(1)}$ and $E_1^{(1)}$, the first-order corrections for the ground state and for the first excited state; use (9.4) for $\hat{H}^0$.

9.2 Consider the one-dimensional potential-energy function

$$V = \infty \quad \text{for } x < 0 \quad \text{and} \quad x > a$$
$$V = 0 \quad \text{for } 0 \leqslant x \leqslant \tfrac{1}{4}a \quad \text{and} \quad \tfrac{3}{4}a \leqslant x \leqslant a$$
$$V = k \quad \text{for } \tfrac{1}{4}a < x < \tfrac{3}{4}a$$

where k is a constant. Treat the system as a perturbed particle in a box, and find the first-order energy and wave-function corrections.

9.3 Assume that the charge of the proton is distributed uniformly throughout the volume of a sphere of radius 10^{-13} cm. Use perturbation theory to find the shift in the ground-state hydrogen-atom energy due to the finite proton size. The potential energy experienced by the electron when it has penetrated the nucleus and is at distance r from the nuclear center is $-eQ/4\pi\varepsilon_0 r$, where Q is the amount of proton charge within the sphere of radius r (*Halliday and Resnick*, Section 28-8). The evaluation of the integral is simplified by noting that the exponential factor in ψ is essentially equal to 1 within the nucleus.

9.4 Prove that the expression (9.46) for $E_n^{(2)}$ does not depend on the choice $a_n = 0$.

9.5 When Hylleraas began his calculations on helium, it was not known whether the isolated hydride ion H⁻ was a stable entity. Calculate the ground-state energy of H⁻ predicted by the trial function (9.77). Compare the result with the ground-state energy of the hydrogen atom, -13.60 eV, and show that this simple variation function (erroneously) indicates H⁻ is unstable with respect to ionization into a hydrogen atom and an electron. (More complicated variational functions give a ground-state energy of -14.35 eV.)

9.6 Show that the secular equation (9.114) can be written as

$$\det[H_{mi} - \delta_{mi}(E_j^{(0)} + E_j^{(1)})] = 0$$

where $H_{mi} = \int \psi_m^{(0)*} \hat{H} \psi_i^{(0)} \, d\tau$.

9.7 By considering Eq. (9.108) for the case $m > n$, derive Eq. (9.119).

9.8 Consider the perturbation treatment of the helium configurations $1s3s$, $1s3p$, and $1s3d$. Without setting up the secular equation, but simply by analogy to the results of Section 9.7, write down the 18 correct zeroth-order wave functions. How many energy levels correspond to each of these three configurations, and what is the degeneracy of each energy level? The levels of which configuration lie lowest? highest?

9.9 There is more than one way to divide a Hamiltonian $\hat{H}$ into an unperturbed part $\hat{H}^0$ and a perturbation $\hat{H}'$. Instead of the division (9.51) and (9.52), consider the following way of dividing up the helium-atom Hamiltonian:

$$\hat{H}^0 = -\frac{\hbar^2}{2m}\nabla_1^2 - \frac{\hbar^2}{2m}\nabla_2^2 - \left(Z - \frac{5}{16}\right)\frac{e'^2}{r_1} - \left(Z - \frac{5}{16}\right)\frac{e'^2}{r_2}$$

$$\hat{H}' = -\frac{5}{16}\frac{e'^2}{r_1} - \frac{5}{16}\frac{e'^2}{r_2} + \frac{e'^2}{r_{12}}$$

What are the unperturbed wave functions? Calculate $E^{(0)}$ and $E^{(1)}$ for the ground state. (See Section 9.4.)

9.10 Verify the result (9.157) for the integral J_{1s2p}.

9.11 We have considered helium configurations in which only one electron is excited. Get a rough estimate of the energy of the $2s^2$ configuration from Eq. (9.59). Compare this to the ground-state energy of the He^+ ion to show that the $2s^2$ helium configuration is unstable with respect to ionization to He^+ and an electron. If we had obtained a more accurate estimate of the $2s^2$ energy by including the first-order energy correction, would this increase or decrease our estimate of the $2s^2$ energy?

9.12 Most (but not all) of the effect of nuclear motion in helium can be corrected for by replacing the electron's mass m by the reduced mass (6.67) in the expression for the energy. The energy of helium is proportional to what power of m? [See Eq. (9.87).] Use of μ instead of m multiplies the energies calculated on the basis of infinite nuclear mass by what factor?

9.13 For a hydrogen atom perturbed by a uniform applied electric field in the z direction, the perturbation Hamiltonian is

$$\hat{H}' = e\mathscr{E}z = e\mathscr{E}r\cos\theta$$

where $\mathscr{E}$ is the magnitude of the electric field. Consider the effect of $\hat{H}'$ on the $n = 2$ energy level, which is fourfold degenerate. Since $\hat{H}'$ commutes with the angular-momentum operator $\hat{L}_z$, the ideas of Section 9.6 lead us to set up the secular determinant using the complex hydrogen-atom orbitals $2s$, $2p_1$, $2p_0$, and $2p_{-1}$, which are eigenfunctions of $\hat{L}_z$. Set up the secular determinant using the fact that matrix elements of $\hat{H}'$ between states with different values of the quantum number m will vanish; also use parity considerations to show that certain other integrals are zero (see Problem 7.16d). Evaluate the nonzero integrals, and find the first-order energy corrections and the correct zeroth-order wave functions.

9.14 For helium the first-order perturbation energy correction is e'^2/r_{12} averaged over the correct unperturbed wave function. Show that if we evaluate $\langle e'^2/r_{12}\rangle$ using the incorrect zeroth-order functions $1s(1)2s(2)$ or $1s(2)2s(1)$, we get J_{1s2s} in each case. Now show that when we use the correct functions (9.148) and (9.149) to evaluate $\langle e'^2/r_{12}\rangle$, we get $J_{1s2s} \pm K_{1s2s}$ (as found from the secular equation). The exchange-integral contribution to the energy thus arises from the indistinguishability of electrons in the same atom.

9.15 Calculate $\langle r_1\rangle$ for the He trial function (9.77). (To save time, use the result of Problem 6.4a.)

9.16 (a) Set f in (7.66) equal to $\hat{B}S$ and operate on each side of the resulting equation with $\hat{A}$. Then multiply by R^* and integrate over all space, thereby obtaining the sum rule

$$\sum_i \langle R|\hat{A}|\varphi_i\rangle\langle\varphi_i|\hat{B}|S\rangle = \langle R|\hat{A}\hat{B}|S\rangle$$

where the functions φ_i form a complete orthonormal set, the functions R and S are any two well-behaved functions, the operators $\hat{A}$ and $\hat{B}$ are linear, and the sum is over all members of the complete set. [For other sum rules, see A. Dalgarno, *Rev. Mod. Phys.*, **35**, 522 (1963).] (b) An approximate way to evaluate $E_n^{(2)}$ in (9.46) is to replace $E_n^{(0)} - E_k^{(0)}$ by ΔE, where ΔE is some sort of average excitation energy for the problem, whose value can be roughly estimated from the spacings of the unperturbed levels. Use the sum rule of (a) to show that this replacement leads to

$$E_n^{(2)} \approx \frac{1}{\Delta E}[\langle n|(\hat{H}')^2|n\rangle - (\langle n|\hat{H}'|n\rangle)^2]$$

where n stands for $\psi_n^{(0)}$.

9.17 True or false: (a) Any linear combination of solutions of the time-dependent Schroedinger equation is a solution of this equation. (b) Any linear combination of solutions of the time-independent Schroedinger equation is a solution of this equation.

9.18 Evaluate $\lim_{s \to 0} (e^{as} - 1)/s$. [$s$ corresponds to $\omega_{mn} \pm \omega$ in (9.173).]

9.19 Evaluate $\langle \psi_m^0 | Qx | \psi_n^0 \rangle$ for the particle in a one-dimensional box.

9.20 Find the selection rules for a charged particle in a three-dimensional box exposed to unpolarized radiation.

10 *ELECTRON SPIN AND THE PAULI PRINCIPLE*

10.1 *ELECTRON SPIN*

All chemists are familiar with the yellow color imparted to a flame by sodium atoms. Examining the emission spectrum of sodium, one finds that the strongest yellow line (the so-called D line) is actually two closely spaced lines. The sodium D line arises from a transition from the excited configuration $1s^2 2s^2 2p^6 3p$ to the ground state. The doublet nature of this and other lines in the Na spectrum indicates a doubling of the expected number of states available to the valence electron.

To explain this *fine structure* of atomic spectra, Uhlenbeck and Goudsmit proposed in 1925 that the electron has an *intrinsic* angular momentum in addition to the orbital angular momentum due to its motion about the nucleus. If one pictures the electron as a sphere of charge spinning about one of its diameters, one can see how such an intrinsic angular momentum can arise. Hence we have the term *spin angular momentum* or, more simply, *spin*. However, electron "spin" is not a classical effect, and the picture of an electron rotating about one of its axes should not be considered to represent physical reality. The intrinsic angular momentum is real, but there is no easily visualizable model that can explain its origin properly. We cannot hope to obtain a proper understanding of microscopic particles based on models taken from our experience in the macroscopic world. Other elementary particles besides the electron have spin angular momentum.

In 1928 Dirac developed the *relativistic* quantum mechanics of an electron, and in his treatment electron spin arises naturally. Dirac's theory also indicated the existence of positively charged electrons, positrons, although Dirac did not fully realize this in 1928. Positrons were discovered in 1932. The positron is the antiparticle of the electron. Nowadays, speculation on the possible existence of galaxies composed of antimatter is commonplace.

In the nonrelativistic quantum mechanics to which we are confining ourselves, electron spin must be introduced as an additional hypothesis. We have learned that each physical property has its corresponding linear Hermitian operator in quantum mechanics. For such properties as orbital angular momentum, we can construct the quantum-mechanical operator from the classical expression by replacing p_x, p_y, p_z by the appropriate operators. The inherent spin angular momentum of a microscopic particle has no analog in classical mechanics, so that we cannot use this method to construct operators for spin. For our purposes, we will simply use symbols for the spin operators, without giving an explicit form for them.

Analogous to the orbital angular-momentum operators $\hat{L}^2$, $\hat{L}_x$, $\hat{L}_y$, $\hat{L}_z$, we have the spin angular-momentum operators $\hat{S}^2$, $\hat{S}_x$, $\hat{S}_y$, $\hat{S}_z$. The operator $\hat{S}^2$ is for the square of the magnitude of the total spin angular momentum of a particle; $\hat{S}_z$ is the operator for the z component of the particle's spin angular momentum. We have

$$\hat{S}^2 = \hat{S}_x^2 + \hat{S}_y^2 + \hat{S}_z^2 \tag{10.1}$$

We hypothesize that the spin angular-momentum operators obey the same commutation relations as the orbital angular-momentum operators. Analogous to Eqs. (5.70) and (5.72), we have

$$[\hat{S}_x, \hat{S}_y] = i\hbar\hat{S}_z, \qquad [\hat{S}_y, \hat{S}_z] = i\hbar\hat{S}_x, \qquad [\hat{S}_z, \hat{S}_x] = i\hbar\hat{S}_y \tag{10.2}$$

From (10.1) and (10.2), it follows, by the same operator algebra used to obtain (5.73) and (5.74), that

$$[\hat{S}^2, \hat{S}_x] = [\hat{S}^2, \hat{S}_y] = [\hat{S}^2, \hat{S}_z] = 0 \tag{10.3}$$

Since Eqs. (10.1) and (10.2) are of the form of Eqs. (5.149) and (5.150), it follows from the work of Section 5.4 (which depended only on the commutation relations and not the specific forms of the operators) that the eigenvalues of $\hat{S}^2$ are [Eq. (5.192)]

$$s(s+1)\hbar^2, \qquad s = 0, \tfrac{1}{2}, 1, \tfrac{3}{2}, \ldots \tag{10.4}$$

and the eigenvalues of $\hat{S}_z$ are [Eq. (5.191)]

$$m_s\hbar, \qquad m_s = -s, -s+1, \ldots, s-1, s \tag{10.5}$$

The quantum number s is called the *spin* of the particle. Although nothing in our work in Section 5.4 restricts electrons to a single value for s, experiment shows that all electrons do have a single value for s, namely $s = \tfrac{1}{2}$. (Protons and neutrons also have spin $\tfrac{1}{2}$; pi-mesons[1] have $s = 0$.) Thus the magnitude of the total intrinsic (spin) angular momentum of an electron is

$$[\tfrac{1}{2}(\tfrac{3}{2})\hbar^2]^{1/2} = \tfrac{1}{2}\sqrt{3}\,\hbar \tag{10.6}$$

---

[1] Photons have spin quantum number $s = 1$. However, Eq. (10.5) does not hold for photons. Photons travel at speed c in vacuum; because of their relativistic nature it turns out that photons can have either $m_s = +1$ or $m_s = -1$, but $m_s = 0$ is not allowed (see *Merzbacher*, Chapter 22). These two m_s values correspond to left-circularly polarized and right-circularly polarized electromagnetic radiation.

Corresponding to $s = \frac{1}{2}$, we have two possible eigenvalues of $\hat{S}_z$: $+\frac{1}{2}\hbar$ and $-\frac{1}{2}\hbar$. The electron spin eigenfunctions corresponding to these eigenvalues of $\hat{S}_z$ are denoted by α and β:

$$\hat{S}_z\alpha = +\tfrac{1}{2}\hbar\alpha \tag{10.7}$$

$$\hat{S}_z\beta = -\tfrac{1}{2}\hbar\beta \tag{10.8}$$

Since $\hat{S}_z$ commutes with $\hat{S}^2$, we can take the eigenfunctions of $\hat{S}_z$ to be eigenfunctions of $\hat{S}^2$ also:

$$\hat{S}^2\alpha = \tfrac{3}{4}\hbar^2\alpha, \qquad \hat{S}^2\beta = \tfrac{3}{4}\hbar^2\beta \tag{10.9}$$

$\hat{S}_z$ does not commute with $\hat{S}_x$ or $\hat{S}_y$, so that α and β are not eigenfunctions of these operators. The terms *spin up* and *spin down* refer to $m_s = +\frac{1}{2}$ and $m_s = -\frac{1}{2}$, respectively. See Fig. 10.1. We will later show that the two possibilities for the quantum number m_s give the doubling of lines in the spectra of the alkali metals.

    The wave functions we have dealt with previously are functions of the spatial coordinates of the particle: $\psi = \psi(x, y, z)$. We might ask: What is the variable for the spin eigenfunctions α and β? Sometimes one talks of a spin coordinate ω, without really specifying what this coordinate is. Most often, one takes the spin quantum number m_s as being the variable on which the spin eigenfunctions depend. This procedure is quite unusual as compared to the spatial wave functions, but because we have only two possible electronic spin eigenfunctions and eigenvalues, this is a convenient choice. We have

$$\alpha = \alpha(m_s), \qquad \beta = \beta(m_s) \tag{10.10}$$

    As usual, we want our eigenfunctions to be normalized. The three variables of a one-particle space wave function range continuously from $-\infty$ to $+\infty$, so that normalization means

$$\int_{-\infty}^{\infty} \int_{-\infty}^{\infty} \int_{-\infty}^{\infty} |\psi(x, y, z)|^2 \, dx \, dy \, dz = 1$$

The variable m_s of the electronic spin eigenfunctions takes on only the two discrete values $+\frac{1}{2}$ and $-\frac{1}{2}$, so that normalization of the one-particle spin eigenfunctions means

$$\sum_{m_s = -1/2}^{1/2} |\alpha(m_s)|^2 = 1, \qquad \sum_{m_s = -1/2}^{1/2} |\beta(m_s)|^2 = 1 \tag{10.11}$$

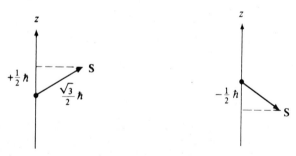

Figure 10.1 *The two possible orientations of the electronic spin vector with respect to the z axis*

Since the eigenfunctions α and β correspond to different eigenvalues of the Hermitian operator $\hat{S}_z$, they are orthogonal:

$$\sum_{m_s=-1/2}^{1/2} \alpha^*(m_s)\beta(m_s) = 0 \tag{10.12}$$

To satisfy (10.11) and (10.12), we can take

$$\alpha(m_s) = 1 \qquad \text{for } m_s = +\tfrac{1}{2}$$

$$\alpha(m_s) = 0 \qquad \text{for } m_s = -\tfrac{1}{2}$$

$$\beta(+\tfrac{1}{2}) = 0, \qquad \beta(-\tfrac{1}{2}) = 1$$

which may be written more compactly using Kronecker deltas:

$$\alpha(m_s) = \delta_{m_s,\,1/2}, \quad \beta(m_s) = \delta_{m_s,\,-1/2} \tag{10.13}$$

When we consider the complete wave function for an electron including both space and spin variables, we will normalize it according to

$$\sum_{m_s=-1/2}^{1/2} \int_{-\infty}^{\infty} \int_{-\infty}^{\infty} \int_{-\infty}^{\infty} |\psi(x, y, z, m_s)|^2 \, dx \, dy \, dz = 1 \tag{10.14}$$

The notation

$$\int |\psi(x, y, z, m_s)|^2 \, d\tau$$

will denote summation over the spin variable and integration over the space variables, as in (10.14). The symbol $\int dv$ will denote integration over spatial variables.

10.2   LADDER OPERATORS FOR ELECTRON SPIN

The spin angular-momentum operators obey the general angular-momentum commutation relations of Section 5.4, and it is often helpful to use spin angular-momentum ladder operators.

From (5.152) and (5.153), the raising and lowering operators for spin angular momentum are

$$\hat{S}_+ = \hat{S}_x + i\hat{S}_y \tag{10.15}$$

$$\hat{S}_- = \hat{S}_x - i\hat{S}_y \tag{10.16}$$

and we have [Eqs. (5.154) and (5.155)]

$$\hat{S}_+\hat{S}_- = \hat{S}^2 - \hat{S}_z^2 + \hbar\hat{S}_z \tag{10.17}$$

$$\hat{S}_-\hat{S}_+ = \hat{S}^2 - \hat{S}_z^2 - \hbar\hat{S}_z \tag{10.18}$$

The spin functions α and β are eigenfunctions of $\hat{S}_z$ with eigenvalues $+\tfrac{1}{2}\hbar$ and $-\tfrac{1}{2}\hbar$, respectively. Since $\hat{S}_+$ is the raising operator, the function $\hat{S}_+\beta$ is an eigenfunction of $\hat{S}_z$ with eigenvalue $+\tfrac{1}{2}\hbar$. The most general eigenfunction of $\hat{S}_z$ with this eigenvalue is an arbitrary constant times α. Hence

$$\hat{S}_+\beta = c\alpha \tag{10.19}$$

where c is some constant. To find c, we use Eq. (10.11):

$$1 = \sum_{m_s} \alpha^*(m_s)\alpha(m_s) = \sum \left(\frac{\hat{S}_+\beta}{c}\right)^*\left(\frac{\hat{S}_+\beta}{c}\right)$$

$$|c|^2 = \sum (\hat{S}_+\beta)^*\hat{S}_+\beta = \sum (\hat{S}_+\beta)^*(\hat{S}_x + i\hat{S}_y)\beta$$

$$|c|^2 = \sum (\hat{S}_+\beta)^*\hat{S}_x\beta + i\sum (\hat{S}_+\beta)^*\hat{S}_y\beta \qquad (10.20)$$

We now use the Hermitian property of $\hat{S}_x$ and $\hat{S}_y$. For an operator $\hat{A}$ that acts on functions of a continuous variable like x, the Hermitian property is

$$\int f^*(x)\,\hat{A}g(x)\,dx = \int g(x)[\hat{A}f(x)]^*\,dx$$

For an operator such as $\hat{S}_x$ that acts on functions of the variable m_s, which takes on discrete values, the Hermitian property is

$$\sum_{m_s} f^*(m_s)\,\hat{S}_x g(m_s) = \sum_{m_s} g(m_s)\,[\hat{S}_x f(m_s)]^* \qquad (10.21)$$

Taking $f = \hat{S}_+\beta$ and $g = \beta$, we can write (10.20) as

$$c^*c = \sum \beta[\hat{S}_x\hat{S}_+\beta]^* + i\sum \beta[\hat{S}_y\hat{S}_+\beta]^*$$

Taking the complex conjugate of this equation, we have

$$cc^* = \sum \beta^*\hat{S}_x\hat{S}_+\beta - i\sum \beta^*\hat{S}_y\hat{S}_+\beta$$

$$|c|^2 = \sum \beta^*(\hat{S}_x - i\hat{S}_y)\hat{S}_+\beta = \sum \beta^*\hat{S}_-\hat{S}_+\beta$$

$$|c|^2 = \sum \beta^*(\hat{S}^2 - \hat{S}_z^2 - \hbar\hat{S}_z)\beta$$

$$|c|^2 = \sum \beta^*\left(\frac{3}{4}\hbar^2 - \frac{\hbar^2}{4} + \frac{\hbar^2}{2}\right)\beta = \hbar^2\sum \beta^*\beta = \hbar^2$$

$$|c| = \hbar$$

Choosing the phase of c as zero, we have $c = \hbar$, and (10.19) reads

$$\hat{S}_+\beta = \hbar\alpha \qquad (10.22)$$

A similar calculation gives

$$\hat{S}_-\alpha = \hbar\beta \qquad (10.23)$$

Since α is the eigenfunction with the highest possible value of m_s, the operator $\hat{S}_+$ acting on α must annihilate it [Eq. (5.181)]:

$$\hat{S}_+\alpha = 0 \qquad (10.24)$$

Likewise

$$\hat{S}_-\beta = 0 \qquad (10.25)$$

From these last four equations, we get

$$(\hat{S}_+ + \hat{S}_-)\beta = \hbar\alpha \qquad (10.26)$$

$$(\hat{S}_+ - \hat{S}_-)\beta = \hbar\alpha \qquad (10.27)$$

Use of Eqs. (10.15) and (10.16) in Eqs. (10.26) and (10.27) gives

$$\hat{S}_x\beta = \tfrac{1}{2}\hbar\alpha, \qquad \hat{S}_y\beta = -\tfrac{1}{2}i\hbar\alpha \tag{10.28}$$

Similarly we find

$$\hat{S}_x\alpha = \tfrac{1}{2}\hbar\beta, \qquad \hat{S}_y\alpha = \tfrac{1}{2}i\hbar\beta \tag{10.29}$$

10.3   SPIN AND THE HYDROGEN ATOM

The wave function specifying the state of an electron depends not only on the coordinates x, y, and z but also on the spin state of the electron. What effect does this have on the wave functions and energy levels of the hydrogen atom?

To a very good approximation, the Hamiltonian for a system of electrons does not involve the spin variables but is a function only of spatial coordinates and derivatives with respect to spatial coordinates. As a result, we can separate the wave function of a single electron into a product of space and spin parts:

$$\psi(x, y, z)g(m_s) \tag{10.30}$$

where $g(m_s)$ is either one of the functions α and β, depending on whether $m_s = \tfrac{1}{2}$ or $-\tfrac{1}{2}$. Since the Hamiltonian operator has no effect on the spin function, we have

$$\hat{H}[\psi(x, y, z)g(m_s)] = g(m_s)\hat{H}\psi(x, y, z) = E[\psi(x, y, z)g(m_s)]$$

so that we get the same energies as previously found without taking spin into account. The only difference spin makes is to double the possible number of states; instead of the state $\psi(x, y, z)$, we have the two possible states $\psi(x, y, z)\alpha$ and $\psi(x, y, z)\beta$. When we take spin into account, the degeneracy of the hydrogen-atom energy levels is $2n^2$ rather than n^2, as given by Eq. (6.113).

10.4   THE PAULI PRINCIPLE

Suppose we have a system of several identical particles. In classical mechanics the identity of the particles leads to no special consequences. For example, consider n identical billiard balls rolling about on a billiard table. It is perfectly possible to follow the motion of any individual ball, say by taking a motion picture of the system. We can say that ball number one is moving along a certain path, ball two is on another definite path, and so on, the paths being determined by Newton's laws of motion. Thus although the balls are identical, we can distinguish among them by specifying the path each one takes. The identity of the balls has no special effect on their motions.

In quantum mechanics the uncertainty principle tells us that we cannot follow the exact path taken by a microscopic "particle." If the microscopic particles of our system all have different masses or charges or spins, etc., we can use one of these properties to distinguish the particles from one another. But if they are all identical, then the one way we had in classical mechanics of distinguishing them, namely by specifying their paths, is lost in quantum mechanics because of the uncertainty principle. Therefore the wave function of a system of interacting identical particles must not distinguish among the particles. For example, in the perturbation treatment of the helium-atom excited states in Chapter 9, we saw that the function $1s(1)2s(2)$, which says that electron 1 is in the $1s$ orbital and electron 2 is in the $2s$ orbital, was not a correct zeroth-order wave function. Rather we had to use the functions (9.161) and (9.162), which do not specify which electron is in which orbital. (If the identical particles are well separated from each other, so that their wave functions do not overlap, they may be regarded as distinguishable.)

We now derive the restrictions on the wave function due to the requirement of indistinguishability of identical particles in quantum mechanics. We consider a system of n identical microscopic particles. The wave function depends on the space and spin variables of all the particles. For particle 1 these variables are x_1, y_1, z_1, m_{s1}. We will use the symbol q_1 to stand for all four of these variables. Thus

$$\psi = \psi(q_1, q_2, \ldots, q_n)$$

We define the *permutation operator* $\hat{P}_{12}$ as the operator that interchanges all the coordinates of particles 1 and 2:

$$\hat{P}_{12} f(q_1, q_2, q_3, \ldots, q_n) = f(q_2, q_1, q_3, \ldots, q_n)$$

As an example, the effect of $\hat{P}_{12}$ on the function that has electron 1 in a $1s$ orbital with spin up and electron 2 in a $3s$ orbital with spin down is

$$\hat{P}_{12}[1s(1)\alpha(1)3s(2)\beta(2)] = 1s(2)\alpha(2)3s(1)\beta(1)$$

What are the eigenvalues of $\hat{P}_{12}$? Applying $\hat{P}_{12}$ twice has no net effect:

$$\hat{P}_{12}\hat{P}_{12} f(q_1, q_2, \ldots, q_n) = \hat{P}_{12} f(q_2, q_1, \ldots, q_n) = f(q_1, q_2, \ldots, q_n)$$

so that $\hat{P}_{12}^2 = \hat{1}$. We showed in Section 7.5 that an operator whose square is the unit operator has the eigenvalues $+1$ and -1. If h is an eigenfunction of $\hat{P}_{12}$ with eigenvalue $+1$, we have

$$\hat{P}_{12}h(q_1, q_2, \ldots, q_n) = +1 \cdot h(q_1, q_2, \ldots, q_n)$$

$$h(q_2, q_1, \ldots, q_n) = h(q_1, q_2, \ldots, q_n) \qquad (10.31)$$

and we say that h is *symmetric* with respect to interchange of particles 1 and 2. For eigenvalue -1 we have

$$h(q_2, q_1, \ldots, q_n) = -h(q_1, q_2, \ldots, q_n) \qquad (10.32)$$

If (10.32) holds, h is *antisymmetric* with respect to interchange of particles 1 and 2. There is no necessity for an arbitrary function $f(q_1, q_2, \ldots, q_n)$ to be either symmetric or antisymmetric with respect to interchange of 1 and 2. However, we can write an arbitrary function as the sum of functions symmetric and antisymmetric with respect to 1–2 interchange:

$$f(q_1, q_2, \ldots, q_n) = \tfrac{1}{2}[f(q_1, q_2, \ldots, q_n) + f(q_2, q_1, \ldots, q_n)]$$

$$+ \tfrac{1}{2}[f(q_1, q_2, \ldots, q_n) - f(q_2, q_1, \ldots, q_n)]$$

This equation shows that the eigenfunctions of $\hat{P}_{12}$ form a complete set.

Do not confuse the property of being symmetric or antisymmetric with respect to particle interchange with the property of being even or odd with respect to inversion. The function $(x_1 + x_2)$ is symmetric with respect to 1–2 interchange and is an odd function of x_1 and x_2. The function $(x_1^2 + x_2^2)$ is symmetric with respect to 1–2 interchange and is an even function of x_1 and x_2.

The operator $\hat{P}_{ij}$ is defined by

$$\hat{P}_{ij} f(q_1, \ldots, q_i, \ldots, q_j, \ldots, q_n) = f(q_1, \ldots, q_j, \ldots, q_i, \ldots, q_n) \quad \textbf{(10.33)}$$

The eigenvalues of $\hat{P}_{ij}$ are, like those of $\hat{P}_{12}$, $+1$ and -1.

We now consider the wave function of a system of identical microscopic particles. Since the particles are indistinguishable, the way we label them cannot affect the state of the system. Thus the two wave functions

$$\psi(q_1, \ldots, q_i, \ldots, q_j, \ldots, q_n) \quad \text{and} \quad \psi(q_1, \ldots, q_j, \ldots, q_i, \ldots, q_n)$$

must correspond to the same state of the system. Two wave functions that correspond to the same state can differ at most by a multiplicative constant. Hence

$$\psi(q_1, \ldots, q_j, \ldots, q_i, \ldots, q_n) = c\psi(q_1, \ldots, q_i, \ldots, q_j, \ldots, q_n)$$

$$\hat{P}_{ij}\psi(q_1, \ldots, q_i, \ldots, q_j, \ldots, q_n) = c\psi(q_1, \ldots, q_i, \ldots, q_j, \ldots, q_n) \quad \textbf{(10.34)}$$

Equation (10.34) states that ψ is an eigenfunction of $\hat{P}_{ij}$. But we know that the only possible eigenvalues of $\hat{P}_{ij}$ are 1 and -1. We conclude that the wave function for a system of n identical particles must be symmetric or antisymmetric with respect to interchange of any two of the identical particles, i and j. Since the n particles are all identical, we could not have the wave function symmetric with respect to some interchanges and antisymmetric with respect to other interchanges. Thus the wave function of n identical particles must be either symmetric with respect to every possible interchange or antisymmetric with respect to every possible interchange of two particles.

We have shown that there are two possible cases for the wave function of a system of identical particles, the symmetric and the antisymmetric cases. Experimental evidence (such as the periodic table of the elements to be discussed later) shows that for electrons only the antisymmetric case occurs. Thus we have an additional postulate of quantum mechanics which states that *the wave function of a system of electrons must be antisymmetric with respect to interchange of any two electrons.* This important postulate is called the *Pauli principle* or the *exclusion principle.* The Pauli exclusion principle was originally stated by Pauli in a more restricted form to be discussed later.

Pauli has shown that relativistic quantum field theory indicates that particles with half-integral spin ($s = \frac{1}{2}, \frac{3}{2}$, etc.) require antisymmetric wave functions, while particles of integral spin ($s = 0$, 1, etc.) require symmetric wave functions. Experimental evidence leads to the same conclusion. Particles requiring antisymmetric wave functions, such as electrons, are called *fermions* (after E. Fermi), while particles requiring symmetric wave functions, such as pi-mesons, are called *bosons* (after S. N. Bose).

The Pauli principle has an interesting consequence for a system of identical fermions. The antisymmetry requirement means that

$$\psi(q_1, q_2, q_3, \ldots, q_n) = -\psi(q_2, q_1, q_3, \ldots, q_n) \tag{10.35}$$

Consider the value of the wave function when electrons 1 and 2 have the same coordinates, i.e., when

$$x_1 = x_2, \qquad y_1 = y_2, \qquad z_1 = z_2, \qquad m_{s1} = m_{s2}$$

Putting $q_2 = q_1$ in (10.35), we have

$$\psi(q_1, q_1, q_3, \ldots, q_n) = -\psi(q_1, q_1, q_3, \ldots, q_n)$$

$$2\psi = 0$$

$$\psi(q_1, q_1, q_3, \ldots, q_n) = 0 \tag{10.36}$$

Thus two electrons with the same spin have zero probability of being found at the same point in three-dimensional space. (By "the same spin," we mean the same value of m_s.) Since ψ is a continuous function, Eq. (10.36) means that the probability of finding two electrons with the same spin close to each other in space is quite small. Thus the Pauli antisymmetry principle forces electrons of like spin to keep apart from one another; to describe this, one often speaks of a *Pauli repulsion* between such electrons. This "repulsion" is not a real physical force, but a reflection of the fact that the electronic wave function must be antisymmetric with respect to exchange.

10.5 *THE HELIUM ATOM*

We now reconsider the helium atom from the standpoint of electron spin and the Pauli principle. In the perturbation treatment of helium in Section 9.3, we found the zeroth-order wave function for the ground state to be $1s(1)1s(2)$. To take spin into account, we must multiply this spatial function by a spin eigenfunction. We therefore consider the possible spin eigenfunctions for two electrons. We will use the notation $\alpha(1)\alpha(2)$ to indicate a state where electron 1 has spin up and electron two has spin up; $\alpha(1)$ stands for $\alpha(m_{s1})$. Since each electron has two possible spin states, we have at first sight the four possible spin functions:

$$\alpha(1)\alpha(2), \qquad \beta(1)\beta(2), \qquad \alpha(1)\beta(2), \qquad \alpha(2)\beta(1)$$

There is nothing wrong with the first two functions, but the third and fourth functions violate the principle of indistinguishability of identical particles. For example, the third function says that electron 1 has spin up and electron 2 has spin down, which *does* distinguish between electrons 1 and 2. More formally, if we apply the permutation operator $\hat{P}_{12}$ to these functions we find that the first two functions are symmetric with respect to interchange of the two electrons, but the third and fourth functions are neither symmetric nor antisymmetric and so are unacceptable.

What now? Recall that we ran into essentially the same situation in treating the helium excited states (Section 9.7), where we started with the functions $1s(1)2s(2)$ and $2s(1)1s(2)$. We found that these two functions, which distinguished between electrons 1 and 2, were not the correct zeroth-order functions and that the correct zeroth-order functions were

$$2^{-1/2}[1s(1)2s(2) \pm 2s(1)1s(2)]$$

This result suggests pretty strongly that instead of $\alpha(1)\beta(2)$ and $\beta(1)\alpha(2)$, we use the spin functions

$$2^{-1/2}[\alpha(1)\beta(2) \pm \beta(1)\alpha(2)]$$

These two functions are the normalized linear combinations of $\alpha(1)\beta(2)$ and $\beta(1)\alpha(2)$ that are eigenfunctions of $\hat{P}_{12}$. To demonstrate that they are normalized, we have

$$\sum_{m_{s1}} \sum_{m_{s2}} \frac{1}{\sqrt{2}}[\alpha(1)\beta(2) \pm \beta(1)\alpha(2)]^* \frac{1}{\sqrt{2}}[\alpha(1)\beta(2) \pm \beta(1)\alpha(2)]$$

$$= \tfrac{1}{2}\sum_{m_{s1}}|\alpha(1)|^2 \sum_{m_{s2}}|\beta(2)|^2 \pm \tfrac{1}{2}\sum_{m_{s1}}\alpha^*(1)\beta(1) \sum_{m_{s2}}\beta^*(2)\alpha(2)$$

$$\pm \tfrac{1}{2}\sum_{m_{s1}}\beta^*(1)\alpha(1)\sum_{m_{s2}}\alpha^*(2)\beta(2) + \tfrac{1}{2}\sum_{m_{s1}}|\beta(1)|^2 \sum_{m_{s2}}|\alpha(2)|^2 = 1$$

where we have used the orthonormality relations (10.11) and (10.12).

Therefore the four normalized two-electron spin eigenfunctions are

symmetric:
$$\begin{cases} \alpha(1)\alpha(2) & \textbf{(10.37)} \\ \beta(1)\beta(2) & \textbf{(10.38)} \\ [\alpha(1)\beta(2) + \beta(1)\alpha(2)]/\sqrt{2} & \textbf{(10.39)} \end{cases}$$

antisymmetric:
$$[\alpha(1)\beta(2) - \beta(1)\alpha(2)]/\sqrt{2} \qquad \textbf{(10.40)}$$

We now include spin in the zeroth-order ground-state wave function. The function $1s(1)1s(2)$ is symmetric with respect to exchange. According to the Pauli principle, the overall wave function including spin must be antisymmetric with respect to interchange of the two electrons. Hence we must multiply the symmetric space function $1s(1)1s(2)$ by an antisymmetric spin function. We have only one antisymmetric two-electron spin function, so that the ground-state zeroth-order wave function for the helium atom including spin is

$$\psi^{(0)} = 1s(1)1s(2) \cdot 2^{-1/2}[\alpha(1)\beta(2) - \beta(1)\alpha(2)] \qquad \textbf{(10.41)}$$

The function $\psi^{(0)}$ is an eigenfunction of $\hat{P}_{12}$ with eigenvalue -1, as the Pauli principle requires.

To a very good approximation, the Hamiltonian does not contain spin terms, so that the energy is unaffected by inclusion of the spin factor in the ground-state wave function. Also, the ground state of helium is still nondegenerate when spin is considered. Since the ground-state helium wave function consists of a *symmetric* space function times an antisymmetric spin function, in doing variational calculations with functions of the form (9.91) we should include only even values of j, so that the factor $(r_1 - r_2)^j$ will be symmetric with respect to interchange of 1 and 2.

To further demonstrate that the spin factor has no effect on the value of the energy, we will assume we are doing a variational calculation for the helium ground state using the trial function

$$\varphi = f(r_1, r_2, r_{12})2^{-1/2}[\alpha(1)\beta(2) - \beta(1)\alpha(2)]$$

where f is a normalized function symmetric in the coordinates of the two electrons. The variational integral is

$$\int \varphi^* \hat{H} \varphi \, d\tau = \sum_{m_{s1}} \sum_{m_{s2}} \int \int f^*(r_1, r_2, r_{12}) \frac{1}{\sqrt{2}} [\alpha(1)\beta(2) - \beta(1)\alpha(2)]^*$$

$$\times \hat{H} f(r_1, r_2, r_{12}) \frac{1}{\sqrt{2}} [\alpha(1)\beta(2) - \beta(1)\alpha(2)] \, dv_1 \, dv_2$$

Since $\hat{H}$ has no effect on the spin functions, the variational integral becomes

$$\int \int f^* \hat{H} f \, dv_1 \, dv_2 \sum_{m_{s1}} \sum_{m_{s2}} \tfrac{1}{2} |\alpha(1)\beta(2) - \beta(1)\alpha(2)|^2$$

The spin function (10.40) is normalized, so that the variational integral reduces to

$$\int \varphi^* \hat{H} \varphi \, d\tau = \int \int f^* \hat{H} f \, dv_1 \, dv_2$$

which is the expression we used before we introduced spin.

Now consider the excited states of helium. We found the lowest excited state to have the zeroth-order space wave function

$$2^{-1/2}[1s(1)2s(2) - 2s(1)1s(2)]$$

Since this space function is antisymmetric, we must multiply it by a symmetric spin function. We can use any one of the three symmetric two-electron spin functions, so that instead of the nondegenerate level previously found, we have a triply degenerate level with the three zeroth-order wave functions

$$2^{-1/2}[1s(1)2s(2) - 2s(1)1s(2)]\alpha(1)\alpha(2) \tag{10.42}$$

$$2^{-1/2}[1s(1)2s(2) - 2s(1)1s(2)]\beta(1)\beta(2) \tag{10.43}$$

$$2^{-1/2}[1s(1)2s(2) - 2s(1)1s(2)]2^{-1/2}[\alpha(1)\beta(2) + \beta(1)\alpha(2)] \tag{10.44}$$

For the next excited state, the requirement of antisymmetry of the overall wave function leads to the zeroth-order wave function

$$2^{-1/2}[1s(1)2s(2) + 2s(1)1s(2)]2^{-1/2}[\alpha(1)\beta(2) - \beta(1)\alpha(2)] \tag{10.45}$$

The same considerations apply for the $1s\,2p$ states.

10.6   *THE LITHIUM ATOM*

So far, we have not seen any very spectacular consequences of electron spin and the Pauli principle. In the hydrogen and helium atoms, the spin factors in the wave functions and the antisymmetry requirement simply affect the degeneracy of the levels but do not (except for very small effects to be considered later) affect the previously obtained energies. For lithium, the story is quite different.

The natural perturbation approach to the lithium atom is to take the interelectronic repulsions as a perturbation on the remaining terms in the Hamiltonian. By the same steps used in the treatment of helium, the unperturbed wave functions are products of three hydrogenlike functions. For the ground state, we have the zeroth-order wave function

$$\psi^{(0)} = 1s(1)\,1s(2)\,1s(3) \tag{10.46}$$

and the zeroth-order (unperturbed) energy is [Eq. (9.59)]

$$E^{(0)} = -\left(\frac{1}{1^2} + \frac{1}{1^2} + \frac{1}{1^2}\right)\left(\frac{Z^2 e'^2}{2a_0}\right) = -27\left(\frac{e'^2}{2a_0}\right)$$

$$E^{(0)} = -27(13.606)\,\text{eV} = -367.4\,\text{eV}$$

The first-order correction to the energy is given by (9.30). The perturbation consists of the interelectronic repulsions, so that

$$E^{(1)} = \int |1s(1)|^2 |1s(2)|^2 |1s(3)|^2 \frac{e'^2}{r_{12}}\,dv + \int |1s(1)|^2 |1s(2)|^2 |1s(3)|^2 \frac{e'^2}{r_{23}}\,dv$$

$$+ \int |1s(1)|^2 |1s(2)|^2 |1s(3)|^2 \frac{e'^2}{r_{13}}\,dv$$

The way we label the dummy integration variables in these definite integrals cannot affect their value. If we interchange the labels 1 and 3 on the variables in the second integral, it is converted to the first integral. Hence these two integrals are equal. Interchange of the labels 2 and 3 in the third integral shows it to be equal to the first integral also. Hence

$$E^{(1)} = 3 \iint |1s(1)|^2 |1s(2)|^2 \frac{e'^2}{r_{12}}\,dv_1\,dv_2 \int |1s(3)|^2\,dv_3$$

The integral over electron 3 gives 1 (normalization). The integral over electrons 1 and 2 was evaluated in the perturbation treatment of helium, and we have [Eqs. (9.63) and (9.67)]

$$E^{(1)} = 3\left(\frac{5Z}{4}\right)\left(\frac{e'^2}{2a_0}\right) = 153.1\,\text{eV}$$

$$E^{(0)} + E^{(1)} = -214.3\,\text{eV}$$

Since we can use the zeroth-order perturbation wave function as a trial variation function (recall the discussion at the beginning of Section 9.4), the value of $E^{(0)} + E^{(1)}$ must be, according to the variation principle, equal to or greater than the true ground-state energy. The experimental value of the lithium ground-state energy is obtained by adding up the three ionization energies, which gives[2]

$$-(5.39 + 75.64 + 122.45)\,\text{eV} = -203.5\,\text{eV}$$

---

[2] C. E. Moore, "Ionization Potentials and Ionization Limits," publication NSRDS-NBS 34 of the National Bureau of Standards (1970).

We thus have $E^{(0)} + E^{(1)}$ as less than the true ground-state energy, which is a violation of the variation principle. Moreover, the supposed configuration $(1s)^3$ for the Li ground state is in complete disagreement with the low value of the first ionization potential and with all chemical evidence. If we continued on in this manner, we would have a $(1s)^Z$ ground-state configuration for the element of atomic number Z. We would not get the well-known periodic behavior of the elements.

Of course, our error is failure to consider spin and the Pauli principle. The hypothetical zeroth-order wave function $1s(1)\,1s(2)\,1s(3)$ is symmetric with respect to interchange of any two electrons. If we are to satisfy the Pauli principle, we must multiply this symmetric space function by an antisymmetric spin function. It is easy to construct completely symmetric spin functions for three electrons, such as $\alpha(1)\alpha(2)\alpha(3)$. However, try as we may, it is impossible to construct a completely antisymmetric spin function for three electrons.

Let us consider how we can systematically construct an antisymmetric function for three electrons. We will use f, g, and h to stand for three functions of electronic coordinates, without specifying whether we are considering space coordinates or spin coordinates or both. We start with the function

$$f(1)\,g(2)\,h(3) \tag{10.47}$$

which is certainly not antisymmetric. The antisymmetric function we desire must be converted into its negative by each of the permutation operators $\hat{P}_{12}$, $\hat{P}_{13}$, and $\hat{P}_{23}$. Applying each of these operators in turn to $f(1)\,g(2)\,h(3)$, we get the functions

$$f(2)\,g(1)\,h(3) \tag{10.48}$$

$$f(3)\,g(2)\,h(1) \tag{10.49}$$

$$f(1)\,g(3)\,h(2) \tag{10.50}$$

We might seek to construct the antisymmetric function as a linear combination of the four functions (10.47)–(10.50), but this could not succeed; the application of $\hat{P}_{12}$ to (10.49) and (10.50) gives

$$f(3)\,g(1)\,h(2) \tag{10.51}$$

$$f(2)\,g(3)\,h(1) \tag{10.52}$$

which are not included in (10.47)–(10.50), so that we must include all six functions (10.47)–(10.52) in our desired antisymmetric linear combination. These six functions are the six $(3 \cdot 2 \cdot 1)$ possible permutations of the three electrons among the three functions f, g, and h. If $f(1)g(2)h(3)$ is a solution of the Schroedinger equation with eigenvalue E, then because of the identity of the particles, each of the functions (10.48)–(10.52) is also a solution with the same eigenvalue E (exchange degeneracy), and any linear combination of these functions is an eigenfunction with eigenvalue E.

The antisymmetric linear combination will have the form

$$c_1 f(1)\,g(2)\,h(3) + c_2 f(2)\,g(1)\,h(3) + c_3 f(3)\,g(2)\,h(1) + c_4 f(1)\,g(3)\,h(2)$$
$$+ c_5 f(3)\,g(1)\,h(2) + c_6 f(2)\,g(3)\,h(1) \tag{10.53}$$

Since $f(2)g(1)h(3) = \hat{P}_{12}f(1)g(2)h(3)$, in order to have (10.53) be an eigenfunction of $\hat{P}_{12}$ with eigenvalue -1, we must have

$$c_2 = -c_1$$

Likewise $f(3)g(2)h(1) = \hat{P}_{13}f(1)g(2)h(3)$ and $f(1)g(3)h(2) = \hat{P}_{23}f(1)g(2)h(3)$ so that $c_3 = -c_1$ and $c_4 = -c_1$. Since $f(3)g(1)h(2) = \hat{P}_{12}f(3)g(2)h(1)$, we must have $c_5 = -c_3 = c_1$. Similarly, we find $c_6 = c_1$. We thus arrive at the linear combination

$$c_1[f(1)g(2)h(3) - f(2)g(1)h(3) - f(3)g(2)h(1) - f(1)g(3)h(2)$$
$$+ f(3)g(1)h(2) + f(2)g(3)h(1)] \tag{10.54}$$

which is easily verified to be antisymmetric with respect to 1–2, 1–3, and 2–3 interchange.[3]

Let us assume f, g, and h to be orthonormal and choose c_1 so that (10.54) is normalized. Multiplying (10.54) by its complex conjugate, we get many terms, but because of the assumed orthogonality the integrals of all products involving two different terms of (10.54) vanish. For example,

$$\int [f(1)g(2)h(3)]^* f(2)g(1)h(3)\, d\tau$$

$$= \int f^*(1)g(1)\, d\tau_1 \int g^*(2)f(2)\, d\tau_2 \int h^*(3)h(3)\, d\tau_3 = 0 \cdot 0 \cdot 1 = 0$$

Integrals involving the product of a term of (10.54) with its own complex conjugate are equal to one, because f, g, and h are normalized. Therefore

$$1 = \int |(10.54)|^2\, d\tau = |c_1|^2(1 + 1 + 1 + 1 + 1 + 1)$$

$$c_1 = 1/\sqrt{6}$$

We could work with (10.54) as it stands, but its properties are most easily determined if we recognize it as simply the expansion [Eq. (8.42)] of the following third-order determinant:

$$\frac{1}{\sqrt{6}} \begin{vmatrix} f(1) & g(1) & h(1) \\ f(2) & g(2) & h(2) \\ f(3) & g(3) & h(3) \end{vmatrix} \tag{10.55}$$

(For the reader familiar with Problem 8.8, this should come as no surprise.) The antisymmetry property is easily seen to hold for (10.55), because interchange of two electrons amounts to interchanging two rows of the determinant, which multiplies it by -1.

We stated that it is impossible to construct an antisymmetric spin function for three electrons. We now use (10.55) to prove this statement. The functions f, g,

---

[3] Taking all signs as plus in (10.54), we get a completely symmetric function.

and h may each be either α or β. We might take $f = \alpha$, $g = \beta$, $h = \alpha$, so that (10.55) becomes

$$\frac{1}{\sqrt{6}} \begin{vmatrix} \alpha(1) & \beta(1) & \alpha(1) \\ \alpha(2) & \beta(2) & \alpha(2) \\ \alpha(3) & \beta(3) & \alpha(3) \end{vmatrix} \qquad (10.56)$$

Although (10.56) is antisymmetric, we must reject it because it is equal to zero. The first and third columns of the determinant are identical, so (Section 8.3) the determinant vanishes. No matter how we choose f, g, and h, at least two columns of the determinant must be equal, so that we cannot construct a nonzero antisymmetric three-electron spin function.

We now use (10.55) to construct the zeroth-order ground-state wave function for lithium, including both space and spin variables. The functions f, g, and h will now involve both space and spin variables. We choose

$$f(1) = 1s(1)\alpha(1) \qquad (10.57)$$

We call a function like (10.57) a *spin-orbital*. A spin-orbital is the product of a one-electron space orbital and a one-electron spin function. If we were to take $g(1) = 1s(1)\alpha(1)$, this would make the first and second columns of (10.55) identical, and the wave function would vanish. This is a particular case of the original form of the Pauli exclusion principle: *No two electrons can occupy the same spin-orbital.* Another way of stating this is to say that no two electrons in an atom can have the same values for all their quantum numbers. These versions of the Pauli principle are consequences of the more general antisymmetry requirement and are less satisfying, since they are based on approximate (zeroth-order) wave functions. We therefore take

$$g(1) = 1s(1)\beta(1)$$

which puts two electrons with opposite spin in the $1s$ orbital. For the spin-orbital h, we cannot use either $1s(1)\alpha(1)$ or $1s(1)\beta(1)$, since these choices make the determinant vanish. We take $h(1) = 2s(1)\alpha(1)$, which gives the familiar Li ground-state configuration $1s^2 2s$ and the zeroth-order wave function

$$\psi^{(0)} = \frac{1}{\sqrt{6}} \begin{vmatrix} 1s(1)\alpha(1) & 1s(1)\beta(1) & 2s(1)\alpha(1) \\ 1s(2)\alpha(2) & 1s(2)\beta(2) & 2s(2)\alpha(2) \\ 1s(3)\alpha(3) & 1s(3)\beta(3) & 2s(3)\alpha(3) \end{vmatrix} \qquad (10.58)$$

Note especially that (10.58) is *not* simply a product of space and spin parts (as we found for H and He), but is a linear combination of terms, each of which is a product of space and spin parts.

We could just as well have taken $h(1) = 2s(1)\beta(1)$, so that the ground state of lithium is, like hydrogen, doubly degenerate, corresponding to the two possible orientations of the spin of the $2s$ electron. We might use the usual orbital diagrams

$1s$	$2s$		$1s$	$2s$
↑↓	↑	and	↑↓	↓

to indicate this. Each space orbital such as $1s$ or $2p_0$ can hold two electrons of opposite spin. A spin-orbital such as $2s\alpha$ can hold one electron.

Although the $1s^2 2p$ configuration will have the same unperturbed energy $E^{(0)}$ as the $1s^2 2s$ configuration, when we take electron repulsion into account by calculating $E^{(1)}$ and higher corrections, we find that the $1s^2 2s$ configuration lies lower for the same reason as in helium.

We close this section by discussing some points about the original form of the Pauli exclusion principle, which we restate as follows: *In a system of identical fermions, no two particles can occupy the same state.* If we have a system of n interacting particles (e.g., an atom), there is a single wave function (involving $4n$ variables) for the entire system. Because of the interactions between the particles, the wave function cannot be written as the product of wave functions of the individual particles; hence, strictly speaking, we cannot talk of the states of individual particles, only the state of the whole system. If, however, the interactions between the particles are not too large, then as an initial approximation we can neglect them and write the zeroth-order wave function of the system as a product of wave functions of the individual particles. In this zeroth-order wave function, no two fermions can have the same wave function (state).

Since bosons require a wave function symmetric with respect to interchange, there is no restriction on the number of bosons in a given state.

10.7 SLATER DETERMINANTS

Slater pointed out in 1929 that a determinant of the form (10.58) satisfies the antisymmetry requirement for a many-electron atom. A determinant like (10.58) is called a *Slater determinant*. All the elements in a given column of a Slater determinant involve the same spin-orbital, while elements in the same row all involve the same electron. Since interchanging rows and columns does not affect the value of a determinant, we could write the Slater determinant in another, equivalent form. For the sake of consistency, we will always write Slater determinants in the form (10.58).

Consider how the zeroth-order helium wave functions which we found previously can be written as Slater determinants. For the ground-state configuration $(1s)^2$, we have the spin-orbitals $1s\alpha$ and $1s\beta$, which give the Slater determinant

$$\frac{1}{\sqrt{2}} \begin{vmatrix} 1s(1)\alpha(1) & 1s(1)\beta(1) \\ 1s(2)\alpha(2) & 1s(2)\beta(2) \end{vmatrix} = 1s(1)\,1s(2)\frac{1}{\sqrt{2}}[\alpha(1)\beta(2) - \beta(1)\alpha(2)] \quad \textbf{(10.59)}$$

which agrees with (10.41). For the states corresponding to the excited configuration $1s\,2s$, we have the possible spin-orbitals $1s\alpha$, $1s\beta$, $2s\alpha$, $2s\beta$, which give the four Slater determinants

$$D_1 = \frac{1}{\sqrt{2}} \begin{vmatrix} 1s(1)\alpha(1) & 2s(1)\alpha(1) \\ 1s(2)\alpha(2) & 2s(2)\alpha(2) \end{vmatrix} \qquad D_2 = \frac{1}{\sqrt{2}} \begin{vmatrix} 1s(1)\alpha(1) & 2s(1)\beta(1) \\ 1s(2)\alpha(2) & 2s(2)\beta(2) \end{vmatrix}$$

$$D_3 = \frac{1}{\sqrt{2}} \begin{vmatrix} 1s(1)\beta(1) & 2s(1)\alpha(1) \\ 1s(2)\beta(2) & 2s(2)\alpha(2) \end{vmatrix} \qquad D_4 = \frac{1}{\sqrt{2}} \begin{vmatrix} 1s(1)\beta(1) & 2s(1)\beta(1) \\ 1s(2)\beta(2) & 2s(2)\beta(2) \end{vmatrix}$$

Comparison with (10.42)–(10.45) shows that the $1s2s$ zeroth-order wave functions are related to these four Slater determinants as follows:

$$2^{-1/2}[1s(1)\,2s(2) - 2s(1)\,1s(2)]\alpha(1)\,\alpha(2) = D_1 \qquad (10.60)$$

$$2^{-1/2}[1s(1)\,2s(2) - 2s(1)\,1s(2)]\beta(1)\,\beta(2) = D_4 \qquad (10.61)$$

$$2^{-1/2}[1s(1)\,2s(2) - 2s(1)\,1s(2)]2^{-1/2}[\alpha(1)\,\beta(2) + \beta(1)\,\alpha(2)] = 2^{-1/2}(D_2 + D_3) \qquad (10.62)$$

$$2^{-1/2}[1s(1)\,2s(2) + 2s(1)\,1s(2)]2^{-1/2}[\alpha(1)\,\beta(2) - \beta(1)\,\alpha(2)] = 2^{-1/2}(D_2 - D_3) \qquad (10.63)$$

(To get a zeroth-order function that is an eigenfunction of the spin and orbital angular-momentum operators, we sometimes have to take a linear combination of the Slater determinants of a configuration; see Chapter 11.)

Next consider some notations used for Slater determinants. Instead of writing α and β for spin functions, one often puts a bar over the space function to indicate the spin function β, while a space function without a bar implies the spin factor α. With this notation (10.58) is written as

$$\psi^{(0)} = \frac{1}{\sqrt{6}} \begin{vmatrix} 1s(1) & \overline{1s}(1) & 2s(1) \\ 1s(2) & \overline{1s}(2) & 2s(2) \\ 1s(3) & \overline{1s}(3) & 2s(3) \end{vmatrix} \qquad (10.64)$$

Given the spin-orbitals occupied by the electrons, we can readily construct the Slater determinant. Thus it is redundant to write out the full determinant. Instead, a shorthand notation which simply specifies the spin-orbitals is often used. In this notation (10.64) is written as

$$\psi^{(0)} = |1s\overline{1s}\,2s| \qquad (10.65)$$

where the vertical lines indicate formation of the determinant and multiplication by $1/\sqrt{6}$.

We showed that the factor $1/\sqrt{6}$ normalizes a third-order Slater determinant constructed of orthonormal functions. The expansion of an nth-order determinant has $n!$ terms (Problem 8.10). For an nth-order Slater determinant of orthonormal spin-orbitals, the same reasoning used in the third-order case shows that the normalization constant is $1/\sqrt{n!}$. We always include a factor $1/\sqrt{n!}$ in defining a Slater determinant of order n.

10.8 PERTURBATION TREATMENT OF THE LITHIUM GROUND STATE

Let us carry out a perturbation treatment of the ground state of the lithium atom.

We take

$$\hat{H}^0 = -\frac{\hbar^2}{2m}\nabla_1^2 - \frac{\hbar^2}{2m}\nabla_2^2 - \frac{\hbar^2}{2m}\nabla_3^2 - \frac{Ze'^2}{r_1} - \frac{Ze'^2}{r_2} - \frac{Ze'^2}{r_3} \qquad (10.66)$$

$$\hat{H}' = \frac{e'^2}{r_{12}} + \frac{e'^2}{r_{23}} + \frac{e'^2}{r_{13}} \tag{10.67}$$

We found in Section 10.6 that to satisfy the Pauli principle the ground-state configuration must be $1s^2 2s$; the correct zeroth-order wave function is (10.58):

$$\psi^{(0)} = 6^{-1/2}[1s(1)\,1s(2)\,2s(3)\,\alpha(1)\,\beta(2)\,\alpha(3) - 1s(1)\,2s(2)\,1s(3)\,\alpha(1)\,\alpha(2)\,\beta(3)$$

$$- 1s(1)\,1s(2)\,2s(3)\,\beta(1)\,\alpha(2)\,\alpha(3) + 1s(1)\,2s(2)\,1s(3)\,\beta(1)\,\alpha(2)\,\alpha(3)$$

$$+ 2s(1)\,1s(2)\,1s(3)\,\alpha(1)\,\alpha(2)\,\beta(3) - 2s(1)\,1s(2)\,1s(3)\,\alpha(1)\,\beta(2)\,\alpha(3)] \tag{10.68}$$

What is $E^{(0)}$? Each term in (10.68) contains the product of two $1s$ hydrogenlike functions and one $2s$ hydrogenlike function, multiplied by a spin factor. $\hat{H}^0$ is the sum of three hydrogenlike Hamiltonians, one for each electron, and does not involve spin. Thus $\psi^{(0)}$ is a linear combination of terms, each of which is an eigenfunction of $\hat{H}^0$ with eigenvalue $E_{1s}^{(0)} + E_{1s}^{(0)} + E_{2s}^{(0)}$, where these are hydrogenlike energies; hence $\psi^{(0)}$ is an eigenfunction of $\hat{H}^0$ with eigenvalue $E_{1s}^{(0)} + E_{1s}^{(0)} + E_{2s}^{(0)}$. Therefore [Eq. (6.107)]

$$E^{(0)} = -\left(\frac{1}{1^2} + \frac{1}{1^2} + \frac{1}{2^2}\right)\left(\frac{Z^2 e'^2}{2a_0}\right) = -\frac{81}{4}(13.606 \text{ eV}) = -275.5 \text{ eV} \tag{10.69}$$

To find $E^{(1)}$, we must evaluate the integral (9.30). We begin by grouping together terms in $\psi^{(0)}$ that have the same spin factor:

$$\psi^{(0)} = 6^{-1/2}[1s(1)\,2s(2)\,1s(3) - 1s(1)\,1s(2)\,2s(3)]\,\beta(1)\,\alpha(2)\,\alpha(3)$$

$$+ 6^{-1/2}[1s(1)\,1s(2)\,2s(3) - 2s(1)\,1s(2)\,1s(3)]\,\alpha(1)\,\beta(2)\,\alpha(3)$$

$$+ 6^{-1/2}[2s(1)\,1s(2)\,1s(3) - 1s(1)\,2s(2)\,1s(3)]\,\alpha(1)\,\alpha(2)\,\beta(3) \tag{10.70}$$

$$\psi^{(0)} = a\beta(1)\,\alpha(2)\,\alpha(3) + b\alpha(1)\,\beta(2)\,\alpha(3) + c\alpha(1)\,\alpha(2)\,\beta(3) = A + B + C \tag{10.71}$$

where the space function multiplying the spin function $\beta(1)\,\alpha(2)\,\alpha(3)$ in (10.70) is called a and where $A = a\beta(1)\,\alpha(2)\,\alpha(3)$, with similar definitions for b, c, B, and C. We have

$$E^{(1)} = \int |\psi^{(0)}|^2 H'\, d\tau$$

$$E^{(1)} = \int |A|^2 H'\, d\tau + \int |B|^2 H'\, d\tau + \int |C|^2 H'\, d\tau + \int A^*BH'\, d\tau$$

$$+ \int B^*CH'\, d\tau + \int A^*CH'\, d\tau + \int AB^*H'\, d\tau$$

$$+ \int BC^*H'\, d\tau + \int AC^*H'\, d\tau \tag{10.72}$$

Because of the orthogonality of the different spin functions in A, B, and C, the last six integrals in (10.72) are zero. For example, the integral $\int A^*BH'\, d\tau$ involves summations over spins, as follows:

$$\sum_{m_{s1}}\sum_{m_{s2}}\sum_{m_{s3}} [\beta(1)\,\alpha(2)\,\alpha(3)]^*[\alpha(1)\,\beta(2)\,\alpha(3)]$$

$$= \sum_{m_{s1}} \beta^*(1)\,\alpha(1) \sum_{m_{s2}} \alpha^*(2)\,\beta(2) \sum_{m_{s3}} \alpha^*(3)\,\alpha(3) = 0 \cdot 0 \cdot 1 = 0$$

Since the spin functions are normalized, summation over spins in the first three integrals in (10.72) gives unity. Therefore

$$E^{(1)} = \int\int\int a^2 H'\, dv_1\, dv_2\, dv_3 + \int\int\int b^2 H'\, dv_1\, dv_2\, dv_3 + \int\int\int c^2 H'\, dv_1\, dv_2\, dv_3 \tag{10.73}$$

where spin is no longer involved. By relabeling dummy integration variables, we can prove that the three integrals in (10.73) are equal to one another. Using orthonormality of the $1s$ and $2s$ orbitals and relabeling of integration variables, we can then show that (Problem 10.12)

$$E^{(1)} = 2 \int\int 1s^2(1)\, 2s^2(2) \frac{e'^2}{r_{12}}\, dv_1\, dv_2 + \int\int 1s^2(1)\, 1s^2(2) \frac{e'^2}{r_{12}}\, dv_1\, dv_2$$

$$- \int\int 1s(1)\, 2s(2)\, 1s(2)\, 2s(1) \frac{e'^2}{r_{12}}\, dv_1\, dv_2 \tag{10.74}$$

These integrals are Coulomb and exchange integrals:

$$E^{(1)} = 2J_{1s2s} + J_{1s1s} - K_{1s2s} \tag{10.75}$$

We have [Eqs. (9.63), (9.67), and (9.157)]

$$J_{1s1s} = \frac{5}{8}\frac{Ze'^2}{a_0}, \qquad J_{1s2s} = \frac{17}{81}\frac{Ze'^2}{a_0}, \qquad K_{1s2s} = \frac{16}{729}\frac{Ze'^2}{a_0} \tag{10.76}$$

$$E^{(1)} = \frac{5965}{972}\left(\frac{e'^2}{2a_0}\right) = 83.5\ \text{eV} \tag{10.77}$$

The energy through first order is $-192.0\ \text{eV}$, as compared to the true ground-state energy of lithium, $-203.5\ \text{eV}$. To improve on this result, we must calculate higher-order wave-function and energy corrections. This will mix into the wave function contributions from Slater determinants involving configurations besides $1s^2 2s$ (configuration interaction).

10.9   VARIATION TREATMENTS OF THE LITHIUM GROUND STATE

The zeroth-order perturbation wave function (10.58) uses the full nuclear charge $(Z = 3)$ for both the $1s$ and $2s$ orbitals of lithium. We expect that the $2s$ electron, which is partially shielded from the nucleus by the two $1s$ electrons, will see an effective nuclear charge that is considerably less than 3. Even the $1s$ electrons partially shield each other (recall the treatment of the helium ground state). This reasoning suggests the introduction of two variational parameters b_1 and b_2 into (10.58).

Instead of using the $Z = 3$ $1s$ function in Table 6.2, we take

$$f \equiv \frac{1}{\pi^{1/2}}\left(\frac{b_1}{a_0}\right)^{3/2} e^{-b_1 r/a_0} \tag{10.78}$$

where b_1 is a variational parameter representing an effective nuclear charge for the $1s$ electrons. Instead of the $Z = 3$ $2s$ function in Table 6.2, we use

$$g = \frac{1}{4(2\pi)^{1/2}}\left(\frac{b_2}{a_0}\right)^{3/2}\left(2 - \frac{b_2 r}{a_0}\right) e^{-b_2 r/2a_0} \tag{10.79}$$

Our trial variation function is then

$$\varphi = \frac{1}{\sqrt{6}} \begin{vmatrix} f(1)\alpha(1) & f(1)\beta(1) & g(1)\alpha(1) \\ f(2)\alpha(2) & f(2)\beta(2) & g(2)\alpha(2) \\ f(3)\alpha(3) & f(3)\beta(3) & g(3)\alpha(3) \end{vmatrix} \tag{10.80}$$

Note that using different charges b_1 and b_2 for the 1s and 2s orbitals destroys their orthogonality, so that (10.80) is not normalized. The best values of the variational parameters are found by setting

$$\frac{\partial W}{\partial b_1} = 0, \qquad \frac{\partial W}{\partial b_2} = 0$$

where the variational integral W is given by the left side of Eq. (8.13). The results are[4]

$$b_1 = 2.686, \qquad b_2 = 1.776 \qquad\qquad\qquad \textbf{(10.81)}$$

$$W = -7.3922 \left(\frac{e'^2}{a_0}\right) = -201.2 \text{ eV}$$

which is considerably closer to the true value -203.5 eV than the result -192.0 eV obtained in the last section using simple perturbation theory. The value of b_2 indicates substantial, but not complete, screening of the 2s electron by the 1s electrons.

We might try other forms for the orbitals besides (10.78) and (10.79) to improve the trial function. However, no matter what orbital functions we try, if we restrict ourselves to a trial function of the form of (10.80), we can never reach the true ground-state energy. To do this, we must either introduce r_{12}, r_{23}, and r_{13} into the trial function or use a linear combination of several Slater determinants corresponding to various configurations (configuration interaction).

10.10   *SPIN MAGNETIC MOMENT*

Recall that the orbital angular momentum $\mathbf{L}$ of an electron has a magnetic moment $-(e/2m)\mathbf{L}$ associated with it [Eq. (6.152)]. It is natural to suppose that there is also a magnetic moment $\boldsymbol{\mu}_S$ associated with the electronic spin angular momentum $\mathbf{S}$. We might guess that $\boldsymbol{\mu}_S$ would be $-e/2m$ times $\mathbf{S}$. Spin is a relativistic phenomenon, however, and we cannot expect $\boldsymbol{\mu}_S$ to be related to $\mathbf{S}$ in exactly the same way that $\boldsymbol{\mu}_L$ is related to $\mathbf{L}$. In fact, Dirac's relativistic treatment of the electron gave the result that (in SI units)

$$\boldsymbol{\mu}_S = -g_e \frac{e}{2m}\mathbf{S} = -\frac{e}{m}\mathbf{S} \qquad\qquad\qquad \textbf{(10.82)}$$

where Dirac found an electron g factor equal to 2. The magnitude of the spin magnetic moment of an electron is (in SI units)

$$|\boldsymbol{\mu}_S| = g_e \frac{e}{2m}|\mathbf{S}| = \sqrt{3}\,\frac{e\hbar}{2m} \qquad\qquad\qquad \textbf{(10.83)}$$

Theoretical and experimental work[5] subsequent to Dirac's treatment has shown that g_e is slightly greater than 2:

$$g_e = 2\left(1 + \frac{\alpha}{2\pi} + \cdots\right) = 2.0023$$

---

[4]   E. B. Wilson, Jr., *J. Chem. Phys.*, **1**, 211 (1933).
[5]   For a review see P. Kusch, *Physics Today*, Feb. 1966, p. 23.

where the dots indicate terms involving higher powers of α and where the *fine-structure constant* α is defined as

$$\alpha \equiv \frac{e^2}{4\pi\varepsilon_0 \hbar c} \equiv \frac{e'^2}{\hbar c} \tag{10.84}$$

The constant α is a dimensionless combination of the three fundamental constants e', $\hbar$, and c and is of great theoretical interest. α is small; its numerical value is 0.007297; its reciprocal, $1/\alpha$, is approximately 137.

The ferromagnetism of iron is due to the electron's magnetic dipole moment.

The two possible orientations of an electron's spin and its associated spin magnetic moment with respect to an axis produce two energy levels in an externally applied magnetic field. In electron-spin-resonance (ESR) spectroscopy, one observes transitions between these two levels. ESR spectroscopy is applicable to species such as free radicals and transition-metal ions that have one or more unpaired electron spins and hence have a nonzero total electron spin and spin magnetic moment.

Certain nuclei have nonzero spins and spin magnetic moments. In nuclear-magnetic-resonance (NMR) spectroscopy, one observes transitions between nuclear-spin energy levels for a sample in an applied magnetic field. The proton (spin $\frac{1}{2}$) is the nucleus most commonly studied.

PROBLEMS

10.1 If electrons had a spin of zero, what would be the zeroth-order (interelectronic repulsions neglected) wave functions for the ground state and first excited state of lithium?

10.2 (a) Show that $\hat{P}_{12}$ commutes with the Hamiltonian for the lithium atom. (b) Show that $\hat{P}_{12}$ and $\hat{P}_{23}$ do not commute with each other.

10.3 The antisymmetrization operator $\hat{A}$ is defined as the operator that antisymmetrizes a product of n one-electron functions and multiplies them by $(n!)^{-1/2}$. For $n = 2$, we have

$$\hat{A}f(1)g(2) = \frac{1}{\sqrt{2}} \begin{vmatrix} f(1) & g(1) \\ f(2) & g(2) \end{vmatrix}$$

(a) For $n = 2$, express $\hat{A}$ in terms of $\hat{P}_{12}$. (b) For $n = 3$, express $\hat{A}$ in terms of $\hat{P}_{12}$, $\hat{P}_{13}$, and $\hat{P}_{23}$.

10.4 A *permanent* is defined by the same expansion as a determinant except that all terms are given a plus sign. Thus the second-order permanent is

$$\begin{vmatrix} \overset{+}{a} & \overset{+}{b} \\ c & d \end{vmatrix} = ad + bc$$

Can you think of a use for permanents in quantum mechanics?

10.5 Show that $\hat{P}_{12}$ is Hermitian.

10.6 If we had incorrectly used as the zeroth-order Li ground-state wave function the nonantisymmetric function $1s(1)1s(2)2s(3)$, what would $E^{(1)}$ be calculated to be?

10.7 Verify that the spin functions (10.13) satisfy Eqs. (10.11) and (10.12).

10.8 Calculate the angle that the spin vector **S** makes with the z axis for an electron with spin function α.

10.9 Which of the following functions are (a) completely symmetric? (b) completely antisymmetric?

(1) $f(1)g(2)\alpha(1)\alpha(2)$

(2) $f(1)f(2)[\alpha(1)\beta(2) - \beta(1)\alpha(2)]$

(3) $f(1)f(2)f(3)\beta(1)\beta(2)\beta(3)$

(4) $[f(1)g(2) - g(1)f(2)][\alpha(1)\beta(2) - \beta(1)\alpha(2)]$

(5) $r_{12}^2 e^{-a(r_1 + r_2)}$

(6) $e^{-a(r_1 - r_2)}$

10.10 Verify Eqs. (10.23) and (10.29).

10.11 (a) If the spin component S_x of an electron is measured, what possible values can result? (b) The functions α and β form a complete set, so that any one-electron spin function can be written as a linear combination of them. Use Eqs. (10.28)–(10.29) to construct the two normalized eigenfunctions of $\hat{S}_x$ with eigenvalues $+\frac{1}{2}\hbar$ and $-\frac{1}{2}\hbar$. (c) Suppose a measurement of S_z for an electron gives the value $+\frac{1}{2}\hbar$; if a measurement of S_x is then carried out, give the probabilities for each possible outcome.[6] (d) Do the same as in (b) for $\hat{S}_y$ instead of $\hat{S}_x$.

10.12 Derive Eq. (10.74) from (10.73).

10.13 Let Y_{jm} be the *normalized* eigenfunction of the generalized angular-momentum operators (Section 5.4) $\hat{M}^2$ and $\hat{M}_z$:

$$\hat{M}^2 Y_{jm} = j(j+1)\hbar^2 Y_{jm}, \qquad \hat{M}_z Y_{jm} = m\hbar Y_{jm}$$

From Section 5.4, the effect of $\hat{M}_+$ on Y_{jm} is to increase the $\hat{M}_z$ eigenvalue by $\hbar$:

$$\hat{M}_+ Y_{jm} = A Y_{j,m+1}$$

where A is a constant. Use the same procedure that led to Eqs. (10.22) and (10.23) to show that

$$\hat{M}_+ Y_{jm} = [j(j+1) - m(m+1)]^{1/2}\hbar Y_{j,m+1} \tag{10.85}$$

$$\hat{M}_- Y_{jm} = [j(j+1) - m(m-1)]^{1/2}\hbar Y_{j,m-1} \tag{10.86}$$

(b) Show that (10.85) and (10.86) are consistent with (10.22) and (10.23). (c) With $\mathbf{M} = \mathbf{L}$, the function Y_{jm} is the spherical harmonic $Y_l^m(\theta, \varphi)$. Verify (10.85) directly for $l = 2$, $m = -1$. [Actually, for consistency with the phase choice of Eqs. (10.85) and (10.86), one must add the factor $(-i)^{m+|m|}$ to the definition (5.146) of the spherical harmonics; this introduces a minus sign for odd positive values of m.]

10.14 A muon has the same charge and spin as an electron, but a heavier mass. What would be the ground-state configuration of a lithium atom with two electrons and one muon?

10.15 Show that α and β are each eigenfunctions of $\hat{S}_x^2$ (but not of $\hat{S}_x$).

---

[6] In the Stern–Gerlach experiment, a beam of particles is sent through an inhomogeneous magnetic field, which splits the beam into several beams each having particles with a different component of magnetic dipole moment in the field direction. For example, a beam of ground-state sodium atoms is split into two beams, corresponding to the two possible orientations of the valence electron's spin. Problem 10.11c corresponds to setting up a Stern–Gerlach apparatus with the field in the z direction and then allowing the $+\frac{1}{2}\hbar$ beam from this apparatus to enter a Stern–Gerlach apparatus that has the field in the x direction.

11 MANY-ELECTRON ATOMS

11.1 THE HARTREE–FOCK SELF-CONSISTENT-FIELD METHOD

For hydrogen the exact wave function is known. For helium and lithium, very accurate wave functions have been calculated by including interelectronic distances in the variation functions. For atoms of higher atomic number, the best approach to finding a good wave function lies in first calculating an approximate wave function using the Hartree–Fock procedure, which we will outline in this section. The Hartree–Fock method is the basis for the use of atomic and molecular orbitals in many-electron systems.

We begin by writing down the atomic Hamiltonian. For an n-electron atom we have,[1] assuming an infinitely heavy point nucleus,

$$\hat{H} = -\frac{\hbar^2}{2m} \sum_{i=1}^{n} \nabla_i^2 - \sum_{i=1}^{n} \frac{Ze'^2}{r_i} + \sum_{i=1}^{n} \sum_{j>i}^{n} \frac{e'^2}{r_{ij}} \tag{11.1}$$

The first sum in (11.1) contains the kinetic-energy operators for the n electrons; the second sum is the potential energy for the attractions between the electrons and the nucleus of charge Z (for a neutral atom $Z = n$); the last sum is the potential energy of the interelectronic repulsions—the restriction $j > i$ avoids counting the same interelectronic repulsion twice, and avoids terms like e'^2/r_{ii}. The Schroedinger equation for the atom is not separable because of the interelectronic repulsion terms, e'^2/r_{ij}. Recalling the perturbation treatment of helium (Section 9.3), we might obtain

---

[1] This Hamiltonian is not complete; it omits spin–orbit and other interactions; these terms are generally small and will be considered later in this chapter.

256

a zeroth-order wave function by neglecting these repulsions. The Schroedinger equation would then separate into n one-electron hydrogenlike equations. The zeroth-order wave function would be a product of n hydrogenlike (one-electron) orbitals:

$$\psi^{(0)} = f_1(r_1, \theta_1, \varphi_1)f_2(r_2, \theta_2, \varphi_2)\cdots f_n(r_n, \theta_n, \varphi_n) \qquad (11.2)$$

where the hydrogenlike orbitals are

$$f = R_{nl}(r)Y_l^m(\theta, \varphi) \qquad (11.3)$$

For the ground state of the atom, we would feed two electrons with opposite spin into each of the lowest orbitals, in accord with the Pauli exclusion principle, giving rise to the ground-state configuration. Although the approximate wave function (11.2) is qualitatively useful, it is gravely lacking in quantitative accuracy. For one thing, all the orbitals use the full nuclear charge Z. Recalling our variational treatments of helium and lithium, we know we can get a better approximation by using different effective atomic numbers for the different orbitals to account for screening of electrons. The use of effective atomic numbers gives considerable improvement, but we are still far from having an accurate wave function. The next step is to use a variation function that has the same form as (11.2), but is not restricted to hydrogenlike or any other particular form of orbitals. Thus we take

$$\varphi = g_1(r_1, \theta_1, \varphi_1)g_2(r_2, \theta_2, \varphi_2)\cdots g_n(r_n, \theta_n, \varphi_n) \qquad (11.4)$$

and we seek to determine the functions $g_1, g_2, \ldots, g_n$ that minimize the variational integral $\int \varphi^* \hat{H} \varphi \, dv / \int \varphi^* \varphi \, dv$. Our task is considerably more difficult than in previous variational calculations where we guessed a trial function that included some parameters and then varied the *parameters*. Here we must vary the *functions* g_i. [After we have found the best possible functions g_i, Eq. (11.4) will still be only an approximate wave function. The many-electron Schroedinger equation is not separable, so that the true wave function cannot be written as the product of n one-electron functions.]

Finding the best possible approximate wave function of the form (11.4) is a formidable computational task for a many-electron atom. To simplify matters somewhat, we approximate the best possible orbitals with orbitals that are the product of a radial factor and a spherical harmonic:

$$g_i = h_i(r_i)Y_{l_i}^{m_i}(\theta_i, \varphi_i) \qquad (11.5)$$

This approximation is generally made in atomic calculations.

The procedure for calculating the g_i's was introduced by Hartree in 1928 and is called the Hartree *self-consistent-field* (SCF) method. Hartree arrived at the SCF procedure by intuitive physical arguments; the proof[2] that Hartree's procedure gives the best possible variation function of the form (11.4) was given in 1930 by Slater and by Fock.

---

[2] For a review of the SCF method, see S. M. Blinder, *Am. J. Phys.*, **33**, 431 (1965).

Hartree's procedure is as follows. We first guess a product wave function

$$\varphi_0 = s_1(r_1, \theta_1, \varphi_1) s_2(r_2, \theta_2, \varphi_2) \cdots s_n(r_n, \theta_n, \varphi_n) \qquad (11.6)$$

where each s_i is a normalized function of r multiplied by a spherical harmonic. A reasonable guess for φ_0 would be a product of hydrogenlike orbitals with effective atomic numbers. For the function (11.6), the probability density of electron i is $|s_i|^2$. We now focus attention on electron 1 and regard electrons 2, 3, ..., n as being smeared out to form a static distribution of electric charge through which electron 1 moves; we are thus averaging out the instantaneous interactions between electron 1 and the other electrons. The potential energy of interaction between two point charges Q_1 and Q_2 is given by (6.66) and (1.34) as $V_{12} = Q_1' Q_2' / r_{12} = Q_1 Q_2 / 4\pi\varepsilon_0 r_{12}$. We now take Q_2 and smear it out into a continuous charge distribution such that ρ_2 is the charge density, the charge per unit volume. The infinitesimal charge in the infinitesimal volume dv_2 is $\rho_2 \, dv_2$, and summing up the interactions between Q_1 and the infinitesimal elements of charge, we have

$$V_{12} = \frac{Q_1}{4\pi\varepsilon_0} \int \frac{\rho_2}{r_{12}} \, dv_2 \qquad (11.7)$$

For electron 2 (with charge $-e$), the charge density of the hypothetical charge cloud is $\rho_2 = -e|s_2|^2$. Hence

$$V_{12} = e'^2 \int \frac{|s_2|^2}{r_{12}} \, dv_2 \qquad (11.8)$$

where $e'^2 = e^2/4\pi\varepsilon_0$. Adding in the interactions with the other electrons, we have

$$V_{12} + V_{13} + \cdots + V_{1n} = \sum_{j=2}^{n} e'^2 \int \frac{|s_j|^2}{r_{1j}} \, dv_j \qquad (11.9)$$

The potential energy of interaction between electron 1 and the other electrons and the nucleus is then

$$V(r_1, \theta_1, \varphi_1) = \sum_{j=2}^{n} e'^2 \int \frac{|s_j|^2}{r_{1j}} \, dv_j - \frac{Ze'^2}{r_1} \qquad (11.10)$$

We now make a further approximation beyond assuming the wave function to be a product of one-electron orbitals. We assume that the effective potential acting on an electron in an atom can be adequately approximated by a function of r only. This *central-field approximation* can be shown to be generally accurate. We therefore average $V(r_1, \theta_1, \varphi_1)$ over the angles to arrive at a potential energy which depends only on r_1:

$$V(r_1) = \frac{\int_0^{2\pi} \int_0^{\pi} V(r_1, \theta_1, \varphi_1) \sin \theta_1 \, d\theta_1 \, d\varphi_1}{\int_0^{2\pi} \int_0^{\pi} \sin \theta \, d\theta \, d\varphi} \qquad (11.11)$$

We now use $V(r_1)$ as the potential energy in a one-electron Schroedinger equation

$$\left[-\frac{\hbar^2}{2m} \nabla_1^2 + V(r_1) \right] t_1(1) = \varepsilon_1 t_1(1) \qquad (11.12)$$

and solve for $t_1(1)$, which will be an improved orbital for electron 1. In (11.12), ε_1 is the energy of the orbital of electron 1 at this stage of the approximation. Since the potential energy in (11.12) is spherically symmetric, the angular factor in $t_1(1)$ is a spherical harmonic involving quantum numbers l_1 and m_1 (Section 6.1). The radial factor $R(r_1)$ in t_1 is the solution of a one-dimensional Schroedinger equation of the form (6.21). We get a set of solutions $R(r_1)$, where the number of nodes k interior to the boundary points ($r = 0$ and ∞) starts at zero for the lowest energy and increases by one for each higher energy (Section 4.2). We now *define* the quantum number n as $n = l + 1 + k$, $k = 0, 1, 2, \ldots$. We thus have $1s$, $2s$, $2p$, etc., orbitals (with orbital energy ε increasing with n) just as in hydrogenlike atoms, and the number of interior radial nodes ($n - l - 1$) is the same as in hydrogenlike atoms (Section 6.5). However, since $V(r_1)$ is not a simple Coulomb potential, the radial factor $R(r_1)$ is not a hydrogenlike function. Of the set of solutions $R(r_1)$, we take the one that corresponds to the orbital we are improving. For example, if electron 1 is a $1s$ electron in the Be $1s^2 2s^2$ configuration, then $V(r_1)$ is calculated from the guessed orbitals of one $1s$ electron and two $2s$ electrons, and we use the radial solution of (11.12) with $k = 0$ to find an improved $1s$ orbital.

We now go to electron 2 and regard it as moving in a charge cloud of density

$$-e[|t_1(1)|^2 + |s_3(3)|^2 + |s_4(4)|^2 + \cdots + |s_n(n)|^2]$$

due to the other electrons. We calculate an effective potential energy $V(r_2)$ and solve a one-electron Schroedinger equation for electron 2 to obtain an improved orbital $t_2(2)$. We continue this process until we have a set of improved orbitals for all n electrons. Then we go back to electron 1 and repeat the process. We continue to calculate improved orbitals until there is no further change from one iteration to the next. The final set of orbitals gives the Hartree self-consistent-field wave function.

How do we get the energy of the atom in the SCF approximation? It seems natural to simply take the sum of the orbital energies of the electrons, $\varepsilon_1 + \varepsilon_2 + \cdots + \varepsilon_n$, and some texts give this expression. This is wrong. In calculating the orbital energy ε_1, we solved a one-electron Schroedinger equation like (11.12). The potential energy in (11.12) includes, in an average way, the energy of the repulsions between electrons 1 and 2, 1 and 3, ..., 1 and n. When we solve for ε_2, we solve a Schroedinger equation whose potential energy includes repulsions between electrons 2 and 1, 2 and 3, ..., 2 and n. If we take $\Sigma \varepsilon_i$, we will count each interelectronic repulsion twice. To obtain the total energy E of the atom, we must take

$$E = \sum_{i=1}^{n} \varepsilon_i - \sum_{i=1}^{n} \sum_{j>i}^{n} \iint \frac{e'^2 |g_i(i)|^2 |g_j(j)|^2}{r_{ij}} \, dv_i \, dv_j$$

$$E = \sum_{i=1}^{n} \varepsilon_i - \sum_{i=1}^{n} \sum_{j>i}^{n} J_{ij} \tag{11.13}$$

where we have subtracted the average repulsions of the electrons from the sum of the orbital energies and where we have used the notation J_{ij} for Coulomb integrals [Eq. (9.144)].

The set of orbitals belonging to a given principal quantum number n constitutes a *shell*. The $n = 1, 2, 3, \ldots$ shells are the $K, L, M, \ldots$ shells, respectively. The orbitals belonging to a given n and a given l constitute a *subshell*. Consider the sum of the Hartree probability densities for the electrons in a filled subshell. Using (11.5), we have

$$2 \sum_{m=-l}^{l} |h_{n,l}(r)|^2 |Y_l^m(\theta, \varphi)|^2 = 2|h_{n,l}(r)|^2 \sum_{m=-l}^{l} |Y_l^m(\theta, \varphi)|^2 \qquad \textbf{(11.14)}$$

where the factor 2 comes from the pair of electrons in each orbital. Now it is a result of the spherical-harmonic addition theorem (*Merzbacher*, Section 9.7) that the sum on the right side of (11.14) equals $(2l + 1)/4\pi$. Hence the sum of the probability densities is $[(2l + 1)/2\pi]|h_{n,l}(r)|^2$, which is independent of the angles. A closed subshell gives a spherically symmetric probability density, a result called *Unsöld's theorem*. For a half-filled subshell, the factor 2 is omitted from (11.14), and here also we get a spherically symmetric probability density.

The alert reader may have realized that there is something fundamentally wrong with the simple Hartree product wave function (11.4). Although we have paid some attention to spin and the Pauli principle by putting no more than two electrons in each spatial orbital, any approximation to the true wave function should include spin explicitly and should be antisymmetric to interchange of electrons (Chapter 10). Hence instead of the spatial orbitals, we must use spin-orbitals and must take an antisymmetric linear combination of products of spin-orbitals. This was pointed out by Fock (and by Slater) in 1930, and an SCF calculation that uses antisymmetrized spin-orbitals is called a *Hartree–Fock calculation*. We have seen that a Slater determinant of spin-orbitals provides the proper antisymmetry. For example, to carry out a Hartree–Fock calculation for the lithium ground state, we start with the function (10.80), where f and g are guesses for the $1s$ and $2s$ orbitals. We then carry out the SCF iterative process until we get no further improvement in f and g; this gives the lithium ground-state Hartree–Fock wave function. The differential equations for finding the Hartree–Fock orbitals have the same general form as (11.12), namely,

$$\hat{F}f_i = \varepsilon_i f_i, \qquad i = 1, 2, \ldots, n \qquad \textbf{(11.15)}$$

where f_i is the ith spin-orbital, the operator $\hat{F}$, called the *Fock* (or *Hartree–Fock*) *operator*, is the effective Hartree–Fock Hamiltonian, and the eigenvalue ε_i is the energy of spin-orbital i. However, the Hartree–Fock operator $\hat{F}$ has extra terms as compared to the effective Hartree Hamiltonian given by the bracketed terms in (11.12). The Hartree–Fock expression for the total energy of the atom involves exchange integrals K_{ij} in addition to the Coulomb integrals that occur in the Hartree expression (11.13). See Section 13.16.

Originally Hartree–Fock calculations were done numerically, and the resulting orbitals were given as tables of the radial functions for various values of r. In 1951 Roothaan proposed representing the Hartree–Fock orbitals as linear combinations of a complete set of known functions, called *basis functions*. Thus for lithium we would write the Hartree–Fock $1s$ and $2s$ spatial orbitals as

$$f = \sum_i b_i \chi_i, \qquad g = \sum_i c_i \chi_i$$

where the χ_i are some complete set of functions and where the b's and c's are the expansion coefficients to be determined by the SCF iterative procedure. Since the χ_i form a complete set, these expansions are valid.

The most commonly used set of basis functions is the set of *Slater-type orbitals* (STO's), which have the normalized form

$$\frac{[2\zeta/a_0]^{n+1/2}}{[(2n)!]^{1/2}} r^{n-1} e^{-\zeta r/a_0} Y_l^m(\theta, \varphi) \tag{11.16}$$

The set of all such functions with n, l, and m being integers but with ζ having all possible positive values forms a complete set. The parameter ζ is called the *orbital exponent*. To get a truly accurate representation of the Hartree–Fock orbitals, we would have to include an infinite number of Slater orbitals in the expansions; in practice, one can get very accurate results by using only a few judiciously chosen Slater orbitals.

Clementi and Roetti[3] performed analytic Hartree–Fock calculations for the ground state and some excited states of the first 54 elements of the periodic table. For example, consider the Hartree–Fock ground-state wave function of helium, which has the form [*cf.* Eq. (10.59)]

$$f(1)f(2) \cdot 2^{-1/2}[\alpha(1)\beta(2) - \alpha(2)\beta(1)]$$

Clementi and Roetti expressed the $1s$ orbital function f as the following combination of five $1s$ Slater-type orbitals:

$$f = \pi^{-1/2} \sum_{i=1}^{5} c_i \left(\frac{\zeta_i}{a_0}\right)^{3/2} e^{-\zeta_i r/a_0}$$

where the expansion coefficients c_i are $c_1 = 0.76838$, $c_2 = 0.22346$, $c_3 = 0.04082$, $c_4 = -0.00994$, $c_5 = 0.00230$ and where the orbital exponents ζ_i are $\zeta_1 = 1.41714$, $\zeta_2 = 2.37682$, $\zeta_3 = 4.39628$, $\zeta_4 = 6.52699$, $\zeta_5 = 7.94252$. [Note that the largest term in the expansion has an orbital exponent that is similar to the orbital exponent (9.86) for the simple trial function (9.77).] The Hartree–Fock energy is $-77.9\,\mathrm{eV}$, as compared to the true energy $-79.0\,\mathrm{eV}$.

For Li, Clementi and Roetti used a basis set consisting of two $1s$ STO's (with different orbital exponents) and four $2s$ STO's (with different orbital exponents). The Li $1s$ and $2s$ Hartree–Fock orbitals were each expressed as a linear combination of all six of these basis functions.

Electron densities calculated from Hartree–Fock wave functions are quite accurate. Figure 11.1 compares the radial distribution function of argon (found by integrating the electron density over the angles θ and φ and multiplying the result by r^2) calculated by the Hartree–Fock method with the radial distribution function found by electron diffraction. (Recall from Section 6.5 that the radial distribution function is proportional to the probability of finding an electron in a thin spherical shell at a distance r from the nucleus.) Note the electronic shell structure in Fig. 11.1. The high nuclear charge causes the inner-shell electrons to have much smaller

---

[3] E. Clementi and C. Roetti, *At. Data Nucl. Data Tables*, **14**, 177 (1974).

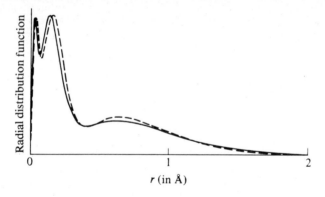

Figure 11.1 *Radial distribution function in* Ar *as a function of r. The broken line is the result of a Hartree–Fock calculation. The solid line is the result of electron-diffraction data.* [*From L. S. Bartell and L. O. Brockway,* Phys. Rev., **90**, *833* (*1953*). *Used by permission.*]

average distances from the nucleus than in hydrogen or helium. Thus there is only a moderate increase in atomic size as we go down a given group in the periodic table. Calculations show that the radius of a sphere containing 98 percent of the Hartree–Fock electron probability density gives an atomic radius in good agreement with the empirically determined van der Waals radius. [See C. W. Kammeyer and D. R. Whitman, *J. Chem. Phys.*, **56**, 4419 (1972).]

Although the radial distribution function of an atom shows the shell structure, the electron probability density integrated over the angles and plotted versus r does not oscillate. Rather, for ground-state atoms this probability density is a maximum at the nucleus (because of the s electrons) and continually decreases as r increases. Similarly, in molecules the maxima in electron probability density usually occur at the nuclei; see, for example, Fig. 13.6. [For further discussion see H. Weinstein, P. Politzer, and S. Srebnik, *Theor. Chim. Acta*, **38**, 159 (1975).]

Accurate representation of a many-electron atomic orbital (AO) requires a linear combination of several Slater-type orbitals. For rough calculations it is convenient to have simple approximations for AO's. We might use hydrogenlike orbitals with effective nuclear charges, but Slater suggested an even simpler method—namely, to approximate an AO by a single function of the form (11.16) with the orbital exponent ζ taken as

$$\zeta = (Z - s)/n \qquad (11.17)$$

where Z is the atomic number, n is the orbital's principal quantum number, and s is a screening constant calculated by a set of rules (see Problem 15.27). A Slater orbital replaces the polynomial in r in a hydrogenlike orbital with a single power of r; hence a Slater orbital does not have the proper number of radial nodes and does not represent well the inner part of an orbital.

An enormous amount of computation is required to carry out a

Hartree–Fock SCF calculation for a many-electron atom. Hartree did several SCF calculations in the 1930s, when electronic computers were not in existence. Fortunately, Hartree's father, a retired engineer, enjoyed numerical computation as a hobby and helped his son. Nowadays computers have replaced Hartree's father.

11.2 ORBITALS AND THE PERIODIC TABLE

The orbital concept and the Pauli exclusion principle allow us to understand the periodic table of the elements. An orbital is a one-electron wave function. We have used orbitals to obtain approximate wave functions for many-electron atoms, writing the wave function as a Slater determinant of one-electron orbitals. In the crudest approximation we neglect all interelectronic repulsions and obtain hydrogenlike orbitals. The best possible orbitals are the Hartree–Fock SCF functions. We build up the periodic table by feeding electrons into these orbitals, each of which can hold a pair of electrons with opposite spin.

Figure 11.2 is a diagram of theoretically calculated AO energies for neutral atoms; these energies were calculated by the Thomas–Fermi–Dirac method, which uses ideas of statistical mechanics to get approximate atomic wave functions.[4] The orbital energies of Fig. 11.2 generally agree well with both Hartree–Fock and experimentally determined orbital energies.[5]

Orbital energies change with changing atomic number Z; as Z increases, the orbital energies decrease, because of the increased attraction between the nucleus and the electrons. This decrease is most rapid for the inner orbitals, which are less well shielded from the nucleus.

For $Z > 1$, orbitals with the same value of n but different l have different energies. Thus for the $n = 3$ orbital energies, we have

$$E_{3s} < E_{3p} < E_{3d}, \qquad Z > 1$$

The splitting of these levels, which are degenerate in the hydrogen atom, arises from the interelectronic repulsions. (Recall the perturbation treatment of helium in Section 9.7.) In the limit $Z \to \infty$, orbitals with the same value of n are again degenerate, because the interelectronic repulsions become insignificant in comparison with the electron–nucleus attractions.

The relative positions of certain orbitals change with changing Z. Thus in hydrogen the $3d$ orbital lies below the $4s$ orbital, but for Z in the range 7 through 20 the $4s$ is below the $3d$. For large values of Z, the $3d$ is again lower. At $Z = 19$, the $4s$ is lower; hence $_{19}$K has the ground-state configuration $1s^2 2s^2 2p^6 3s^2 3p^6 4s$. Recall that s orbitals are more penetrating than p or d orbitals; this allows the $4s$ orbital to lie below the $3d$ orbital for some values of Z. Note the sudden drop in the $3d$ energy which starts at $Z = 21$, when filling of the $3d$ orbital begins. The electrons of the $3d$

---

[4] For the Thomas-Fermi-Dirac method, see Chapter 5 of *Bethe and Jackiw*.

[5] See the figures on pp. 146, 147, 325, and 326 of J. C. Slater, *Quantum Theory of Matter*, Second Edition, McGraw-Hill, New York, 1968.

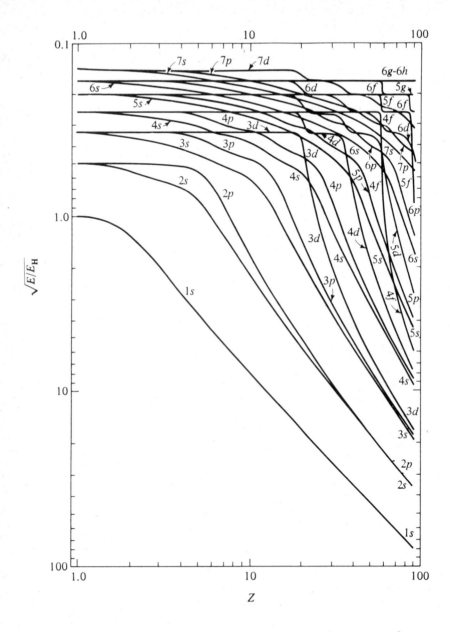

Figure 11.2 *Atomic-orbital energies as a function of atomic number for neutral atoms, as calculated by Latter.* [*Figure redrawn by M. Kasha from R. Latter,* Phys. Rev., **99**, *510 (1955). Used by permission.*] *Note the logarithmic scales. E_H is the ground-state hydrogen-atom energy,* $-13.6\,\text{eV}$.

orbital do not shield each other very well; hence the sudden drop in $3d$ energy. Similar drops occur for other orbitals.

Since $3d$ lies below $4s$ for Z above 20, one might wonder why the ground-state configuration of, say, $_{23}V$ is $\ldots 3d^3 4s^2$, rather than $\ldots 3d^5$. It is true that the $3d^5$ configuration has a lower sum of orbital energies than the $3d^3 4s^2$. However, as we saw in Eq. (11.13) for the Hartree method and will see in Section 13.16 for the Hartree–Fock method, the energy of an atom is not given by the sum of the orbital energies of the electrons. Despite the lower sum of orbital energies for $3d^5$, the energy corresponding to the $3d^5$ configuration turns out to be higher than the energy of the $3d^3 4s^2$ configuration. (For the ion V^{2+}, the $3d^3$ configuration has both a lower sum of orbital energies and a lower energy than either of the configurations $3d 4s^2$ or $3d^2 4s$, so the ground-state of V^{2+} has the $3d^3$ configuration.)

Figure 11.2 shows that the separation between ns and np orbitals is much less than that between np and nd orbitals, giving the familiar $ns^2 np^6$ stable octet.

The orbital concept[6] is the basis for most qualitative discussions of the chemistry of atoms and molecules. The use of orbitals, however, is an approximation; to reach the true wave function, we must go beyond a Slater determinant of spin-orbitals.

11.3 ELECTRON CORRELATION

Energies calculated by the Hartree–Fock method are typically in error by about 1 percent. On an absolute basis this is not much, but for the chemist it is too large. For example, the total energy of the carbon atom is about 1000 eV, and 1 percent of this is 10 eV. Chemical bond energies run about 100 kcal/mole, which is 5 eV/molecule. Attempting to calculate a bond energy by taking the difference between Hartree–Fock molecular and atomic energies, which are in error by several eV for light atoms, is an unreliable procedure. We must seek a way of improving on Hartree–Fock wave functions and energies. (Our discussion will be applicable to molecules as well as atoms.)

A Hartree–Fock scf wave function takes into account the interactions between electrons only in an average way. Actually we must consider the instantaneous interactions between electrons. Since electrons repel each other, they tend to keep out of each other's way. For example, in helium, if one electron is close to the nucleus at a given instant, it is energetically more favorable for the other electron to be far from the nucleus at that instant. One sometimes speaks of a *Coulomb hole* surrounding each electron in an atom—this is a region in which the probability of finding another electron is small. The motions of electrons are correlated with each other, and we speak of *electron correlation*. We must find a way to introduce the instantaneous electron correlation into the wave function.

Actually a Hartree–Fock wave function does have some instantaneous electron correlation built into it. A Hartree–Fock function satisfies the antisym-

───────────

   [6] Two review articles on atomic orbitals are R. S. Berry, *J. Chem. Educ.*, **43**, 283 (1966); I. Cohen and T. Bustard, *J. Chem. Educ.*, **43**, 187.

metry requirement of the Pauli principle; therefore [Eq. (10.36)] it vanishes when two electrons with the same spin have the same spatial coordinates. For a Hartree–Fock function, there is little probability of finding electrons of the same spin in the same region of space, so that a Hartree–Fock function has some correlation of the motions of electrons with the same spin. This causes the Hartree–Fock energy to be lower than the Hartree energy. (One sometimes refers to a *Fermi hole* around each electron in a Hartree–Fock wave function, thereby indicating a region in which the probability of finding another electron with the same spin is small.)

The *correlation energy* is the difference between the exact nonrelativistic energy and the Hartree–Fock energy. For helium the correlation energy is 1.1 eV (Section 11.1).

We have already indicated the two main ways in which we may provide for instantaneous electron correlation. The first method is to introduce the interelectronic distances r_{ij} into the wave function (Section 9.4). This method is only practicable for systems with a few electrons.

The second method is configuration interaction. We found (Sections 9.3 and 10.5) the zeroth-order wave function for the helium $1s^2$ ground state to be $1s(1)1s(2)[\alpha(1)\beta(2) - \beta(1)\alpha(2)]/\sqrt{2}$. We remarked that first-order and higher-order corrections to the wave function will mix in contributions from excited configurations, giving rise to *configuration interaction* (CI), also called *configuration mixing* (CM).

The most common way to do a configuration-interaction calculation on an atom (or molecule) uses the variation method. One starts by choosing a basis set of one-electron functions χ_i. In principle this basis set should be complete. In practice one is limited to a basis set of finite size; one hopes that a good choice of basis functions will give a good approximation to a complete set. For atomic calculations, STO's [Eq. (11.16)] are usually chosen.

The SCF atomic (or molecular) orbitals φ_i are written as linear combinations of the basis-set members, and the Hartree–Fock equations (Section 13.16) are solved to give the coefficients in these linear combinations. The number of atomic (or molecular) orbitals obtained equals the number of basis functions used. The lowest-energy orbitals are the occupied orbitals for the ground state. The remaining unoccupied orbitals are called virtual orbitals.

Using the set of occupied and virtual spin-orbitals, one can form antisymmetric many-electron functions that have different orbital occupancies. For example, for He, one can form functions that correspond to the electron configurations $1s^2$, $1s2s$, $1s2p$, $2s^2$, $2s2p$, $2p^2$, $1s3s$, etc. Moreover, more than one function can correspond to a given electron configuration; recall the functions (10.42)–(10.45) corresponding to the He $1s2s$ configuration. Each such many-electron function Φ_i is a Slater determinant or a linear combination of a few Slater determinants; use of more than one Slater determinant is required for certain open-shell functions like (10.62) and (10.63). Each Φ_i is called a *configuration state function* or a *configuration function* or simply a "configuration." (This last name is unfortunate, since it leads to confusion between an electron configuration like $1s^2$ and a configuration function like $|1s1s|$.)

As we saw in perturbation theory, the true atomic (or molecular) wave function ψ contains contributions from configurations other than the one that makes the main contribution to ψ, so one expresses ψ as a linear combination of the configuration functions Φ_i:

$$\psi = \sum_i c_i \Phi_i \tag{11.18}$$

We then regard (11.18) as a linear variation function (Section 8.5). Variation of the coefficients c_i to minimize the variational integral leads to the determinantal equation

$$\det(H_{ij} - ES_{ij}) = 0 \tag{11.19}$$

[We shall see in Section 11.5 that each atomic ψ is an eigenfunction of the operators $\hat{S}^2$ and $\hat{S}_z$ for the square of the magnitude of the total electronic spin and the z component of the total electron spin. Therefore only configuration functions that have the same $\hat{S}^2$ and $\hat{S}_z$ eigenvalues as the state under consideration need be included in the expansion (11.18).]

Because the many-electron configuration functions Φ_i are ultimately based on a one-electron basis set that is a complete set, the set of all possible configuration functions is a complete set for the many-electron problem—any antisymmetric many-electron function can be expressed as a linear combination of the Φ_i functions. [For a proof of this, see P.-O. Löwdin, *Adv. Chem. Phys.*, **2**, 207 (1959); the proof is on pages 260–261.] Therefore, if one starts with a complete one-electron basis set and includes all possible configuration functions, a CI calculation will give the exact atomic (or molecular) wave function ψ for the state under consideration. (In practice one is limited to a finite, incomplete basis set, rather than an infinite, complete basis set. Moreover, even with a modest-size basis set, the number of possible configuration functions is extremely large, and one usually does not include all possible configuration functions. Part of the "art" of the CI method is choosing those configurations that will make the greatest contributions.)

Because it generally takes a very large number of configuration functions to give a truly accurate wave function, configuration-interaction calculations for systems with more than a few electrons are very difficult.

In summary, to do a CI calculation, one chooses a one-electron basis set χ_i, determines one-electron atomic (or molecular) orbitals φ_i as linear combinations of the basis set, forms many-electron configuration functions Φ_i from the orbitals φ_i, expresses ψ as a linear combination of these configuration functions, solves (11.19) for the energy, and solves the associated simultaneous linear equations for the expansion coefficients c_i.

As an example, consider the ground state of beryllium. The Hartree–Fock SCF method would find the best forms for the $1s$ and $2s$ orbitals in the Slater determinant $|1s\overline{1s}\,2s\overline{2s}|$ and use this for the ground-state wave function. [We are using the notation of Eq. (10.65).] Going beyond the Hartree–Fock method, we would include contributions from excited configuration functions in a linear variation function for the ground state. Bunge did a configuration-interaction calculation[7] for the Be

---

[7] C. F. Bunge, *Phys. Rev.*, **168**, 92 (1968).

ground state, using a linear combination of 180 configuration functions. The Hartree–Fock energy is $-14.573\ (e'^2/a_0)$, Bunge's CI result is $-14.664\ (e'^2/a_0)$, and the experimental energy is $-14.667\ (e'^2/a_0)$. Bunge was able to account for almost all the correlation energy.

11.4   *ADDITION OF ANGULAR MOMENTA*

For a many-electron atom, the operators for individual angular momenta of the electrons do not commute with the Hamiltonian, but their sum does. Hence we want to learn how to add angular momenta.

Suppose we have a system with two angular-momentum vectors $\mathbf{M}_1$ and $\mathbf{M}_2$. They might be the orbital angular-momentum vectors of two electrons in an atom, or they might be the spin angular-momentum vectors of two electrons, or one might be the spin and the other the orbital angular momentum of a single electron. The eigenvalues of $\hat{M}_1^2$, $\hat{M}_2^2$, $\hat{M}_{z1}$, and $\hat{M}_{z2}$ are $j_1(j_1+1)\hbar^2$, $j_2(j_2+1)\hbar^2$, $m_1\hbar$, and $m_2\hbar$, where the quantum numbers obey the usual restrictions. The components of $\hat{\mathbf{M}}_1$ and $\hat{\mathbf{M}}_2$ obey the angular-momentum commutation relations [Eq. (5.149)]

$$[\hat{M}_{x1}, \hat{M}_{y1}] = i\hbar\, \hat{M}_{z1}, \text{ etc.} \tag{11.20}$$

$$[\hat{M}_{x2}, \hat{M}_{y2}] = i\hbar\, \hat{M}_{z2}, \text{ etc.} \tag{11.21}$$

We define the total angular momentum $\mathbf{M}$ of the system as the vector sum

$$\mathbf{M} = \mathbf{M}_1 + \mathbf{M}_2 \tag{11.22}$$

$\mathbf{M}$ is a vector with three components:

$$\mathbf{M} = M_x\mathbf{i} + M_y\mathbf{j} + M_z\mathbf{k} \tag{11.23}$$

The vector equation (11.22) gives the three scalar equations

$$M_x = M_{x1} + M_{x2}, \qquad M_y = M_{y1} + M_{y2}, \qquad M_z = M_{z1} + M_{z2} \tag{11.24}$$

For the operator $\hat{M}^2$, we have

$$\hat{M}^2 = \hat{\mathbf{M}} \cdot \hat{\mathbf{M}} = \hat{M}_x^2 + \hat{M}_y^2 + \hat{M}_z^2 \tag{11.25}$$

$$\hat{M}^2 = (\hat{\mathbf{M}}_1 + \hat{\mathbf{M}}_2) \cdot (\hat{\mathbf{M}}_1 + \hat{\mathbf{M}}_2)$$

$$= \hat{M}_1^2 + \hat{M}_2^2 + \hat{\mathbf{M}}_1 \cdot \hat{\mathbf{M}}_2 + \hat{\mathbf{M}}_2 \cdot \hat{\mathbf{M}}_1 \tag{11.26}$$

If $\hat{\mathbf{M}}_1$ and $\hat{\mathbf{M}}_2$ refer to different electrons, they will commute with each other, since each will affect only functions of the coordinates of one electron and not the other. Even if $\hat{\mathbf{M}}_1$ and $\hat{\mathbf{M}}_2$ are the orbital and spin angular momenta of the same electron, they will commute, as one will affect only functions of the spatial coordinates while the other will affect functions of the spin coordinates. Thus Eq. (11.26) becomes

$$\hat{M}^2 = \hat{M}_1^2 + \hat{M}_2^2 + 2\hat{\mathbf{M}}_1 \cdot \hat{\mathbf{M}}_2 \tag{11.27}$$

$$\hat{M}^2 = \hat{M}_1^2 + \hat{M}_2^2 + 2(\hat{M}_{x1}\hat{M}_{x2} + \hat{M}_{y1}\hat{M}_{y2} + \hat{M}_{z1}\hat{M}_{z2}) \tag{11.28}$$

We now show that the components of the total angular momentum obey the usual angular-momentum commutation relations. We have

$$[\hat{M}_x, \hat{M}_y] = [\hat{M}_{x1} + \hat{M}_{x2}, \hat{M}_{y1} + \hat{M}_{y2}]$$

$$= [\hat{M}_{x1}, \hat{M}_{y1} + \hat{M}_{y2}] + [\hat{M}_{x2}, \hat{M}_{y1} + \hat{M}_{y2}]$$

$$= [\hat{M}_{x1}, \hat{M}_{y1}] + [\hat{M}_{x1}, \hat{M}_{y2}] + [\hat{M}_{x2}, \hat{M}_{y1}] + [\hat{M}_{x2}, \hat{M}_{y2}]$$

Since all components of $\hat{\mathbf{M}}_1$ commute with all components of $\hat{\mathbf{M}}_2$, we have

$$[\hat{M}_x, \hat{M}_y] = [\hat{M}_{x1}, \hat{M}_{y1}] + [\hat{M}_{x2}, \hat{M}_{y2}] = i\hbar\hat{M}_{z1} + i\hbar\hat{M}_{z2}$$

$$[\hat{M}_x, \hat{M}_y] = i\hbar\hat{M}_z \tag{11.29}$$

Cyclic permutation of x, y, and z gives

$$[\hat{M}_y, \hat{M}_z] = i\hbar\hat{M}_x, \qquad [\hat{M}_z, \hat{M}_x] = i\hbar\hat{M}_y \tag{11.30}$$

The same commutator algebra used to derive (5.151) gives

$$[\hat{M}^2, \hat{M}_x] = [\hat{M}^2, \hat{M}_y] = [\hat{M}^2, \hat{M}_z] = 0 \tag{11.31}$$

Thus we can simultaneously quantize M^2 and one of its components, say M_z. Since the components of the total angular momentum obey the angular-momentum commutation relations, the work of Section 5.4 shows that the eigenvalues of $\hat{M}^2$ are

$$J(J+1)\hbar^2, \qquad J = 0, \tfrac{1}{2}, 1, \tfrac{3}{2}, 2, \ldots \tag{11.32}$$

and the eigenvalues of $\hat{M}_z$ are

$$M_J\hbar, \qquad M_J = -J, -J+1, \ldots, J-1, J \tag{11.33}$$

We want to find out how the total angular-momentum quantum numbers J and M_J are related to the quantum numbers j_1, j_2, m_1, m_2 of the two angular momenta we are adding in (11.22). We also want the eigenfunctions of $\hat{M}^2$ and $\hat{M}_z$. These eigenfunctions are characterized by the quantum numbers J and M_J, and using ket notation (Section 7.3) we write them as $|JM_J\rangle$. Similarly, let $|j_1m_1\rangle$ denote the eigenfunctions of $\hat{M}_1^2$ and $\hat{M}_{z1}$ and $|j_2m_2\rangle$ denote the eigenfunctions of $\hat{M}_2^2$ and $\hat{M}_{z2}$. Now it is readily shown (Problem 11.4a) that

$$[\hat{M}_x, \hat{M}_1^2] = [\hat{M}_y, \hat{M}_1^2] = [\hat{M}_z, \hat{M}_1^2] = [\hat{M}^2, \hat{M}_1^2] = 0 \tag{11.34}$$

with similar equations with $\hat{M}_2^2$ replacing $\hat{M}_1^2$. Hence we can have simultaneous eigenfunctions of all four operators $\hat{M}_1^2, \hat{M}_2^2, \hat{M}^2, \hat{M}_z$, and the eigenfunctions $|JM_J\rangle$ can be more fully written as $|j_1j_2JM_J\rangle$. If we take the complete set of functions $|j_1m_1\rangle$ for particle 1 and the complete set $|j_2m_2\rangle$ for particle 2 and form all possible products of the form $|j_1m_1\rangle|j_2m_2\rangle$, we will have a complete set of functions for the two particles. Each unknown eigenfunction $|j_1j_2JM_J\rangle$ can then be expanded using this complete set:

$$|j_1j_2JM_J\rangle = \sum c(j_1j_2JM_J; m_1m_2)|j_1m_1\rangle|j_2m_2\rangle \tag{11.35}$$

where the expansion coefficients are the $c(j_1 \cdots m_2)$'s. Since the function being expanded is an eigenfunction of $\hat{M}_1^2$ with eigenvalue $j_1(j_1+1)\hbar^2$, we only include terms in the sum that have the same j_1 value as in the function $|j_1j_2JM_J\rangle$. [See the reasoning following Eq. (7.66).] Likewise, only terms with the same j_2 value as in $|j_1 j_2 J M_J\rangle$ are included in the sum. Hence the sum only goes over the m_1 and m_2 values. Furthermore, using $\hat{M}_z = \hat{M}_{z1} + \hat{M}_{z2}$, we can prove (Problem 11.4b) that the coefficient c vanishes unless

$$m_1 + m_2 = M_J \tag{11.36}$$

To explicitly obtain the total angular-momentum eigenfunctions, we must evaluate the coefficients in (11.35). These are called *Clebsch-Gordan* or *Wigner* or *vector addition coefficients*. We omit their evaluation (see *Merzbacher*, Section 16.6); *Anderson* gives twelve pages of tables of these coefficients.

Thus each total angular-momentum eigenfunction $|j_1 j_2 J M_J\rangle$ is a linear combination of those product functions $|j_1 m_1\rangle|j_2 m_2\rangle$ whose m values satisfy $m_1 + m_2 = M_J$. We now find the relation between J and the j_1, j_2 values. Before discussing the general case, we consider the case with $j_1 = 1, j_2 = 2$. The possible values of m_1 are $-1, 0, +1$, and the possible values of m_2 are $-2, -1, 0, +1, +2$. If we describe the system by the quantum numbers j_1, j_2, m_1, m_2, then the total number of possible states is fifteen, corresponding to three possibilities for m_1 and five for m_2. Instead, we can describe the system using the quantum numbers j_1, j_2, J, M_J, and we must have the same number of states in this description. Let us tabulate the fifteen possible values of M_J using (11.36):

$m_1 = -1$	0	$+1$	
-3	-2	-1	$-2 = m_2$
-2	-1	0	-1
-1	0	$+1$	0
0	$+1$	$+2$	$+1$
$+1$	$+2$	$+3$	$+2$

The number of times each value of M_J occurs is

value of M_J	3	2	1	0	-1	-2	-3
number of occurrences	1	2	3	3	3	2	1

The highest value of M_J is $+3$; since M_J ranges from $-J$ to $+J$, the highest value of J must be 3. Corresponding to $J = 3$, there are seven values of M_J ranging from -3 to $+3$. Eliminating these seven values, we are left with

value of M_J	2	1	0	-1	-2
number of occurrences	1	2	2	2	1

The highest remaining value, $M_J = 2$, must correspond to $J = 2$; for $J = 2$, we have five values of M_J, which when eliminated leave

value of M_J	1	0	-1
number of occurrences	1	1	1

These remaining values of M_J clearly correspond to $J = 1$. Thus for $j_1 = 1, j_2 = 2$, the possible values of J are 3, 2, and 1.

Now we consider the general case. We have $2j_1 + 1$ values of m_1 and $2j_2 + 1$ values of m_2; hence there are $(2j_1 + 1)(2j_2 + 1)$ possible states. The highest possible values of m_1 and m_2 are j_1 and j_2, respectively, so that the highest possible value of M_J is $j_1 + j_2$. Since M_J ranges from $-J$ to $+J$, the highest value of J must also be $j_1 + j_2$. The second-highest value of M_J is $j_1 + j_2 - 1$, which arises in two ways: $m_1 = j_1 - 1, m_2 = j_2$ and $m_1 = j_1, m_2 = j_2 - 1$. Linear combinations of these two states must give one state with $J = j_1 + j_2, M_J = j_1 + j_2 - 1$ and one state with $J = j_1 + j_2 - 1, M_J = j_1 + j_2 - 1$. Continuing on in this manner, we find the possible values of J to be $j_1 + j_2, j_1 + j_2 - 1, j_1 + j_2 - 2, \ldots, j_1 + j_2 - b$, where $j_1 + j_2 - b$ is the lowest value of J. We determine b by the requirement that the total number of states be $(2j_1 + 1)(2j_2 + 1)$. For the state with $J = j_1 + j_2 - k$, there are $2(j_1 + j_2 - k) + 1$ possible values of M_J. Hence

$$\sum_{k=0}^{b} [2(j_1 + j_2 - k) + 1] = (2j_1 + 1)(2j_2 + 1)$$

$$\sum_{k=0}^{b} (2j_1 + 2j_2 + 1) - 2 \sum_{k=0}^{b} k = (2j_1 + 1)(2j_2 + 1) \tag{11.37}$$

The first sum has $b + 1$ terms, each of which is the same; hence its value is $(b + 1)(2j_1 + 2j_2 + 1)$. From Problem 6.25, the second sum equals $\frac{1}{2}b(b + 1)$. Equation (11.37) becomes

$$b^2 - 2(j_1 + j_2)b + 4j_1 j_2 = 0$$

The roots of this quadratic equation are $b = 2j_1$ and $b = 2j_2$. Corresponding to these roots, we have the minimum values of J:

$$J_{\min} = j_2 - j_1 \qquad J_{\min} = j_1 - j_2 \tag{11.38}$$

If $j_1 = j_2$, then $J_{\min} = 0$. If $j_1 \neq j_2$, then one of the values in (11.38) is negative and must be rejected [Eq. (11.32)]. Thus

$$J_{\min} = |j_2 - j_1| \tag{11.39}$$

To summarize, we have shown that the addition of two angular momenta characterized by quantum numbers j_1 and j_2 results in a total angular momentum whose quantum number J has the possible values

$$J = j_1 + j_2, j_1 + j_2 - 1, \ldots, |j_1 - j_2| \tag{11.40}$$

For example, if $j_1 = 2, j_2 = 3$, we have the possibilities $J = 5, 4, 3, 2, 1$. For $j_1 = 3$, $j_2 = 3/2$, we have $J = 9/2, 7/2, 5/2, 3/2$.

To add more than two angular momenta, we apply (11.40) repeatedly. Thus to add three angular momenta with $j_1 = 1, j_2 = 2, j_3 = 3$, we first add j_1 and j_2 to give the possible values 3, 2, 1. Adding j_3 to each of these values, we get the following possibilities for the total angular-momentum quantum number

$$6, 5, 4, 3, 2, 1, 0; \quad 5, 4, 3, 2, 1; \quad 4, 3, 2 \tag{11.41}$$

We have one set of states with total angular-momentum quantum number 6, two sets of states with angular momentum 5, etc.

11.5   *ANGULAR MOMENTUM IN MANY-ELECTRON ATOMS*

We define the total orbital angular momentum **L** of an n-electron atom as the vector sum of the orbital angular momenta of the individual electrons:

$$\mathbf{L} = \sum_{i=1}^{n} \mathbf{L}_i \qquad (11.42)$$

Although the individual orbital angular momenta of the electrons do not commute with the atomic Hamiltonian (11.1), it can be shown[8] that the total orbital angular-momentum operator does commute with the Hamiltonian. We may therefore characterize an atomic state by a quantum number L, where $L(L + 1)\hbar^2$ is the square of the magnitude of the total electronic orbital angular momentum.

As an example, consider an excited state of carbon with the electron configuration $1s^2 2s^2 2p\,3d$. The s electrons have zero orbital angular momentum; combining the orbital angular momenta of the p and d electrons, we find that L can be 3 or 2 or 1. [The Hartree–Fock central-field approximation has each electron moving in a central-field potential, $V = V(r)$; hence within this approximation, the individual electronic orbital angular momenta are constant, giving rise to a wave function composed of a single configuration which specifies the individual orbital angular momenta. When we go beyond the SCF central-field approximation, we mix in other configurations, so that we no longer specify precisely the individual orbital angular momenta; even so, we can still use the rule (11.40) for finding the possible values of the total orbital angular momentum.]

We indicate the total orbital angular momentum of an atom by a letter, as follows:

L	0	1	2	3	4	5	6	7	8
letter	S	P	D	F	G	H	I	K	L

The total orbital angular momentum is designated by a capital letter, while lowercase letters are used for orbital angular momenta of individual electrons. The configuration $1s^2 2s^2 2p\,3d$ gives rise to P, D, and F states.

We define the total electronic spin angular momentum as the vector sum of the spins of the individual electrons of the atom:

$$\mathbf{S} = \sum_{i=1}^{n} \mathbf{S}_i \qquad (11.43)$$

The total electronic spin angular-momentum operator commutes with the Hamiltonian (11.1). We may characterize an atomic state by a quantum number S,

---

[8] *Pilar*, Section 12-1. This statement is true provided spin–orbit interaction (Section 11.6) is neglected.

where $S(S+1)\hbar^2$ is the square of the magnitude of the total electronic spin angular momentum.[9] For example, for two electrons, the individual spin quantum numbers s_1 and s_2 are both $1/2$, and the rule (11.40) gives the possibilities $S = 0, 1$, corresponding to the two spins being antiparallel or parallel.

The quantity $2S + 1$ is written as a left superscript to the letter designating L. As an example, we again consider the configuration $1s^2 2s^2 2p3d$. Consider first the two $1s$ electrons. To satisfy the exclusion principle, one of these electrons must have $m_s = +\frac{1}{2}$, while the other has $m_s = -\frac{1}{2}$. If M_S is the quantum number specifying the z component of the total spin of the $1s$ electrons, then the only possible value of M_S is zero [Eq. (11.36)]. This single value of M_S clearly means that the total spin of the two $1s$ electrons is zero. Thus although in general we have the two possibilities $S = 0$, $S = 1$ when adding the spins of two electrons, the restriction imposed by the Pauli principle leaves $S = 0$ as the only possibility in this case. Likewise, the spins of the $2s$ electrons add up to 0. The Pauli principle imposes no restriction on the $2p$ and $3d$ electrons, and we get both possibilities, $S = 0, S = 1$, for the sum of their spins. The total atomic spin quantum number is 0 or 1 for this configuration, so that $2S + 1$ is 1 or 3. We thus have the possibilities

$$^1P, \; ^3P, \; ^1D, \; ^3D, \; ^1F, \; ^3F \qquad (11.44)$$

for the $1s^2 2s^2 2p3d$ configuration. The symbols in (11.44) are called *term symbols*. Atomic states arising from the same electron configuration and having the same value of L and of S are said to belong to the same *term*. States belonging to different terms have different energies, since the amount of interelectronic repulsion is different for each term. For example, see the discussion of helium at the end of this section.

The total electronic angular momentum $\mathbf{J}$ of the atom is the vector sum of the total electronic orbital and spin angular momenta:

$$\mathbf{J} = \mathbf{L} + \mathbf{S} \qquad (11.45)$$

The operator for the total electronic angular momentum commutes with the Hamiltonian, and we may characterize an atomic state by a quantum number J, which has the possible values [Eq. (11.40)]

$$L + S, \; L + S - 1, \; \ldots, \; |L - S| \qquad (11.46)$$

The value of J is written as a subscript on the term symbol. Thus for a ^3P term, $L = 1$ and $S = 1$, so that J can be 2, 1, or 0, giving the possibilities $^3P_2$, $^3P_1$, and $^3P_0$. States belonging to the same term and having the same value of J are said to belong to the same *level*. As far as the Hamiltonian (11.1) is concerned, all states belonging to the same term have the same energy, but when we include spin-orbit interaction in the Hamiltonian (Section 11.6), we find that the value of J affects the energy slightly; hence the energies of different levels belonging to the same term are slightly different (see Fig. 11.6).

_____

[9] The proof often given is that since $\hat{H}$ does not involve spin, it must commute with $\hat{S}$. This proof is not quite adequate; see *Bethe and Jackiw*, pp. 103–104.

The quantity $2S + 1$ is called the *multiplicity* of the term. If $L \geqslant S$, the possible values of J range from $L + S$ to $L - S$ and are $2S + 1$ in number; in this case the multiplicity gives the number of levels arising from a given term. For $L < S$, the values of J range from $S + L$ to $S - L$ and are $2L + 1$ in number; in this case the multiplicity is greater than the number of levels. Thus if $L = 0$ and $S = 1$, we designate the term as ^3S although there is only one possible value for J: $J = 1$. For $2S + 1 = 1, 2, 3, 4, 5, 6, \ldots$, we use the words *singlet, doublet, triplet, quartet, quintet, sextet,* ... to designate the multiplicity. We read $^3P_1$ as "triplet P one."

We now show how to derive the terms that arise from a given electron configuration. We first consider configurations that contain only completely filled subshells. In such configurations, for each electron with $m_s = +\frac{1}{2}$ there is an electron with $m_s = -\frac{1}{2}$. Let the quantum number specifying the z component of the total electronic spin angular momentum be M_S. The only possible value for M_S is zero ($M_S = \Sigma m_s = 0$); hence S must be zero. For each electron in a closed subshell with magnetic quantum number m, there is an electron with magnetic quantum number $-m$; e.g., for a $2p^6$ configuration we have two electrons with $m = +1$, two with $m = -1$, and two with $m = 0$. Denoting the quantum number specifying the z component of the total electronic orbital angular momentum by M_L, we have $M_L = \Sigma m = 0$; we conclude that L must be zero. In summary, a configuration of closed subshells gives rise to only one term: ^1S. For configurations consisting of closed subshells and open subshells, the closed subshells make no contribution to L or S and may be ignored in finding the terms. (Note that our conclusions are consistent with Unsöld's theorem.)

We now consider two electrons in different subshells; such electrons are called *nonequivalent*. Nonequivalent electrons have different values of n or l or both, and we need not worry about any restrictions imposed by the exclusion principle when we derive the terms. We simply find the possible values of L from l_1 and l_2 according to (11.40); combining s_1 and s_2 gives $S = 0, 1$. We previously worked out the pd case, which gives the terms in (11.44). If we have more than two nonequivalent electrons, we combine the individual l's to find the values of L, and we combine the individual s's to find the values of S. For example, consider a pdf configuration. The possible values of L are given by (11.41). Combining three angular momenta, each of which is $\frac{1}{2}$, gives $S = \frac{3}{2}, \frac{1}{2}, \frac{1}{2}$. Each of the three possibilities in (11.41) with $L = 3$ may be combined with each of the two possibilities for $S = \frac{1}{2}$, giving six ^2F terms. Continuing in this manner, we find that the following terms arise from a pdf configuration: $^2S(2)$, $^2P(4)$, $^2D(6)$, $^2F(6)$, $^2G(6)$, $^2H(4)$, $^2I(2)$, ^4S, $^4P(2)$, $^4D(3)$, $^4F(3)$, $^4G(3)$, $^4H(2)$, ^4I, where the number of times each term occurs is in parentheses.

Now consider two electrons in the same subshell (*equivalent* electrons). Equivalent electrons have the same value of n and of l, and the situation is complicated by the necessity to avoid giving two electrons the same four quantum numbers; hence not all the terms derived for nonequivalent electrons are possible. As an example, consider the terms arising from two equivalent p electrons, an np^2 configuration. (The carbon ground-state configuration is $1s^2 2s^2 2p^2$.) We list the possible values of m and m_s for the two electrons in Table 11.1, which also gives M_L and M_S.

Note that certain combinations are missing from this table. For example, we

Table 11.1 *Quantum numbers for two equivalent p electrons*

m_1	m_{s1}	m_2	m_{s2}	$M_L = \sum m$	$M_S = \sum m_s$
1	$\frac{1}{2}$	1	$-\frac{1}{2}$	2	0
1	$\frac{1}{2}$	0	$\frac{1}{2}$	1	1
1	$\frac{1}{2}$	0	$-\frac{1}{2}$	1	0
1	$-\frac{1}{2}$	0	$\frac{1}{2}$	1	0
1	$-\frac{1}{2}$	0	$-\frac{1}{2}$	1	-1
1	$\frac{1}{2}$	-1	$\frac{1}{2}$	0	1
1	$\frac{1}{2}$	-1	$-\frac{1}{2}$	0	0
1	$-\frac{1}{2}$	-1	$\frac{1}{2}$	0	0
1	$-\frac{1}{2}$	-1	$-\frac{1}{2}$	0	-1
0	$\frac{1}{2}$	0	$-\frac{1}{2}$	0	0
0	$\frac{1}{2}$	-1	$\frac{1}{2}$	-1	1
0	$\frac{1}{2}$	-1	$-\frac{1}{2}$	-1	0
0	$-\frac{1}{2}$	-1	$\frac{1}{2}$	-1	0
0	$-\frac{1}{2}$	-1	$-\frac{1}{2}$	-1	-1
-1	$\frac{1}{2}$	-1	$-\frac{1}{2}$	-2	0

do not have $m_1 = 1, m_{s1} = \frac{1}{2}, m_2 = 1, m_{s2} = \frac{1}{2}$, as this violates the exclusion principle. Another missing combination is $m_1 = 1, m_{s1} = -\frac{1}{2}, m_2 = 1, m_{s2} = \frac{1}{2}$. This combination differs from $m_1 = 1, m_{s1} = \frac{1}{2}, m_2 = 1, m_{s2} = -\frac{1}{2}$ solely by interchange of electrons 1 and 2. Each row in Table 11.1 stands for a Slater determinant, which when expanded contains terms for all possible electron interchanges among the spin-orbitals. Two rows that differ from each other solely by interchange of two electrons correspond to the same Slater determinant, and we include only one of them in the table.

The highest value of M_L in Table 11.1 is 2, which must correspond to a term with $L = 2$, a D term. The $M_L = 2$ value occurs in conjunction with $M_S = 0$, indicating that $S = 0$ for the D term; thus we have a ^1D term corresponding to the five states

$$
\begin{array}{cccccc}
M_L = 2 & 1 & 0 & -1 & -2 \\
M_S = 0 & 0 & 0 & 0 & 0
\end{array}
\tag{11.47}
$$

The highest value of M_S in Table 11.1 is 1, indicating a term with $S = 1$, a triplet term. $M_S = 1$ occurs in conjunction with $M_L = 1, 0, -1$, which indicates a P term. Hence we have a ^3P term corresponding to the nine states

$$
\begin{array}{ccccccccc}
M_L = 1 & 1 & 1 & 0 & 0 & 0 & -1 & -1 & -1 \\
M_S = 1 & 0 & -1 & 1 & 0 & -1 & 1 & 0 & -1
\end{array}
\tag{11.48}
$$

Elimination of the states of (11.47) and (11.48) from Table 11.1 leaves only a single state, which has $M_L = 0$, $M_S = 0$, corresponding to a ^1S term. Thus a p^2 configuration gives rise to the terms ^1S, ^3P, ^1D. (In contrast, two nonequivalent p electrons give rise to six terms: ^1S, ^3S, ^1P, ^3P, ^1D, ^3D.)

Table 11.2 *Terms arising from various electron configurations*

Configuration	Terms
	(a) *Equivalent electrons*[1]
s^2; p^6; d^{10}	^1S
p; p^5	^2P
p^2; p^4	^3P, ^1D, ^1S
p^3	^4S, ^2D, ^2P
d; d^9	^2D
d^2; d^8	^3F, ^3P, ^1G, ^1D, ^1S
d^3; d^7	^4F, ^4P, ^2H, ^2G, ^2F, $^2D(2)$, ^2P
d^4; d^6	$\begin{cases} ^5D,\ ^3H,\ ^3G,\ ^3F(2),\ ^3D,\ ^3P(2) \\ ^1I,\ ^1G(2),\ ^1F,\ ^1D(2),\ ^1S(2) \end{cases}$
d^5	$\begin{cases} ^6S,\ ^4G,\ ^4F,\ ^4D,\ ^4P,\ ^2I,\ ^2H,\ ^2G(2) \\ ^2F(2),\ ^2D(3),\ ^2P,\ ^2S \end{cases}$
	(b) *Nonequivalent electrons*
ss	^1S, ^3S
sp	^1P, ^3P
sd	^1D, ^3D
pp	^3D, ^1D, ^3P, ^1P, ^3S, ^1S

[1] "Equivalent" electrons have the same values of n and l.

Table 11.2a lists the terms arising from various configurations of equivalent electrons. These results may be derived in the same manner that we found the p^2 terms, but this procedure can become quite involved; to derive the terms of the f^7 configuration would require a table with 3432 rows. A more efficient method[10] uses group theory. Note that the terms for a subshell containing N electrons are the same as the terms for a subshell that is N electrons short of being full; e.g., the terms for p^2 and p^4 are the same. We can divide the electrons of a closed subshell into two groups and find the terms for each group; because a closed subshell gives only a ^1S term, the terms for each of these two groups must be the same. One sometimes speaks of the absence of an electron from a subshell as a "hole." Table 11.2b gives the terms arising from some nonequivalent electron configurations.

To deal with a configuration containing both equivalent and nonequivalent electrons, we first find separately the terms from the nonequivalent electrons and the terms from the equivalent electrons. We then take all possible combinations of the L and S values of these two sets of terms. For example, consider an sp^3 configuration. From the s electron, we get a ^2S term. From the three equivalent p electrons, we get

---

[10] See R. F. Curl, Jr., and J. E. Kilpatrick, *Am. J. Phys.*, **28**, 357 (1960).

the terms ^2P, ^2D, and ^4S. Combining the L and S values of these terms, we have as the terms of an sp^3 configuration

$$^3P, \, {}^1P, \, {}^3D, \, {}^1D, \, {}^5S, \, {}^3S$$

We now consider the terms and levels for the hydrogen and helium atoms. For hydrogen we have one electron; hence $L = l$ and $S = s = \frac{1}{2}$. The possible values of J are $L \pm \frac{1}{2}$ except for $L = 0$, where we can have only $J = \frac{1}{2}$. Each configuration gives rise to a single term, which is composed of one or two levels. The ground-state configuration $1s$ gives the term ^2S, which is composed of the single level $^2S_{1/2}$; the level is twofold degenerate, corresponding to the two possible orientations $M_J = +\frac{1}{2}$ and $-\frac{1}{2}$. The $2s$ configuration also gives a $^2S_{1/2}$ level. For the $2p$ configuration, we have the levels $^2P_{3/2}$ and $^2P_{1/2}$; the $^2P_{3/2}$ level is fourfold degenerate ($M_J = -\frac{3}{2}$, $-\frac{1}{2}, \frac{1}{2}, \frac{3}{2}$), and the $^2P_{1/2}$ level is twofold degenerate. All told, we have 8 states with quantum number $n = 2$; omitting spin-orbit interaction, these states all have the same energy. Previously we discussed the degeneracy in terms of possible values for m and m_s; now we are representing things in terms of the quantum numbers J and M_J. We get the same number of states no matter how we view things.

For helium the ground-state configuration $1s^2$ is a closed subshell and gives the single level $^1S_0$; we have $M_J = 0$, and the level is nondegenerate. The $1s2s$ excited configuration gives rise to two terms, ^1S and ^3S, each of which has one level. The $^1S_0$ level is nondegenerate, while the $^3S_1$ level is threefold degenerate ($M_J = 1, 0, -1$). (Although in this case the multiplicity is equal to the degeneracy, this is certainly not true in general.) In Section 10.5 we found the lower of the two energy levels of the $1s2s$ helium configuration to be triply degenerate, corresponding to the three zeroth-order wave functions (10.42)–(10.44). This must be the $^3S_1$ level. The nondegenerate $^1S_0$ level has the zeroth-order wave function (10.45).

Since $S = 1$ for the $^3S_1$ level, we should find the three spin functions in (10.42)–(10.44) to be eigenfunctions of $\hat{S}^2$ with eigenvalue $2\hbar^2$, and eigenfunctions of $\hat{S}_z$ with eigenvalues $\hbar, 0, -\hbar$ ($M_S = 1, 0, -1$). The spin function in (10.45) should be an eigenfunction of $\hat{S}^2$ and $\hat{S}_z$ with eigenvalue zero in each case, since $S = 0$ here. We now verify these assertions. From Eq. (11.43) the total electronic spin operator is the sum of the spin operators for each electron:

$$\hat{\mathbf{S}} = \hat{\mathbf{S}}_1 + \hat{\mathbf{S}}_2 \tag{11.49}$$

Taking the z components of (11.49), we have

$$\hat{S}_z = \hat{S}_{z1} + \hat{S}_{z2} \tag{11.50}$$

$$\hat{S}_z \alpha(1)\alpha(2) = \hat{S}_{z1}\alpha(1)\alpha(2) + \hat{S}_{z2}\alpha(1)\alpha(2)$$

$$= \alpha(2)\hat{S}_{z1}\alpha(1) + \alpha(1)\hat{S}_{z2}\alpha(2)$$

$$= \tfrac{1}{2}\hbar\alpha(1)\alpha(2) + \tfrac{1}{2}\hbar\alpha(1)\alpha(2)$$

$$= \hbar\alpha(1)\alpha(2) \tag{11.51}$$

where Eq. (10.7) has been used. Similarly we find

$$\hat{S}_z \beta(1)\beta(2) = -\hbar\beta(1)\beta(2) \tag{11.52}$$

$$\hat{S}_z[\alpha(1)\beta(2) + \beta(1)\alpha(2)] = 0 \qquad (11.53)$$

$$\hat{S}_z[\alpha(1)\beta(2) - \beta(1)\alpha(2)] = 0 \qquad (11.54)$$

Consider now $\hat{S}^2$. We have [Eq. (11.28)]

$$\hat{S}^2 = (\hat{\mathbf{S}}_1 + \hat{\mathbf{S}}_2)\cdot(\hat{\mathbf{S}}_1 + \hat{\mathbf{S}}_2)$$

$$= \hat{S}_1^2 + \hat{S}_2^2 + 2(\hat{S}_{x1}\hat{S}_{x2} + \hat{S}_{y1}\hat{S}_{y2} + \hat{S}_{z1}\hat{S}_{z2}) \qquad (11.55)$$

$$\hat{S}^2\alpha(1)\alpha(2) = \alpha(2)\hat{S}_1^2\alpha(1) + \alpha(1)\hat{S}_2^2\alpha(2) + 2\hat{S}_{x1}\alpha(1)\hat{S}_{x2}\alpha(2)$$

$$+ 2\hat{S}_{y1}\alpha(1)\hat{S}_{y2}\alpha(2) + 2\hat{S}_{z1}\alpha(1)\hat{S}_{z2}\alpha(2) \qquad (11.56)$$

Using Eqs. (10.7)–(10.9) and (10.28) and (10.29), we find

$$\hat{S}^2\alpha(1)\alpha(2) = 2\hbar^2\alpha(1)\alpha(2)$$

showing that $\alpha(1)\alpha(2)$ is an eigenfunction of $\hat{S}^2$ corresponding to $S = 1$. Similarly we find

$$\hat{S}^2\beta(1)\beta(2) = 2\hbar^2\beta(1)\beta(2)$$

$$\hat{S}^2[\alpha(1)\beta(2) + \beta(1)\alpha(2)] = 2\hbar^2[\alpha(1)\beta(2) + \beta(1)\alpha(2)]$$

$$\hat{S}^2[\alpha(1)\beta(2) - \beta(1)\alpha(2)] = 0$$

Thus the spin eigenfunctions in (10.42)–(10.45) correspond to the following values for the total spin quantum numbers:

		S	M_S	
	$\alpha(1)\alpha(2)$	1	1	(11.57)
triplet	$2^{-1/2}[\alpha(1)\beta(2) + \beta(1)\alpha(2)]$	1	0	(11.58)
	$\beta(1)\beta(2)$	1	−1	(11.59)
singlet	$\{2^{-1/2}[\alpha(1)\beta(2) - \beta(1)\alpha(2)]$	0	0	(11.60)

[In the notation of Section 11.4, we are dealing with the addition of two angular momenta with quantum numbers $j_1 = \frac{1}{2}$ and $j_2 = \frac{1}{2}$ to give eigenfunctions with total angular-momentum quantum numbers $J = 1$ and $J = 0$; the coefficients in (11.57)–(11.60) correspond to the coefficients c in (11.35) and are examples of Clebsch–Gordan coefficients.]

Figure 11.3 shows the vector addition of $\mathbf{S}_1$ and $\mathbf{S}_2$ to form $\mathbf{S}$. It might seem surprising that the spin function (11.58), which has the z components of the spins of the two electrons pointing in opposite directions, could have total spin quantum number $S = 1$. Figure 11.3 shows how this is possible.

In Section 10.7 we showed that two of the four zeroth-order wave functions for the $1s2s$ helium configuration could be written as single Slater determinants, but the remaining two functions had to be expressed as linear combinations of two Slater determinants. Since $\hat{L}^2$ and $\hat{S}^2$ commute with the Hamiltonian (11.1), the zeroth-order functions should be eigenfunctions of $\hat{L}^2$ and $\hat{S}^2$. The Slater determinants D_2 and D_3 of Section 10.7 are not eigenfunctions of these operators and hence are not suitable zeroth-order functions. We have just shown that the

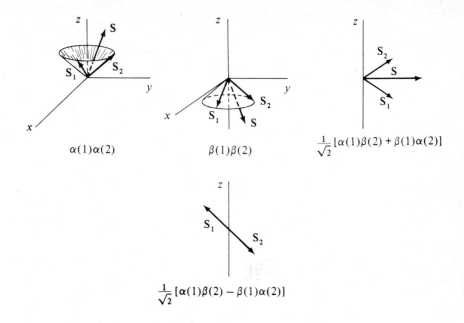

Figure 11.3 *Vector addition of the spins of two electrons. For $\alpha(1)\alpha(2)$ and $\beta(1)\beta(2)$, the projections of $\mathbf{S}_1$ and $\mathbf{S}_2$ in the xy plane make an angle of $90°$ with each other (Problem 11.8c).*

linear combinations (10.62) and (10.63) are eigenfunctions of $\hat{S}^2$, and we can also show them to be eigenfunctions of $\hat{L}^2$.

For a configuration of closed subshells (e.g., the helium ground state), we can write only a single Slater determinant; this determinant is an eigenfunction of $\hat{L}^2$ and $\hat{S}^2$ and is the correct zeroth-order function for the nondegenerate $^1S_0$ level. A configuration with one electron outside closed subshells (e.g., the lithium ground state) gives rise to only one term; the Slater determinants for such a configuration differ from one another only in the m and m_s values of this electron and are the correct zeroth-order functions for the states of the term. Except for these two kinds of configurations (and a configuration that is one electron short of having closed subshells), we generally have to take linear combinations of the Slater determinants to obtain the correct zeroth-order functions, which are eigenfunctions of $\hat{L}^2$ and $\hat{S}^2$. The correct linear combinations can be found by solving the secular equation of degenerate perturbation theory, or by operator techniques. Tabulations of the correct combinations for various configurations are available.[11] Hartree–Fock calculations of atomic term energies use these linear combinations and find the best possible orbital functions for the Slater determinants.

To decide which of the terms arising from a given atomic configuration is lowest in energy, we use the empirical *Hund's rule: The term with the highest*

[11] *Slater, Atomic Structure*, Vol. II.

multiplicity is lowest in energy; if there is more than one term with the highest multiplicity, then the term with the highest multiplicity and the largest value of L lies lowest. For example, the ground-state configuration of carbon $1s^2 2s^2 2p^2$ gives the terms ^1S, ^3P, and ^1D. Hund's rule tells us that ^3P is the lowest term, in agreement with experiment. For a d^2 configuration (Table 11.2), the ^3F term lies lowest according to Hund's rule. Hund's rule gives only the lowest term, and should not be used to decide the order of the remaining terms. Thus for the $1s^2 2s 2p^3$ configuration of carbon, the experimental order of the terms is

$$^5S < {}^3D < {}^3P < {}^1D < {}^3S < {}^1P$$

The ^3S term lies above the ^1D term, even though ^3S has the higher multiplicity. For the $1s^2 2s^2 2p^6 3s^2 3p^6 3d^2 4s^2$ titanium configuration, the lowest singlet term is ^1D and not ^1G, even though ^1G has the larger value of L.

Hund's rule works very well for the ground-state configuration, but occasionally fails for an excited-state configuration.

The traditional explanation of Hund's rule is as follows: Electrons with the same spin tend to keep out of each other's way (recall the idea of Fermi holes), thereby minimizing the Coulombic repulsion between them. The term that has the greatest number of parallel spins (i.e., the greatest value of S) will therefore be lowest in energy. For example, the ^3S term of the helium $1s2s$ configuration has an antisymmetric space function which vanishes when the spatial coordinates of the two electrons are equal; hence the ^3S term is lower than the ^1S term.

This traditional explanation turns out to be wrong in many cases. It is true that the probability that the two electrons are very close together is smaller for the He ^3S $1s2s$ term than for the ^1S $1s2s$ term; however, calculations with accurate wave functions also show that the probability that the two electrons are very far apart is also less for the ^3S term. The net result is that the average distance between the two electrons is slightly less for the ^3S term than for the ^1S term and the interelectronic repulsion is slightly greater for the ^3S term. The reason the ^3S term lies below the ^1S term is because of a substantially greater electron–nucleus attraction in the ^3S term as compared to the ^1S term. Similar results are found for singlet and triplet terms of Be and C. [For details see J. Katriel and R. Pauncz, *Adv. Quantum Chem.*, **10**, 143 (1977).] The following explanation of these results has been proposed [I. Shim and J. P. Dahl, *Theor. Chim. Acta*, **48**, 165 (1978)]. The Pauli-principle "repulsion" between electrons of like spin makes the average angle between the radius vectors of the two electrons larger for the ^3S term than for the ^1S term; this allows the electrons to get closer to the nucleus in the ^3S term, making the electron–nucleus attraction greater for the ^3S term.

It is not necessary to consult Table 11.2a to find the lowest term of a partly filled subshell configuration. We simply put the electrons in the orbitals so as to give the greatest number of parallel spins. Thus for a d^3 configuration, we have

$$\begin{array}{cccccc} & \uparrow & \uparrow & \uparrow & & \\ m: & +2 & +1 & 0 & -1 & -2 \end{array}$$

The lowest term thus has three parallel spins, so $S = \frac{3}{2}$, giving a quartet term. The maximum value of M_L is 3, corresponding to $L = 3$, an F term. Hund's rule thus predicts ^4F as the lowest term of a d^3 configuration.

An important concept is the parity of atomic states. Consider the atomic Hamiltonian (11.1). We showed in Section 7.5 that the parity operator $\hat{\Pi}$ commutes with the kinetic-energy operator. The quantity $1/r_i$ is given by

$$r_i^{-1} = [x_i^2 + y_i^2 + z_i^2]^{-1/2}$$

Replacement of each coordinate by its negative leaves $1/r_i$ unchanged. Also

$$r_{ij}^{-1} = [(x_i - x_j)^2 + (y_i - y_j)^2 + (z_i - z_j)^2]^{-1/2}$$

so that inversion has no effect on $1/r_{ij}$. Thus $\hat{\Pi}$ commutes with the atomic Hamiltonian, and we can choose atomic eigenfunctions to have definite parity.

For a one-electron atom, the spatial wave function has the form

$$\psi = R(r) Y_l^m(\theta, \varphi)$$

The radial function is unchanged on inversion, so that the parity is determined by the angular factor. In Problem 7.16 we showed that Y_l^m is an even function when l is even and is an odd function when l is odd. Thus the states of one-electron atoms have even or odd parity according to whether l is even or odd.

Now consider an n-electron atom. In the Hartree–Fock central-field approximation, we write the wave function as a Slater determinant (or linear combination of Slater determinants) of spin-orbitals. The wave function is the sum of terms, the spatial factor in each term having the form

$$R_1(r_1) \cdots R_n(r_n) Y_{l_1}^{m_1}(\theta_1, \varphi_1) \cdots Y_{l_n}^{m_n}(\theta_n, \varphi_n)$$

The parity of this product is determined by the spherical-harmonic factors; we see that the product is an even or odd function according to whether $l_1 + l_2 + \cdots + l_n$ is an even or odd number. Thus the sum of the l values of the electrons determines the parity. For example, the configuration $1s^2 2s 2p^3$ has $\Sigma \, l_i = 3$, and all states arising from this configuration have odd parity. (Our argument was based on the SCF approximation to the wave function, but the conclusions are valid for the true wave function.)

11.6   SPIN–ORBIT INTERACTION

Up to now we have used an atomic Hamiltonian [Eq. (11.1)] that does not involve electron spin. In reality, the existence of spin adds an additional term, usually small, to the Hamiltonian. This term, called the *spin–orbit interaction*, causes the fine structure of atomic spectra. Spin–orbit interaction is a relativistic effect and is properly derived using Dirac's relativistic treatment of the electron. We will give a rough "derivation."

If we imagine ourselves riding on an electron in an atom, from our viewpoint the nucleus is moving around the electron (as the sun appears to move about the earth). This apparent motion of the nucleus gives rise to a magnetic field that interacts with the intrinsic (spin) magnetic moment of the electron, giving the spin–

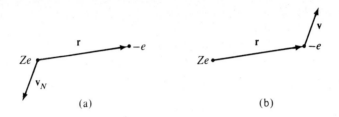

Figure 11.4 (a) Nucleus viewed from an electron   (b) Laboratory
view of things

orbit interaction term in the Hamiltonian. The electron's spin magnetic moment is
proportional to its spin **S**, and the magnetic field arising from the apparent nuclear
motion is proportional to the electron's orbital angular momentum **L**; hence the
spin–orbit interaction term is proportional to **L** · **S**. The dot product of **L** and **S**
depends on the relative orientation of these two vectors. The total angular
momentum **J** = **L** + **S** also depends on the relative orientation of **L** and **S**, and
therefore the spin–orbit interaction energy depends on J [Eq. 11.74)].

From the viewpoint of the electron, the magnetic field produced by the
apparent nuclear motion is given by Eq. (6.148) as (all equations in this section are in
SI units)

$$\mathbf{B} = \frac{\mu_0 Ze}{4\pi r^3} \mathbf{v}_N \times \mathbf{r} \tag{11.61}$$

where $\mathbf{v}_N$ is the apparent velocity of the nucleus and **r** is the vector from the nucleus to
the electron (Fig. 11.4). We have $\mathbf{v}_N = -\mathbf{v}$, where **v** is the actual velocity of the
electron. Hence [Eq. (5.37)] $\mathbf{B} = (\mu_0 Ze/4\pi r^3)\mathbf{r} \times \mathbf{v}$. The electric field **E** of the nucleus
at the electron is $Ze\mathbf{r}/4\pi\varepsilon_0 r^3$, so $\mathbf{B} = \mu_0\varepsilon_0 \mathbf{E} \times \mathbf{v}$. Maxwell's electromagnetic theory of
light shows that (*Halliday and Resnick*, Section 41-8) the speed of light is
$c = (\varepsilon_0\mu_0)^{-1/2}$, so

$$\mathbf{B} = (1/c^2)\mathbf{E} \times \mathbf{v} \tag{11.62}$$

For the energy of the spin–orbit interaction between **B** and the spin magnetic
moment $\boldsymbol{\mu}_S$ of the electron, we have [Eq. (6.155)]

$$E_{\text{S.O.}} = -\mathbf{B} \cdot \boldsymbol{\mu}_S = -(1/c^2)(\mathbf{E} \times \mathbf{v}) \cdot \boldsymbol{\mu}_S \tag{11.63}$$

The product of the electric field at the electron and its charge gives the force on the
electron:

$$\mathbf{F} = -e\mathbf{E} = -\nabla V \tag{11.64}$$

where Eq. (5.46) has been used. Hence

$$E_{\text{S.O.}} = -(1/c^2 e)(\nabla V \times \mathbf{v}) \cdot \boldsymbol{\mu}_S \tag{11.65}$$

At this point we must break the news that (11.65) is wrong. It turns out that there is
an additional relativistic effect[12] (discovered by L. H. Thomas in 1926), which when

---

[12] For a discussion see W. H. Furry, *Am. J. Phys.*, **23**, 517 (1955).

added in gives for the spin–orbit interaction energy

$$E_{\text{S.O.}} = -(1/2c^2 e)(\nabla V \times \mathbf{v}) \cdot \boldsymbol{\mu}_S \qquad (11.66)$$

If we assume the validity of the central-field approximation (Section 11.1), we can accurately approximate the potential energy of an electron in an atom by a function of r only: $V = V(r)$, so that [Eqs. (5.46) and (6.4)]

$$\nabla V = \frac{dV}{dr}\frac{\mathbf{r}}{r} \qquad (11.67)$$

Substitution of (11.67), (10.82), and $\mathbf{v} = \mathbf{p}/m$ (where $\mathbf{p}$ is the linear momentum of the electron) into (11.66) gives

$$E_{\text{S.O.}} = \frac{1}{2m^2 c^2}\frac{1}{r}\frac{dV}{dr}(\mathbf{r} \times \mathbf{p}) \cdot \mathbf{S} \qquad (11.68)$$

The quantity $\mathbf{r} \times \mathbf{p}$ is the electronic orbital angular momentum $\mathbf{L}$, and introduction of operators into (11.68) gives the following term in the Hamiltonian due to spin–orbit interaction:

$$\hat{H}_{\text{S.O.}} = \frac{1}{2m^2 c^2}\frac{1}{r}\frac{dV}{dr}\hat{\mathbf{L}} \cdot \hat{\mathbf{S}} \qquad (11.69)$$

Equation (11.69) is for one electron; for the many-electron case, we sum over the individual spin–orbit interactions to get

$$\hat{H}_{\text{S.O.}} = \frac{1}{2m^2 c^2}\sum_i \frac{1}{r_i}\frac{dV(r_i)}{dr_i}\hat{\mathbf{L}}_i \cdot \hat{\mathbf{S}}_i = \sum_i \xi_i(r_i)\hat{\mathbf{L}}_i \cdot \hat{\mathbf{S}}_i \qquad (11.70)$$

where the definition of ξ_i is clear.

An exact calculation of the effect of the spin–orbit interaction requires that we find the eigenvalues and eigenfunctions of the Hamiltonian $\hat{H} + \hat{H}_{\text{S.O.}}$, where $\hat{H}$ is the Hamiltonian of Eq. (11.1). Such a calculation is too difficult, and we will use an approximate treatment. The effect of $\hat{H}_{\text{S.O.}}$ is generally small compared to the terms in $\hat{H}$, so that we may use first-order perturbation theory to calculate the energy correction. We have [Eq. (9.30)]

$$E_{\text{S.O.}} = \int \psi^* \hat{H}_{\text{S.O.}}\psi \, d\tau = \langle H_{\text{S.O.}} \rangle \qquad (11.71)$$

where ψ is an eigenfunction of $\hat{H}$. For a one-electron atom,

$$E_{\text{S.O.}} = \int \psi^* \xi(r)\hat{\mathbf{L}} \cdot \hat{\mathbf{S}}\psi \, d\tau \qquad (11.72)$$

Now

$$\mathbf{J} \cdot \mathbf{J} = (\mathbf{L} + \mathbf{S}) \cdot (\mathbf{L} + \mathbf{S}) = L^2 + S^2 + 2\mathbf{L} \cdot \mathbf{S}$$

$$\mathbf{L} \cdot \mathbf{S} = \tfrac{1}{2}(J^2 - L^2 - S^2)$$

$$\hat{\mathbf{L}} \cdot \hat{\mathbf{S}}\psi = \tfrac{1}{2}(\hat{J}^2 - \hat{L}^2 - \hat{S}^2)\psi = \tfrac{1}{2}[J(J+1) - L(L+1) - S(S+1)]\hbar^2\psi$$

since the unperturbed ψ is an eigenfunction of $\hat{L}^2$, $\hat{S}^2$, and $\hat{J}^2$. Therefore

$$E_{\text{S.O.}} = \tfrac{1}{2}\langle \xi \rangle \hbar^2 [J(J+1) - L(L+1) - S(S+1)] \qquad (11.73)$$

For a many-electron atom, it can be shown (*Bethe and Jackiw*, p. 163) that the spin–orbit interaction energy is

$$E_{\text{s.o.}} = \tfrac{1}{2} A \hbar^2 [J(J+1) - L(L+1) - S(S+1)] \tag{11.74}$$

where A is a constant for a given term; i.e., A depends on L and S but not on J. Equation (11.74) shows that when we include the spin–orbit interaction, the energy of an atomic state depends on its total electronic angular momentum J. Thus each atomic term is split into levels, each level having a different value of J. For example, the $1s^2 2s^2 2p^6 3p$ configuration of sodium has the single term ^2P, which is composed of the two levels $^2P_{3/2}$ and $^2P_{1/2}$. The splitting of these levels gives the observed *fine structure* of the sodium D line (Fig. 11.5). The levels of a given term are said to form its *multiplet structure*.

What about the order of levels within a given term? Since L and S are the same for such levels, their relative energies are determined, according to Eq. (11.74), by $AJ(J+1)$. If A is positive, the level with the lowest value of J lies lowest, and the multiplet is said to be *regular*; if A is negative, the level with the highest value of J lies lowest, and the multiplet is said to be *inverted*. The following rule usually applies to a configuration with only one partly filled subshell: If this subshell is less than half filled, the multiplet is regular; if this subshell is more than half filled, the multiplet is inverted. (A few exceptions exist.) For example, according to this rule, the multiplets of the $1s^2 2s^2 2p^2$ carbon configuration are regular; previously we used Hund's rule to conclude that the lowest term was ^3P, and we now conclude that $^3P_0$ is the lowest level of this term. For a subshell that is half-filled, Hund's rule tells us that the lowest-lying term has $L = 0$. For example, we have for the lowest term of a d^5 configuration

$$\begin{array}{ccccc} \uparrow & \uparrow & \uparrow & \uparrow & \uparrow \\ \hline \end{array}$$
$$m: \quad +2 \quad +1 \quad 0 \quad -1 \quad -2$$

With $L = 0$, there is only one possible value of J, and the lowest term has but one level. Thus we don't need a rule to find the lowest level of the lowest term of a half-filled subshell.

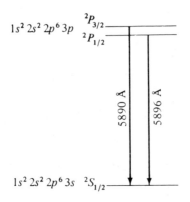

Figure 11.5 *Fine structure of the sodium D line*

The tables[13] of C. E. Moore and others give the spectroscopically determined atomic energy levels for all atoms except the actinides. Also given are the levels of various atomic ions. Spectroscopists use the symbol I to indicate a neutral atom, the symbol II to indicate a singly ionized atom, etc. Thus C III refers to the C^{2+} ion.

11.7   THE ATOMIC HAMILTONIAN

The Hamiltonian of an atom can be divided into three parts:

$$\hat{H} = \hat{H}^0 + \hat{H}_{rep} + \hat{H}_{S.O.} \tag{11.75}$$

where $\hat{H}^0$ is the sum of hydrogenlike Hamiltonians

$$\hat{H}^0 = \sum_{i=1}^{n} \left[-\frac{\hbar^2}{2m} \nabla_i^2 - \frac{Ze'^2}{r_i} \right] \tag{11.76}$$

$\hat{H}_{rep}$ consists of the interelectronic repulsions

$$\hat{H}_{rep} = \sum_{i=1}^{n} \sum_{j>i} \frac{e'^2}{r_{ij}} \tag{11.77}$$

and $\hat{H}_{S.O.}$ is the spin–orbit interaction

$$\hat{H}_{S.O.} = \sum_{i=1}^{n} \xi_i \hat{\mathbf{L}}_i \cdot \hat{\mathbf{S}}_i \tag{11.78}$$

If we consider just $\hat{H}^0$, all atomic states corresponding to the same electronic *configuration* are degenerate. Adding in $\hat{H}_{rep}$, we lift the degeneracy between states with different L or S or both, thus splitting each configuration into *terms*. Next, we add in $\hat{H}_{S.O.}$, which splits each term into *levels*; each level is composed of *states* with the same value of J and is $(2J+1)$-fold degenerate, corresponding to the possible values of M_J.

We can remove the degeneracy of the levels by applying an external magnetic field (the *Zeeman effect*). If **B** is the applied field, we have the additional term in the Hamiltonian [Eq. (6.155)]

$$\hat{H}_B = -\hat{\boldsymbol{\mu}} \cdot \mathbf{B} = -(\hat{\boldsymbol{\mu}}_L + \hat{\boldsymbol{\mu}}_S) \cdot \mathbf{B} \tag{11.79}$$

where we have included both the orbital and the spin magnetic moments. Using Eqs. (6.152), (6.154), and (10.82), we have

$$\hat{H}_B = \beta_e \hbar^{-1} (\hat{\mathbf{L}} + 2\hat{\mathbf{S}}) \cdot \mathbf{B} = \beta_e \hbar^{-1} (\hat{\mathbf{J}} + \hat{\mathbf{S}}) \cdot \mathbf{B} = \beta_e B \hbar^{-1} (\hat{J}_z + \hat{S}_z) \tag{11.80}$$

where β_e is the Bohr magneton; we have taken the z axis along the direction of the

    [13] C. E. Moore, *Atomic Energy Levels*, National Bureau of Standards Circular 467, vols. I, II, and III, 1949, 1952, and 1958, Washington D.C.; these have been reprinted as Natl. Bur. Stand. Publ. NSRDS-NBS 35, 1971. W. C. Martin et al., *Atomic Energy Levels—The Rare-Earth Elements*, Natl. Bur. Stand. Publ. NSRDS-NBS 60, Washington, D.C., 1978. Revisions of Moore's tables are found in C. E. Moore, Natl. Bur. Stand. Publ. NSRDS-NBS 3, Sections 1–9, 1965–1980 and in papers in the *Journal of Physical and Chemical Reference Data*.

field. If the external field is reasonably weak, its effect will be less than that of the spin–orbit interaction, and we can calculate the effect of the field by using first-order perturbation theory on the levels. Since

$$\hat{J}_z \psi = M_J \hbar \psi$$

we have for the energy of interaction with the applied field

$$E_B = \beta_e B M_J + \beta_e B \hbar^{-1} \langle S_z \rangle$$

Evaluation[14] of the average value of S_z gives the final result

$$E_B = \beta_e g B M_J \tag{11.81}$$

where g (*the Landé g factor*) is given by

$$g = 1 + \frac{[J(J+1) - L(L+1) + S(S+1)]}{2J(J+1)} \tag{11.82}$$

Thus the external field splits each level into $2J + 1$ states, each state having a different value of M_J.

Figure 11.6 illustrates what happens when we consider successive interactions in an atom, using the $1s2p$ configuration of helium as the example.

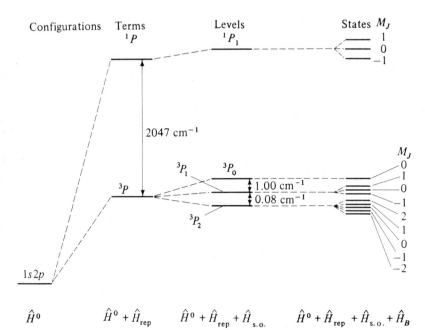

Figure 11.6 *Effect of inclusion of successive terms in the atomic Hamiltonian for the $1s2p$ helium configuration. ($\hat{H}_B$ is not part of the atomic Hamiltonian but is due to an applied magnetic field.)*

---

[14] See *Bethe and Jackiw*, p. 168, or *Landau and Lifshitz*, p. 428.

We have based our discussion on a scheme in which we first added the individual electronic orbital angular momenta to form a total orbital angular-momentum vector, and did the same for the spins: $\mathbf{L} = \Sigma \mathbf{L}_i$, $\mathbf{S} = \Sigma \mathbf{S}_i$. We then combined $\mathbf{L}$ and $\mathbf{S}$ to get $\mathbf{J}$. This scheme is known as *Russell–Saunders* coupling (or *L–S* coupling) and is appropriate where the spin–orbit interaction energy is small compared to the interelectronic repulsion energy. Although $\hat{L}$ and $\hat{S}$ commute with $\hat{H}^0 + \hat{H}_{\text{rep}}$, when we include $\hat{H}_{\text{S.O.}}$ in the Hamiltonian, $\hat{L}$ and $\hat{S}$ no longer commute with $\hat{H}$. If the spin–orbit interaction is small, then $\hat{L}$ and $\hat{S}$ "almost" commute with $\hat{H}$, and L–S coupling is valid.

As the atomic number increases, the average speed v of the electrons increases; as v/c increases, relativistic effects such as the spin–orbit interaction increase. For atoms with very high atomic number, the spin–orbit interaction exceeds the interelectronic repulsion energy, and we can no longer consider $\hat{L}$ and $\hat{S}$ to commute with $\hat{H}$; the operator $\hat{J}$, however, still commutes with $\hat{H}$. In this case we first add in $\hat{H}_{\text{S.O.}}$ to $\hat{H}^0$ and then consider $\hat{H}_{\text{rep}}$. This corresponds to first combining the spin and orbital angular momenta of each electron to give a total angular momentum $\mathbf{j}_i$ for each electron: $\mathbf{j}_i = \mathbf{L}_i + \mathbf{S}_i$. We then add the $\mathbf{j}_i$ to get the total electronic angular momentum: $\mathbf{J} = \Sigma \mathbf{j}_i$. This scheme is called *j–j* coupling. For most heavy atoms, the situation is intermediate between *j–j* and *L–S* coupling, and computations are complex.

There are several other effects which should be included in the atomic Hamiltonian. The finite size of the nucleus and the effect of nuclear motion cause slight changes[15] in the energy. There is a small relativistic term due to the interaction between the spin magnetic moments of the electrons (*spin–spin interaction*). We should also take into account the relativistic change of electronic mass with velocity; this effect is of appreciable magnitude for inner-shell electrons of heavy atoms, where average electronic speeds are not negligible in comparison with the speed of light.

If the nucleus has a nonzero spin, there is an interaction between the nuclear spin magnetic moment and the electronic spin and orbital magnetic moments, giving rise to atomic *hyperfine structure*. The nuclear spin angular momentum $\mathbf{I}$ adds vectorially to the total electronic angular momentum $\mathbf{J}$ to give the total angular momentum $\mathbf{F}$ of the atom: $\mathbf{F} = \mathbf{I} + \mathbf{J}$. For example, consider the ground state of the hydrogen atom. The spin of a proton is $\frac{1}{2}$, so that $I = \frac{1}{2}$; also $J = \frac{1}{2}$. Hence the quantum number F can be 0 or 1, corresponding to the proton and electron spins being antiparallel or parallel. The transition $F = 1 \to 0$ gives a line at 1420 megahertz, the famous 21-centimeter line emitted by hydrogen atoms in outer space. In 1951 Ewen and Purcell (who is also codiscoverer of nuclear magnetic resonance) stuck a horn-shaped antenna out the window of a Harvard physics laboratory and detected this line. The frequency of the hyperfine splitting in the ground state of hydrogen is probably the most accurately measured physical constant[16]: $1420.405751786 \pm 0.000000002$ MHz.

---

[15] See *Bethe and Salpeter*, pp. 102, 166, 351; *Landau and Lifshitz*, p. 462.
[16] R. Vessot et al., *IEEE Trans. Instr. Meas.*, **IM-15**, 165 (1966).

11.8 *THE CONDON–SLATER RULES*

In the Hartree–Fock approximation, the wave function of an atom (or molecule) is a Slater determinant or a linear combination of a few Slater determinants [e.g., Eq. (10.62)]. A configuration-interaction wave function such as (11.18) is a linear combination of many Slater determinants. To evaluate the energy and other properties of atoms and molecules using Hartree–Fock or configuration-interaction wave functions, one must be able to evaluate integrals of the form $\langle D'|\hat{A}|D\rangle$, where D and D' are Slater determinants of orthonormal spin-orbitals and $\hat{A}$ is an operator.

Each spin-orbital u_i is a product of a spatial orbital θ_i and a spin function σ_i, where σ_i is either α or β. We have $u_i = \theta_i\sigma_i$ and $\langle u_i(1)|u_j(1)\rangle = \delta_{ij}$, where $\langle u_i(1)|u_j(1)\rangle$ involves a sum over the spin coordinate of electron 1 and an integration over its spatial coordinates. If u_i and u_j have different spin functions, then (10.12) ensures the orthogonality of u_i and u_j; if u_i and u_j have the same spin function, their orthogonality is due to the orthogonality of the spatial orbitals θ_i and θ_j.

For an n-electron system, D is

$$D = \frac{1}{\sqrt{n!}} \begin{vmatrix} u_1(1) & u_2(1) & \cdots & u_n(1) \\ u_1(2) & u_2(2) & \cdots & u_n(2) \\ \vdots & \vdots & \vdots & \vdots \\ u_1(n) & u_2(n) & \cdots & u_n(n) \end{vmatrix} \tag{11.83}$$

An example with $n = 3$ is Eq. (10.58). D' has the same form as D except that u_1, $u_2, \ldots, u_n$ are replaced by $u'_1, u'_2, \ldots, u'_n$.

We shall assume that the columns of D and D' are arranged so as to have as many as possible of their left-hand columns match. For example, if we were working with the Slater determinants $|1s\overline{1s}\,2s\,3p_0|$ and $|1s\overline{1s}\,3p_0\,4s|$, we would interchange the third and fourth columns of the first determinant (thereby multiplying it by -1) and let $D = |1s\overline{1s}\,3p_0\,2s|$ and $D' = |1s\overline{1s}\,3p_0\,4s|$.

The operator $\hat{A}$ typically has the form

$$\hat{A} = \sum_{i=1}^{n} \hat{f}_i + \sum_{j>i}^{n-1}\sum_{i=1} \hat{g}_{ij} \tag{11.84}$$

where the *one-electron operator* $\hat{f}_i$ involves only coordinate and momentum operators of electron i and where the *two-electron operator* $\hat{g}_{ij}$ involves coordinate and momentum operators of electrons i and j. For example, if $\hat{A}$ is the Hamiltonian operator, then $\hat{f}_i = -(\hbar^2/2m)\nabla_i^2 - Ze'^2/r_i$ and $\hat{g}_{ij} = e'^2/r_{ij}$.

Condon and Slater showed that the n-electron integral $\langle D'|A|D\rangle$ could be reduced to sums of certain one- and two-electron integrals. The derivation of these Condon–Slater formulas uses the determinant expression of Problem 8.8 together with the orthonormality of the spin-orbitals. (See *Parr*, pp. 23–27 for the derivation.) Table 11.3 gives the Condon–Slater formulas.

In Table 11.3 each matrix element of $\hat{g}_{12}$ involves summation over the spin

Table 11.3 *The Condon–Slater rules*

D and D' differ by	$\left\langle D' \left\| \sum_{i=1}^{n} \hat{f}_i \right\| D \right\rangle$	$\left\langle D' \left\| \sum_{j>i}^{n-1} \sum_{i=1} \hat{g}_{ij} \right\| D \right\rangle$
no spin-orbitals	$\sum_{i=1}^{n} \langle u_i(1)|\hat{f}_1|u_i(1)\rangle$	$\sum_{j>i}^{n-1} \sum_{i=1} [\langle u_i(1)u_j(2)|\hat{g}_{12}|u_i(1)u_j(2)\rangle \\ \qquad - \langle u_i(1)u_j(2)|\hat{g}_{12}|u_j(1)u_i(2)\rangle]$
one spin-orbital $u'_n \neq u_n$	$\langle u'_n(1)|\hat{f}_1|u_n(1)\rangle$	$\sum_{j=1}^{n-1} [\langle u'_n(1)u_j(2)|\hat{g}_{12}|u_n(1)u_j(2)\rangle \\ \qquad - \langle u'_n(1)u_j(2)|\hat{g}_{12}|u_j(1)u_n(2)\rangle]$
two spin-orbitals $u'_n \neq u_n,$ $u'_{n-1} \neq u_{n-1}$	0	$\langle u'_n(1)u'_{n-1}(2)|\hat{g}_{12}|u_n(1)u_{n-1}(2)\rangle \\ \qquad - \langle u'_n(1)u'_{n-1}(2)|\hat{g}_{12}|u_{n-1}(1)u_n(2)\rangle$
three or more spin-orbitals	0	0

coordinates of electrons 1 and 2 and integration over the full range of the spatial coordinates of electrons 1 and 2; each matrix element of $\hat{f}_1$ involves summation over the spin coordinate of electron 1 and integration over its spatial coordinates. The variables in the sums and definite integrals are dummy variables.

If the operators $\hat{f}_i$ and $\hat{g}_{ij}$ do not involve spin, the expressions in Table 11.3 can be further simplified. We have $u_i = \theta_i \sigma_i$ and

$$\langle u_i(1)|\hat{f}_1|u_i(1)\rangle = \int \theta_i^*(1)\hat{f}_1\theta_i(1)\,dv_1 \sum_{m_{s1}} \sigma_i^*(1)\sigma_i(1)$$

$$= \int \theta_i^*(1)\hat{f}_1\theta_i(1)\,dv_1 = \langle \theta_i(1)|\hat{f}_1|\theta_i(1)\rangle$$

since σ_i is normalized. Using this result and the orthonormality of σ_i and σ_j, we get for the case $D = D'$ (Problem 11.23)

$$\left\langle D \left| \sum_{i=1}^{n} \hat{f}_i \right| D \right\rangle = \sum_{i=1}^{n} \langle \theta_i(1)|\hat{f}_1|\theta_i(1)\rangle \qquad \textbf{(11.85)}$$

$$\left\langle D \left| \sum_{j>i}^{n-1} \sum_{i=1} \hat{g}_{ij} \right| D \right\rangle = \sum_{j>i}^{n-1} \sum_{i=1} [\langle \theta_i(1)\theta_j(2)|\hat{g}_{12}|\theta_i(1)\theta_j(2)\rangle$$

$$- \delta_{m_{s,i} m_{s,j}} \langle \theta_i(1)\theta_j(2)|\hat{g}_{12}|\theta_j(1)\theta_i(2)\rangle] \qquad \textbf{(11.86)}$$

where $\delta_{m_{s,i} m_{s,j}}$ is 0 or 1 according to whether $m_{s,i} \neq m_{s,j}$ or $m_{s,i} = m_{s,j}$. Similar equations hold for the other integrals.

Let us apply these equations to evaluate $\langle D|\hat{H}|D\rangle$, where $\hat{H}$ is the Hamiltonian of an n-electron atom with spin–orbit interaction neglected and where

D is a Slater determinant of n spin-orbitals. We have $\hat{H} = \sum_i \hat{f}_i + \sum_{j>i} \sum_i \hat{g}_{ij}$, where $\hat{f}_i = -(\hbar^2/2m)\nabla_i^2 - Ze'^2/r_i$ and $\hat{g}_{ij} = e'^2/r_{ij}$. Introducing the Coulomb and exchange integrals of Eq. (9.146) and using (11.86) and (11.85), we have

$$\langle D|\hat{H}|D\rangle = \sum_{i=1}^{n} \langle \theta_i(1)|\hat{f}_1|\theta_i(1)\rangle + \sum_{j>i}^{n-1} \sum_{i=1} (J_{ij} - \delta_{m_{s,i} m_{s,j}} K_{ij}) \quad \textbf{(11.87)}$$

$$J_{ij} = \langle \theta_i(1)\theta_j(2)|e'^2/r_{12}|\theta_i(1)\theta_j(2)\rangle,$$
$$K_{ij} = \langle \theta_i(1)\theta_j(2)|e'^2/r_{12}|\theta_j(1)\theta_i(2)\rangle \qquad\qquad \textbf{(11.88)}$$

$$\hat{f}_1 = -(\hbar^2/2m)\nabla_1^2 - Ze'^2/r_1 \qquad\qquad\qquad\qquad \textbf{(11.89)}$$

The Kronecker delta in (11.87) results from the orthonormality of the one-electron spin functions.

As an example, consider Li. The SCF approximation to the ground-state ψ is the Slater determinant $D = |1s\overline{1s}2s|$. The spin-orbitals are $u_1 = 1s\alpha$, $u_2 = 1s\beta$, and $u_3 = 2s\alpha$. The spatial orbitals are $\theta_1 = 1s$, $\theta_2 = 1s$, and $\theta_3 = 2s$. We have $J_{12} = J_{1s1s}$ and $J_{13} = J_{23} = J_{1s2s}$. Since $m_{s1} \neq m_{s2}$ and $m_{s2} \neq m_{s3}$, the only exchange integral that appears in the energy expression is $K_{13} = K_{1s2s}$. We only get exchange integrals between spin-orbitals with the same spin. Equation (11.87) gives the SCF energy as

$$E = \langle D|\hat{H}|D\rangle = 2\langle 1s(1)|\hat{f}_1|1s(1)\rangle + \langle 2s(1)|\hat{f}_1|2s(1)\rangle + J_{1s1s} + 2J_{1s2s} - K_{1s2s}$$

The terms involving $\hat{f}_1$ are hydrogenlike energies, and their sum equals $E^{(0)}$ in Eq. (10.69). The remaining terms equal $E^{(1)}$ in Eqs. (10.74) and (10.75). As noted at the beginning of Section 9.4, $E^{(0)} + E^{(1)}$ equals the variational integral $\langle D|\hat{H}|D\rangle$, so the Condon–Slater rules have been checked in this case.

For an atom with closed subshells only (e.g., ground-state Be with a $1s^2 2s^2$ configuration), the n electrons reside in $n/2$ different spatial orbitals, so that $\theta_1 = \theta_2$, $\theta_3 = \theta_4$, etc. Let $\varphi_1 \equiv \theta_1 = \theta_2$, $\varphi_2 \equiv \theta_3 = \theta_4$, $\ldots$, $\varphi_{n/2} \equiv \theta_{n-1} = \theta_n$. If one rewrites Eq. (11.87) using the φ's instead of the θ's, one finds (Problem 11.24) for the SCF energy

$$E = \langle D|\hat{H}|D\rangle = 2\sum_{i=1}^{n/2} \langle \varphi_i(1)|\hat{f}_1|\varphi_i(1)\rangle + \sum_{j=1}^{n/2} \sum_{i=1}^{n/2} (2J_{ij} - K_{ij}) \quad \textbf{(11.90)}$$

where $\hat{f}_1$ is given by (11.89) and where J_{ij} and K_{ij} have the forms in (11.88) but with θ_i and θ_j replaced by φ_i and φ_j. Each sum in (11.90) goes over all the $n/2$ different spatial orbitals.

For example, consider the $1s^2 2s^2$ configuration. We have $n = 4$ and the two different spatial orbitals are $1s$ and $2s$. The double sum in (11.90) equals $2J_{1s1s} - K_{1s1s} + 2J_{1s2s} - K_{1s2s} + 2J_{2s1s} - K_{2s1s} + 2J_{2s2s} - K_{2s2s}$. From the definition (11.88), it follows that $J_{ii} = K_{ii}$. The labels 1 and 2 in (11.88) are dummy variables, and interchanging them can have no effect on the integrals. Interchanging 1 and 2 in J_{ij} converts it to J_{ji}; therefore $J_{ij} = J_{ji}$. The same reasoning gives $K_{ij} = K_{ji}$. Thus

$$J_{ii} = K_{ii}, \qquad J_{ij} = J_{ji}, \qquad K_{ij} = K_{ji} \qquad \textbf{(11.91)}$$

Use of (11.91) gives the Coulomb- and exchange-integrals expression for the $1s^2 2s^2$ configuration as $J_{1s1s} + J_{2s2s} + 4J_{1s2s} - 2K_{1s2s}$. Between the two electrons in the

$1s$ orbital, there is only one Coulombic interaction, and we get the term J_{1s1s}. Each $1s$ electron interacts with two $2s$ electrons, for a total of four $1s$-$2s$ interactions, and we get the term $4J_{1s2s}$. As noted earlier, exchange integrals occur only between spin-orbitals of the same spin. There is an exchange integral between the $1s\alpha$ and $2s\alpha$ spin-orbitals and an exchange integral between the $1s\beta$ and $2s\beta$ spin-orbitals, which gives the $-2K_{1s2s}$ term.

We close this section by noting that the magnitude of the exchange integrals is generally much less than the magnitude of the Coulomb integrals [e.g., see Eq. (9.157)].

PROBLEMS

11.1 How many electrons can be put in each of the following: (a) a shell with principal quantum number n; (b) a subshell with quantum numbers n and l; (c) an orbital; (d) a spin-orbital?

11.2 Give the designations for the ground-state energy levels of the first 10 elements; e.g., $^3P_0$ for carbon.

11.3 What terms does the electron configuration $1s^2 2s^2 2p^6 3s^2 3p5g$ give rise to?

11.4 (a) Verify Eq. (11.34). (b) Apply $\hat{M}_z = \hat{M}_{z1} + \hat{M}_{z2}$ to (11.35), combine terms, and use the linear independence of the functions involved to prove (11.36).

11.5 Verify Eqs. (11.52)–(11.54) and the four equations following (11.56).

11.6 How many states belong to a term that has quantum numbers L and S?

11.7 Draw a diagram similar to Fig. 11.6 for the carbon $1s^2 2s^2 2p^2$ configuration. (The ^1S term is the highest.)

11.8 (a) Calculate the angle in Fig. 11.3 between the z axis and $\mathbf{S}$ for the spin function $\alpha(1)\alpha(2)$. (b) Calculate the angle between $\mathbf{S}_1$ and $\mathbf{S}_2$ for each of the functions (11.57)–(11.60). [*Hint:* One approach is to use the law of cosines. A second approach is to use $\mathbf{S} \cdot \mathbf{S} = (\mathbf{S}_1 + \mathbf{S}_2) \cdot (\mathbf{S}_1 + \mathbf{S}_2)$.] (c) If a vector $\mathbf{A}$ has components (A_x, A_y, A_z), what are the components of the projection of $\mathbf{A}$ in the xy plane? Use the answer to this question to find the angle between the projections of $\mathbf{S}_1$ and $\mathbf{S}_2$ in the xy plane for the function $\alpha(1)\alpha(2)$.

11.9 Which STO's have the same form as hydrogenlike AO's?

11.10 Does Fig. 11.6 contain a violation of the rule given in Section 11.6 for determining whether a multiplet is regular or inverted?

11.11 Find the $\hat{S}^2$ and $\hat{S}_z$ eigenvalues for the spin function

$$3^{-1/2}[\alpha(1)\alpha(2)\beta(3) + \alpha(1)\beta(2)\alpha(3) + \beta(1)\alpha(2)\alpha(3)]$$

11.12 Estimate the nonrelativistic $1s$ orbital energy in Ar; check with Fig. 11.2.

11.13 At what atomic number does the second crossing of the $3d$ and $4s$ orbital energies occur in Fig. 11.2? Take account of the logarithmic scale. (Atomic spectral data show that the crossing actually occurs between 20 and 21.)

11.14 If $R(r_1)$ is the radial factor in the function t_1 in the Hartree differential equation (11.12), write down the differential equation satisfied by R.

11.15 Verify the terms in Table 11.2b.

11.16 Give the level designations for the ground states of elements 21 through 30. Which of these elements has the most degenerate ground level?

11.17 How many states belong to the carbon configurations: (a) $1s^2 2s^2 2p^2$; (b) $1s^2 2s^2 2p3p$?

11.18 Use Eq. (11.73) to calculate the separation between the $^2P_{3/2}$ and $^2P_{1/2}$ levels of the hydrogen-atom $2p$ configuration. (Because of other relativistic effects, the result will not agree accurately with experiment.)

11.19 Use Eq. (11.81) to calculate the energy separation between the $M_J = \frac{1}{2}$ and $M_J = -\frac{1}{2}$ states of the $2p\ ^2P_{1/2}$ hydrogen-atom level, if a magnetic field of 0.200 T is applied.

11.20 Consult Moore's table of atomic energy levels (footnote at the end of Section 11.6), and find a configuration of the neutral carbon atom for which Hund's rule does not correctly predict the lowest term.

11.21 Explain why it would be incorrect to calculate the experimental ground-state energy of lithium by taking $E_{2s} + 2E_{1s}$, where E_{2s} is the experimental energy needed to remove the $2s$ electron from lithium and E_{1s} is the experimental energy needed to remove a $1s$ electron from lithium.

11.22 Which of the first 10 elements have ground states of odd parity?

11.23 Derive Eqs. (11.85) and (11.86).

11.24 For a closed-subshell configuration, (a) show that the double sum in (11.87) equals

$$\sum_{\substack{j > i \\ j > i \ i = 1}}^{n/2} \sum_{i=1}^{n/2} (4J_{ij} - 2K_{ij}) + \sum_{i=1}^{n/2} J_{ii}$$

where the Coulomb and exchange integrals are defined in terms of the $n/2$ different spatial orbitals φ_i; (b) use (11.91) and the result of part (a) to derive (11.90).

11.25 Use the Condon–Slater rules to prove the orthonormality of two n-electron Slater determinants of orthonormal spin-orbitals.

11.26 Consider the double sum in Eq. (11.1). How many terms are there in this sum that have $i = n$? Could this sum be written as $\sum_{i=1}^{n-1} \sum_{j > i} e'^2 / r_{ij}$?

12 *MOLECULAR SYMMETRY*

12.1 *SYMMETRY ELEMENTS AND OPERATIONS*

The quantum-mechanical treatment of molecules is difficult; hence it is heartening to know that we can often obtain qualitative information about molecular wave functions and properties from the symmetry of the molecule. By the symmetry of a molecule, we will mean the symmetry of the framework formed by the nuclei held fixed in their equilibrium positions. (Our starting point for molecular quantum mechanics will be the Born–Oppenheimer approximation, which regards the nuclei as fixed when solving for the electronic wave function; see Section 13.1.) It should be noted that the symmetry of a molecule can be different in different electronic states. For example, HCN is linear in its ground electronic state, but nonlinear in certain excited states. Unless we specify otherwise, we will be considering the symmetry of the ground electronic state.

By a *symmetry operation* we mean a transformation of a body such that the final position is physically indistinguishable from the initial position, and such that the distances between all pairs of points in the body are preserved. For example, consider the trigonal-planar molecule BF_3 (Fig. 12.1a), where for convenience we have numbered the fluorine nuclei. If we rotate the molecule counterclockwise through 120° about an axis through the boron nucleus and perpendicular to the plane of the molecule, the new position will be as in Fig. 12.1b. Since in reality the fluorine nuclei are physically indistinguishable from one another, we have carried out a symmetry operation. The axis about which we rotated is an example of a *symmetry element*. Symmetry elements and symmetry operations are related but different things which are often confused. A symmetry *element* is a geometrical entity (point, line, or plane) with respect to which a symmetry *operation* is carried out.

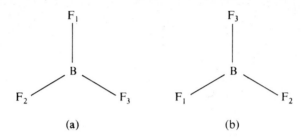

(a)   (b)

Figure 12.1 (a) The BF_3 molecule   (b) BF_3 after a 120° rotation

We say that a body has an *n-fold axis of symmetry* (also called an *n-fold proper axis* or an *n-fold rotation axis*) if rotation about this axis by $2\pi/n$ radians (where n is an integer) gives a configuration physically indistinguishable from the original position; n is called the *order* of the axis. For example, BF_3 has a threefold axis of symmetry perpendicular to the molecular plane. The symbol for an *n-fold rotation axis* is C_n. The threefold axis in BF_3 is a C_3 axis. To denote the operation of counterclockwise rotation by $2\pi/n$ radians, we use the symbol $\hat{C}_n$. The "hat" distinguishes symmetry operations from symmetry elements. BF_3 has three more rotation axes; each B—F bond is a twofold symmetry axis (Fig. 12.2).

A second kind of symmetry element is a *plane of symmetry*. A molecule has a plane of symmetry if reflection of all the nuclei through that plane gives a configuration physically indistinguishable from the original one. The symbol for a symmetry plane is σ. (*Spiegel* is the German word for mirror.) The symbol for the operation of reflection is $\hat{\sigma}$. BF_3 has four symmetry planes. The plane of the molecule is a symmetry plane, since nuclei lying in a reflection plane undergo no change in position when a reflection is carried out. The plane passing through the B and F_1 nuclei and perpendicular to the plane of the molecule is a symmetry plane, since reflection in this plane merely interchanges F_2 and F_3. It might be thought that this reflection is the same symmetry operation as rotation by 180° about the C_2 axis passing through B and F_1, which also interchanges F_2 and F_3. This is not so; the reflection carries points lying above the plane of the molecule into points that also lie

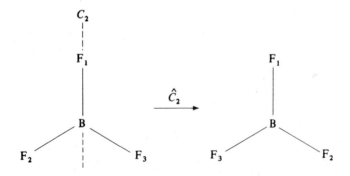

Figure 12.2 A C_2 axis in BF_3

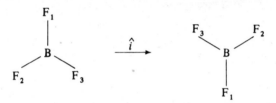

Figure 12.3 *Effect of inversion in* BF_3

above the molecular plane, whereas the $\hat{C}_2$ rotation carries points lying above the plane of the molecule into points below the molecular plane. Two symmetry operations are equal only when they represent the same transformation of three-dimensional space. The remaining two symmetry planes in BF_3 pass through B—F_2 and B—F_3 and are perpendicular to the molecular plane.

The third kind of symmetry element is a *center of symmetry*, symbolized by i (no connection with $\sqrt{-1}$). A molecule has a center of symmetry if the operation of inverting all the nuclei through the center gives a configuration indistinguishable from the original one. If we set up a Cartesian coordinate system, the operation of inversion through the origin (symbolized by $\hat{i}$) carries a nucleus originally at (x, y, z) to $(-x, -y, -z)$. Does BF_3 have a center of symmetry? With the origin at the boron nucleus, inversion gives the result shown in Fig. 12.3. Since we get a configuration that is physically distinguishable from the original one, BF_3 does not have a center of symmetry. For SF_6 inversion through the sulfur nucleus is shown in Fig. 12.4, and it is clear that SF_6 has a center of symmetry. (An operation such as $\hat{i}$, $\hat{C}_n$, etc., may or may not be a symmetry operation; thus $\hat{i}$ is a symmetry operation in SF_6 but not in BF_3.)

The fourth and final kind of symmetry element is an *n-fold alternating axis of symmetry* (also called an *improper axis* or a *rotation-reflection axis*), symbolized by S_n. A body has an S_n axis if rotation by $2\pi/n$ radians (n integral) about the axis, followed by reflection in a plane perpendicular to the axis, carries the body into a position physically indistinguishable from the original one. Clearly if a body has a C_n axis and also has a plane of symmetry perpendicular to this axis, then the C_n axis is also an S_n axis. Thus the C_3 axis in BF_3 is also an S_3 axis. It is possible to have an S_n axis that is not a C_n axis. An example is CH_4. In Fig. 12.5 we have first carried out a 90° proper rotation ($\hat{C}_4$) about what we assert is an S_4 axis. As can be seen, this

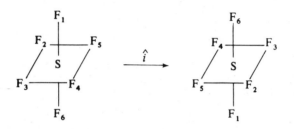

Figure 12.4 *Effect of inversion in* SF_6

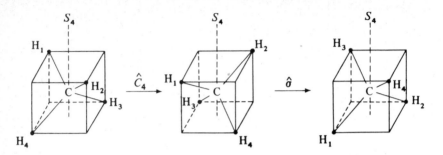

Figure 12.5 *An S_4 axis in methane*

operation does not result in an equivalent configuration. When we follow the $\hat{C}_4$ operation by reflection in the plane perpendicular to the axis and passing through the carbon atom, we do get a configuration equivalent to the one existing before we performed the rotation and reflection; hence we have an S_4 axis. The S_4 axis is not a C_4 axis, although it is a C_2 axis. There are two other S_4 axes in methane, each one perpendicular to a pair of faces of the cube in which the tetrahedral molecule is inscribed.

The operation of rotation by $2\pi/n$ radians about an axis, followed by reflection in a plane perpendicular to the axis, is denoted by $\hat{S}_n$. An $\hat{S}_1$ operation is a $360°$ rotation about an axis, followed by a reflection in a plane perpendicular to the axis. Since a $360°$ rotation restores the body to its original position, an $\hat{S}_1$ operation is the same as reflection in a plane, $\hat{S}_1 = \hat{\sigma}$; any plane of symmetry has an S_1 axis perpendicular to it.

Consider now the $\hat{S}_2$ operation. We choose the coordinate system so that the S_2 axis is the z axis (Fig. 12.6). Rotation by $180°$ about the S_2 axis changes the x and y coordinates of a point to $-x$ and $-y$, respectively, and leaves the z coordinate unaffected. Reflection in the xy plane then converts the z coordinate to $-z$. The net effect of the $\hat{S}_2$ operation is to bring a point originally at (x, y, z) to $(-x, -y, -z)$, which amounts to an inversion through the origin: $\hat{S}_2 = \hat{\imath}$. Any axis passing through a center of symmetry is an S_2 axis. Reflection in a plane and inversion are special cases of the $\hat{S}_n$ operation.

The $\hat{S}_n$ operation may seem an arbitrary kind of operation, but it must be

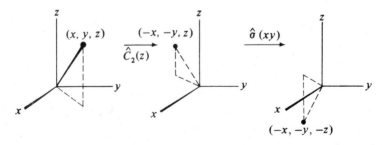

Figure 12.6 *The $\hat{S}_2$ operation*

included as one of the kinds of symmetry operations. Thus the transformation from the first to the third CH_4 configuration in Fig. 12.5 certainly meets the definition of a symmetry operation, but it is neither a proper rotation nor a reflection nor an inversion.

Performing a symmetry operation on a molecule gives a nuclear configuration physically indistinguishable from the original one; hence the center of mass must have the same position in space before and after a symmetry operation. For the operation $\hat{C}_n$, the only points that do not move are those on the C_n axis; therefore a C_n symmetry axis must pass through the molecular center of mass. Similarly a center of symmetry must coincide with the center of mass; a plane of symmetry and an alternating axis of symmetry must pass through the center of mass. The center of mass is the common intersection of all the symmetry elements of the molecule.

In discussing the symmetry of a molecule, we often place it in a Cartesian coordinate system with the molecular center of mass at the origin. The rotational axis of highest order is made the z axis. A plane of symmetry containing this axis is designated σ_v (for *vertical*); a plane of symmetry perpendicular to this axis is designated σ_h (for *horizontal*).

What is the relation between the symmetry operations of a molecule and quantum mechanics? To classify the states of a quantum-mechanical system, we consider those operators that commute with the Hamiltonian and with each other. For example, we classified the states of many-electron atoms using the quantum numbers L, S, J, and M_J, which correspond to the operators $\hat{L}^2$, $\hat{S}^2$, $\hat{J}^2$, and $\hat{J}_z$, all of which commute with each other and with the Hamiltonian (omitting spin–orbit interaction). The symmetry operations discussed in this chapter act on *points* in three-dimensional space, transforming each point to a corresponding point. All the quantum-mechanical operators we have discussed act on *functions*, transforming each function to a corresponding function. Corresponding to each symmetry operation $\hat{R}$, we define an operator $\hat{O}_R$ that acts on functions in the following manner. Let $\hat{R}$ bring a point originally at (x, y, z) to the location (x', y', z'):

$$\hat{R}(x, y, z) \rightarrow (x', y', z') \tag{12.1}$$

The operator $\hat{O}_R$ is defined so that the function $\hat{O}_R f$ has the same value at (x', y', z') that the function f has at (x, y, z):

$$\hat{O}_R f(x', y', z') = f(x, y, z) \tag{12.2}$$

We shall illustrate this definition with an example. Let $\hat{R}$ be a counterclockwise 90° proper rotation about the z axis: $\hat{R} = \hat{C}_4(z)$. Let f be a $2p_x$ hydrogen orbital:

$$f = 2p_x = Nxe^{-k(x^2+y^2+z^2)^{1/2}}$$

where N and k are constants. A contour showing where f has its maximum positive value is a distorted ellipsoid of revolution about the positive x axis (Section 6.6); likewise, f has its maximum negative value on an ellipsoid lying on the negative x axis. Let us say that these ellipsoids are "centered" about the points $(a, 0, 0)$ and $(-a, 0, 0)$. The operator $\hat{C}_4(z)$ has the following effect (Fig. 12.7):

$$\hat{C}_4(z)(x, y, z) \rightarrow (-y, x, z) \tag{12.3}$$

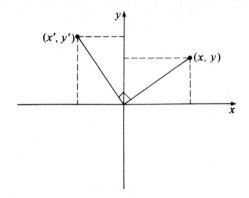

Figure 12.7 *Effect of a $\hat{C}_4(z)$ rotation. A little trigonometry shows that $x' = -y$, $y' = x$.*

For example, the point originally at $(a, 0, 0)$ is moved to $(0, a, 0)$, while the point at $(-a, 0, 0)$ is moved to $(0, -a, 0)$. From (12.2) the function $\hat{O}_{C_4(z)}2p_x$ must have its contours of maximum positive and negative values centered about $(0, a, 0)$ and $(0, -a, 0)$, respectively. We conclude that

$$\hat{O}_{C_4(z)}2p_x = 2p_y \tag{12.4}$$

(See Fig. 12.8.)

For the inversion operation, we have

$$\hat{i}(x, y, z) \rightarrow (-x, -y, -z)$$

so that (12.2) reads

$$\hat{O}_i f(-x, -y, -z) = f(x, y, z)$$

We now carry out the following renaming of the variables:

$$\bar{x} = -x, \qquad \bar{y} = -y, \qquad \bar{z} = -z$$

Hence

$$\hat{O}_i f(\bar{x}, \bar{y}, \bar{z}) = f(-\bar{x}, -\bar{y}, -\bar{z})$$

The point $(\bar{x}, \bar{y}, \bar{z})$ is a general point in space, and we can drop the bars to get

$$\hat{O}_i f(x, y, z) = f(-x, -y, -z)$$

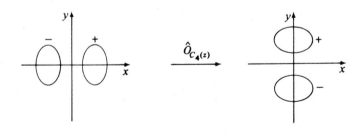

Figure 12.8 *Effect of $\hat{O}_{C_4(z)}$ on a p_x orbital*

We conclude that $\hat{O}_i$ is the parity operator:

$$\hat{O}_i = \hat{\Pi} \tag{12.5}$$

The wave function of an n-particle system is a function of $4n$ variables, and we extend the definition (12.2) of $\hat{O}_R$ to read

$$\hat{O}_R f(x_1', y_1', z_1', m_{s1}, \ldots, x_n', y_n', z_n', m_{sn}) = f(x_1, y_1, z_1, m_{s1}, \ldots, x_n, y_n, z_n, m_{sn})$$

Note that $\hat{O}_R$ does not affect the spin coordinates. Thus in looking at the parity of atomic states in Section 11.5, we looked at the spatial factors in each term of the expansion of the Slater determinant and omitted consideration of the spin factors, since these are unaffected by $\hat{\Pi}$.

When a system is characterized by the symmetry operations $\hat{R}_1, \hat{R}_2, \ldots,$ then the corresponding operators $\hat{O}_{R_1}, \hat{O}_{R_2}, \ldots$ commute with the Hamiltonian.[1] For example, if the nuclear framework of a molecule has a center of symmetry, then the parity operator $\hat{\Pi}$ commutes with the Hamiltonian for the electronic motion. We can then choose the electronic states (wave functions) as even or odd, according to the eigenvalue of $\hat{\Pi}$. Of course, all the symmetry operations may not commute among themselves (see the next section); hence the wave functions cannot in general be chosen as eigenfunctions of all the operators $\hat{O}_R$. (Since the correspondence between the symmetry transformation $\hat{R}$ and the operator $\hat{O}_R$ is so close, one often doesn't distinguish between them.)

There is a close connection between symmetry and the *constants of the motion* (these are properties whose operators commute with the Hamiltonian $\hat{H}$). For a system whose Hamiltonian is invariant (i.e., doesn't change) under any translation of spatial coordinates, the linear-momentum operator $\hat{p}$ will commute with $\hat{H}$, and $\hat{p}$ can be assigned a definite value in a stationary state. (An example is the free particle.) For a system with $\hat{H}$ invariant under any rotation of coordinates, the operators for the angular-momentum components commute with $\hat{H}$, and the total angular momentum and one of its components are specifiable. (An example is an atom.) A linear molecule has axial symmetry, rather than the spherical symmetry of an atom; here only the axial component of angular momentum can be specified (Chapter 13).

12.2 PRODUCTS OF SYMMETRY OPERATIONS

Symmetry operations are operators that cause transformations of three-dimensional space, and (as with any operators) we define the product of two such operators as meaning successive application of the operators, the operator on the right of the product being applied first. Clearly the product of any two symmetry operations of a molecule must also be a symmetry operation.

As an example, consider BF_3. The product of the $\hat{C}_3$ operator with itself, $\hat{C}_3\hat{C}_3 = \hat{C}_3^2$, rotates the molecule $240°$ counterclockwise. If we take $\hat{C}_3\hat{C}_3\hat{C}_3 = \hat{C}_3^3$, we have a $360°$ rotation, which restores the molecule to its original position. We

---

[1] For a careful discussion, see *Schonland*, Sections 7.1 and 7.2.

define the *identity operation* $\hat{E}$ as the operation that does nothing to a body. We have $\hat{C}_3^3 = \hat{E}$. (The symbol comes from the German word *Einheit*, meaning unity. The symbol $\hat{I}$ is sometimes used instead of $\hat{E}$.) The operator $\hat{O}_E$ is the unit operator: $\hat{O}_E = \hat{1}$.

Now consider a molecule with a sixfold axis of symmetry, e.g., C_6H_6. The operation $\hat{C}_6$ is a 60° rotation, and $\hat{C}_6^2$ is a 120° rotation; hence $\hat{C}_6^2 = \hat{C}_3$. Also $\hat{C}_6^3 = \hat{C}_2$. We conclude that a C_6 symmetry axis is also a C_3 and a C_2 axis. In general, a C_n axis is also a C_m axis if n/m is an integer.

Since two successive reflections in the same plane bring all nuclei back to their original positions, we have $\hat{\sigma}^2 = \hat{E}$. Also $\hat{\imath}^2 = \hat{E}$. More generally, $\hat{\sigma}^n = \hat{E}, \hat{\imath}^n = \hat{E}$ for even n, while $\hat{\sigma}^n = \hat{\sigma}, \hat{\imath}^n = \hat{\imath}$ for odd n.

Do symmetry operators always commute? Consider SF_6. We examine the products of a $\hat{C}_4$ rotation about the z axis and a $\hat{C}_2$ rotation about the x axis. Figure 12.9 shows that $\hat{C}_4(z)\hat{C}_2(x) \neq \hat{C}_2(x)\hat{C}_4(z)$. Thus symmetry operations do not always commute. [Note that we describe symmetry operations with respect to a fixed coordinate system; our convention is that the symmetry elements do not move with

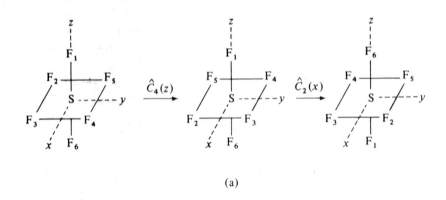

(a)

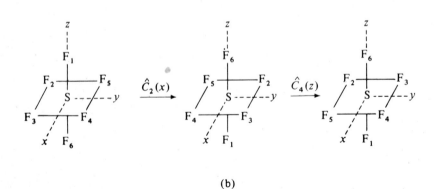

(b)

Figure 12.9 *Products of symmetry operations in* SF_6: (a) $\hat{C}_2(x)\hat{C}_4(z)$; (b) $\hat{C}_4(z)\hat{C}_2(x)$

the molecule when we perform a symmetry operation but remain fixed in space. For example, the $C_2(x)$ axis does not move when we perform the $\hat{C}_4(z)$ operation.]

12.3   SYMMETRY AND DIPOLE MOMENTS

As an application of symmetry, we consider molecular dipole moments. Since a symmetry operation produces a configuration physically indistinguishable from the original one, the direction of the dipole-moment vector must remain unchanged after a symmetry operation. (This is a nonrigorous, unsophisticated argument.) Hence if we have a proper axis of symmetry, the dipole moment must lie along this axis. If we have two or more noncoincident symmetry axes, the molecule cannot have a dipole moment, since the dipole moment cannot lie on two different axes. CH_4, which has four noncoincident C_3 axes, has no dipole moment. If there is a plane of symmetry, the dipole moment must lie in this plane. If there are several symmetry planes, the dipole moment must lie along the line of intersection of these planes. In H_2O the dipole moment lies on the C_2 axis, which is also the intersection of the two symmetry planes (Fig. 12.10). A molecule with a center of symmetry cannot have a dipole moment, since inversion reverses the direction of a vector. A monatomic molecule has a center of symmetry; hence atoms do not have dipole moments. (There is one exception to this statement—see Problem 13.30.) Thus we can use symmetry to discover whether a molecule has a dipole moment; in many cases symmetry considerations also tell us along what line the dipole moment lies.

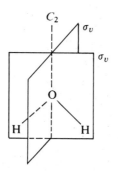

Figure 12.10 *The symmetry elements of* H_2O

12.4   SYMMETRY AND OPTICAL ACTIVITY

The ability of certain molecules to rotate the plane of polarization of plane-polarized light is well known. Experimental evidence and a quantum-mechanical treatment[2] both show that the optical rotary powers of two molecules that are

---

[2] *Kauzmann*, pp. 703–713.

mirror images of each other are equal in magnitude but opposite in sign. A simple way to see this is from the fact that physical processes occur in the same manner for a system and for its mirror image. Reflection of an optical-rotation experiment in a mirror converts the molecules to their mirror images and converts a clockwise optical rotation to a counterclockwise rotation of equal magnitude. (Interestingly, for the physical processes that occur in elementary-particle decay, nature *does* distinguish between a system and its mirror image; Yang and Lee won the 1957 Nobel Prize in physics for suggesting this possibility. See M. Gardner, *The Ambidextrous Universe*, Mentor–New American Library, New York, 1969.) Hence if a molecule is its own mirror image, it is optically inactive: $R = -R$, $2R = 0$, $R = 0$, where R is the optical rotary power. If a molecule is not superimposable on its mirror image, it may be optically active. (If the conformation of the mirror image differs from that of the original molecule only by rotation about a bond with a low rotational barrier, then we will not have optical activity.)

What is the connection between symmetry and optical activity? Consider the $\hat{S}_n$ operation. It consists of a rotation ($\hat{C}_n$) and a reflection ($\hat{\sigma}$). The reflection part of the $\hat{S}_n$ operation converts the molecule to its mirror image, and if the $\hat{S}_n$ operation is a symmetry operation for the molecule, then the $\hat{C}_n$ rotation will superimpose the molecule and its mirror image:

$$\text{molecule} \xrightarrow{\hat{C}_n} \text{rotated molecule} \xrightarrow{\hat{\sigma}} \text{rotated mirror image}$$

We conclude that a molecule with an S_n axis is optically inactive; if the molecule has no S_n axis, it may be optically active.

Since $\hat{S}_1 = \hat{\sigma}$ and $\hat{S}_2 = \hat{\imath}$, a molecule with either a plane or a center of symmetry is optically inactive (as you probably learned in organic chemistry). However, an S_n axis of any order rules out optical activity. As an example consider the molecule of Fig. 12.11, which has two mutually perpendicular rings with one atom in common. The molecule has neither a plane nor a center of symmetry but is optically inactive because of the presence of an S_4 axis. As a check, it is easily verified that the mirror image generated by reflection in a plane perpendicular to the S_4 axis is superimposable on the original molecule after a $\hat{C}_4$ rotation.

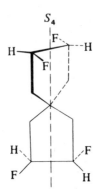

Figure 12.11 A molecule that is optically inactive because of an S_4 axis

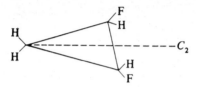

Figure 12.12 *An optically active molecule with a C_2 axis*

A molecule can have a symmetry element and still be optically active; if a C_n axis is present and there is no S_n axis, we can have optical activity. An example is the molecule in Fig. 12.12, which has a C_2 axis but no S_n axis. As a check, we readily verify that this molecule is not superimposable on its mirror image.

We have made no reference to asymmetric carbon atoms. There are optically active organic compounds that have no asymmetric carbons; there are optically inactive molecules that have two or more asymmetric carbons.

12.5 SYMMETRY POINT GROUPS

We now systematically consider the possible combinations of symmetry elements. We cannot have arbitrary combinations of symmetry elements in a molecule. For example, suppose we have a molecule with one and only one C_3 axis. Any symmetry operation must send this axis into itself. We cannot, therefore, have a plane of symmetry at an arbitrary angle to the C_3 axis; any plane of symmetry must either contain this axis or be perpendicular to it. (In BF_3 there are three σ_v planes and one σ_h plane.) The only possibility for a C_n axis noncoincident with the C_3 axis is a C_2 axis perpendicular to the C_3 axis; the corresponding $\hat{C}_2$ operation will send the C_3 axis into itself. Since $\hat{C}_3$ and $\hat{C}_3^2$ are symmetry operations, if we have one C_2 axis perpendicular to the C_3 axis, we must have a total of three such axes (as in BF_3).

The set of all the symmetry operations of a molecule forms a mathematical group. A *group* is a set of entities and a rule for forming the product of these entities, such that certain requirements are met. (The rule of combination of symmetry operations is successive performance of them.) We will not make explicit use of the mathematics of group theory. Note that it is the symmetry *operations*, not the symmetry elements, that constitute the group. For any symmetry operation of a molecule, the point that is the center of mass remains fixed. Hence the symmetry groups of isolated molecules are called *point groups*. For a crystal of infinite extent, we can have symmetry operations (e.g., translations) that leave no point fixed, giving rise to *space groups*. We omit consideration of space groups.

Any molecule can be classified as belonging to one of the symmetry point groups that we now list. For convenience we have divided the point groups into four divisions. (Script letters denote point groups.)

1. *Groups with no C_n axis:* $\mathscr{C}_1$, $\mathscr{C}_s$, $\mathscr{C}_i$

(a) $\mathscr{C}_1$

If a molecule has no symmetry elements at all, it belongs to this group. The only symmetry operation is $\hat{E}$ (which is a $\hat{C}_1$ rotation). CHFClBr belongs to point group $\mathscr{C}_1$.

(b) $\mathscr{C}_s$

A molecule whose only symmetry element is a plane of symmetry belongs to this group. The symmetry operations are $\hat{E}$ and $\hat{\sigma}$. An example is HOCl (Fig. 12.13).

(c) $\mathscr{C}_i$

A molecule whose only symmetry element is a center of symmetry belongs to this group. The symmetry operations are $\hat{i}$ and $\hat{E}$.

II. *Groups with a single C_n axis: $\mathscr{C}_n$, $\mathscr{C}_{nh}$, $\mathscr{C}_{nv}$, $\mathscr{S}_{2n}$*

(a) $\mathscr{C}_n$, $n = 2, 3, 4, \ldots$

A molecule whose only symmetry element is a C_n axis belongs to this group. The symmetry operations are $\hat{C}_n$, $\hat{C}_n^2$, $\ldots$, $\hat{C}_n^{n-1}$, $\hat{E}$. Examples of molecules belonging to $\mathscr{C}_2$ are shown in Figs. 12.14 and 12.12.

(b) $\mathscr{C}_{nh}$, $n = 2, 3, 4, \ldots$

If we add a plane of symmetry perpendicular to the C_n axis, we have a molecule belonging to this group. Since $\hat{\sigma}_h\hat{C}_n = \hat{S}_n$, the C_n axis is also an S_n axis. If n is even, the C_n axis is also a C_2 axis, and we have the symmetry operation

$$\hat{\sigma}_h\hat{C}_2 = \hat{S}_2 = \hat{i}$$

Thus for n even, a molecule belonging to $\mathscr{C}_{nh}$ has a center of symmetry. (The group $\mathscr{C}_{1h}$ is the group $\mathscr{C}_s$ discussed previously.) Examples of molecules belonging to groups $\mathscr{C}_{2h}$ and $\mathscr{C}_{3h}$ are shown in Fig. 12.14.

(c) $\mathscr{C}_{nv}$, $n = 2, 3, 4, \ldots$

A molecule in this group has a C_n axis and n vertical symmetry planes (passing through the C_n axis). (Group $\mathscr{C}_{1v}$ is the group $\mathscr{C}_s$.) Water, with a C_2 axis and two vertical symmetry planes, belongs to $\mathscr{C}_{2v}$. Ammonia belongs to $\mathscr{C}_{3v}$. (See Fig. 12.14.)

(d) $\mathscr{S}_n$, $n = 4, 6, 8, \ldots$

$\mathscr{S}_n$ is the group of symmetry operations associated with an S_n axis. First consider the case of odd n. We have $\hat{S}_n = \hat{\sigma}_h\hat{C}_n$. The operation $\hat{C}_n$ affects the x and y coordinates only, while the $\hat{\sigma}_h$ operation affects the z coordinate only;

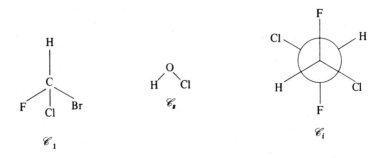

Figure 12.13 *Molecules with no C_n axis*

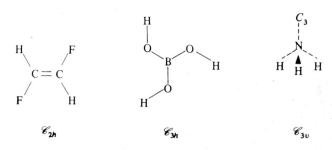

Figure 12.14   *Molecules with a single C_n axis*

hence these operations commute, and we have

$$\hat{S}_n^n = (\hat{\sigma}_h \hat{C}_n)^n = \hat{\sigma}_h \hat{C}_n \hat{\sigma}_h \hat{C}_n \cdots \hat{\sigma}_h \hat{C}_n = \hat{\sigma}_h^n \hat{C}_n^n$$

Now $\hat{C}_n^n = \hat{E}$, and for odd n $\hat{\sigma}_h^n = \hat{\sigma}_h$. Hence the symmetry operation $\hat{S}_n^n$ equals $\hat{\sigma}_h$ for odd n, and the group $\mathscr{S}_n$ has a horizontal symmetry plane if n is odd. Also

$$\hat{S}_n^{n+1} = \hat{S}_n^n \hat{S}_n = \hat{\sigma}_h \hat{S}_n = \hat{\sigma}_h \hat{\sigma}_h \hat{C}_n = \hat{C}_n, \qquad n \text{ odd}$$

so that we have a C_n axis if n is odd. We conclude that the group $\mathscr{S}_n$ is identical to the group $\mathscr{C}_{nh}$ if n is odd.

Now consider even values of n. Since $\hat{S}_2 = \hat{\imath}$, the group $\mathscr{S}_2$ is identical to $\mathscr{C}_i$. Thus it is only for $n = 4, 6, 8, \ldots$ that we get new groups. The S_{2n} axis is also a C_n axis:

$$\hat{S}_{2n}^2 = \hat{\sigma}_h^2 \hat{C}_{2n}^2 = \hat{E}\hat{C}_n = \hat{C}_n$$

The spiran of Fig. 12.11 belongs to the group $\mathscr{S}_4$.

III. *Groups with one C_n axis and n C_2 axes: $\mathscr{D}_n$, $\mathscr{D}_{nh}$, $\mathscr{D}_{nd}$*

(a) $\mathscr{D}_n$, $n = 2, 3, 4, \ldots$

A molecule with a C_n axis and n C_2 axes perpendicular to the C_n axis (and no symmetry planes) belongs to group $\mathscr{D}_n$. The angle between adjacent C_2 axes is π/n radians. For the group $\mathscr{D}_2$, we have three mutually perpendicular C_2 axes, and the symmetry operations are $\hat{E}$, $\hat{C}_2(x)$, $\hat{C}_2(y)$, $\hat{C}_2(z)$.

(b) $\mathscr{D}_{nh}$, $n = 2, 3, 4, \ldots$

This is the group of a molecule with a C_n axis, n C_2 axes, and a σ_h symmetry plane perpendicular to the C_n axis. As in $\mathscr{C}_{nh}$, the C_n axis is also an S_n axis. If n is

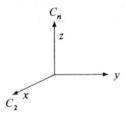

Figure 12.15 *Two of the symmetry axes in a $\mathscr{D}_{nh}$ molecule*

even, the C_n axis is a C_2 and an S_2 axis, and we have a center of symmetry. Molecules in $\mathscr{D}_{nh}$ also have n vertical planes of symmetry, each such plane passing through the C_n axis and a C_2 axis. We now prove this assertion. We set up a coordinate system with the C_n axis as the z axis and let one of the C_2 axes be the x axis (Fig. 12.15). This makes the xy plane the σ_h symmetry plane. Looking at the effect of the product $\hat{\sigma}(xy)\hat{C}_2(x)$ on a point originally at (x, y, z), we have

$$(x, y, z) \xrightarrow{\hat{C}_2(x)} (x, -y, -z) \xrightarrow{\hat{\sigma}(xy)} (x, -y, z)$$

We also have

$$(x, y, z) \xrightarrow{\hat{\sigma}(xz)} (x, -y, z)$$

Since $\hat{\sigma}(xy)\hat{C}_2(x)$ and $\hat{\sigma}(xz)$ both bring a point originally at (x, y, z) to the final position $(x, -y, z)$, they are equal:

$$\hat{\sigma}(xy)\hat{C}_2(x) = \hat{\sigma}(xz)$$

$\hat{C}_2(x)$ and $\hat{\sigma}(xy)$ are symmetry operations, and their product must be a symmetry operation; hence the xz plane is a symmetry plane. The same argument holds for any C_2 axis, so that we have $n \, \sigma_v$ planes. BF_3 belongs to $\mathscr{D}_{3h}$; $PtCl_4^{2-}$ belongs to $\mathscr{D}_{4h}$; benzene belongs to $\mathscr{D}_{6h}$ (Fig. 12.16).

(c) $\mathscr{D}_{nd}$, $n = 2, 3, 4, \ldots$

A molecule with a C_n axis, $n \, C_2$ axes, and n vertical planes of symmetry, which pass through the C_n axis and bisect the angles between adjacent C_2 axes, belongs to this group. The n vertical planes are called *diagonal* planes and are symbolized by σ_d. The C_n axis can be shown to be an S_{2n} axis. The staggered conformation of ethane is an example of group $\mathscr{D}_{3d}$ (Fig. 12.16). [The symmetry of molecules with internal rotation (e.g., ethane) actually requires special consideration, which we omit.[3]]

IV. *Groups with more than one C_n axis, $n > 2$: $\mathscr{T}_d$, $\mathscr{T}$, $\mathscr{T}_h$, $\mathcal{O}_h$, $\mathcal{O}$, $\mathscr{I}_h$, $\mathscr{I}$, $\mathscr{K}_h$*

These groups are related to the symmetries of the Platonic solids, solids bounded by congruent regular polygons and having congruent polyhedral angles. There are five such solids: the tetrahedron has four triangular faces, the cube has six square faces, the octahedron has eight triangular faces, the

^{3} See H. C. Longuet-Higgins, *Mol. Phys.*, **6**, 445 (1963).

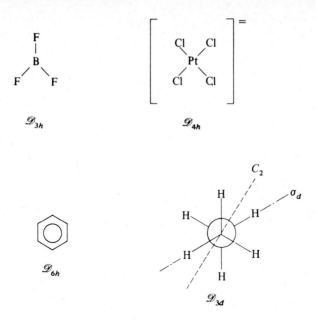

Figure 12.16 *Molecules with a C_n axis and n C_2 axes*

pentagonal dodecahedron has twelve pentagonal faces, the icosahedron has twenty triangular faces. (The pentagonal dodecahedron should not be confused with the triangular dodecahedron; the latter has twelve triangular faces and is not a Platonic solid.)

(a) $\mathcal{T}_d$

The symmetry operations of a regular tetrahedron constitute this group. The prime example is CH_4. The symmetry elements of methane are four C_3 axes (each C—H bond), three S_4 axes which are also C_2 axes (Fig. 12.5), and six symmetry planes, each such plane containing two C—H bonds. (The number of combinations of 4 things taken 2 at a time is $4!/2!2! = 6$.)

(b) $\mathcal{O}_h$

The symmetry operations of a cube or a regular octahedron constitute this group. The cube and octahedron are said to be *dual* to each other—if we connect the midpoints of adjacent faces of a cube, we get an octahedron, and vice versa. Hence the cube and octahedron have the same symmetry elements and operations. A cube has six faces, eight vertices, and twelve edges. It has these symmetry elements: a center of symmetry, three C_4 axes passing through the centers of opposite faces of the cube (these are also S_4 and C_2 axes), four C_3 axes passing through opposite corners of the cube (these are also S_6 axes), six C_2 axes connecting the midpoints of pairs of opposite edges, three planes of symmetry parallel to pairs of opposite faces, six planes of symmetry passing through pairs of opposite edges. Octahedral molecules such as SF_6 belong to $\mathcal{O}_h$.

(c) $\mathcal{I}_h$

The symmetry operations of a regular pentagonal dodecahedron or ico-

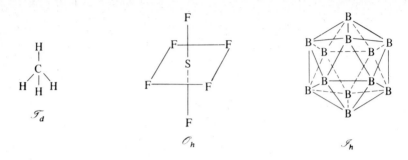

Figure 12.17 *Molecules with more than one C_n axis, $n > 2$. (For the $B_{12}H_{12}^{2-}$ ion, the hydrogen atoms have been omitted for clarity.)*

sahedron (which are dual to each other) constitute this group. The $B_{12}H_{12}^{2-}$ ion belongs to group $\mathscr{I}_h$; the twelve boron atoms lie at the vertices of a regular icosahedron (Fig. 12.17).

(d) $\mathscr{K}_h$

This is the group of symmetry operations of a sphere. (*Kugel* is the German word for sphere.) An atom belongs to this group.

For completeness we mention the remaining groups related to the Platonic solids; these groups are chemically unimportant. The groups $\mathscr{T}$, $\mathscr{O}$, and $\mathscr{I}$ are the groups of symmetry proper *rotations* of a tetrahedron, cube, and icosahedron, respectively; these groups do not have the symmetry reflections and improper rotations of these solids or the inversion operation of the cube and icosahedron. The group $\mathscr{T}_h$ contains the symmetry rotations of a tetrahedron, the inversion operation, and certain reflections and improper rotations.

What groups do linear molecules belong to? A rotation by any angle about the internuclear axis of a linear molecule is a symmetry operation. A regular polygon of n sides has a C_n axis, and taking the limit as $n \to \infty$ we get a circle, which has a C_∞ axis. The internuclear axis of a linear molecule is a C_∞ axis. Any plane containing this axis is a symmetry plane. If the linear molecule does not have a center of symmetry (e.g., CO, HCN), it belongs to the group $\mathscr{C}_{\infty v}$. If the linear molecule has a center of symmetry (e.g., H_2, C_2H_2), then it also has a σ_h symmetry plane and an infinite number of C_2 axes perpendicular to the molecular axis; hence it belongs to $\mathscr{D}_{\infty h}$.

How do we find what point group a molecule belongs to? One way is to find all the symmetry elements and then compare with the above list of groups. A more systematic procedure[4] is given in Fig. 12.18; the relationship of this procedure to our four divisions of point groups should be clear.

We begin by checking whether or not the molecule is linear; linear molecules are classified in $\mathscr{D}_{\infty h}$ or $\mathscr{C}_{\infty v}$ according to whether or not there is a center of symmetry. If the molecule is nonlinear, we look for two or more rotational axes of threefold or higher order; if these are present, the molecule is classified in one of the

---

[4] J. B. Calvert, *Am. J. Phys.*, **31**, 569 (1963).

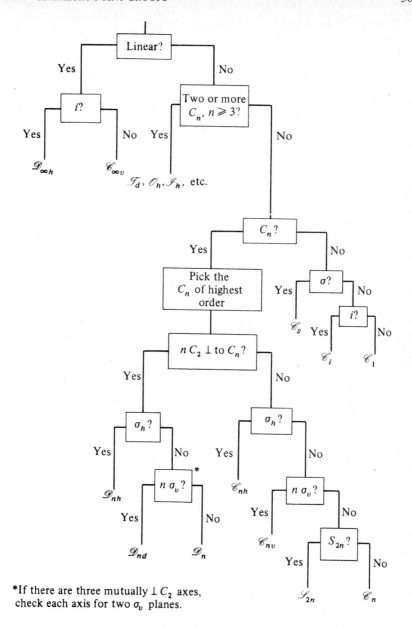

*If there are three mutually $\perp C_2$ axes,
check each axis for two σ_v planes.

Figure 12.18 *How to determine the point group of a molecule*

groups related to the symmetry of the regular polyhedra (division IV). If these axes
are not present, we look for any C_n axis at all. If there is no C_n axis, the molecule
belongs to one of the groups $\mathscr{C}_s$, $\mathscr{C}_i$, $\mathscr{C}_1$ (division I). If there is at least one C_n axis, we
pick the C_n axis of highest order as the main symmetry axis before proceeding to the
next step. (Sometimes there will be three mutually perpendicular C_2 axes; in these

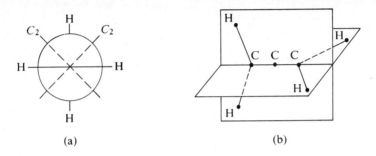

Figure 12.19 *Two views of allene. The* C=C=C *axis is perpendicular to the plane of the paper in* (a).

cases we may pick any one of these axes as the main axis.) We next check for $n C_2$ axes at right angles to the main C_n axis. If these are present, we have one of the division III groups; if these are absent, we have one of the division II groups. If we find the $n C_2$ axes, we look for a symmetry plane perpendicular to the main C_n axis; if it is present, the group is $\mathcal{D}_{nh}$. If it is absent, we check for n planes of symmetry containing the main C_n axis (if we have three mutually perpendicular C_2 axes, we must try each axis as the main axis in looking for the two σ_v planes—the three C_2 axes are equivalent in the groups $\mathcal{D}_{nh}$ and $\mathcal{D}_n$, but not in $\mathcal{D}_{nd}$). If we find $n \sigma_v$ planes, the group is $\mathcal{D}_{nd}$; otherwise it is $\mathcal{D}_n$. If we do not have $n C_2$ axes perpendicular to the main C_n axis, we classify the molecule in one of the groups $\mathscr{C}_{nh}$, $\mathscr{C}_{nv}$, $\mathscr{S}_{2n}$, or $\mathscr{C}_n$, by looking first for a σ_h plane, then for $n \sigma_v$ planes, and finally, if these are absent, checking whether or not the C_n axis is an S_{2n} axis. The procedure of Fig. 12.18 does not locate all symmetry elements; after classifying a molecule, check that all the required symmetry elements are indeed present. Although the above procedure might seem involved, it is really quite simple and is easily memorized.

The most common error students make in classifying a molecule is to miss the $n C_2$ axes perpendicular to the C_n axis of a molecule belonging to $\mathcal{D}_{nd}$. For example, it is easy to see that the C=C=C axis of allene is a C_2 axis, but the other two C_2 axes (Fig. 12.19) are often overlooked. Molecules with two equal halves "staggered" with respect to each other generally belong to $\mathcal{D}_{nd}$. Models may be helpful for those with visualization difficulties.

PROBLEMS

12.1 Two people play the following game. Each in turn places a penny on the surface of a large chessboard. The pennies can be put anywhere on the board, so long as they do not overlap previously placed pennies. A penny may overlap more than one square. Once placed, a penny cannot be moved. When one of the players finds there is no room to place another penny on the board, she loses. With best play, will the person placing the first penny or her opponent win? Give the winning strategy.

12.2 Give all the symmetry elements of each of the following molecules: (a) H_2S; (b) NH_3; (c) CHF_3; (d) $HOCl$; (e) 1,3,5-trichlorobenzene; (f) CH_2F_2; (g) CHFClBr.

12.3 For SF_6, which of the following pairs of operations commute? (a) $\hat{C}_4(z)$, $\hat{\sigma}_h$; (b) $\hat{C}_4(z)$, $\hat{\sigma}(yz)$; (c) $\hat{C}_2(z)$, $\hat{C}_2(x)$; (d) $\hat{\sigma}_h$, $\hat{\sigma}(yz)$; (e) $\hat{\imath}$, $\hat{\sigma}_h$.

12.4 What information does symmetry give about the dipole moment of each of the molecules in Problem 12.2?

12.5 Verify directly that the molecule in Fig. 12.11 is superimposable on its mirror image, while the molecule in Fig. 12.12 is not superimposable on its mirror image.

12.6 (a) What Platonic solid is dual to the regular tetrahedron? (b) How many vertices does a pentagonal dodecahedron have?

12.7 (a) Does H_2O_2 have an S_n axis? (b) Is it optically active? Explain.

12.8 For which point groups can a molecule have a dipole moment?

12.9 For which point groups can a molecule be optically active?

12.10 (a) For what values of n does the presence of an S_n axis imply the presence of a plane of symmetry? (b) For what values of n does the presence of an S_n axis imply the presence of a center of symmetry? (c) The group $\mathscr{D}_{nd}$ has an S_{2n} axis; for what values of n does it have a center of symmetry?

12.11 Give the point group of each of the following molecules. (a) CH_4; (b) CH_3F; (c) CH_2F_2; (d) CHF_3; (e) SF_6; (f) SF_5Br; (g) *trans*-SF_4Br_2; (h) CDH_3.

12.12 Give the point group of: (a) $CH_2{=}CH_2$; (b) $CH_2{=}CHF$; (c) $CH_2{=}CF_2$; (d) *cis*-$CHF{=}CHF$; (e) *trans*-$CHF{=}CHF$.

12.13 Give the point group of: (a) benzene; (b) fluorobenzene; (c) *o*-difluorobenzene; (d) *m*-difluorobenzene; (e) *p*-difluorobenzene; (f) 1,3,5-trifluorobenzene; (g) 1,4-difluoro-2,5-dibromobenzene; (h) naphthalene; (i) 2-chloronaphthalene.

12.14 Give the point group of: (a) HCN; (b) H_2S; (c) CO_2; (d) CO; (e) C_2H_2; (f) CH_3OH; (g) ND_3; (h) OCS; (i) P_4; (j) PCl_3; (k) PCl_5; (l) $B_{12}Cl_{12}{}^{2-}$; (m) UF_6; (n) Ar.

12.15 Give the point group of: (a) $FeF_6{}^{3-}$; (b) IF_5; (c) $CH_2{=}C{=}CH_2$; (d) C_8H_8, cubane; (e) $C_6H_6Cr(CO)_3$; (f) B_2H_6; (g) XeF_4; (h) F_2O; (i) spiropentane; (j) $B_{10}H_{10}{}^{2-}$ (for the structure, see E. L. Muetterties and W. H. Knoth, *Chem. Eng. News*, May 9, 1966, p. 88).

12.16 The structure of ferrocene, $C_5H_5FeC_5H_5$, is an iron atom sandwiched midway between two parallel regular pentagons. For the eclipsed conformation, the vertices of the two pentagons are aligned; for the staggered conformation, one pentagon is rotated $2\pi/10$ radians with respect to the other. Electron diffraction results show that the gas-phase equilibrium conformation is the eclipsed one, with a quite low barrier to internal rotation of the rings. [A. Haaland and J. E. Nilsson, *Acta Chem. Scand.*, **22**, 2653 (1968).] What is the point group of (a) eclipsed ferrocene; (b) staggered ferrocene?

12.17 What is the point group of the tris(ethylenediamine)cobalt(III) complex ion? (Each $NH_2CH_2CH_2NH_2$ group occupies two adjacent positions of the octahedral coordination sphere.)

12.18 Give the point group of: (a) a square-based pyramid; (b) a right circular cone; (c) a square lamina; (d) a square lamina with the top and bottom sides painted different colors;

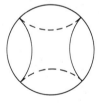

Figure 12.20 A baseball. The dashed and solid parts of the seam are in different hemispheres.

(e) a right circular cylinder; (f) a right circular cylinder with the two ends painted different colors; (g) a right circular cylinder with a stripe painted parallel to the axis; (h) a snowflake; (i) a doughnut; (j) a baseball (Fig. 12.20); (k) a $2p_z$ orbital; (l) a human being (ignore internal organs and slight external left–right asymmetries).

12.19 (a) What are the eigenvalues of $\hat{O}_{C_4}$? (b) Is this operator Hermitian?

12.20 Do the same as in Problem 12.19 for $\hat{O}_{C_2}$.

12.21 To what function is a $2p_z$ hydrogenlike orbital converted by: (a) $\hat{O}_{C_4(z)}$; (b) $\hat{O}_{C_4(y)}$?

12.22 Consider the following transformation:

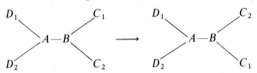

where C_1 and C_2 are identical and D_1 and D_2 are identical. Does it meet the definition of a symmetry operation (Section 12.1)? If so, express it in terms of some combination of the four kinds of symmetry operations discussed.

12.23 It is common to use rotation-inversion axes (rather than rotation-reflection axes) to classify the symmetry of crystals. Any S_n axis is equivalent to a rotation-inversion axis (symbolized by $\bar{p}$) whose order p may differ from n. A rotation-inversion operation consists of rotation by $2\pi/p$ radians followed by inversion. Show that

$$\hat{S}_n(z) = \hat{\imath}\,[\hat{C}_n(z)\,\hat{C}_2(z)]$$

Thus we have the following correspondence:

S_n	1	2	3	4
$\bar{p}$	2	$\bar{1}$	$\bar{6}$	$\bar{4}$

Give the next three pairs of entries in this table.

13

ELECTRONIC STRUCTURE OF DIATOMIC MOLECULES

13.1 THE BORN–OPPENHEIMER APPROXIMATION

We now begin the study of molecular quantum mechanics. If we assume the nuclei and electrons to be point masses and neglect spin–orbit and other relativistic interactions (Sections 11.6 and 11.7), then the molecular Hamiltonian is

$$\hat{H} = -\frac{\hbar^2}{2}\sum_\alpha \frac{1}{m_\alpha}\nabla_\alpha^2 - \frac{\hbar^2}{2m}\sum_i \nabla_i^2 + \sum_\alpha \sum_{\beta>\alpha} \frac{Z_\alpha Z_\beta e'^2}{r_{\alpha\beta}} - \sum_\alpha \sum_i \frac{Z_\alpha e'^2}{r_{i\alpha}} + \sum_j \sum_{i>j} \frac{e'^2}{r_{ij}}$$

(13.1)

where α and β refer to nuclei and i and j refer to electrons. The first term in (13.1) is the operator for the kinetic energy of the nuclei; the second term is the operator for the kinetic energy of the electrons; the third term represents the repulsions between the nuclei, $r_{\alpha\beta}$ being the distance between nuclei α and β of atomic numbers Z_α and Z_β; the fourth term represents the attractions between the electrons and the nuclei, $r_{i\alpha}$ being the distance between electron i and nucleus α; the last term represents the repulsions between the electrons, r_{ij} being the distance between electrons i and j.

As an example, consider H_2. Let α and β be the two protons, let 1 and 2 be the two electrons, and let M be the proton mass; we have

$$\hat{H} = -\frac{\hbar^2}{2M}\nabla_\alpha^2 - \frac{\hbar^2}{2M}\nabla_\beta^2 - \frac{\hbar^2}{2m}\nabla_1^2 - \frac{\hbar^2}{2m}\nabla_2^2 + \frac{e'^2}{r_{\alpha\beta}} - \frac{e'^2}{r_{1\alpha}} - \frac{e'^2}{r_{1\beta}} - \frac{e'^2}{r_{2\alpha}} - \frac{e'^2}{r_{2\beta}} + \frac{e'^2}{r_{12}}$$

(13.2)

The wave functions and energies of a molecule are found from the Schroedinger equation:

$$\hat{H}\psi(q_i, q_\alpha) = E\psi(q_i, q_\alpha)$$

(13.3)

313

where q_i and q_α symbolize the electronic and nuclear coordinates, respectively. The molecular Hamiltonian (13.1) is formidable enough to strike terror in the heart of any quantum chemist; fortunately there exists a highly accurate, simplifying approximation. The key lies in the fact that nuclei are much heavier than electrons: $m_\alpha \gg m$. Hence the electrons move much faster than the nuclei, and to a good approximation as far as the electrons are concerned, we can regard the nuclei as fixed while the electrons carry out their motions. Speaking classically, during the time of a cycle of electronic motion, the change in nuclear configuration is negligible. Thus considering the nuclei as fixed, we omit the nuclear kinetic-energy terms from (13.1) to obtain the Schroedinger equation for electronic motion:

$$(\hat{H}_{el} + V_{NN})\psi_{el} = U\psi_{el} \tag{13.4}$$

where the *purely electronic Hamiltonian* $\hat{H}_{el}$ is

$$\hat{H}_{el} = -\frac{\hbar^2}{2m}\sum_i \nabla_i^2 - \sum_\alpha \sum_i \frac{Z_\alpha e'^2}{r_{i\alpha}} + \sum_j \sum_{i>j} \frac{e'^2}{r_{ij}} \tag{13.5}$$

The electronic Hamiltonian including nuclear repulsion is $\hat{H}_{el} + V_{NN}$. The nuclear-repulsion term V_{NN} is given by

$$V_{NN} = \sum_\alpha \sum_{\beta > \alpha} \frac{Z_\alpha Z_\beta e'^2}{r_{\alpha\beta}} \tag{13.6}$$

The energy U in (13.4) is the electronic energy including the energy of nuclear repulsion. The internuclear distances $r_{\alpha\beta}$ in (13.4) are not variables, but are each fixed at some constant value. Of course, there are an infinite number of possible nuclear configurations, and for each of these we may solve the electronic Schroedinger equation (13.4) to get a set of electronic wave functions (and corresponding electronic energies); each member of the set corresponds to a different molecular electronic state. The electronic wave functions and energies thus depend parametrically on the nuclear configuration:

$$\psi_{el} = \psi_{el,n}(q_i; q_\alpha)$$

$$U = U_n(q_\alpha)$$

where n symbolizes the electronic quantum numbers.

The variables in the electronic Schroedinger equation (13.4) are the electronic coordinates. The quantity V_{NN} is independent of these coordinates and is a constant for a given nuclear configuration. Now it is easily proved (Problem 4.20) that the omission of a constant term C from the Hamiltonian does not affect the wave functions, and simply decreases each energy eigenvalue by C. Hence if V_{NN} is omitted from (13.4), we get

$$\hat{H}_{el}\psi_{el} = E_{el}\psi_{el} \tag{13.7}$$

where the *purely electronic energy* E_{el} is related to the electronic energy including internuclear repulsion by

$$U = E_{el} + V_{NN} \tag{13.8}$$

We can thus omit the internuclear repulsion from the electronic Schroedinger equation and simply add it to E_{el} after solving (13.7).

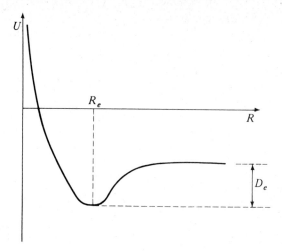

Figure 13.1 *Electronic energy of a diatomic molecule bound state, nuclear repulsion included*

For the hydrogen molecule, with the two protons at a fixed distance $r_{\alpha\beta} = R$, the purely electronic Hamiltonian is given by (13.2) with the first, second, and fifth terms omitted; the nuclear repulsion V_{NN} equals e'^2/R. The purely electronic Hamiltonian involves the six electronic coordinates $x_1, y_1, z_1, x_2, y_2, z_2$ and involves the nuclear coordinates as parameters.

The electronic Schroedinger equation (13.4) can be dealt with by approximate methods to be discussed later. If we plot the electronic energy including nuclear repulsion for a bound state of a diatomic molecule against the internuclear distance R, we find a curve like Fig. 13.1. At $R = 0$, the internuclear repulsion causes U to go to infinity. The internuclear separation at the minimum in this curve is called the *equilibrium internuclear distance* R_e. The difference between the limiting value of U at infinite internuclear separation and its value at R_e is called the *equilibrium dissociation energy* D_e:

$$D_e \equiv U(\infty) - U(R_e) \qquad (13.9)$$

When nuclear motion is considered, one finds that the equilibrium dissociation energy D_e differs from the molecular ground-state dissociation energy D_0. The lowest state of nuclear motion has zero rotational energy [as shown by Eq. (6.58)] but has a nonzero vibrational energy—the zero-point energy. If we use the harmonic-oscillator approximation for the vibration of a diatomic molecule (Section 4.3), then this zero-point energy is $\frac{1}{2}h\nu$. This zero-point energy raises the energy for the ground state of nuclear motion $\frac{1}{2}h\nu$ above the minimum in the $U(R)$ curve, so D_0 is less than D_e and $D_0 \approx D_e - \frac{1}{2}h\nu$. (Different electronic states of the same molecule have different values of R_e, D_e, D_0, and ν.)

No doubt you are familiar with tables of bond energies. It might be thought that D_0 of the ground electronic state is identical with the tabulated bond energy, but this is not so. Bond energies[1] are usually tabulated as the standard enthalpy

---

[1] See S. W. Benson, *J. Chem. Educ.*, **42**, 502 (1965).

change ΔH° for the dissociation reaction at $25^\circ C$. For $H_2 \rightarrow 2H$, we have $\Delta H^\circ_{298} = 104.16$ kcal/mole. For an ideal-gas reaction at temperature T, the standard enthalpy change is related to the standard internal-energy change ΔU° [not to be confused with $U(R)$] by $\Delta H^\circ = \Delta U^\circ + RT\,\Delta n$, where Δn is the change in the number of moles of gas and R is the gas constant. For the dissociation of H_2, $\Delta n = 1$, and we find $\Delta U^\circ_{298} = 103.57$ kcal/mole, which is an energy change of 4.491 eV/molecule. This is not D_0, however. The ground electronic state D_0 is the energy change for the dissociation reaction with all species in their ground states of electronic and nuclear motion, i.e., at 0 K. At 298 K, we have translational energy for the atoms and the molecule, and rotational energy for the molecule; for H_2 the energy differences between the ground and first excited electronic states and the ground and first excited vibrational states are too large to have appreciable population of excited electronic or vibrational states at room temperature. (Recall the form of the Boltzmann distribution law.) When we allow for the energy change on raising the atoms and molecules from absolute zero to 298 K, we find (Problem 13.2) that for H_2, $D_0 = 4.478$ eV.

For some electronic states of a molecule, the solution of the electronic Schroedinger equation gives a $U(R)$ curve with no minimum. For this case we do not have a stable bound state, and the molecule will dissociate. (Examples are some of the states in Fig. 13.4.)

It occasionally happens that an electronic state shows two minima in its potential-energy curve. We shall not consider this case.

Assuming that we have solved the electronic Schroedinger equation (which is quite an assumption), we next consider nuclear motions. According to our picture, the electrons move much faster than the nuclei; when the nuclei change their configuration slightly, say from q'_α to q''_α, the electrons immediately adjust to the change, with the electronic wave function changing from $\psi_{el}(q_i; q'_\alpha)$ to $\psi_{el}(q_i; q''_\alpha)$ and the electronic energy changing from $U(q'_\alpha)$ to $U(q''_\alpha)$. Thus as the nuclei move, the electronic energy varies smoothly as a function of the parameters defining the nuclear configuration, and $U(q_\alpha)$ becomes, in effect, the potential energy for the nuclear motion. The electrons act like a spring connecting the nuclei; as the internuclear distance changes, the energy stored in the spring changes. Hence the Schroedinger equation for nuclear motion is

$$\hat{H}_N \psi_N = E\psi_N \tag{13.10}$$

$$\hat{H}_N = -\frac{\hbar^2}{2} \sum_\alpha \frac{1}{m_\alpha} \nabla^2_\alpha + U(q_\alpha) \tag{13.11}$$

The variables in the nuclear Schroedinger equation are the nuclear coordinates, symbolized by q_α. The energy eigenvalue E in (13.10) is the *total* energy of the molecule, since the Hamiltonian (13.11) includes operators for both nuclear energy and electronic energy. E is of course simply a number and does not depend on any coordinates. Note that for each electronic state of a molecule, one must solve a different nuclear Schroedinger equation, since U differs from state to state. In this chapter we will concentrate on the electronic Schroedinger equation (13.4).

The approximation of separating electronic and nuclear motion is called the

Born–Oppenheimer approximation[2] and is basic to quantum chemistry. Born and Oppenheimer's mathematical treatment showed that the true molecular wave function is adequately approximated as

$$\psi(q_i, q_\alpha) = \psi_{el}(q_i; q_\alpha)\psi_N(q_\alpha) \qquad (13.12)$$

if $(m/m_\alpha)^{1/4} \ll 1$. The Born–Oppenheimer approximation introduces little error for the ground electronic states of diatomic molecules. Corrections for excited electronic states are larger than for the ground state, but still are generally negligible as compared to the errors introduced by the approximations used to solve the electronic Schroedinger equation of a many-electron molecule. Hence we will not worry about corrections to the Born–Oppenheimer approximation.

A rough justification of the Born–Oppenheimer approximation is as follows. Using the definitions (13.5) and (13.6), we write the molecular Hamiltonian $\hat{H}$ in (13.1) as $\hat{H} = -\frac{1}{2}\hbar^2\Sigma_\alpha m_\alpha^{-1}\nabla_\alpha^2 + \hat{H}_{el} + V_{NN}$. We substitute the assumed solution (13.12) into the exact molecular Schroedinger equation (13.3); using the result of Problem 5.7 and the fact that $\hat{H}_{el}$ does not affect ψ_N, we get

$$-\frac{\hbar^2}{2}\sum_\alpha \frac{1}{m_\alpha}[\psi_{el}\nabla_\alpha^2\psi_N + \psi_N\nabla_\alpha^2\psi_{el} + 2\nabla_\alpha\psi_{el}\cdot\nabla_\alpha\psi_N] + \psi_N\hat{H}_{el}\psi_{el} + V_{NN}\psi_{el}\psi_N$$
$$= E\psi_{el}\psi_N$$

We shall show below that the second and third terms in the brackets are substantially smaller than the first term. Neglecting these two terms and using (13.4), we get

$$-\frac{\hbar^2}{2}\psi_{el}\sum_\alpha \frac{1}{m_\alpha}\nabla_\alpha^2\psi_N + U\psi_{el}\psi_N = E\psi_{el}\psi_N$$

Division by ψ_{el} yields the Schroedinger equation for nuclear motion, Eqs. (13.10) and (13.11), thereby justifying the Born–Oppenheimer approximation.

To complete things we must show that the neglected terms are small. The nuclear coordinates q_α enter into the electronic Schroedinger equation (13.4) through the dependence of the distances $r_{i\alpha}$ in (13.5) on the nuclear coordinates. Each $r_{i\alpha}$ is a function of $x_i - x_\alpha$, $y_i - y_\alpha$, and $z_i - z_\alpha$. Hence we expect the nuclear coordinates to enter into ψ_{el} as $x_i - x_\alpha$, etc. Therefore $\partial\psi_{el}/\partial x_\alpha = -\partial\psi_{el}/\partial x_i$ and $\partial^2\psi_{el}/\partial x_\alpha^2 = \partial^2\psi_{el}/\partial x_i^2$. Hence $\nabla_\alpha^2\psi_{el} = \nabla_i^2\psi_{el}$ and $\nabla_\alpha\psi_{el} = -\nabla_i\psi_{el}$. Using (5.44), we see that the three terms in brackets become (when multiplied by $-\hbar^2$) $\psi_{el}\hat{p}_\alpha^2\psi_N + \psi_N\hat{p}_i^2\psi_{el} - 2(\hat{p}_i\psi_{el})\cdot(\hat{p}_\alpha\psi_N)$. We can expect that the ratio of the third term to the first will be of the order of magnitude $p_ip_\alpha/p_\alpha^2 = p_i/p_\alpha$, where p_i and p_α are average molecular electronic and nuclear linear momenta. The ratio of the second to the first term will be of the order of magnitude p_i^2/p_α^2.

To estimate the average momentum p_α of a nucleus, we use the harmonic-oscillator vibrational levels $(v + \frac{1}{2})h\nu$. The result of Problem 4.1 says that for a $v = 0$ harmonic oscillator the average kinetic energy is half the total energy. Therefore for $v = 0$ we have $p_\alpha^2/2m_\alpha \approx h\nu/4$ and $p_\alpha \approx (h\nu m_\alpha/2)^{1/2}$. Vibrational frequencies for relatively light diatomic molecules are of the order of magnitude 10^{14} Hz, so we get $p_\alpha \approx (0.2\text{ eV})^{1/2}m_\alpha^{1/2}$.

───────────

[2] The American physicist J. Robert Oppenheimer (1904–1967) was a graduate student of Born in 1927. During World War II, Oppenheimer was director of the Los Alamos laboratory that developed the atomic bomb. In 1954 Oppenheimer was judged a security risk and his security clearance was revoked by the U.S. Atomic Energy Commission. In 1963 the Atomic Energy Commission gave him the Enrico Fermi award.

We can estimate the energy of an electron in a molecule by taking the energy needed to ionize the molecule. A typical molecular ionization energy is 10 eV. The virial theorem (Section 14.1 and Problem 6.8) indicates that in atoms and molecules an electron's kinetic energy is of the same order of magnitude as minus the total energy of the electron, so we have $p_i^2/2m_i \approx 10 \text{ eV}$ and $p_i \approx (20 \text{ eV})^{1/2} m_i^{1/2}$. Then $p_i/p_\alpha \approx (20 \text{ eV}/0.2 \text{ eV})^{1/2}(m_i/m_\alpha)^{1/2} \approx 0.2$ or less, since m_α/m_i is 2000 or greater for relatively light molecules. Therefore the neglected second and third terms are substantially smaller than the first term.

13.2 ATOMIC UNITS

Most quantum chemists report the results of their calculations using a system of units called *atomic units*.[3]

First consider the cgs Gaussian system of units. The hydrogenlike-atom Hamiltonian in these units is (assuming infinite nuclear mass)

$$-(\hbar^2/2m)\nabla^2 - Ze'^2/r \qquad (13.13)$$

The system of atomic units that is based on Gaussian units is defined as follows. The unit of mass is the electron's mass m, rather than the gram; the unit of charge is the proton's charge e', rather than the statcoulomb; the unit of angular momentum is $\hbar$, rather than the g cm^2/sec. (The atomic unit of mass used in quantum chemistry should not be confused with the quantity 1 amu, which is one-twelfth the mass of a ^{12}C atom.) When we switch to atomic units, $\hbar$, m, and e' each have a numerical value of 1. Hence to change a formula from cgs Gaussian units to atomic units, we simply set these quantities equal to 1. Thus in atomic units the hydrogenlike Hamiltonian is $-\frac{1}{2}\nabla^2 - Z/r$, where r is now measured in atomic units of length, rather than in centimeters as in (13.13). The ground-state energy of the hydrogenlike atom is given by (6.107) as $-\frac{1}{2}Z^2(e'^2/a_0)$. Since [Eq. (6.125)] $a_0 = \hbar^2/me'^2$, the numerical value of a_0 in atomic units is 1, and the ground-state energy of the hydrogenlike atom has the numerical value (neglecting nuclear motion) $-\frac{1}{2}Z^2$ in atomic units.

The atomic unit of energy, e'^2/a_0, is called the *hartree*:

$$1 \text{ hartree} = e'^2/a_0 = 27.212 \text{ eV} \qquad (13.14)$$

The ground-state energy of the hydrogen atom is $-\frac{1}{2}$ hartree if nuclear motion is neglected. (Unfortunately, some workers report energies in rydbergs, where 1 rydberg $= e'^2/2a_0 = \frac{1}{2}$ hartree.)

The atomic unit of length is called the *bohr*:

$$1 \text{ bohr} = a_0 = 0.52918 \text{ Å} \qquad (13.15)$$

Now suppose we start with SI units. The hydrogenlike Hamiltonian in SI units is $-(\hbar^2/2m)\nabla^2 - Ze^2/4\pi\varepsilon_0 r$. The system of atomic units based on SI units is defined as follows. The units of mass, charge, and angular momentum are defined as the electron's mass m, the proton's charge e, and $\hbar$, respectively (rather than the

---

[3] H. Shull and G. G. Hall, *Nature*, **184**, 1559 (1959).

kilogram, the coulomb, and the $kg\,m^2/s$); the unit of permittivity is $4\pi\varepsilon_0$, rather than the $C^2\,N^{-1}\,m^{-2}$. (In Gaussian units, charge is expressible in terms of mass, length, and time, whereas in SI units this is not so. So in SI atomic units, we define four quantities as compared to three for Gaussian atomic units.) When we switch to atomic units, $\hbar, m, e,$ and $4\pi\varepsilon_0$ each have a numerical value of 1. In SI atomic units, the hydrogenlike Hamiltonian is $-\frac{1}{2}\nabla^2 - Z/r$; the Bohr radius $a_0 = 4\pi\varepsilon_0\hbar^2/me^2$ has the numerical value 1; the hydrogenlike ground-state energy is $-\frac{1}{2}Z^2$.

Use of atomic units saves time by eliminating $m, e',$ and $\hbar$ from equations. Also, if the results of a quantum-mechanical calculation are reported in eV, ergs, or joules, the value given depends on the currently accepted values of the physical constants; reporting results in atomic units avoids this.

Atomic units will be used in Chapters 13 and 15.

13.3 *THE HYDROGEN MOLECULE ION*

We now begin the study of the electronic energies of molecules. We will use the Born–Oppenheimer approximation, keeping the nuclei fixed while we solve, as best we can, the Schroedinger equation for the motion of the electrons. We will usually be considering an isolated molecule, ignoring intermolecular interactions. Our results will be most applicable to molecules in the gas phase at low pressure.

We start with diatomic molecules, the simplest of which is H_2^+, the hydrogen molecule ion, consisting of two protons and one electron. (H_2^+ was discovered by J. J. Thomson in cathode rays.) Just as the one-electron hydrogen atom serves as a starting point in the discussion of many-electron atoms, the one-electron hydrogen molecular ion furnishes many ideas useful for discussing many-electron diatomic molecules. The electronic Schroedinger equation for H_2^+ is separable, and we can obtain exact solutions for the eigenfunctions and eigenvalues.

Figure 13.2 shows H_2^+. The nuclei are at a and b; R is the internuclear distance; r_a and r_b are the distances from the electron to nuclei a and b. Since the nuclei are fixed, we have a one-particle problem whose Hamiltonian is [Eq. (13.5)]

$$-\frac{\hbar^2}{2m}\nabla^2 - \frac{e'^2}{r_a} - \frac{e'^2}{r_b}$$

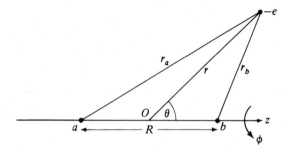

Figure 13.2 *The* H_2^+ *molecule*

The first term is the electronic kinetic-energy operator; the second and third terms are the attractions between the electron and the nuclei. In atomic units the purely electronic Hamiltonian for $H_2{}^+$ is

$$\hat{H}_{el} = -\tfrac{1}{2}\nabla^2 - \frac{1}{r_a} - \frac{1}{r_b} \tag{13.16}$$

In Fig. 13.2 the coordinate origin is on the internuclear axis, midway between the nuclei, with the z axis lying along the internuclear axis. We might attempt to solve the electronic Schroedinger equation using the spherical polar coordinates (r, θ, φ) of the electron, expressing r_a and r_b in these coordinates (Problem 13.5). We would find that the Schroedinger equation is not separable in spherical polar coordinates. However, Burrau showed in 1927 that separation of variables is possible using *confocal elliptic coordinates* ξ, η, φ. The coordinate φ is the angle of rotation of the electron about the internuclear (z) axis, the same as in spherical polar coordinates. The coordinates ξ (xi) and η (eta) are defined by

$$\xi = \frac{r_a + r_b}{R}, \qquad \eta = \frac{r_a - r_b}{R} \tag{13.17}$$

The ranges of these coordinates are (Problem 13.5)

$$0 \leqslant \varphi \leqslant 2\pi, \qquad 1 \leqslant \xi \leqslant \infty, \qquad -1 \leqslant \eta \leqslant 1 \tag{13.18}$$

(Instead of ξ and η, some people use λ and μ, respectively; still another notation for these coordinates is μ and v.)

We must put the Hamiltonian (13.16) into these coordinates. We have

$$r_a = \tfrac{1}{2}R(\xi + \eta), \qquad r_b = \tfrac{1}{2}R(\xi - \eta) \tag{13.19}$$

We also need the expression for the Laplacian in confocal elliptic coordinates. One way to find this is to express ξ, η, and φ in terms of x, y, and z, the Cartesian coordinates of the electron, and then use the chain rule to find $\partial/\partial x$, $\partial/\partial y$, and $\partial/\partial z$ in terms of $\partial/\partial\xi$, $\partial/\partial\eta$, and $\partial/\partial\varphi$. We then form the operator (3.55). [We used this procedure to derive Eq. (6.6).] We omit the long, tiresome details and simply state the result[4]:

$$\nabla^2 = \frac{4}{R^2(\xi^2 - \eta^2)}\left[(\xi^2 - 1)\frac{\partial^2}{\partial\xi^2} + 2\xi\frac{\partial}{\partial\xi} + (1 - \eta^2)\frac{\partial^2}{\partial\eta^2} - 2\eta\frac{\partial}{\partial\eta} \right.$$
$$\left. + \left(\frac{1}{\xi^2 - 1} + \frac{1}{1 - \eta^2}\right)\frac{\partial^2}{\partial\varphi^2} \right] \tag{13.20}$$

The $H_2{}^+$ electronic Hamiltonian (13.16) in atomic units is

$$\hat{H}_{el} = -\frac{2}{R^2(\xi^2 - \eta^2)}\left[(\xi^2 - 1)\frac{\partial^2}{\partial\xi^2} + 2\xi\frac{\partial}{\partial\xi} + (1 - \eta^2)\frac{\partial^2}{\partial\eta^2} - 2\eta\frac{\partial}{\partial\eta} \right.$$
$$\left. + \left(\frac{1}{\xi^2 - 1} + \frac{1}{1 - \eta^2}\right)\frac{\partial^2}{\partial\varphi^2} \right] - \frac{2}{R(\xi + \eta)} - \frac{2}{R(\xi - \eta)} \tag{13.21}$$

---

[4] For a systematic discussion see *Margenau and Murphy*, Chapter 5.

We could now form the electronic Schroedinger equation (13.7) and carry out a separation of variables. However, we can save time by considering the electronic angular momentum. For the hydrogen atom, the operators $\hat{L}^2$ and $\hat{L}_z$ both commute with the Hamiltonian. For H_2^+ we no longer have spherical symmetry, and we would find that $[\hat{L}^2, \hat{H}_{el}] \neq 0$. However, we do have axial symmetry, and we can show that $\hat{L}_z$ commutes with $\hat{H}_{el}$. $\hat{L}_z$ is given by (5.96) and involves only φ; hence in calculating the commutator, we need only consider the part of $\hat{H}_{el}$ that involves φ; we have (in atomic units)

$$[\hat{L}_z, \hat{H}_{el}] = \left[-i\frac{\partial}{\partial\varphi}, \ -\frac{2}{R^2(\xi^2 - \eta^2)}\left(\frac{1}{\xi^2 - 1} + \frac{1}{1 - \eta^2}\right)\frac{\partial^2}{\partial\varphi^2} \right] = 0$$

Therefore the electronic wave functions can be chosen to be eigenfunctions of $\hat{L}_z$; the eigenfunctions of $\hat{L}_z$ are [Eq. (5.114)]

$$\text{constant} \cdot (2\pi)^{-1/2} e^{im\varphi} \tag{13.22}$$

$$m = 0, \ \pm 1, \ \pm 2, \ \pm 3, \ldots \tag{13.23}$$

The z component of electronic orbital angular momentum in H_2^+ is $m\hbar$ in cgs units or m in atomic units. The total electronic orbital angular momentum is not a constant for H_2^+.

The "constant" in (13.22) is a constant only as far as $\partial/\partial\varphi$ is concerned, so that the H_2^+ wave functions have the form

$$\psi_{el} = F(\xi, \eta)(2\pi)^{-1/2} e^{im\varphi} \tag{13.24}$$

We now try a separation of variables:

$$\psi_{el} = L(\xi)M(\eta)(2\pi)^{-1/2} e^{im\varphi} \tag{13.25}$$

Substitution of (13.25) into the electronic Schroedinger equation (13.7) leads to an equation in which the variables are separable; we are led to the two ordinary differential equations (Problem 13.3)

$$(\xi^2 - 1)L'' + 2\xi L' + \left(A + 2R\xi + \tfrac{1}{2}E_{el}R^2\xi^2 - \frac{m^2}{\xi^2 - 1}\right)L = 0 \tag{13.26}$$

$$(1 - \eta^2)M'' - 2\eta M' - \left(A + \tfrac{1}{2}E_{el}R^2\eta^2 + \frac{m^2}{1 - \eta^2}\right)M = 0 \tag{13.27}$$

where A is the separation constant.

Solving these equations is involved, and we simply sketch the results. The solution for $M(\eta)$ is an infinite series of associated Legendre functions. $L(\xi)$ also involves an infinite series. The values of the coefficients in the series for L and M are available. The requirement that the wave function be well behaved leads to the conclusion that for a fixed value of R, only certain values of E_{el} are allowed; this gives a series of electronic states. There is no algebraic formula for E_{el}; it must be calculated numerically for each desired value of R for each state. Tables of E_{el} are available.[5]

For the ground electronic state, the quantum number m is zero. At $R = \infty$, the H_2^+ ground state is dissociated into a proton and a ground-state hydrogen atom; hence $E_{el}(\infty) = -\frac{1}{2}$ hartree. At $R = 0$, the two protons have come together to form the He^+ ion with ground-state energy: $-\frac{1}{2}(2)^2$ hartrees $= -2$ hartrees. Addition of the internuclear repulsion $1/R$ (in atomic units) to $E_{el}(R)$ gives the $U(R)$ potential-energy curve for nuclear motion. Plots of the ground-state $E_{el}(R)$ and $U(R)$, as found from solution of the electronic Schroedinger equation, are shown in Fig. 13.3. At $R = \infty$, the internuclear repulsion is 0, and U is $-\frac{1}{2}$ hartree.

The $U(R)$ curve is found to have a minimum at [see J. L. Schaad and W. V. Hicks, *J. Chem. Phys.*, **53**, 851 (1970)]

$$R_e = 1.9972 \text{ bohrs} \approx 2.00 \text{ bohrs} = 1.06 \text{ Å}$$

indicating that the H_2^+ ground electronic state is a stable bound state. The calculated value of E_{el} at 1.9972 bohrs is -1.1033 hartrees; addition of the

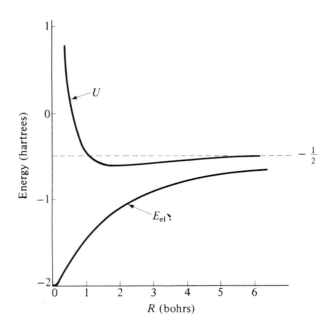

Figure 13.3 *Electronic energy with (U) and without (E_{el}) nuclear repulsion for the H_2^+ ground state*

[5] D. R. Bates, K. Ledsham, and A. L. Stewart, *Phil. Trans. Roy. Soc.*, **A246**, 215 (1953). The wave functions given by Bates et al. are not normalized. For the normalization constants, see E. M. Roberts, M. R. Foster, and F. F. Selig, *J. Chem. Phys.*, **37**, 485 (1962). For accurate ground-state energy values, see H. Wind, *J. Chem. Phys.*, **42**, 2371 (1965).

internuclear repulsion $1/R$ gives $U(R_e) = -0.6026$ hartree, compared to -0.5000 hartree at $R = \infty$. Thus the ground-state binding energy is

$$D_e = 0.1026 \text{ hartree} = 2.79 \text{ eV} = 64.4 \text{ kcal/mole} \qquad (13.28)$$

The binding energy is only 17 percent of the total energy at the equilibrium internuclear distance. Thus a reasonably small error in the total energy can correspond to a large error in the binding energy. For heavier molecules the situation is even worse, since chemical binding energies are of the same order of magnitude (100 kcal/mole) for most diatomic molecules, but the total electronic energy increases markedly for heavier molecules.

Note that the single electron in $H_2{}^+$ is sufficient to give a stable bound state. Figure 13.4 shows the $U(R)$ curves for the first several electronic energy levels

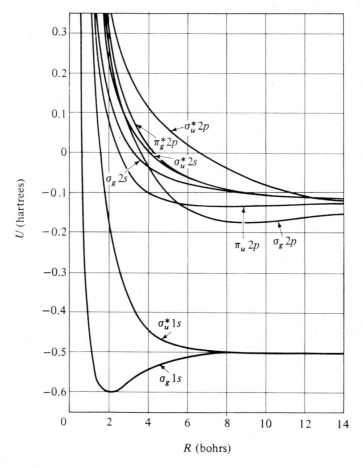

Figure 13.4 *Potential-energy curves for several* $H_2{}^+$ *electronic states.* [*Curves taken from J. C. Slater,* Quantum Theory of Molecules and Solids, *vol. 1, McGraw-Hill, New York, 1963. Used by permission.*]

of $H_2{}^+$, as calculated from the solutions to the Schroedinger equation. Each $H_2{}^+$ state has a definite value for m. Of course, since this is a three-dimensional problem, the value of the single quantum number m does not specify the state uniquely; we have two other quantum numbers, and these specify which solutions of the two differential equations (13.27) and (13.26) are used. These quantum numbers give the number of nodes in the ξ and η factors of the wave function; we shall not make any further reference to them.

Note that m enters Eqs. (13.27) and (13.26) as m^2; hence the electronic energy depends only on $|m|$. Each electronic energy level with $m \neq 0$ is doubly degenerate, corresponding to states with quantum numbers $+|m|$ and $-|m|$. In the standard notation[6] for diatomic molecules, the absolute value of m is called λ:

$$\lambda = |m|$$

(Some texts define λ as identical to m.) Somewhat similar to the s, p, d, f notation for hydrogen-atom states, a letter code is used to specify λ, the absolute value (in atomic units) of the component along the molecular axis of the electron's orbital angular momentum:

λ	0	1	2	3	4
letter	σ	π	δ	φ	γ

Thus the lowest $H_2{}^+$ electronic state is a σ state.

Besides classifying the states of $H_2{}^+$ according to λ, we can also classify them according to their parity (Section 7.5). From Fig. 13.11, inversion of the electron's coordinates through the origin O changes φ to $\varphi + \pi$, r_a to r_b, and r_b to r_a; this leaves the potential-energy part of the electronic Hamiltonian (13.16) unchanged. We previously showed the kinetic-energy operator to be invariant under inversion. Hence the parity operator commutes with the Hamiltonian (13.16), and the $H_2{}^+$ electronic wave functions can be classified as either even or odd. For even electronic wave functions, we use the subscript g (from the German word *gerade*, meaning even); for odd wave functions, we use u (from *ungerade*).

We further classify the $H_2{}^+$ electronic energy levels according to the state of the hydrogen atom obtained on dissociation, writing this state following the symbol for λ. Thus the two lowest levels dissociate to a $1s$ hydrogen atom. The significance of the star in Fig. 13.4 will be explained later.

For completeness we must take spin into account by multiplying each spatial $H_2{}^+$ electronic wave function by α or β, depending on whether the component of electron spin along the internuclear axis is $+\frac{1}{2}$ or $-\frac{1}{2}$ (in atomic units). Inclusion of spin doubles the degeneracy of all levels.

The Schroedinger equation for any one-electron diatomic molecule is separable in elliptical coordinates; thus the electronic energies for the heteronuclear diatomic molecule HeH^{2+} are available.[7]

In the rest of this chapter, we will drop the subscript el from the electronic wave function, Hamiltonian, and energy; it will be understood in Chapters 13 and 15 that ψ means ψ_{el}.

---

[6] F. A. Jenkins, *J. Opt. Soc. Am.*, **43**, 425 (1953).

[7] D. R. Bates and T. R. Carson, *Proc. Roy. Soc.*, **A234**, 207 (1956).

13.4   APPROXIMATE TREATMENTS OF THE H_2^+ GROUND ELECTRONIC STATE

For a many-electron atom, the self-consistent-field (SCF) method is used to construct an approximate wave function as a Slater determinant of (one-electron) spin-orbitals. The one-electron spatial part of a spin-orbital is an atomic orbital (AO). We took each AO as a product of a spherical harmonic and a radial factor; as an initial approximation to the radial factors, we can use hydrogenlike radial functions with effective nuclear charges.

For many-electron molecules, which (unlike H_2^+) cannot be solved exactly, we want to use many of the ideas of the SCF treatment of atoms. We will write an approximate molecular electronic wave function as a Slater determinant of (one-electron) spin-orbitals. The one-electron spatial part of a molecular spin-orbital will be called a *molecular orbital* (MO). Because of the Pauli principle, each MO can hold no more than two electrons, just as for AO's. What kind of functions do we use for the MO's? Ideally the analytic form of each MO is found by an SCF calculation. Such calculations are difficult for molecules, and accurate SCF molecular wave functions first became available in the 1960s. We seek a simple approximation for the MO's which will enable us to gain some qualitative understanding of chemical bonding. Just as we took the angular part of each AO to be the same kind of function (a spherical harmonic) as in the one-electron hydrogenlike atom, we will take the angular part of each diatomic MO to be $(2\pi)^{-1/2}e^{im\varphi}$, as in H_2^+. However, the ξ and η factors in the H_2^+ wave functions are complicated functions and do not readily lend themselves to use in MO calculations. We therefore seek simpler functions that will provide reasonably accurate approximations to the H_2^+ wave functions and that can be used to construct molecular orbitals for many-electron diatomic molecules. With this discussion as motivation for looking at approximate solutions in a case where the Schroedinger equation is exactly solvable, we consider approximate treatments of H_2^+.

We will use the variation method, writing down some function containing several parameters and varying them to minimize the variational integral. This will give an approximation to the ground-state wave function and an upper bound to the ground-state energy. By use of the factor $e^{im\varphi}$ in the trial function, we can get an upper bound to the energy of the lowest H_2^+ level for any given value of m (see Section 8.2). By using linear variation functions, we can get approximations for excited states.

The H_2^+ ground state has $m = 0$, and the wave function depends only on ξ and η. We could try any well-behaved function of these coordinates as a trial variation function; we will, however, use a more systematic approach based on the idea of a molecule as being formed from the interaction of atoms.

Consider what the H_2^+ wave function would look like for large values of the internuclear separation R. When the electron is near nucleus a, nucleus b is so far away that we essentially have a hydrogen atom with origin at a; thus when r_a is small, the ground-state H_2^+ electronic wave function should resemble the ground-state hydrogen-atom wave function of Eq. (6.120). We have $Z = 1$, and the Bohr radius a_0 has the numerical value 1 in atomic units; hence (6.120) becomes

$$\pi^{-1/2}e^{-r_a}$$

(13.29)

Similarly we conclude that when the electron is near nucleus b, the H_2^+ ground-state wave function will be approximated by

$$\pi^{-1/2}e^{-r_b} \tag{13.30}$$

These considerations suggest that we try as a variation function

$$c_1\pi^{-1/2}e^{-r_a}+c_2\pi^{-1/2}e^{-r_b} \tag{13.31}$$

where c_1 and c_2 are variational parameters. When the electron is near nucleus a, the variable r_a is small and r_b is large, so that the first term in (13.31) predominates, giving a function resembling (13.29). The function (13.31) is a linear variation function, and we are led to solve a secular equation, which has the form (8.87), where the subscripts 1 and 2 refer to the functions (13.29) and (13.30).

We can also approach the problem using perturbation theory. We take the unperturbed system as the H_2^+ molecule with $R = \infty$. For $R = \infty$, the electron can be bound to nucleus a with wave function (13.29), or it can be bound to nucleus b with wave function (13.30). In either case the energy is $-\frac{1}{2}$ hartree, so that we have a doubly degenerate unperturbed energy level. Bringing the nuclei in from infinity gives rise to a perturbation which splits the doubly degenerate unperturbed level into two levels; this is illustrated by the $U(R)$ curves for the two lowest H_2^+ electronic states, which both dissociate to a ground-state hydrogen atom (see Fig. 13.4). The correct zeroth-order wave functions for the perturbed levels are linear combinations of the form (13.31), and we are led to a secular equation of the form (8.87) with W replaced by $E^{(0)} + E^{(1)}$ (see Problem 9.6).

Before proceeding with the solution of (8.87), let us improve the trial function (13.31). Consider the limiting behavior of the H_2^+ ground-state electronic wave function as R goes to zero. In this limit we get the He^+ ion, which has the ground-state wave function [put $Z = 2$ in (6.120)]

$$2^{3/2}\pi^{-1/2}e^{-2r} \tag{13.32}$$

From Fig. 13.2 we see that as R goes to zero, both r_a and r_b go to r; hence as R goes to zero, the trial function (13.31) goes to $(c_1 + c_2)\pi^{-1/2}e^{-r}$. Comparing with (13.32), we see that our trial function has the wrong limiting behavior at $R = 0$; it should go to e^{-2r}, not e^{-r}. We can fix things by multiplying r_a and r_b in the exponentials by a variational parameter k, which will be some function of R: $k = k(R)$. For the correct limiting behavior at zero and at infinite internuclear distance, we have

$$k(0) = 2, \qquad k(\infty) = 1$$

for the ground state of H_2^+. Physically k is some sort of effective nuclear charge, which increases as the nuclei come together. We thus take as the trial function

$$c_a 1s_a + c_b 1s_b \tag{13.33}$$

where the c's are variational parameters and where

$$1s_a = k^{3/2}\pi^{-1/2}e^{-kr_a}, \qquad 1s_b = k^{3/2}\pi^{-1/2}e^{-kr_b} \tag{13.34}$$

The factor $k^{3/2}$ normalizes $1s_a$ and $1s_b$ [see Eq. (6.120)]. The molecular-orbital function (13.33) is a *linear combination of atomic orbitals*, an LCAO-MO. The trial function (13.33) was first used by Finkelstein and Horowitz in 1928.

For the function (13.33), the secular equation (8.87) is

$$\begin{vmatrix} H_{aa} - WS_{aa} & H_{ab} - WS_{ab} \\ H_{ba} - WS_{ba} & H_{bb} - WS_{bb} \end{vmatrix} = 0 \tag{13.35}$$

The integrals H_{aa} and H_{bb} are

$$H_{aa} = \int 1s_a^* \hat{H} 1s_a \, dv, \qquad H_{bb} = \int 1s_b^* \hat{H} 1s_b \, dv \tag{13.36}$$

where the $H_2{}^+$ electronic Hamiltonian operator $\hat{H}$ is given by (13.16). These two integrals are called *Coulomb integrals*. [Their form differs considerably from that of the Coulomb integral (9.144); the designation of (13.36) as Coulomb integrals is not really appropriate.] We can relabel the variables in a definite integral without affecting its value; changing a to b and b to a changes $1s_a$ to $1s_b$ but leaves $\hat{H}$ unaffected (this would not be true for a heteronuclear diatomic molecule). Hence $H_{aa} = H_{bb}$. We have

$$H_{ab} = \int 1s_a^* \hat{H} 1s_b \, dv, \qquad H_{ba} = \int 1s_b^* \hat{H} 1s_a \, dv \tag{13.37}$$

Since $\hat{H}$ is Hermitian and the functions in these integrals are real, we conclude that $H_{ab} = H_{ba}$. The integral H_{ab} is called a *resonance* (or *bond*) *integral*. Since $1s_a$ and $1s_b$ are normalized and real, we have

$$S_{aa} = \int 1s_a^* 1s_a \, dv = 1 = S_{bb} \tag{13.38}$$

$$S_{ab} = \int 1s_a^* 1s_b \, dv = S_{ba} \tag{13.39}$$

The secular equation (13.35) becomes

$$\begin{vmatrix} H_{aa} - W & H_{ab} - S_{ab}W \\ H_{ab} - S_{ab}W & H_{aa} - W \end{vmatrix} = 0 \tag{13.40}$$

$$H_{aa} - W = \pm (H_{ab} - S_{ab}W) \tag{13.41}$$

$$W_1 = \frac{H_{aa} + H_{ab}}{1 + S_{ab}}, \qquad W_2 = \frac{H_{aa} - H_{ab}}{1 - S_{ab}} \tag{13.42}$$

These two roots are upper bounds for the energies of the ground and first excited electronic states of $H_2{}^+$.

We now find the coefficients in (13.33) for each of the roots of the secular equation. From Eq. (8.86) we have

$$(H_{aa} - W)c_a + (H_{ab} - S_{ab}W)c_b = 0 \tag{13.43}$$

Substituting in W_1 from (13.42) [or using (13.41)], we get

$$c_a/c_b = 1 \tag{13.44}$$

$$\varphi_1 = c_a(1s_a + 1s_b) \tag{13.45}$$

We fix c_a by normalization:

$$|c_a|^2 \int (1s_a^2 + 1s_b^2 + 2 \cdot 1s_a \, 1s_b) \, dv = 1$$

$$|c_a| = \frac{1}{[2 + 2S_{ab}]^{1/2}} \tag{13.46}$$

The normalized trial function corresponding to the energy W_1 is thus

$$\varphi_1 = \frac{1s_a + 1s_b}{\sqrt{2}(1 + S_{ab})^{1/2}} \tag{13.47}$$

For the root W_2, we find $c_b = -c_a$ and

$$\varphi_2 = \frac{1s_a - 1s_b}{\sqrt{2}(1 - S_{ab})^{1/2}} \tag{13.48}$$

Equations (13.47) and (13.48) come as no surprise. Since the nuclei are identical, we expect $|\varphi|^2$ to remain unchanged on interchanging a and b; in other words, we expect no polarity in the bond.

    Let us now evaluate H_{aa}, H_{ab}, and S_{ab}. [This can be skimmed, if desired; begin reading again after Eq. (13.58).] We begin with the *overlap integral* S_{ab}. From (13.34) and (13.17), we have

$$1s_a \, 1s_b = k^3 \pi^{-1} e^{-k(r_a + r_b)} = k^3 \pi^{-1} e^{-kR\xi}$$

We need the expression for the volume element dv in elliptic coordinates; this turns out to be (*Eyring, Walter, and Kimball*, Appendix III)

$$dv = \tfrac{1}{8} R^3 (\xi^2 - \eta^2) \, d\xi \, d\eta \, d\varphi \tag{13.49}$$

We have

$$S_{ab} = \frac{k^3 R^3}{8\pi} \int_0^{2\pi} d\varphi \int_{-1}^1 \left[\int_1^\infty e^{-kR\xi}(\xi^2 - \eta^2) \, d\xi \right] d\eta \tag{13.50}$$

Using the Appendix integral (A.7), we find

$$\int_1^\infty e^{-kR\xi} \xi^2 \, d\xi - \int_1^\infty e^{-kR\xi} \eta^2 \, d\xi = \frac{2e^{-kR}}{k^3 R^3} \left(1 + kR + \frac{k^2 R^2}{2} \right) - \eta^2 \frac{e^{-kR}}{kR}$$

Substituting in (13.50) and doing the eta integration, we get

$$S_{ab} = e^{-kR}[1 + kR + \tfrac{1}{3} k^2 R^2] \tag{13.51}$$

Now for H_{aa}. Adding and subtracting k/r_a in (13.16), we get

$$\hat{H} = \left[-\tfrac{1}{2} \nabla^2 - \frac{k}{r_a} \right] + \frac{(k-1)}{r_a} - \frac{1}{r_b} = \hat{H}_a + \frac{(k-1)}{r_a} - \frac{1}{r_b} \tag{13.52}$$

where $\hat{H}_a$ is the Hamiltonian operator for a hydrogenlike atom of nuclear charge k located at a. Since $1s_a$ is a 1s hydrogenlike function for nuclear charge k, we have (Section 13.2)

$$\hat{H}_a 1s_a = -\tfrac{1}{2} k^2 1s_a \tag{13.53}$$

Hence

$$H_{aa} = \int 1s_a^* \left[\hat{H}_a + \frac{(k-1)}{r_a} - \frac{1}{r_b} \right] 1s_a \, dv$$

$$H_{aa} = -\tfrac{1}{2} k^2 + (k-1) \int \frac{1s_a^* 1s_a}{r_a} \, dv - \int \frac{1s_a^* 1s_a}{r_b} \, dv \tag{13.54}$$

The first integral in (13.54) can be done in elliptical coordinates, but it is faster to use spherical polar coordinates with origin at a:

$$\int \frac{1s_a^2}{r_a} dv = \frac{k^3}{\pi} \int_0^\infty \frac{e^{-2kr_a}}{r_a} r_a^2 \, dr_a \int_0^{2\pi} d\varphi \int_0^\pi \sin\theta \, d\theta = k \qquad (13.55)$$

where Eq. (A.4) has been used. Equations (13.49), (13.34), and (13.19) give for the second integral in (13.54)

$$\int \frac{1s_a^2}{r_b} dv = \frac{k^3 R^2}{4\pi} \int_0^{2\pi} d\varphi \int_{-1}^1 e^{-kR\eta} \left[\int_1^\infty e^{-kR\xi} (\xi + \eta) \, d\xi \right] d\eta$$

$$\int \frac{1s_a^2}{r_b} dv = \frac{1}{R} - e^{-2kR} \left(\frac{1}{R} + k \right) \qquad (13.56)$$

Therefore

$$H_{aa} = \tfrac{1}{2}k^2 - k - \frac{1}{R} + e^{-2kR} \left(k + \frac{1}{R} \right) \qquad (13.57)$$

The evaluation of H_{ab} is left as an exercise. We find

$$H_{ab} = -\tfrac{1}{2}k^2 S_{ab} + k(k-2)(1+kR)e^{-kR} \qquad (13.58)$$

Substituting the values for the integrals into (13.42), we get

$$W_{1,2} = -\tfrac{1}{2}k^2 + \frac{k^2 - k - R^{-1} + R^{-1}(1+kR)e^{-2kR} \pm k(k-2)(1+kR)e^{-kR}}{1 \pm e^{-kR}(1+kR+k^2 R^2/3)} \qquad (13.59)$$

where the upper signs are for W_1.

The final task is to vary the parameter k, setting

$$\frac{\partial W_1}{\partial k} = 0, \qquad \frac{\partial W_2}{\partial k} = 0$$

For the procedure used, see Problem 13.9. The results[8] are that for the $1s_a + 1s_b$ function (13.47), the parameter k increases monotonically from 1 to 2 as R decreases from ∞ to 0; for the $1s_a - 1s_b$ function (13.48), k decreases almost monotonically from 1.0 to 0.4 as R decreases from ∞ to 0. Since k is positive and never greater than 2, and since the overlap integral S_{ab} is positive, we see from (13.58) that the resonance integral H_{ab} is always negative. We conclude from (13.42) that the root W_1 corresponds to the ground electronic state $\sigma_g 1s$ of H_2^+. For the ground state, one finds

$$k(R_e) = 1.24$$

We might ask why the variational parameter k for the $\sigma_u^* 1s$ state goes to 0.4, rather than to 2, as R goes to zero. The answer is that this state of H_2^+ does not go to the ground state ($1s$) of He^+ as R goes to zero. The $\sigma_u^* 1s$ state has odd parity and must correlate with an odd state of He^+. The lowest odd states of He^+ are the $2p$ states (Section 11.5); since the $\sigma_u^* 1s$ state has zero electronic orbital angular momentum along the internuclear (z) axis, this state must go to an atomic $2p$ state with $m = 0$, i.e., to the $2p_0 = 2p_z$ state.

---

[8] C. A. Coulson, *Trans. Faraday Soc.*, **33**, 1479 (1937).

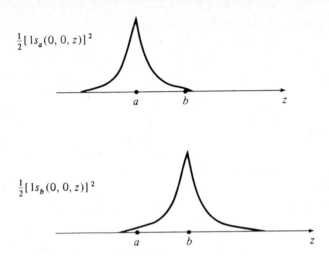

$\frac{1}{2}[1s_a(0, 0, z)]^2$

$\frac{1}{2}[1s_b(0, 0, z)]^2$

Figure 13.5 *Atomic charge probability densities for* $H_2{}^+$. *Note the cusps at the nuclei.*

Having found $k(R)$ for each root, we can calculate the variational approximation to the energies from (13.59). To find the $U(R)$ curves, we add the internuclear repulsion $1/R$ to W. The calculated ground-state $U(R)$ curve has a minimum at 2.02 bohrs, in good agreement with the true value 2.00 bohrs; the total energy at this distance is found to be 15.95 eV, giving a binding energy of 2.35 eV, as compared to the true value 2.79 eV [Eq. (13.28)]. (If we omit varying k but simply set it equal to 1, we get $R_e = 2.50$ bohrs, $D_e = 1.76$ eV.)

Now consider the appearance of the trial functions for the $\sigma_g 1s$ and $\sigma_u^* 1s$ states at intermediate values of R. Figure 13.5 shows the values of the functions $(1s_a)^2$ and $(1s_b)^2$ at points on the internuclear axis (see also Fig. 6.6). For the $\sigma_g 1s$ function $1s_a + 1s_b$, we get a buildup of electronic probability density between the nuclei, as shown in Fig. 13.6. It is especially significant that the buildup of charge between the nuclei is greater than that obtained by simply taking the sum of the separate atomic charge densities. The charge probability density for an electron in a $1s_a$ atomic orbital is $-(1s_a)^2$. (The electron's charge is -1 in atomic units.) If we add the charge

$\dfrac{[1s_a(0, 0, z) + 1s_b(0, 0, z)]^2}{2(1 + S_{ab})}$

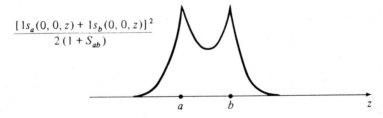

Figure 13.6 *Charge probability density for the* LCAO-MO *function* $1s_a + 1s_b$

probability density for half an electron in a $1s_a$ atomic orbital and half an electron in a $1s_b$ atomic orbital, we get

$$-\tfrac{1}{2}[1s_a^2 + 1s_b^2] \tag{13.60}$$

However, in quantum mechanics we do not add the separate atomic charge probability densities; instead, we add the wave functions, as in (13.47). The charge probability density for the H_2^+ ground state is then

$$-\varphi_1^2 = -\frac{1}{2(1+S_{ab})}[1s_a^2 + 1s_b^2 + 2\cdot 1s_a 1s_b] \tag{13.61}$$

The difference between (13.61) and (13.60) is

$$-\frac{1}{2(1+S_{ab})}[2\cdot 1s_a 1s_b - S_{ab}(1s_a^2 + 1s_b^2)] \tag{13.62}$$

Putting $R = 2.00$ and $k = 1.24$ in Eq. (13.51), we find that the overlap integral S_{ab} equals 0.46 at R_e. (It might be thought that because of the orthogonality of different AO's, the integral S_{ab} should be zero. However, the AO's $1s_a$ and $1s_b$ are eigenfunctions of *different* Hamiltonian operators, one for a hydrogen atom at a and one for a hydrogen atom at b; hence the orthogonality theorem does not apply.) Consider now the relative magnitudes of the two terms in brackets in (13.62) for points on the molecular axis. To the left of nucleus a, the function $1s_b$ is very small; to the right of nucleus b, the function $1s_a$ is very small. Hence in the regions outside the nuclei, the product $1s_a 1s_b$ is small, and the second term in (13.62) is dominant. This gives a subtraction of electronic charge density in the regions outside the nuclei, as compared to the sum of the densities of the individual atoms. Now consider the region between the nuclei. At the midpoint of the internuclear axis (and anywhere on the plane perpendicular to the axis and bisecting it), we have $1s_a = 1s_b$, and the bracketed terms in (13.62) become $2(1s_a)^2 - 0.92(1s_a)^2 \approx 1s_a^2$, which is positive. We thus get a buildup of charge probability density between the nuclei in the molecule, as compared to the sum of the densities of the individual atoms. This buildup of electronic charge between the nuclei allows the electron to feel the attractions of both nuclei at the same time, which lowers its potential energy. The greater the overlap in the internuclear region between the atomic orbitals forming the bond, the

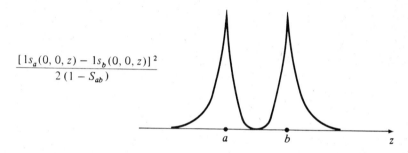

$$\frac{[1s_a(0,0,z) - 1s_b(0,0,z)]^2}{2(1 - S_{ab})}$$

Figure 13.7 *Charge probability density for the* LCAO-MO *function* $1s_a - 1s_b$

greater the charge buildup in this region, giving the familiar *principle of maximum overlap*. [Our argument has been based on the approximate wave function (13.47). If we plot Fig. 13.6 using the true wave function, we get a very similar graph, except that the true wave function is somewhat larger between the nuclei and somewhat smaller outside the internuclear region than (13.47).]

The preceding discussion seems to attribute the bonding in H_2^+ mainly to the lowering in the average electronic potential energy that results from having the shared electron interact with two nuclei instead of one. This, however, is an incomplete picture. Calculations on H_2^+ by Feinberg and Ruedenberg show that the decrease in electronic potential energy due to the sharing is of the same order of magnitude as the nuclear repulsion energy $1/R$ and hence is insufficient by itself to give binding. Two other effects also contribute to the bonding. The increase in atomic orbital exponent ($k = 1.24$ at R_e versus 1.0 at ∞) causes charge to accumulate near the nuclei (as well as in the internuclear region), and this further lowers the electronic potential energy. Moreover, the buildup of charge in the internuclear region makes $\partial \psi / \partial z$ zero at the midpoint of the molecular axis and small in the region close to this point; hence the z component of the average electronic kinetic energy [which can be expressed as $\frac{1}{2} \int |\partial \psi / \partial z|^2 \, d\tau$; Problem 7.3] is lowered as compared to the atomic $\langle T_z \rangle$. (However, the *total* average electronic kinetic energy is raised; see Section 14.2.) For further details, see M. J. Feinberg and K. Ruedenberg, *J. Chem. Phys.*, **54**, 1495 (1971). Wilson and Goddard have also emphasized the important role of kinetic energy in bonding. [C. W. Wilson and W. A. Goddard, *Chem. Phys. Letters*, **5**, 45 (1970); *Theor. Chim. Acta*, **26**, 195 (1972); W. A. Goddard and C. W. Wilson, *Theor. Chim. Acta*, **26**, 211 (1972).] Further study is needed before the origin of the covalent bond can be considered a settled question.

The $\sigma_u^* 1s$ trial function $1s_a - 1s_b$ is proportional to $e^{-r_a} - e^{-r_b}$. On a plane perpendicular to the internuclear axis and midway between the nuclei, we have $r_a = r_b$, so that this plane is a nodal plane for the $\sigma_u^* 1s$ function. We do not get a buildup of charge between the nuclei for this state, and the $U(R)$ curve has no minimum. We say that the $\sigma_g 1s$ orbital is *bonding* and the $\sigma_u^* 1s$ orbital is *antibonding*. (See Fig. 13.7.)

Reflection of the electron's coordinates in the σ_h symmetry plane perpendicular to the molecular axis and midway between the nuclei converts r_a to r_b and r_b to r_a and leaves φ unchanged [Eq. (13.75)]. The operator $\hat{O}_{\sigma_h}$ (Section 12.1) commutes with the electronic Hamiltonian (13.16) and with the parity (inversion) operator. Hence we can choose the H_2^+ wave functions to be eigenfunctions of this reflection operator as well as of the parity operator. Since the square of this reflection operator is the unit operator, its eigenvalues must be $+1$ and -1 (Section 7.5). States of H_2^+ for which the wave function changes sign upon reflection in this plane (eigenvalue -1) are indicated by a star as a superscript to the letter that specifies λ; states whose wave functions are unchanged on reflection in this plane are left unstarred. Since orbitals with eigenvalue -1 for this reflection have a nodal plane between the nuclei, starred orbitals are antibonding.

Instead of using graphs, we can make contour diagrams of the orbitals (Section 6.6); see Fig. 13.8.

We might compare the preceding treatment of H_2^+ with the perturbation

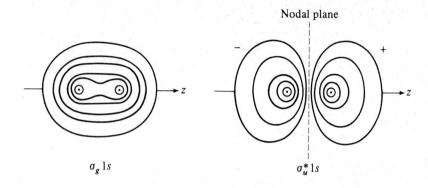

Figure 13.8 *Contours of constant $|\psi|$ for the $\sigma_g 1s$ and $\sigma_u^* 1s$ MO's. The three-dimensional contour surfaces are generated by rotating these figures about the z axis. Note the resemblance of the anti-bonding MO contours to those of a $2p_z$ AO.*

treatment of the helium $1s2s$ levels (Section 9.7). There we started with the degenerate functions $1s(1)2s(2)$ and $1s(2)2s(1)$. Because of the symmetry of the Hamiltonian with respect to interchange of the identical electrons, we found the correct zeroth-order functions to be (9.161) and (9.162). For H_2^+ we started with the degenerate functions $1s_a$ and $1s_b$. Because of the symmetry of the electronic Hamiltonian with respect to the identical nuclei, we found the correct zeroth-order functions to be (13.47) and (13.48).

Sometimes one speaks of a resonance "phenomenon" in H_2^+, saying that the electron resonates back and forth between the $1s$ orbital on nucleus a and the $1s$ orbital on nucleus b. Just as with "exchange" in helium, there is no physical reality to such a "phenomenon." The electron in the ground state of H_2^+ is in neither the $1s_a$ orbital nor the $1s_b$ orbital, but is in a molecular orbital that (in the zeroth-order approximation) is a linear combination of both AO's. A state where the electron "resonated" between different orbitals would be a nonstationary state, whereas we are dealing with the stationary states of the molecule.

Sometimes one attributes the binding in H_2^+ to the resonance integral H_{ab}, since in the approximate treatment we have given, it provides most of the binding energy. This viewpoint is misleading; in the exact treatment of Section 13.3, there arose no such resonance integral. The resonance integral simply arises out of the nature of the LCAO approximation we used.

The resonance integral H_{ab} is often called an exchange integral; however, the following distinction is worth preserving. An exchange integral such as K_{1s2s} in Eq. (9.145) involves two functions that differ from each other in the interchange of two electrons. A resonance integral such as H_{ab} in (13.37) involves only a single electron and two functions that differ from each other in the spatial location of this electron.

In summary, we have formed the two H_2^+ MO's (13.47) and (13.48), one bonding and one antibonding, from the AO's $1s_a$ and $1s_b$. The MO energies are given by Eq. (13.42) as

$$W_{1,2} = H_{aa} \pm \frac{H_{ab} - H_{aa}S_{ab}}{1 \pm S_{ab}} \tag{13.63}$$

where $H_{aa} = \langle 1s_a | \hat{H} | 1s_a \rangle$, with $\hat{H}$ being the purely electronic Hamiltonian of $H_2{}^+$. The integral H_{aa} would be the molecule's purely electronic energy if the electron's wave function in the molecule were $1s_a$. In a sense H_{aa} is the energy of the $1s_a$ orbital in the molecule; in the limit $R = \infty$, H_{aa} becomes the $1s$ AO energy in the H atom. In the molecule, H_{aa} is substantially lower than the electronic energy of an H atom because the electron is attracted to both nuclei. A diagram of MO formation from AO's is given in Fig. 13.21. To get $U(R)$, the electronic energy including nuclear repulsion, we must add $1/R$ to (13.63).

We have described the $H_2{}^+$ states according to the state of the hydrogen atom obtained on dissociation; this is a *separated-atoms description*. Alternatively, we can use the state of the atom formed as the internuclear distance goes to zero; this is a *united-atom description*. We have seen that for the two lowest electronic states of $H_2{}^+$, the united-atom states are the $1s$ and $2p_0$ states of He^+. The united-atom designation is put on the left of the symbol for λ. The $\sigma_g 1s$ state thus has the united-atom designation $1s\sigma_g$. The $\sigma_u^* 1s$ state has the united-atom designation $2p\sigma_u^*$. It is not necessary to write this state as $2p_0 \sigma_u^*$, because the fact that it is a σ state tells us that it correlates with the united-atom $2p_0$ state. For the united-atom description, the subscripts g and u are redundant, in that molecular states correlating with $s, d, g, \ldots$ atomic states must be g, while states correlating with $p, f, h, \ldots$ atomic states must be u. From the separated-atoms states, we cannot tell whether the molecular wave function is g or u. Thus from the $1s$ separated-atoms state, we formed both a g and a u function for $H_2{}^+$.

Before constructing approximate molecular orbitals for other $H_2{}^+$ states, we consider how the trial function (13.47) can be improved. From the viewpoint of perturbation theory, (13.47) is the correct zeroth-order wave function. We know that the perturbation of molecule formation will mix in other hydrogen-atom states besides $1s$. Dickinson in 1933 used a trial function with some $2p_0$ character mixed in (since the ground state of $H_2{}^+$ is a σ state, it would not do to mix in $2p_{\pm 1}$ functions); he took

$$\varphi = [1s_a + c(2p_0)_a] + [1s_b + c(2p_0)_b] \tag{13.64}$$

where c is a variational parameter and where (Table 6.2)

$$1s_a = k^{3/2} \pi^{-1/2} e^{-kr_a}$$

$$(2p_0)_a = (2p_z)_a = \frac{\zeta^{5/2}}{4(2\pi)^{1/2}} r_a e^{-\zeta r_a/2} \cos \theta_a$$

with k and ζ being two other variational parameters. We have similar expressions for $1s_b$ and $(2p_0)_b$. The angles θ_a and θ_b refer to two sets of spherical polar coordinates, one set at each nucleus; see Fig. 13.9. The definitions of θ_a and θ_b correspond to using a right-handed coordinate system on atom a and a left-handed system on atom b; this is the usual convention. Naturally c goes to zero as R goes to either zero or infinity.

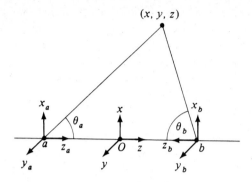

Figure 13.9 *Coordinate systems for a homonuclear diatomic molecule*

The function $(1s + c2p_0)$ is a *hybridized* atomic orbital. Since the $2p_0$ function is positive in one lobe and negative in the other, its inclusion leads to further building up of charge between the nuclei, giving a greater binding energy. The hybridization allows for the polarization of the $1s_a$ and $1s_b$ atomic orbitals that occurs on molecule formation. The function (13.64) gives a $U(R)$ curve with a minimum at 2.01 bohrs. At this distance the parameters have the values [F. Weinhold, *J. Chem. Phys.*, **54**, 530 (1971)]

$$k = 1.246, \qquad \zeta = 2.965, \qquad c = 0.138$$

The binding energy is calculated to be 2.73 eV, in good agreement with the true value 2.79 eV.

Besides the variational functions discussed in this section, many others[9] have been used for the H_2^+ ground state.

One final point. The approximate wave functions in this chapter are written in atomic units. In rewriting these functions in ordinary units, one must remember that wave functions are not dimensionless. A one-particle wave function ψ has units of length$^{-3/2}$. This follows from the fact that $|\psi|^2 \, dx \, dy \, dz$ is a probability and probabilities are dimensionless. The AO's $1s_a$ and $1s_b$ that occur in the functions (13.48) and (13.47) are given by (13.34) in atomic units; in ordinary units $1s_a = (k/a_0)^{3/2}\pi^{-1/2}e^{-kr_a/a_0}$.

13.5 MOLECULAR ORBITALS FOR H_2^+ EXCITED STATES

In the last section, we used the approximate functions (13.47) and (13.48) for the two lowest H_2^+ electronic states. Now we construct approximate functions for further excited states, so as to build up a supply of H_2^+-like molecular orbitals. We

[9] For a partial listing, see M. Geller, A. A. Frost, and P. G. Lykos, *J. Chem. Phys.*, **36**, 2693 (1962); F. Weinhold and A. B. Chinen, *J. Chem. Phys.*, **56**, 3798 (1972).

will then use these MO's to give a qualitative discussion of many-electron diatomic molecules, just as we used hydrogenlike AO's to discuss many-electron atoms.

To get approximations to higher H_2^+ MO's, we can use the method of the linear variation function. We saw that it was natural to take our variation functions for H_2^+ as linear combinations of hydrogenlike atomic-orbital functions, giving LCAO-MO's. To get approximate MO's for higher states, we add in more AO's to the linear combination. Thus to get approximate wave functions for the six lowest H_2^+ σ states, we use a linear combination of the three lowest $m = 0$ hydrogenlike functions on each atom:

$$\varphi = c_1 1s_a + c_2 2s_a + c_3 (2p_0)_a + c_4 1s_b + c_5 2s_b + c_6 (2p_0)_b$$

As found in the last section for the function (13.33), the symmetry of the homonuclear diatomic molecule makes the coefficients of the atom b orbitals equal to ± 1 times the corresponding atom a orbital coefficients:

$$\varphi = [c_1 1s_a + c_2 2s_a + c_3 (2p_0)_a] \pm [c_1 1s_b + c_2 2s_b + c_3 (2p_0)_b] \quad \textbf{(13.65)}$$

where the upper sign goes with the even (g) states.

Consider the relative magnitudes of the coefficients in (13.65). For the two electronic states that dissociate into a $1s$ hydrogen atom, we expect that c_1 will be considerably greater than c_2 or c_3, since c_2 and c_3 vanish in the limit of R going to infinity. Thus the Dickinson function (13.64) had the $2p_0$ coefficient about one-seventh the $1s$ coefficient at R_e. (This function does not include a $2s$ term, but if it did, we no doubt would find its coefficient to be small compared to the $1s$ coefficient.) As a first approximation, we therefore set c_2 and c_3 equal to zero, taking

$$\varphi = c_1 (1s_a \pm 1s_b) \quad \textbf{(13.66)}$$

as an approximation for the wave functions of these two states (as we already have done). The same argument for the two states that dissociate to a $2s$ hydrogen atom gives as approximate wave functions for them

$$\varphi = c_2 (2s_a \pm 2s_b) \quad \textbf{(13.67)}$$

since c_1 and c_3 will be small for these states. The functions (13.67) are only an approximation to what we would find if we carried out the linear variation treatment; to find rigorous upper bounds to the energies of these two H_2^+ states, we must use the trial function (13.65) and solve the appropriate sixth-order determinantal equation.

From the viewpoint of perturbation theory, taking the separated atoms as the unperturbed problem, the functions (13.66) and (13.67) are the correct zeroth-order wave functions.

In general we have two H_2^+ states correlating with each separated-atoms state, and rough approximations to the wave functions of these two states will be the LCAO functions $f_a + f_b$ and $f_a - f_b$, where f is a hydrogenlike wave function. The functions (13.66) give the $\sigma_g 1s$ and $\sigma_u^* 1s$ states. Similarly the functions (13.67) give the $\sigma_g 2s$ and $\sigma_u^* 2s$ molecular orbitals. The outer contour lines for these orbitals are like those for the corresponding MO's made from $1s$ AO's. However, since the $2s$ AO has a nodal sphere while the $1s$ AO does not, each of these MO's has one more nodal surface than the corresponding $\sigma_g 1s$ or $\sigma_u^* 1s$ MO.

Next we have the combinations

$$(2p_0)_a \pm (2p_0)_b = (2p_z)_a \pm (2p_z)_b \qquad (13.68)$$

giving the $\sigma_g 2p$ and $\sigma_u^* 2p$ molecular orbitals (Fig. 13.10); these are σ MO's even though they correlate with $2p$ separated AO's, since they have $m = 0$.

[The preceding discussion is oversimplified. For the hydrogen atom, the $2s$ and $2p$ AO's are degenerate, so that we can expect the correct zeroth-order functions for the $\sigma_g 2s$, $\sigma_u^* 2s$, $\sigma_g 2p$, and $\sigma_u^* 2p$ MO's of H_2^+ to each be mixtures of $2s$ and $2p$ AO's rather than containing only $2s$ or only $2p$ character. For molecules that dissociate into many-electron atoms, the separated-atoms $2s$ and $2p$ AO's are not degenerate but do lie close together in energy; hence the first-order corrections to the wave functions will mix substantial $2s$ character into the $\sigma 2p$ MO's and substantial $2p$ character into the $\sigma 2s$ MO's. Thus the designation of an MO as $\sigma 2s$ or $\sigma 2p$ should not be taken too literally. For H_2^+ and H_2, the united-atom designations of the MO's are preferable to the separated-atoms designations, but we shall use mostly the latter.]

For the other two $2p$ atomic orbitals, we can use either the $2p_{+1}$ and $2p_{-1}$ complex functions or the $2p_x$ and $2p_y$ real functions. If we want MO's that are eigenfunctions of $\hat{L}_z$, we will choose the complex p orbitals. We then have the four MO's

$$(2p_{+1})_a + (2p_{+1})_b \qquad (13.69)$$

$$(2p_{+1})_a - (2p_{+1})_b \qquad (13.70)$$

$$(2p_{-1})_a + (2p_{-1})_b \qquad (13.71)$$

$$(2p_{-1})_a - (2p_{-1})_b \qquad (13.72)$$

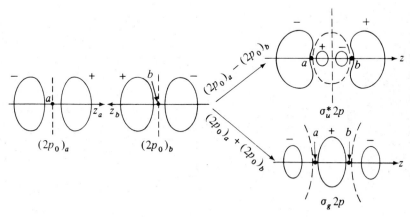

Figure 13.10 *Formation of $\sigma_g 2p$ and $\sigma_u^* 2p$ MO's from $2p_z$ AO's. The dashed lines indicate nodal surfaces. The signs on the contours give the sign of the wave function. The contours are symmetric about the z axis. (Because of substantial $2s$–$2p$ hybridization, these contours are not accurate representations of true MO shapes; for accurate contours see the reference to Fig. 13.19.)*

From Eq. (6.131) we have, since $\varphi_a = \varphi_b = \varphi$,

$$(2p_{+1})_a + (2p_{+1})_b = \tfrac{1}{8}\pi^{-1/2}[r_a e^{-r_a/2} \sin \theta_a + r_b e^{-r_b/2} \sin \theta_b]e^{i\varphi} \quad (13.73)$$

Since $\lambda = |m| = 1$, this is a π orbital. The inversion operation amounts to the coordinate transformation (Fig. 13.11)

$$r_a \rightarrow r_b, \qquad r_b \rightarrow r_a, \qquad \varphi \rightarrow \varphi + \pi \qquad\qquad (13.74)$$

We have

$$e^{i(\varphi+\pi)} = (\cos \pi + i \sin \pi) \, e^{i\varphi} = -e^{i\varphi}$$

From Fig. 13.11 we see that inversion converts θ_a to θ_b and vice versa. Thus inversion converts (13.73) to its negative, meaning it is a u orbital. Reflection in the plane perpendicular to the axis and midway between the nuclei causes the following transformations (Problem 13.14):

$$r_a \rightarrow r_b, \qquad r_b \rightarrow r_a, \qquad \varphi \rightarrow \varphi, \qquad \theta_a \rightarrow \theta_b, \qquad \theta_b \rightarrow \theta_a \qquad (13.75)$$

This leaves (13.73) unchanged, so that we have an unstarred (bonding) orbital. The designation of (13.73) is then $\pi_u 2p_{+1}$.

    The function (13.73) is complex; taking its absolute value, we can plot the orbital contours of constant probability density (Section 6.6). Since $|e^{i\varphi}| = 1$, the probability density is independent of φ, giving a density that is symmetric about the z (internuclear) axis. Figure 13.12 shows a cross section of this orbital in a plane containing the nuclei; the three-dimensional shape is found by rotating this figure about the z axis, creating a sort of fat doughnut. (Sometimes this orbital is pictured as having the shape of a right circular cylinder; this is inaccurate, since the probability density is not constant as we move parallel to the z axis.)

    The molecular orbital (13.71) differs from (13.73) only in having $e^{i\varphi}$ replaced

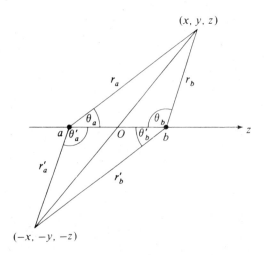

Figure 13.11 *The effect of inversion of the electron's coordinates in* $H_2{}^+$. *We have* $r'_a = r_b$, $r'_b = r_a$, *and* $\varphi' = \varphi + \pi$.

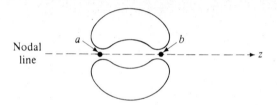

Figure 13.12 *Cross section of the* $\pi_u 2p_{+1}$ *(or* $\pi_u 2p_{-1}$*) molecular orbital. To obtain the three-dimensional contour surface, rotate the figure about the z axis. The z axis is a nodal line for this* MO *(as it is for the* $2p_{+1}$ AO*).*

by $e^{-i\varphi}$ and is designated $\pi_u 2p_{-1}$. The coordinate φ enters the Hamiltonian (13.21) as $\partial^2/\partial\varphi^2$. We have

$$\frac{\partial^2 e^{i\varphi}}{\partial\varphi^2} = \frac{\partial^2 e^{-i\varphi}}{\partial\varphi^2}$$

so that the states (13.69) and (13.71) have the same energy. Recall (Section 13.3) that the $\lambda = 1$ energy levels are doubly degenerate, corresponding to $m = \pm 1$. Since $|e^{i\varphi}| = |e^{-i\varphi}|$, the $\pi_u 2p_{+1}$ and $\pi_u 2p_{-1}$ MO's have the same shapes, just as the $2p_{+1}$ and $2p_{-1}$ AO's have the same shapes.

The functions (13.70) and (13.72) give the $\pi_g^* 2p_{+1}$ and $\pi_g^* 2p_{-1}$ MO's. These functions do not give charge buildup between the nuclei, as can be seen from Fig. 13.13.

Let us now consider the more familiar alternative of using the $2p_x$ and $2p_y$ AO's to make our MO's. The linear combination

$$(2p_x)_a + (2p_x)_b \tag{13.76}$$

gives the $\pi_u 2p_x$ MO (Fig. 13.14). This MO is not symmetric about the internuclear axis but builds up electronic charge probability density in two lobes, one above and one below the yz plane, which is a nodal plane for this function. The wave function has

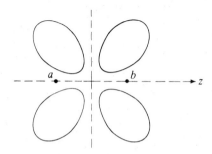

Figure 13.13 *Cross section of the* $\pi_g^* 2p_{+1}$ *(or* $\pi_g^* 2p_{-1}$*)* MO. *To obtain the three-dimensional contour surface, rotate about the z axis. The z axis and the xy plane are nodes.*

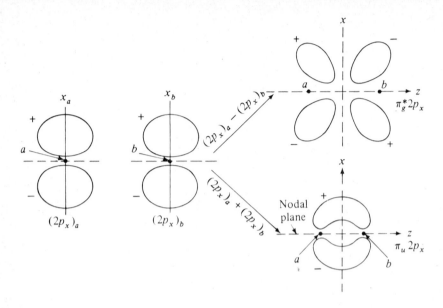

Figure 13.14 *Formation of the $\pi_u 2p_x$ and $\pi_g^* 2p_x$ MO's. Since the coordinate φ has the value zero in the xz plane, the cross sections of these MO's in the xz plane are the same as for the corresponding $\pi_u 2p_{+1}$ and $\pi_g^* 2p_{+1}$ MO's. However, the $\pi 2p_x$ MO's are not symmetrical about the z axis; rather, they consist of blobs of probability density above and below the nodal yz plane.*

opposite signs on each side of this plane. The linear combination

$$(2p_x)_a - (2p_x)_b \qquad (13.77)$$

gives the $\pi_g^* 2p_x$ MO (Fig. 13.14). Since the $2p_y$ functions differ from the $2p_x$ functions solely by a rotation of $90°$ about the internuclear axis, they give molecular orbitals differing from those of Fig. 13.14 by a $90°$ rotation about the z axis. The linear combinations

$$(2p_y)_a + (2p_y)_b \qquad (13.78)$$

$$(2p_y)_a - (2p_y)_b \qquad (13.79)$$

give the $\pi_u 2p_y$ and $\pi_g^* 2p_y$ molecular orbitals. The MO's (13.76) and (13.78) have the same energy; the MO's (13.77) and (13.79) have the same energy. (Note that the g $\pi 2p$ MO's are antibonding, while the u $\pi 2p$ MO's are bonding.)

　　　Just as the $2p_x$ and $2p_y$ AO's are linear combinations of the $2p_{+1}$ and $2p_{-1}$ AO's [Eqs. (6.135) and (6.139)], the $\pi_u 2p_x$ and $\pi_u 2p_y$ MO's are linear combinations of the $\pi_u 2p_{+1}$ and $\pi_u 2p_{-1}$ MO's. We can use any linear combination of the eigenfunctions of a degenerate energy level and still have an energy eigenfunction. Just as the $2p_{+1}$ and $2p_{-1}$ AO's are eigenfunctions of $\hat{L}_z$ and the $2p_x$ and $2p_y$ AO's are not, the $\pi_u 2p_{+1}$ and $\pi_u 2p_{-1}$ MO's are eigenfunctions of $\hat{L}_z$ and the $\pi_u 2p_x$ and $\pi_u 2p_y$ MO's are not. Although the molecular orbitals $\pi_u 2p_{+1}, \pi_u 2p_{-1}, \pi_u 2p_x, \pi_u 2p_y$ all have the same energy, they represent different states of the system, since they have different wave functions. The question arises: Given an H_2^+ molecule in the $\pi_u 2p$ energy level, what is its wave function (state)? The answer is that the state depends on the past history of the system. Thus if we have just measured L_z for a $\pi_u 2p$ H_2^+ molecule and found the result $+\hbar$, then we know that the molecule is in the state $\pi_u 2p_{+1}$. This assertion holds just after the measurement, but not just before. If the state just before the measurement were either $\pi_u 2p_x$ or $\pi_u 2p_y$, then (since these states are superpositions of the $\pi_u 2p_{+1}$ and $\pi_u 2p_{-1}$ states with coefficients equal in magnitude) we would have had a 50 percent chance of finding each of the results $+\hbar$ and $-\hbar$; the measurement of L_z changes the state to either $\pi_u 2p_{+1}$ or $\pi_u 2p_{-1}$, depending on what the outcome of the measurement is (reduction of the wave function—Section 7.8). For the H_2^+ $\pi_u 2p$ energy level, we can use the pair of real MO's (13.76) and (13.78), or the pair of complex MO's (13.69) and (13.71), or any two linearly independent linear combinations of these functions. (It might be argued that since $\hat{L}_z$ commutes with the Hamiltonian for H_2^+, the $\pi_u 2p_{+1}$ and $\pi_u 2p_{-1}$ molecular orbitals are preferable, since they are eigenfunctions of $\hat{L}_z$. However, for degenerate levels there is no necessity that the molecule be in an eigenstate of $\hat{L}_z$; linear combinations that are not eigenfunctions of $\hat{L}_z$ are just as valid stationary states as those that are eigenfunctions of $\hat{L}_z$. For nondegenerate (σ) levels there is no element of choice in the wave function and the wave functions must be eigenfunctions of $\hat{L}_z$.)

We have shown the correlation of the H_2^+ MO's with the separated-atoms AO's. We can also show how they correlate with the united-atom AO's. As R goes to zero, the $\sigma_u^* 1s$ MO (Fig. 13.8) resembles more and more the $2p_z$ AO, with which it correlates. Similarly the $\pi_u 2p$ MO's correlate with p united-atom states, while the $\pi_g^* 2p$ MO's correlate with d united-atom states.

We can treat a molecule by applying perturbation theory to the corresponding united atom, taking the perturbation $\hat{H}'$ as the difference between the molecular and united-atom Hamiltonians:

$$\hat{H}' = \hat{H}_{mol} - \hat{H}_{UA}$$

It has been shown[10] that the expression for the purely electronic energy of a diatomic molecule can be written as

$$E_{el} = E_{UA} + aR^2 + bR^3 + cR^4 + \cdots \tag{13.80}$$

where E_{UA} is the corresponding united-atom energy, R is the internuclear distance, and a, b, c,... are constants. Note the absence of a term linear in R. The matrix elements of $\hat{H}'$ between the various united-atom states turn out to be proportional to R^2 for small values of R; since the second-order perturbation energy correction $E^{(2)}$ [Eq. (9.46)] involves the squares of H'_{ij}, it was thought that $E^{(2)}$ was proportional to R^4 for small R, so that a and b could be evaluated from the first-order energy correction $E^{(1)}$. However, this is not so.[11] Although the individual terms in Eq. (9.46) are proportional to R^4, when the infinite summation over bound states and integration over continuum states is performed, the resulting $E^{(2)}$ is proportional to R^3. (See Problem 13.12.) The series (13.80) converges slowly.

[10] W. A. Bingel, *J. Chem. Phys.*, **30**, 1250 (1959).
[11] I. N. Levine, *J. Chem. Phys.*, **40**, 3444 (1964); **41**, 2044 (1964); W. Byers Brown and E. Steiner, *J. Chem. Phys.*, **44**, 3934 (1966). This last reference shows the presence of an $R^5 \ln R$ term in (13.80).

13.6   MOLECULAR-ORBITAL CONFIGURATIONS OF HOMONUCLEAR DIATOMIC MOLECULES

We now use the H_2^+ MO's developed in the last section to discuss many-electron homonuclear diatomic molecules. If we ignore the interelectronic repulsions, the zeroth-order wave function is a Slater determinant of H_2^+-like one-electron spin-orbitals. We approximate the spatial part of the H_2^+ spin-orbitals by the LCAO-MO's of the last section. Treatments that go beyond this crude first approximation will be discussed later.

The sizes and energies of the MO's vary with varying internuclear distance for each molecule and vary as we go from one molecule to another. Thus we saw how the orbital exponent k in the H_2^+ trial function (13.47) varied with R. As we go to molecules with higher nuclear charge, the parameter k for the $\sigma_g 1s$ MO will increase, giving a more compact MO. We want to consider the order of the MO energies. Because of the variation of these energies with R and variations from molecule to molecule, numerous crossings occur, just as for atomic-orbital energies (Fig. 11.2). Hence we cannot give a definitive order. However, the following is the order in which the MO's fill as we go across the periodic table:

$$\sigma_g 1s < \sigma_u^* 1s < \sigma_g 2s < \sigma_u^* 2s < \pi_u 2p_{+1} = \pi_u 2p_{-1} < \sigma_g 2p < \pi_g^* 2p_{+1}$$

$$= \pi_g^* 2p_{-1} < \sigma_u^* 2p$$

Each bonding orbital fills before the corresponding antibonding orbital. The $\pi_u 2p$ orbitals are close in energy to the $\sigma_g 2p$ orbital, and it was formerly believed that the $\sigma_g 2p$ MO filled first.

Besides the separated-atoms designation, there are other ways of referring to these MO's; see Table 13.1. The second column of this table gives the united-atom designations. The nomenclature of the third column uses $1\sigma_g$ for the lowest σ_g MO, $2\sigma_g$ for the second lowest σ_g MO, etc.

A diagram showing how the MO's correlate with the separated-atoms and

Table 13.1 *Molecular-orbital nomenclature for homonuclear diatomic molecules*

Separated-Atoms Description	United-Atom Description	Numbering by Symmetry
$\sigma_g 1s$	$1s\sigma_g$	$1\sigma_g$
$\sigma_u^* 1s$	$2p\sigma_u^*$	$1\sigma_u$
$\sigma_g 2s$	$2s\sigma_g$	$2\sigma_g$
$\sigma_u^* 2s$	$3p\sigma_u^*$	$2\sigma_u$
$\pi_u 2p$	$2p\pi_u$	$1\pi_u$
$\sigma_g 2p$	$3s\sigma_g$	$3\sigma_g$
$\pi_g^* 2p$	$3d\pi_g^*$	$1\pi_g$
$\sigma_u^* 2p$	$4p\sigma_u^*$	$3\sigma_u$

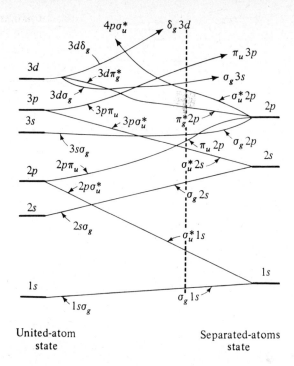

Figure 13.15 *Correlation diagram for homonuclear diatomic* MO'*s. (This diagram doesn't hold for* H_2^+.) *The dashed vertical line corresponds to the order in which the* MO'*s fill.*

united-atom states is shown in Fig. 13.15. Because of the variation of MO energies from molecule to molecule, this diagram is not quantitative.

Recall (Problem 7.16 and Figure 6.12) that s, d, g, ... united-atom AO's are even functions and hence correlate with *gerade* (g) MO's, whereas p, f, h, ... AO's are odd functions and hence correlate with *ungerade* (u) MO's.

A useful principle in drawing orbital correlation diagrams is the *noncrossing rule*, which states that for correlation diagrams of many-electron diatomic molecules, the energies of MO's with the same symmetry cannot cross. For diatomic MO's the word *symmetry* refers to whether the orbital is g or u and whether it is σ, π, δ, ... For example, two σ_g MO's cannot cross on a correlation diagram. From the noncrossing rule, we conclude that the lowest MO of a given symmetry type must correlate with the lowest united-atom AO of that symmetry, and similarly for higher orbitals. [A similar noncrossing rule holds for potential-energy curves $U(R)$ for different electronic states of a many-electron diatomic molecule.] The proof of the noncrossing rule is a bit subtle; see C. A. Mead, *J. Chem. Phys.*, **70**, 2276 (1979) for a thorough discussion.

Just as we discussed atoms by filling in the AO's, giving rise to atomic configurations such as $1s^2 2s^2$, we will discuss homonuclear diatomic molecules by filling in the MO's, giving rise to molecular electronic configurations such as

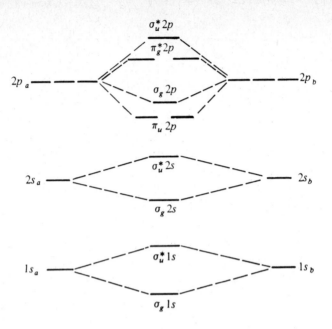

Figure 13.16 *Homonuclear diatomic* MO's *formed from* 1s, 2s, *and* 2p AO's.

$(\sigma_g 1s)^2 (\sigma_u^* 1s)^2$. (Recall that with a single atomic configuration, there is associated a hierarchy of terms, levels, and states; the same is true for a molecular configuration—see Section 13.7.)

Figure 13.16 shows the homonuclear diatomic MO's formed from the 1s, 2s, and 2p AO's.

For H_2^+ we have the ground-state configuration $\sigma_g 1s$, which gives a one-electron bond. For excited states the electron is in one of the higher MO's.

For H_2 we put the two electrons in the $\sigma_g 1s$ MO with opposite spins, giving the ground-state configuration $(\sigma_g 1s)^2$. The two bonding electrons give a single bond. The ground-state dissociation energy D_e is 4.75 eV. In the crudest approximation, the H_2 dissociation energy would be twice the H_2^+ dissociation energy, or 5.6 eV [Eq. (13.28)]. Because of interelectronic repulsion, the true binding energy is somewhat less than this.

Now consider He_2. Two electrons go in the $\sigma_g 1s$ MO, thereby filling it. The other two go in the next highest MO, $\sigma_u^* 1s$. The ground-state configuration is $(\sigma_g 1s)^2 (\sigma_u^* 1s)^2$. With two bonding and two antibonding electrons, we expect no net bonding, in agreement with the well-known fact that the ground electronic state of He_2 is unstable, showing no significant minimum in the potential-energy curve. However, if an electron is excited from the antibonding $\sigma_u^* 1s$ MO to a higher MO that is bonding, we will have three bonding electrons and only one antibonding electron, so that there should be bonding excited states, with a significant minimum in each $U(R)$ curve. Indeed, about two dozen such bound excited states of He_2 have been spectroscopically observed in gas discharge tubes. Of course, such excited states decay to the ground electronic state, and then the molecule dissociates.

The repulsion of two $1s^2$ helium atoms can be ascribed to the Pauli repulsion between electrons with parallel spins (Section 10.4). Each helium atom has a pair of electrons with opposite spin, and each pair tends to exclude the other pair from occupying the same region of space.

Removal of an antibonding electron from He_2 gives the $He_2{}^+$ ion, with ground-state configuration $(\sigma_g 1s)^2 (\sigma_u^* 1s)$; we have one net bonding electron. Ground-state properties of this molecule are quite close to those for $H_2{}^+$; see Table 13.2.

For Li_2 we have the ground-state configuration $(\sigma_g 1s)^2 (\sigma_u^* 1s)^2 (\sigma_g 2s)^2$ with two net bonding electrons, leading to the description of the molecule as containing an Li—Li single bond. Experimentally Li_2 is found to be a stable species. In Li_2 the orbital exponent of the $1s$ AO's is considerably greater than in $H_2{}^+$ or H_2, because of the increase in the nuclear charges from 1 to 3. This shrinks the $1s_a$ and $1s_b$ AO's in closer to the corresponding nuclei; there is thus only very slight overlap between these two AO's, and the integrals S_{ab} and H_{ab} are very small for these AO's. As a result the energies of the $\sigma_g 1s$ and $\sigma_u^* 1s$ MO's in Li_2 are nearly equal to each other and to the energy of a $1s$ Li AO. (For very small R, the $1s_a$ and $1s_b$ AO's do overlap appreciably and their energies then differ considerably.) The Li_2 ground-state configuration is frequently written as $KK(\sigma_g 2s)^2$ to indicate the negligible change in inner-shell orbital energies on molecule formation, which is in accord with the chemist's usual idea of bonding involving only the valence electrons. The orbital exponent of the $2s$ AO's in Li_2 is not much greater than 1, because these electrons are screened from the nucleus by the $1s$ electrons.

Be_2 has the ground-state configuration $KK(\sigma_g 2s)^2 (\sigma_u^* 2s)^2$ with no net bonding electrons.

B_2 has the ground-state configuration $KK(\sigma_g 2s)^2 (\sigma_u^* 2s)^2 (\pi_u 2p)^2$ with two net bonding electrons, indicating a stable ground state, as is found experimentally. The bonding electrons are π electrons, which is at variance with the notion that single bonds are always σ bonds. We have two degenerate $\pi_u 2p$ MO's. Recall that when we had an atomic configuration such as $1s^2 2s^2 2p^2$, we obtained several terms, which because of interelectronic repulsions had different energies. We saw that the term with the highest total spin was generally the lowest (Hund's rule). With the molecular configuration of B_2 given above, we will also have a number of terms. Since the lower (σ) MO's are all filled, their electrons must be paired and contribute nothing to the total spin. If the two $\pi_u 2p$ electrons are both in the same MO (e.g., both in $\pi_u 2p_{+1}$) their spins must be paired (antiparallel), giving a total molecular electronic spin of zero. If, however, we have one electron in the $\pi_u 2p_{+1}$ MO and the other in the $\pi_u 2p_{-1}$ MO, their spins can be parallel, giving a net spin of 1; by Hund's rule, this term will be lowest, and the ground state of B_2 will have $2S + 1 = 3$. Experimental evidence as to the nature of the B_2 ground electronic state is not conclusive, but a highly accurate configuration-interaction calculation [M. Dupuis and B. Liu, J. Chem. Phys., **68**, 2902 (1978)] confirms that the triplet term of the $\ldots(\pi_u 2p)^2$ configuration is in fact the ground state.

C_2 has the ground-state configuration $KK(\sigma_g 2s)^2 (\sigma_u^* 2s)^2 (\pi_u 2p)^4$ with four net bonding electrons, indicating a stable ground state with a double bond. As mentioned, the $\pi_u 2p$ and $\sigma_g 2p$ MO's have nearly the same energy in many molecules. It was formerly believed that the ground-state configuration of C_2 was $KK(\sigma_g 2s)^2(\sigma_u^* 2s)^2(\pi_u 2p)^3(\sigma_g 2p)$. This configuration would allow two of the electrons to have their spins parallel, giving a triplet state; because of the minimization of the repulsion between these two electrons, the triplet state of this configuration was thought to be lower than the $(\pi_u 2p)^4$ configuration, which can give only a singlet state. (Recall that the ground-state configuration of Cr is $3d^5 4s$, rather than $3d^4 4s^2$.) However, experimental evidence[12] shows $(\pi_u 2p)^4$ to be the ground-state configuration of gas-phase C_2.

The N_2 ground-state electronic configuration is

$$KK(\sigma_g 2s)^2(\sigma_u^* 2s)^2(\pi_u 2p)^4(\sigma_g 2p)^2$$

The six net bonding electrons indicate a triple bond, in accord with the Lewis structure $:N\!\equiv\!N:$.

The O_2 ground-state configuration is

$$KK(\sigma_g 2s)^2(\sigma_u^* 2s)^2(\sigma_g 2p)^2(\pi_u 2p)^4(\pi_g^* 2p)^2$$

Spectroscopic evidence indicates that in O_2 (and in F_2) the $\sigma_g 2p$ MO is lower in energy than the $\pi_u 2p$ MO. The four net bonding electrons give a double bond. The $\pi_g^* 2p_{+1}$ and $\pi_g^* 2p_{-1}$ MO's have the same energy, and by putting one electron in each we get a triplet state. By Hund's rule this is the ground state. This explanation of the paramagnetism of O_2 was one of the early triumphs of molecular-orbital theory.

For F_2 the ground-state configuration is

$$KK(\sigma_g 2s)^2(\sigma_u^* 2s)^2(\sigma_g 2p)^2(\pi_u 2p)^4(\pi_g^* 2p)^4$$

The two net bonding electrons give a single bond.

For Ne_2 we have

$$KK(\sigma_g 2s)^2(\sigma_u^* 2s)^2(\sigma_g 2p)^2(\pi_u 2p)^4(\pi_g^* 2p)^4(\sigma_u^* 2p)^2$$

There are no net bonding electrons, and thus there is no chemical bonding, in agreement with experiment.

We can go on to describe homonuclear diatomic molecules of higher atomic number. Thus the lowest electron configuration of Na_2 is $KKLL(\sigma_g 3s)^2$.

Table 13.2 lists D_e and R_e for the ground states of some homonuclear diatomic molecules; it also lists the bond order, which is one-half the difference between the number of bonding and antibonding electrons. The dissociation energy increases and the bond length decreases as the bond order increases. The spectroscopic designations in this table will be explained in the next section.

Bonding MO's produce charge buildup between the nuclei, whereas antibonding MO's produce charge depletion between the nuclei. Hence it is usually true that removal of an electron from a bonding MO decreases D_e, whereas removal of an

[12] E. A. Ballik and D. A. Ramsay, *J. Chem. Phys.*, **31**, 1128 (1959); *Astrophys. J.*, **137**, 61, 84 (1963).

Table 13.2 *Properties of homonuclear diatomic molecules in their ground electronic states*

Molecule	Ground Term	Bond Order	D_e/eV	R_e/Å
$H_2{}^+$	$^2\Sigma_g^+$	$\frac{1}{2}$	2.79	1.06
H_2	$^1\Sigma_g^+$	1	4.75	0.741
$He_2{}^+$	$^2\Sigma_u^+$	$\frac{1}{2}$	2.5	1.08
He_2	$^1\Sigma_g^+$	0	0.0009	3.0
Li_2	$^1\Sigma_g^+$	1	1.07	2.67
Be_2	$^1\Sigma_g^+$	0	0.1	2.5
B_2	$^3\Sigma_g^-$	1	3	1.59
C_2	$^1\Sigma_g^+$	2	6.3	1.24
$N_2{}^+$	$^2\Sigma_g^+$	$2\frac{1}{2}$	8.85	1.12
N_2	$^1\Sigma_g^+$	3	9.91	1.10
$O_2{}^+$	$^2\Pi_g$	$2\frac{1}{2}$	6.78	1.12
O_2	$^3\Sigma_g^-$	2	5.21	1.21
F_2	$^1\Sigma_g^+$	1	1.66	1.41
Ne_2	$^1\Sigma_g^+$	0	0.0036	3.1

Data from K. P. Huber and G. Herzberg, *Constants of Diatomic Molecules* (vol. IV of *Molecular Spectra and Molecular Structure*), Van Nostrand, New York, 1979; and B. Liu and A. D. McLean, *J. Chem. Phys.*, **72**, 3418 (1980).

electron from an antibonding MO increases D_e. (Note that as R decreases in Fig. 13.15, the energies of bonding MO's decrease, while the energies of antibonding MO's increase.) For example, the highest filled MO in N_2 is bonding, and Table 13.2 shows that in going from the ground state of N_2 to that of $N_2{}^+$, the dissociation energy decreases (and the bond length increases). In contrast, the highest filled MO of O_2 is antibonding, and in going from O_2 to $O_2{}^+$, the dissociation energy increases (and R_e decreases). The designation of bonding or antibonding is not relevant to the effect of the electrons on the total energy of the molecule. Energy is always required to ionize a stable molecule, no matter which electron is removed. Hence both bonding and antibonding electrons in a stable molecule decrease the total molecular energy.

If the interaction between two ground-state He atoms were strictly repulsive (as predicted by MO theory), the atoms in He gas would not attract one another at all and the gas would never liquefy. Of course, helium gas can be liquefied. Configuration-interaction calculations and direct experimental evidence from scattering experiments show that as two He atoms approach each other, there is an initial weak attraction, with the potential energy reaching a minimum at 3.0 Å of 0.0009 eV below the separated-atoms energy; at distances less than 3.0 Å, the force becomes increasingly repulsive because of overlap of the electron probability densities. The initial attraction (called a *London* or *dispersion force*) can be explained as follows. At a given instant of time, the two electrons in one He atom will probably not be symmetrically located on opposite sides of the nucleus, so the atom will have an instantaneous dipole moment; this instantaneous dipole moment induces an instantaneous dipole moment in the other atom, and the two instantaneous dipole moments produce an attractive force. The dispersion force results from instantaneous correlation between the motions of the electrons in one atom and the motions of the electrons in the second atom; therefore a configuration-interaction calculation is needed.

(Besides dispersion forces, the other kinds of important attractive intermolecular forces are due to permanent dipole moments and are the dipole–dipole and dipole–induced-dipole forces; see *Levine, Physical Chemistry*, Section 22.10. Except for highly polar molecules, the dispersion force is greater than the dipole forces. The general term for all kinds of intermolecular forces is *van der Waals forces*. The dispersion force clearly increases as the number of electrons increases, so molecular boiling points tend to increase as the molecular weight increases.)

The slight minimum in the potential-energy curve at relatively large intermolecular separation produced by the dispersion force can be sufficiently deep to allow the existence at low temperatures of molecules bound by the dispersion interaction. Such species are called *van der Waals molecules*. For example, argon gas at 100 K has a small concentration of Ar_2 van der Waals molecules for which $D_e = 0.012\,eV$ and $R_e = 3.76\,\text{Å}$; Ar_2 has seven bound vibrational levels ($v = 0$, ..., 6). For He_2 calculations show that the zero-point vibrational energy is approximately equal to the depth of the potential well, and whether a bound He_2 van der Waals molecule exists is uncertain; calculations based on the experimentally determined potential-energy curve of He_2 strongly suggest that He_2 does have a single *very* weakly bound state [see R. Feltgren et al., *J. Chem. Phys.*, **76**, 2360 (1982) and Y. H. Uang and W. C. Stwalley, *J. Chem. Phys.*, **76**, 5069 (1982)].

Examples of diatomic van der Waals molecules and their R_e and D_e values include Ne_2, 3.1 Å, 0.0036 eV; HeNe, 3.2 Å, 0.0012 eV; Ca_2, 4.28 Å, 0.13 eV; Mg_2, 3.89 Å, 0.053 eV. Observed polyatomic van der Waals molecules include $(O_2)_2$, H_2–N_2, Ar–HCl, and $(Cl_2)_2$. Note that for van der Waals bonding, R_e is significantly greater and D_e very substantially less than the corresponding values for chemically bound molecules. For more on van der Waals molecules, see B. L. Blaney and G. E. Ewing, *Ann. Rev. Phys. Chem.*, **27**, 553 (1976).

13.7   MOLECULAR ELECTRONIC TERMS

We now consider the terms arising from a given molecular electronic configuration.

For atoms each set of degenerate atomic orbitals constitutes an atomic subshell. For example, the $2p_{+1}$, $2p_0$, and $2p_{-1}$ AO's constitute the $2p$ subshell. An atomic electronic configuration is defined by giving the number of electrons in each subshell; e.g., $1s^2 2s^2 2p^4$. For molecules each set of degenerate molecular orbitals constitutes a molecular subshell. For example, the $\pi_u 2p_{+1}$ and $\pi_u 2p_{-1}$ MO's constitute the $\pi_u 2p$ subshell. Each σ subshell consists of one MO, while each

$\pi, \delta, \varphi, \ldots$ subshell consists of two MO's; σ subshells are filled with two electrons, while non-σ subshells hold up to four electrons. We define a molecular electronic configuration by giving the number of electrons in each subshell, e.g.,

$$(\sigma_g 1s)^2 (\sigma_u^* 1s)^2 (\sigma_g 2s)^2 (\sigma_u^* 2s)^2 (\pi_u 2p)^3$$

For H_2^+ we saw that although $\hat{L}^2$ does not commute with $\hat{H}$, the operator $\hat{L}_z$ does commute with $\hat{H}$. For a many-electron diatomic molecule, one finds that the operator for the axial component of the total electronic orbital angular momentum commutes with $\hat{H}$. The component of electronic orbital angular momentum along the molecular axis has the possible values $M_L \hbar$, where $M_L = 0, \pm 1, \pm 2, \pm \ldots$. To calculate M_L we simply add algebraically the m's of the individual electrons. Analogous to the symbol λ for the one-electron molecule, Λ is defined as

$$\Lambda \equiv |M_L| \qquad (13.81)$$

The following code is used to indicate the value of Λ:

Λ	0	1	2	3	4
letter	Σ	Π	Δ	Φ	Γ

For $\Lambda \neq 0$, there are two possible values of M_L: $+\Lambda$ and $-\Lambda$; as in H_2^+, the electronic energy depends on M_L^2, so that there is a double degeneracy associated with the two values of M_L. (Note that lowercase letters refer to individual electrons, while capital letters refer to the whole molecule.)

Just as in atoms, the individual electronic spins add vectorially to give a total electronic spin S, whose magnitude has the possible values $\sqrt{S(S+1)}\hbar$, with $S = 0, \frac{1}{2}, 1, \frac{3}{2}, \ldots$. The component of S along an axis has the possible values $M_S \hbar$, where $M_S = S, S-1, \ldots, -S$. As in atoms, the quantity $(2S+1)$ is called the *multiplicity* and is written as a left superscript to the code letter for Λ. Diatomic electronic states that have the same value for Λ and the same value for S are said to belong to the same electronic *term*. We now consider how the terms belonging to a given electronic configuration are derived. (We are assuming Russell–Saunders coupling, which holds for molecules composed of atoms of not-too-high atomic number.)

A filled molecular subshell consists of one or two filled molecular orbitals. The Pauli principle requires that for two electrons in the same molecular orbital, one have $m_s = +\frac{1}{2}$ and the other have $m_s = -\frac{1}{2}$. Hence the quantum number M_S, which is the algebraic sum of the individual m_s values, must be zero for a filled-subshell molecular configuration; therefore we must have $S = 0$ for a configuration containing only filled molecular subshells. A filled σ subshell has two electrons with $m = 0$, so that M_L is zero. A filled π subshell has two electrons with $m = +1$ and two electrons with $m = -1$, so that M_L (which is the algebraic sum of the m's) is zero. The same situation holds for filled $\delta, \varphi, \ldots$ subshells. Thus a closed-subshell molecular configuration has both S and Λ equal to zero and gives rise to a single $^1\Sigma$ term. An example is the ground electronic configuration of H_2. (Recall that a filled-subshell atomic configuration gives only a ^1S term.) In deriving molecular terms, we need only consider electrons outside filled subshells.

A single σ electron has $s = \frac{1}{2}$, so S must be $\frac{1}{2}$, and we get a $^2\Sigma$ term. An example is the ground electronic configuration of $H_2{}^+$. A single π electron gives a $^2\Pi$ term. And so on.

Now consider more than one electron. Electrons that are in different subshells are called *nonequivalent*, just as for atoms. For such electrons we do not have to worry about giving two of them the same set of quantum numbers, and the terms are easily derived. Consider two nonequivalent σ electrons, a $\sigma\sigma$ configuration. Since both m's are zero, we have $M_L = 0$. Each s is $\frac{1}{2}$, so S can be 1 or 0. We thus have the terms $^1\Sigma$ and $^3\Sigma$. Similarly a $\sigma\pi$ configuration gives $^1\Pi$ and $^3\Pi$ terms.

For a $\pi\delta$ configuration, we have singlet and triplet terms. The π electron can have $m = \pm 1$, and the δ electron can have $m = \pm 2$. The possible values for M_L are thus $+3$, -3, $+1$, and -1. This gives $\Lambda = 3$ or 1, and we have the terms $^1\Pi$, $^3\Pi$, $^1\Phi$, $^3\Phi$. (In atoms we add the *vectors* $\mathbf{L}_i$ to get the total $\mathbf{L}$; hence a *pd* atomic configuration gives P, D, and F terms. In molecules, however, we add the z *components* of the orbital angular momenta; this is an algebraic rather than a vectorial addition, so a $\pi\delta$ molecular configuration gives Π and Φ terms and no Δ terms.)

For a $\pi\pi$ configuration of two nonequivalent electrons, each electron has $m = \pm 1$, and we have the M_L values 2, -2, 0, 0. The values of Λ are 2, 0, and 0; the terms are $^1\Delta$, $^3\Delta$, $^1\Sigma$, $^3\Sigma$, $^1\Sigma$, and $^3\Sigma$. The values $+2$ and -2 correspond to the two degenerate states of the same Δ term. However, Σ terms are nondegenerate (apart from spin degeneracy), and the two values of M_L that are zero indicate two different Σ terms (which become four Σ terms when we consider spin).

Consider the forms of the wave functions for the terms. We shall call the two π subshells π and π', and shall use a subscript to indicate the m value. For the Δ terms, both electrons have $m = +1$ or both have $m = -1$. For $M_L = +2$, we might write as the spatial factor in the wave function $\pi_{+1}(1)\pi'_{+1}(2)$ or $\pi_{+1}(2)\pi'_{+1}(1)$. However, these functions are neither symmetric nor antisymmetric with respect to exchange of the indistinguishable electrons and are unacceptable. Instead, we must take the two linear combinations (we will not bother with normalization constants)

$$^1\Delta: \quad \pi_{+1}(1)\pi'_{+1}(2) + \pi_{+1}(2)\pi'_{+1}(1) \qquad (13.82)$$

$$^3\Delta: \quad \pi_{+1}(1)\pi'_{+1}(2) - \pi_{+1}(2)\pi'_{+1}(1) \qquad (13.83)$$

Similarly, with both electrons having $m = -1$, we have the spatial factors

$$^1\Delta: \quad \pi_{-1}(1)\pi'_{-1}(2) + \pi_{-1}(2)\pi'_{-1}(1) \qquad (13.84)$$

$$^3\Delta: \quad \pi_{-1}(1)\pi'_{-1}(2) - \pi_{-1}(2)\pi'_{-1}(1) \qquad (13.85)$$

The functions (13.82) and (13.84) are symmetric with respect to exchange; they therefore go with the antisymmetric two-electron spin factor (11.60), which has $S = 0$; thus (13.82) and (13.84) are the spatial factors in the wave functions for the two states of the doubly degenerate $^1\Delta$ term. The antisymmetric functions (13.83) and (13.85) must go with the symmetric two-electron spin functions (11.57), (11.58), and (11.59), giving the six states of the $^3\Delta$ term. These states all have the same energy (if we neglect spin–orbit interaction).

Now consider the wave functions of the Σ terms. These have one electron with $m = +1$ and one electron with $m = -1$. We start with the four functions

$$\pi_{+1}(1)\pi'_{-1}(2) \qquad \pi_{+1}(2)\pi'_{-1}(1) \qquad \pi_{-1}(1)\pi'_{+1}(2) \qquad \pi_{-1}(2)\pi'_{+1}(1)$$

Combining them to get symmetric and antisymmetric functions, we have

$^1\Sigma^+$: $\quad \pi_{+1}(1)\pi'_{-1}(2) + \pi_{+1}(2)\pi'_{-1}(1) + \pi_{-1}(1)\pi'_{+1}(2) + \pi_{-1}(2)\pi'_{+1}(1)$

$^1\Sigma^-$: $\quad \pi_{+1}(1)\pi'_{-1}(2) + \pi_{+1}(2)\pi'_{-1}(1) - \pi_{-1}(1)\pi'_{+1}(2) - \pi_{-1}(2)\pi'_{+1}(1)$

$^3\Sigma^+$: $\quad \pi_{+1}(1)\pi'_{-1}(2) - \pi_{+1}(2)\pi'_{-1}(1) + \pi_{-1}(1)\pi'_{+1}(2) - \pi_{-1}(2)\pi'_{+1}(1)$ $\qquad$ **(13.86)**

$^3\Sigma^-$: $\quad \pi_{+1}(1)\pi'_{-1}(2) - \pi_{+1}(2)\pi'_{-1}(1) - \pi_{-1}(1)\pi'_{+1}(2) + \pi_{-1}(2)\pi'_{+1}(1)$

The first two functions in (13.86) are symmetric; they therefore go with the antisymmetric singlet spin function (11.60). Clearly these two spatial functions have different energies. The last two functions in (13.86) are antisymmetric and hence are the spatial factors in the wave functions of the two $^3\Sigma$ terms. These four functions are found to have eigenvalue $+1$ or -1 with respect to reflection of electronic coordinates in the xz σ_v symmetry plane containing the molecular (z) axis (Problem 13.31); the superscripts $+$ and $-$ refer to this eigenvalue.

Examination of the Δ terms (13.82)–(13.85) shows that they are not eigenfunctions of $\hat{O}_{\sigma_v}$. Since there is a twofold degeneracy (apart from spin degeneracy) associated with these terms, there is no necessity that their wave functions be eigenfunctions of this operator; however, since $\hat{O}_{\sigma_v}$ commutes with the Hamiltonian, we can *choose* the eigenfunctions to be eigenfunctions of $\hat{O}_{\sigma_v}$. Thus we can combine the functions (13.82) and (13.84), which belong to a degenerate energy level, as follows:

$$(13.82) + (13.84) \qquad \text{and} \qquad (13.82) - (13.84)$$

These two linear combinations are eigenfunctions of $\hat{O}_{\sigma_v}$ with eigenvalues $+1$ and -1, and we could refer to them as $^1\Delta^+$ and $^1\Delta^-$ states. Since they have the same energy, there is no point in using the $+$ and $-$ superscripts. Thus the $+$ and $-$ designations are used only for Σ terms. However, when one considers the interaction between the molecular rotational angular momentum and the electronic orbital angular momentum, there is a very slight splitting (called Λ-type doubling) of the two states of a $^1\Delta$ term; it turns out that the correct zeroth-order wave functions for this perturbation are the linear combinations that are eigenfunctions of $\hat{O}_{\sigma_v}$, so in this case there is a point to distinguishing between Δ^+ and Δ^- states. Note that the linear combinations $(13.82) \pm (13.84)$, which are eigenfunctions of $\hat{O}_{\sigma_v}$, are not eigenfunctions of $\hat{L}_z$ but are superpositions of $\hat{L}_z$ eigenfunctions with eigenvalues $+2$ and -2.

We can distinguish $+$ and $-$ terms for one-electron configurations. The wave function of a single σ electron has no phi factor and hence must correspond to a Σ^+ term. For a π electron, the MO's that are eigenfunctions of $\hat{L}_z$ are the π_{+1} and π_{-1} functions (whose probability densities are each symmetric about the z axis— Fig. 13.12). The π_{+1} and π_{-1} functions are not eigenfunctions of $\hat{O}_{\sigma_v}$, but the linear combinations $\pi_{+1} + \pi_{-1} = \pi_x$ and $\pi_{+1} - \pi_{-1} = \pi_y$ are; the π_x and π_y MO's (whose probability densities are not symmetric about the z axis—Fig. 13.14) are the correct zeroth-order functions if the perturbation of the electronic wave functions due to molecular rotation is considered. The π_x and π_y MO's have eigenvalues $+1$ and -1, respectively, for reflection in the xz plane and eigenvalues -1 and $+1$, respectively, for reflection in the yz plane. (The operators $\hat{L}_z$ and $\hat{O}_{\sigma_v}$ do not commute—Problem 13.33. Hence we cannot have all the eigenfunctions of $\hat{H}$ being eigenfunctions of both these operators as well. However, since each of these operators commutes with the electronic Hamiltonian and since there is no element of choice in the wave function of a nondegenerate level, all the σ MO's must be eigenfunctions of both $\hat{L}_z$ and $\hat{O}_{\sigma_v}$.)

Electrons in the same molecular subshell are called *equivalent*. For equivalent electrons, there are fewer terms than for the corresponding nonequivalent electron configuration, because of the Pauli principle. Thus for a π^2 configuration of two equivalent π electrons, four of the eight functions (13.82)–(13.86) vanish; the remaining functions give a $^1\Delta$ term, a $^1\Sigma^+$ term, and a $^3\Sigma^-$ term. Alternatively, we can make a table similar to Table 11.1, and use it to derive the terms for equivalent electrons.

Table 13.3 lists terms arising from various electron configurations. A filled subshell always gives the single term $^1\Sigma^+$. A π^3 configuration gives the same result as a π configuration, since we can view things in terms of "holes," instead of electrons.

For homonuclear diatomic molecules, a g or u right subscript is added to the term symbol to show the parity of the electronic states belonging to the term. Terms arising from an electron configuration that has an odd number of electrons in molecular orbitals of odd parity are odd; all other terms are even. (This is the same rule as for atoms.)

The term symbols given in Table 13.2 are readily derived from the MO configurations. For example, O_2 has a π^2 configuration, which gives the three terms $^1\Sigma_g^+$, $^3\Sigma_g^-$, and $^1\Delta_g$. Hund's rule tells us that $^3\Sigma_g^-$ is the lowest term, as listed. The $v=0$ levels of the $^1\Delta_g$ and $^1\Sigma_g^+$ O_2 terms lie 0.98 eV and 1.6 eV, respectively, above the $v=0$ level of the ground $^3\Sigma_g^-$ term. Singlet O_2 is a reaction intermediate in many organic, biochemical, and inorganic reactions. [See H. H. Wasserman and R. W.

Table 13.3 *Molecular electronic terms*

Configuration	Terms
$\sigma\sigma$	$^1\Sigma^+$, $^3\Sigma^+$
$\sigma\pi$	$^1\Pi$, $^3\Pi$
$\pi\pi$	$^1\Sigma^+$, $^3\Sigma^+$, $^1\Sigma^-$, $^3\Sigma^-$, $^1\Delta$, $^3\Delta$
$\pi\delta$	$^1\Pi$, $^3\Pi$, $^1\Phi$, $^3\Phi$
σ	$^2\Sigma^+$
σ^2; π^4; δ^4	$^1\Sigma^+$
π; π^3	$^2\Pi$
π^2	$^1\Sigma^+$, $^3\Sigma^-$, $^1\Delta$
δ; δ^3	$^2\Delta$
δ^2	$^1\Sigma^+$, $^3\Sigma^-$, $^1\Gamma$

Murray, eds., *Singlet Oxygen*, Academic Press, New York, 1979; B. Ranby and J. F. Rabeck, eds., *Singlet Oxygen*, Wiley, New York, 1978.]

Nearly all stable diatomic molecules have a $^1\Sigma^+$ ground term ($^1\Sigma_g^+$ for homonuclear diatomics). Exceptions include B_2 and O_2, and NO, which has a $^2\Pi$ ground term.

Spectroscopists prefix the ground term of a molecule by the symbol X. Excited terms of the same multiplicity as the ground term are designated as A, B, $C, \ldots$, while excited terms of different multiplicity from the ground term are designated as $a, b, c, \ldots$. Exceptions are C_2 and N_2, where the ground terms are $^1\Sigma_g^+$ but the letters $A, B, C, \ldots$ are used for excited triplet terms.

Just as for atoms, spin–orbit interaction can split a molecular term into closely spaced energy levels, giving a multiplet structure to the term. The projection of the total electronic spin S on the molecular axis is $M_S\hbar$; in molecules the quantum number M_S is called Σ (not to be confused with the symbol meaning $\Lambda = 0$):

$$\Sigma = S, S-1, \ldots, -S$$

The axial components of electronic orbital and spin angular momenta add, giving as the total axial component of electronic angular momentum $(\Lambda + \Sigma)\hbar$. (Recall that Λ is the absolute value of M_L; we consider Σ to be positive when it has the same direction as Λ, and negative when it has the opposite direction as Λ.) The possible values of $\Lambda + \Sigma$ are

$$\Lambda + S, \Lambda + S - 1, \ldots, \Lambda - S$$

The value of $\Lambda + \Sigma$ is written as a right subscript to the term symbol to distinguish the energy levels of the term. Thus a $^3\Delta$ term has $\Lambda = 2$ and $S = 1$ and gives rise to the levels $^3\Delta_3$, $^3\Delta_2$, and $^3\Delta_1$. In a sense $\Lambda + \Sigma$ is the analog in molecules of the quantum number J in atoms. However, $\Lambda + \Sigma$ is the quantum number of the z *component* of total electronic angular momentum and therefore can take on negative values. Thus a $^4\Pi$ term has the four levels $^4\Pi_{5/2}$, $^4\Pi_{3/2}$, $^4\Pi_{1/2}$, and $^4\Pi_{-1/2}$. The absolute value of $\Lambda + \Sigma$ is called Ω:

$$\Omega \equiv |\Lambda + \Sigma| \tag{13.87}$$

It is the value of $\Lambda + \Sigma$, and not the value of Ω, which is written as a subscript on the term symbol.

The spin–orbit interaction energy in diatomic molecules can be shown to be well approximated by $A\Lambda\Sigma$, where A depends on Λ and on the internuclear distance R but not on Σ. The spacing between levels of the multiplet is thus constant. When A is positive, the level with the lowest value of $\Lambda + \Sigma$ lies lowest, and the multiplet is *regular*. When A is negative, the multiplet is *inverted*. Note that for $\Lambda \neq 0$, the multiplicity $(2S+1)$ always equals the number of multiplet components; this is not always true for atoms.

Each energy level of a multiplet with $\Lambda \neq 0$ is doubly degenerate, corresponding to the two values for M_L. Thus a $^3\Delta$ term has six different wave functions [Eqs. (13.83), (13.85), (11.57)–(11.59)] and therefore six different molecular electronic states. Spin–orbit interaction splits the $^3\Delta$ term into three levels, each doubly degenerate. (The double degeneracy of the levels is removed by the Λ-type doubling mentioned previously.)

For Σ terms ($\Lambda = 0$), the spin–orbit interaction is very small (zero in the first approximation), and the quantum numbers Σ and Ω are not defined.

A $^1\Sigma$ term always corresponds to a single nondegenerate energy level.

13.8   *THE HYDROGEN MOLECULE*

The hydrogen molecule is the simplest molecule containing an electron-pair bond and is probably the most studied molecule in quantum chemistry. Many of the ideas applied in treating H_2 are used in discussing more complicated molecules. We will devote most of our attention to the ground electronic state.

The purely electronic Hamiltonian for H_2 is

$$\hat{H} = -\tfrac{1}{2}\nabla_1^2 - \tfrac{1}{2}\nabla_2^2 - \frac{1}{r_{a1}} - \frac{1}{r_{a2}} - \frac{1}{r_{b1}} - \frac{1}{r_{b2}} + \frac{1}{r_{12}} \tag{13.88}$$

where 1 and 2 are the electrons and a and b are the nuclei (Fig. 13.17). Just as in the helium atom, the $1/r_{12}$ interelectronic-repulsion term prevents the Schroedinger equation from being separable. We therefore use approximation methods.

We start with the molecular-orbital approach. The ground-state electronic configuration of H_2 is $(\sigma_g 1s)^2$, and we can write an approximate wave function as the Slater determinant

$$\frac{1}{\sqrt{2}} \begin{vmatrix} \sigma_g 1s(1)\alpha(1) & \sigma_g 1s(1)\beta(1) \\ \sigma_g 1s(2)\alpha(2) & \sigma_g 1s(2)\beta(2) \end{vmatrix} = \sigma_g 1s(1)\sigma_g 1s(2)2^{-1/2}[\alpha(1)\beta(2) - \beta(1)\alpha(2)]$$

$$= f(1)f(2)2^{-1/2}[\alpha(1)\beta(2) - \beta(1)\alpha(2)] \tag{13.89}$$

which is similar to (10.41) for the helium atom. To save time we write f instead of $\sigma_g 1s$. As we saw in Section 10.5, omission of the spin factor does not affect the variational integral for a two-electron problem. Hence we want to choose f so as to minimize

$$\frac{\int\int f^*(1) f^*(2) \hat{H} f(1) f(2)\, dv_1\, dv_2}{\int\int |f(1)|^2 |f(2)|^2\, dv_1\, dv_2} \tag{13.90}$$

where the integration is over the space coordinates of the two electrons. Ideally f should be found by an SCF calculation. For simplicity we can use an H_2^+-like MO. (The H_2 Hamiltonian becomes the sum of two H_2^+ Hamiltonians if we omit the $1/r_{12}$ term.) We saw in Section 13.4 that the function [Eq. (13.47)]

$$\frac{k^{3/2}}{(2\pi)^{1/2}(1+S_{ab})^{1/2}} (e^{-kr_a} + e^{-kr_b})$$

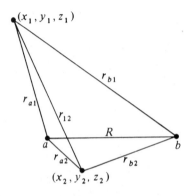

Figure 13.17 The hydrogen molecule

tal D_e/hc was $(38{,}293 \pm 1)\,\text{cm}^{-1}$, and the $4\,\text{cm}^{-1}$ discrepancy caused much consternation among theoreticians, who kept looking for additional theoretical corrections, without success. Finally a reinvestigation [G. Herzberg, *J. Mol. Spectry.*, **33**, 147 (1970)] of the H$_2$ spectrum gave a new experimental value in full agreement with the Kolos–Wolniewicz value, thereby vindicating the theoreticians.

13.9   *THE VALENCE-BOND TREATMENT OF* H$_2$

The first quantum-mechanical treatment of the hydrogen molecule was by Heitler and London in 1927. Their ideas have been extended to give a general theory of chemical bonding, known as the *valence-bond* (VB) theory, or the HLSP theory, after Heitler, London, Slater, and Pauling. The valence-bond method is more closely related to the chemist's idea of molecules as consisting of atoms held together by localized bonds than is the molecular-orbital method. The VB method views molecules as composed of atomic cores (nuclei plus inner-shell electrons) and bonding valence electrons. For H$_2$ both electrons are valence electrons.

The first step in the Heitler–London treatment of the H$_2$ ground state is to approximate the molecule as two ground-state hydrogen atoms. The wave function for two such noninteracting atoms is

$$f_1 = 1s_a(1)1s_b(2) \qquad\qquad (13.97)$$

where a and b refer to the nuclei and where 1 and 2 refer to the electrons. Of course, the function

$$f_2 = 1s_a(2)\,1s_b(1) \qquad\qquad (13.98)$$

is also a valid wave function. This then suggests the trial variation function

$$c_1 f_1 + c_2 f_2 = c_1\,1s_a(1)1s_b(2) + c_2\,1s_a(2)1s_b(1) \qquad\qquad (13.99)$$

This is a linear variation function, which leads to the determinantal equation (8.87), where

$$H_{11} = \int f_1^* \hat{H} f_1 \, dv, \qquad S_{11} = \int f_1^* f_1 \, dv, \qquad \text{etc.} \qquad (13.100)$$

We can also consider the problem using perturbation theory (as Heitler and London did). A ground-state hydrogen molecule dissociates to two neutral ground-state hydrogen atoms. We therefore take as our unperturbed problem two ground-state hydrogen atoms at infinite separation. One possible zeroth-order (unperturbed) wave function is $1s_a(1)1s_b(2)$; however, electron 2 could just as well be bound to nucleus a, giving the unperturbed wave function $1s_a(2)1s_b(1)$. These two unperturbed wave functions belong to a doubly degenerate energy level (exchange degeneracy). Under the perturbation of molecule formation, the doubly degenerate level is split into two levels, and the *correct* zeroth-order wave functions are linear combinations of the two unperturbed wave functions:

$$c_1\,1s_a(1)1s_b(2) + c_2\,1s_a(2)1s_b(1)$$

This leads to a 2×2 secular determinant that is the same as (8.87), except that W is replaced by $E^{(0)} + E^{(1)}$; see Problem 9.6.

We now solve (8.87). The Hamiltonian is Hermitian, all functions are real, and f_1 and f_2 are normalized; hence

$$H_{12} = H_{21}, \qquad S_{12} = S_{21}, \qquad S_{11} = S_{22} = 1 \qquad \text{(13.101)}$$

Consider H_{11} and H_{22}:

$$H_{11} = \langle 1s_a(1)1s_b(2)|\hat{H}|1s_a(1)1s_b(2) \rangle$$

$$H_{22} = \langle 1s_a(2)1s_b(1)|\hat{H}|1s_a(2)1s_b(1) \rangle$$

Interchange of the coordinate labels 1 and 2 in H_{22} converts H_{22} to H_{11}, since this relabeling leaves $\hat{H}$ unchanged. Hence $H_{11} = H_{22}$. The secular equation (8.87) becomes

$$\begin{vmatrix} H_{11} - W & H_{12} - WS_{12} \\ H_{12} - WS_{12} & H_{11} - W \end{vmatrix} = 0 \qquad \text{(13.102)}$$

This equation has the same form as Eq. (13.40), and by analogy to Eqs. (13.42), (13.47), and (13.48) the approximate energies and wave functions are

$$W_1 = \frac{H_{11} + H_{12}}{1 + S_{12}}, \qquad W_2 = \frac{H_{11} - H_{12}}{1 - S_{12}} \qquad \text{(13.103)}$$

$$\varphi_1 = \frac{f_1 + f_2}{\sqrt{2}(1 + S_{12})^{1/2}}, \qquad \varphi_2 = \frac{f_1 - f_2}{\sqrt{2}(1 - S_{12})^{1/2}} \qquad \text{(13.104)}$$

For S_{12} we have

$$S_{12} = \int f_1^* f_2 \, dv = \iint 1s_a(1)1s_b(2)1s_a(2)1s_b(1) \, dv_1 \, dv_2$$

$$S_{12} = \langle 1s_a(1)|1s_b(1) \rangle \langle 1s_a(2)|1s_b(2) \rangle = S_{ab}^2 \qquad \text{(13.105)}$$

where the overlap integral S_{ab} is defined by (13.39).

The numerators of (13.104) are

$$f_1 \pm f_2 = 1s_a(1)1s_b(2) \pm 1s_a(2)1s_b(1) \qquad \text{(13.106)}$$

From our previous discussion in this chapter, we know that the ground state of H_2 is a $^1\Sigma$ state with the antisymmetric spin factor (11.60) and a symmetric spatial factor. Hence φ_1 must be the ground state. The Heitler–London ground-state wave function is

$$\frac{1s_a(1)1s_b(2) + 1s_a(2)1s_b(1)}{\sqrt{2}(1 + S_{ab}^2)^{1/2}} \frac{1}{\sqrt{2}}[\alpha(1)\beta(2) - \alpha(2)\beta(1)] \qquad \text{(13.107)}$$

The Heitler–London wave functions for the three states of the lowest $^3\Sigma$ term are

$$\frac{1s_a(1)1s_b(2) - 1s_a(2)1s_b(1)}{\sqrt{2}(1 - S_{ab}^2)^{1/2}} \begin{cases} \alpha(1)\alpha(2) \\ 2^{-1/2}[\alpha(1)\beta(2) + \beta(1)\alpha(2)] \\ \beta(1)\beta(2) \end{cases} \qquad \text{(13.108)}$$

Now consider the ground-state energy expression. We divide the molecular electronic Hamiltonian into the sum of two hydrogen-atom Hamiltonians plus perturbing terms:

$$\hat{H} = \hat{H}_a(1) + \hat{H}_b(2) + \hat{H}' \tag{13.109}$$

$$\hat{H}_a(1) = -\tfrac{1}{2}\nabla_1^2 - \frac{1}{r_{a1}}, \qquad \hat{H}_b(2) = -\tfrac{1}{2}\nabla_2^2 - \frac{1}{r_{b2}} \tag{13.110}$$

$$\hat{H}' = -\frac{1}{r_{b1}} - \frac{1}{r_{a2}} + \frac{1}{r_{12}} \tag{13.111}$$

We then have

$$H_{11} = \langle 1s_a(1)1s_b(2)|\hat{H}_a(1) + \hat{H}_b(2) + \hat{H}'|1s_a(1)1s_b(2)\rangle \tag{13.111}$$

For the integral involving $\hat{H}_a(1)$, we have

$$\langle 1s_a(1)1s_b(2)|\hat{H}_a(1)|1s_a(1)1s_b(2)\rangle = \langle 1s_a(1)|\hat{H}_a(1)|1s_a(1)\rangle \langle 1s_b(2)|1s_b(2)\rangle$$

The Heitler–London calculation does not introduce an effective nuclear charge into the $1s$ function; hence $1s_a(1)$ is an eigenfunction of $\hat{H}_a(1)$ with eigenvalue $-\tfrac{1}{2}$ hartree, the hydrogen-atom ground-state energy. The $1s$ function is normalized, and we conclude that the $\hat{H}_a(1)$ integral equals $-\tfrac{1}{2}$ in atomic units. Similarly the $\hat{H}_b(2)$ integral in (13.111) equals $-\tfrac{1}{2}$. Defining the Coulomb integral Q as

$$Q \equiv \langle 1s_a(1)1s_b(2)|\hat{H}'|1s_a(1)1s_b(2)\rangle \tag{13.112}$$

we have

$$H_{11} = Q - 1 \tag{13.113}$$

For H_{12} we have

$$H_{12} = H_{21} = \langle 1s_a(2)1s_b(1)|\hat{H}_a(1) + \hat{H}_b(2) + \hat{H}'|1s_a(1)1s_b(2)\rangle \tag{13.114}$$

The $\hat{H}_a(1)$ integral is easily evaluated as

$$\langle 1s_a(2)|1s_b(2)\rangle \langle 1s_b(1)|\hat{H}_a(1)|1s_a(1)\rangle = -\tfrac{1}{2}S_{ab}^2$$

Defining the exchange integral A as

$$A \equiv \langle 1s_a(2)1s_b(1)|\hat{H}'|1s_a(1)1s_b(2)\rangle \tag{13.115}$$

we have

$$H_{12} = A - S_{ab}^2 \tag{13.116}$$

Substitution into (13.103) gives

$$W_1 = -1 + \frac{Q+A}{1+S_{ab}^2} \tag{13.117}$$

$$W_2 = -1 + \frac{Q-A}{1-S_{ab}^2} \tag{13.118}$$

The quantity -1 hartree in these expressions is the energy of two ground-state hydrogen atoms. To obtain the $U(R)$ potential-energy curves, we add the internuclear repulsion $1/R$ to these expressions.

Many of the integrals needed to evaluate W_1 and W_2 have been evaluated in the treatment of H_2^+ in Section 13.4; the only new integrals are those involving $1/r_{12}$. The hardest one is the two-center two-electron exchange integral:

$$\iint 1s_a(1)1s_b(2)\frac{1}{r_{12}}1s_a(2)1s_b(1)\,dv_1\,dv_2$$

(*Two-center* means that the integrand contains functions centered on two different nuclei, a and b; *two-electron* means that the coordinates of two electrons occur in the integrand.) This must be evaluated using an expansion for $1/r_{12}$ in confocal elliptic coordinates, similar to the expansion (9.64) in spherical polar coordinates. (Heitler and London only approximated this integral. Its exact evaluation was given by Sugiura.) We omit the integral evaluations.[15] The results of the Heitler–London treatment are $D_e = 3.20$ eV, $R_e = 0.80$ Å. The agreement with the experimental values $D_e = 4.75$ eV, $R_e = 0.74$ Å is only fair. In this treatment, most of the binding energy is provided by the exchange integral A. However, as discussed for helium, "exchange" is not a real physical process. The Heitler–London treatment of H_2 bears some resemblance to the Heisenberg treatment of the helium $1s2s$ configuration given in Section 9.7. However, for H_2 the exchange integral is negative, whereas for helium it is positive.

Consider some improvements on the Heitler–London function (13.107). One obvious step is the introduction of an orbital exponent ζ in the $1s$ function. This was done by Wang in 1928. The optimum value of ζ is 1.166 at R_e, and the binding energy is improved to 3.78 eV. Recall that Dickinson in 1933 improved the Finkelstein–Horowitz H_2^+ trial function by mixing in some $2p_z$ character into the atomic orbitals (hybridization). In 1931 Rosen used this idea to improve the Heitler–London–Wang function. He took the trial function

$$\varphi = \varphi_a(1)\varphi_b(2) + \varphi_a(2)\varphi_b(1)$$

where the atomic orbital φ_a is given by $\varphi_a = e^{-\zeta r_a}(1 + cz_a)$, with a similar expression for φ_b. This allows for the polarization of the AO's on molecule formation. The result is a binding energy of 4.04 eV. Another improvement, the use of ionic structures, will be considered in the next section.

13.10 PRELIMINARY COMPARISON OF THE MO AND VB THEORIES

Let us compare the molecular-orbital and valence-bond treatments of the hydrogen-molecule ground state.

If φ_a symbolizes an atomic orbital centered on nucleus a, the spatial factor of the (unnormalized) LCAO-MO wave function for the H_2 ground state is

$$[\varphi_a(1) + \varphi_b(1)][\varphi_a(2) + \varphi_b(2)] \tag{13.119}$$

In the simplest treatment, φ is a $1s$ AO. The function (13.119) equals

$$\varphi_a(1)\varphi_a(2) + \varphi_b(1)\varphi_b(2) + \varphi_a(1)\varphi_b(2) + \varphi_b(1)\varphi_a(2) \tag{13.120}$$

---

[15] See *Slater, Quantum Theory of Molecules and Solids*, vol. 1, Appendix 6.

What is the physical significance of the terms? The last two terms have each electron in an atomic orbital centered on a different nucleus; these are covalent terms, corresponding to equal sharing of the electrons between the atoms. The first two terms have both electrons in AO's centered on the same nucleus; these are ionic terms, corresponding to the chemical structures

$$H^- \; H^+ \quad \text{and} \quad H^+ \; H^- \tag{13.121}$$

The covalent and ionic terms occur with equal weight, so that according to this simple MO function, there is a 50–50 chance as to whether the H_2 ground state dissociates to two neutral hydrogen atoms or to a proton and a hydride ion. Actually, the H_2 ground state dissociates to two neutral hydrogen atoms. Thus the simple MO function gives the wrong limiting value of the energy as R goes to infinity.

How can we remedy this? Since H_2 is nonpolar, chemical intuition tells us that ionic terms should make substantially less contribution to the wave function than covalent terms. The simplest procedure is to omit the ionic terms of the MO function (13.120). This gives

$$\varphi_a(1)\varphi_b(2) + \varphi_b(1)\varphi_a(2) \tag{13.122}$$

We immediately recognize (13.122) as being the Heitler–London function (13.107).

Although interelectronic repulsion causes the electrons to avoid each other, there is some probability of finding both electrons near the same nucleus, corresponding to an ionic structure. Therefore instead of simply dropping the ionic terms from (13.120), we might try

$$\varphi_{\text{VB imp}} = \varphi_a(1)\varphi_b(2) + \varphi_b(1)\varphi_a(2) + \delta[\varphi_a(1)\varphi_a(2) + \varphi_b(1)\varphi_b(2)] \tag{13.123}$$

where $\delta(R)$ is a variational parameter and where the subscript indicates an improved VB function. In the language of valence-bond theory, this trial function represents *ionic–covalent resonance*. Of course, the ground-state wave function of H_2 does not undergo a time-dependent change back and forth from a covalent function corresponding to the structure H—H to ionic functions corresponding to the structures (13.121). Rather (in the particular approximation we are considering), the wave function is a time-independent mixture of covalent and ionic functions. Since H_2 dissociates to neutral atoms, we know that $\delta(\infty) = 0$. A variational calculation done by Weinbaum in 1933 using $1s$ AO's with an orbital exponent gave the result that at R_e the parameter δ has the value 0.26; the orbital exponent was found to be 1.19, and the dissociation energy was calculated as 4.02 eV, a modest improvement over the Heitler–London–Wang value of 3.78 eV. With δ equal to zero in (13.123), we get the VB function (13.122); with δ equal to 1, we get the LCAO-MO function (13.120). The optimum value of δ turns out to be closer to zero than to 1, and, in fact, the Heitler–London–Wang VB function gives a better dissociation energy than the LCAO-MO function.

Let us compare the improved valence-bond trial function (13.123) with the simple LCAO-MO function improved by configuration interaction. The LCAO-MO CI trial function (13.95) has the (unnormalized) form

$$\varphi_{\text{MO, imp}} = [\varphi_a(1) + \varphi_b(1)][\varphi_a(2) + \varphi_b(2)] + \gamma[\varphi_a(1) - \varphi_b(1)][\varphi_a(2) - \varphi_b(2)]$$

Since we have not yet normalized this function, there is no harm in multiplying it by the constant $1/(1-\gamma)$. Doing so and rearranging terms, we get

$$\varphi_{\text{MO, imp}} = \varphi_a(1)\varphi_b(2) + \varphi_b(1)\varphi_a(2) + \frac{1+\gamma}{1-\gamma}[\varphi_a(1)\varphi_a(2) + \varphi_b(1)\varphi_b(2)]$$

There is also no harm done if we define a new constant δ as

$$\delta = (1+\gamma)/(1-\gamma)$$

We see then that this improved MO function and the improved VB function (13.123) are *identical*. Weinbaum viewed his H_2 calculation as a valence-bond calculation with inclusion of ionic terms. We have shown that we can just as well view the Weinbaum calculation as an MO calculation with configuration interaction. (This was the viewpoint adopted in Section 13.8.)

The MO function (13.120) underestimates electron correlation, in that it says that structures with both electrons on the same atom are just as likely as structures with each electron on a different atom. The VB function (13.122) overestimates electron correlation, in that it has no contribution from structures with both electrons on the same atom. In MO theory electron correlation is introduced by configuration interaction. In VB theory electron correlation is reduced by ionic–covalent resonance. The simple VB method is more reliable at large R than the simple MO method, since the latter predicts the wrong dissociation products.

To fix further the differences between the MO and VB approaches, consider how each method divides the H_2 electronic Hamiltonian into unperturbed and perturbation Hamiltonians. For the MO method, we write

$$\hat{H} = \left[\left(-\tfrac{1}{2}\nabla_1^2 - \frac{1}{r_{a1}} - \frac{1}{r_{b1}}\right) + \left(-\tfrac{1}{2}\nabla_2^2 - \frac{1}{r_{a2}} - \frac{1}{r_{b2}}\right)\right] + \frac{1}{r_{12}} \quad (13.124)$$

where the unperturbed Hamiltonian consists of the bracketed terms. In MO theory the unperturbed Hamiltonian for H_2 is the sum of two $H_2{}^+$ Hamiltonians, one for each electron. Accordingly the zeroth-order MO wave function is a product of two $H_2{}^+$-like wave functions, one for each electron. Since the $H_2{}^+$ functions are rather complicated, we approximate the $H_2{}^+$-like MO's as LCAO's. The effect of the $1/r_{12}$ perturbation is taken into account in an average way through use of self-consistent-field molecular orbitals; to take instantaneous electron correlation into account, we use configuration interaction, getting a kind of calculation called LCAO-MO SCF CI.

For the valence-bond method, the terms in the Hamiltonian are grouped in either of two ways:

$$\hat{H} = \left[\left(-\tfrac{1}{2}\nabla_1^2 - \frac{1}{r_{a1}}\right) + \left(-\tfrac{1}{2}\nabla_2^2 - \frac{1}{r_{b2}}\right)\right] - \frac{1}{r_{a2}} - \frac{1}{r_{b1}} + \frac{1}{r_{12}} \quad (13.125)$$

$$\hat{H} = \left[\left(-\tfrac{1}{2}\nabla_1^2 - \frac{1}{r_{b1}}\right) + \left(-\tfrac{1}{2}\nabla_2^2 - \frac{1}{r_{a2}}\right)\right] - \frac{1}{r_{a1}} - \frac{1}{r_{b2}} + \frac{1}{r_{12}} \quad (13.126)$$

The unperturbed system is two hydrogen atoms. We have two zeroth-order functions consisting of products of hydrogen-atom wave functions, and these belong to a degenerate level. The correct ground-state zeroth-order function is the linear combination (13.107).

The majority of quantitative molecular calculations are done using the MO method, because it is computationally much simpler than the VB method. The MO method was developed by Hund, Mulliken, and Lennard-Jones in the late 1920s. Originally it was used largely for qualitative descriptions of molecules, but the electronic digital computer has made possible the calculation of accurate MO functions (Section 13.17).

13.11 *ELECTRON PROBABILITY DENSITY*

We now consider how the wave function of a system is related to the electron probability density. We want the probability of finding an electron in the rectangular volume element at point (x, y, z) in space with edges dx, dy, dz. The electronic wave function ψ is a function of the space and spin coordinates of the n electrons. (For simplicity the parametric dependence on the nuclear configuration will not be explicitly indicated.) We know that

$$|\psi(x_1, \ldots, z_n, m_{s1}, \ldots, m_{sn})|^2 \, dx_1 \, dy_1 \, dz_1 \cdots dx_n \, dy_n \, dz_n \quad (13.127)$$

is the probability of simultaneously finding electron 1 with spin m_{s1} in the volume element $dx_1 \, dy_1 \, dz_1$ at (x_1, y_1, z_1), electron 2 with spin m_{s2} in the volume element $dx_2 \, dy_2 \, dz_2$ at (x_2, y_2, z_2), etc. Since we are not interested in what spin the electron we find at (x, y, z) has, we sum the probability (13.127) over all possible spin states of all electrons, to give the probability of simultaneously finding each electron in the appropriate volume element with no regard for spin:

$$\sum_{m_{s1}} \cdots \sum_{m_{sn}} |\psi|^2 \, dx_1 \cdots dz_n \quad (13.128)$$

Suppose we ask for the probability of finding electron 1 in the volume element $dx \, dy \, dz$ at (x, y, z). For this probability we do not care where electrons 2 through n are; we therefore add the probabilities for all possible locations for these electrons. This amounts to integrating (13.128) over the coordinates of electrons 2, 3, ..., n:

$$\left[\sum_{\text{all } m_s} \int \cdots \int |\psi(x, y, z, x_2, y_2, z_2, \ldots, x_n, y_n, z_n, m_{s1}, \ldots, m_{sn})|^2 \, dx_2 \cdots dz_n \right] dx \, dy \, dz$$
$$(13.129)$$

where there is a $(3n - 3)$-fold integration over x_2 through z_n.

Now suppose we ask for the probability of finding electron 2 in the volume element $dx \, dy \, dz$ at (x, y, z). By analogy to (13.129), this is

$$\left[\sum_{\text{all } m_s} \int \cdots \int |\psi(x_1, y_1, z_1, x, y, z, x_3, \ldots, z_n, m_{s1}, \ldots, m_{sn})|^2 \right.$$
$$\left. \times \, dx_1 \, dy_1 \, dz_1 \, dx_3 \cdots dz_n \right] dx \, dy \, dz \quad (13.130)$$

Of course electrons do not come with labels, and because of this indistinguishability (Chapter 10) we know that the probabilities (13.129) and (13.130) must be equal. This equality is readily proved. ψ is antisymmetric with respect to electron exchange, and therefore $|\psi|^2$ is unchanged by such an exchange. Interchanging the space and spin coordinates of electrons 1 and 2 in ψ in (13.130) and doing some relabeling of dummy variables, we see that (13.130) is equal to (13.129). Thus (13.129) gives the probability of finding any one particular electron in the volume element. Since we have n electrons, the probability of finding *an* electron in the volume element is n times (13.129). (In drawing this conclusion, we are assuming that the probability of finding more than one electron in the infinitesimal volume element is negligible compared to the probability of finding one electron; this is certainly valid, since the probability of finding two electrons will involve the product of six infinitesimal quantities as compared to the product of three infinitesimal quantities for the probability of finding one electron.)

Thus the probability density ρ for finding an electron in the neighborhood of point (x, y, z) is

$$\rho(x, y, z) = n \sum_{\text{all } m_s} \int \cdots \int |\psi(x, y, z, x_2, \ldots, z_n, m_{s1}, \ldots, m_{sn})|^2 \, dx_2 \cdots dz_n$$

$$(13.131)$$

[Experimental determination of ρ in molecules involves measurement of electron diffraction intensities or x-ray diffraction intensities; see D. A. Kohl and L. S. Bartell, *J. Chem. Phys.*, **51**, 2891, 2896 (1969); P. Coppens and E. D. Stevens, *Adv. Quantum Chem.*, **10**, 1 (1977).] The atomic units of ρ are electrons/bohr^3.

To illustrate (13.131), we calculate the electron density for the simple VB and MO ground-state H_2 functions. The wave function is a product of a spatial factor and the spin function (11.60). (For more than two electrons, the wave function cannot be factored into a simple product of space and spin parts; see Chapter 10.) Summation of (11.60) over m_{s1} and m_{s2} gives unity (Section 10.5). Thus (13.131) becomes for H_2

$$\rho(x, y, z) = 2 \int \int \int |\varphi(x, y, z, x_2, y_2, z_2)|^2 \, dx_2 \, dy_2 \, dz_2 \qquad (13.132)$$

where φ is the spatial factor. The valence-bond function (13.107) gives

$$|\varphi(1,2)|^2 = \frac{[1s_a(1)]^2 [1s_b(2)]^2 + [1s_b(1)]^2 [1s_a(2)]^2 + 2 \cdot 1s_a(1) \, 1s_a(2) \, 1s_b(1) \, 1s_b(2)}{2(1 + S_{ab}^2)}$$

$$\rho_{VB} = \frac{2}{2(1 + S_{ab}^2)} \left[1s_a^2 \int [1s_b(2)]^2 \, dv_2 + 1s_b^2 \int [1s_a(2)]^2 \, dv_2 \right.$$

$$\left. + 2 \cdot 1s_a 1s_b \int 1s_a(2) \, 1s_b(2) \, dv_2 \right]$$

$$\rho_{VB} = \frac{1s_a^2 + 1s_b^2 + 2S_{ab} \, 1s_a 1s_b}{1 + S_{ab}^2} \qquad (13.133)$$

The MO function (13.92) gives (Problem 13.20b)

$$\rho_{MO} = \frac{1s_a^2 + 1s_b^2 + 2 \cdot 1s_a 1s_b}{1 + S_{ab}} \qquad (13.134)$$

Let us compare ρ_{MO} and ρ_{VB}. Both functions pile up more charge between the nuclei than is given by the sum of the atomic charge densities ($1s_a^2 + 1s_b^2$). (Recall the discussion of H_2^+ in Section 13.4.) Midway between the nuclei, we have $1s_a = 1s_b$, and $\rho_{MO} - \rho_{VB}$ is

$$\frac{2(S_{ab}-1)^2}{(1 + S_{ab})(1 + S_{ab}^2)} 1s_a^2 \qquad (13.135)$$

The numerator of (13.135) is zero for S_{ab} equal to 1 and is positive for all other values of S_{ab}. Since the overlap integral S_{ab} is less than 1, the MO function piles up more charge between the nuclei than the VB function. The VB function gives a lower energy (higher dissociation energy) than the MO function, so that the MO function must pile up too much charge between the nuclei. This is unfavorable because of increased interelectronic repulsion. Recall that the MO function underestimates electron correlation.

The MO probability density (13.134) is just twice the probability density for an H_2^+-like $\sigma_g 1s$ molecular orbital [Eq. (13.61)]. We can prove that for a molecular-orbital trial function, the probability density is found by multiplying the probability-density function of each MO by the number of electrons occupying it and summing the results:

$$\rho(x, y, z) = \sum_j n_j |\varphi_j|^2 \qquad (13.136)$$

where the sum is over all the different spatial MO's and where n_j ($= 0$ or 1 or 2) is the number of electrons in the orbital φ_j. [We used (13.136) in Eq. (11.14).]

13.12 DIPOLE MOMENTS

We now show how to calculate molecular dipole moments from wave functions.

The classical expression for the electric dipole moment $\mathbf{d}_{cl}$ of a set of discrete charges Q_i is

$$\mathbf{d}_{cl} = \sum_i Q_i \mathbf{r}_i \qquad (13.137)$$

where $\mathbf{r}_i$ is the position vector from the origin to the ith charge. The electric dipole moment is a vector; its x component is

$$d_{x,cl} = \sum_i Q_i x_i \qquad (13.138)$$

with similar expressions for the other components. (The symbol $\boldsymbol{\mu}$ is commonly used for the electric dipole moment, but this can be confused with the symbol for magnetic dipole moment.) For a continuous charge distribution with charge density $\rho_Q(x, y, z)$, $\mathbf{d}_{cl}$ is found by summing over the infinitesimal elements of charge $dQ_i = \rho_Q(x, y, z) \, dx \, dy \, dz$:

$$\mathbf{d}_{cl} = \int \rho_Q(x, y, z) \mathbf{r} \, dx \, dy \, dz \qquad (13.139)$$

$$\mathbf{r} = x\mathbf{i} + y\mathbf{j} + z\mathbf{k} \qquad (13.140)$$

Now consider the quantum-mechanical definition of the electric dipole moment. Suppose we apply a uniform external electric field $\mathbf{E}$ to an atom or molecule and ask for the effect on the energy of the system. To form the Hamiltonian operator, we first need the classical expression for the energy. The definition of an electric field is the force $\mathbf{F}$ it exerts on a charge Q, divided by Q: $\mathbf{E} = \mathbf{F}/Q$. We take the z direction as the direction of the applied field: $\mathbf{E} = \mathscr{E}_z \mathbf{k}$. The potential energy V is [Eq. (4.26)]

$$\frac{dV}{dz} = -F_z = -Q\mathscr{E}_z \tag{13.141}$$

$$V = -Q\mathscr{E}_z z \tag{13.142}$$

This is the potential energy of a single charge in the field. For a system of charges, we have

$$V = -\mathscr{E}_z \sum_i Q_i z_i \tag{13.143}$$

where z_i is the z coordinate of charge Q_i. The extension of (13.143) to the case where the electric field points in an arbitrary direction follows from (4.26) and is

$$V = -\mathscr{E}_x \sum_i Q_i x_i - \mathscr{E}_y \sum_i Q_i y_i - \mathscr{E}_z \sum_i Q_i z_i \tag{13.144}$$

$$V = -\mathbf{E} \cdot \mathbf{d}_{cl} \tag{13.145}$$

This is the classical-mechanical expression for the energy of an electric dipole in a uniform applied electric field.

To calculate the quantum-mechanical expression, we use perturbation theory. The perturbation operator $\hat{H}'$ corresponding to (13.145) is

$$\hat{H}' = -\mathbf{E} \cdot \hat{\mathbf{d}} \tag{13.146}$$

where the *electric dipole-moment operator* $\hat{\mathbf{d}}$ is given by

$$\hat{\mathbf{d}} = \sum_i Q_i \hat{\mathbf{r}}_i = \mathbf{i}\hat{d}_x + \mathbf{j}\hat{d}_y + \mathbf{k}\hat{d}_z \tag{13.147}$$

$$\hat{d}_x = \sum_i Q_i x_i, \qquad \hat{d}_y = \sum_i Q_i y_i, \qquad \hat{d}_z = \sum_i Q_i z_i \tag{13.148}$$

The first-order correction to the energy is [Eq. (9.30)]

$$E^{(1)} = -\mathbf{E} \cdot \int \psi^{(0)*} \hat{\mathbf{d}} \psi^{(0)} \, d\tau \tag{13.149}$$

where $\psi^{(0)}$ is the unperturbed wave function. Comparison of (13.149) and (13.145) shows that the quantum-mechanical quantity that corresponds to $\mathbf{d}_{cl}$ is the integral

$$\mathbf{d} = \int \psi^{(0)*} \hat{\mathbf{d}} \psi^{(0)} \, d\tau \tag{13.150}$$

The quantity $\mathbf{d}$ in (13.150) is the quantum-mechanical *electric dipole moment* of the system.

An objection to taking (13.150) as the dipole moment is that we considered only the first-order energy correction. If we had included $E^{(2)}$ in (13.149), the comparison with (13.145) would not have given (13.150) as the dipole moment. Actually (13.150) is the dipole moment of the system in the absence of an applied electric field and is the *permanent* electric dipole moment. Application of the field distorts the wave function from $\psi^{(0)}$, giving rise to an *induced* electric dipole moment in addition to the permanent dipole moment; the induced dipole moment corresponds[16] to the energy correction $E^{(2)}$. The induced dipole

---

[16] For the details, see *Merzbacher*, Section 17.4.

moment $\mathbf{d}_{ind}$ is related to the applied electric field by

$$\mathbf{d}_{ind} = \alpha \mathbf{E} \tag{13.151}$$

where α is the *polarizability* of the atom or molecule. The polarizability is important in discussions of van der Waals forces[17] and determines the index of refraction and dielectric constant of a gas.

The shift in the energy of a quantum-mechanical system caused by an applied electric field is called the *Stark effect* (after a German physicist). The *first-order* (or *linear*) Stark effect is given by (13.149), and from (13.150) it vanishes for a system with no permanent electric dipole moment. The *second-order* (or *quadratic*) Stark effect is given by the energy correction $E^{(2)}$ and is proportional to the square of the applied field.

The electric dipole-moment operator (13.147) is an odd function of the coordinates. If the wave function in (13.150) is either even or odd, then the integrand in (13.150) is an odd function, and the integral over all space vanishes. We conclude that the permanent electric dipole moment $\mathbf{d}$ is zero for states of definite parity.

The permanent electric dipole moment of a molecule whose electronic state is ψ_{el} is

$$\mathbf{d} = \int \psi_{el}^* \hat{\mathbf{d}} \psi_{el} \, d\tau_{el} \tag{13.152}$$

We have seen that the electronic wave functions of homonuclear diatomic molecules can be classified as g or u, according to their parity. Hence a homonuclear diatomic molecule has a zero permanent electric dipole moment, a not too astonishing result. The same holds true for any molecule with a center of symmetry.[18] The electric dipole-moment operator for a molecule includes summation over both the electronic and nuclear charges:

$$\hat{\mathbf{d}} = \sum_i (-e\,\mathbf{r}_i) + \sum_\alpha Z_\alpha e \mathbf{r}_\alpha \tag{13.153}$$

where $\mathbf{r}_\alpha$ is the vector from the origin to the nucleus of atomic number Z_α and where $\mathbf{r}_i$ is the vector to electron i. Since both the dipole-moment operator (13.153) and the electronic wave function depend on the parameters defining the nuclear configuration, the molecular electronic dipole moment $\mathbf{d}$ depends on the nuclear configuration. To indicate this, the quantity (13.152) can be called the dipole-moment *function* of the molecule. In writing (13.152), we ignore the nuclear motion; when the dipole moment of a molecule is experimentally determined, what is measured is the quantity (13.152) averaged over the zero-point vibrations (assuming the temperature is not high enough for there to be appreciable population of higher vibrational levels). We might use $\mathbf{d}_0$ and $\mathbf{d}_e$ to indicate the dipole moment averaged over zero-point vibrations and the dipole moment at the equilibrium nuclear configuration, respectively.

Since the second sum in (13.153) is independent of the electronic coordinates, we have

$$\mathbf{d} = \int \psi_{el}^* \sum_i (-e\mathbf{r}_i) \psi_{el} \, d\tau_{el} + \sum_\alpha Z_\alpha e \mathbf{r}_\alpha \int \psi_{el}^* \psi_{el} \, d\tau_{el} \tag{13.154}$$

$$\mathbf{d} = -e \int |\psi_{el}|^2 \sum_i \mathbf{r}_i \, d\tau_{el} + e \sum_\alpha Z_\alpha \mathbf{r}_\alpha \tag{13.155}$$

Because of the indistinguishability of the electrons, we can write (13.155) as

$$\mathbf{d} = -en \int |\psi_{el}|^2 \mathbf{r}_1 \, d\tau_{el} + e \sum_\alpha Z_\alpha \mathbf{r}_\alpha \tag{13.156}$$

where n is the number of electrons in the molecule and where $\mathbf{r}_1$ is the position vector of

---

[17] See *Kauzmann*, Chapter 13.

[18] Of course, there are other cases where $\mathbf{d}$ vanishes; see Section 12.3.

electron 1. Introducing the electronic probability density (13.131), we write

$$\mathbf{d} = -e \iiint \rho(x, y, z)\mathbf{r} \, dx \, dy \, dz + e \sum_\alpha Z_\alpha \mathbf{r}_\alpha \qquad (13.157)$$

The formula (13.157) is sometimes derived by considering the electronic charge as being smeared out into a continuous charge distribution whose charge density is found from (13.131); the classical expression (13.139) is then used to give the contribution of the electronic "charge cloud" to the molecular dipole moment. As mentioned in Section 6.6, the electrons are *not* smeared out into a charge cloud. However, the quantum-mechanical expression (13.150) for the electric dipole moment involves a spatial average over the electrons' positions. To determine $\langle \psi | \mathbf{r} | \psi \rangle$, we make a single measurement of $\mathbf{r}$ on a large number of identical noninteracting systems, each in the same state ψ, and average the result. Mathematically the result is the same as if we considered the electrons in a single system to be smeared out into a charge cloud and calculated $\langle \mathbf{r} \rangle$ on this basis.

13.13 MO AND VB WAVE FUNCTIONS FOR HOMONUCLEAR DIATOMIC MOLECULES

The molecular-orbital approximation puts the electrons of a molecule in molecular orbitals, which extend over the whole molecule. As an approximation to the molecular orbitals, we usually use linear combinations of atomic orbitals. The valence-bond method puts the electrons of a molecule in atomic orbitals and constructs the molecular wave function by allowing for "exchange" of the valence electron pairs between the atomic orbitals of the bonding atoms. We have compared the two methods for H_2; we now consider other homonuclear diatomic molecules.

We begin with the ground state of He_2. The separated helium atoms each have the ground-state configuration $1s^2$; for this closed-subshell configuration, we do not have any unpaired electrons to form valence bonds, and the VB wave function is simply the antisymmetrized product of the atomic-orbital functions:

$$\frac{1}{\sqrt{24}} \begin{vmatrix} 1s_a(1) & \overline{1s_a}(1) & 1s_b(1) & \overline{1s_b}(1) \\ 1s_a(2) & \overline{1s_a}(2) & 1s_b(2) & \overline{1s_b}(2) \\ 1s_a(3) & \overline{1s_a}(3) & 1s_b(3) & \overline{1s_b}(3) \\ 1s_a(4) & \overline{1s_a}(4) & 1s_b(4) & \overline{1s_b}(4) \end{vmatrix} \qquad (13.158)$$

The $1s$ function in this wave function is a helium-atom $1s$ function, which ideally is an SCF atomic function but can be approximated by a hydrogenlike function with an effective nuclear charge. The subscripts a and b refer to the two atoms, and the bar indicates spin function β. In the shorthand notation of (10.65), the wave function (13.158) is

$$|1s_a \, \overline{1s_a} \, 1s_b \, \overline{1s_b}| \qquad (13.159)$$

The valence-bond wave function for He_2 has each electron paired with another electron in an orbital on the *same* atom and hence predicts no bonding.

In the MO approach, He_2 has the ground-state configuration $(\sigma_g 1s)^2 (\sigma_u^* 1s)^2$; with no net bonding electrons, no bonding is predicted, in agreement with the VB method. The MO approximation to the wave function is

$$|\sigma_g 1s \overline{\sigma_g 1s} \, \sigma_u^* 1s \overline{\sigma_u^* 1s}| \qquad (13.160)$$

The simplest way to approximate the (unnormalized) MO's is to take them as linear combinations of the helium-atom AO's:

$$\sigma_g 1s = 1s_a + 1s_b, \qquad \sigma_u^* 1s = 1s_a - 1s_b$$

With this approximation (13.160) becomes

$$|(1s_a + 1s_b)\overline{(1s_a + 1s_b)}(1s_a - 1s_b)\overline{(1s_a - 1s_b)}| \qquad (13.161)$$

We can add or subtract one column of the determinant from another column without changing the determinant's value. If we add column 1 to column 3 and column 2 to column 4, we simplify (13.161) to

$$4|(1s_a + 1s_b)\overline{(1s_a + 1s_b)}\, 1s_a \overline{1s_a}|$$

We now subtract column 3 from column 1 and column 4 from column 2 to get

$$4|1s_b \overline{1s_b}\, 1s_a \overline{1s_a}| \qquad (13.162)$$

The interchange of columns 1 and 3 and of columns 2 and 4 multiplies the determinant by $(-1)^2$, so that (13.162) is equal to

$$4|1s_a \overline{1s_a}\, 1s_b \overline{1s_b}| \qquad (13.163)$$

which is identical (after normalization) to the valence-bond function (13.159). This result is easily generalized to the statement that the simple VB and simple LCAO-MO methods give the same approximate wave functions for diatomic molecules formed from separated atoms with completely filled atomic subshells. For example, the two methods give the same wave function for the Be_2 ground state. We could now substitute the trial function (13.163) into the variational integral and calculate the repulsive curve for the interaction of two ground-state helium atoms.

Before going on to Li_2, let us express the Heitler–London valence-bond functions for H_2 as Slater determinants. The ground-state Heitler–London function (13.107) can be written as

$$\frac{1}{2}(1 + S_{ab}^2)^{-1/2} \left\{ \begin{vmatrix} 1s_a(1)\alpha(1) & 1s_b(1)\beta(1) \\ 1s_a(2)\alpha(2) & 1s_b(2)\beta(2) \end{vmatrix} - \begin{vmatrix} 1s_a(1)\beta(1) & 1s_b(1)\alpha(1) \\ 1s_a(2)\beta(2) & 1s_b(2)\alpha(2) \end{vmatrix} \right\}$$

$$= (2 + 2S_{ab}^2)^{-1/2} \{|1s_a \overline{1s_b}| - |\overline{1s_a} 1s_b|\} \qquad (13.164)$$

In each Slater determinant, the electron on atom a is paired with an electron of opposite spin on atom b, corresponding to the Lewis structure H—H. The Heitler–London functions (13.108) for the lowest H_2 triplet state can also be written as Slater determinants. Omitting normalization constants, we write the Heitler–London H_2 functions as

$$\text{Singlet:} \quad |1s_a \overline{1s_b}| - |\overline{1s_a} 1s_b| \qquad (13.165)$$

$$\text{Triplet:} \begin{cases} |1s_a\,1s_b| \\ |1s_a\,\overline{1s_b}| + |\overline{1s_a}\,1s_b| \\ |\overline{1s_a}\,\overline{1s_b}| \end{cases} \tag{13.166}$$

Now consider Li_2. The ground-state configuration of Li is $1s^2\,2s$, and the Lewis structure of Li_2 is Li—Li, with the two $2s$ Li electrons paired and the $1s$ electrons remaining in the inner shell of each atom. The part of the valence-bond wave function involving the $1s$ electrons will be like the He_2 function (13.159), while the part of the VB wave function involving the $2s$ electrons (which form the bond) will be like the Heitler–London H_2 function (13.165). Of course, because of the indistinguishability of the electrons, there is complete electronic democracy, and we must allow every electron to be in every orbital; hence we write the ground-state valence-bond function for Li_2 using 6×6 Slater determinants:

$$|1s_a\,\overline{1s_a}\,1s_b\,\overline{1s_b}\,2s_a\,\overline{2s_b}| - |1s_a\,\overline{1s_a}\,1s_b\,\overline{1s_b}\,\overline{2s_a}\,2s_b| \tag{13.167}$$

We have written down (13.167) simply by analogy to (13.159) and (13.165)—for a fuller justification of it, we should show that it is an eigenfunction of the spin operators $\hat{S}^2$ and $\hat{S}_z$ with eigenvalue zero for each operator, which corresponds to a singlet state. This can be shown, but we omit doing so. To save space (13.167) is sometimes written as

$$|1s_a\,\overline{1s_a}\,1s_b\,\overline{1s_b}\,\overline{2s_a}\,2s_b| \tag{13.168}$$

where the curved line indicates the pairing (bonding) of the $2s_a$ and $2s_b$ AO's.

The MO wave function for the Li_2 ground state is

$$|\sigma_g1s\,\overline{\sigma_g1s}\,\sigma_u^*1s\,\overline{\sigma_u^*1s}\,\sigma_g2s\,\overline{\sigma_g2s}| \tag{13.169}$$

If we approximate the two lowest MO's by $1s_a \pm 1s_b$ and carry out the same manipulations we did for the He_2 MO function, we can write (13.169) as

$$|1s_a\,\overline{1s_a}\,1s_b\,\overline{1s_b}\,\sigma_g2s\,\overline{\sigma_g2s}| \tag{13.170}$$

[Recall the notation $KK(\sigma_g2s)^2$ for the Li_2 ground-state configuration.] From (13.170) it might be thought that the contribution of the σ_g1s and σ_u^*1s electrons to the probability density in Li_2 is the same as that of the sum of two $1s_a$ and two $1s_b$ electrons, but this is not quite so. The theorem stated at the end of Section 13.11 that allows the addition of orbital probability densities applies only to orthogonal orbitals, and the $1s_a$ and $1s_b$ orbitals are not orthogonal. However, as mentioned earlier, the overlap between these AO's in Li_2 is slight, so that they are nearly orthogonal.

Now consider the VB treatment of the N_2 ground state. The lowest configuration of N is $1s^2\,2s^2\,2p^3$; Hund's rule gives the ground level as $^4S_{3/2}$, with one electron in each of the three $2p$ AO's. We can thus pair the two $2p_x$ electrons, the two $2p_y$ electrons, and the two $2p_z$ electrons to form a triple bond; the Lewis structure is $:N\!\equiv\!N:$. How is this Lewis structure translated into the VB wave function? In the VB method, opposite spins are given to orbitals bonded together; we have three such pairs of orbitals, and two ways to give opposite spins to the electrons of each bonding pair of AO's. Hence there are $2^3 = 8$ possible Slater determinants that we can write. We begin with

$$D_1 = |1s_a\overline{1s_a}\,2s_a\overline{2s_a}\,1s_b\overline{1s_b}\,2s_b\overline{2s_b}\,2p_{xa}\overline{2p_{xb}}\,2p_{ya}\overline{2p_{yb}}\,2p_{za}\overline{2p_{zb}}|$$

In all eight determinants, the first eight columns will remain unchanged, and to save space we write D_1 as

$$D_1 = |\cdots 2p_{xa}\overline{2p_{xb}}\,2p_{ya}\overline{2p_{yb}}\,2p_{za}\overline{2p_{zb}}| \tag{13.171}$$

Reversing the spins of the electrons in $2p_{xa}$ and $2p_{xb}$, we get

$$D_2 = |\cdots \overline{2p_{xa}}\,2p_{xb}\,2p_{ya}\overline{2p_{yb}}\,2p_{za}\overline{2p_{zb}}| \tag{13.172}$$

There are six other determinants formed by interchanges of spins within the three pairs of bonding orbitals, and the VB wave function is a linear combination of eight determinants (Problem 13.29). The following rule (proved in *Pilar*, pp. 553–554) gives a VB wave function that is an eigenfunction of $\hat{S}^2$ with eigenvalue 0 (as is desired for the ground state): the coefficient of each determinant is $+1$ or -1 according to whether the number of spin interchanges required to generate the determinant from D_1 is even or odd, respectively. Thus D_2 has coefficient -1. [Compare also (13.165).] Clearly the single-determinant ground-state N_2 MO function is easier to handle than the eight-determinant VB function.

The VB method places great emphasis on pairing of electrons. In treating O_2, whose ground state is a triplet, the VB method runs into difficulties. It *is* possible to give a VB explanation of why O_2 has a triplet ground state, but the reasoning is rather involved [see B. J. Moss et al., *J. Chem. Phys.*, **63**, 4632 (1975)] in contrast to the very simple MO explanation.

13.14 EXCITED STATES OF H_2

We have concentrated mostly on the ground electronic states of diatomic molecules; in this section we consider some of the excited states of H_2. Figure 13.18 gives the potential-energy curves for some of the H_2 electronic energy levels.

The lowest MO configuration is $(\sigma_g 1s)^2$; this closed-subshell configuration gives only a nondegenerate $^1\Sigma_g^+$ level, designated $X^1\Sigma_g^+$. The LCAO-MO function is (13.89).

The next lowest MO configuration is $(\sigma_g 1s)(\sigma_u^* 1s)$, which gives rise to the terms $^1\Sigma_u^+$ and $^3\Sigma_u^+$ (Table 13.3). There is no axial electronic orbital angular momentum, so that each term corresponds to one level; spectroscopists have named these electronic levels $B^1\Sigma_u^+$ and $b^3\Sigma_u^+$. By Hund's rule the $b^3\Sigma_u^+$ is the lower of the two levels. The LCAO-MO functions for these levels are [see Eqs. (10.42)–(10.45)]

$$b^3\Sigma_u^+: \quad 2^{-1/2}[\sigma_g 1s(1)\sigma_u^* 1s(2) - \sigma_g 1s(2)\sigma_u^* 1s(1)]\begin{cases} \alpha(1)\alpha(2) \\ 2^{-1/2}[\alpha(1)\beta(2)+\alpha(2)\beta(1)] \\ \beta(1)\beta(2) \end{cases}$$

$$B^1\Sigma_u^+: \quad 2^{-1/2}[\sigma_g 1s(1)\sigma_u^* 1s(2) + \sigma_g 1s(2)\sigma_u^* 1s(1)]2^{-1/2}[\alpha(1)\beta(2)-\alpha(2)\beta(1)]$$

where $\sigma_g 1s \approx N(1s_a + 1s_b)$ and $\sigma_u^* 1s \approx N'(1s_a - 1s_b)$. The $b^3\Sigma_u^+$ level is triply degenerate; the $B^1\Sigma_u^+$ level is nondegenerate. [The Heitler–London wave functions for the $b^3\Sigma_u^+$ level are

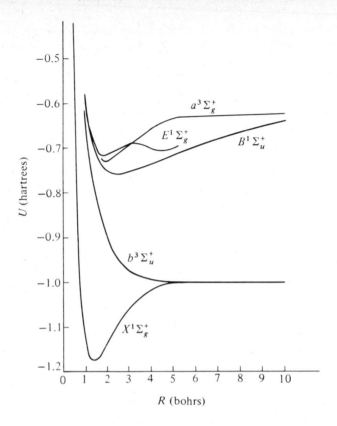

Figure 13.18 *Potential-energy curves for some states of* H_2. [See W. Kolos and L. Wolniewicz, J. Chem. Phys., **43**, 2429 (1965); **45**, 509 (1966); J. Gerhauser and H. S. Taylor, J. Chem. Phys., **42**, 3621 (1965).]

given by (13.108).] Both these levels have one bonding and one antibonding electron, and we would expect the potential-energy curves for both levels to be repulsive. However, the $B^1\Sigma_u^+$ level actually has a minimum in its $U(R)$ curve; the stability of this state should caution us against drawing too hasty conclusions from very approximate wave functions.

We expect the next lowest configuration to be $(\sigma_g 1s)(\sigma_g 2s)$, giving rise to $^1\Sigma_g^+$ and $^3\Sigma_g^+$ levels; these levels of H_2 are designated $E^1\Sigma_g^+$ and $a^3\Sigma_g^+$. By Hund's rule the triplet lies lower. Both levels are expected to be stable, as is observed. The E state is especially interesting, as it shows two substantial minima in its $U(R)$ curve.

Although the $\sigma_u^* 2s$ MO fills before the two $\pi_u 2p$ MO's in a progression across the periodic table, the $\pi_u 2p$ MO's are much lower than the $\sigma_u^* 2s$ MO for H_2. The configuration $(\sigma_g 1s)(\pi_u 2p)$ gives rise to the terms $^1\Pi_u$ and $^3\Pi_u$, the triplet lying lower. These terms are designated $C^1\Pi_u$ and $c^3\Pi_u$. The c term has three levels associated with it: $c^3\Pi_{2u}$, $c^3\Pi_{1u}$, and $c^3\Pi_{0u}$; however, these levels lie so close together that they are generally not resolved in spectroscopic work. The C level shows a slight hump in the potential-energy curve at large R. Each level is twofold degenerate, giving a total of eight electronic states arising from the $(\sigma_g 1s)(\pi_u 2p)$ configuration.

13.15 *THE VALENCE-ELECTRON APPROXIMATION*

Suppose we want to treat K_2, which has thirty-eight electrons. In the MO method, we would start by writing down a 38×38 Slater determinant of molecular orbitals. We would then approximate the MO's by functions containing variational parameters and go on to minimize the variational integral. Clearly the large number of electrons makes this a formidable task. One way to simplify the problem is to divide the electrons into two groups: the thirty-six *core* electrons and the two $4s$ *valence* electrons, which provide bonding; we then attempt to treat the valence electrons separately from the core, taking the molecular energy as the sum of core- and valence-electron energies. This approach, introduced in the 1930s, is called the *valence-electron approximation.*

The simplest approach is to regard the core electrons as point charges coinciding with the nucleus. For K_2 this would give a Hamiltonian for the two valence electrons that is identical with the electronic Hamiltonian for H_2. If we then go ahead and minimize the variational integral for the valence electrons in K_2, with no restrictions on the valence-electron trial functions, we will clearly be in trouble; such a procedure will cause the valence-electrons' MO to "collapse" to the $\sigma_g 1s$ MO, since the core electrons are considered absent. To avoid this collapse one can impose the constraint that the variational functions used for the valence electrons be orthogonal to the orbitals of the core electrons. Of course, the task of keeping the valence orbitals orthogonal to the core orbitals means more work. A somewhat different approach is to drop the approximation of treating the core electrons as coinciding with the nucleus, and to treat them as a charge distribution that provides some sort of effective repulsive potential for the motion of the valence electrons. This leads to an effective Hamiltonian for the valence electrons, which is then used in the variational integral. The valence-electron approximation is widely used in approximate treatments of polyatomic molecules (Sections 15.9–15.14).

13.16 *THE HARTREE–FOCK METHOD FOR MOLECULES*

A very important development in quantum chemistry has been the computation of accurate self-consistent-field wave functions for many diatomic and polyatomic molecules. The principles of molecular SCF calculations are essentially the same as for atomic SCF calculations (Section 11.1). We shall restrict ourselves to consideration of closed-subshell configurations; for open subshells, the formulas are more complicated.

The molecular Hartree–Fock wave function is written as an antisymmetrized product (Slater determinant) of spin-orbitals, each spin-orbital being a product of a spatial orbital φ_i and a spin function (either α or β).

The expression for the Hartree–Fock molecular energy E_{HF} is given by the variation theorem as $E_{HF} = \langle D | \hat{H}_{el} + V_{NN} | D \rangle$, where D is the Slater-determinant Hartree–Fock wave function and $\hat{H}_{el}$ and V_{NN} are given by (13.5) and (13.6). Since V_{NN} does not involve electronic coordinates and D is normalized, we have $\langle D | V_{NN} | D \rangle = V_{NN} \langle D | D \rangle = V_{NN}$. The operator $\hat{H}_{el}$ is the sum of one-electron operators $\hat{f}_i$ and two-electron operators $\hat{g}_{ij}$; we have $\hat{H}_{el} = \Sigma_i \hat{f}_i + \Sigma_j \Sigma_{i>j} \hat{g}_{ij}$, where $\hat{f}_i = -\frac{1}{2}\nabla_i^2 - \Sigma_\alpha Z_\alpha/r_{i\alpha}$ and $\hat{g}_{ij} = 1/r_{ij}$. The Hamiltonian $\hat{H}_{el}$ is the same as the Hamiltonian $\hat{H}$ for an atom except that $\Sigma_\alpha Z_\alpha/r_{i\alpha}$ replaces Z/r_i in $\hat{f}_i$. Hence Eq. (11.90) can be used to give $\langle D|\hat{H}_{el}|D \rangle$. Therefore the Hartree–Fock energy of a diatomic or polyatomic molecule with only closed subshells is

$$E_{HF} = 2\sum_{i=1}^{n/2} h_i + \sum_{i=1}^{n/2}\sum_{j=1}^{n/2} (2J_{ij} - K_{ij}) + V_{NN} \tag{13.173}$$

$$h_i \equiv \langle \varphi_i(1)| -\tfrac{1}{2}\nabla_1^2 - \Sigma_\alpha Z_\alpha/r_{1\alpha}|\varphi_i(1)\rangle \tag{13.174}$$

$$J_{ij} \equiv \langle \varphi_i(1)\varphi_j(2)|1/r_{12}|\varphi_i(1)\varphi_j(2)\rangle, \qquad K_{ij} \equiv \langle \varphi_i(1)\varphi_j(2)|1/r_{12}|\varphi_j(1)\varphi_i(2)\rangle \tag{13.175}$$

where the sums are over the $n/2$ occupied spatial orbitals φ_i of the n-electron molecule. In the Coulomb integrals J_{ij} and the exchange integrals K_{ij}, the integration goes over the spatial coordinates of electrons 1 and 2.

The Hartree–Fock method looks for those orbitals φ_i that minimize the variational integral E_{HF}. Of course, each MO is taken to be normalized: $\langle \varphi_i(1)|\varphi_i(1)\rangle = 1$. Moreover, for computational convenience one takes the MO's to be orthogonal: $\langle \varphi_i(1)|\varphi_j(1)\rangle = 0$ for $i \neq j$. [It might be thought that a lower energy could be obtained if the orthogonality restriction were omitted, but this is not so. A closed-subshell antisymmetric wave function is a Slater determinant, and one can use the properties of determinants to show that a Slater determinant of nonorthogonal orbitals is equal to a Slater determinant in which the orbitals have been orthogonalized by the Schmidt or some other procedure; see Section 15.6 and F. W. Bobrowicz and W. A. Goddard, Chapter 4, Section 3.1 of Schaefer, *Methods of Electronic Structure Theory*. (In effect the Pauli antisymmetry requirement removes nonorthogonalities from the orbitals.)]

The derivation of the equation that determines the orthonormal φ_i's that minimize E_{HF} is rather complicated and will be omitted.[19] One finds that the closed-subshell Hartree–Fock MO's satisfy

$$\hat{F}(1)\varphi_i(1) = \varepsilon_i\varphi_i(1) \tag{13.176}$$

where ε_i is the orbital energy and where the (*Hartree–*) *Fock operator* $\hat{F}$ is (in atomic units)

$$\hat{F}(1) = -\tfrac{1}{2}\nabla_1^2 - \sum_\alpha \frac{Z_\alpha}{r_{1\alpha}} + \sum_{j=1}^{n/2} [2\hat{J}_j(1) - \hat{K}_j(1)] \tag{13.177}$$

where the *Coulomb operator* $\hat{J}_j$ and the *exchange operator* $\hat{K}_j$ are defined by

$$\hat{J}_j(1)\varphi_i(1) = \varphi_i(1)\int |\varphi_j(2)|^2 \frac{1}{r_{12}}\,dv_2 \tag{13.178}$$

$$\hat{K}_j(1)\varphi_i(1) = \varphi_j(1)\int \frac{\varphi_j^*(2)\varphi_i(2)}{r_{12}}\,dv_2 \tag{13.179}$$

---

[19] For a derivation see *Parr*, pp. 21–32; *Pilar*, Chapter 13.

The first term on the right of (13.177) is the operator for the kinetic energy of one electron; the second term is the potential-energy operators for the attractions between one electron and the nuclei. (These two terms form the one-electron *core Hamiltonian*; the core Hamiltonian omits interactions with all other electrons.) The Coulomb operator $\hat{J}_j(1)$ is the potential energy of interaction between electron 1 and a smeared-out electron with electronic density $-|\varphi_j(2)|^2$; the factor 2 occurs because there are two electrons in each spatial orbital. The exchange operator has no simple physical interpretation but arises from the requirement that the wave function be antisymmetric with respect to electron exchange; the exchange operators are absent from the Hartree equations (11.12). The Hartree–Fock MO's φ_i in (13.176) are eigenfunctions of the same operator $\hat{F}$, the eigenvalues being the orbital energies ε_i.

The orthogonality of the MO's greatly simplifies MO calculations, causing many integrals to vanish. In contrast, the VB method uses atomic orbitals, and AO's centered on different atoms are not orthogonal. MO calculations are mathematically simpler than VB calculations, and the large majority of molecular calculations are MO calculations.

The true Hamiltonian operator and wave function involve the coordinates of all n electrons. The Hartree–Fock Hamiltonian operator $\hat{F}$ is a one-electron operator (i.e., it involves the coordinates of only one electron), and (13.176) is a one-electron differential equation. This has been indicated in (13.176) by writing $\hat{F}$ and φ_i as functions of the coordinates of electron 1; of course, the coordinates of any electron could have been used. The operator $\hat{F}$ is peculiar in that it depends on its own eigenfunctions [see Eqs. (13.177)–(13.179)], which are not known initially. Hence the Hartree–Fock equations must be solved by an iterative process.

To obtain the expression for the orbital energies ε_i, we multiply (13.176) by $\varphi_i^*(1)$ and integrate over all space; using the fact that φ_i is normalized and using the result of Problem 13.24, we obtain $\varepsilon_i = \int \varphi_i^*(1)\hat{F}(1)\varphi_i(1)\,dv_1$ and

$$\varepsilon_i = \langle \varphi_i(1) | -\tfrac{1}{2}\nabla_1^2 - \sum_\alpha Z_\alpha/r_{1\alpha} | \varphi_i(1) \rangle$$
$$+ \sum_j [2\langle \varphi_i(1) | \hat{J}_j(1) | \varphi_i(1) \rangle - \langle \varphi_i(1) | \hat{K}_j(1) | \varphi_i(1) \rangle]$$

$$\varepsilon_i = h_i + \sum_{j=1}^{n/2} (2J_{ij} - K_{ij}) \tag{13.180}$$

Summation of (13.180) over the occupied orbitals gives

$$2\sum_{i=1}^{n/2} \varepsilon_i = 2\sum_{i=1}^{n/2} h_i + 2\sum_{i=1}^{n/2}\sum_{j=1}^{n/2} (2J_{ij} - K_{ij})$$

Solving this equation for $2\sum_i h_i$ and substituting the result into (13.173), we obtain the Hartree–Fock energy as

$$E_{\text{HF}} = 2\sum_{i=1}^{n/2} \varepsilon_i - \sum_{i=1}^{n/2}\sum_{j=1}^{n/2} (2J_{ij} - K_{ij}) + V_{NN} \tag{13.181}$$

Since there are two electrons per MO, the quantity $2\sum_i \varepsilon_i$ is the sum of the orbital energies. Subtraction of the double sum in (13.181) avoids counting each interelectronic repulsion twice, as discussed in Section 11.1.

A key development that helped make feasible the calculation of accurate SCF wave functions for molecules was Roothaan's 1951 proposal to expand the spatial orbitals as linear combinations of a complete set of basis functions χ_k:

$$\varphi_i = \sum_k c_{ki} \chi_k \tag{13.182}$$

Substitution of this expansion into (13.176) gives

$$\sum_k c_{ki} \hat{F} \chi_k = \varepsilon_i \sum_k c_{ki} \chi_k$$

Multiplication by χ_j^* and integration gives

$$\sum_k c_{ki} (F_{jk} - \varepsilon_i S_{jk}) = 0, \qquad j = 1, 2, 3, \ldots \tag{13.183}$$

where

$$F_{jk} = \langle \chi_j | \hat{F} | \chi_k \rangle, \qquad S_{jk} = \langle \chi_j | \chi_k \rangle \tag{13.184}$$

The equations (13.183) are a set of simultaneous linear homogeneous equations in the unknowns c_{ki}. For a nontrivial solution, we must have

$$\det (F_{jk} - \varepsilon_i S_{jk}) = 0 \tag{13.185}$$

This is a secular equation whose roots give the orbital energies. The (*Hartree–Fock–*) *Roothaan equations* (13.183) must be solved by an iterative process, since the F_{jk} integrals depend on the orbitals φ_i (through the dependence of $\hat{F}$ on the φ_i's), which in turn depend on the unknown coefficients c_{ki}.

One starts with guesses for the occupied-MO expressions as linear combinations of the basis functions, as in (13.182). This initial set of MO's is used to compute the Fock operator $\hat{F}$ from (13.177)–(13.179). The matrix elements (13.184) are computed, and the secular equation (13.185) is solved to give an initial set of ε_i's; these ε_i's are used to solve (13.183) for an improved set of coefficients, giving an improved set of MO's, which are then used to compute an improved $\hat{F}$, etc. One keeps going until no further improvement in MO coefficients and energies occurs from one cycle to the next. The calculations are done using a computer. (The most efficient way to solve the Roothaan equations is to use matrix-algebra methods; see Section 15.21.)

Now consider the basis functions used. Generally each MO is written as a linear combination of one-electron functions (orbitals) centered on each atom. For diatomic molecules the functions most frequently used for the AO's are Slater functions, Eq. (11.16). To have a complete set of AO basis functions, it is necessary to use an infinite number of Slater orbitals, but the true molecular Hartree–Fock wave function can be closely approximated with a reasonably small number of Slater orbitals, if these are carefully chosen. Molecular SCF calculations can be divided into two broad classes: those that use a minimal basis set of AO's and those that use an extended basis set of AO's. A *minimal basis set* consists of only inner-shell and valence-shell AO's, while an *extended basis set* uses higher-shell AO's in addition to inner- and valence-shell AO's. Of course, minimal-basis-set SCF calculations are easier than extended-basis-set calculations, but the latter are more accurate.

We have used the terms scf wave function and Hartree–Fock wave function interchangeably. In practice the term *scf wave function* is applied to any wave function obtained by iterative solution of the Roothaan equations, whether or not the basis set is large enough to give a really accurate approximation to the Hartree–Fock scf wave function. There is only one true Hartree–Fock scf wave function, which is the best possible wave function that can be written as a Slater determinant of spin-orbitals. Some of the extended-basis-set calculations are believed to approach the true Hartree–Fock wave function quite closely; such functions are called "near Hartree–Fock wave functions" or, less cautiously, "Hartree–Fock wave functions."

13.17 *SCF WAVE FUNCTIONS FOR DIATOMIC MOLECULES*

Having discussed the equations used to calculate molecular scf wave functions, we now consider some examples of such wave functions for diatomic molecules.[20]

scf wave functions using a minimal basis set were calculated by Ransil for several light diatomic molecules.[21] As an example, the scf mo's for the ground state of Li_2 [mo configuration $(1\sigma_g)^2(1\sigma_u)^2(2\sigma_g)^2$] at $R = R_e$ are

$$1\sigma_g = 0.706(1s_a + 1s_b) + 0.009(2s_{\perp a} + 2s_{\perp b}) + 0.0003(2p\sigma_a + 2p\sigma_b)$$

$$1\sigma_u = 0.709(1s_a - 1s_b) + 0.021(2s_{\perp a} - 2s_{\perp b}) + 0.003(2p\sigma_a - 2p\sigma_b) \quad \textbf{(13.186)}$$

$$2\sigma_g = -0.059(1s_a + 1s_b) + 0.523(2s_{\perp a} + 2s_{\perp b}) + 0.114(2p\sigma_a + 2p\sigma_b)$$

The ao functions in these equations are sto's, except for $2s_\perp$. A Slater-type $2s$ ao has no radial nodes and is not orthogonal to a $1s$ sto. The Hartree–Fock $2s$ ao has one radial node $(n - l - 1 = 1)$ and is orthogonal to the $1s$ ao. We can form an orthogonalized $2s$ orbital with the proper number of nodes by taking the following normalized linear combination of $1s$ and $2s$ sto's of the same atom (Schmidt orthogonalization):

$$2s_\perp = (1 - S^2)^{-1/2}(2s - S \cdot 1s) \quad \textbf{(13.187)}$$

where S is the overlap integral $\langle 1s|2s \rangle$. Ransil expressed the Li_2 orbitals using the (nonorthogonal) $2s$ sto, but since the orthogonalized $2s_\perp$ function gives a better representation of the $2s$ ao, we have rewritten the orbitals using $2s_\perp$. This changes the $1s$ and $2s$ coefficients, but the actual orbital is, of course, unchanged; see Problem 13.28. The notation $2p\sigma$ for an ao indicates that the p orbital points along the molecular axis; i.e., a $2p\sigma$ ao is a $2p_z$ ao. (The $2p_x$ and $2p_y$ ao's are called $2p\pi$ ao's.) The optimum orbital exponents for the orbitals in (13.186) are

$$\zeta_{1s} = 2.689, \qquad \zeta_{2s} = 0.634, \qquad \zeta_{2p\sigma} = 0.761$$

(Since six ao's were used as basis functions, the Roothaan equations yielded

[20] For a summary of calculations on diatomic molecules, see *Mulliken and Ermler*.

[21] B. J. Ransil, *Rev. Mod. Phys.*, **32**, 245 (1960).

approximations for the six lowest MO's of ground-state Li_2; only three of these MO's are occupied. The expressions for the other three can be found in Ransil's paper.) Our previous simple expressions for these MO's were

$$1\sigma_g = \sigma_g 1s = 2^{-1/2}(1s_a + 1s_b)$$

$$1\sigma_u = \sigma_u^* 1s = 2^{-1/2}(1s_a - 1s_b)$$

$$2\sigma_g = \sigma_g 2s = 2^{-1/2}(2s_a + 2s_b)$$

Comparison of these with (13.186) shows the simple LCAO functions to be reasonable first approximations to the minimal-basis-set SCF MO's. The approximation is best for the $1\sigma_g$ and $1\sigma_u$ MO's, whereas the $2\sigma_g$ MO has a substantial $2p\sigma$ AO contribution in addition to the $2s$ AO contributions. For this reason the notation of the third column of Table 13.1 is preferable to the separated-atoms MO notation. The substantial amount of $2s$–$2p\sigma$ hybridization is to be expected, since the $2s$ and $2p$ AO's are close in energy [see Eq. (9.42)]; the hybridization allows for the polarization of the $2s$ AO's in forming the molecule.

Let us compare the $3\sigma_g$ MO of the F_2 ground state at R_e as calculated by Ransil using a minimal basis set and as calculated by Wahl[22] using an extended basis set:

$$3\sigma_{g,\,min} = 0.038(1s_a + 1s_b) - 0.184(2s_a + 2s_b) + 0.648(2p\sigma_a + 2p\sigma_b)$$

$$\zeta_{1s} = 8.65, \qquad \zeta_{2s} = 2.58, \qquad \zeta_{2p\sigma} = 2.49$$

$$3\sigma_{g,\,ext} = 0.048(1s_a + 1s_b) + 0.003(1s_a' + 1s_b') - 0.257(2s_a + 2s_b)$$

$$+ 0.582(2p\sigma_a + 2p\sigma_b) + 0.307(2p\sigma_a' + 2p\sigma_b') + 0.085(2p\sigma_a'' + 2p\sigma_b'')$$

$$- 0.056(3s_a + 3s_b) + 0.046(3d\sigma_a + 3d\sigma_b) + 0.014(4f\sigma_a + 4f\sigma_b)$$

$$\zeta_{1s} = 8.27, \qquad \zeta_{1s'} = 13.17, \qquad \zeta_{2s} = 2.26$$

$$\zeta_{2p\sigma} = 1.85, \qquad \zeta_{2p\sigma'} = 3.27, \qquad \zeta_{2p\sigma''} = 5.86$$

$$\zeta_{3s} = 4.91, \qquad \zeta_{3d\sigma} = 2.44, \qquad \zeta_{4f\sigma} = 2.83$$

Just as several STO's are needed to give an accurate representation of Hartree–Fock AO's (Section 11.1), one needs more than one STO of a given n and l in the linear combination of STO's that is to accurately represent the Hartree–Fock MO. The primed and double-primed AO's in the extended-basis-set function are STO's with different orbital exponents. The $3d\sigma$ and $4f\sigma$ AO's are AO's with quantum number $m = 0$, i.e., the $3d_0$ and $4f_0$ AO's. The total energies found are -197.877 and -198.768 hartrees for the minimal and extended calculations, respectively. The experimental energy of F_2 at R_e is -199.670 hartrees, so that the error for the minimal calculation is twice that of the extended calculation. The extended-basis-set calculation is believed to give a wave function quite close to the true Hartree–Fock wave function; therefore the correlation energy in F_2 is about 24 eV.

In discussing H_2^+ and H_2, we saw how hybridization (the mixing of different AO's of the same atom) improves molecular wave functions. There is a tendency to

22   A. C. Wahl, *J. Chem. Phys.*, **41**, 2600 (1964).

think of hybridization as occurring only for certain molecular geometries. The SCF calculations make clear that *all* MO's are hybridized to some extent. Thus any diatomic-molecule σ MO is a linear combination of $1s, 2s, 2p_0, 3s, 3p_0, 3d_0, \dots$ AO's of the separated atoms.

To aid in deciding which AO's contribute to a given diatomic MO, we use two rules. First, only σ-type AO's ($s, p\sigma, d\sigma$, etc.) can contribute to a σ MO; only π-type AO's ($p\pi, d\pi$, etc.) can contribute to a π MO; etc. Second, only AO's of reasonably similar energy make substantial contributions to a given MO. (For examples, see the minimal- and extended-basis-set MO's quoted above.)

Wahl has plotted the contours of the near Hartree–Fock molecular orbitals of homonuclear diatomic molecules from H_2 through F_2. Figure 13.19 shows these plots for H_2 and Li_2.

Of course, Hartree–Fock wave functions are only approximations to the true wave functions. It is possible to prove that a Hartree–Fock wave function gives a very good approximation to the electron probability density $\rho(x, y, z)$. A molecular property that involves only one-electron operators can be expressed as an integral involving ρ; consequently such properties are accurately calculated using Hartree–Fock wave functions. The prime example is the molecular dipole moment [Eq. (13.157)], for which the success of the near Hartree–Fock wave functions has

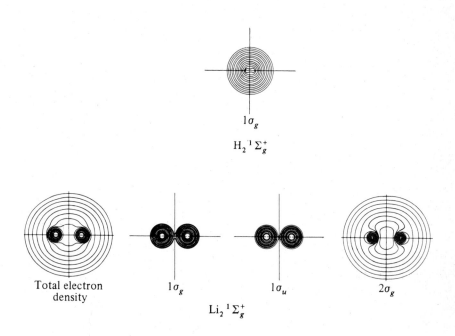

Figure 13.19 *Hartree–Fock MO electron-density contours for the ground states of* H_2 *and* Li_2, *as calculated by Wahl.* [*A. C. Wahl,* Science, *151, 961 (1966); Scientific American,* April 1970, p. 54; Atomic and Molecular Structure: 4 Wall Charts, McGraw-Hill, New York, 1970.]

been impressive. As an example, for LiH the dipole moment calculated[23] with a near-Hartree–Fock ψ is 6.00 D (debyes), compared to the experimental value 5.83 D. (One debye $= 10^{-18}$ statcoul cm.) For NaCl the calculated[24] and experimental d's are 9.18 and 9.02 D. An error of, say, 0.2 D is typical in such calculations, but where the dipole moment is small, the percent error can become large. An extreme example is CO. Here the experimental moment is 0.11 D with the polarity C^-O^+, but the near-Hartree–Fock moment is 0.27 D with the erroneous polarity C^+O^-. However, a wave function improved by configuration interaction gives[23] 0.12 D with the correct polarity.

Molecular Hartree–Fock calculations do not give good values for dissociation energies. For example, an extended-basis-set calculation[25] gives $D_e = 5.3$ eV for N_2, as compared to the true value 9.9 eV. (To calculate the Hartree–Fock D_e, the molecular energy at the minimum in the $U(R)$ Hartree–Fock curve is subtracted from the sum of the Hartree–Fock energies of the separated atoms.) For F_2 the true D_e is 1.66 eV, while the Hartree–Fock D_e is -1.4 eV; in other words, the Hartree–Fock calculation predicts the separated atoms to be more stable than the fluorine molecule. A related defect of Hartree–Fock molecular wave functions is that the energy approaches the wrong limit for large values of R; recall the MO discussion of H_2.

In nearly all calculations of Hartree–Fock wave functions, electrons that are paired in an AO or MO (e.g., the two 1s electrons in Li) are given precisely the same spatial orbital function. This produces the *conventional* or *restricted* Hartree–Fock wave function. For atoms and molecules with partially filled subshells, one finds that a slightly lower variational energy can be obtained if paired electrons are allowed to have different spatial orbital functions. This produces the *unrestricted* Hartree–Fock wave function (see Schaefer, *The Electronic Structure of Atoms and Molecules*, pp. 22–27, 95–101). The term *Hartree–Fock* without an adjective denotes the restricted Hartree–Fock method. We shall not consider the unrestricted Hartree–Fock method.

13.18 CI WAVE FUNCTIONS

To overcome the deficiencies of the Hartree–Fock wave function (e.g., improper behavior as $R \to \infty$ and incorrect D_e values), one must introduce configuration interaction (CI), thus going beyond the Hartree–Fock approximation. Recall (Section 11.3) that in a molecular CI calculation one begins with a set of basis functions χ_i, does an SCF calculation to find SCF occupied and virtual (unoccupied) MO's, uses these MO's to form configuration (state) functions Φ_i, writes the molecular wave function ψ as a linear combination $\sum_i c_i \Phi_i$ of the configuration functions, and uses the variation method to find the c_i's. In calculations on diatomic molecules, the basis functions are usually Slater-type AO's, some centered on one atom, the remainder on the second atom.

[23] S. Green, *J. Chem. Phys.*, **54**, 827 (1971).
[24] R. L. Matcha, *J. Chem. Phys.*, **48**, 335 (1968).
[25] P. E. Cade, K. D. Sales, and A. C. Wahl, *J. Chem. Phys.*, **44**, 1973 (1966).

The number of MO's produced equals the number of basis functions used. The type of MO's produced depends on the type of basis functions used. For example, if one includes only s AO's in the basis set, one will get only σ MO's, and no $\pi, \delta, \ldots$ MO's.

The configuration functions in a CI calculation are classified as *singly excited*, *doubly excited*, *triply excited*, $\ldots$ according to whether $1, 2, 3, \ldots$ electrons are excited from occupied to unoccupied (virtual) orbitals. For example, the H_2 configuration function $2^{-1/2}|\sigma_u^* 1s \overline{\sigma_u^* 1s}|$ used in Eq. (13.95) is doubly excited (another term sometimes used is *doubly substituted*).

In the CI expansion $\psi = \Sigma_i c_i \Phi_i$, one includes only configuration functions that have the same symmetry properties (symmetry eigenvalues) as the state that is being approximated by the expansion. [This follows from the discussion in the paragraph after Eq. (7.66).] For example, the ground state of H_2 is a $^1\Sigma_g^+$ state, and a CI calculation of the H_2 ground state would include only configuration functions that correspond to $^1\Sigma_g^+$ terms. A $\sigma_g \sigma_u$ electron configuration would have states of odd parity (u) and would not be included in ψ; a $\sigma_g \pi_g$ configuration would produce only Π terms (Table 13.3) and would not be included; for a π_g^2 or π_u^2 configuration, only the configuration function corresponding to the $^1\Sigma_g^+$ term (Table 13.3) would contribute to ψ.

The number of possible configuration functions with the proper symmetry increases very rapidly as the number of electrons and the number of basis functions increase; for n electrons and b basis functions, the number of configuration functions turns out to be roughly proportional to b^n. A CI calculation that includes all possible configuration functions with proper symmetry is called a *full* CI calculation. Because of the large number of configuration functions, full CI calculations are out of the question except for very small molecules (small n) and small basis sets (small b).

One must therefore decide which types of configuration functions are likely to make the largest contributions to ψ, and one must include only these. We expect the unexcited configuration function (the SCF wave function) to make the largest contribution. The question then is which types of excited configurations make significant contributions to ψ. To answer this we consider the effects of instantaneous electron correlation as a perturbation on the Hartree–Fock wave function $\psi^{(0)}$. It turns out that the first-order correction to the unperturbed (Hartree–Fock) wave function involves only configuration functions Φ_i for which the matrix element $\langle \Phi_i | \hat{H} | \psi^{(0)} \rangle$ is nonzero, where $\hat{H}$ is the molecular electronic Hamiltonian. [Recall Eq. (9.42).] The Hamiltonian $\hat{H}$ is the sum of one-electron and two-electron operators, and the Condon–Slater rules (Table 11.3) show that $\langle \Phi_i | \hat{H} | \psi^{(0)} \rangle$ is zero whenever Φ_i differs by three or more orbitals from the Hartree–Fock $\psi^{(0)}$. Moreover, *Brillouin's theorem* (Pilar, p. 364) states that $\langle \Phi_i | \hat{H} | \psi^{(0)} \rangle = 0$ when Φ_i is a singly excited configuration function and $\psi^{(0)}$ is a closed-shell SCF wave function. Therefore for a closed-shell state, the only nonzero Hamiltonian matrix elements between the unexcited SCF wave function and excited configuration functions are those for double excitations.

Thus we expect the most important correction to the Hartree–Fock wave function to come from doubly excited configuration functions. Although singly excited configuration functions are of lesser importance than double excitations in affecting the wave function, it turns out (see I. Shavitt, in *Schaefer, Methods of*

Electronic Structure Theory, p. 255) that single excitations have a significant effect on one-electron properties. [A *one-electron property* is one calculated as $\langle\psi|\hat{B}|\psi\rangle$, where the operator $\hat{B}$ is a sum of terms, each of which involves only a single particle; an example is the dipole moment—Eqs. (13.152) and (13.153).] Therefore one usually includes single excitations in a CI calculation, and the most common type of CI calculation includes the singly and doubly excited configuration functions. (It turns out that the second-order correction to the Hartree–Fock function includes single, triple, and quadruple excitations.)

In the majority of calculations (e.g., molecular dissociation, excitation of valence-shell electrons to produce excited electronic states), one is looking at energy changes in processes affecting primarily the valence-shell electrons, so one expects the correlation energies involving the inner-shell electrons to change only slightly. Hence one often makes the further approximation of considering only configuration functions involving excitation of valence-shell electrons.

The CI procedure just discussed calculates SCF occupied and virtual orbitals from the basis functions and uses these SCF MO's to form configuration functions. With this procedure the rate of convergence is quite slow, and very large numbers of configuration functions must be included. For example, an accurate dissociation energy for LiH was calculated by CI that included all single and double excitations; for this small, four-electron molecule, the CI wave function consisted of 939 configuration functions. [C. F. Bender and E. R. Davidson, *Phys. Rev.*, **183**, 23 (1969).] For medium-sized molecules, thousands or tens of thousands of configuration functions may be required for accurate results. A major reason for the very slow convergence is that the excited (virtual) SCF orbitals have much of their probability density at large distances from the nuclei, whereas the ground-state wave function has most of its probability density reasonably near the nuclei.

Actually there is no necessity to use SCF MO's in a CI calculation. Any set of MO's calculated from the basis set will produce the same final wave function provided a full CI calculation is carried out. Moreover, if the non-SCF MO's are well chosen, they can produce much faster convergence to the true wave function than is obtained with SCF MO's, thereby allowing substantially fewer configuration functions to be included in ψ. Two approaches that use this idea are the multiconfiguration SCF method and the method of natural orbitals.

In the *multiconfiguration* SCF (MCSCF) *method*, one writes the molecular wave function as a linear combination of configuration functions Φ_i and varies not only the expansion coefficients c_i in $\psi = \sum_i c_i \Phi_i$ but also the forms of the molecular orbitals that compose the configuration functions. (This is done by varying the expansion coefficients that relate the MO's to the basis functions χ_i.) The optimum orbitals are found by an iterative process (somewhat similar to the usual SCF iterative process used to find the Hartree–Fock wave function); see A. C. Wahl and G. Das, Chapter 3 in *Schaefer, Methods of Electronic Structure Theory*. By optimizing the orbitals, one can get good results with inclusion of relatively few configuration functions. Because the orbitals are varied, the amount of calculation required in the MCSCF procedure is very great, and the method has been applied mainly to diatomic (and a few triatomic) molecules.

Wahl, Das, and co-workers have developed an MCSCF approach to the calculation of diatomic D_e's, an approach which they call the *optimized valence configuration* (OVC) *method*. Since the atomic cores remain relatively unchanged on molecule formation, the OVC method includes only configuration functions with valence electrons excited. To correctly take into account the changes that occur as the molecule forms, one must add those configuration functions needed to give proper dissociation of the molecule to Hartree–Fock atoms in their correct states, together with a few other configuration functions that make important contributions in the molecule near R_e. (The method does involve some trial and error to find the right configuration functions to include and has been criticized on this account.) Although the OVC CI ψ gives only a part (about 25 percent for F_2) of the molecular correlation energy (which is the error in the Hartree–Fock ψ; Section 11.3), it includes that part of the correlation energy that varies with R and hence gives a good $U(R)$ curve and a good D_e. As noted above, one of the outstanding failures of the Hartree–Fock method is the F_2 D_e. An OVC ψ for F_2 containing only six configuration functions gives $D_e = 1.82$ eV, and when contributions from further configuration functions are included by a perturbation calculation, the final result $D_e = 1.67$ eV is obtained[26]; this is in excellent agreement with the experimental value 1.66 eV. The OVC method has been applied with success to several other diatomics. (See A. C. Wahl and G. Das, Chapter 3 in *Schaefer, Methods of Electronic Structure Theory*.)

Another alternative to the use of SCF MO's in CI calculations is provided by *natural orbitals*. The definition of natural orbitals is rather technical and can be skimmed over if desired.

Natural orbitals are defined as follows. For an n-electron molecule whose wave function is ψ, the spinless first-order reduced density matrix $\rho(1, 1')$ is defined as

$$\rho(1, 1') = n \sum_{\text{all } m_s} \int \cdots \int \psi^*(x'_1, y'_1, z'_1, x_2, \ldots, z_n, m_{s1}, \ldots, m_{sn})$$

$$\times \psi(x_1, y_1, z_1, x_2, \ldots, z_n, m_{s1}, \ldots, m_{sn}) \, dx_2 \cdots dz_n$$

$\rho(1, 1')$ is a function of $x_1, y_1, z_1, x'_1, y'_1, z'_1$. For $x'_1 = x_1, y'_1 = y_1, z'_1 = z_1$, we see that $\rho(1, 1')$ reduces to the ordinary electron probability density (13.131). For a wave function that is a linear combination of Slater determinants, it turns out that $\rho(1, 1')$ can be expressed as

$$\rho(1, 1') = \sum_i \sum_j a_{ij} \varphi_i^*(1') \varphi_j(1)$$

where the φ's are all the MO's that appear in the Slater determinants and where the a_{ij}'s are a set of numbers (also called a density matrix); $\varphi_i(1')$ and $\varphi_j(1)$ are functions of x'_1, y'_1, z'_1 and x_1, y_1, z_1, respectively. As we shall see in Section 15.6, one can take linear combinations of MO's to form a new set of MO's. It can be shown that there exists a particular set of MO's θ_i, called the *natural orbitals*, that have the property that when ψ is expressed in terms of the θ_i's, one finds that $\rho(1, 1')$ has the form

$$\rho(1, 1') = \sum_i b_i \theta_i^*(1') \theta_i(1)$$

where the numbers b_i (called the *occupation numbers*) give the relative importance of each natural orbital in the overall wave function.

---

[26] G. Das and A. C. Wahl, *J. Chem. Phys.*, **56**, 3532 (1972).

It turns out that a CI calculation using natural orbitals converges substantially faster than one using SCF orbitals, so one needs to include far fewer configuration functions to obtain a given level of accuracy. Unfortunately, the natural orbitals are defined in terms of the final CI wave function and cannot be calculated until a CI calculation using SCF orbitals has been completed. Hence several schemes have been devised to calculate approximate natural orbitals and use these in CI calculations.

For example, the *iterative natural-orbital* (INO) *method* starts by calculating a CI wave function using a manageable number of configurations; from this CI wave function, one calculates approximate natural orbitals and uses these orbitals to get an improved CI wave function; this process is then repeated to get further improvement. As an example, an INO calculation on the LiH ground state using only 45 configuration functions obtained 87 percent of the correlation energy and gave a slightly lower energy than the ordinary CI wave function of 939 configuration functions mentioned earlier in this section; see C. F. Bender and E. R. Davidson, *J. Phys. Chem.*, **70**, 2675 (1966).

In calculating a diatomic-molecule potential-energy curve $U(R)$, one must repeat the CI calculation at several values of R. Usually one configuration function makes the predominant contribution at R_e, but as R increases, the contribution of excited configuration functions increases.

Although calculation of SCF wave functions for closed-shell reasonably small molecules is essentially a routine procedure, each CI calculation of a molecule presents its own special problems. To obtain reliable results, one must use sound judgment in choosing the basis set, the molecular orbitals, the configuration functions to be included, and the procedure to be used.

Numerous CI calculations have been performed on diatomic molecules to yield potential-energy curves, dipole moments, and dissociation energies for ground and excited states. For discussions of results, see *Mulliken and Ermler, Diatomic Molecules* and *Schaefer, The Electronic Structure of Atoms and Molecules.*

13.19 *MO TREATMENT OF HETERONUCLEAR DIATOMIC MOLECULES*

The treatment of heteronuclear diatomic molecules is similar to that of homonuclear diatomic molecules. We first consider the MO description.

Suppose the two atoms have atomic numbers that differ only slightly; an example is CO. We could consider CO as being formed from the isoelectronic molecule N_2 by a gradual transfer of nuclear charge from one nucleus to the other. During this hypothetical transfer, the original N_2 MO's would slowly vary to give finally the CO MO's. We therefore expect the CO molecular orbitals to bear some resemblance to those of N_2. For a heteronuclear diatomic molecule such as CO, the symbols used for the MO's are similar to those for homonuclear diatomics; however, for a heteronuclear diatomic, the electronic Hamiltonian (13.5) is not invariant with respect to inversion of the electronic coordinates (i.e., $\hat{H}_{el}$ does not commute with $\hat{\Pi}$) so that the g, u property of the MO's disappears. The correlation between the N_2 and

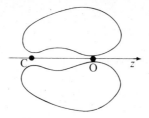

Figure 13.20 *Cross section of a contour of the $1\pi_{\pm 1}$ MO's in carbon monoxide*

CO subshell designations is

N_2	$1\sigma_g$	$1\sigma_u$	$2\sigma_g$	$2\sigma_u$	$1\pi_u$	$3\sigma_g$	$1\pi_g$	$3\sigma_u$
CO	1σ	2σ	3σ	4σ	1π	5σ	2π	6σ

MO's of the same symmetry are numbered in order of increasing energy; because of the absence of the g, u property, the numbers of corresponding homonuclear and heteronuclear MO's differ. Figure 13.20 is a sketch of a contour of the CO $1\pi_{\pm 1}$ MO's, as determined by an extended-basis-set SCF calculation.[27] Note its resemblance to the contour of Fig. 13.12, which is for the $1\pi_{u,\pm 1}$ MO's of a homonuclear diatomic molecule.

The ground-state configuration of CO is $1\sigma^2\, 2\sigma^2\, 3\sigma^2\, 4\sigma^2\, 1\pi^4\, 5\sigma^2$, as compared to the N_2 configuration $(1\sigma_g)^2\,(1\sigma_u)^2\,(2\sigma_g)^2\,(2\sigma_u)^2\,(1\pi_u)^4\,(3\sigma_g)^2$.

As in homonuclear diatomics, the heteronuclear diatomic MO's are approximated as linear combinations of atomic orbitals. The coefficients are found by solving the Roothaan equations (13.183). For example, a minimal-basis-set SCF calculation[28] using Slater AO's (with nonoptimized exponents given by Slater's rules) gives for the CO 5σ, 1π, and 2π MO's at $R = R_e$

$$5\sigma = 0.027\,(1s_C) + 0.011\,(1s_O) + 0.739\,(2s_{\perp C}) + 0.036\,(2s_{\perp O})$$
$$- 0.566(2p\sigma_C) - 0.438\,(2p\sigma_O)$$
$$1\pi = 0.469\,(2p\pi_C) + 0.771\,(2p\pi_O)$$
$$2\pi = 0.922\,(2p\pi_C) - 0.690\,(2p\pi_O)$$

(The expressions for the π MO's are simpler than those for the σ MO's because s and $p\sigma$ AO's cannot contribute to π MO's.) For comparison the corresponding MO's in N_2 are found[28] to be (at $R = R_e$)

$$3\sigma_g = 0.030(1s_a + 1s_b) + 0.395\,(2s_{\perp a} + 2s_{\perp b}) - 0.603\,(2p\sigma_a + 2p\sigma_b)$$
$$1\pi_u = 0.624\,(2p\pi_a + 2p\pi_b)$$
$$1\pi_g = 0.835\,(2p\pi_a - 2p\pi_b)$$

[27] W. M. Huo, *J. Chem. Phys.*, **43**, 624 (1965).
[28] B. J. Ransil, *Rev. Mod. Phys.*, **32**, 245 (1960).

The resemblance of CO and N_2 MO's is apparent. The 1σ MO in CO is found to be nearly the same as a $1s$ oxygen-atom AO; the 2σ MO in CO is essentially a carbon-atom $1s$ AO.

In general, for a heteronuclear diatomic molecule AB where the valence AO's of each atom are of s and p type and where the valence AO's of A do not differ greatly in energy from the valence AO's of B, we can expect the Figure 13.16 pattern of

$$\sigma s < \sigma^* s < \pi p < \sigma p < \pi^* p < \sigma^* p$$

valence-shell MO's formed from s and p valence-shell AO's to hold reasonably well. Figure 13.16 would be modified in that each valence AO of the more electronegative atom would lie below the corresponding valence AO of the other atom.

When the valence-shell AO energies of B lie very substantially below those of A, the s and $p\sigma$ valence AO's of B lie below the s valence-shell AO of A, and this affects which AO's contribute to each MO. Consider the molecule BF, for example. A minimal-basis-set calculation [Ransil, *Rev. Mod. Phys.*, **32**, 245 (1960)] gives the 1σ MO as essentially $1s_F$ and the 2σ MO as essentially $1s_B$. The 3σ MO is predominantly $2s_F$, with small amounts of $2s_B$, $2p\sigma_B$, and $2p\sigma_F$. The 4σ MO is predominantly $2p\sigma_F$, with significant amounts of $2s_B$ and $2s_F$ and a small amount of $2p\sigma_B$; this is quite different from N_2, where the corresponding MO is formed predominantly from the $2s$ AO's on each N. The 1π MO is a bonding combination of $2p\pi_B$ and $2p\pi_F$. The 5σ MO is predominantly $2s_B$, with a substantial contribution from $2p\sigma_B$ and a significant contribution from $2p\sigma_F$; this is unlike the corresponding MO in N_2, where the largest contributions are from $2p\sigma$ MO's on each atom. The 2π MO is an antibonding combination of $2p\pi_B$ and $2p\pi_F$. The 6σ MO has important contributions from $2p\sigma_B$, $2s_B$, $2s_F$, and $2p\sigma_F$.

We see from Fig. 11.2 that the $2p_F$ AO lies well below the $2s_B$ AO. This causes the $2p\sigma_F$ AO to contribute substantially to lower-lying MO's and the $2s_B$ AO to contribute substantially to higher-lying MO's, as compared to what happens in N_2. (This effect occurs in CO, although to a lesser extent. Note the very substantial contribution of $2s_C$ to the 5σ MO; also, the 4σ MO in CO has a very substantial contribution from $2p\sigma_O$.)

For a diatomic molecule AB where each atom has s and p valence-shell AO's (this excludes H and transition elements) and where the A and B valence AO's differ widely in energy, we may expect the pattern of valence MO's to be $\sigma < \sigma < \pi < \sigma < \pi < \sigma$, but it is not so easy to guess which AO's contribute to the various MO's or the bonding or antibonding character of the MO's. By feeding the valence electrons into these MO's, we can come up with a plausible guess as to the number of unpaired electrons and the ground term of the AB molecule (Problem 13.35).

Diatomic hydrides are a special case, since H has only a $1s$ valence AO. Consider HF as an example. The ground-state configurations of the atoms are $1s$ for H and $1s^2 2s^2 2p^5$ for F. We expect the filled $1s$ and $2s$ F subshells to take little part in the bonding. The four $2p\pi$ fluorine electrons are nonbonding (there are no π valence AO's on H). The hydrogen $1s$ AO and the fluorine $2p\sigma$ AO have the same symmetry (σ) and have rather similar energies (Fig. 11.2), and a linear combination of these two AO's will form a σ MO for the bonding electron pair:

$$\varphi = c_1(1s_H) + c_2(2p\sigma_F)$$

where the contributions of $1s_F$ and $2s_F$ to this MO have been neglected. Since fluorine is more electronegative than hydrogen, we expect that $c_2 > c_1$. (In addition the $1s_H$ and $2p\sigma_F$ AO's form an antibonding MO, which is unoccupied in the ground state.)

The picture of HF just given is only a crude qualitative approximation. A minimal-basis-set SCF calculation[29] using Slater orbitals with optimized exponents gives for the HF MO's

$$1\sigma = 1.000(1s_F) + 0.012(2s_{\perp F}) + 0.002(2p\sigma_F) - 0.003(1s_H)$$

$$2\sigma = -0.018(1s_F) + 0.914(2s_{\perp F}) + 0.090(2p\sigma_F) + 0.154(1s_H)$$

$$3\sigma = -0.023(1s_F) - 0.411(2s_{\perp F}) + 0.711(2p\sigma_F) + 0.516(1s_H)$$

$$1\pi_{+1} = (2p\pi_{+1})_F$$

$$1\pi_{-1} = (2p\pi_{-1})_F$$

The ground-state MO configuration of HF is $1\sigma^2 2\sigma^2 3\sigma^2 1\pi^4$. The 1σ MO is virtually identical with the $1s$ fluorine AO. The 2σ MO is pretty close to the $2s$ fluorine AO. The 1π MO's are required by symmetry to be the same as the corresponding fluorine π AO's. The bonding 3σ MO has its largest contribution from the $2p\sigma$ fluorine and $1s$ hydrogen AO's, as would be expected from the discussion of the preceding paragraph; however, there is a substantial contribution to this MO from the $2s$ fluorine AO. (Since a single $2s$ function is only an approximation to the $2s$ AO of F, one cannot use this calculation to say exactly how much $2s$ AO character the 3σ HF molecular orbital has.)

Accurate MO expressions require the solution of the Roothaan equations for their determination. For qualitative discussion (but not quantitative work), it is useful to have simple approximations for heteronuclear diatomic MO's. In the crudest approximation, we can take each valence MO of a heteronuclear diatomic molecule as a linear combination of two AO's φ_a and φ_b, one on each atom. (As the discussions of CO and BF show, this approximation is often quite inaccurate.) From the two AO's, we can form two MO's:

$$c_1\varphi_a + c_2\varphi_b \qquad \text{and} \qquad c_1'\varphi_a + c_2'\varphi_b$$

The lack of symmetry in the heteronuclear diatomic causes the coefficients to be unequal in magnitude. The coefficients are determined by solving the secular equation [see Eq. (13.35)]

$$\begin{vmatrix} H_{aa} - W & H_{ab} - WS_{ab} \\ H_{ab} - WS_{ab} & H_{bb} - W \end{vmatrix} = 0$$

$$(H_{aa} - W)(H_{bb} - W) - (H_{ab} - WS_{ab})^2 = 0 \qquad \textbf{(13.188)}$$

where $\hat{H}$ is some sort of effective one-electron Hamiltonian. Suppose that $H_{aa} > H_{bb}$, and let $f(W)$ be defined as the left side of (13.188). The overlap integral S_{ab} is less than 1 (except at $R = 0$). [A rigorous proof of this follows from the Schwarz inequality, Eq. (3-114) in *Margenau and Murphy*.] The coefficient of W^2 in $f(W)$ is $(1 - S_{ab}^2)$

---

[29] B. J. Ransil, *Rev. Mod. Phys.*, **32**, 245 (1960).

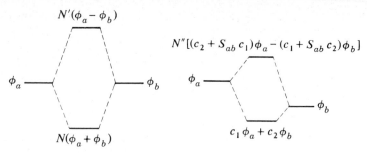

Figure 13.21 *Formation of bonding and antibonding MO's from AO's in the homonuclear and heteronuclear cases. (See Problem 13.27.)*

> 0; hence we have $f(\infty) = f(-\infty) = +\infty > 0$. For $W = H_{aa}$ or H_{bb}, the first product in (13.188) vanishes; hence $f(H_{aa}) < 0$ and $f(H_{bb}) < 0$. The roots of (13.188) occur where $f(W)$ equals 0; hence by continuity, one root must be between $+\infty$ and H_{aa} and the other between H_{bb} and $-\infty$. Therefore the orbital energy of one MO is less than both H_{aa} and H_{bb} (the energies of the two AO's in the molecule—Section 13.4), and the energy of the other MO is greater than both H_{aa} and H_{bb}. One bonding and one antibonding MO are formed from the two AO's. Figure 13.21 shows the formation of bonding and antibonding MO's from two AO's, for the homonuclear and heteronuclear cases. These figures are gross oversimplifications, since a given MO has contributions from many AO's, not just two.

The coefficients c_1 and c_2 in the bonding heteronuclear MO in Fig. 13.21 are both positive, so as to build up charge between the nuclei. For the antibonding heteronuclear MO, the coefficients of φ_a and φ_b have opposite signs, causing charge depletion between the nuclei.

13.20 VB TREATMENT OF HETERONUCLEAR DIATOMIC MOLECULES

We will illustrate the valence-bond approach to the ground states of heteronuclear diatomic molecules by considering HF. Adopting the valence-electron approximation (Section 13.15), we will look at the wave function for the bonding electrons only. We expect a single bond to be formed by the pairing of the hydrogen $1s$ electron and the unpaired fluorine $2p\sigma$ electron. The Heitler–London function corresponding to this pairing is [Eq. (13.164)]

$$\varphi_{\text{cov}} = (2 + 2S^2)^{-1/2}\{|1s_H\overline{2p\sigma_F}| - |\overline{1s_H}2p\sigma_F|\} \qquad (13.189)$$

This function is essentially covalent, the electrons being shared by the two atoms. However, the high electronegativity of fluorine leads us to include a contribution from an ionic structure as well. [Actually the "covalent" function (13.189) does give some polarity to the bond for heteronuclear diatomic molecules; in the electron-density expression corresponding to (13.133), the AO's φ_a and φ_b are of different size.] An ionic valence-bond function has the form $\varphi_a(1)\varphi_a(2)$ [Eq. (13.123)]. Introduction of the required antisymmetric spin factor gives as the valence-bond function for an ionic structure in HF

$$\varphi_{\text{ion}} = |2p\sigma_F \overline{2p\sigma_F}| \qquad \textbf{(13.190)}$$

The wave function for the bonding electrons is then written as

$$\varphi = c_1 \varphi_{\text{cov}} + c_2 \varphi_{\text{ion}} \qquad \textbf{(13.191)}$$

The optimum values of c_1 and c_2 are determined by the variation method, using an effective Hamiltonian; this leads to the usual secular equation. We have ionic-covalent "resonance," involving the structures H—F and $H^+ F^-$; the true molecular structure is intermediate between the covalent and ionic structures. A term $c_3|\overline{1s_H 1s_H}|$ corresponding to the ionic structure $H^- F^+$ could also be included in the wave function, but this should make a negligible contribution for HF. For molecules that are less ionic, both ionic structures might well be included.

For a molecule like NaCl, which is highly ionic, we expect the valence-bond function to have $c_2 \gg c_1$. It might be thought that NaCl would dissociate to Na^+ and Cl^- ions, but this is not accurate. The ionization potential of Na is 5.1 eV, while the electron affinity of Cl is only 3.6 eV. Hence in the gas phase, the neutral separated ground-state atoms Na + Cl are more stable than the ground-state separated ions $Na^+ + Cl^-$. (Of course, in aqueous solution the ions are more stable because of the hydration energy, which makes the separated ions more stable than even the diatomic NaCl molecule.) If the nuclei are slowly pulled apart, a gas-phase NaCl molecule will dissociate to neutral atoms; this means that as R increases from R_e, the ratio c_2/c_1 in (13.191) must decrease, becoming zero at $R = \infty$. For intermediate values of R, the Coulombic attraction between the ions is greater than the 1.5 eV difference between the ionization potential and electron affinity, and the molecule is largely ionic; for very large R, the Coulombic attraction between the ions is less than 1.5 eV, and the molecule is largely covalent. However, if the nuclei in NaCl are pulled apart very rapidly, then the electrons will not have a chance to adjust their wave function from the ionic to the covalent wave function, and both bonding electrons will go with the chlorine nucleus, giving dissociation into ions.

Cesium has the lowest ionization potential, 3.9 eV. Chlorine has the highest electron affinity, 3.6 eV. (The electron affinity of fluorine is 3.45 eV.) Thus even for CsCl and CsF, the separated ground-state neutral atoms are more stable than the separated ground-state ions. There are, however, cases of excited states of diatomic molecules whose potential-energy curves correspond to dissociation to ions.

PROBLEMS

13.1 Give the numerical value in atomic units of each of the following quantities: (a) proton mass; (b) electron charge; (c) Planck's constant; (d) He^+ ground-state energy, assuming infinite nuclear mass; (e) one second; (f) c (speed of light); (g) hydrogen-atom ground-state energy, taking internal nuclear motion into account; (h) one debye.

13.2 Use results of statistical mechanics to show that the standard change in internal energy $\Delta U°$ for the gas-phase dissociation of a diatomic molecule to atoms at 298 K is $RT/2$ greater than $\Delta U°$ at 0 K if there is negligible population of excited vibrational and electronic states (as is usually the case).

13.3 Derive Eqs. (13.26) and (13.27) from (13.25) and (13.21).

13.4 In confocal elliptic coordinates, what is the shape of surfaces of constant (a) ξ; (b) η; (c) φ?

13.5 (a) From Fig. 13.2 show that

$$R\xi = [r^2 \sin^2 \theta + (z + \tfrac{1}{2}R)^2]^{1/2} + [r^2 \sin^2 \theta + (z - \tfrac{1}{2}R)^2]^{1/2}$$

and that η is given by a similar expression with the plus sign between the brackets replaced by a minus sign. (b) Verify the ranges given for the coordinates in Eq. (13.18).

13.6 Use the D_0 value of H_2 (4.478 eV) and the D_0 value of H_2^+ (2.651 eV) to calculate the first ionization energy of H_2 (i.e., the energy needed to remove an electron from H_2).

13.7 The infrared absorption spectrum of $^1H^{35}Cl$ has its strongest band at 8.65 $\times 10^{13}$ Hz. For this molecule, $D_0 = 4.43$ eV. (a) Find D_e for $^1H^{35}Cl$. (b) Find D_0 for $^2H^{35}Cl$.

13.8 Verify Eq. (13.58).

13.9 (a) Show that the variational integral W_1 for the ground state of H_2^+ [Eq. (13.59)] can be written as

$$W_1 = k^2 F(t) + kG(t)$$

where $t \equiv kR$ and where F and G are certain functions of t. (b) Show that the minimization condition $\partial W_1 / \partial k = 0$ leads to

$$k = -\frac{G(t) + tG'(t)}{2F(t) + tF'(t)}$$

Using this equation, we can find k for a given value of t; we then use $R = t/k$ to find the value of R corresponding to our value of k.

13.10 Consider the trial function (13.48) for the first excited state of H_2^+. Find the limit of this function as R goes to zero. Compare the result to the $2p_z$ hydrogenlike function.

13.11 Calculate the energy at $R = 2.00$ bohrs for the H_2^+ trial function $e^{-k\xi}$. Calculate the percent error in (a) the electronic energy including nuclear repulsion; (b) the purely electronic energy; (c) the dissociation energy.

13.12 Evaluate the geometric series $\sum_{n=0}^{\infty} R^4 e^{-Rn}$, and show that the infinite sum behaves as R^3 for small values of R, even though each term in the sum behaves as R^4 for small R.

13.13 Write a computer program to calculate (at $R = R_e$) values of the H_2^+ MO's (13.47) and (13.48) at points in a plane containing the nuclei. Use the output to make contour plots of the MO's. Label the contours in units of electrons/bohr³ (probability density).

13.14 Verify Eq. (13.75) for a σ_h reflection.

13.15 Which species of each pair has the greater D_e? (a) Li_2 or Li_2^+; (b) C_2 or C_2^+; (c) O_2 or O_2^+; (d) F_2 or F_2^+.

13.16 How many independent electronic wave functions correspond to each of the following diatomic-molecule terms: (a) $^1\Sigma^-$; (b) $^3\Sigma^+$; (c) $^3\Pi$; (d) $^1\Phi$; (e) $^6\Delta$?

13.17 Give the levels belonging to each of the terms in Problem 13.16.

13.18 The ground state of H_2 has $^1\Sigma_g^+$ symmetry. What restriction does this impose on the values of m, n, j, and k in (13.96)?

13.19 (a) Show that the simple MO wave function for the $b^3\Sigma_u^+$ level of H_2 is the same as the Heitler–London VB function for this level. (b) Show that the simple MO wave function for the $B^1\Sigma_u^+$ level of H_2 contains only ionic terms.

13.20 (a) Use (13.131) to find the electron probability density for the simple MO wave function of the $b^3\Sigma_u^+$ level of H_2. Verify that (13.136) holds for this case. (b) Verify Eq. (13.134).

13.21 Show that the dipole moment (13.137) of a system of charges is independent of the choice of the coordinate origin, *provided* the system has no net charge.

13.22 Prove that the one-electron Hartree–Fock operator (13.177) is Hermitian.

13.23 Explain the origin of the extra terms in the molecular Hartree–Fock operator (13.177) as compared to the atomic Hartree operator of (11.12) and (11.10).

13.24 Verify that the Coulomb and exchange integrals J_{ij} and K_{ij} can be written in terms of the Coulomb and exchange operators of Section 13.16 as

$$J_{ij} = \langle \varphi_i(1)|\hat{J}_j(1)|\varphi_i(1)\rangle, \qquad K_{ij} = \langle \varphi_i(1)|\hat{K}_j(1)|\varphi_i(1)\rangle$$

13.25 Verify that for a closed-subshell molecule

$$E_{\text{HF}} = \sum_{i=1}^{n/2} (h_i + \varepsilon_i) + V_{NN}$$

13.26 In applying quantum chemistry to chemical kinetics, which would be the more accurate approximation, the simple MO or the simple VB method?

13.27 (a) Use orthogonality (Section 8.5 and Problem 8.15) to derive the expression given in Fig. 13.21 for the heteronuclear antibonding MO. Then do the same for the homonuclear antibonding MO. (b) Verify that the functions (13.47) and (13.48) are orthogonal.

13.28 (a) Verify that the $2s_{\perp}$ AO in (13.187) is orthogonal to the $1s$ AO and is normalized. (b) Let an MO φ have the form $\varphi = a(1s) + b(2s) + \cdots$ when expressed using a nonorthogonal $2s$ STO and the form $\varphi = c(1s) + d(2s_{\perp}) + \cdots$ when expressed using an orthogonalized $2s$ orbital. Show that $c = a + Sb$ and $d = b(1 - S^2)^{1/2}$, where $S = \langle 1s|2s\rangle$. (c) Let ζ_1 and ζ_2 be the orbital exponents of $1s$ and $2s$ STO's, respectively. Show that $S = 24\zeta_1^{3/2}\zeta_2^{5/2}/3^{1/2}(\zeta_1 + \zeta_2)^4$.

13.29 Write down abbreviated expressions for the remaining six determinants of the N_2 VB function of Section 13.13. Use the rule given in that section to find the coefficient of each determinant in the wave function.

13.30 (a) Explain why the permanent dipole moment of a many-electron atom is always zero. (b) Explain why the permanent electric dipole moment of H can be nonzero for certain excited states. (c) Show qualitatively that two of the four correct zeroth-order functions of Problem 9.13 give nonzero permanent dipole moments.

13.31 Show that the four functions of (13.86) have the indicated eigenvalues with respect to a $\sigma_v(xz)$ reflection of electronic coordinates. (Start by showing that this reflection converts φ to $-\varphi$ and leaves r_a and r_b unchanged.)

13.32 True or false: (a) If $|\psi_1| = |\psi_2|$, then ψ_1 and ψ_2 must represent the same state. (b) The probability density for the wave function (13.191) is the sum of the probability densities for the functions $c_1\varphi_{\text{cov}}$ and $c_2\varphi_{\text{ion}}$.

13.33 Show that for a diatomic molecule $[\hat{L}_z, \hat{O}_{\sigma_v}] \neq 0$.

13.34 Use simple MO theory and Table 13.3 to predict the bond order, the number of unpaired electrons, and the ground term for each of the following molecules: (a) S_2; (b) S_2^+; (c) S_2^-; (d) N_2^+; (e) N_2^-; (f) F_2^+; (g) F_2^-; (h) Ne_2^+; (i) Na_2^+; (j) Na_2^-; (k) H_2^-; (l) C_2^+; (m) C_2; (n) C_2^-.

13.35 Use simple MO theory to predict the number of unpaired electrons and the ground term of each of the following: (a) BF; (b) BN; (c) BeS; (d) BO; (e) NO; (f) CF; (g) CP; (h) NBr; (i) LiO; (j) ClO; (k) BrCl. Compare your results with the experimentally observed ground terms: (a) $^1\Sigma^+$; (b) $^3\Pi$; (c) $^1\Sigma^+$; (d) $^2\Sigma^+$; (e) $^2\Pi$; (f) $^2\Pi$; (g) $^2\Sigma^+$; (h) $^3\Sigma^-$; (i) $^2\Pi$; (j) $^2\Pi$; (k) $^1\Sigma^+$.

13.36 (a) Verify that if anharmonicity is taken into account by inclusion of the $v_e x_e$ term in the vibrational energy, then $D_e = D_0 + \frac{1}{2}hv_e - \frac{1}{4}hv_e x_e$. (b) For the $^7\text{Li}^1\text{H}$ ground electronic state, one has $D_0 = 2.4287$ eV, $v_e/c = 1405.65\,\text{cm}^{-1}$, and $v_e x_e/c = 23.20\,\text{cm}^{-1}$, where c is the speed of light. (These last two quantities are usually designated ω_e and $\omega_e x_e$ in the literature.) Calculate D_e for $^7\text{Li}^1\text{H}$.

13.37 For NaCl, $R_e = 2.36$ A. The ionization potential of Na is 5.14 eV, and the electron affinity of Cl is 3.61 eV. Use the simple model of NaCl as a pair of spherical ions in contact to estimate D_e and the dipole moment d of NaCl. Compare with the experimental values $D_e = 4.25$ eV and $d = 9.0$ D. [One debye (D) $= 10^{-18}$ statC cm.]

13.38 A common way of approximating the potential-energy curve $U(R)$ of a bound state of a diatomic molecule is by the Morse function $U(R) = D_e[1 - e^{-a(R - R_e)}]^2$, where a is a parameter whose value depends on the molecule and the electronic state. For convenience, the Morse function takes the zero of energy at $R = R_e$. (A slight defect of the Morse function is that it approaches a large finite value, rather than infinity, as R goes to zero. This defect is unimportant, since the probability of R being very near zero is extremely small.) For the Morse function, (a) verify that $U(\infty) - U(R_e) = D_e$; (b) verify that $U'(R_e) = 0$, as must be true, since U is a minimum at R_e; (c) use $k = d^2U/dR^2|_{R=R_e}$ (Section 4.3) to show that $a = (k/2D_e)^{1/2}$.

13.39 Use data in Problem 13.36 and the result of Problem 13.38 to calculate the Morse function a of the $^7\text{Li}^1\text{H}$ ground electronic state.

13.40 For a CI calculation on the ground electronic state of H_2, which of the following configurations will produce configuration functions that will contribute to the wave function? (a) $(\sigma_g 1s)(\sigma_g 2s)$; (b) $(\sigma_g 1s)(\sigma_u^* 2s)$; (c) $(\sigma_u^* 1s)^2$; (d) $(\pi_u 2p)(\pi_g^* 2p)$; (e) $(\pi_u 2p)(\sigma_u^* 2p)$; (f) $(\pi_u 2p)^2$; (g) $(\pi_u 2p)(\pi_u 3p)$.

14

THE VIRIAL THEOREM AND THE HELLMANN–FEYNMAN THEOREM

14.1 THE VIRIAL THEOREM

In this chapter we discuss two theorems that aid in the understanding of chemical bonding. We begin with the virial theorem.

Consider a system in the stationary state ψ and let $\hat{H}$ be its time-independent Hamiltonian:

$$\hat{H}\psi = E\psi \tag{14.1}$$

Let $\hat{A}$ be a linear operator that does not involve the time. We consider the following integral:

$$\int \psi^*[\hat{H}, \hat{A}]\psi \, d\tau \tag{14.2}$$

$$= \int \psi^*(\hat{H}\hat{A} - \hat{A}\hat{H})\psi \, d\tau = \int \psi^*\hat{H}\hat{A}\psi \, d\tau - E\int \psi^*\hat{A}\psi \, d\tau \tag{14.3}$$

Since $\hat{H}$ is Hermitian, we have

$$\langle \psi|\hat{H}|\hat{A}\psi \rangle = \langle \hat{A}\psi|\hat{H}|\psi \rangle^* \tag{14.4}$$

Using (14.4) and (14.1) in (14.3), we get

$$\int \psi^*[\hat{H}, \hat{A}]\psi \, d\tau = E\int (\hat{A}\psi)\psi^* \, d\tau - E\int \psi^*\hat{A}\psi \, d\tau$$

$$\int \psi^*[\hat{H}, \hat{A}]\psi \, d\tau = 0 \tag{14.5}$$

393

Equation (14.5) has been christened the *hypervirial theorem*.[1]

We now derive the virial theorem from (14.5). We choose $\hat{A}$ to be

$$\sum_i \hat{q}_i \hat{p}_i = -i\hbar \sum_i q_i \frac{\partial}{\partial q_i} \tag{14.6}$$

where the sum runs over the $3n$ Cartesian coordinates of the n particles. (Particle 1 has Cartesian coordinates q_1, q_2, q_3 and corresponding linear-momentum components p_1, p_2, p_3. Throughout this chapter the symbol q will indicate a *Cartesian coordinate*.) We want to evaluate $[\hat{H}, \hat{A}]$. Using (5.5), (5.7), (5.12), and (5.13), we get

$$\left[\hat{H}, \sum_i \hat{q}_i \hat{p}_i \right] = \sum_i [\hat{H}, \hat{q}_i \hat{p}_i] = \sum_i \hat{q}_i [\hat{H}, \hat{p}_i] + \sum_i [\hat{H}, \hat{q}_i] \hat{p}_i$$

$$= i\hbar \sum_i q_i \frac{\partial V}{\partial q_i} - i\hbar \sum_i \frac{1}{m_i} \hat{p}_i^2 = i\hbar \sum_i q_i \frac{\partial V}{\partial q_i} - 2i\hbar \hat{T}$$

where $\hat{T}$ and $\hat{V}$ are the kinetic- and potential-energy operators for the system. Application of (14.5) gives

$$\left\langle \psi \left| \sum_i q_i \frac{\partial V}{\partial q_i} \right| \psi \right\rangle = 2 \langle \psi | \hat{T} | \psi \rangle \tag{14.7}$$

Using $\langle B \rangle$ for the quantum-mechanical average value of the quantity B, we write (14.7) as

$$\left\langle \sum_i q_i \frac{\partial V}{\partial q_i} \right\rangle = 2 \langle T \rangle \tag{14.8}$$

Equation (14.8) is the quantum-mechanical *virial theorem*. Note that its validity is restricted to stationary states. [The name of the theorem comes from a similar theorem in classical mechanics, which has the same form as (14.8) except that the averages are time averages. The word *vires* is Latin for "forces"; the derivatives of the potential energy give the negatives of the force components.]

For certain systems the form of V causes the virial theorem to take on a particularly simple form. To discuss these systems, we introduce the idea of a homogeneous function. A function $f(x_1, x_2, \ldots, x_j)$ of several variables is said to be *homogeneous of degree n* if it satisfies

$$f(sx_1, sx_2, \ldots, sx_j) = s^n f(x_1, x_2, \ldots, x_j) \tag{14.9}$$

where s is an arbitrary parameter. For example, the function

$$g = \frac{1}{x^3} + \frac{1}{y^3} + \frac{1}{z^3} + \frac{x}{y^2 z^2} \tag{14.10}$$

is homogeneous of degree -3, since

$$g(sx, sy, sz) = \frac{1}{s^3 x^3} + \frac{1}{s^3 y^3} + \frac{1}{s^3 z^3} + \frac{sx}{s^2 y^2 s^2 z^2} = s^{-3} g(x, y, z)$$

---

[1] For some of its applications, see J. O. Hirschfelder, *J. Chem. Phys.*, **33**, 1462 (1960); J. H. Epstein and S. T. Epstein, *Am. J. Phys.*, **30**, 266 (1962).

Euler's theorem on homogeneous functions states that if $f(x_1, \ldots, x_j)$ is homogeneous of degree n, then

$$\sum_{k=1}^{j} x_k \frac{\partial f}{\partial x_k} = nf \tag{14.11}$$

The theorem is proved as follows. Let

$$u_1 = sx_1, u_2 = sx_2, \ldots, u_j = sx_j \tag{14.12}$$

From the chain rule, the derivative of the left side of (14.9) is

$$\frac{\partial f(u_1, \ldots, u_j)}{\partial s} = \frac{\partial f}{\partial u_1}\frac{\partial u_1}{\partial s} + \frac{\partial f}{\partial u_2}\frac{\partial u_2}{\partial s} + \cdots + \frac{\partial f}{\partial u_j}\frac{\partial u_j}{\partial s}$$

$$= x_1 \frac{\partial f}{\partial u_1} + x_2 \frac{\partial f}{\partial u_2} + \cdots + x_j \frac{\partial f}{\partial u_j} \tag{14.13}$$

For the partial derivative of (14.9) with respect to s, we then have

$$\sum_{k=1}^{j} x_k \frac{\partial f(u_1, \ldots, u_j)}{\partial u_k} = ns^{n-1} f(x_1, \ldots, x_j) \tag{14.14}$$

Now let $s = 1$, so that $u_i = x_i$; Eq. (14.14) then gives (14.11). This completes the proof. As an example of Euler's theorem, let $f = x^2 + y^2 + z^2$. We have

$$\sum_{k} x_k \frac{\partial f}{\partial x_k} = x \cdot 2x + y \cdot 2y + z \cdot 2z = 2f \tag{14.15}$$

Now we return to the virial theorem (14.8). If V happens to be a homogeneous function of degree n when expressed in Cartesian coordinates, then Euler's theorem gives

$$\sum_{i} q_i \frac{\partial V}{\partial q_i} = nV \tag{14.16}$$

and the virial theorem (14.8) simplifies to

$$2 \langle T \rangle = n \langle V \rangle \tag{14.17}$$

Since (Problem 6.7)

$$\langle T \rangle + \langle V \rangle = E \tag{14.18}$$

we can write (14.17) in two other forms·

$$\langle V \rangle = \frac{2E}{n+2} \tag{14.19}$$

$$\langle T \rangle = \frac{nE}{n+2} \tag{14.20}$$

Consider some examples. For the one-dimensional harmonic oscillator, we have $V = kx^2/2$, which is homogeneous of degree 2. Hence

$$\langle T \rangle = \langle V \rangle = E/2 = \tfrac{1}{2}hv(v + \tfrac{1}{2}) \tag{14.21}$$

This was verified for the ground state in Problem 4.1.

For the hydrogen atom, V in Cartesian coordinates is

$$V = -\frac{e'^2}{(x^2 + y^2 + z^2)^{1/2}} \tag{14.22}$$

which is a homogeneous function of degree -1. Hence

$$2\langle T \rangle = -\langle V \rangle \tag{14.23}$$

a relation verified for the ground state in Problem 6.8. For any hydrogen-atom stationary state, we have

$$\langle V \rangle = 2E \tag{14.24}$$

$$\langle T \rangle = -E \tag{14.25}$$

For a many-electron atom, the potential energy is

$$V = -Ze'^2 \sum_{i=1}^{n} \frac{1}{(x_i^2 + y_i^2 + z_i^2)^{1/2}} + \sum_i \sum_{j>i} \frac{e'^2}{[(x_i - x_j)^2 + (y_i - y_j)^2 + (z_i - z_j)^2]^{1/2}} \tag{14.26}$$

Replacing each of the $3n$ coordinates by s times the coordinate, we find that V is homogeneous of degree -1. Hence the relations (14.23)–(14.25) are valid for any atom.

Now consider molecules. We shall assume the validity of the Born–Oppenheimer approximation and write the molecular wave function ψ as [Eq. (13.12)]

$$\psi = \psi_{el}(q_i; q_\alpha)\psi_N(q_\alpha) \tag{14.27}$$

where q_i and q_α symbolize the electronic and nuclear coordinates, respectively. The electronic wave function is found by solving the electronic Schroedinger equation (13.7):

$$\hat{H}_{el}\psi_{el}(q_i; q_\alpha) = E_{el}(q_\alpha)\psi_{el}(q_i; q_\alpha) \tag{14.28}$$

where E_{el} is the electronic energy and where

$$\hat{H}_{el} = \hat{T}_{el} + \hat{V}_{el} \tag{14.29}$$

$$\hat{T}_{el} = -\frac{\hbar^2}{2m} \sum_i \left(\frac{\partial^2}{\partial x_i^2} + \frac{\partial^2}{\partial y_i^2} + \frac{\partial^2}{\partial z_i^2}\right) \tag{14.30}$$

$$\hat{V}_{el} = -\sum_\alpha \sum_i \frac{Z_\alpha e'^2}{[(x_i - x_\alpha)^2 + (y_i - y_\alpha)^2 + (z_i - z_\alpha)^2]^{1/2}}$$

$$+ \sum_i \sum_{j>i} \frac{e'^2}{[(x_i - x_j)^2 + (y_i - y_j)^2 + (z_i - z_j)^2]^{1/2}} \tag{14.31}$$

Let the system be in the electronic stationary state ψ_{el}. If we put the subscript el on $\hat{H}$ and ψ in (14.1) and regard the variables of ψ_{el} to be the electronic coordinates q_i (with the nuclear coordinates q_α being parameters), then the derivation of the virial theorem (14.8) is seen to be valid for the electronic kinetic- and potential-energy operators, and we have

$$2\langle\psi_{el}|\hat{T}_{el}|\psi_{el}\rangle = \left\langle\psi_{el}\left|\sum_i q_i\frac{\partial V_{el}}{\partial q_i}\right|\psi_{el}\right\rangle \tag{14.32}$$

Viewed as a function of the electronic coordinates, V_{el} is *not* a homogeneous function, since

$$[(sx_i - x_\alpha)^2 + (sy_i - y_\alpha)^2 + (sz_i - z_\alpha)^2]^{-1/2}$$
$$\neq s^{-1}[(x_i - x_\alpha)^2 + (y_i - y_\alpha)^2 + (z_i - z_\alpha)^2]^{-1/2}$$

Thus the virial theorem for the average electronic kinetic and potential energies of a molecule will not have the simple form (14.23), which holds for atoms. We can, however, view V_{el} as a function of both the electronic and the nuclear Cartesian coordinates; from this viewpoint V_{el} is a homogeneous function of degree -1, since

$$\frac{1}{[(sx_i - sx_\alpha)^2 + (sy_i - sy_\alpha)^2 + (sz_i - sz_\alpha)^2]^{1/2}} = \frac{1}{s[(x_i - x_\alpha)^2 + (y_i - y_\alpha)^2 + (z_i - z_\alpha)^2]^{1/2}}$$

Therefore, considering V_{el} as a function of both electronic and nuclear coordinates and applying Euler's theorem (14.11), we have

$$\sum_i q_i\frac{\partial V_{el}}{\partial q_i} + \sum_\alpha q_\alpha\frac{\partial V_{el}}{\partial q_\alpha} = -V_{el} \tag{14.33}$$

Using (14.33) in (14.32), we get

$$2\langle\psi_{el}|\hat{T}_{el}|\psi_{el}\rangle = -\langle\psi_{el}|\hat{V}_{el}|\psi_{el}\rangle - \left\langle\psi_{el}\left|\sum_\alpha q_\alpha\frac{\partial V_{el}}{\partial q_\alpha}\right|\psi_{el}\right\rangle \tag{14.34}$$

which contains an additional term as compared to the atomic virial theorem (14.23).

Consider this extra term. We have

$$\left\langle\psi_{el}\left|\sum_\alpha q_\alpha\frac{\partial V_{el}}{\partial q_\alpha}\right|\psi_{el}\right\rangle = \sum_\alpha q_\alpha\int\psi_{el}^*\frac{\partial V_{el}}{\partial q_\alpha}\psi_{el}\,d\tau_{el} \tag{14.35}$$

where the nuclear coordinate q_α has been taken outside the integral over electronic coordinates. In Section 14.3 we will show that

$$\int\psi_{el}^*\frac{\partial V_{el}}{\partial q_\alpha}\psi_{el}\,d\tau_{el} = \frac{\partial E_{el}}{\partial q_\alpha} \tag{14.36}$$

[Equation (14.36) is an example of the Hellmann–Feynman theorem.] Using these last two equations in the molecular electronic virial theorem (14.34), we get

$$2\langle\psi_{el}|\hat{T}_{el}|\psi_{el}\rangle = -\langle\psi_{el}|\hat{V}_{el}|\psi_{el}\rangle - \sum_\alpha q_\alpha\frac{\partial E_{el}}{\partial q_\alpha} \tag{14.37}$$

where the q_α are the nuclear *Cartesian* coordinates. We abbreviate (14.37) as

$$2\langle T_{el}\rangle = -\langle V_{el}\rangle - \sum_\alpha q_\alpha\frac{\partial E_{el}}{\partial q_\alpha} \tag{14.38}$$

Using

$$\langle T_{el} \rangle + \langle V_{el} \rangle = E_{el} \qquad (14.39)$$

we can eliminate either $\langle T_{el} \rangle$ or $\langle V_{el} \rangle$ from (14.38).

Now consider a diatomic molecule. The electronic energy is a function of the internuclear distance R: $E_{el} = E_{el}(R)$. The summation in (14.38) is over the nuclear Cartesian coordinates x_a, y_a, z_a, x_b, y_b, z_b. We have

$$\frac{\partial E_{el}}{\partial x_a} = \frac{dE_{el}}{dR}\frac{\partial R}{\partial x_a}, \qquad \frac{\partial E_{el}}{\partial x_b} = \frac{dE_{el}}{dR}\frac{\partial R}{\partial x_b} \qquad (14.40)$$

$$R = [(x_a - x_b)^2 + (y_a - y_b)^2 + (z_a - z_b)^2]^{1/2} \qquad (14.41)$$

$$\frac{\partial R}{\partial x_a} = \frac{x_a - x_b}{R}, \qquad \frac{\partial R}{\partial x_b} = \frac{x_b - x_a}{R} \qquad (14.42)$$

with similar equations for the y and z coordinates. The sum in (14.38) becomes

$$\sum_{\alpha} q_{\alpha} \frac{\partial E_{el}}{\partial q_{\alpha}} = \frac{1}{R}\frac{dE_{el}}{dR}[x_a(x_a - x_b) + x_b(x_b - x_a) + y_a(y_a - y_b)$$

$$+ y_b(y_b - y_a) + z_a(z_a - z_b) + z_b(z_b - z_a)]$$

$$\sum_{\alpha} q_{\alpha} \frac{\partial E_{el}}{\partial q_{\alpha}} = R\frac{dE_{el}}{dR} \qquad (14.43)$$

The virial theorem (14.38) for a diatomic molecule reads

$$2\langle T_{el} \rangle = -\langle V_{el} \rangle - R\frac{dE_{el}}{dR} \qquad (14.44)$$

Using (14.39), we have the two alternative forms

$$\langle T_{el} \rangle = -E_{el} - R\frac{dE_{el}}{dR} \qquad (14.45)$$

$$\langle V_{el} \rangle = 2E_{el} + R\frac{dE_{el}}{dR} \qquad (14.46)$$

In deriving the molecular electronic virial theorem, we omitted the internuclear repulsion

$$V_{NN} = \sum_{\beta}\sum_{\alpha > \beta} \frac{Z_{\alpha}Z_{\beta}e'^2}{[(x_{\alpha} - x_{\beta})^2 + (y_{\alpha} - y_{\beta})^2 + (z_{\alpha} - z_{\beta})^2]^{1/2}} \qquad (14.47)$$

from the electronic Hamiltonian (14.29)–(14.31). Let

$$V = V_{el} + V_{NN} \qquad (14.48)$$

where V_{el} is given by (14.31). We can rewrite the electronic Schroedinger equation (14.28) as [Eq. (13.4)]

$$(\hat{T}_{el} + \hat{V})\psi_{el} = U(q_{\alpha})\psi_{el} \qquad (14.49)$$

where

$$U(q_{\alpha}) = E_{el}(q_{\alpha}) + V_{NN} \qquad (14.50)$$

$U(q_\alpha)$ is the potential-energy function for nuclear motion. Now consider what happens to the right side of Eq. (14.37) when we add in V_{NN} to V_{el} and E_{el}. We have

$$-\int \psi_{el}^*(\hat{V}_{el} + V_{NN})\psi_{el}\, d\tau_{el} - \sum_\alpha q_\alpha \frac{\partial U}{\partial q_\alpha}$$

$$= -\langle \psi_{el}|\hat{V}_{el}|\psi_{el}\rangle - V_{NN} - \sum_\alpha q_\alpha \frac{\partial E_{el}}{\partial q_\alpha} - \sum_\alpha q_\alpha \frac{\partial V_{NN}}{\partial q_\alpha} \quad \textbf{(14.51)}$$

V_{NN} is a homogeneous function of the nuclear Cartesian coordinates of degree -1, so that (Euler's theorem)

$$\sum_\alpha q_\alpha \frac{\partial V_{NN}}{\partial q_\alpha} = -V_{NN} \quad \textbf{(14.52)}$$

and (14.51) becomes

$$-\langle \psi_{el}|\hat{V}_{el} + V_{NN}|\psi_{el}\rangle - \sum_\alpha q_\alpha \frac{\partial U}{\partial q_\alpha} = -\langle \psi_{el}|\hat{V}_{el}|\psi_{el}\rangle - \sum_\alpha q_\alpha \frac{\partial E_{el}}{\partial q_\alpha} \quad \textbf{(14.53)}$$

Substitution of (14.53) into (14.37) gives

$$2\langle \psi_{el}|\hat{T}_{el}|\psi_{el}\rangle = -\langle \psi_{el}|\hat{V}_{el} + V_{NN}|\psi_{el}\rangle - \sum_\alpha q_\alpha \frac{\partial U}{\partial q_\alpha}$$

$$2\langle T_{el}\rangle = -\langle V \rangle - \sum_\alpha q_\alpha \frac{\partial U}{\partial q_\alpha} \quad \textbf{(14.54)}$$

so that the molecular electronic virial theorem holds whether or not we include the internuclear repulsion. Corresponding to Eqs. (14.44)–(14.46) for diatomic molecules, we have

$$2\langle T_{el}\rangle = -\langle V \rangle - R(dU/dR) \quad \textbf{(14.55)}$$

$$\langle T_{el}\rangle = -U - R(dU/dR) \quad \textbf{(14.56)}$$

$$\langle V \rangle = 2U + R(dU/dR) \quad \textbf{(14.57)}$$

The potential energy V is given by Eqs. (14.48), (14.47), and (14.31); we see that the zero of energy is taken with all particles (electrons and nuclei) at infinite separation from each other. Thus the function $U(R)$ in (14.55)–(14.57) does not go to zero at $R = \infty$ but goes to the sum of the energies of the separated atoms, which is negative.

The molecular electronic virial theorem was first derived by Slater.

For polyatomic molecules (14.54) can be written as

$$2\langle T_{el}\rangle = -\langle V \rangle - \sum_\alpha \sum_{\beta > \alpha} R_{\alpha\beta}(\partial U/\partial R_{\alpha\beta})$$

where the sum runs over either all internuclear distances or the bond lengths only; proofs are given in R. G. Parr and J. E. Brown, *J. Chem. Phys.*, **49**, 4849 (1968) and B. Nelander, *J. Chem. Phys.*, **51**, 469 (1969).

The true wave functions for a system with V a homogeneous function of the coordinates must satisfy the form of the virial theorem (14.17). We now ask: What determines whether an *approximate* wave function for such a system satisfies (14.17)? The answer is that by inserting a variational parameter as a multiplier of each Cartesian coordinate and choosing this parameter to minimize the variational integral, we can make any trial variation function satisfy the virial theorem. (For the proof, see *Kauzmann*, p. 229.) This process is called *scaling*, and the variational parameter multiplying each coordinate is called a *scale factor*. For a molecular trial function, the scaling parameter must be inserted in front of the nuclear Cartesian coordinates, as well as in front of the electronic coordinates.

Consider some examples. The zeroth-order perturbation wave function (9.60) for the heliumlike atom has no scale factor and so does not satisfy the virial theorem. If we were to calculate $\langle T \rangle$ and $\langle V \rangle$ for (9.60), we would find $2\langle T \rangle \neq -\langle V \rangle$; see Problem 14.7. The Heitler–London trial function for H_2, Eq. (13.107), has no scale factor and does not satisfy the virial theorem. The Heitler–London–Wang function, which uses a variationally determined orbital exponent, satisfies the virial theorem. Hartree–Fock wave functions satisfy the virial theorem—note the scale factor in the Slater basis functions (11.16).

14.2 THE VIRIAL THEOREM AND CHEMICAL BONDING

We now use the virial theorem to examine the changes in electronic kinetic and potential energy that occur when a covalent chemical bond is formed in a diatomic molecule. The prime requisite for formation of a stable bond is that the $U(R)$ curve have a substantial minimum. At this minimum we have

$$\left.\frac{dU}{dR}\right|_{R_e} = 0 \qquad (14.58)$$

and Eqs. (14.55)–(14.57) become

$$2\langle T_{el}\rangle|_{R_e} = -\langle V \rangle|_{R_e} \qquad (14.59)$$

$$\langle T_{el}\rangle|_{R_e} = -U(R_e) \qquad (14.60)$$

$$\langle V \rangle|_{R_e} = 2U(R_e) \qquad (14.61)$$

These equations resemble those for atoms [Eqs. (14.23)–(14.25)]. At $R = \infty$ we have the separated atoms, and the atomic virial theorem gives

$$2\langle T_{el}\rangle|_\infty = -\langle V \rangle|_\infty \qquad (14.62)$$

$$\langle T_{el}\rangle|_\infty = -U(\infty) \qquad (14.63)$$

$$\langle V \rangle|_\infty = 2U(\infty) \qquad (14.64)$$

$U(\infty)$ is the sum of the energies of the two separated atoms. Therefore

$$\langle T_{el}\rangle|_{R_e} - \langle T_{el}\rangle|_\infty = U(\infty) - U(R_e) \qquad (14.65)$$

$$\langle V \rangle|_{R_e} - \langle V \rangle|_\infty = 2[U(R_e) - U(\infty)] \qquad (14.66)$$

For bonding we have

$$U(R_e) < U(\infty)$$

Thus the average molecular potential energy at R_e is *less* than the sum of the potential energies of the separated atoms, while the average molecular kinetic energy is *greater* at R_e than at ∞. The decrease in potential energy is twice the increase in kinetic energy, and results from allowing the electrons to feel the attractions of both nuclei and from the increase in orbital exponents in the molecule (Section 13.4). The dissociation energy (13.9) is

$$D_e = \tfrac{1}{2}[\langle V \rangle|_\infty - \langle V \rangle|_{R_e}] \tag{14.67}$$

Consider the behavior of the average potential and kinetic energies for large R. The interactions between atoms at large distances are called *van der Waals forces*. For two neutral atoms, at least one of which is in an S state, quantum-mechanical perturbation theory shows[2] that the van der Waals force of attraction is proportional to $1/R^7$, and the potential energy behaves like

$$U(R) \approx U(\infty) - \frac{A}{R^6}, \qquad R \text{ large} \tag{14.68}$$

where A is a positive constant. This expression was first derived by London, and van der Waals forces between neutral atoms are called *London forces* (or dispersion forces); London forces[3] can be viewed as due to the polarization of one atom by the instantaneous dipole moment of the other.

Using (14.68), (14.56), (14.57), (14.63), and (14.64), we have

$$\langle V \rangle \approx \langle V \rangle|_\infty + \frac{4A}{R^6}, \qquad R \text{ large} \tag{14.69}$$

$$\langle T_{el} \rangle \approx \langle T_{el} \rangle|_\infty - \frac{5A}{R^6}, \qquad R \text{ large} \tag{14.70}$$

Hence as R decreases from infinity, the average potential energy at first increases, while the average kinetic energy at first decreases. The combination of these conclusions with our conclusions about $\langle V \rangle|_{R_e}$ and $\langle T_{el} \rangle|_{R_e}$ shows that $\langle V \rangle$ must go through a maximum somewhere between R_e and infinity and $\langle T_{el} \rangle$ must go through a minimum in this region.

For R much less than R_e, we can use (13.80) and (13.8) to write

$$U(R) \approx \frac{Z_a Z_b e'^2}{R} + E_{UA} + aR^2, \qquad R \text{ small} \tag{14.71}$$

where E_{UA} is the united-atom energy. The virial theorem then gives

$$\langle T_{el} \rangle \approx -E_{UA} - 3aR^2, \qquad R \text{ small} \tag{14.72}$$

$$\langle V \rangle \approx \frac{Z_a Z_b e'^2}{R} + 2E_{UA} + 4aR^2, \qquad R \text{ small} \tag{14.73}$$

---

[2] See *Landau and Lifshitz*, Section 89; *Kauzmann*, Chapter 13.
[3] Recall the discussion near the end of Sec. 13.6.

The virial theorem holds for the united atom:

$$\langle T_{\text{el}} \rangle|_0 = -E_{\text{UA}}, \qquad \langle V_{\text{el}} \rangle|_0 = 2E_{\text{UA}} \tag{14.74}$$

and we have

$$\langle T_{\text{el}} \rangle \approx \langle T_{\text{el}} \rangle|_0 - 3aR^2, \qquad\qquad R \text{ small} \tag{14.75}$$

$$\langle V \rangle \approx \frac{Z_a Z_b e'^2}{R} + \langle V_{\text{el}} \rangle|_0 + 4aR^2, \quad R \text{ small} \tag{14.76}$$

The average value of $\langle V \rangle$ goes to infinity as R goes to zero, because of the internuclear repulsion.

Having found the general behavior of $\langle V \rangle$ and $\langle T_{\text{el}} \rangle$ as functions of R, we now draw Fig. 14.1. This figure is not for any particular molecule but resembles the known curves[4] for H_2 and $H_2{}^+$.

What explanation can we give for the change in average kinetic and potential energy with R? Consider $H_2{}^+$. The electronic potential-energy function is

$$V_{\text{el}} = -\frac{e'^2}{r_a} - \frac{e'^2}{r_b} \tag{14.77}$$

If we plot the values of V_{el} for points on the molecular axis for a large value of R, we get a curve like Fig. 14.2, which resembles two hydrogen-atom potential-energy curves (Fig. 6.5) placed side by side. We have seen that the overlapping of the $1s$ AO's occurring in molecule formation leads to a buildup of charge probability density between the nuclei for the ground state. However, Fig. 14.2 shows that the potential energy is relatively *high* in the region midway between the nuclei when R is large. Thus we get an initial increase in $\langle V \rangle$ as R decreases from infinity. Now consider the kinetic energy. The uncertainty principle (5.18) gives $(\Delta x)^2 (\Delta p_x)^2 \geq \hbar^2/4$. For a stationary state, $\langle p_x \rangle$ is zero [see for example (3.125)] and (5.15) gives $(\Delta p_x)^2 = \langle p_x^2 \rangle$. Hence a small value of $(\Delta x)^2$ means a large value of $\langle p_x^2 \rangle$ and a large value of the average kinetic energy (which equals $\langle p^2 \rangle/2m$). Thus a compact ψ_{el} corresponds to a large electronic kinetic energy. In the separated atoms, the wave function is concentrated in two rather small regions about each nucleus (Fig. 6.6). In the initial stages of molecule formation, the buildup of probability density between the nuclei corresponds to having a wave function that is less compact than it was in the separated atoms. Thus as R decreases from infinity, the electronic kinetic energy initially decreases. The energies E_{el} of the two lowest $H_2{}^+$ states have been indicated in Fig. 14.2. For large R the region between the nuclei is classically forbidden, but it is accessible according to quantum mechanics (tunneling).

Now consider what happens as R decreases further. Plotting (14.77) for an intermediate value of R, we find that now the region between the nuclei is a region of *low* potential energy, since an electron in this region feels substantial attractions from both nuclei. (See Fig. 14.3.) Hence at intermediate values of R, the overlap charge buildup between the nuclei (and the increase in orbital exponents) lowers the potential energy. For intermediate values of R, the wave function has become more

---

[4] W. Kolos and L. Wolniewicz, *J. Chem. Phys.*, **41**, 3663 (1964); *Slater, Quantum Theory of Molecules and Solids*, vol. 1, p. 36.

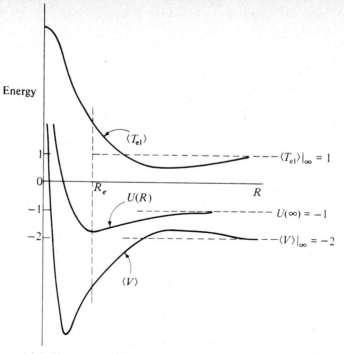

Figure 14.1 *Variation of the average potential and kinetic energies of a diatomic molecule. The unit of energy is taken as the electronic kinetic energy of the separated atoms.*

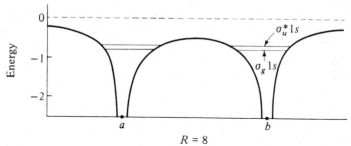

Figure 14.2 *Potential energy along the internuclear axis for electronic motion in H_2^+ for large internuclear separation. Atomic units are used.*

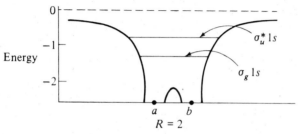

Figure 14.3 *Potential energy along the internuclear axis for electronic motion in H_2^+ for intermediate internuclear distance*

compact as compared to large R (in part because of the increase in orbital exponents), which gives an increase in $\langle T_{el} \rangle$ as R is reduced. In fact, we see from Fig. 14.1 and Eq. (14.65) that $\langle T_{el} \rangle$ is *greater* at R_e in the molecule than in the separated atoms; hence the molecular wave function at R_e is more compact than the separated-atoms wave functions.

For very small R, the average potential energy goes to infinity, because of the internuclear repulsion. *However*, for $R = R_e$ Fig. 14.1 shows that $\langle V \rangle$ is still decreasing sharply with decreasing R, and it is the increase in $\langle T_{el} \rangle$, and not the nuclear repulsion, that causes the $U(R)$ curve to turn up as R becomes less than R_e. The squeezing of the molecular wave function into a smaller region with the associated increase in $\langle T_{el} \rangle$ is more important than the internuclear repulsion in causing the initial repulsion between the atoms.

14.3   *THE HELLMANN–FEYNMAN THEOREM*

Consider a system with a time-independent Hamiltonian $\hat{H}$ that involves parameters. An obvious example is the molecular electronic Hamiltonian (13.5), which depends parametrically on the nuclear coordinates. However, the Hamiltonian of any system contains parameters. For example, the one-dimensional harmonic-oscillator Hamiltonian is

$$-\frac{\hbar^2}{2m}\frac{d^2}{dx^2} + \tfrac{1}{2}kx^2 \tag{14.78}$$

The force constant k is a parameter, as is the mass m. Although $\hbar$ is a constant, there is nothing to prevent us from considering it as a parameter also. The stationary-state energies E_n are functions of the same parameters as $\hat{H}$. For example, for the harmonic oscillator

$$E_n = (v + \tfrac{1}{2})h\nu = (v + \tfrac{1}{2})\hbar(k/m)^{1/2} \tag{14.79}$$

The stationary-state wave functions also depend on the parameters in $\hat{H}$. We now investigate how E_n varies with each of the parameters. More specifically, let λ be one of the parameters; we ask for $\partial E_n/\partial \lambda$, where the partial derivative is taken with all other parameters held constant.

We begin with the Schroedinger equation

$$\hat{H}\psi_n = E_n\psi_n \tag{14.80}$$

where the ψ_n are the normalized stationary-state eigenfunctions. Because of normalization, we have

$$E_n = \int \psi_n^* \hat{H} \psi_n \, d\tau \tag{14.81}$$

$$\frac{\partial E_n}{\partial \lambda} = \frac{\partial}{\partial \lambda} \int \psi_n^* \hat{H} \psi_n \, d\tau \tag{14.82}$$

The integral in (14.81) is a definite integral over all space, and its value depends parametrically on λ, since $\hat{H}$ and ψ_n depend on λ. Provided the integrand is well behaved, we can evaluate the integral's derivative with respect to a parameter by differentiating under the integral sign. (We used this procedure in Problems 7.18 and 8.2.) Therefore

$$\frac{\partial E_n}{\partial \lambda} = \int \frac{\partial}{\partial \lambda} (\psi_n^* \hat{H} \psi_n) \, d\tau \tag{14.83}$$

$$\frac{\partial E_n}{\partial \lambda} = \int \frac{\partial \psi_n^*}{\partial \lambda} \hat{H} \psi_n \, d\tau + \int \psi_n^* \frac{\partial}{\partial \lambda} (\hat{H} \psi_n) \, d\tau \tag{14.84}$$

We have

$$\frac{\partial}{\partial \lambda} (\hat{H} \psi_n) = \frac{\partial}{\partial \lambda} (\hat{T} \psi_n) + \frac{\partial}{\partial \lambda} (\hat{V} \psi_n) \tag{14.85}$$

The potential-energy operator is just multiplication by V, so

$$\frac{\partial}{\partial \lambda} (\hat{V} \psi_n) = \frac{\partial V}{\partial \lambda} \psi_n + V \frac{\partial \psi_n}{\partial \lambda} \tag{14.86}$$

The parameter λ will occur in the kinetic-energy operator as part of the factor multiplying one or more of the derivatives with respect to the coordinates. For example, taking λ as the mass of the particle, we have for a one-particle problem

$$\hat{T} = -\frac{\hbar^2}{2\lambda} \left(\frac{\partial^2}{\partial x^2} + \frac{\partial^2}{\partial y^2} + \frac{\partial^2}{\partial z^2} \right) \tag{14.87}$$

$$\frac{\partial}{\partial \lambda} (\hat{T} \psi) = -\frac{\hbar^2}{2} \frac{\partial}{\partial \lambda} \left[\frac{1}{\lambda} \left(\frac{\partial^2 \psi}{\partial x^2} + \frac{\partial^2 \psi}{\partial y^2} + \frac{\partial^2 \psi}{\partial z^2} \right) \right]$$

$$= \frac{\hbar^2}{2\lambda^2} \left(\frac{\partial^2 \psi}{\partial x^2} + \frac{\partial^2 \psi}{\partial y^2} + \frac{\partial^2 \psi}{\partial z^2} \right) - \frac{\hbar^2}{2\lambda} \left(\frac{\partial^2}{\partial x^2} + \frac{\partial^2}{\partial y^2} + \frac{\partial^2}{\partial z^2} \right) \left(\frac{\partial \psi}{\partial \lambda} \right)$$

since we can change the order of the partial differentiations without affecting the result. We can write this last equation as

$$\frac{\partial}{\partial \lambda} (\hat{T} \psi_n) = \left(\frac{\partial \hat{T}}{\partial \lambda} \right) \psi_n + \hat{T} \left(\frac{\partial \psi_n}{\partial \lambda} \right) \tag{14.88}$$

where $(\partial \hat{T} / \partial \lambda)$ is found by differentiating $\hat{T}$ with respect to λ just as if it were a function instead of an operator. Although we got (14.88) by considering a specific $\hat{T}$ and λ, the same arguments show it to be generally valid. Combining (14.88) and (14.86), we write

$$\frac{\partial}{\partial \lambda} (\hat{H} \psi_n) = \left(\frac{\partial \hat{H}}{\partial \lambda} \right) \psi_n + \hat{H} \left(\frac{\partial \psi_n}{\partial \lambda} \right) \tag{14.89}$$

Equation (14.84) becomes

$$\frac{\partial E_n}{\partial \lambda} = \int \frac{\partial \psi_n^*}{\partial \lambda} \hat{H} \psi_n \, d\tau + \int \psi_n^* \frac{\partial \hat{H}}{\partial \lambda} \psi_n \, d\tau + \int \psi_n^* \hat{H} \frac{\partial \psi_n}{\partial \lambda} \, d\tau \tag{14.90}$$

For the first integral in (14.90), we have

$$\int \frac{\partial \psi_n^*}{\partial \lambda} \hat{H} \psi_n \, d\tau = E_n \int \frac{\partial \psi_n^*}{\partial \lambda} \psi_n \, d\tau \tag{14.91}$$

The Hermitian property of $\hat{H}$ and (14.80) give for the last integral in (14.90)

$$\int \psi_n^* \hat{H} \frac{\partial \psi_n}{\partial \lambda} d\tau = \int \frac{\partial \psi_n}{\partial \lambda} (\hat{H} \psi_n)^* d\tau = E_n \int \psi_n^* \frac{\partial \psi_n}{\partial \lambda} d\tau$$

Therefore

$$\frac{\partial E_n}{\partial \lambda} = \int \psi_n^* \frac{\partial \hat{H}}{\partial \lambda} \psi_n d\tau + E_n \int \frac{\partial \psi_n^*}{\partial \lambda} \psi_n d\tau + E_n \int \psi_n^* \frac{\partial \psi_n}{\partial \lambda} d\tau \quad \textbf{(14.92)}$$

The wave function is normalized; hence

$$\int \psi_n^* \psi_n d\tau = 1$$

$$\frac{\partial}{\partial \lambda} \int \psi_n^* \psi_n d\tau = 0$$

$$\int \frac{\partial \psi_n^*}{\partial \lambda} \psi_n d\tau + \int \psi_n^* \frac{\partial \psi_n}{\partial \lambda} d\tau = 0 \quad \textbf{(14.93)}$$

Using (14.93) in (14.92), we obtain

$$\frac{\partial E_n}{\partial \lambda} = \int \psi_n^* \frac{\partial \hat{H}}{\partial \lambda} \psi_n d\tau \quad \textbf{(14.94)}$$

Equation (14.94) is called the *generalized Hellmann–Feynman theorem*, although it seems to have first been given by Pauli.[5]

Consider some examples. If we take λ to be the force constant k in the harmonic-oscillator Hamiltonian (14.78), then (14.94) gives

$$\frac{\partial}{\partial k} \left[(v + \tfrac{1}{2}) \hbar \left(\frac{k}{m} \right)^{1/2} \right] = \int \psi_n^* (\tfrac{1}{2} x^2) \psi_n d\tau$$

$$\int \psi_n^* x^2 \psi_n d\tau = (v + \tfrac{1}{2}) \frac{h\nu}{k} \quad \textbf{(14.95)}$$

We have found an expression for the average value of x^2 for any harmonic-oscillator stationary state, without evaluating any integrals. This result was also obtained from the virial theorem; see Eq. (14.21). For a third derivation, see *Eyring, Walter, and Kimball*, p. 79.

Application of the Hellmann–Feynman theorem to the hydrogenlike atom, with Z as the parameter, gives (Problem 14.6a)

$$\int r^{-1} |\psi|^2 d\tau = \left\langle \frac{1}{r} \right\rangle = \frac{Z}{n^2} \left(\frac{1}{a_0} \right) \quad \textbf{(14.96)}$$

This result was also obtained from the virial theorem; see Eq. (14.24).

The first-order perturbation theory result Eq. (9.30) is a special case of the

---

[5] For a discussion of the origin of the Hellmann–Feynman and related theorems, see J. I. Musher, *Am. J. Phys.*, **34**, 267 (1966).

Hellmann–Feynman theorem. Differentiation of Eqs. (9.7) and (9.18) gives

$$\frac{\partial E_n}{\partial \lambda} = E_n^{(1)} + 2\lambda E_n^{(2)} + \cdots \tag{14.97}$$

$$\frac{\partial \hat{H}}{\partial \lambda} = \hat{H}' \tag{14.98}$$

The Hellmann–Feynman theorem (14.94) gives

$$E_n^{(1)} + 2\lambda E_n^{(2)} + \cdots = \int \psi_n^* \hat{H}' \psi_n \, d\tau \tag{14.99}$$

If we set λ equal to zero, then ψ_n goes over to $\psi_n^{(0)}$ [Eq. (9.17)], and we get the result (9.30).[6]

The Hellmann–Feynman theorem (with E_n being the Hartree–Fock energy) is obeyed by Hartree–Fock (as well as exact) wave functions.[7]

14.4   THE ELECTROSTATIC THEOREM

Hellmann and Feynman independently applied Eq. (14.94) to molecules, taking λ as a nuclear Cartesian coordinate. We now consider their results.

[Hellmann was a German physical chemist. The Nazis had him dismissed from his position, and he then settled in Russia. He was executed as a "German spy" in the Stalinist purges of the late 1930s. His results appear in his textbook *Einführung in die Quantenchemie*, Deuticke, Leipzig, 1937, p. 285. The American physicist Richard Feynman is noted for his contributions to quantum field theory. His paper "Forces in Molecules"—*Phys. Rev.*, **56**, 340 (1939)—consists of work he did as an undergraduate at MIT.]

As usual, we are using the Born–Oppenheimer approximation, solving the electronic Schroedinger equation for a fixed nuclear configuration [Eq. (14.49)]:

$$\hat{H}\psi_{el} = (\hat{T}_{el} + \hat{V})\psi_{el} = U\psi_{el} \tag{14.100}$$

The potential energy V is

$$\hat{V} = \hat{V}_{el} + \hat{V}_{NN} \tag{14.101}$$

The operators $\hat{T}_{el}$, $\hat{V}_{el}$, and $\hat{V}_{NN}$ are given by (14.30), (14.31), and (14.47). The Hamiltonian $\hat{H}$ depends on the nuclear Cartesian coordinates as parameters. If x_δ is the x coordinate of nucleus δ, the generalized Hellmann–Feynman theorem (14.94) gives

$$\frac{\partial U}{\partial x_\delta} = \int \psi_{el}^* \frac{\partial \hat{H}}{\partial x_\delta} \psi_{el} \, d\tau_{el} \tag{14.102}$$

---

[6] For further discussion of the relation between perturbation theory and the Hellmann–Feynman theorem, see S. T. Epstein, *Am. J. Phys.*, **22**, 613 (1954).

[7] For a proof see R. E. Stanton, *J. Chem. Phys.*, **36**, 1298 (1962).

The kinetic-energy part of $\hat{H}$ is independent of the nuclear Cartesian coordinates, as can be seen from (14.30). Hence

$$\frac{\partial U}{\partial x_\delta} = \int \psi_{el}^* \frac{\partial V}{\partial x_\delta} \psi_{el} \, d\tau_{el} \qquad (14.103)$$

[If we had omitted the internuclear repulsion V_{NN} from V, we would have obtained Eq. (14.36), which was used in deriving the molecular electronic virial theorem.] We have

$$\frac{\partial V}{\partial x_\delta} = \frac{\partial V_{el}}{\partial x_\delta} + \frac{\partial V_{NN}}{\partial x_\delta} \qquad (14.104)$$

From (14.31) we get

$$\frac{\partial V_{el}}{\partial x_\delta} = -\sum_i \frac{Z_\delta (x_i - x_\delta) e'^2}{r_{i\delta}^3} \qquad (14.105)$$

where $r_{i\delta}$ is the distance from electron i to nucleus δ. To evaluate $\partial V_{NN}/\partial x_\delta$, we need consider only those internuclear repulsion terms that involve nucleus δ; hence

$$\frac{\partial V_{NN}}{\partial x_\delta} = \frac{\partial}{\partial x_\delta} \sum_{\alpha \neq \delta} \frac{Z_\alpha Z_\delta e'^2}{[(x_\alpha - x_\delta)^2 + (y_\alpha - y_\delta)^2 + (z_\alpha - z_\delta)^2]^{1/2}}$$

$$= \sum_{\alpha \neq \delta} Z_\alpha Z_\delta e'^2 \frac{(x_\alpha - x_\delta)}{R_{\alpha\delta}^3} \qquad (14.106)$$

where $R_{\alpha\delta}$ is the distance between nuclei α and δ. Noting that (14.106) does not involve the electronic coordinates and that ψ_{el} is normalized, we see that (14.103) becomes

$$\frac{\partial U}{\partial x_\delta} = -\sum_i Z_\delta e'^2 \int |\psi_{el}|^2 \frac{(x_i - x_\delta)}{r_{i\delta}^3} \, d\tau_{el} + \sum_{\alpha \neq \delta} Z_\alpha Z_\delta e'^2 \frac{(x_\alpha - x_\delta)}{R_{\alpha\delta}^3} \qquad (14.107)$$

Consider the integral in (14.107). The factor multiplying $|\psi_{el}|^2$ depends only on the coordinates of electron i, and this integral equals (recall that $\int d\tau$ implies a summation over all spin variables, as well as integration over all space coordinates)

$$\int_{-\infty}^{\infty} \int_{-\infty}^{\infty} \int_{-\infty}^{\infty} \left[\sum_{\text{all } m_s} \int' |\psi_{el}|^2 \, dv_{el}' \right] \frac{(x_i - x_\delta)}{r_{i\delta}^3} \, dx_i \, dy_i \, dz_i$$

where by $\int' dv_{el}'$ we mean an integration over the $3n - 3$ spatial coordinates of all electrons except electron i. From Eq. (13.131) we see that the quantity in brackets is equal to ρ/n, where ρ is the electron probability density of the n-electron molecule. Hence (14.107) simplifies to[8]

$$\frac{\partial U}{\partial x_\delta} = -\frac{1}{n} \sum_{i=1}^{n} Z_\delta e'^2 \int \rho(x_i, y_i, z_i) \frac{(x_i - x_\delta)}{r_{i\delta}^3} \, dv_i + \sum_{\alpha \neq \delta} Z_\alpha Z_\delta e'^2 \frac{(x_\alpha - x_\delta)}{R_{\alpha\delta}^3} \qquad (14.108)$$

The integration variables in each of the n definite integrals in (14.108) are

---

[8] Recall that we used this kind of simplification in Section 13.12 on dipole moments.

dummy variables, and the integrals have the same value no matter what the value of i is. Thus each term in the first sum in (14.108) has the same value, and since there are n terms in this sum, we get

$$\frac{\partial U}{\partial x_\delta} = -Z_\delta e'^2 \iiint \rho(x, y, z)\frac{(x - x_\delta)}{r_\delta^3}\, dx\, dy\, dz + \sum_{\alpha \neq \delta} Z_\alpha Z_\delta e'^2 \frac{(x_\alpha - x_\delta)}{R_{\alpha\delta}^3} \qquad (14.109)$$

where we have dropped the superfluous subscript i. The variable r_δ is the distance between nucleus δ and point (x, y, z) in space:

$$r_\delta = [(x - x_\delta)^2 + (y - y_\delta)^2 + (z - z_\delta)^2]^{1/2}$$

What is the significance of (14.109)? In the Born–Oppenheimer approximation, the function $U(x_\alpha, y_\alpha, z_\alpha, x_\beta, \ldots)$ is the potential-energy function for nuclear motion, the nuclear Schroedinger equation being

$$\left[-\frac{\hbar^2}{2}\sum_\alpha \frac{1}{m_\alpha}\nabla_\alpha^2 + U \right]\psi_N = E\psi_N \qquad (14.110)$$

The quantity $-\partial U/\partial x_\delta$ can thus be viewed [see Eq. (5.46)] as the x component of the effective force on nucleus δ due to the other nuclei and the electrons. In addition to (14.109), we have two corresponding equations for $\partial U/\partial y_\delta$ and $\partial U/\partial z_\delta$; if $\mathbf{F}_\delta$ is the effective force on nucleus δ, we have

$$\mathbf{F}_\delta = -\mathbf{i}\frac{\partial U}{\partial x_\delta} - \mathbf{j}\frac{\partial U}{\partial y_\delta} - \mathbf{k}\frac{\partial U}{\partial z_\delta} \qquad (14.111)$$

$$\mathbf{F}_\delta = -Z_\delta e'^2 \iiint \rho(x, y, z)\frac{\mathbf{r}_\delta}{r_\delta^3}\, dx\, dy\, dz + e'^2 \sum_{\alpha \neq \delta} Z_\alpha Z_\delta \frac{\mathbf{R}_{\alpha\delta}}{R_{\alpha\delta}^3} \qquad (14.112)$$

where $\mathbf{r}_\delta$ is the vector from point (x, y, z) to nucleus δ:

$$\mathbf{r}_\delta = \mathbf{i}(x_\delta - x) + \mathbf{j}(y_\delta - y) + \mathbf{k}(z_\delta - z) \qquad (14.113)$$

and where $\mathbf{R}_{\alpha\delta}$ is the vector from nucleus α to nucleus δ:

$$\mathbf{R}_{\alpha\delta} = \mathbf{i}(x_\delta - x_\alpha) + \mathbf{j}(y_\delta - y_\alpha) + \mathbf{k}(z_\delta - z_\alpha) \qquad (14.114)$$

Equation (14.112) has a simple physical interpretation. Let us imagine the electrons smeared out into a charge distribution whose density is $-e\rho(x, y, z)$. The force on nucleus δ exerted by the infinitesimal element of electronic charge $-e\rho\, dx\, dy\, dz$ is [Eq. (6.64)]

$$-Z_\delta e'^2\frac{\mathbf{r}_\delta}{r_\delta^3}\rho\, dx\, dy\, dz \qquad (14.115)$$

and integration of (14.115) shows that the total force exerted on δ by this hypothetical electron smear is given by the first term on the right of (14.112). The second term on the right of (14.112) is clearly the Coulomb's law force on nucleus δ due to the electrostatic repulsions of the other nuclei.

Thus the effective force acting on a nucleus in a molecule can be calculated by simple electrostatics as the sum of the Coulombic forces exerted by the other nuclei and by a hypothetical electron cloud whose charge density $-e\rho(x, y, z)$ is found by solving the electronic Schroedinger equation. This statement is called the *electrostatic theorem*. It is frequently called the Hellmann–Feynman theorem, but to avoid confusion with the generalized Hellmann–Feynman theorem (of which it is a special case), the term "electrostatic theorem" is widely used. The electron probability density depends on the parameters defining the nuclear configuration:

$$\rho = \rho(x, y, z; x_\alpha, y_\alpha, z_\alpha, x_\beta, \dots)$$

It is quite reasonable that the electrostatic theorem follows from the Born–Oppenheimer approximation, since the rapid motion of the electrons allows the electronic wave function and probability density to adjust immediately to changes in nuclear configuration; the rapid motion of the electrons causes the sluggish nuclei to "see" the electrons as a charge cloud, rather than as discrete particles. The fact that the effective forces on the nuclei are electrostatic affirms that there are no "mysterious quantum-mechanical forces" acting in molecules.

Let us consider the implications of the electrostatic theorem for chemical bonding in diatomic molecules. We take the internuclear axis as the z axis (Fig. 14.4). By symmetry the x and y components of the effective forces on the two nuclei are zero. For the z force component on nucleus a, we have [Eq. (14.112)]

$$F_{z, a} = -Z_a e'^2 \iiint \rho \frac{(z_a - z)}{r_a^3} \, dx \, dy \, dz - \frac{Z_a Z_b e'^2}{R^2} \qquad (14.116)$$

where R is the internuclear distance. From Fig. 14.4 we see that $r_a \cos\theta_a = -z_a + z$ (z_a is negative). Hence

$$F_{z, a} = Z_a e'^2 \iiint \rho \frac{\cos\theta_a}{r_a^2} \, dx \, dy \, dz - \frac{Z_a Z_b e'^2}{R^2} \qquad (14.117)$$

Similarly we find

$$F_{z, b} = -Z_b e'^2 \iiint \rho \frac{\cos\theta_b}{r_b^2} \, dx \, dy \, dz + \frac{Z_a Z_b e'^2}{R^2} \qquad (14.118)$$

Using Eq. (14.42), we have

$$F_{z, a} = -\frac{dU(R)}{dR} \frac{\partial R}{\partial z_a} = -\frac{dU}{dR} \frac{(z_a - z_b)}{R} = \frac{dU}{dR} \frac{\partial R}{\partial z_b} = -F_{z, b} \qquad (14.119)$$

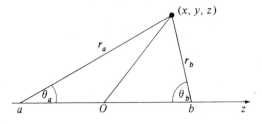

Figure 14.4 *Coordinate system for a diatomic molecule. The origin is at O.*

so that the effective forces on nuclei a and b are equal in magnitude and opposite in direction.

From (14.117) the z component of the effective force on nucleus a due to the element of electronic charge in the region about (x, y, z) is

$$e'^2 Z_a \rho \frac{\cos \theta_a}{r_a^2} \, dx \, dy \, dz \qquad (14.120)$$

Similarly the z component of force on nucleus b due to this charge is

$$-e'^2 Z_b \rho \frac{\cos \theta_b}{r_b^2} \, dx \, dy \, dz \qquad (14.121)$$

A positive value of (14.120) or (14.121) corresponds to a force in the $+z$ direction, i.e., to the right in Fig. 14.4. When the force on nucleus a is algebraically greater than the force on nucleus b, then the element of electronic charge tends to draw a toward b; hence electronic charge that is binding is located in the region where

$$e'^2 Z_a \rho \frac{\cos \theta_a}{r_a^2} \, dx \, dy \, dz > -e'^2 Z_b \rho \frac{\cos \theta_b}{r_b^2} \, dx \, dy \, dz \qquad (14.122)$$

Since the probability density ρ is nonnegative, division by ρ preserves the direction of the inequality sign, and the binding region of space is where

$$Z_a \frac{\cos \theta_a}{r_a^2} + Z_b \frac{\cos \theta_b}{r_b^2} > 0 \qquad (14.123)$$

When the force on b is algebraically greater than that on a, the electronic charge element tends to draw b away from a; the antibinding region of space is thus characterized by a negative value for the left side of (14.123). The surfaces for which the left side of (14.123) equals zero divide space into the binding and antibinding regions. (This concept of *binding* and *antibinding* regions was proposed[9] by Berlin.) Figures 14.5 and 14.6 show this division for a homonuclear and a heteronuclear diatomic molecule. As might be expected, the binding region for a homonuclear diatomic molecule lies between the nuclei. Charge in this region tends to draw the nuclei together. We have seen time and again that bonding leads to a transfer of charge probability density into the region between the nuclei, because of the overlap between the bonding AO's. Electronic charge that is "behind" the nuclei (to the left of nucleus a or to the right of nucleus b in Fig. 14.5) exerts a greater attraction on the nucleus that is nearer to it than on the other nucleus and thus tends to pull the nuclei apart.

Bader, Henneker, and Cade have[10] taken the electron probability densities for homonuclear diatomic molecules as calculated from Hartree–Fock functions and subtracted off the probability densities for the corresponding separated atoms, as calculated from atomic SCF wave functions. They then plotted contours of this

---

[9] T. Berlin, *J. Chem. Phys.*, **19**, 208 (1951). [An extension of Berlin's ideas to polyatomic molecules is given in T. Koga et al., *J. Am. Chem. Soc.*, **100**, 7522 (1978).]

[10] R. F. W. Bader, W. H. Henneker, and P. E. Cade, *J. Chem. Phys.*, **46**, 3341 (1967).

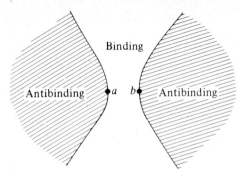

Figure 14.5 *Cross section of binding and antibinding regions in a homonuclear diatomic molecule. To obtain the three-dimensional regions, rotate the figure about the internuclear axis.*

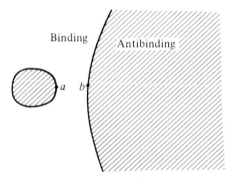

Figure 14.6 *Binding and antibinding regions for a heteronuclear diatomic molecule with $Z_b > Z_a$*

density difference $\Delta\rho$ for Li_2 through F_2. A contour with positive $\Delta\rho$ corresponds to a buildup in charge over that of the separated atoms. Not surprisingly they found a charge buildup in the central part of the binding region, between the nuclei. However, they also found that molecule formation is accompanied by a charge buildup in most of the volume of the antibinding regions also. (The buildup is at the expense of charge probability density in the regions near the nuclei and containing the boundaries between binding and antibinding regions; see Fig. 3 of the paper.) Of course, the charge buildup between the nuclei is such that its contribution to the attractive force between the nuclei exceeds the Hellmann–Feynman repulsive force due to the charge buildup in the antibinding region. At $R = R_e$, the effective force on each nucleus is zero ($\partial U/\partial R|_{R_e} = 0$), and the hypothetical electron smear must exert a net attractive Hellmann–Feynman force to counterbalance the internuclear repulsion. As mentioned (Section 13.17), Hartree–Fock wave functions provide accurate electron probability-density functions. However, the difference between the molecular and atomic ρ's is sensitive to small errors in the ρ's, and one might therefore question the accuracy of Hartree–Fock $\Delta\rho$ maps. For H_2 and for Li_2, when a configuration-interaction wave function is used in place of the Hartree–Fock ψ, one finds little change in the $\Delta\rho$ maps, indicating that the Hartree–Fock functions probably give accurate $\Delta\rho$'s [R. F. W. Bader and A. K. Chandra, *Can. J. Chem.*, **46**, 953 (1968)].

In discussing things from the Hellmann–Feynman viewpoint, we seem to be considering chemical bonding solely in terms of potential energy, whereas the virial-theorem discussion talked about both potential and kinetic energy. For the purposes of the Hellmann–Feynman discussion, we are imagining the electrons to be smeared out into a continuous charge distribution; hence we make no reference to electronic kinetic energy. The use of the electrostatic theorem to explain chemical bonding has been criticized by some quantum chemists on the grounds that it hides the important role of kinetic energy in bonding. [See the references cited after Eq. (13.62).]

PROBLEMS

14.1 Which of the following functions are homogeneous? Give the degree of homogeneity. (a) $x + 3yz$; (b) 179; (c) x^2/yz^3; (d) $(ax^3 + bxy^2)^{1/2}$.

14.2 Let 1 and 2 be two states of an atom, with $E_2 > E_1$. For which state is the average electronic kinetic energy larger?

14.3 Show that the hypervirial theorem follows from Eq. (7.151).

14.4 The $U(R)$ curve for a repulsive state can be roughly approximated by the function $ae^{-bR} - c$, where a, b, and c are positive constants with $a > c$. (This function omits the van der Waals minimum and fails to go to infinity at $R = 0$.) Sketch U, $\langle T_{el} \rangle$, and $\langle V \rangle$ as functions of R for this function.

14.5 Prove that $\partial \langle V \rangle / \partial R$ must be nonnegative at $R = R_e$, i.e., that $\langle V \rangle$ cannot be increasing with decreasing R as we go through the minimum in the $U(R)$ curve (Fig. 14.1). State and prove the corresponding theorem for $\langle T_{el} \rangle$.

14.6 (a) Use the generalized Hellmann–Feynman theorem to calculate $\langle 1/r \rangle$ for the hydrogen-atom bound states. (b) We now want an expression for $\langle 1/r^2 \rangle$, and we proceed as follows. Equation (6.70) can be written as $\hat{H}_r R(r) = ER(r)$, where the "radial Hamiltonian" is given by deleting the function R from the left side of (6.70). Since $\hat{H}_r$ does not affect the angles θ and φ, we can multiply the above equation by Y_l^m to get $\hat{H}_r \psi = E\psi$. If we view the quantum number l as a parameter in $\hat{H}_r$, then the generalized Hellmann–Feynman theorem gives

$$\left\langle \psi \left| \frac{\partial \hat{H}_r}{\partial l} \right| \psi \right\rangle = \frac{\partial E}{\partial l}$$

Use this equation and Eq. (6.104) to show that for hydrogen-atom bound states

$$\left\langle \frac{1}{r^2} \right\rangle = \frac{2Z^2}{(2l+1)n^3} \left(\frac{1}{a_0^2} \right)$$

14.7 (a) Calculate $\langle T \rangle$ and $\langle V \rangle$ for the helium-atom trial function (9.77). All the needed integrals were evaluated in Chapter 9. (b) Verify that the virial theorem is satisfied for $\zeta = Z - 5/16$ but not for $\zeta = Z$.

14.8 A particle is subject to the potential energy $V = ax^4 + by^4 + cz^4$. If its ground-state energy is 10 eV, calculate $\langle T \rangle$ and $\langle V \rangle$ for the ground state.

14.9 Let ψ be the complete wave function for a molecule, with the Born–Oppenheimer approximation of $\psi = \psi_{el}\psi_N$ not necessarily holding. Is it true that

$$2\langle\psi|\hat{T}_{el}+\hat{T}_N|\psi\rangle = -\langle\psi|\hat{V}|\psi\rangle$$

where $\hat{T}_{el}$ and $\hat{T}_N$ are the kinetic-energy operators for the electrons and nuclei and $\hat{V}$ is the complete potential-energy operator? Justify your answer.

14.10 The R_e values for the ground electronic states of HF, HCl, HBr, and HI are 0.9, 1.3, 1.4, and 1.6 Å. The surface enclosing the antibinding region "behind" the proton in these molecules intersects the internuclear axis at two points, one of which is the proton location. Calculate the distance between these two points of intersection for each of the hydrogen halides.

15 ELECTRONIC STRUCTURE OF POLYATOMIC MOLECULES

15.1 INTRODUCTION

The search for accurate electronic wave functions of polyatomic molecules uses mainly the MO method. The presence of several nuclei causes greater computational difficulties than for diatomic molecules. Moreover, the electronic wave function of a diatomic molecule is a function of only one parameter—the internuclear distance; in contrast, the electronic wave function of a polyatomic molecule depends simultaneously on several parameters—the bond distances and bond angles. A full theoretical treatment of a polyatomic molecule involves calculation of the electronic wave function for a range of each of these parameters; the equilibrium bond distances and angles are then found as those values that minimize the electronic energy (including nuclear repulsion). Such a calculation is a formidable task, and the majority of polyatomic-molecule calculations have calculated ψ_{el} for only the experimentally determined bond distances and angles.

Molecular quantum-mechanical calculations can be divided into two classes: *ab initio* and *semiempirical*. Semiempirical calculations generally use a simpler Hamiltonian than the correct molecular Hamiltonian and incorporate into the calculation experimental data or parameters that can be adjusted to fit experimental data. (Certain semiempirical theories adjust their parameters to fit results of ab initio calculations, rather than experimental data; see Section 15.14.) The prime example of a semiempirical method is the Hückel MO treatment of conjugated hydrocarbons (Section 15.11), which uses a one-electron Hamiltonian and takes the bond integrals as adjustable parameters rather than quantities to be calculated theoretically. In contrast, an ab initio calculation uses the correct Hamiltonian for the system and attempts a solution without the use of experimental data (other than molecular geometry). A Hartree–Fock SCF calculation seeks the antisymmetrized product φ of one-electron functions that minimizes $\int \varphi^* \hat{H} \varphi \, d\tau$, where $\hat{H}$ is the true Hamiltonian, and is thus an ab initio calculation.

15.2 CLASSIFICATION OF ELECTRONIC TERMS OF POLYATOMIC MOLECULES

For polyatomic molecules the operator $\hat{S}^2$ for the square of the total electronic spin angular momentum commutes with the electronic Hamiltonian, and, as for diatomic molecules, the electronic terms of polyatomic molecules are classified as singlets, doublets, triplets, etc., according to the value of $2S + 1$. (The commutation of $\hat{S}^2$ and $\hat{H}_{el}$ holds provided spin–orbit interaction is omitted from the Hamiltonian; for molecules containing heavy atoms, spin–orbit interaction is considerable, and S is not a good quantum number.)

For linear polyatomic molecules, the operator $\hat{L}_z$ for the axial component of the total electronic orbital angular momentum commutes with the electronic Hamiltonian, and the same term classifications are used as for diatomic molecules; we have such possibilities as $^1\Sigma^+$, $^1\Sigma^-$, $^3\Sigma^+$, $^1\Pi$, etc. For linear polyatomic molecules with a center of symmetry, the g, u classification is added.

For nonlinear polyatomic molecules, there is no orbital angular-momentum operator that commutes with the electronic Hamiltonian, and the angular-momentum classification of electronic terms cannot be used. Operators that do commute with the electronic Hamiltonian are the symmetry operators $\hat{O}_R$ of the molecule (Section 12.1), and the electronic states of polyatomic molecules are classified according to the behavior of the electronic wave function on application of these operators. Consider H_2O as an example.

In its equilibrium configuration, water belongs to group $\mathscr{C}_{2v}$ with the symmetry operations

$$\hat{E} \qquad \hat{C}_2(z) \qquad \hat{\sigma}_v(xz) \qquad \hat{\sigma}_v(yz) \qquad\qquad \textbf{(15.1)}$$

We follow the standard convention[1] and take the molecular plane as the yz plane (Fig. 15.1). We readily find that each of the symmetry operations commutes with the other three, so that the electronic wave functions can be chosen as simultaneous eigenfunctions of all four symmetry operators. Since $\hat{O}_E$ is the unit operator, we have $\hat{O}_E \psi_{el} = \psi_{el}$. Each of the remaining symmetry operators satisfies $\hat{O}_R^2 = \hat{1}$, so that each has as its eigenvalues $+1$ and -1 (Eq. 7.87). Therefore each electronic wave function of water is an eigenfunction of $\hat{O}_E$ with eigenvalue $+1$, and an eigenfunction of each of the other three symmetry operators with the eigenvalues $+1$ or -1.

How many different combinations of these symmetry eigenvalues are there for water? At first sight, we might think there are $(1)(2)(2)(2) = 8$ possible sets. However, certain sets can be ruled out, as we now show. Let the product of the

---

[1] R. S. Mulliken, *J. Chem. Phys.*, **23**, 1997 (1955).

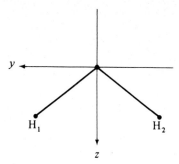

Figure 15.1 *Coordinate axes for the water molecule. The x axis is perpendicular to the molecular plane.*

symmetry operations $\hat{R}$ and $\hat{S}$ be the symmetry operation $\hat{T}$:

$$\hat{R}\hat{S} = \hat{T} \tag{15.2}$$

Let ψ_{el} be an eigenfunction of $\hat{O}_R$, $\hat{O}_S$, and $\hat{O}_T$ with eigenvalues r, s, and t, respectively. Since the symmetry operators multiply the same way the symmetry operations do, we have

$$t\psi_{el} = \hat{O}_T\psi_{el} = \hat{O}_R[\hat{O}_S\psi_{el}] = s\hat{O}_R\psi_{el} = rs\psi_{el} \tag{15.3}$$

Dividing by ψ_{el}, we have $rs = t$. Hence the eigenvalues of the symmetry operators must multiply in the same way the symmetry operators do. Now consider water. Let us examine the set of symmetry eigenvalues

$$1 \qquad -1 \qquad -1 \qquad -1 \tag{15.4}$$

where the eigenvalues are listed in the order corresponding to (15.1). We have $\hat{C}_2(z)\hat{\sigma}_v(xz) = \hat{\sigma}_v(yz)$; the set (15.4), however, has the product of the $\hat{O}_{C_2}$ and $\hat{O}_{\sigma_v(xz)}$ eigenvalues as $(-1)(-1) = 1$, which differs from the $\hat{O}_{\sigma_v(yz)}$ eigenvalue in (15.4). The set (15.4) must be discarded. Of the eight possible symmetry-eigenvalue combinations, only four are found to multiply properly; these four sets are (Problem 15.3)

	$\hat{E}$	$\hat{C}_2(z)$	$\hat{\sigma}_v(xz)$	$\hat{\sigma}_v(yz)$	
A_1	1	1	1	1	
A_2	1	1	-1	-1	(15.5)
B_1	1	-1	1	-1	
B_2	1	-1	-1	1	

In this table the sets have been labeled A_1, A_2, B_1, and B_2. The letter A or B indicates whether the symmetry eigenvalue for the highest-order $\hat{C}_n$ or $\hat{S}_n$ operation of the molecule [$\hat{C}_2(z)$ for water] is $+1$ or -1, respectively. The subscripts 1 and 2 distinguish sets having the same letter label. Each possible set of symmetry eigenvalues in (15.5) is called a *symmetry species* (or *symmetry type*). (The group-theory term is *irreducible representation*.) The symmetry species with all symmetry eigenvalues $+1$ (A_1 for H_2O) is called the *totally symmetric* species.

Each molecular electronic term of H_2O is designated by giving the symmetry species of the electronic wave functions of the term, with the spin multiplicity $(2S + 1)$ as a left superscript. For example, an electronic state of water with two electrons unpaired and with the electronic wave function unchanged by all four symmetry operators belongs to a $^3A_1$ term. (The subscript 1 is not an angular-momentum eigenvalue, but is part of the symmetry-species label.)

We now consider the *orbital degeneracy* of molecular electronic terms; this is degeneracy connected with the electrons' spatial (orbital) motion, as distinguished from spin degeneracy. Thus $^1\Pi$ and $^3\Pi$ terms of linear molecules are orbitally degenerate, while $^1\Sigma$ and $^3\Sigma$ terms are orbitally nondegenerate. Consider an operator $\hat{F}$ that commutes with the molecular electronic Hamiltonian and that does not involve spin; we have

$$\hat{F}\hat{H}_{el}\psi_{el,i} = \hat{F}E_{el}\psi_{el,i}$$

$$\hat{H}_{el}[\hat{F}\psi_{el,i}] = E_{el}[\hat{F}\psi_{el,i}] \tag{15.6}$$

so that $\hat{F}\psi_{el,i}$ is an eigenfunction of $\hat{H}_{el}$ with eigenvalue E_{el}. Let the orbital degeneracy of the term to which $\psi_{el,i}$ belongs be n. It follows from (15.6) that $\hat{F}\psi_{el,i}$ must be some linear combination of the n orbitally degenerate wave functions of the term that have the same value of S_z as $\psi_{el,i}$:

$$\hat{F}\psi_{el,i} = \sum_{j=1}^{n} c_{ij,F}\psi_{el,j} \tag{15.7}$$

where the c's are certain constants. (These arguments were previously given in Section 7.4.) For a level that is not orbitally degenerate ($n = 1$), Eq. (15.7) reduces to

$$\hat{F}\psi_{el,i} = c_F\psi_{el,i} \tag{15.8}$$

so that here $\psi_{el,i}$ *must* be an eigenfunction of $\hat{F}$. For $n > 1$, we can *choose* n linear combinations of the $\psi_{el,j}$'s that are eigenfunctions of $\hat{F}$, but there is no *necessity* for the eigenfunctions of a degenerate level to be eigenfunctions of $\hat{F}$. If we have two operators $\hat{F}$ and $\hat{G}$ that commute with $\hat{H}_{el}$, but not with each other, we can pick the linear combinations to be eigenfunctions of $\hat{F}$ or of $\hat{G}$, but not in general of both $\hat{F}$ and $\hat{G}$ simultaneously.

Now consider symmetry operators. For water all the symmetry operators commute among themselves, so that each electronic wave function is simultaneously an eigenfunction of all the symmetry operators. If all the electronic wave functions were orbitally nondegenerate, then it would automatically follow from (15.8) that they were simultaneously eigenfunctions of all the symmetry operators. We therefore suspect (but have not proved) that the electronic wave functions of water are all orbitally nondegenerate. This statement is in fact correct. The letters A and B designate symmetry species of orbitally nondegenerate electronic terms. For any molecule all of whose symmetry operators commute with one another, the electronic wave functions are each simultaneous eigenfunctions of all symmetry operators, and the only symmetry species are nondegenerate A and B species.

For some point groups, the symmetry operators do not all commute; an example is $\mathscr{C}_h$ (Fig. 12.9). When the symmetry operators do not all commute, some of the electronic terms are orbitally degenerate; a symmetry operator applied to an electronic wave function of an orbitally degenerate term converts it to a linear combination of the wave functions of the term [Eq. (15.7)]. The effects of the symmetry operator $\hat{O}_R$ on the wave functions of an n-fold orbitally degenerate electronic term are specified by the n^2 numbers

$$c_{ij,R}, \qquad i = 1, \ldots, n, \qquad j = 1, \ldots, n \tag{15.9}$$

The n^2 numbers (15.9) when arranged in an $n \times n$ square array form a *matrix*. If there are h symmetry operations in the molecular point group, then the h matrices of coefficients in (15.7) constitute the *symmetry species* (or *symmetry type*) of the degenerate-term wave functions in (15.7). The following letter labels are used for the symmetry species, according to the orbital degeneracy n:

$$
\begin{array}{c|c|c|c|c|c}
n & 1 & 2 & 3 & 4 & 5 \\
\hline
\textit{Letter} & A, B & E & T & G & H
\end{array}
\tag{15.10}
$$

Numerical subscripts distinguish different symmetry species having the same letter designation. For molecules with a center of symmetry, a g or u subscript is added, depending on whether the wave function has the eigenvalue $+1$ or -1 for inversion of all electronic spatial coordinates. The possible symmetry species can be found in a systematic way using group theory (see *Schonland*). As an example, group theory shows the possible symmetry species of a $\mathcal{D}_{6h}$ molecule to be

$$
\begin{aligned}
A_{1g} \quad A_{2g} \quad B_{1g} \quad B_{2g} \quad E_{1g} \quad E_{2g} \\
A_{1u} \quad A_{2u} \quad B_{1u} \quad B_{2u} \quad E_{1u} \quad E_{2u}
\end{aligned}
\tag{15.11}
$$

We shall not give the numbers $c_{ij,R}$ that specify these symmetry species, except to note that A_{1g} is the totally symmetric symmetry species, with all c's equal to 1.

It is an empirical fact that for most molecules in their electronic ground states, the wave function belongs to the (nondegenerate) totally symmetric species; also the electronic spins are usually all paired in the ground state, so that the ground state is a singlet. For water the ground electronic state is $^1A_1$; for benzene it is $^1A_{1g}$.

We have been using the equilibrium-geometry point groups of H_2O and C_6H_6. For nonequilibrium nuclear configurations, the symmetry is in general less than that of $\mathcal{C}_{2v}$ or $\mathcal{D}_{6h}$; for reasonably small departures from equilibrium, however, the symmetry behavior should be given to a good approximation by the symmetry species of the equilibrium-geometry point group. We also note that excited states sometimes differ in their equilibrium point group from the ground electronic state. For example, the point group of NH_3 is $\mathcal{D}_{3h}$ for several excited electronic states.[2]

Corresponding to a given molecular electronic term, there are in general several electronic states. The interactions between electronic spin and electronic orbital motion and between electronic and nuclear motions split the energies of these states. These splittings are usually small.

---

[2] For properties of molecular electronic states, see G. Herzberg, *Electronic Spectra of Polyatomic Molecules*, Van Nostrand, New York, 1966, Appendix VI.

15.3   *THE SCF MO TREATMENT OF POLYATOMIC MOLECULES*

The purely electronic nonrelativistic Hamiltonian for a polyatomic molecule is (in atomic units)

$$\hat{H}_{el} = -\frac{1}{2} \sum_i \nabla_i^2 - \sum_i \sum_\alpha \frac{Z_\alpha}{r_{i\alpha}} + \sum_i \sum_{j>i} \frac{1}{r_{ij}} \qquad (15.12)$$

If interelectronic repulsions are neglected, the zeroth-order wave function is the product of one-electron spatial functions (molecular orbitals). Allowance for electron spin and the Pauli principle gives a zeroth-order wave function that is an antisymmetrized product of molecular spin-orbitals, each spin-orbital being a product of a spatial MO and a spin function. The best possible variation function that has the form of an antisymmetrized product of spin-orbitals is the Hartree–Fock SCF function. [Improvements beyond the Hartree–Fock stage require configuration interaction (CI).] The MO's are usually expressed as linear combinations of basis functions, the coefficients being found by solution of the Roothaan equations (Section 13.16). If a sufficiently large basis set is used, the MO's are accurate approximations to the Hartree–Fock MO's. If a minimal basis set is used, the MO's are only rough approximations to the Hartree–Fock MO's but are still referred to as SCF molecular orbitals.

How are polyatomic MO's classified? As might be expected, the MO's of a polyatomic molecule show the same kinds of possible symmetry behavior as the overall electronic wave function does[3] (Section 15.2); the MO's are therefore classified according to the symmetry species of the molecular point group. For example, the MO's of H_2O have the possible symmetry species a_1, a_2, b_1, and b_2 of (15.5). (Lowercase letters are used for MO symmetry species.) To distinguish MO's of the same symmetry species, one numbers them in order of increasing energy. Thus the lowest three a_1 MO's of water are called $1a_1$, $2a_1$, and $3a_1$. (This nomenclature is similar to that of the third column in Table 13.1.)

Each MO holds two electrons of opposite spin. MO's having the same energy constitute a *subshell* (often simply called a *shell*). A subshell that consists of a single MO of symmetry species a or b holds two electrons; a subshell that consists of two e MO's having the same energy holds four electrons; and so on. For water there are no degenerate symmetry species, and each subshell holds two electrons. For benzene there are some doubly degenerate symmetry species [see (15.11)], so that some of the benzene MO's occur in pairs having the same energy. Specification of the number of electrons in each subshell specifies the molecular electronic *configuration*. As for atoms and diatomic molecules, a given electron configuration of a polyatomic molecule gives rise in general to several electronic terms. The systematic method for finding the possible terms uses group theory and will not be discussed here.[4] [As an example, an $(e_{1g})^2$ configuration of a $\mathscr{D}_{6h}$ molecule gives the terms $^1A_{1g}$, $^1E_{2g}$, and $^3A_{2g}$.] *A closed-subshell configuration gives rise to a single nondegenerate term, whose*

---

[3] For the proof see C. C. J. Roothaan, *Rev. Mod. Phys.*, **23**, 161 (1951).

[4] For tables of the possible terms arising from various configurations, see G. Herzberg, *Electronic Spectra of Polyatomic Molecules*, pp. 330–334, 570–573.

multiplicity is 1 and whose symmetry species is the totally symmetric one. Most polyatomic molecules have a closed-subshell ground state; here, the MO wave function is a single Slater determinant. For states arising from open-subshell configurations, the MO wave function may require a linear combination of several Slater determinants.

We now consider the kinds of basis functions used in molecular SCF calculations. For diatomic molecules the basis functions are usually taken as atomic orbitals, some centered on one atom, the remainder centered on the other atom, with each AO being represented as a *linear combination* of one or more Slater-type orbitals (11.16); the real form of the STO's is used for nonlinear molecules. We have an LC-STO MO. For polyatomic molecules the LC-STO method involves one or more STO's centered on each of the atoms. The presence of more than two atoms causes difficulties in evaluating the needed integrals. For a triatomic molecule, one must deal with three-center as well as one- and two-center integrals. For a molecule with four or more atoms, one has four-center integrals as well, but the number of centers occurring in any one integral does not exceed four, as we now demonstrate. Let $\varphi_1, \varphi_2, \ldots, \varphi_n$ be the spatial MO's of electrons $1, 2, \ldots, n$. The Hamiltonian (15.12) contains only one-electron and two-electron terms. A typical two-electron integral is

$$\int\int\cdots\int \varphi_1^* \varphi_2^* \cdots \varphi_n^* \frac{1}{r_{12}} \varphi_1 \varphi_2 \cdots \varphi_n \, dv_1 \, dv_2 \cdots dv_n = \int\int |\varphi_1|^2 |\varphi_2|^2 \frac{1}{r_{12}} \, dv_1 \, dv_2$$

$$(15.13)$$

Each MO is expanded in terms of a set of basis functions $\chi_1, \chi_2, \ldots, \chi_m$, and we have

$$\varphi_1 = c_1 \chi_1(1) + c_2 \chi_2(1) + \cdots + c_m \chi_m(1) \tag{15.14}$$

$$\varphi_2 = d_1 \chi_1(2) + d_2 \chi_2(2) + \cdots + d_m \chi_m(2) \tag{15.15}$$

where the c's and d's are certain coefficients. Substituting into the integral (15.13), we obtain (15.13) as the sum of integrals involving the basis functions; a typical such integral is

$$c_i^* c_j d_k^* d_l \int\int \chi_i^*(1)\chi_j(1)\chi_k^*(2)\chi_l(2) \frac{1}{r_{12}} \, dv_1 \, dv_2 \tag{15.16}$$

If it happens that the basis functions $\chi_i, \chi_j, \chi_k, \chi_l$ are each centered on a different nucleus, then (15.16) is a four-center integral.

With m different choices for each basis function in (15.16), we can expect that the number of electron-repulsion integrals over the basis functions will increase as m^4. A detailed investigation that takes account of the permutational symmetry shows that one must evaluate $(m^4 + 2m^3 + 3m^2 + 2m)/8$ two-electron integrals over the basis functions. For accurate SCF calculations on typical small to medium-size molecules, one might use 20 to 100 STO basis functions, producing 20,000 to 13,000,000 two-electron integrals. Computer evaluation of three- and four-center integrals over STO basis functions is quite time-consuming; this has led to two other choices of basis functions. These are the one-center expansion method and the Gaussian-orbital method.

The *one-center expansion* (OCE) *method* takes each MO as a linear combination of functions (usually STO's), *all* of which are centered at the same point in space. For example, a one-center MO calculation of CH_4 takes each MO as a linear combination of functions, all of which are centered on the carbon atom. With the OCE method, all integrals are one-center, but the method has drawbacks. The presence of a nucleus at a point in space introduces a cusp (Fig. 6.6) into the electronic wave function at that point. The representation of an MO as a linear combination of orbitals all centered at a single nucleus makes it difficult to get the proper cusp behavior at the other nuclei. A one-center wave function for CH_4 is inaccurate in the regions near the hydrogen nuclei, unless very many basis functions are used. The OCE method is clearly most applicable to molecules with the formula AH_n; here, most of the electron probability density is located near atom A and the hydrogen atoms introduce only small cusps into the wave function. The OCE method is related to the united-atom viewpoint; the corresponding united atom for CH_4 is Ne, and a united-atom perturbation treatment would express the methane wave function as an expansion of neon ground-state and excited-state (one-center) wave functions. Because of generally discouraging results and the limited applicability of the method, there is currently not much interest in the OCE method.

To simplify molecular integral evaluation, Boys proposed in 1950 the use of *Gaussian-type orbitals* (GTO's) instead of STO's for the atomic orbitals in an LCAO wave function. A *Gaussian function* centered on nucleus a has the form

$$Nr^l e^{-\zeta r_a^2} Y_l^m(\theta, \varphi) \qquad (15.17)$$

which is in contrast to hydrogenlike and Slater functions, whose exponential factor is $\exp(-\zeta r_a)$. As with Slater orbitals, the real form of the spherical harmonics (Problem 15.31) is used in molecular calculations. Thus any s AO (whether $1s$ or $2s$ or ...) is represented by a linear combination of several Gaussians with different orbital exponents, each Gaussian having the form $\exp(-\zeta r_a^2)$; any atomic p_x orbital is represented by a linear combination of Gaussians, each of the form $x \exp(-\zeta r_a^2)$; and so on. [It is possible to use higher powers of r in (15.17), but this has been found to be of no advantage.] The behavior of the Gaussian exponential factor is shown in Fig. 4.1a, where the origin is at nucleus a. A Gaussian function does not have the desired cusp at the nucleus and hence gives a poor representation of an AO for small values of r_a. To get an accurate representation of an AO, we must therefore use a linear combination of several Gaussians. Therefore an LC-GTO SCF MO calculation involves evaluation of very many more integrals than the corresponding LC-STO SCF MO calculation, since the number of two-electron integrals is proportional to the fourth power of the number of basis functions. However, Gaussian integral evaluation takes much less computer time than Slater integral evaluation. This is because the product of two Gaussian functions centered at two different points is equal to a single Gaussian centered at a third point. Thus all three- and four-center two-electron repulsion integrals are reduced to two-center integrals. [A modification of the LC-GTO method is the *Gaussian-lobe method*, which uses $1s$ Gaussian functions placed at points in space to simulate s, p, d, etc., orbitals; for example, a p_x orbital (which has two lobes) can be simulated by two Gaussians centered on the x axis, one on each side of the nucleus.]

A device widely used with GTO basis functions is that of a *contracted* Gaussian set. Here, not all the coefficients of the basis functions are independently varied;

rather, the coefficient ratios between certain basis functions are kept fixed at predetermined values. This gives substantial savings in computation time with but little loss in accuracy.

Let us discuss some of the terminology used to describe basis sets. A *minimal basis set* consists of one STO for each inner-shell and valence-shell AO of each atom (Section 13.16). For example, for C_2H_2 a minimal basis set consists of $1s$, $2s$, $2p_x$, $2p_y$, and $2p_z$ AO's on each carbon and a $1s$ STO on each hydrogen; there are five STO's on each C and one on each H, for a total of twelve basis functions. This set contains two s-type STO's and one set of p-type STO's on each carbon and one s-type STO on each hydrogen; such a set is denoted by $(2s\, 1p)$ for the carbon functions and $(1s)$ for the hydrogen functions, a notation which is further abbreviated to $(2s\, 1p/1s)$.

A *double-zeta basis set* is obtained by replacing each STO of a minimal basis set by two STO's that differ in their orbital exponents ζ (zeta). (Recall that a single STO is not an accurate representation of an AO; use of two STO's gives substantial improvement.) For example, for C_2H_2 a double-zeta set consists of two $1s$ STO's on each H, two $1s$ STO's, two $2s$ STO's, two $2p_x$, two $2p_y$, and two $2p_z$ STO's on each carbon, for a total of twenty-four basis functions; this is a $(4s\, 2p/2s)$ basis set. (A *split-valence basis set* uses two STO's for each valence-shell AO but only one STO for each inner-shell AO.)

AO's are distorted (or polarized) upon molecule formation. To allow for this polarization, one must introduce basis-function STO's whose l quantum numbers are greater than the maximum l in the corresponding free atom. For example, for a hydrocarbon one would add to the double-zeta basis set the following *polarization functions*: $2p_x$, $2p_y$, and $2p_z$ on each hydrogen and a set of five $3d$ functions on each carbon; this gives a $(4s\, 2p\, 1d/2s\, 1p)$ *double-zeta plus polarization basis set*. (Higher polarization functions can also be included.)

A common practice in molecular calculations is to use a contracted Gaussian function to replace each STO. A *contracted Gaussian* has the form $\sum_i a_i f_i$, where the functions f_i are Gaussians (called *primitive Gaussians*) and where the coefficients a_i are chosen to give a good fit to the STO and are subsequently held fixed in the molecular calculation. Pople and co-workers have done extensive SCF calculations with this approach, using basis sets they call STO-3G, 4-31G, and 6-31G*.

An STO-NG basis set uses a minimal basis set of one STO per AO; the orbital exponents are fixed at values found to work well in calculations on small molecules. To avoid the time-consuming integrals that occur with STO's, each STO is approximated as a linear combination of N Gaussian functions, where the coefficients in the linear combination and the Gaussian orbital exponents are chosen to give the best least-squares fit to the STO. Most commonly, N = 3. Since a linear combination of three Gaussians is only an approximation to an STO, the STO-3G method gives results not quite as good as a minimal-basis-set STO calculation. As noted earlier, a minimal basis set of STO's for a hydrocarbon is denoted by $(2s\, 1p/1s)$. Since each STO is replaced by a linear combination of three primitive Gaussians (which is one contracted Gaussian), the STO-3G basis set for a hydrocarbon is denoted by $(6s\, 3p/3s)$ contracted to $[2s\, 1p/1s]$, where parentheses indicate the uncontracted (primitive) Gaussians and brackets indicate the contracted Gaussians.

The 4-31G basis set is a split-valence set using contracted Gaussians. Each inner-shell $1s$ AO is taken as a linear combination of four Gaussians with fixed coefficients. For each valence-shell $2s$ or $2p$ AO, one uses a double-zeta approach, taking one $2s$ (or $2p$) function that is a fixed linear combination of three Gaussians and a second $2s$ (or $2p$) function that consists of a single Gaussian with orbital exponent different from those in the linear combination of three Gaussians. The $1s$ valence AO of each hydrogen is treated like the $2s$ valence AO of C, N, or O. The 4-31G basis set is somewhat inferior to a double-zeta basis set.

The 6-31G* basis set improves on the 4-31G set by using a linear combination of six Gaussians (rather than four) for each inner-shell $1s$ AO and by adding a single set of d-type Gaussian polarization functions on each nonhydrogen atom.

Whether Gaussian or Slater basis sets are a better choice depends on the kind of calculation. For calculations on atoms, Slater functions are the obvious choice since atomic (one-center) integrals are no more difficult for STO's than for GTO's. Most SCF calculations on linear molecules have used STO's, since linear geometry allows special numerical methods to be used. For nonlinear molecules, the much faster rate of evaluation of GTO integrals makes GTO's generally a better choice for SCF calculations, despite the larger number of integrals that must be evaluated with GTO's. However, for a configuration-interaction calculation, evaluation of integrals takes up a substantially smaller fraction of the total computation time than it does for an SCF calculation, so that STO's are about as good a choice as GTO's in CI calculations.

15.4   THE SCF MO TREATMENT OF WATER

For a minimal-basis-set MO treatment of H_2O (Fig. 15.1), we start with the O$1s$, O$2s$, O$2p_x$, O$2p_y$, and O$2p_z$ inner-shell and valence oxygen AO's and the $H_1 1s$ and $H_2 1s$ valence AO's of the hydrogen atoms. Linear combinations of these seven basis AO's give LCAO approximations to the seven lowest MO's of water. As stated in the last section, the MO's can be chosen so that upon application of the molecular symmetry operators each MO transforms according to one of the symmetry species of the molecular point group. To aid in choosing the right linear combinations of AO's, we examine the symmetry behavior of the AO's.

The $1s$ and $2s$ oxygen AO's are spherically symmetric; rotation about the $C_2(z)$ axis and reflection in the xz or yz plane has no effect on them; they thus belong to the totally symmetric symmetry species a_1 of (15.5). The effect of a $\hat{C}_2(z)$ rotation on the oxygen $2p_y$ AO is shown in Fig. 15.2; this rotation sends O$2p_y$ into its negative [see also Fig. 12.8]. Reflection in the molecular plane sends O$2p_y$ into itself, while reflection in the xz plane sends it into its negative. The symmetry eigenvalues of O$2p_y$ are 1, -1, -1, and 1; the symmetry species of this AO is b_2. Similarly, the symmetry species of O$2p_x$ and O$2p_z$ are found to be b_1 and a_1, respectively.

Now for the hydrogen $1s$ AO's. Reflection in the yz plane leaves each of them unchanged. However, rotation by 180° about the z axis sends $H_1 1s$ into $H_2 1s$ and vice versa (Fig. 15.3); the result is the same for reflection in the xz plane. The $H_1 1s$ and $H_2 1s$ functions are *not* eigenfunctions of $\hat{O}_{C_2(z)}$ and $\hat{O}_{\sigma_v(xz)}$, and they

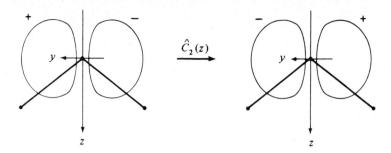

Figure 15.2 *The effect of $\hat{C}_2(z)$ on the $2p_y$ oxygen AO in water*

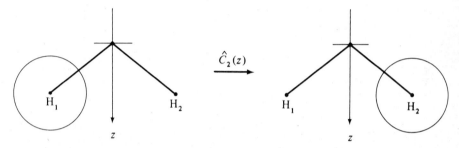

Figure 15.3 *The effect of a $\hat{C}_2(z)$ rotation on the $H_1 1s$ AO in water*

therefore do not transform according to any of the symmetry species of water.

As a preliminary step in finding the MO's of a molecule, it is helpful (but not essential) to construct linear combinations of the original basis AO's such that each linear combination does transform according to one of the molecular symmetry species. Such linear combinations are called *symmetry orbitals*. (Synonyms are *symmetry-adapted orbitals* and *group orbitals*.[5]) The symmetry orbitals are used as the basis functions f_k in the expansion (13.182). *The use of basis functions that transform according to the molecular symmetry species simplifies the calculation by putting the secular determinant in block-diagonal form*; this will be illustrated below.

Each oxygen AO transforms according to one of the symmetry species of water and can serve as a symmetry orbital. We must, however, construct two symmetry orbitals from the two hydrogen $1s$ AO's, neither of which belongs to a symmetry species of water. Consider the following linear combinations:

$$H_1 1s + H_2 1s \tag{15.18}$$

$$H_1 1s - H_2 1s \tag{15.19}$$

We have

$$\hat{O}_{C_2(z)}(H_1 1s + H_2 1s) = H_2 1s + H_1 1s \tag{15.20}$$

$$\hat{O}_{C_2(z)}(H_1 1s - H_2 1s) = H_2 1s - H_1 1s \tag{15.21}$$

---

[5] The term *group orbital* is also applied to an orbital having the local symmetry of a group of atoms in the molecule, rather than the symmetry of the molecule. For example, in toluene one can take linear combinations of the three methyl hydrogen $1s$ orbitals to form three group orbitals, each belonging to a $\mathscr{C}_{3v}$ symmetry species.

Thus (15.18) and (15.19) are eigenfunctions of $\hat{O}_{C_{2(z)}}$ with eigenvalues $+1$ and -1, respectively. Examination of the effects of the other three symmetry operators shows (15.18) and (15.19) to belong to the symmetry species a_1 and b_2, respectively. We shall not bother to normalize the symmetry orbitals (15.18) and (15.19). The seven basis symmetry orbitals are then

Symmetry Orbital	Symmetry Species	
$f_1 = H_1 1s + H_2 1s$	a_1	
$f_2 = O1s$	a_1	
$f_3 = O2s$	a_1	
$f_4 = O2p_z$	a_1	**(15.22)**
$f_5 = H_1 1s - H_2 1s$	b_2	
$f_6 = O2p_y$	b_2	
$f_7 = O2p_x$	b_1	

Now consider the secular determinant in (13.185). We assert that

$$F_{jk} = 0 \qquad (15.23)$$

whenever f_j and f_k belong to different symmetry species. This result follows from the theorem (Section 7.4) that $\langle f_j | \hat{B} | f_k \rangle = 0$ if f_j and f_k are eigenfunctions of a Hermitian operator $\hat{A}$ with different eigenvalues, where $\hat{A}$ commutes with $\hat{B}$. The symmetry orbitals of water are eigenfunctions of the symmetry operators, each of which commutes with the electronic Hamiltonian and with the Fock operator $\hat{F}$. Symmetry orbitals f_j and f_k that belong to different symmetry species differ in at least one symmetry eigenvalue. Hence (15.23) follows. Moreover, since two eigenfunctions of a Hermitian operator that correspond to different eigenvalues are orthogonal, we have

$$S_{jk} = 0 \qquad (15.24)$$

whenever f_j and f_k belong to different symmetry species. From (15.23) and (15.24), it follows that using symmetry orbitals as basis functions puts the secular determinant of water in block-diagonal form, each block corresponding to a different symmetry species; the blocks are 4×4, 2×2, and 1×1. [For molecules with degenerate symmetry species $(E, T, \text{etc.})$, the symmetry orbitals of the degenerate species are not necessarily eigenfunctions of the symmetry operators; nevertheless, the symmetry orbitals still put the secular determinant in block-diagonal form, as can be shown using group theory.]

The set of simultaneous equations (13.183) then breaks up into one set of four simultaneous equations, one set of two simultaneous equations, and one set of one "simultaneous" equation (Section 9.6). The first set contains matrix elements involving only the four a_1 symmetry orbitals. Therefore four of the lowest seven water MO's are linear combinations of the four a_1 symmetry orbitals; these four MO's must have a_1 symmetry. Similarly, we have two MO's of b_2 symmetry and one MO of b_1 symmetry. The symmetry orbitals are *not* (in general) the MO's, but each MO must be a linear combination of those symmetry orbitals having the same symmetry species as the MO. The forms of the lowest MO's of water are then

$$\varphi_1 = c_{11}f_1 + c_{12}f_2 + c_{13}f_3 + c_{14}f_4$$

$$\varphi_2 = c_{21}f_1 + c_{22}f_2 + c_{23}f_3 + c_{24}f_4$$

$$\varphi_3 = c_{31}f_1 + c_{32}f_2 + c_{33}f_3 + c_{34}f_4$$

$$\varphi_4 = c_{41}f_1 + c_{42}f_2 + c_{43}f_3 + c_{44}f_4 \tag{15.25}$$

$$\varphi_5 = c_{55}f_5 + c_{56}f_6$$

$$\varphi_6 = c_{65}f_5 + c_{66}f_6$$

$$\varphi_7 = f_7$$

The next step in the SCF MO calculation is to choose explicit forms for the seven AO's. The orbital energies and the coefficients of the symmetry orbitals are then found using Roothaan's equations.

Pitzer and Merrifield[6] did an H_2O minimal-basis-set calculation, representing each AO by a single STO. They optimized the orbital exponents, finding 1.27 for H$1s$, 7.66 for O$1s$, 2.25 for O$2s$, 2.21 for O$2p$. (To optimize the exponents, one must repeat the entire SCF iterative calculation for several different sets of orbital exponents to locate the set that gives the minimum energy. Since orbital-exponent optimization is time-consuming, it is only feasible for rather small molecules.) The orbital energies in hartrees are found to be the following: $1a_1$, -20.56; $2a_1$, -1.28; $1b_2$, -0.62; $3a_1$, -0.47; $1b_1$, -0.40. The ground-state electronic configuration of this ten-electron molecule is

$$(1a_1)^2 (2a_1)^2 (1b_2)^2 (3a_1)^2 (1b_1)^2 \tag{15.26}$$

The ground state has a closed-subshell configuration and is a $^1A_1$ state.

The five lowest SCF MO's found by Pitzer and Merrifield at the experimental geometry are

$$1a_1 = 1.000(O1s) + 0.015(O2s_\perp) + 0.003(O2p_z) - 0.004(H_1 1s + H_2 1s)$$

$$2a_1 = -0.027(O1s) + 0.820(O2s_\perp) + 0.132(O2p_z) + 0.152(H_1 1s + H_2 1s)$$

$$1b_2 = 0.624(O2p_y) + 0.424(H_1 1s - H_2 1s) \tag{15.27}$$

$$3a_1 = -0.026(O1s) - 0.502(O2s_\perp) + 0.787(O2p_z) + 0.264(H_1 1s + H_2 1s)$$

$$1b_1 = O2p_x$$

The $O2s_\perp$ orbital in (15.27) is an orthogonalized orbital [Eq. (13.187)]:

$$O2s_\perp = 1.028[O2s - 0.2313(O1s)] \tag{15.28}$$

where $O2s$ is the ordinary $2s$ STO:

$$O2s = 2.25^{5/2} \pi^{-1/2} 3^{-1/2} r_O \exp(-2.25r_O)$$

where r_O is the distance to the oxygen nucleus. The MO approximation to the ground-

⁶ R. M. Pitzer and D. P. Merrifield, *J. Chem. Phys.*, **52**, 4782 (1970); S. Aung, R. M. Pitzer, and S. I. Chan, *J. Chem. Phys.*, **49**, 2071 (1968).

state wave function of H_2O is the 10×10 Slater determinant

$$|1a_1 \overline{1a_1} 2a_1 \overline{2a_1} 1b_2 \overline{1b_2} 3a_1 \overline{3a_1} 1b_1 \overline{1b_1}| \qquad \textbf{(15.29)}$$

Consider the nature of the MO's. The lowest MO, $1a_1$, is essentially a pure nonbonding $1s$ oxygen AO, which is hardly surprising.

The oxygen part of the $2a_1$ MO is mostly $O2s$ with some $O2p_z$ mixed in. This mixing in (hybridization) of $O2p_z$ adds to the value of the $O2s$ AO along the positive z axis (which points toward the hydrogens—Fig. 15.1) and subtracts from $O2s$ along the negative z axis. The combination of the hybridized $O2s$ and $O2p_z$ orbitals with the $H_1 1s$ and $H_2 1s$ orbitals in the $2a_1$ MO then gives electron probability-density buildup in the region enclosed by the three nuclei. Therefore the $2a_1$ MO contributes to the bonding in water.

Consider the $1b_2$ MO. The $2p_y$ oxygen AO has its positive lobe on the H_1 side of the molecule, so that the positive lobe of $O2p_y$ adds to $H_1 1s$ in the $1b_2$ MO, giving electron charge buildup between the H_1 and O nuclei. Similarly, the negative lobe of $O2p_y$ adds to $-H_2 1s$, giving charge buildup between O and H_2 in this MO. Hence $1b_2$ is a bonding MO.

The hybridization of the $2s$ and $2p_z$ oxygen AO's in $3a_1$ builds up electron probability density in the region around the negative z axis, away from the hydrogens, giving this MO substantial lone-pair character. On the positive side of the z axis, the oxygen $2s$ and $2p_z$ AO's tend to cancel each other, and we get little bonding overlap between the oxygen hybrid and the hydrogen AO's. Alternatively, we can look separately at the·overlap between $O2s_1$ and the H's (which is negative or antibonding) and the overlap between $O2p_z$ and the hydrogens (which is positive or bonding); because of the approximate cancellation of these overlaps (see Section 15.5), the $3a_1$ MO has little bonding character and is best described as mainly a lone-pair MO. (It is sometimes erroneously stated that $3a_1$ and $1b_2$ are the main bonding MO's in water.)

The $1b_1$ MO is a nonbonding lone-pair $2p_x$ oxygen AO.

Figure 15.4 shows the shapes of the bonding MO's $2a_1$ and $1b_2$. Note that their symmetry eigenvalues are as given by (15.5). [For accurately plotted H_2O MO contours, see T. H. Dunning, R. M. Pitzer, and S. Aung, *J. Chem. Phys.*, **57**, 5044 (1972).]

Each of the SCF bonding MO's in (15.27) is delocalized over the entire molecule

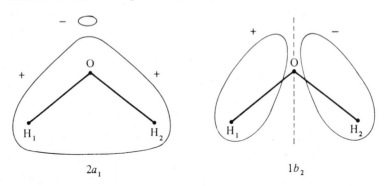

$$2a_1 \qquad\qquad\qquad 1b_2$$

Figure 15.4 *Sketches of the two main bonding MO's of water*

and does not resemble a chemical bond. Hence the reader may be wondering about the relation of these MO's to the picture presented in most freshman-chemistry books, where one bonding MO points along the O—H_1 bond and the other points along the O—H_2 bond. We shall discuss this point in Section 15.6.

The unoccupied $4a_1$ and $2b_2$ MO's of water calculated by Pitzer and Merrifield (for nonoptimized, Slater-rule exponents) are

$$4a_1 = 0.08(O1s) + 0.84(O2s_\perp) + 0.70(O2p_z) - 0.75(H_1 1s + H_2 1s)$$
$$2b_2 = 0.99(O2p_y) - 0.89(H_1 1s - H_2 1s)$$

(15.30)

For these two MO's, the opposite signs of the oxygen and hydrogen AO coefficients give charge depletion between the nuclei. These MO's are antibonding.

The unoccupied orbitals (15.30) are called *virtual orbitals*. Because they were calculated for the electron configuration (15.26), they are not accurate representations of the SCF orbitals actually occupied in H_2O excited electronic states. When the electron configuration changes, the interorbital electronic interactions change, thereby changing the forms of all the SCF orbitals. We can, however, use the virtual orbitals and their calculated energies as rough approximations to the higher SCF orbitals and energies.

The formation of the H_2O MO's from the separated-atoms AO's is illustrated schematically in Fig. 15.5; the dotted lines indicate which AO's contribute signifi-

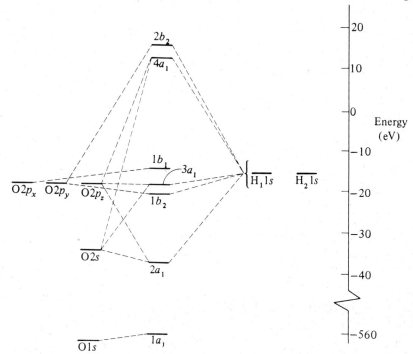

Figure 15.5 *Formation of the H_2O MO's from the minimal basis AO's. The five lowest MO's are filled in the ground state. (Note the break in the scale.)*

cantly to each MO. Note that only AO's of roughly the same energy combine to a significant degree in a given MO. [This fact is explained by the $1/(E_n^{(0)} - E_m^{(0)})$ term in Eq. (9.36).] We can thus get a qualitative idea of the MO's without doing any calculations by using the rules that only symmetry orbitals of the same symmetry species combine and that only AO's of comparable energy (Fig. 11.2) contribute significantly to a given MO.

Table 15.1 lists some of the many calculations on the H_2O ground state. The experimental equilibrium electronic energy of H_2O is -76.480 hartrees. All the calculations listed are nonrelativistic. The only significant relativistic correction will occur in the $1s$ inner-shell electrons of oxygen (see Section 11.7); since this shell remains essentially unchanged on molecule formation, one can use the relativistic energy correction for the O atom (which has been calculated to be -0.05 hartree) as the relativistic correction for H_2O. Thus we can add -0.05 to all the calculated energies listed, or, alternatively, we can compare the nonrelativistic calculated energies with the exact nonrelativistic H_2O energy $-76.48 + 0.05 = -76.43$ hartrees.

All the SCF MO calculations listed are multicenter except for Moccia's one-center calculation. The size and quality of the basis set used determine how closely the Hartree–Fock limit is approached. Montgomery, Bruner, and Knight used the variational perturbation method discussed at the end of Section 9.3, taking the unperturbed problem as an O^{2-} ion with ten noninteracting electrons; the perturbation is then the sum of interelectronic repulsions and electron–proton attractions; the energy was calculated through second order. Rosenberg, Ermler, and Shavitt's CI calculations used a thirty-nine-function STO basis set consisting of $(5s\,4p\,2d)$ on oxygen and $(3s1p)$ on each hydrogen; they included all single and double excitations, a total of 4120 configuration functions for geometries with equal bond lengths and 7966 configuration functions for unequal bond lengths; they obtained 75 percent of the correlation energy. These workers did SCF and CI calculations at thirty-six geometries to generate the potential-energy surface in the region of the equilibrium geometry; they also determined the variation of one-electron properties like the dipole moment with geometry in the region of the equilibrium geometry. Lucchese, Conrad, and Schaefer used a basis set of fifty-five contracted Gaussians and included 10705 configuration functions. The other three calculations will be discussed later in this chapter.

Although the SCF MO calculations give good geometries and dipole moments, they give poor binding energies. For the minimal-basis Pitzer–Merrifield calculation, the difference between the energy of two isolated hydrogen atoms and one oxygen atom (as calculated with the minimal basis set) and the calculated H_2O energy gives a binding energy of only 4.5 eV, compared to the experimental value 10.1 eV. (With nonoptimized exponents, the minimal basis set gives 3.4 eV.) The near-Hartree–Fock calculations of Dunning, Pitzer, and Aung give a binding energy of 6.9 eV. This rather large error in dissociation energy is typical of Hartree–Fock calculations, as we saw for diatomics. When H_2O is formed from $2H + O$, two new electron pairs are formed. Two electrons paired in the same MO move through the same region of space, and hence the correlation energy of such a pair is high. The Hartree–Fock calculation does not take into account this extra correlation in the

Table 15.1 *Calculations for* H_2O

Reference[1]	Method	Total Energy[2] (hartrees)	Dipole Moment	Predicted Bond Angle	Predicted Bond Length
Lathan et al.	SCF STO-3G	-74.97 (21; 7)	1.69 D	100°	0.990 Å
Pitzer, Merrifield	SCF STO	-75.70 (7)	1.92 D	100°	0.990 Å
Lathan et al.	SCF 4-31G	-75.91 (28; 13)	2.52 D	111°	0.951 Å
Moccia	SCF 1-center	-75.92 (28)	2.08 D	107°	0.960 Å
Aung, Pitzer, Chan	SCF STO	-76.00 (26)	2.03 D	—	—
Harihan, Pople	SCF 6-31G*	-76.01 (36; 19)	2.19 D	105.5°	0.948 Å
Dunning, Pitzer, Aung	SCF GTO	-76.062 (56; 41)	2.08 D	106.6°	0.941 Å
Rosenberg, Ermler, Shavitt	SCF STO	-76.064 (39)	2.00 D	106.1°	0.940 Å
Clementi, Popkie	SCF GTO	-76.066	—	—	—
Estimated Hartree–Fock energy		-76.068			
Montgomery, Bruner, Knight	Varn.-Pert.	-75.99	—	—	—
Peterson, Pfeiffer	Ab initio VB	-76.02	—	106.5°	0.968 Å
Rosenberg, Ermler, Shavitt	CI, STO	-76.34	1.92 D	104.9°	0.953 Å
Lucchese, Conrad, Schaefer	CI, GTO	-76.35	—	—	—
Meyer	PNO-CI	-76.37	—	—	—
Meyer	PNO-CEPA	-76.38	1.84 D	105.1°	0.955 Å
Experimental values		-76.48	1.85 D	104.5°	0.957 Å

[1] W. J. Lathan, W. J. Hehre, L. A. Curtiss, and J. A. Pople, *J. Am. Chem. Soc.*, **93**, 6377 (1971); R. M. Pitzer and D. P. Merrifield, *J. Chem. Phys.*, **52**, 4782 (1970); W. J. Lathan et al., *op. cit.*; R. Moccia, *J. Chem. Phys.*, **40**, 2186 (1964); S. Aung, R. M. Pitzer, and S. I. Chan, *J. Chem. Phys.*, **49**, 2071 (1968); P. C. Harihan and J. A. Pople, *Mol. Phys.*, **27**, 209 (1974); T. H. Dunning, R. M. Pitzer, and S. Aung, *J. Chem. Phys.*, **57**, 5044 (1972); B. J. Rosenberg and I. Shavitt, *J. Chem. Phys.*, **63**, 2162 (1975), and B. J. Rosenberg, W. C. Ermler, and I. Shavitt, *J. Chem. Phys.*, **65**, 4072 (1976); E. Clementi and H. Popkie, *J. Chem. Phys.*, **57**, 1077 (1972); H. E. Montgomery, B. L. Bruner, and R. E. Knight, *J. Chem. Phys.*, **52**, 4407 (1970); C. Peterson and G. V. Pfeiffer, *Theor. Chim. Acta*, **26**, 321 (1972); Rosenberg and Shavitt, *op. cit.*, and Rosenberg, Ermler, and Shavitt, *op. cit.*; R. R. Lucchese, M. P. Conrad, and H. F. Schaefer, *J. Chem. Phys.*, **68**, 5292 (1978); W. Meyer, *Int. J. Quantum Chem.*, **5S**, 341 (1971); W. Meyer in *Schaefer, Methods of Electronic Structure Theory*, Chapter 11.

[2] This is the total electronic energy (including nuclear repulsion) calculated at or near the true structural values. Number of basis functions in parentheses; the notation (46; 28) means 46 Gaussians contracted to 28.

molecule (as compared to the separated atoms) and hence gives too small a binding energy.

Hartree–Fock orbital energies have experimental as well as theoretical significance. In 1933 Koopmans gave arguments that indicate that the energy required to remove an electron from a closed-shell atom or molecule is reasonably well approximated by minus the orbital energy ε of the AO or MO from which the electron is removed. A partial justification of this result is the fact that if we neglect the change in the form of the MO's that occurs when the molecule is ionized, then the difference between the Hartree–Fock energies of the ion and the neutral closed-shell molecule can be shown to equal the orbital energy of the MO from which the electron was removed (see Problem 15.14).

The energy needed to remove an electron from an MO of a molecule can be found experimentally using photoelectron spectroscopy; here, one uses photons of known energy to knock electrons out of gas-phase molecules and measures the kinetic energies of the emitted electrons. A comparison of minus the near-Hartree–Fock orbital energies (Rosenberg and Shavitt, Table 15.1) with the experimentally observed ionization energies from the various H_2O MO's follows, where the energies are in electron volts and the experimental values are in parentheses: $1a_1$, 559.5 (539.7); $2a_1$, 36.7 (32.2); $1b_2$, 19.5 (18.5); $3a_1$, 15.9 (14.7); $1b_1$, 13.8 (12.6). Koopmans' theorem ionization potentials are somewhat inaccurate because of (a) neglect of the change in the forms of the MO's that occurs on ionization and (b) neglect of the change in correlation energy between the neutral molecule and the ion.

15.5   POPULATION ANALYSIS

A widely used method to analyze LCAO-MO SCF wave functions is population analysis, introduced by Mulliken. Suppose that the MO φ_i has the form $\varphi_i = c_1\chi_1 + c_2\chi_2 + \cdots + c_m\chi_m$, where the c's are constants and where the χ's are AO's. For simplicity, we shall assume the c's and χ's are real. The probability density associated with an electron in this MO is $|\varphi_i|^2 = c_1^2\chi_1^2 + c_2^2\chi_2^2 + \cdots + 2c_1c_2\chi_1\chi_2 + 2c_1c_3\chi_1\chi_3 + 2c_2c_3\chi_2\chi_3 + \cdots$. Integrating this equation over three-dimensional space and using the fact that φ_i and the χ_i's are normalized, we get

$$1 = c_1^2 + c_2^2 + c_3^2 + \cdots + 2c_1c_2S_{12} + 2c_1c_3S_{13} + 2c_2c_3S_{23} + \cdots \quad (15.31)$$

where the S's are overlap integrals: $S_{12} = \int \chi_1\chi_2 \, dv$, etc.

Mulliken proposed two ways of apportioning the probability density in (15.31). In one approach the molecule's electron probability density is divided into *net* populations in the various AO's and *overlap* populations for the various pairs of AO's. One says that an electron in MO φ_i contributes c_1^2 to the net atomic population of the AO χ_1, c_2^2 to the net atomic population of χ_2, etc., and contributes $2c_1c_2S_{12}$ to the overlap population between χ_1 and χ_2, $2c_1c_3S_{13}$ to the overlap population between χ_1 and χ_3, etc. If the AO's χ_i and χ_j are centered on the same atom, then S_{ij} vanishes (provided we use orthogonalized STO's like $2s_\perp$) and the overlap population is zero. Summing the contributions of all the electrons, we obtain the Mulliken net population for each AO and overlap population for each AO pair.

For example, consider the Pitzer–Merrifield minimal-basis H_2O MO's in (15.27). For the H_2O equilibrium geometry, the overlap integrals are found to be (Problem 15.15)

$$\langle H_1 1s | O1s \rangle = \langle H_2 1s | O1s \rangle = 0.054, \quad \langle H_1 1s | O2s_\perp \rangle = \langle H_2 1s | O2s_\perp \rangle = 0.471$$

$$\langle H_1 1s | O2p_y \rangle = -\langle H_2 1s | O2p_y \rangle = 0.319, \quad \langle H_1 1s | O2p_z \rangle = \langle H_2 1s | O2p_z \rangle = 0.247$$

$$\langle H_1 1s | H_2 1s \rangle = 0.238$$

An electron in the MO $2a_1$ contributes $(-0.027)^2 = 0.0007$ to the net population of the $O1s$ AO, $(0.820)^2 = 0.672$ to the net population of $O2s_\perp$, etc., and contributes $2(-0.027)(0.152)(0.054) = -0.0004$ to the overlap population between $O1s$ and $H_1 1s$, $2(0.820)(0.152)(0.471) = 0.117$ to the overlap population between $O2s_\perp$ and $H_1 1s$, etc. Note from (15.31) that the net and overlap contributions of one electron in a given MO add to 1.

Since there are two electrons in each of the MO's (15.27), the net population of the $O1s$ AO is $2(1.000)^2 + 2(-0.027)^2 + 2(-0.026)^2 = 2.00$. Net populations for the remaining AO's are $O2p_y$—0.78, $O2s_\perp$—1.85, $O2p_z$—1.27, $O2p_x$—2.00, $H_1 1s$—0.54_5, $H_2 1s$—0.54_5.

Summation of the AO overlap contributions over all pairs of AO's in a given MO φ_i gives the total overlap population for that MO; if this overlap population is substantially positive, the MO is bonding; if it is substantially negative, the MO is antibonding; if it is zero or near-zero, the MO is nonbonding.

For example, for H_2O we have the following contributions to the total overlap population of the $3a_1$ MO: $O1s$ overlap with the two hydrogen $1s$ AO's contributes $2(2)(-0.026)(0.264 + 0.264)(0.054) = -0.003$; $O2s_\perp$ overlap with the H's contributes $2(2)(-0.502)(0.264)2(0.471) = -0.499$; $O2p_z$ overlap with the H's contributes 0.411; and overlap between the hydrogen AO's contributes $2(2)(0.264)(0.264)(0.238) = 0.066$. Summing, we get a total overlap population of -0.02 for the $3a_1$ MO. This is essentially zero, indicating a nonbonding (lone-pair) MO. The total overlap population for $2a_1$ is found to be 0.53, and that for $1b_2$ is 0.50 (Problem 15.16). These are bonding MO's. For the inner-shell $1a_1$ MO, we get 0.00.

Mulliken's second method divides the probability density into *gross* populations in the individual AO's, with no probability density assigned to overlap regions. This is done by dividing the overlap population $2c_i c_j S_{ij}$ equally between the AO's χ_i and χ_j. The contribution of one electron in the MO φ_i to the gross atomic population of AO χ_1 is then taken as [see (15.31)] $c_1^2 + c_1 c_2 S_{12} + c_1 c_3 S_{13} + \cdots + c_1 c_m S_{1m}$, with similar expressions for the other AO's.

For example, consider the $2a_1$ H_2O MO in (15.27). The contribution of the two electrons in this MO to the gross population of the $O2s_\perp$ AO is

$$2[(0.820)^2 + (0.820)(0.152)(0.471) + (0.820)(0.152)(0.471)] = 1.58$$

One finds these other contributions to the gross population of $O2s_\perp$: 0.00 from $1a_1$, 0.25 from $3a_1$, zero from $1b_2$ and $1b_1$. Addition of these contributions gives a total gross AO population (or occupation number) of 1.83 for $O2s_\perp$. Carrying out the calculation for the other AO's, one finds (Problem 15.16) the following total gross AO populations (where the contributions are listed in the order $1a_1, 2a_1, 1b_2, 3a_1, 1b_1$):

$O1s$—$2.00 + 0.00 + 0 + 0.00 + 0 = 2.00$; $O2s_\perp$—1.83; $O2p_x$—$0 + 0 + 0 + 0 + 2 = 2$; $O2p_y$—$0 + 0 + 1.12 + 0 + 0 = 1.12$; $O2p_z$—$0 + 0.05_5 + 0 + 1.44_5 + 0 = 1.50$; H_11s— $0.00 + 0.18_4 + 0.44_2 + 0.150 + 0 = 0.77_6$; H_21s—0.77_6.

Addition of these gross AO populations on each atom gives the total gross atomic population for that atom. The total gross atomic populations are then 8.45 for O and 0.77_6 for each H. (Of course, the total gross atomic populations add to 10, the number of electrons.) Since the atomic numbers are 8 and 1, the population analysis indicates a net charge of $-0.45e$ on oxygen and 0.22_4e on each hydrogen.

One should not put too much reliance on numbers calculated by population analysis. Mulliken's assignment of half the overlap probability density to each AO is rather arbitrary and occasionally leads to unphysical results (see *Mulliken and Ermler, Diatomic Molecules*, pp. 36–38, 88–89); moreover, one finds that a small change in basis set can produce a large change in calculated population-analysis charges on atoms. Several other methods have been proposed to assign charges to atoms in molecules; see P. Politzer et al., *Theor. Chim. Acta*, **38**, 101 (1975). Most of these methods give results in semiquantitative agreement with Mulliken population-analysis values.

15.6 LOCALIZED MO'S

The idea of a chemical *bond* between a pair of atoms in a molecule is fundamental to chemistry. The experimental evidence supporting this concept is substantial. One can assign bond energies to various kinds of bonds and obtain good estimates for the heats of formation of most molecules by adding the energies of the individual bonds. Other molecular properties (e.g., magnetic susceptibility, dipole moment) can also be analyzed as the sum of contributions from individual bonds (and lone pairs). Examination of the infrared spectrum of a compound containing an —O—H group shows a characteristic band near 3600 cm^{-1}; this O—H stretching vibrational band occurs at nearly the same frequency no matter whether it is HOH or HOCl or CH$_3$OH which is observed. The length of an O—H bond is nearly constant from molecule to molecule, about 0.96 Å.

The MO picture of water presented in Section 15.4 appears to be gravely deficient, in that it is seemingly inconsistent with the existence of individual bonds in the molecule. Each of the bonding MO's in (15.27) is delocalized over the entire molecule; if one were to compare the bonding MO's for HOH and HOCl, one would find them to be quite different. Yet we know that the OH parts of these two molecules are similar. Actually, MO theory *can* explain the observed near invariance of a given kind of chemical bond, as we now show.

The MO approximation to the ground state of water is a Slater determinant of the form

$$|\overline{\varphi_1}\,\overline{\varphi_1}\,\overline{\varphi_2}\,\overline{\varphi_2}\,\overline{\varphi_3}\,\overline{\varphi_3}\,\overline{\varphi_4}\,\overline{\varphi_4}\,\overline{\varphi_5}\,\overline{\varphi_5}| \qquad (15.32)$$

Putting (15.32) into words, we say that two electrons are in each of the orthonormal MO's $\varphi_1, \varphi_2, \varphi_3, \varphi_4,$ and φ_5. However, this MO description is not unique. The addition of a multiple of one column of a determinant to another column leaves the determinant unchanged in value. Adding column 7 to column 9 and column 8 to column 10 in (15.32), we get

$$|\varphi_1\overline{\varphi_1}\varphi_2\overline{\varphi_2}\varphi_3\overline{\varphi_3}\varphi_4\overline{\varphi_4}(\varphi_4+\varphi_5)\overline{(\varphi_4+\varphi_5)}| \qquad (15.33)$$

This determinant leads to the description of the molecule as having the MO's $\varphi_1, \varphi_2,$ $\varphi_3, \varphi_4,$ and $(\varphi_4+\varphi_5)$ each doubly occupied. Despite the different verbal descriptions, the wave functions (15.32) and (15.33) are identical.

[In discussing SCF calculations, we used an orthogonalized $2s$ AO of the form $a(1s)+b(2s)$, where $2s$ is a (nodeless) $2s$ Slater-type orbital. This procedure is justified by use of the freedom of adding a multiple of one column of a determinant to another; the determinantal wave function for the oxygen atom is the same whether it is the $2s$ or the orthogonalized $2s_\perp$ AO that is used.]

An objection that can be raised to (15.33) is that the MO $(\varphi_4+\varphi_5)$ is neither normalized nor orthogonal to φ_4. Consider, however, the Slater determinant

$$|\varphi_1\overline{\varphi_1}\varphi_2\overline{\varphi_2}\varphi_3\overline{\varphi_3}(b\varphi_4+c\varphi_5)\overline{(b\varphi_4+c\varphi_5)}(c\varphi_4-b\varphi_5)\overline{(c\varphi_4-b\varphi_5)}| \quad (15.34)$$

where b and c are any two real constants such that

$$b^2+c^2=1 \qquad (15.35)$$

We now show (15.34) and (15.32) to be the same wave function. Multiplication of columns 7 and 8 of (15.32) by b and columns 9 and 10 by $-b^{-1}$ gives

$$|\cdots b\varphi_4\overline{(b\varphi_4)}(-b^{-1}\varphi_5)\overline{(-b^{-1}\varphi_5)}| \qquad (15.36)$$

This step multiplies (15.32) by $b^2(-b^{-1})^2=1$. Next, $-bc$ times column 9 is added to column 7 and $-bc$ times column 10 is added to column 8 to give

$$|\cdots(b\varphi_4+c\varphi_5)\overline{(b\varphi_4+c\varphi_5)}(-b^{-1}\varphi_5)\overline{(-b^{-1}\varphi_5)}| \qquad (15.37)$$

Finally, c/b times column 7 is added to column 9, c/b times column 8 is added to column 10, and (15.35) is used; this gives

$$|\cdots(b\varphi_4+c\varphi_5)\overline{(b\varphi_4+c\varphi_5)}(c\varphi_4-b\varphi_5)\overline{(c\varphi_4-b\varphi_5)}| \qquad (15.38)$$

This completes the proof. The orthonormality of the MO's $(c\varphi_4-b\varphi_5)$ and $(b\varphi_4+c\varphi_5)$ readily follows from (15.35) and the orthonormality of φ_4 and φ_5. Thus we can describe the molecule as having the orthonormal MO's

$$\varphi_1, \varphi_2', \varphi_3, b\varphi_4+c\varphi_5, c\varphi_4-b\varphi_5 \qquad (15.39)$$

each doubly occupied.

There are an infinite number of other MO descriptions consistent with (15.32): Let d and e be any two real constants such that $d^2+e^2=1$. If the procedure used to go from (15.32) to (15.34) is applied to columns 5, 6, 7, and 8 of (15.34), we end up with

$$|\varphi_1\overline{\varphi_1}\varphi_2\overline{\varphi_2}(d\varphi_3+be\varphi_4+ce\varphi_5)\overline{(d\varphi_3+be\varphi_4+ce\varphi_5)}$$

$$\times(e\varphi_3-bd\varphi_4-cd\varphi_5)\overline{(e\varphi_3-bd\varphi_4-cd\varphi_5)}(c\varphi_4-b\varphi_5)\overline{(c\varphi_4-b\varphi_5)}|$$

We can thus describe the electronic configuration as having the orthonormal MO's

$$\varphi_1, \varphi_2, (d\varphi_3 + be\varphi_4 + ce\varphi_5), (e\varphi_3 - bd\varphi_4 - cd\varphi_5), (c\varphi_4 - b\varphi_5)$$

each doubly occupied. Continuing on in this manner, we can derive MO's, each of which is a linear combination of all the original MO's φ_1, φ_2, φ_3, φ_4, φ_5. [For the general conditions that must be satisfied by the coefficients in these linear combinations, see *Pilar*, p. 345.]

Thus for a given closed-subshell electronic state of a molecule, there are many possible MO descriptions. The delocalized MO's (15.27) of water are uniquely determined as the solutions of the SCF equations [Eq. (13.176)]

$$\hat{F}\varphi_i = \varepsilon_i\varphi_i, \qquad i = 1, 2, \ldots \tag{15.40}$$

Hence the reader may be wondering how we can have other sets of MO's that also minimi˙ the variational integral. Actually, Eq. (15.40) is not the most general equation satisfied by the MO's that minimize the variational integral; instead, one finds in deriving the SCF equations that the Hartree–Fock MO's must satisfy[7]

$$\hat{F}\varphi_i = \sum_{j=1}^{N} \lambda_{ji}\varphi_j, \qquad i = 1, \ldots, N \tag{15.41}$$

where the sum runs over the occupied MO's. The λ_{ji}'s are certain constants (Lagrangian multipliers) which can be chosen arbitrarily, subject to the MO orthonormality requirement. Different choices of the λ_{ji}'s give different sets of MO's, but each such set minimizes the variational integral. For a closed-subshell configuration, one possible choice of the λ_{ji}'s can be shown to be

$$\lambda_{ji} = \varepsilon_i\delta_{ij} \tag{15.42}$$

This choice reduces (15.41) to (15.40). The delocalized MO's [(15.27) for water] that satisfy (15.40) are called the *canonical* SCF MO's. The canonical MO's are eigenfunctions of the Fock operator, which commutes with the molecular symmetry operators, and hence the canonical MO's each transform according to one of the possible molecular symmetry species. A set of MO's like (15.39) satisfies (15.41) but not (15.40) and does not necessarily transform according to the molecular symmetry species.

For an open-subshell configuration, one can also find canonical delocalized MO's that satisfy (15.40) and that transform according to the molecular symmetry species; however, the form of $\hat{F}$ is more complicated than for the closed-subshell case. [See C. C. J. Roothaan, *Rev. Mod. Phys.*, **32**, 179 (1960).]

Of the possible MO sets formed as linear combinations of the delocalized canonical MO's of water, a set that would appeal to a chemist is one for which the charge probability density of each bonding MO is localized in the region of one of the O—H bonds. There are many ways of taking linear combinations of the delocalized MO's to get such *localized* MO's. A natural requirement that the two localized bonding

---

[7] For the proof of (15.41) and (15.42), see, for example, S. M. Blinder, *Am. J. Phys.*, **33**, 431 (1965), Appendix D; *Pilar*, Chapter 13.

MO's in water should satisfy is that they be equivalent to each other; i.e., each localized bonding MO of water should be transformed into the other by the $\hat{C}_2(z)$ symmetry operation (Fig. 15.7). Localized MO's that are permuted among one another by a symmetry operation that permutes equivalent chemical bonds are called *equivalent orbitals*. The concept of localized equivalent orbitals goes back to papers by Hund (1932) and Coulson (1942), but the systematic procedure for deriving equivalent MO's from the canonical MO's was given by Lennard-Jones in 1949.[8] As we shall see, equivalent MO's reconcile MO theory with the chemist's intuitive picture of chemical bonding.

For water, chemical intuition (H—Ö—H) suggests the equivalent MO's to be a pair of equivalent bonding orbitals $b(OH_1)$ and $b(OH_2)$, an inner-shell 1s oxygen orbital $i(O)$ (which is equivalent to itself), and two lone-pair equivalent MO's $l_1(O)$ and $l_2(O)$ on oxygen. Actually, the symmetry requirement that the MO's be equivalent is not sufficient to determine the equivalent MO's of water uniquely, and many possible equivalent MO's can be constructed.

The concept of equivalent orbitals has two shortcomings. First, it cannot be used to define localized MO's in molecules having no equivalent chemical bonds. For HOCl the O—H and O—Cl bonds are not equivalent, and the localized bonding MO's cannot be expected to be equivalent. Second, even for symmetric molecules, the choice of equivalent MO's is not unique in many cases (e.g., HOH). Hence we want a definition that will specify a unique set of localized MO's for any molecule.

The most "localized" MO's are those that are most separated from one another; by this we mean that set of MO's for which the interelectronic repulsion between the different MO's viewed as "charge clouds" is a minimum. The energy of repulsion between the charge clouds of electron 1 in the MO φ_i and electron 2 in the MO φ_j is a Coulomb integral of the form (9.144). The total interorbital charge-cloud repulsion energy is then

$$4 \sum_i \sum_{j>i} \int \int |\varphi_i(1)|^2 |\varphi_j(2)|^2 \frac{1}{r_{12}} dv_1 \, dv_2 \qquad (15.43)$$

where the sums run over the occupied MO's. (The factor 4 comes from the four interorbital repulsions of the electrons in each MO pair.) We define the localized MO's as those orthonormal MO's that minimize (15.43). [It turns out that minimization of (15.43) implies the maximization of the total *intra*orbital electron repulsion and the minimization of the magnitude of the (interorbital) exchange energy.] This definition was originally suggested by Lennard-Jones and Pople and has been applied to many molecules by Edmiston and Ruedenberg.[9] We shall call these MO's the *energy-localized* MO's. For molecules with symmetry, we expect that the energy-localized MO's will also be equivalent MO's. [An apparent exception was found by

---

[8] J. E. Lennard-Jones, *Proc. Roy. Soc.*, **A198**, 1, 14 (1949); G. G. Hall and J. E. Lennard-Jones, *Proc. Roy. Soc.*, **A202**, 155 (1950); J. E. Lennard-Jones and J. A. Pople, *Proc. Roy. Soc.*, **A202**, 166 (1950).

[9] C. Edmiston and K. Ruedenberg, *Rev. Mod. Phys.*, **35**, 457 (1963); *J. Chem. Phys.*, **43**, S97 (1965); P. O. Löwdin, ed., *Quantum Theory of Atoms, Molecules, and the Solid State*, Academic Press, New York, 1966, pp. 263–280.

Edmiston and Ruedenberg for C_2, but calculations by von Niessen using several improved C_2 wave functions always produced equivalent localized MO's; W. von Niessen, *Theor. Chim. Acta*, **27**, 9 (1972).] Energy-localized MO's are thus a generalization of equivalent MO's.

Liang and Taylor calculated the energy-localized MO's for water, starting with the canonical MO's (15.27) of Pitzer and Merrifield. In terms of the canonical (delocalized) MO's, Liang and Taylor find[10]

$$i(O) = 0.99(1a_1) - 0.12(2a_1) \qquad\qquad\qquad +0.06(3a_1)$$

$$b(OH_1) = 0.05(1a_1) + 0.57(2a_1) + 0.71(1b_2) + 0.42(3a_1)$$

$$b(OH_2) = 0.05(1a_1) + 0.57(2a_1) - 0.71(1b_2) + 0.42(3a_1) \qquad\qquad \textbf{(15.44)}$$

$$l_1(O) = 0.08(1a_1) + 0.42(2a_1) \qquad\qquad -0.57(3a_1) - 0.71(1b_1)$$

$$l_2(O) = 0.08(1a_1) + 0.42(2a_1) \qquad\qquad -0.57(3a_1) + 0.71(1b_1)$$

The lone-pair $1b_1$ MO $(2p_{xO})$ is equally divided between the equivalent lone-pair localized orbitals. ($\sqrt{\frac{1}{2}} = 0.71$.) The inner-shell $1a_1$ MO contributes substantially to only the $i(O)$ MO. The bonding $1b_2$ MO is equally divided between the two bonding localized MO's. The bonding $2a_1$ MO makes substantial contributions to the two bonding localized MO's and smaller, but still substantial, contributions to the lone-pair localized MO's. The largely lone-pair $3a_1$ MO makes substantial contributions to the lone-pair localized MO's and lesser contributions to the bonding localized MO's.

In terms of the AO's, the energy-localized MO's of water are

$$i(O) = -0.007(H_1 1s) - 0.007(H_2 1s) + 0.99(O1s) - 0.12(O2s_\perp)$$
$$+0.03(O2p_z)$$

$$b(OH_1) = \quad 0.50(H_1 1s) - 0.10(H_2 1s) + 0.02(O1s) + 0.25(O2s_\perp)$$
$$+0.407(O2p_z) + 0.441(O2p_y)$$

$$b(OH_2) = -0.10(H_1 1s) + 0.50(H_2 1s) + 0.02(O1s) + 0.25(O2s_\perp)$$
$$+0.407(O2p_z) - 0.441(O2p_y) \qquad\qquad \textbf{(15.45)}$$

$$l_1(O) = -0.09(H_1 1s) - 0.09(H_2 1s) + 0.09(O1s) + 0.63(O2s_\perp)$$
$$-0.39(O2p_z) - 0.71(O2p_x)$$

$$l_2(O) = -0.09(H_1 1s) - 0.09(H_2 1s) + 0.09(O1s) + 0.63(O2s_\perp)$$
$$-0.39(O2p_z) + 0.71(O2p_x)$$

The MO wave function

$$|i(O)\overline{i(O)}b(OH_1)\overline{b(OH_1)}b(OH_2)\overline{b(OH_2)}l_1(O)\overline{l_1(O)}l_2(O)\overline{l_2(O)}| \quad \textbf{(15.46)}$$

is identical to (15.29).

---

[10] J. H. Liang, Ph.D. thesis, Ohio State U., 1970. I am grateful to Professor William J. Taylor for making this work available to me. See also F. Franks, ed., *Water*, Vol. 1, Plenum, New York, 1972, p. 42.

Let us analyze these equivalent MO's for water. The $i(O)$ MO is nearly a pure $1s$ inner-shell oxygen AO.

To define the angle between the two localized bonding MO's, we draw the line from O to H_1 along which the electron probability density in $b(OH_1)$ is a maximum, and we draw a similar line from O to H_2. The angle between these lines where they intersect at the O nucleus defines the angle between the localized bonding MO's. (If this angle differs significantly from the angle defined by straight lines between the nuclei, the bonds are said to be *bent*.) The angle between the localized bonding MO's is determined mainly by the oxygen $2p_y$ and $2p_z$ AO contributions (with a small influence by the hydrogen $1s$ AO's). For $b(OH_1)$ the $O2p_yO2p_z$ contribution contains the factor $0.407z_O + 0.441y_O$. Let us carry out a rotation of coordinates in the zy plane by an angle $\alpha = \arctan(0.441/0.407) = 47\frac{1}{2}°$, as in Fig. 15.6. The relation between coordinates in the unrotated system and the rotated $z'y'$ system is given by the well-known formulas

$$z' = z \cos\alpha + y \sin\alpha$$
$$y' = -z \sin\alpha + y \cos\alpha \tag{15.47}$$

From (15.47) we have

$$0.407 z_O + 0.441 y_O = 0.600 z'_O \tag{15.48}$$

Multiplication by the exponential factor of the $2p$ oxygen AO then gives

$$0.407(O2p_z) + 0.441(O2p_y) = 0.60(O2p_{z'}) \tag{15.49}$$

In other words, the hybridized $2p_z2p_y$ AO on the left of (15.49) is the same function as 0.60 times a $2p$ AO inclined at an angle of $47\frac{1}{2}°$ with the z axis. Thus the bonding $2p_z2p_y$ hybrid AO's of oxygen in $b(OH_1)$ and $b(OH_2)$ point in the general direction of the hydrogen atoms. The contribution from the hydrogen atoms to $b(OH_1)$ is mostly from $H_1 1s$, and the overlap between $H_1 1s$ and the $2p_z2p_y$ oxygen hybrid then forms the $O-H_1$ chemical bond. The angle between the two hybrid oxygen AO's in $b(OH_1)$ and $b(OH_2)$ is $95°$, and this is approximately the angle between the localized bonding MO's.

Energy-localized H_2O bonding MO's calculated from an extended-basis-set SCF MO wave function were found to have an angle of $96°$ between the oxygen hybrids that contribute to these MO's and an angle of $103°$ between the localized MO's themselves, which is nearly the same as the $104\frac{1}{2}°$ molecular bond angle, indicating that the bonds in water are not bent to any significant degree. [W. von Niessen, *Theor. Chim. Acta*, **29**, 29 (1973).] In contrast, in the strained molecule cyclopropane, the angle between the carbon hybrids contributing to the carbon–carbon bonding

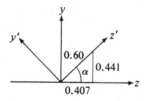

Figure 15.6 *A coordinate rotation in the yz plane*

energy-localized MO's deviates outward by $28°$ from the internuclear lines [M. D. Newton, E. Switkes, and W. N. Lipscomb, *J. Chem. Phys.*, **53**, 2645 (1970)].

The $b(OH_1)$ energy-localized MO is not completely confined to the region of the O—H_1 bond. We see from (15.45) that this MO has a small contribution from the $H_2 1s$ AO. (The ratio of the contributions of two AO's to an MO is given essentially by the square of the ratio of their coefficients.) Moreover, consider the contributions of $O2s_\perp$, $O2p_y$, and $O2p_z$ to $b(OH_1)$. The $2p_y 2p_z$ hybrid has a positive lobe in the region of the OH_1 bond, and the contribution of the $O2s_\perp$ AO reinforces this positive lobe; overlap with $H_1 1s$ then gives the O—H_1 bond. The $2p_y 2p_z$ hybrid has a negative lobe on the side of the oxygen opposite the OH_1 bond; this negative lobe is partly, but not completely, canceled by the $O2s_\perp$ contribution. Thus the $b(OH_1)$ MO has a "tail" on the side of the O away from H_1, this tail being somewhat distorted toward H_2 by the $-0.10H_2 1s$ contribution (Fig. 15.7). (For accurately plotted contours, see F. Franks, ed., *Water*, Vol. 1, Plenum, New York, 1972, p. 46.) Despite this tail, this MO is far more localized than any of the bonding canonical MO's in (15.27). Because $b(OH_1)$ is mostly localized in the OH_1 region, we expect only a small change in its form on going from HOH to, say, HOCl. The observed near invariance of the O—H bond from molecule to molecule is explained by MO theory using localized bonding MO's. (Neglect of the contribution of $H_2 1s$ to the $b(OH_1)$ localized MO gives a two-center orbital called a *bond orbital*.)

Finally, consider the two lone-pair MO's $l_1(O)$ and $l_2(O)$. These two MO's are mainly localized on the oxygen atom and are equivalent to each other. They are directed away from the hydrogen atoms and project above and below the molecular (yz) plane (Fig. 15.8). The $\hat{\sigma}_v(yz)$ operation converts x to $-x$ and hence interchanges the lone-pair MO's. The angle between them is approximately

$$2 \arctan (0.71/0.39) = 122°$$

[The angle between localized lone-pair MO's calculated from a more accurate wave function is $114°$; von Niessen, *Theor. Chim. Acta*, **29**, 29 (1973).]

Edmiston and Ruedenberg calculated energy-localized MO's for several diatomic molecules. Consider Li_2. Chemical intuition (Li—Li) suggests the localized MO's to be an inner-shell $1s$ AO on Li_a, an equivalent inner-shell AO on Li_b, and a bonding MO extending over the molecule. In terms of Ransil's minimal-basis-

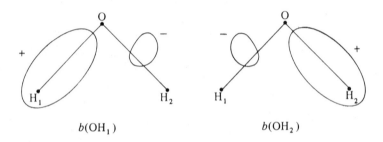

$b(OH_1)$ $b(OH_2)$

Figure 15.7 *Rough sketches of the localized equivalent bonding MO's in water*

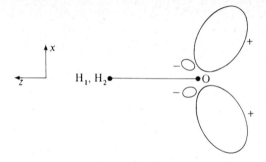

Figure 15.8 *Rough sketches of the localized equivalent lone-pair MO's of water*

set SCF MO's with optimized orbital exponents (Section 13.17), Edmiston and Ruedenberg found the energy-localized MO's of Li_2 to be

$$i(Li_a) = 0.703(1\sigma_g) + 0.707(1\sigma_u) - 0.074(2\sigma_g)$$

$$i(Li_b) = 0.703(1\sigma_g) - 0.707(1\sigma_u) - 0.074(2\sigma_g)$$

$$b(Li_a Li_b) = 0.105(1\sigma_g) \qquad\qquad + 0.995(2\sigma_g)$$

In terms of AO's, we have

$$i(Li_a) = 1.00(1s_a) - 0.02(2s_{\perp a}) - 0.006(2p\sigma_a) - 0.0001(1s_b)$$
$$- 0.05(2s_{\perp b}) - 0.01(2p\sigma_b)$$

$$i(Li_b) = -0.0001(1s_a) - 0.05(2s_{\perp a}) - 0.01(2p\sigma_a) + 1.00(1s_b)$$
$$- 0.02(2s_{\perp b}) - 0.006(2p\sigma_b)$$

$$b(Li_a Li_b) = 0.02(1s_a) + 0.52(2s_{\perp a}) + 0.11(2p\sigma_a) + 0.02(1s_b)$$
$$+ 0.52(2s_{\perp b}) + 0.11(2p\sigma_b) \qquad\qquad \textbf{(15.50)}$$

The bonding MO is very nearly the same as the $2\sigma_g$ MO; the lone-pair MO's are very nearly $1s$ AO's on each atom.

For N_2 we expect (: N≡N :) the localized MO's to be an inner-shell $1s$ AO on each atom, a lone-pair $2s$ AO on each atom, and three bonding MO's spread over the two atoms. The canonical-MO picture is that the triple bond consists of one σ bond and two π bonds, as in Section 13.6. The energy-localized bonding MO's were found by Edmiston and Ruedenberg to be three equivalent banana-shaped bond orbitals spaced 120° apart from one another; the AO's that make significant contributions to the localized bonding MO's are the $2s$, $2p\sigma$, $2p\pi_x$, and $2p\pi_y$ orbitals of each atom. The $i(N_a)$ and $i(N_b)$ localized MO's were found to be nearly pure $1s$ nitrogen AO's. Each of the $l(N_a)$ and $l(N_b)$ localized MO's is a hybrid of the $2s$ and $2p\sigma$ AO's of the relevant nitrogen atom, with the $2s$ AO making the larger contribution to the MO; each lone-pair localized MO is directed away from the other nitrogen atom.

The concept of localized MO's is not as widely applicable as that of delocalized canonical MO's. Delocalized MO's are valid for any molecule. However (as noted in

Section 11.5), the Hartree–Fock wave functions of nonclosed-subshell electronic states are, in most cases, linear combinations of *several* Slater determinants [e.g., see (10.62) and (10.63)], and the above localization procedure is not applicable to the open-subshell orbitals in these wave functions. Thus in a molecule in an excited electronic state with an open-subshell configuration, the electrons in the incompletely filled MO's are delocalized over much of the molecule.

The great success of the concept of chemical bonds between pairs of atoms is a reflection of the fact that the electronic ground states of most molecules have closed-subshell configurations, for which the localized MO description is just as valid as the delocalized MO description.

For a closed-subshell ground-state molecule, those properties that involve only the ground-state wave function can be calculated just as well with either the localized or the delocalized MO description. Such properties include electron probability density, dipole moment, geometry, heat of formation, etc. Properties of a molecule that involve the wave function of the ground state and also the wave function of an open-subshell excited state or the wave function of an open-subshell ion cannot be calculated using a localized MO description. Such properties include the ultraviolet absorption spectrum and molecular ionization energies. (For the special case of conjugated and aromatic molecules, see Section 15.13.)

Aside from justifying chemists' intuitive picture of bonding, the concept of localized MO's is significant in that localized MO's should be approximately transferable from molecule to molecule (which is not true, of course, for delocalized MO's). Having calculated the localized MO for the C—H bond in, say, methane, we expect to find this localized $b(CH)$ MO to be about the same in any hydrocarbon. Rothenberg calculated energy-localized $b(CH)$ MO's in CH_4, staggered and eclipsed C_2H_6, and staggered and eclipsed CH_3OH, and found them to have essentially the same contour shape (which resembles Fig. 15.22); the overlap integrals (calculated for coincident carbon atoms and coincident C—H axes) between pairs of these localized MO's all exceeded 0.996; properties such as the bond dipole moment of the $b(CH)$ MO were found to be nearly the same in all three molecules. Similar results were found for the energy-localized $b(CH)$ MO's in ethylene and acetylene. [See S. Rothenberg, *J. Chem. Phys.*, **51**, 3389 (1969); *J. Am. Chem. Soc.*, **93**, 68 (1971).]

The labor in calculating a wave function increases steeply with increasing numbers of electrons and nuclei. The approximate transferability of localized MO's raises the possibility of building up the MO wave function of a large molecule using localized MO's found from calculations on smaller molecules. Unfortunately, calculation of energy-localized MO's from canonical MO's is quite time-consuming, and most such calculations have been done with minimal (rather than extended) basis sets.

15.7 *THE SCF MO TREATMENT OF METHANE, ETHANE, AND ETHYLENE*

The AO's for a minimal-basis-set SCF calculation of CH_4 are the carbon $1s$, $2s$, $2p_x$, $2p_y$, and $2p_z$ AO's and a $1s$ AO on each hydrogen atom. The point group of methane is $\mathscr{T}_d$; group theory (see *Cotton* or *Schonland*) gives the possible symmetry species as A_1, A_2, E, T_1, and T_2. We shall not worry about specifying the symmetry behavior that corresponds to each symmetry species, but we shall use the species mainly as labels for the MO's. As usual, we set up the coordinate system with the z axis coinciding with the highest-order C_n or S_n axis; for methane this is an S_4 axis. The coordinates of the hydrogen atoms are (q, q, q), $(q, -q, -q)$, $(-q, q, -q)$, and $(-q, -q, q)$, where $2q$ is the edge of the cube in which the molecule is inscribed (Fig. 15.9). Note the equivalence of the x, y, and z axes.

The carbon atom is at the center of the molecule, and the carbon $1s$ and $2s$ AO's are each sent into themselves by every symmetry operation; these AO's transform according to the totally symmetric species A_1. The carbon $2p_x$, $2p_y$, and $2p_z$ AO's are given by x, y, or z times a radial function; their symmetry behavior is the same as that of the functions x, y, and z, respectively. From the formulas for rotation of coordinates [Eq. (15.63)], we see that any proper rotation sends each of the functions x, y, and z into some linear combination of x, y, and z. Any improper rotation is the product of some proper rotation and an inversion (Problem 12.23); the inversion simply converts each coordinate to its negative. Hence the three carbon $2p$ orbitals are sent into linear combinations of one another by each symmetry operation; they must therefore transform according to one of the triply degenerate symmetry species. Further investigation (which we omit) shows the symmetry species of the $2p$ AO's to be T_2.

Just as in water, each $1s$ hydrogen AO in methane does not transform according to any of the molecular symmetry species, and it is convenient to form symmetry orbitals by taking linear combinations of the $1s$ AO's. One obvious symmetry orbital is

$$f_1 = H_1 1s + H_2 1s + H_3 1s + H_4 1s \qquad (15.51)$$

Since each methane symmetry operator permutes the hydrogen $1s$ orbitals among themselves, (15.51) is sent into itself by each symmetry operation and belongs to the totally symmetric species A_1. We need three more symmetry orbitals. The construction of these is not obvious without the use of group theory, and we shall simply write down the results. The remaining three orthogonal (unnormalized)

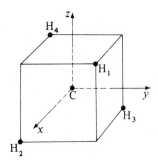

Figure 15.9 *Coordinate axes for methane. The origin is at the center of the cube.*

symmetry orbitals can be taken as

$$f_2 = H_1 1s + H_2 1s - H_3 1s - H_4 1s \tag{15.52}$$

$$f_3 = H_1 1s - H_2 1s + H_3 1s - H_4 1s \tag{15.53}$$

$$f_4 = H_1 1s - H_2 1s - H_3 1s + H_4 1s \tag{15.54}$$

Each of these three functions is transformed into some linear combination of the three functions by each symmetry operation. For example, a $\hat{C}_3$ rotation about the CH_1 bond causes the following permutation of hydrogen atoms:

$$1 \to 1, \ 2 \to 3, \ 3 \to 4, \ 4 \to 2$$

The corresponding $\hat{O}_{C_3}$ operator transforms f_2, f_3, and f_4 into f_3, f_4, and f_2, respectively. These three symmetry orbitals therefore transform according to one of the triply degenerate symmetry species. The function f_2 has positive signs for the hydrogen AO's with a positive x coordinate and negative signs for the hydrogen AO's with a negative x coordinate; we thus expect f_2 to have the same symmetry behavior as the function x. Similarly, f_3 and f_4 behave as y and z, respectively. (As an example, note that a $120°$ counterclockwise rotation about the CH_1 bond has the following effect on the three unit vectors: $\mathbf{i} \to \mathbf{j}, \mathbf{j} \to \mathbf{k}, \mathbf{k} \to \mathbf{i}$; this is the same behavior shown by the functions f_2, f_3, and f_4 for this rotation.) These three symmetry orbitals thus transform according to the species T_2.

The symmetry orbitals are then

Symmetry Orbital	f_1	f_2	f_3	f_4	C1s	C2s	$C2p_x$	$C2p_y$	$C2p_z$
Symmetry Species	a_1	t_2	t_2	t_2	a_1	a_1	t_2	t_2	t_2

The nine lowest MO's therefore consist of three a_1 and six t_2 MO's. The six t_2 MO's belong to triply degenerate levels and thus fall into two different subshells $1t_2$ and $2t_2$; each such subshell contains three MO's of equal orbital energy and each subshell holds six electrons. SCF calculations give the three lowest subshells as $1a_1$, $2a_1$, and $1t_2$, with energies -11.20, -0.93, and -0.54 hartrees, respectively. The ground state of methane has the closed-subshell configuration

$$(1a_1)^2(2a_1)^2(1t_2)^6 \tag{15.55}$$

and is a $^1A_1$ state.

Pitzer[11] did SCF calculations for methane at several CH distances, representing each basis AO by a single STO and optimizing the orbital exponents. He found a minimum in energy for a bond distance of 1.089 Å, as compared to the experimental value 1.085 Å. The MO's found by Pitzer at the experimental equilibrium bond distance are

$$1a_1 = -0.005(H_1 1s + H_2 1s + H_3 1s + H_4 1s) + 1.001(C1s) + 0.025(C2s_\perp)$$

$$2a_1 = 0.186(H_1 1s + H_2 1s + H_3 1s + H_4 1s) - 0.064(C1s) + 0.584(C2s_\perp)$$

---

[11] R. M. Pitzer, *J. Chem. Phys.*, **46**, 4871 (1967).

$$1t_{2x} = 0.318(H_1 1s + H_2 1s - H_3 1s - H_4 1s) + 0.554(C2p_x) \qquad (15.56)$$

$$1t_{2y} = 0.318(H_1 1s - H_2 1s + H_3 1s - H_4 1s) + 0.554(C2p_y)$$

$$1t_{2z} = 0.318(H_1 1s - H_2 1s - H_3 1s + H_4 1s) + 0.554(C2p_z)$$

The $1a_1$ MO is essentially the carbon $1s$ AO. The $2a_1$ MO is a bonding combination of the carbon $2s$ AO and the symmetry orbital (15.51); this MO has charge buildup between the carbon atom and each of the four hydrogen atoms. The $1t_{2x}$ MO is a bonding MO; the function (15.52) is positive on the positive half of the x axis and negative on the negative half of the x axis, and its overlap with $C2p_x$ gives charge buildup in the regions about the x axis. Similarly, the bonding $1t_{2y}$ and $1t_{2z}$ MO's have charge buildup in the regions about the y and z axes, respectively.

We now consider the four localized (equivalent) bonding MO's of methane. Because of the tetrahedral symmetry, each of these orbitals *must* point along a CH bond, since otherwise they would not be equivalent to one another. (This is not true for water, where the equivalence requirement is satisfied by any two bonding MO's that make the same angle with the C_2 axis.) Each localized bonding MO is some linear combination of the five canonical occupied MO's:

$$b(CH_1) = a(1a_1) + b(2a_1) + d(1t_{2x}) + e(1t_{2y}) + f(1t_{2z}) \qquad (15.57)$$

with similar expressions for $b(CH_2)$, $b(CH_3)$, and $b(CH_4)$. (Since $1a_1$ is a nonbonding low-energy inner-shell orbital, we expect $|a| \ll |b|$.) The $1a_1$ and $2a_1$ canonical MO's are directed equally to all four hydrogens, so that varying a or b in (15.57) does not direct $b(CH_1)$ preferentially to any one hydrogen. The $1t_{2x}$, $1t_{2y}$, and $1t_{2z}$ MO's are directed along the x, y, and z axes, respectively, so that by adjusting d, e, and f appropriately, we can get $b(CH_1)$ to be localized mainly in the region of the CH_1 bond. To fix d, e, and f, we use some properties of direction cosines.

If line L passes through the origin and makes the angles α, β, and γ with the x, y, and z axes, respectively, then the quantities

$$l \equiv \cos \alpha, \qquad m \equiv \cos \beta, \qquad n \equiv \cos \gamma \qquad (15.58)$$

are the *direction cosines* of L. If (x_L, y_L, z_L) is a point on L, then clearly

$$x_L = r \cos \alpha, \qquad y_L = r \cos \beta, \qquad z_L = r \cos \gamma \qquad (15.59)$$

where r is the distance from the origin. From $x^2 + y^2 + z^2 = r^2$, it follows that

$$l^2 + m^2 + n^2 = 1 \qquad (15.60)$$

Let the lines L_1 and L_2 go from the origin to (x_1, y_1, z_1) and (x_2, y_2, z_2), respectively; if θ_{12} is the angle between L_1 and L_2, then [Eq. (5.29)]

$$\cos \theta_{12} = \frac{x_1 x_2 + y_1 y_2 + z_1 z_2}{r_1 r_2} \qquad (15.61)$$

$$\cos \theta_{12} = l_1 l_2 + m_1 m_2 + n_1 n_2 \qquad (15.62)$$

Direction cosines are useful in discussing changes in coordinate axes. Let the $x'y'z'$ and the xyz Cartesian coordinate systems have a common origin, and let the $x'y'z'$ axes be obtained from the xyz axes by rotation, or reflection, or inversion, or some combination of these operations. Let the direction cosines of the x' axis with respect to the xyz system be l_1, m_1, n_1; let l_2, m_2, n_2 and l_3, m_3, n_3 be the direction cosines of the y' and z' axes, respectively. Let the vector $\mathbf{s}$ have coordinates (x, y, z) and (x', y', z') in the unrotated and rotated coordinate systems. We have $x' = \mathbf{s} \cdot \mathbf{i}'$, where $\mathbf{i}'$ is a unit vector along the x' axis; since $\mathbf{i}'$ is of unit length, it follows from (15.59) and (15.58) that the coordinates of $\mathbf{i}'$ in the xyz system are l_1, m_1, and n_1. Hence

$$x' = l_1 x + m_1 y + n_1 z$$

$$y' = l_2 x + m_2 y + n_2 z \tag{15.63}$$

$$z' = l_3 x + m_3 y + n_3 z$$

where the y' and z' equations follow from $y' = \mathbf{s} \cdot \mathbf{j}'$, $z' = \mathbf{s} \cdot \mathbf{k}'$. [Equation (15.47) is a special case of (15.63).] Since the angle between any pair of the x', y', and z' axes is $90°$, it follows from (15.62) that

$$l_1 l_2 + m_1 m_2 + n_1 n_2 = 0$$

$$l_1 l_3 + m_1 m_3 + n_1 n_3 = 0 \tag{15.64}$$

$$l_2 l_3 + m_2 m_3 + n_2 n_3 = 0$$

Now we return to the determination of d, e, and f. The MO's t_{2x}, t_{2y}, and t_{2z} are directed along the x, y, and z axes, respectively, and the contributions of the carbon $2p_x$, $2p_y$, and $2p_z$ AO's to these MO's are

$$xe^{-\zeta r}, \qquad ye^{-\zeta r}, \qquad ze^{-\zeta r} \tag{15.65}$$

The contribution of the hydrogen AO's to the t_2 MO's has a more complicated form (the hydrogens are not at the coordinate origin), but we need not explicitly consider the hydrogen part of the MO's; this is because the hydrogen symmetry orbitals (15.52)–(15.54) have the same directional properties as the corresponding carbon $2p$ AO's (15.65) with which each is combined in the $1t_2$ MO's [Eq. (15.56)]. The linear combination (15.57) has as its carbon $2p$ contribution

$$(dx + ey + fz)e^{-\zeta r} \tag{15.66}$$

Let l_1, m_1, and n_1 be the direction cosines of the CH_1 line. We assert that if d, e, and f are chosen as proportional to these direction cosines, then $b(CH_1)$ will be directed toward H_1. To verify this, we set $d : e : f = l_1 : m_1 : n_1$ in (15.66) to get

$$c(l_1 x + m_1 y + n_1 z)e^{-\zeta r} = cx'e^{-\zeta r} \tag{15.67}$$

where c is some constant and the x' axis runs from C to H_1.

Similarly, by picking d, e, and f proportional to the direction cosines of the other CH lines, we form localized orbitals along these bonds. From (15.59) the direction cosines of the CH lines are

$$CH_1: \ 3^{-1/2}, 3^{-1/2}, 3^{-1/2} \qquad\qquad CH_2: \ 3^{-1/2}, -3^{-1/2}, -3^{-1/2}$$

$$CH_3: \ -3^{-1/2}, 3^{-1/2}, -3^{-1/2} \qquad CH_4: \ -3^{-1/2}, -3^{-1/2}, 3^{-1/2} \tag{15.68}$$

To satisfy the equivalence requirement, the values of a and b in (15.57) must be the same for each bonding localized MO. The equivalent localized MO's for methane thus have the forms

$$b(CH_1) = a(1a_1) + b(2a_1) + 3^{-1/2}c(1t_{2x} + 1t_{2y} + 1t_{2z})$$
$$b(CH_2) = a(1a_1) + b(2a_1) + 3^{-1/2}c(1t_{2x} - 1t_{2y} - 1t_{2z})$$
$$b(CH_3) = a(1a_1) + b(2a_1) + 3^{-1/2}c(-1t_{2x} + 1t_{2y} - 1t_{2z})$$
$$b(CH_4) = a(1a_1) + b(2a_1) + 3^{-1/2}c(-1t_{2x} - 1t_{2y} + 1t_{2z})$$

(15.69)

Orthonormality of the bonding localized MO's (15.69) requires that

$$a^2 + b^2 + c^2 = 1 \tag{15.70}$$

$$a^2 + b^2 - \tfrac{1}{3}c^2 = 0 \tag{15.71}$$

Hence

$$c = \tfrac{1}{2}\sqrt{3}, \qquad (a^2 + b^2)^{1/2} = \tfrac{1}{2} \tag{15.72}$$

The equivalence, direction, and orthonormality requirements have fixed all but one parameter (the ratio a/b) in the localized bonding MO's of methane.

The $1t_{2x}$ MO points equally in the $+x$ and $-x$ directions. Similarly, the $1t_{2y}$ and $1t_{2z}$ MO's point equally on both sides of the carbon atom. This is not true of the bonding localized MO's: The $2a_1$ MO is positive in most of the bonding region between the carbon atom and the hydrogen atoms. [It is negative in the region very near the carbon atom, because of the $-0.06(C1s)$ term and the negative portion of the orthogonalized C2s AO—see Eq. (15.28).] The linear combination $(1t_{2x} + 1t_{2y} + 1t_{2z})$ points equally in the $(1, 1, 1)/\sqrt{3}$ and the $(-1, -1, -1)/\sqrt{3}$ directions; if b in (15.69) is taken as positive (as we have taken c), then the $2a_1$ MO adds to the positive half of this linear combination of the $1t_2$ MO's and cancels much of the negative half of this linear combination. With b and c having the same sign, the $b(CH_1)$ MO points mostly in the $(1, 1, 1)/\sqrt{3}$ direction, with only a small "tail" in the $(-1, -1, -1)/\sqrt{3}$ direction.

Pitzer's SCF calculation gives the energy-localized bonding and inner-shell methane MO's as

$$b(CH_1) = 0.055(1a_1) + 0.497(2a_1) + \tfrac{1}{2}(1t_{2x} + 1t_{2y} + 1t_{2z})$$
$$b(CH_2) = 0.055(1a_1) + 0.497(2a_1) + \tfrac{1}{2}(1t_{2x} - 1t_{2y} - 1t_{2z})$$
$$b(CH_3) = 0.055(1a_1) + 0.497(2a_1) + \tfrac{1}{2}(-1t_{2x} + 1t_{2y} - 1t_{2z}) \tag{15.73}$$
$$b(CH_4) = 0.055(1a_1) + 0.497(2a_1) + \tfrac{1}{2}(-1t_{2x} - 1t_{2y} + 1t_{2z})$$
$$i(C) = 0.994(1a_1) - 0.111(2a_1)$$

From (15.56) and (15.73), we have

$$i(C) = 1.002(C1s) - 0.040(C2s_\perp)$$
$$\qquad\qquad - 0.025(H_1\,1s + H_2\,1s + H_3\,1s + H_4\,1s)$$
$$b(CH_1) = 0.024(C1s) + 0.292(C2s_\perp) + 0.569(H_1\,1s) \tag{15.74}$$
$$\qquad\qquad - 0.066(H_2\,1s + H_3\,1s + H_4\,1s)$$
$$\qquad\qquad + 0.277(C2p_x + C2p_y + C2p_z)$$

The carbon AO contribution to each energy-localized bonding MO is approximately an equal mixture of $2s$, $2p_x$, $2p_y$, and $2p_z$ orbitals, i.e., approximately an sp^3 hybrid. (See also the VB discussion of methane in Section 15.17.)

The optimized orbital exponents found by Pitzer are H1s, 1.17; C1s, 5.68; C2s, 1.76; C2p, 1.76. These values may be compared with the values 1.0, 5.7, 1.625, 1.625 given by Slater's rules (Problem 15.27). (Several other SCF MO calculations have been done on methane.)

Next we consider ethane. The most fascinating property of ethane is the barrier to internal rotation about the carbon–carbon single bond. The staggered (equilibrium) conformation of C_2H_6 is more stable than the eclipsed conformation by

$$2.9 \text{ kcal/mole} = 0.13 \text{ eV/molecule} = 0.0046 \text{ hartree/molecule} \quad \textbf{(15.75)}$$

In 1936 J. D. Kemp and Kenneth Pitzer discovered this fact in examining thermodynamic data for ethane. The experimental electronic energy (including nuclear repulsion) of ethane is -80 hartrees $= -2200$ eV. The correlation energy runs one-half to one percent of the total energy of an atom or molecule. Thus we expect the Hartree–Fock energy of ethane to differ by, say, 20 eV from the true molecular electronic energy; this difference is far greater than the barrier height. At first sight, it seems hopeless to expect an SCF MO calculation to give a meaningful result for the ethane barrier.

The minimal-basis-set AO's for ethane are the hydrogen 1s orbitals and the carbon 1s, 2s, and 2p orbitals, a total of $6(1) + 2(5) = 16$ basis AO's. To calculate the barrier in ethane, one must calculate the energy of the staggered and the eclipsed conformations, which requires two separate SCF calculations. One first forms appropriate linear combinations of the hydrogen AO's and of the carbon AO's to get symmetry-adapted orbitals. The Roothaan equations are then solved iteratively to give the orbital coefficients and orbital energies, and the total molecular energy is found. We omit details and simply list the energy results from several SCF MO calculations (Table 15.2). All the calculations listed are many-center ones.

The pioneering ethane calculation is by Russell Pitzer (Kenneth Pitzer's son) and W. N. Lipscomb. They used a minimal basis set of sixteen STO's; the orbital exponents were not optimized but were chosen according to Slater's rules; all integrals were evaluated accurately. The Pitzer calculation is similar to the Pitzer–Lipscomb calculation, except that the orbital exponents used are those that minimize the energy of methane.

The first four calculations listed used the experimentally observed equilibrium geometry for staggered ethane and assumed the geometry of the eclipsed form to be that given by rigidly rotating one methyl group with respect to the other. Veillard varied the bond angles and the CC bond length for each conformation. He found an increase of 0.02 Å in the CC length and a decrease of $\frac{1}{2}°$ in the HCH angle on going from the staggered to the eclipsed form; this change affected the barrier by $\frac{1}{2}$ kcal/mole. Clementi and Popkie's near-Hartree–Fock calculation also uses geometry optimization.

All the SCF MO calculations give good values of the ethane rotational barrier. Why should this be so? The answer lies in the correlation energy. Electrons paired in

Table 15.2 *SCF MO calculations of ethane*

Reference[1]	Basis Functions	Eclipsed Energy (hartrees)	Staggered Energy (hartrees)	Barrier (kcal/mole)
Pitzer and Lipscomb	16 STO's	−78.98593	−78.99115	3.3
Pitzer	16 STO's	−79.09233	−79.09797	3.5
Clementi and Davis	86 GTO's	−79.10247	−79.10824	3.6
Fink and Allen	110 GTO's	−79.14377	−79.14778	2.5
Veillard	120 GTO's	−79.23410	−79.23899	3.07
Clementi and Popkie	150 GTO's	−79.25364	−79.25875	3.21
Experimental values			−79.84	2.90 ± 0.03

[1] R. M. Pitzer and W. N. Lipscomb, *J. Chem. Phys.*, **39**, 1995 (1963); R. M. Pitzer, *J. Chem. Phys.*, **47**, 965 (1967); E. Clementi and D. R. Davis, *J. Chem. Phys.*, **45**, 2593 (1966); W. H. Fink and L. C. Allen, *J. Chem. Phys.*, **46**, 2261 (1967); A. Veillard, *Theor. Chim. Acta*, **18**, 21 (1970); E. Clementi and H. Popkie, *J. Chem. Phys.*, **57**, 4870 (1972). Experimental barrier: E. Hirota et al., **71**, 1183 (1979).

the same localized orbital move through the same region of space. Hence the intraorbital correlation for such a pair should be substantially greater than the interorbital correlation energy for two electrons in different localized MO's. In line with this, it was formerly believed that the total interorbital molecular correlation energy was much less than the total intraorbital correlation energy. However, there are many more interorbital pair correlations than intraorbital correlations, and it is now recognized that the total interorbital correlation is not negligible and in some cases can be of comparable magnitude to the total intraorbital correlation. [See E. Steiner, *J. Chem. Phys.*, **54**, 1114 (1971).] Hence we must consider both kinds of correlation in ethane. Rotation of a methyl group in ethane does not change any of the bonds, and thus intraorbital correlation should be essentially the same in the staggered and eclipsed forms. Moreover, we expect most of the interorbital correlations to be essentially unchanged in the two forms. In particular, correlations between the C—H bonds within each methyl group should remain virtually the same, and so should correlations between the C—C and C—H bonding pairs. It is only the correlations between C—H pairs of different methyl groups that should change, and these make the smallest contributions to interorbital correlation. Thus we expect the total correlation energy to be only very slightly changed from staggered to eclipsed. Thus the energy error in an SCF calculation is about the same for the two forms, and the Hartree–Fock energy difference should yield a good estimate of the barrier. None of the SCF calculations listed in Table 15.2 have actually reached the Hartree–Fock limit, but they still yield good values for the barrier. (Recall that Hartree–Fock calculations give poor values for dissociation energies. This is because the number of electron pairs changes in forming a chemical bond from atoms, thereby changing the correlation energy substantially.)

Rotational barriers in other molecules have been studied by the SCF MO method.[12] Some results follow (values in kcal/mole): CH_3OH, 1.4 calculated versus 1.1 experimental; CH_3NH_2, 2.4 calculated versus 2.0 experimental; CH_3CHO, 1.1 calculated versus 1.2 experimental; CH_3SiH_3, 1.4 calculated versus 1.7 experimental.

Table 15.2 shows that for ethane a reasonably accurate barrier can be gotten with an SCF calculation that has a minimal basis set and does not use geometry optimization, and this is true for most barriers investigated. For H_2O_2 such calculations give quite poor barrier results. However, when a large basis set (including polarization functions) and geometry optimization are used, the H_2O_2 barriers can be calculated by the Hartree–Fock method, as shown by the following results [T. H. Dunning and N. W. Winter, *J. Chem. Phys.*, **63**, 1847 (1975); see also D. Cremer, *J. Chem. Phys.*, **69**, 4440 (1978)]: calculated *cis* barrier (0° dihedral angle) 8.4 kcal/mole versus 7.0–7.6 kcal/mole experimental; calculated *trans* barrier (180° dihedral angle) 1.1 kcal/mole versus 1.1 kcal/mole experimental; calculated equilibrium dihedral angle 114° versus 112° experimental.

The physical origin of the ethane rotational barrier has been attributed to the Pauli-exclusion-principle repulsion (Section 10.4) between the eclipsing localized C—H bonding electron pairs in eclipsed ethane; see O. J. Sovers, C. W. Kern, R. M. Pitzer, and M. Karplus, *J. Chem. Phys.*, **49**, 2592 (1968); R. M. Stevens and M. Karplus, *J. Am. Chem. Soc.*, **94**, 5140 (1972). However, this interpretation has been criticized and a different one offered [T. K. Brunck and F. Weinhold, *J. Am. Chem. Soc.*, **101**, 1700 (1979); C. T. Corcoran and F. Weinhold, *J. Chem. Phys.*, **72**, 2866 (1980)], so the origin of the barrier is a matter of controversy. For discussions of proposed explanations of rotational barriers, see P. W. Payne and L. C. Allen in *Schaefer, Applications of Electronic Structure Theory*, Chapter 2.

Finally, consider the SCF MO treatment of ethylene. The ground-state equilibrium-geometry point group is $\mathscr{D}_{2h}$. The standard choice of coordinate axes is shown in Fig. 15.10. (Unfortunately, the standard convention has often been ignored, and one finds calculations with the molecular plane being the xz or the xy plane; this affects the naming of the MO symmetry species, so that different workers use different symbols for the same MO.)

There are eight symmetry operations for $\mathscr{D}_{2h}$. Each symmetry operation

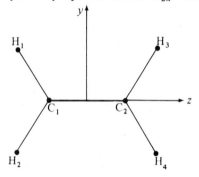

Figure 15.10 *Coordinate axes for ethylene. The x axis is perpendicular to the molecular plane.*

---

[12] For summaries of SCF MO barrier calculations, see P. W. Payne and L. C. Allen in *Schaefer, Applications of Electronic Structure Theory*, Chapter 2.

commutes with every other symmetry operation, so that the electronic wave function must be an eigenfunction of all the symmetry operators, and we have only nondegenerate (A and B) symmetry species. The three rotations, the three reflections, and the inversion operation each have their squares equal to the identity operation, so that these symmetry operations must have the eigenvalues ± 1. All eight symmetry operations can be expressed as the product of one, two, or three of the reflections (each reflection simply converts one coordinate to its negative):

$$\hat{E} = [\hat{\sigma}(xy)]^2, \qquad \hat{\imath} = \hat{\sigma}(xy)\hat{\sigma}(xz)\hat{\sigma}(yz)$$

$$\hat{C}_2(x) = \hat{\sigma}(xy)\hat{\sigma}(xz), \qquad \hat{C}_2(y) = \hat{\sigma}(xy)\hat{\sigma}(yz), \qquad \hat{C}_2(z) = \hat{\sigma}(xz)\hat{\sigma}(yz)$$

Since the symmetry eigenvalues multiply the same way the symmetry operations do, specification of the eigenvalues of the three reflections is sufficient to fix all eight symmetry eigenvalues. There are thus $2^3 = 8$ possible symmetry species. The standard notation for these is

	$\hat{\sigma}(xy)$	$\hat{\sigma}(xz)$	$\hat{\sigma}(yz)$	
A_g	$+1$	$+1$	$+1$	
A_u	-1	-1	-1	
B_{1g}	$+1$	-1	-1	
B_{1u}	-1	$+1$	$+1$	(15.76)
B_{2g}	-1	$+1$	-1	
B_{2u}	$+1$	-1	$+1$	
B_{3g}	-1	-1	$+1$	
B_{3u}	$+1$	$+1$	-1	

The g or u subscript corresponds to eigenvalue $+1$ or -1 for $\hat{\imath}$. For $\mathscr{D}_{2h}$ the designation A is used only for symmetry species with eigenvalue $+1$ for all three $\hat{C}_2$ rotations.

There are fourteen minimal-basis-set AO's. It is easy to set up symmetry orbitals by trial and error; the reader will readily verify the symmetry species of the following (unnormalized) symmetry orbitals:

Symmetry Orbital	Symmetry Species	
$H_1\,1s + H_2\,1s + H_3\,1s + H_4\,1s$	a_g	
$C_1\,1s + C_2\,1s$	a_g	
$C_1\,2s + C_2\,2s$	a_g	
$C_1\,2p_z - C_2\,2p_z$	a_g	
$H_1\,1s + H_2\,1s - H_3\,1s - H_4\,1s$	b_{1u}	
$C_1\,1s - C_2\,1s$	b_{1u}	
$C_1\,2s - C_2\,2s$	b_{1u}	(15.77)
$C_1\,2p_z + C_2\,2p_z$	b_{1u}	
$H_1\,1s - H_2\,1s + H_3\,1s - H_4\,1s$	b_{2u}	
$C_1\,2p_y + C_2\,2p_y$	b_{2u}	
$H_1\,1s - H_2\,1s - H_3\,1s + H_4\,1s$	b_{3g}	
$C_1\,2p_y - C_2\,2p_y$	b_{3g}	
$C_1\,2p_x + C_2\,2p_x$	b_{3u}	
$C_1\,2p_x - C_2\,2p_x$	b_{2g}	

The fourteen lowest MO's thus include four a_g, four b_{1u}, two b_{2u}, two b_{3g}, one b_{3u}, and one b_{2g} MO. In the ground state, eight of these MO's are occupied. SCF calculations[13] yield the following configuration for the ^1A_g ground state:

$$(1a_g)^2(1b_{1u})^2(2a_g)^2(2b_{1u})^2(1b_{2u})^2(3a_g)^2(1b_{3g})^2(1b_{3u})^2 \qquad (15.78)$$

Each canonical MO of a planar molecule is classified as σ or π, according to whether its eigenvalue for reflection in the molecular plane is $+1$ or -1. (This usage is somewhat inconsistent with the $\sigma, \pi, \delta, \ldots$ classification of linear-molecule MO's. For linear molecules the symbols σ and π signify an axial component of electronic orbital angular momentum of 0, and $\pm\hbar$, respectively; for nonlinear molecules one cannot specify the component of $\mathbf{L}$ along an internuclear line. For linear molecules σ MO's are nondegenerate and π MO's are doubly degenerate; for nonlinear molecules the σ–π classification is unrelated to the degeneracy.) For the ground state of water, all the occupied MO's are σ MO's, except the lone-pair $1b_1$ MO, which is a π MO.

For ethylene the only occupied π MO in the ground electronic state is the $1b_{3u}$ MO, the highest occupied MO. Since there is only one b_{3u} symmetry orbital in (15.77), the $1b_{3u}$ MO must be identical (apart from a normalization constant) to this symmetry orbital. One finds (using STO's)

$$1b_{3u} = 0.63\,(C_1\,2p_x + C_2\,2p_x) \qquad (15.79)$$

This bonding MO formed by overlap of the $2p_x$ AO's of carbon resembles the π_u MO of Fig. 13.14. The $1b_{3u}$ pi MO accounts for the planarity of ethylene in its ground state. As one CH_2 group is rotated relative to the other, the overlap between the two carbon $2p_x$ AO's rapidly diminishes, and the energy of the $1b_{3u}$ MO increases; hence the molecule shows a considerable resistance to torsion about the carbon–carbon bond.

The lowest-lying unoccupied MO of ethylene is the $1b_{2g}$ antibonding π MO

$$1b_{2g} = 0.82\,(C_1\,2p_x - C_2\,2p_x) \qquad (15.80)$$

It resembles the π_g MO in Fig. 13.14. The excited ethylene configuration

$$(1a_g)^2 \cdots (1b_{3g})^2(1b_{3u})(1b_{2g}) \qquad (15.81)$$

gives rise to two terms (a singlet and a triplet), and it is likely that these electronic states have a nonplanar configuration, with one CH_2 group rotated $90°$ with respect to the other.

The canonical σ MO's of ethylene each contain contributions from AO's of all six atoms and are delocalized over the whole molecule. By taking appropriate linear combinations of the canonical σ MO's, one can form localized σ MO's; we expect these MO's to consist of the following inner-shell and bonding MO's:

$$i(C_1), \qquad i(C_2) \qquad (15.82)$$

$$b(C_1C_2), \qquad b(C_1H_1), \qquad b(C_1H_2), \qquad b(C_2H_3), \qquad b(C_2H_4) \quad (15.83)$$

The MO's $b(C_1C_2)$ and $1b_{3u}$ constitute the familiar description of the carbon–carbon double bond as one σ and one π bond (Fig. 15.11). We still do not, however, have the

---

[13] U. Kaldor and I. Shavitt, *J. Chem. Phys.*, **48**, 191 (1968).

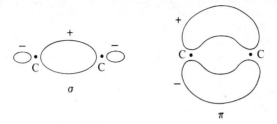

Figure 15.11 *σ–π description of the ethylene double bond. The cross sections are taken in the plane perpendicular to the molecular plane, which is a nodal plane for the π MO.*

equivalent MO's of ethylene, since the π MO $1b_{3u}$ is not equivalent to any of the σ MO's in (15.83), nor is it equivalent to itself—it goes into its negative upon a $\hat{C}_2(z)$ rotation. By adding and subtracting the $b(C_1C_2)$ and $1b_{3u}$ MO's, we can form two equivalent localized carbon–carbon bonding MO's:

$$b_1(C_1C_2) = 2^{-1/2}[b(C_1C_2) + 1b_{3u}] \tag{15.84}$$

$$b_2(C_1C_2) = 2^{-1/2}[b(C_1C_2) - 1b_{3u}] \tag{15.85}$$

The MO's (15.84) and (15.85) lead to the description of the carbon–carbon bond as composed of two "banana" bonds (Fig. 15.12).

There has been considerable controversy as to whether the ethylene double bond is best described as two equivalent bent banana bonds or as a σ and a π bond. Kaldor[14] calculated the energy-localized MO's of ethylene from the minimal-basis-set canonical SCF MO's. Since there is no a priori necessity that the energy-localized MO's be equivalent orbitals, this calculation provides evidence as to which is the "better" description of the carbon–carbon double bond. Kaldor found the energy-

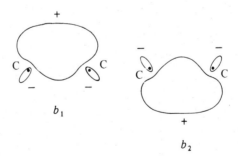

Figure 15.12 *Banana-bond description of the ethylene double bond. (Cross sections as in Fig. 15.11.)*

---

[14] U. Kaldor, *J. Chem. Phys.*, **46**, 1981 (1967).

localized carbon–carbon bond orbitals in ethylene to be the two equivalent banana bonds. For acetylene he found the energy-localized carbon–carbon bond orbitals to be three equivalent banana bonds and not one σ and two π bonds. Of course, although the electron densities in the individual MO's differ for the banana-bond versus σ–π descriptions, the total probability density for the four or six electrons in the double or triple bond is the same in either picture.

The energy-localized C_1H_1 bond MO in ethylene was found by Kaldor to be

$$0.373\,(C_1\,2s) + 0.414\,(C_1\,2p_y) - 0.257\,(C_1\,2p_z) + 0.494\,(H_1\,1s) + \cdots$$

where the dots indicate small contributions from other AO's. From Problem 15.61 the ratio of $2p_y$ to $2p_z$ orbital coefficients in an sp^2 hybrid atomic orbital is $1.73:1$; for $b(C_1H_1)$ this ratio is $1.61:1$, so that the hybridization of carbon is very close to sp^2 in this MO. The angle made by $b(C_1H_1)$ with the C—C axis is $\pi - \arctan(1.61) = 122°$, essentially the same as the experimental angle; hence the localized C—H bond orbitals are not bent in planar ethylene.

Localized MO's for butadiene and benzene are discussed at the end of Section 15.13.

Localized MO's have been calculated for some of the boron hydrides. These compounds have fewer than $2(n-1)$ valence electrons, where n is the number of atoms. Thus there are not enough electrons to form $n-1$ electron-pair bonds between the atoms; from the VB viewpoint such compounds are anomalous and have been termed electron deficient. However, it is only chemical prejudices (and not quantum-mechanical requirements) that lead one to expect $n-1$ electron-pair bonds. The MO treatment of boron hydrides shows all the bonding MO's to be completely filled in the ground state, and the term *electron deficient* is not really appropriate. Calculation of the localized bonding MO's in B_2H_6 shows that there are two localized MO's, each of which extends over three nuclei (two borons and one hydrogen) giving two three-center bonds [E. Switkes et al., *J. Chem. Phys.*, **51**, 2085 (1969)]. These three-center localized MO's resemble the banana orbitals of the isoelectronic molecule ethylene; we can view B_2H_6 as being obtained from C_2H_4 by removing a proton from each carbon nucleus and placing it in the midst of one of the banana orbitals. The higher boron hydrides have three-center bonds involving three boron atoms. (See *Murrell, Kettle, and Tedder*, Chapter 14.)

We noted earlier in this section that each *canonical* MO of a planar molecule is classified as π or σ according to whether or not the molecular plane is a nodal plane for the MO. A *localized* bonding MO of a molecule (planar or nonplanar) is classified as σ or π or δ according to whether this MO has 0 or 1 or 2 nodal planes containing the line joining the nuclei of the two bonded atoms. The bonding localized MO's in water (Fig. 15.7) are σ MO's; the ethylene double-bond localized MO's in Fig. 15.11 consist of one σ and one π MO; the $b(CH_1)$ MO (15.74) in CH_4 is a σ MO. In certain transition-metal compounds (e.g., $Re_2Cl_8^{2-}$), overlap of d AO's produces a δ localized bonding MO, and these compounds have a quadruple bond consisting of one σ, two π, and one δ MO; see F. A. Cotton, *Chem. Soc. Rev.*, **4**, 27 (1975).

15.8 *MOLECULAR GEOMETRY*

The equilibrium geometry of a molecule corresponds to the nuclear configuration that minimizes the molecular electronic energy including nuclear repulsion.

The molecular energy is found to be relatively insensitive to even large changes in molecular geometry. For example, a large-scale CI calculation[15] (using as many as 30,380 configuration functions) of the H_2O potential-energy surface U as a function of molecular geometry yields the following values (in hartrees) of electronic energy including nuclear repulsion as a function of bond angle θ at the equilibrium bond distance:

θ	60°	90°	105°	120°	150°	180°
U	−76.288	−76.342	−76.348	−76.343	−76.314	−76.294

The variation in energy over the range of bond angles is only 0.08 percent, as compared to a correlation energy (Table 15.1) of 0.5 percent. Especially noteworthy is the relative flatness of the surface in the region of the equilibrium geometry; a variation of $\pm 15°$ from the equilibrium 105° angle produces a variation of only 0.008 percent in U. (The data quoted are for CI calculations; for near-Hartree–Fock SCF calculations,[15] the magnitudes of the changes versus geometry are very similar to those for the CI calculations.) Changes in bond lengths also produce relatively small changes in energy. For example, for H_2O at the equilibrium bond angle, CI calculations[15] give the following energies in hartrees as a function of OH bond distance for bonds of equal length: −76.294 at 0.794 Å, −76.348 at 0.957 Å, −76.295 at 1.164 Å. (Note that the small bond bending of 15° has much less effect on the energy than the small bond stretching of 0.2 Å; recall from organic chemistry that infrared vibrational bands corresponding to bond bending occur at lower frequencies than bond-stretching vibrations.)

Despite the smallness of these changes in molecular energy versus molecular geometry as compared to the energy error (the correlation energy) inherent in the SCF method, SCF wave functions give good predictions (0–3 percent error) of equilibrium bond angles and distances. Some SCF 4-31G-basis-set bond angles and lengths follow,[16] where the experimental values are in parentheses. NH_2: 108° (103°) and 1.02 Å (1.02 Å); H_2O: 111° ($104\frac{1}{2}°$) and 0.95 Å (0.96 Å); C_2H_6: 108° (108°) for HCH and 1.53 Å (1.53 Å) for C—C; C_2H_4: 116° ($116\frac{1}{2}°$) for HCH and 1.32 Å (1.33 Å) for C═C; H_2CO: 116° (116°) for HCH and 1.21 Å (1.20 Å) for C═O.

Evidently, the correlation energy remains approximately constant for bond angle and length variations in the region of the equilibrium geometry, and the SCF potential-energy surface, although lying above the true potential-energy surface, is approximately parallel to it in this region.

A molecule of considerable interest to organic chemists is the reaction

---

[15] P. Hennig, W. P. Kraemer, and G. H. F. Diercksen, *Theor. Chim. Acta*, **47**, 233 (1978).

[16] J. A. Pople in *Schaefer, Applications of Electronic Structure Theory*, Chapter 1.

intermediate CH_2, the methylene radical. Experimental and theoretical work are in agreement that the CH_2 ground state is a triplet. The first spectroscopic observations on CH_2 were interpreted as indicating a linear ground state; however, accurate theoretical calculations indicated a bent molecule. Further experimental work showed that indeed the molecule is bent with a bond angle $136° \pm 10°$. An SCF MO calculation with the 4-31G basis set gave $132°$; a large-scale CI calculation yielded $132\frac{1}{2}°$ [C. W. Bauschlicher and I. Shavitt, *J. Am. Chem. Soc.*, **100**, 739 (1978)].

The VSEPR method Accurate SCF MO calculations for large molecules are quite difficult. We want a simple procedure that will give reasonably accurate predictions of bond angles. One such procedure is the *valence-shell electron-pair repulsion* (VSEPR) theory. This theory originated with Sidgwick and Powell in 1940. Gillespie has done much to extend and popularize it.[17] The VSEPR viewpoint has even displaced the traditional VB hybridization explanation (Section 15.17) in freshman-chemistry discussions of bond angles.

We saw in Section 15.6 that the energy-localized MO's for the eight valence-shell electrons of water consist of two bonding orbitals $b(OH_1)$ and $b(OH_2)$ and two lone-pair MO's $l_1 (O)$ and $l_2(O)$; roughly speaking, these four MO's are tetrahedrally disposed about the oxygen atom. (Actually, the bonding MO's are separated by $103°$ and the lone-pair MO's by $114°$.) Similarly, the energy-localized MO's found by Edmiston and Ruedenberg for the valence-shell electrons of NH_3 consist of three bonding MO's making angles of $104\frac{1}{2}°$ with one another, and a lone-pair MO making an angle of $114°$ with each bonding localized MO. Again, these four MO's are approximately tetrahedrally disposed.

To apply the VSEPR method, one draws the Lewis dot structure of the molecule and counts the number of valence electron pairs around the central atom. These valence electron pairs are then assumed to be arranged in space so as to minimize the repulsions between the localized MO's of the various pairs. [Recall that the condition for the energy-localized MO's is the minimization of (15.43).] More important than the Coulombic repulsion between localized pairs is the Pauli repulsion (Section 10.4), which forces electrons with like spins to keep apart; each localized pair contains electrons with both possible spin values and thus excludes other electrons from the space it occupies. Note that the VSEPR method is based on electron pair–electron pair repulsions (where the pairs are bond or lone pairs) rather than ligand atom–ligand atom repulsions.

For four pairs of electrons, the minimum repulsion occurs with the four MO's disposed tetrahedrally about the central atom. The most favorable arrangements for various numbers of electron pairs are (see Fig. 15.13)

2	3	4	5	6	8
straight line	equilateral triangle	tetrahedron	trigonal bipyramid	octahedron	square antiprism

[17] R. J. Gillespie, *J. Chem. Educ.*, **40**, 295 (1963); **47**, 18 (1970); *Can. J. Chem.*, **38**, 818 (1960); *Angew. Chem. Intern. Ed. Engl.*, **6**, 819 (1967); *Molecular Geometry*, Van Nostrand, New York, 1972.

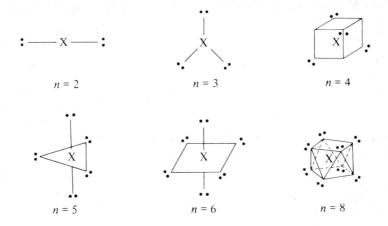

Figure 15.13 *Preferred arrangements for n electron pairs about a central atom* X

Consider some examples. For water there are eight valence electrons, and we have two bonding pairs and two lone pairs arranged tetrahedrally. The bond angle should be close to the $109\frac{1}{2}°$ tetrahedral angle. To explain the deviation of the bond angle ($104\frac{1}{2}°$) from tetrahedral, the VSEPR approach points out that since lone-pair orbitals are under the influence of only one nucleus (rather than two), they are "fatter" (more spread out) than bonding orbitals; hence lone-pair electrons exert greater repulsions on neighboring pairs than bonding pairs exert on neighboring pairs. The angle between the lone pairs of water is thus greater than tetrahedral, while the angle between the bonding pairs is less than tetrahedral. This view is confirmed by the values of 114° and 103° found for the angles between the lone-pair and bonding energy-localized MO's of water. [The "fatness" of lone-pair MO's is illustrated by the greater contribution of the 2s oxygen AO to the lone-pair water MO's than to the bonding MO's—see Eq. (15.45).] For ammonia the four valence pairs are arranged approximately tetrahedrally; since there is only one lone pair instead of two, the bonding pairs are not pushed together as much as in the case of water, and the observed bond angle (107°) is closer to tetrahedral than it is for water.

For IF_5 the dot formula shows six electron pairs around the central atom, I. These are arranged octahedrally. One pair is a lone pair, giving the observed square-pyramidal structure. For XeF_4 the dot formula has six pairs around Xe, which are arranged octahedrally; the two fat lone pairs are directed toward opposite vertices of the octahedron, giving the observed square-planar structure.

ClF_3 has five pairs around Cl. The two fat lone pairs are placed so as to minimize the repulsions. A trigonal bipyramid has three equatorial and two axial positions. An axial electron pair makes an angle of 90° with three neighboring pairs; an equatorial pair makes an angle of 90° with only two neighboring pairs and an angle of 120° with two other neighboring pairs. The repulsions fall off rapidly with increasing angle, so that the most favorable arrangement is with the two lone pairs of ClF_3 in equatorial positions. This gives the molecule a (planar) T structure. The observed slight distortion of $2\frac{1}{2}°$ from the T structure is explained by the fat lone pairs pushing the bonding axial pairs toward the bonding equatorial pair.

For molecules with multiple bonds, Gillespie recommends that for simplicity a double or triple bond be considered as a single electron pair. For example, for $CH_2{=}CH_2$, the expected bond angles are all $120°$, since each carbon is to be considered as surrounded by three electron pairs. The observed H—C—H angle of about $116°$ can be explained by the double-bond electrons pushing together the single-bond pairs. An alternative view (which is more consistent with the energy-localized MO's of ethylene) is to consider the double bond as two electron pairs forming two banana bonds which start out from each carbon atom at roughly the tetrahedral angle (the angle between the energy-localized banana bonds in ethylene is $112°$); with four pairs about each carbon, the expected H—C—H angle is $109\frac{1}{2}°$; the observed angle results from the banana double-bond orbitals exerting less repulsion on the bonding C—H pairs than would two ordinary tetrahedrally directed orbitals.

XeF_6 has seven valence-shell pairs about Xe. The most favorable arrangement for seven electron pairs is not certain, since it depends on the specific form of the effective repulsive forces between the pairs. However, with seven pairs the VSEPR approach predicts that the molecule will certainly not be octahedral. Electron-diffraction results do indicate that XeF_6 is not octahedral, but the departure from the octahedral structure is less than would be expected on the VSEPR model. Although the final word on the XeF_6 structure has yet to be said, it seems likely that XeF_6 is a nonrigid molecule showing large-amplitude motions of the atoms as the molecule interconverts among eight equivalent structures (probably of $\mathscr{C}_{3v}$ symmetry) separated by low-lying potential barriers (a process called pseudo-rotation). See K. S. Pitzer and L. S. Bernstein, *J. Chem. Phys.*, **63**, 3849 (1975); W. J. Ehlhardt and L. L. Lohr, *J. Chem. Phys.*, **67**, 1935 (1977); M. J. Rothman et al., *J. Chem. Phys.*, **73**, 375 (1980).

The molecule IF_7, with seven valence pairs around I, is a somewhat deformed pentagonal bipyramid [W. J. James, H. B. Thompson, and L. S. Bartell, *J. Chem. Phys.*, **53**, 4040 (1970)].

The arrangements of Fig. 15.13 assume that the inner shells of the central atom are spherically symmetric and so do not affect the arrangement of the valence pairs. Recall (Section 11.1) that filled and half-filled subshells give spherically symmetric probability densities. Hence Fig. 15.13 applies to main-group elements (which all have filled inner subshells) and applies to those transition elements with d^0, d^5, or d^{10} subshells. In transition elements with nonspherical d subshells, distortions from the shapes of Fig. 15.13 may be observed.

The VSEPR model has a remarkably high batting average but is not infallible. Thus $TeCl_6^{2-}$ has seven electron pairs about Te, but its observed structure is octahedral. Other exceptions are some of the alkaline earth dihalides (e.g., BaF_2, $SrCl_2$) that are bent, rather than linear. Gillespie has discussed the probable reasons for these exceptions in *J. Chem. Educ.*, **47**, 18 (1970).

Walsh diagrams An alternative approach to predicting molecular geometry is the method of Walsh diagrams, developed by Walsh in 1953.

Consider the molecule AH_2, where A is a second-row element (Li, Be, B, C, N, O, or F). We shall discuss the nature of the AH_2 canonical (delocalized) MO's, aided by our experience with H_2O (Section 15.4). AH_2 might be linear or bent; for the purposes of discussion, we shall use the symmetry-species labels corresponding to a bent geometry, point group $\mathscr{C}_{2v}$; see Eq. (15.5). We choose the coordinate axes as in Fig. 15.1.

We expect the MO's to lie in the following order: inner shell, bonding, nonbonding lone pair, antibonding.

Clearly, the lowest MO is the inner-shell $1a_1$ MO, which is a nearly pure $1s$ AO on atom A; $1a_1 \approx A1s$.

To form the bonding MO's, we combine the $H_1 1s + H_2 1s$ symmetry orbital, which has a_1 symmetry [see (15.22)], with those valence AO's of A that have a_1 symmetry, and we combine the b_2 symmetry orbital $H_1 1s - H_2 1s$ with the valence AO of A that has b_2 symmetry. Overlap of $H_1 1s + H_2 1s$ with the $A2s$ AO [and with a small contribution from $A2p_z$, as in (15.27)] forms the bonding $2a_1$ MO. Overlap of $H_1 1s - H_2 1s$ with $A2p_y$ gives the bonding $1b_2$ MO. The nodal plane in $1b_2$ (Fig. 15.4) and the fact that it is formed from a $2p$ AO as compared to the lower-energy $2s$ AO in $2a_1$ indicate that $1b_2$ lies above $2a_1$.

Next we have the lone-pair MO $3a_1$, composed of a hybrid of $A2s$ and $A2p_z$ and $H_1 1s + H_2 1s$, all of which have a_1 symmetry. This MO contains a positive (bonding) overlap between $A2p_z$ and $H_1 1s + H_2 1s$ [see Eq. (15.27)] but is nonbonding overall, as shown by population analysis (Section 15.5). Next is the lone-pair MO $1b_1$, consisting solely of $A2p_x$, the only valence AO with b_1 symmetry.

Next is the antibonding MO $4a_1$, formed by antibonding interaction between $H_1 1s + H_2 1s$ and an $A2s$-$A2p_z$ hybrid, as in Eq. (15.30). Finally, we have the antibonding $2b_2$ MO, formed by antibonding interaction between the b_2 symmetry orbitals $H_1 1s - H_2 1s$ and $A2p_y$. (See Fig. 15.5.) The $2b_2$ MO lies above the $4a_1$ MO because $A2p_y$ of $2b_2$ lies above the $A2s$-$A2p_z$ hybrid of $4a_1$.

To decide whether AH_2 will be linear or bent, we examine the effect of a change in bond angle on each MO's orbital energy.

We expect the energy of the inner-shell $1a_1$ MO to be essentially unaffected by the bond angle.

For $2a_1$, as the bond angle decreases from $180°$, the overlap between $A2s$ and $H_1 1s + H_2 1s$ is unaffected, since this overlap depends only on the AH distances; the overlap between $H_1 1s + H_2 1s$ and $A2p_z$ (which makes only a small contribution to $2a_1$) increases; and the overlap between $H_1 1s + H_2 1s$ increases. Because H_1 and H_2 are relatively far apart, this increase in H–H overlap will have only a small energy-lowering effect. As a result of the small increases in overlap, we expect the orbital energy ε_{2a_1} to decrease slowly as the bond angle θ decreases from $180°$.

For $1b_2$ a decrease of angle from $180°$ decreases substantially the bonding overlap between $A2p_y$ (whose maximum probability density lies on the y axis in Fig. 15.1) and $H_1 1s - H_2 1s$; also, the antibonding interaction between $H_1 1s$ and $H_2 1s$ is increased, which is a smaller effect. Therefore ε_{1b_2} increases rapidly as θ decreases below $180°$.

For $3a_1$, decreasing θ increases substantially the bonding overlap between $A2p_z$ and $H_1 1s + H_2 1s$ and increases slightly the H–H bonding overlap. Therefore

ε_{3a_1} decreases rapidly as θ decreases from 180°.

The orbital energy of the pure lone pair $1b_1$ MO will be essentially unchanged as θ changes.

For $4a_1$ a decrease in θ will increase the antibonding interaction between $A2p_z$ and $H_1 1s + H_2 1s$, causing ε_{4a_1} to increase rapidly. For $2b_2$, decreasing θ reduces the antibonding overlap between $A2p_y$ and $H_1 1s - H_2 1s$ and decreases ε_{2b_2} rapidly.

Figure 15.14 is a rough sketch of the expected orbital energies versus bond angle θ for the AH_2 valence MO's. The important overlaps for the linear and bent MO's are sketched at each side. Also, the symmetry species of the linear-AH_2 MO's are listed.

Note that for linear AH_2, the $1\pi_{uz}$ MO, which correlates with the $3a_1$ MO of bent AH_2, cannot contain any contributions from $H_1 1s + H_2 1s$ or $A2s$, since these

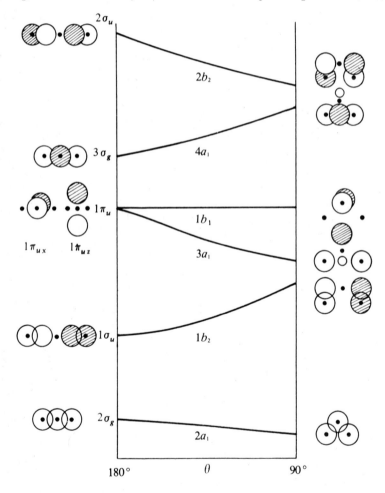

Figure 15.14 AH_2 MO *energies versus bond angle. Shading indicates negative portions of AO's.*

orbitals do not have π_u symmetry; the contributions of these orbitals grow in as the molecule is bent. Similarly, in linear AH_2 there is no contribution of $A2p_z$ to σ_g MO's.

We assume that minimization of the sum of the orbital energies of the valence electrons will determine whether the molecule is linear or bent.

Although we have so far assumed A to be a second-row element, the same sort of diagram will hold for the valence MO's of a main-group element of a later period. We will have more inner-shell MO's, but the nature of the valence MO's will be the same; only the numbering of the valence-shell MO's will be changed.

AH_2 molecules with four valence electrons will have the ground-state valence-electron configuration $(2a_1)^2 (1b_2)^2$. The rapid increase in ε_{1b_2} as the molecule is bent will outweigh the slow decrease in ε_{2a_1}, and we expect a linear geometry. The isolated BeH_2 molecule has not been observed (this compound forms a polymeric solid), but ab initio SCF and CI calculations show that the BeH_2 molecule is linear.

For five valence electrons, $(2a_1)^2 (1b_2)^2 (3a_1)$ is the AH_2 ground-state configuration. It turns out that the combined effect of the two $2a_1$ electrons (which weakly favor bent geometry) and the $3a_1$ electron (which strongly favors bent geometry) outweigh the effect of the two $1b_2$ electrons, and such species are bent. Some experimentally observed angles are $131°$ for BH_2 and $119°$ for AlH_2.

For six valence electrons, the ground-state configuration $(2a_1)^2 (1b_2)^2 (3a_1)^2$ produces bent geometry. The double occupancy of $3a_1$ produces smaller bond angles than for five-valence-electron species. An example is the $93°$ angle observed for SiH_2. However, CH_2 (methylene) does not have the expected ground-state configuration; rather, its ground term is the triplet term of the configuration $(2a_1)^2 (1b_2)^2 (3a_1)(1b_1)$, which lies about 10 kcal/mole (0.4 eV) below the singlet $(2a_1)^2 (1b_2)^2 (3a_1)^2$ configuration [D. Feller and E. R. Davidson, *Chem. Phys. Lett.*, **69**, 201 (1980)]. The CH_2 ground state has a $136°$ bond angle, similar to those of BH_2, which also has one $3a_1$ electron. The CH_2 $(3a_1)^2$ excited state has a $105°$ angle.

For seven valence electrons, the predicted ground-state configuration is $(2a_1)^2 (1b_2)^2 (3a_1)^2 (1b_1)$. The molecule will be bent with an angle similar to those of the six-electron species, since $1b_1$ favors neither linear nor bent geometry. Examples are NH_2 (bond angle $103°$), PH_2 ($92°$), AsH_2 ($91°$), H_2O^+ ($110\frac{1}{2}°$), and CH_2^- ($99°$).

For eight valence electrons and configuration $(2a_1)^2 (1b_2)^2 (3a_1)^2 (1b_1)^2$, the molecule is bent with an angle similar to those of the six- and seven-valence-electron species. Examples are H_2O ($104\frac{1}{2}°$), H_2S ($92°$), H_2Se ($90\frac{1}{2}°$), H_2Te ($90°$), and NH_2^- ($104°$).

The MO's $2a_1$ and $1b_2$ are bonding; $3a_1$ and $1b_1$ are nonbonding; and $4a_1$ and $2b_2$ are antibonding. Therefore BeH_2, BH_2, CH_2, NH_2, and OH_2 each have four net bonding electrons and a total bond order of 2.

Although we considered only AH_2 molecules, similar arguments have been developed by Gimarc, Hoffmann, and others to rationalize and predict the geometry of many other species, including AH_3, AH_4, A_2H_2, A_2H_4, A_2H_6, HAB, AB_2, AB_3, AB_4, AB_5, and AB_6, and to achieve qualitative understanding of reaction mechanisms (see Section 15.19 for an example). This general approach is called *qualitative molecular-orbital theory*; see B. M. Gimarc, *Molecular Structure and Bonding*, Academic Press, New York, 1979; *Lowe*, Chapter 14.

The Walsh-diagram method works quite well, although there are a few exceptions. However, a good theoretical justification for the method is lacking. The method assumes that the equilibrium geometry is determined by minimizing the sum $\sum_i n_i \varepsilon_i$ of orbital energies, where n_i is the number of electrons in MO i. Actually this is false, since the equilibrium geometry is determined by minimization of the molecular electronic energy including nuclear repulsion; in the SCF MO method (which is known to predict ground-state geometries accurately), this energy E_{HF} is equal to [Eq. (13.181)] $\sum_i n_i \varepsilon_i - V_{ee} + V_{NN}$, where V_{ee} is the interelectronic repulsion energy including exchange interactions (and is needed to correct for the fact that the sum of orbital energies counts each interelectronic repulsion twice) and V_{NN} is the internuclear repulsion energy. Various conflicting arguments have been advanced to explain why Walsh diagrams work so well, but the "ultimate theoretical basis of Walsh diagrams is still unclear" [I. Ferguson and N. C. Pyper, *Theor. Chim. Acta*, **55**, 283 (1980)]. Further references on this question include: P. K. Mehrotra and R. Hoffmann, *Theor. Chim. Acta*, **48**, 301 (1978); R. J. Buenker and S. D. Peyerimhoff, *Chem. Rev.*, **74**, 127 (1974); B. M. Gimarc, *Molecular Structure and Bonding*, Academic Press, New York, 1979, Chapter 9.

15.9 SEMIEMPIRICAL MO TREATMENTS OF PLANAR CONJUGATED MOLECULES

The canonical MO's of a planar unsaturated organic molecule can be divided into σ and π MO's (Section 15.7). Semiempirical treatments of planar conjugated organic compounds usually make the approximation of treating the π electrons separately from the σ electrons. This is sometimes justified by arguing that since the π MO's have a node in the molecular plane, the π-electron density is well separated from the σ-electron density. Actually this is false, and the σ and π probability densities overlap appreciably. Coulson has stated that the justification for $\sigma-\pi$ separability lies in the different symmetry of the σ and π orbitals and in the greater polarizability of the π electrons, which makes them more susceptible to perturbations such as those occurring in chemical reactions.

In the *π-electron approximation*, the n_π π electrons are treated separately by incorporating the effects of the σ electrons and the nuclei into some sort of effective π-electron Hamiltonian $\hat{H}_\pi$ (recall the similar valence-electron approximation—Section 13.15):

$$\hat{H}_\pi = \sum_{i=1}^{n_\pi} \hat{H}_{core}(i) + \sum_{i=1}^{n_\pi} \sum_{j>i} \frac{1}{r_{ij}} \tag{15.86}$$

$$\hat{H}_{core}(i) = -\tfrac{1}{2}\nabla_i^2 + V_i \tag{15.87}$$

where V_i is the potential energy of the ith π electron in the field produced by the nuclei and the σ electrons. The variational principle is then applied to find a π-electron wave function ψ_π that minimizes the variational integral $\int \psi_\pi^* \hat{H}_\pi \psi_\pi \, d\tau$. The validity of the π-electron approximation has been discussed by Lykos and Parr.[18]

---

[18] See *Parr*, pp. 41–45, 211–218; *Pilar*, Section 18-1.

Since (15.86) is not the true molecular electronic Hamiltonian, treatments that make the π-electron approximation are semiempirical. The main π-electron MO theories are the free-electron MO method (Section 15.10), the Hückel MO method (Section 15.11), and the Pariser–Parr–Pople method (Section 15.12).

15.10 *THE FREE-ELECTRON MO METHOD*

Perhaps the simplest semiempirical π-electron theory is the *free-electron molecular-orbital* (FE MO) *treatment*, developed about 1950 by Kuhn, Bayliss, Platt, and Simpson. Here the interelectronic repulsions $1/r_{ij}$ are ignored, and the effect of the σ electrons is represented by a particle-in-a-box potential-energy function: $V = 0$ in a certain region, while $V = \infty$ outside this region. With the interelectronic repulsions omitted, $\hat{H}_\pi$ in (15.86) becomes the sum of Hamiltonians for each π electron; hence (Section 6.2)

$$\hat{H}_\pi \psi_\pi = E_\pi \psi_\pi \tag{15.88}$$

$$\psi_\pi = \prod_{i=1}^{n_\pi} \varphi_i \tag{15.89}$$

$$\hat{H}_{\text{core}}(i)\varphi_i = e_i \varphi_i \tag{15.90}$$

$$E_\pi = \sum_{i=1}^{n_\pi} e_i \tag{15.91}$$

The wave function (15.89) takes no account of spin or the Pauli principle. To do so, we must put each electron in a spin-orbital $u_i = \varphi_i \sigma_i$, where σ_i is a spin function (either α or β). The wave function ψ_π is then written as an antisymmetrized product (Slater determinant) of spin-orbitals. Since $\hat{H}_{\text{core}}(i)$ does not involve spin, we have

$$\hat{H}_{\text{core}}(i)u_i = e_i u_i \tag{15.92}$$

Each term in the antisymmetrized product function ψ_π has each electron in a different spin-orbital [see, for example, (10.68)]. When $\hat{H}_\pi$, which is being approximated as the sum of the $\hat{H}_{\text{core}}(i)$'s, acts on each term in ψ_π, it gives the sum of the e_i's. Hence $\hat{H}_\pi \psi_\pi$ equals $\sum_i e_i \psi_\pi$, and (15.91) still holds when spin and the Pauli principle are allowed for.

For conjugated chain molecules, the FE MO approximation takes the box in which the π electrons move as one dimensional. We have

$$e_i = \frac{n_i^2 h^2}{8ml^2}, \qquad n_i = 1, 2, \ldots \tag{15.93}$$

The n_i's are restricted by the Pauli principle—no more than two electrons can be in a given spatial MO. In the ground electronic state, the π electrons fill the $\frac{1}{2}n_\pi$ lowest free-electron π MO's.

The π-electron transition giving the lowest-frequency electronic absorption involves an electron going from the highest occupied to the lowest vacant MO—Fig. 15.15. (The particle-in-a-box selection rule allows the quantum number of an

n

5 ——————

4 ——————

3 ——————

2 ——————

1 ——————

Figure 15.15 *The longest wavelength* FE MO *electronic transition in a conjugated molecule with six* π *electrons*

electron in the FE MO model to change by an odd integer, as shown in Section 9.10; hence the transition of Fig. 15.15 is allowed.) For the wavelength λ of this transition, the FE MO model gives

$$\frac{1}{\lambda} = \frac{1}{hc} \left(e_{n_\pi/2 + 1} - e_{n_\pi/2} \right) = \frac{h}{8mcl^2} \left(n_\pi + 1 \right) \tag{15.94}$$

We shall apply the FE MO model to the conjugated polyenes

$$CH_2 = [CH—CH =]_k CH_2 \tag{15.95}$$

We take l as the zig-zag length along the carbon chain. (Another possibility is to use the direct distance between the end carbons.) Let l_1 and l_2 be the carbon–carbon single- and double-bond lengths in (15.95). For *s-trans*-1,3-butadiene these values are[19] 1.46 and 1.34 Å. The term *s-trans* refers to the configuration about the single bond:

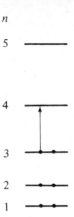

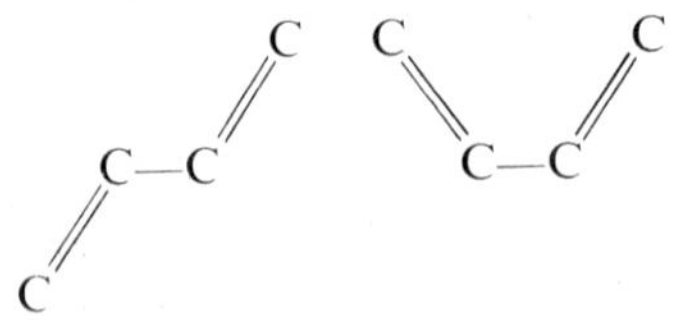

s-trans   *s-cis*

The *s-trans* configuration is the one observed for 1,3-butadiene.[20] Analysis of the ultraviolet and infrared spectra of butadiene condensed on a cold window from a high-temperature gas indicates the presence of a small amount of planar *s-cis*-1,3-butadiene [M. E. Squillacote et al., *J. Am. Chem. Soc.*, **101**, 3657 (1979)]; analysis of

---

[19] K. Kuchitsu, F. Tsutomu, and Y. Morino, *J. Mol. Struct.*, **1**, 463 (1967).
[20] D. R. Lide, *J. Chem. Phys.*, **37**, 2074 (1962).

the Raman spectrum of 1, 3-butadiene indicates the s-cis form to be $2\frac{1}{2}$ to 3 kcal/mole higher in energy than the s-trans form [L. A. Carriera, J. Chem. Phys., **62**, 3851 (1975); J. R. Durig et al., Can. J. Phys., **53**, 1832 (1976)]. Statistical mechanics shows the entropy difference between the s-cis and s-trans forms to be quite small, so this energy difference gives on the order of 1 percent s-cis form present at room temperature.

We shall allow the π electrons to move a bit beyond the ends of the carbon chain by adding in a distance $\frac{3}{4}l_1 + \frac{1}{4}l_2$ at each end. This is about one average bond length and allows for the fact that the MO's are not confined to the regions between the bonding atoms. We have

$$l = kl_1 + (k+1)l_2 + \tfrac{3}{2}l_1 + \tfrac{1}{2}l_2 = (k + \tfrac{3}{2})(l_1 + l_2)$$

$$= \tfrac{1}{2}(n_C + 1)(l_1 + l_2) \tag{15.96}$$

where n_C is the number of carbons. Using (15.94), (15.96), and $n_\pi = n_C$, we have

$$\frac{1}{\lambda} = \frac{h}{2mc(l_1 + l_2)^2(n_C + 1)} = \frac{1}{n_C + 1} \times 155{,}000 \text{ cm}^{-1} \tag{15.97}$$

The observed[21] first electronic absorption bands for the conjugated polyenes (15.95) decrease monotonically from $1/\lambda = 61{,}500 \text{ cm}^{-1}$ to $1/\lambda = 22{,}400 \text{ cm}^{-1}$ as k goes from 0 to 9. The FE MO Eq. (15.97) does correctly predict the trend of increasing wavelength with increasing chain length, but quantitative agreement with experiment is quite poor; the predicted values have a whopping $15{,}000 \text{ cm}^{-1}$ average absolute deviation from observed polyene bands (Problem 15.73). This disagreement is hardly surprising in view of the crudity of the model. [The first excited FE MO configuration gives rise to two electronic terms, a singlet and a triplet. In the FE MO model, neglect of electronic repulsions gives the singlet and triplet terms the same energy. The observed longest wavelength electronic transition is a singlet-singlet transition, since singlet-triplet transitions are forbidden. Strictly speaking, one should compare (15.97) with the average of the singlet-singlet and singlet-triplet energy differences.]

The FE MO model of the polyenes can be improved. We expect the π-electron probability density to be greater in the double-bond regions than in the single-bond regions. To allow for this, Kuhn replaced the $V = 0$ part of the free-electron potential-energy function by a sine curve with minima at the double-bond centers, maxima at the single-bond centers, and length given by (15.96); V goes to infinity at each end of the curve. For

$$V = V_0 \sin \frac{2\pi x}{l_1 + l_2} \tag{15.98}$$

the Schroedinger equation cannot be solved in analytic form. Kuhn used an analog computer to find numerical values of the energy levels; the values for the longest

[21] See H. Suzuki, *Electronic Absorption Spectra*, Academic Press, New York, 1967, p. 204.

wavelength transitions of the polyenes (15.95) are accurately reproduced by the following formula[22]:

$$\frac{1}{\lambda} = \frac{1}{n_C + 1} \times 155{,}000\,\text{cm}^{-1} + \frac{0.83\,V_0}{hc}\left(1 - \frac{1}{n_C}\right) \tag{15.99}$$

With $V_0 = 0$, Eq. (15.99) reduces to (15.97). With $V_0 = 2.4\,\text{eV}$, Eq. (15.99) gives a good fit to the polyene data; the average absolute deviation is only 1300 cm^{-1}.

Although the simple FE MO method fares poorly for the conjugated polyenes, it should work better for polymethine ions with the formula

$$(\text{CH}_3)_2\overset{+}{\text{N}}\!\!=\!\!\text{CH}(\text{---CH}\!\!=\!\!\text{CH})_k\text{---}\ddot{\text{N}}(\text{CH}_3)_2 \tag{15.100}$$

Here the carbon–carbon bond lengths are all equal. (There is an equivalent VB resonance structure with carbon–carbon single and double bonds interchanged and the charge on the other nitrogen.) We shall assume the carbon–carbon bond length to be 1.40 Å as in benzene. We shall not worry about the small difference in carbon–nitrogen and carbon–carbon bond lengths. Adding in an extra bond length at each end of the chain, we set $l = (2k + 4)(1.40\,\text{Å})$ and $n_\pi = 2k + 4$ in (15.94) to get

$$\frac{1}{\lambda} = \frac{h}{8mc(1.96\,\text{Å}^2)}\frac{2k + 5}{(2k + 4)^2} = \frac{(2k + 5)}{(2k + 4)^2} \times 155{,}000\,\text{cm}^{-1} \tag{15.101}$$

The observed[23] $1/\lambda$ values decrease monotonically from 44,600 cm^{-1} to 11,800 cm^{-1} as k goes from 0 to 6. Values calculated from (15.101) are in pretty good agreement with experiment, the average absolute deviation being 2200 cm^{-1} (Problem 15.74).

The FE MO model and its refined versions have been applied to conjugated ring hydrocarbons with reasonable success, but we omit any discussion (see Problem 15.32).

15.11 THE HUCKEL MO METHOD

The most celebrated semiempirical π-electron theory is the *Hückel molecular-orbital* (HMO) *method*, developed in the 1930s. Here the π-electron Hamiltonian (15.86) is approximated by the simpler form

$$\hat{H}_\pi = \sum_{i=1}^{n_\pi} \hat{H}^{\text{eff}}(i) \tag{15.102}$$

where $\hat{H}^{\text{eff}}(i)$ somehow incorporates the effects of the π-electron repulsions in an average way. This sounds rather vague, and in fact the Hückel method does not specify any explicit form for $\hat{H}^{\text{eff}}(i)$. Since the Hückel π-electron Hamiltonian is the sum of one-electron Hamiltonians, a separation of variables is possible, as in FE MO

---

[22] H. Kuhn, *Fortschritte der Chemie Organischer Naturstoffe*, **17**, 404 (1959)—this is a review article with many references; *J. Chem. Phys.*, **17**, 1198 (1949).
[23] See Suzuki, *Electronic Absorption Spectra*, p. 365.

theory. Equations (15.89) and (15.91) hold, where the Hückel MO's satisfy

$$\hat{H}^{\,\text{eff}}(i)\varphi_i = e_i\varphi_i \tag{15.103}$$

Since $\hat{H}^{\,\text{eff}}$ is not specified, there is no point in trying to solve (15.103) directly. Instead, the variational method is used.

The next assumption in the HMO method is to approximate the π MO's as LCAO's. In a minimal-basis-set calculation of a planar conjugated hydrocarbon, the only AO's of π symmetry are the carbon $2p\pi$ orbitals, where by $2p\pi$ we mean the real $2p$ AO's that are perpendicular to the molecular plane. We thus write

$$\varphi_i = \sum_{r=1}^{n_C} c_{ir} f_r \tag{15.104}$$

where f_r is a $2p\pi$ AO on the rth carbon atom and n_C is the number of carbon atoms. (15.104) is a linear variation function. The optimum values of the coefficients for the n_C lowest π MO's satisfy Eq. (8.85):

$$\sum_{s=1}^{n_C} [(H_{rs}^{\text{eff}} - S_{rs}e_i)c_{is}] = 0, \qquad r = 1, 2, \ldots, n_C \tag{15.105}$$

where the e_i's are the roots of the secular equation (8.88)

$$\det (H_{rs}^{\text{eff}} - S_{rs}e_i) = 0 \tag{15.106}$$

The key assumptions in the Hückel theory involve the integrals in (15.106). The integral H_{rr} is assumed to have the same value for every carbon atom in the molecule. (For benzene the six carbons are equivalent, and this is no assumption; for 1,3-butadiene one would expect H_{rr} for an end carbon and a middle carbon to differ slightly.) Moreover, H_{rr} is assumed to be the same for carbon atoms in different planar hydrocarbons. The integral H_{rs} is assumed to have the same value for any two carbon atoms bonded to each other and to vanish for two nonbonded carbons. The integral S_{rr} is equal to 1, since the AO's are normalized. The overlap integral S_{rs} is assumed to vanish for $r \neq s$. We have

$$H_{rr}^{\text{eff}} = \int f_r^*(i)\hat{H}^{\,\text{eff}}(i)f_r(i)\,dv_i \equiv \alpha \tag{15.107}$$

$$H_{rs}^{\text{eff}} = \int f_r^*(i)\hat{H}^{\,\text{eff}}(i)f_s(i)\,dv_i \equiv \beta \quad \text{for } C_r \text{ and } C_s \text{ bonded} \tag{15.108}$$

$$H_{rs}^{\text{eff}} = 0 \quad \text{for } C_r \text{ and } C_s \text{ not bonded together} \tag{15.109}$$

$$S_{rs} = \int f_r^*(i)f_s(i)\,dv_i = \delta_{rs} \tag{15.110}$$

$$f_r \equiv C_r 2p\pi \tag{15.111}$$

(The integrals α and β are called the *Coulomb integral* and *bond integral*, respectively. Often β is called the *resonance integral*; however, the term "resonance" is more appropriate to the VB method than to the MO method.) Since carbons not bonded to each other are well separated in space, the assumption (15.109) is reasonable. However, taking the overlap integral as zero for carbons bonded to each other is a poor assumption; for Slater orbitals, S_{rs} for adjacent carbons ranges between 0.2 and 0.3, depending on the bond distance. Inclusion of overlap will be considered later.

We want each Hückel MO to be normalized. Use of (15.104) and (15.110) gives

$$1 = \langle \varphi_i | \varphi_i \rangle = \langle \Sigma_r c_{ir} f_r | \Sigma_s c_{is} f_s \rangle = \Sigma_r \Sigma_s c_{ir}^* c_{is} \langle f_r | f_s \rangle = \Sigma_r \Sigma_s c_{ir}^* c_{is} \delta_{rs} = \Sigma_r c_{ir}^* c_{ir},$$

$$\sum_{r=1}^{n_C} |c_{ir}|^2 = 1 \qquad (15.112)$$

This is the normalization condition for the ith Hückel MO.

In application of HMO theory to conjugated hydrocarbons, planarity is assumed. Occasionally this assumption does not hold. In gas-phase biphenyl, the two rings are twisted at an angle of 40° with each other, because of steric interference between the *ortho* hydrogens.

From (15.105) it is clear that the order of the HMO secular determinant equals the number of conjugated carbons. (Students sometimes make the error of assuming that this order always equals the number of π electrons.)

Butadiene   To illustrate the HMO method, we consider 1,3-butadiene. The only thing of significance in the simple HMO treatment of a planar hydrocarbon is the topology of the carbon-atom framework; by this we mean which carbons are bonded together. No distinction is made between the *s-cis* and *s-trans* forms of butadiene. The numbering of the carbon atoms is

$$\underset{1}{CH_2}=\underset{2}{CH}-\underset{3}{CH}=\underset{4}{CH_2}$$

The Hückel assumptions give

$$H_{11}^{eff} = H_{22}^{eff} = H_{33}^{eff} = H_{44}^{eff} = \alpha$$

$$H_{12}^{eff} = H_{23}^{eff} = H_{34}^{eff} = \beta$$

$$H_{13}^{eff} = H_{14}^{eff} = H_{24}^{eff} = 0$$

The secular equation (15.106) is

$$\begin{vmatrix} \alpha - e_k & \beta & 0 & 0 \\ \beta & \alpha - e_k & \beta & 0 \\ 0 & \beta & \alpha - e_k & \beta \\ 0 & 0 & \beta & \alpha - e_k \end{vmatrix} = 0$$

We now divide each row of the determinant by β. This divides the determinant by β^4, and since $0/\beta^4 = 0$, we get

$$\begin{vmatrix} x & 1 & 0 & 0 \\ 1 & x & 1 & 0 \\ 0 & 1 & x & 1 \\ 0 & 0 & 1 & x \end{vmatrix} = 0 \qquad (15.113)$$

where

$$x \equiv \frac{\alpha - e_k}{\beta}, \qquad e_k = \alpha - \beta x \qquad (15.114)$$

The secular determinant is a continuant, and Eq. (8.56) gives

$$\prod_{j=1}^{4} \left(x - 2 \cos \frac{j\pi}{5} \right) = 0$$

$$x = 2 \cos (j\pi/5), \qquad j = 1, 2, 3, 4 \tag{15.115}$$

$$x = -1.618, \; -0.618, \; 0.618, \; 1.618 \tag{15.116}$$

Alternatively, we can expand the secular determinant to yield the algebraic equation $x^4 - 3x^2 + 1 = 0$. The substitution $z = x^2$ yields $z^2 - 3z + 1$, and the quadratic formula gives $z = (3 \pm 5^{1/2})/2$. Since $x = \pm z^{1/2}$, we have

$$x = (3 \pm 5^{1/2})^{1/2}/2^{1/2}, \quad -(3 \pm 5^{1/2})^{1/2}/2^{1/2}$$

$$x = (5^{1/2} \pm 1)/2, \quad -(5^{1/2} \pm 1)/2 \tag{15.117}$$

Corresponding to (15.113), the equations for the HMO coefficients of butadiene are

$$
\begin{aligned}
xc_{j1} + c_{j2} \phantom{+ c_{j3}} &= 0 \\
c_{j1} + xc_{j2} + c_{j3} \phantom{+ c_{j4}} &= 0 \\
c_{j2} + xc_{j3} + c_{j4} &= 0 \\
c_{j3} + xc_{j4} &= 0
\end{aligned}
\tag{15.118}
$$

We must now substitute each of the roots (15.116) in turn into (15.118), as discussed in Section 8.5.

Consider first the root $x = -1.618$. The first equation of (15.118) reads $-1.618\,c_1 + c_2 = 0$ (where, for simplicity, the j subscript has been omitted). As discussed in Section 8.4, the solutions c_1, c_2, c_3, c_4 each contain an arbitrary multiplicative constant. Hence we shall solve for $c_2, c_3,$ and c_4 in terms of c_1. We have $c_2 = 1.618\,c_1$. The second equation in (15.118) gives $c_3 = -c_1 - xc_2 = -c_1 + 1.618(1.618\,c_1) = 1.618c_1$. The fourth equation in (15.118) gives $c_4 = -c_3/x = 1.618c_1/1.618 = c_1$.

The normalization condition (15.112) is now used to fix c_1. Taking c_1 to be a real, positive number, we have $1 = c_1^2 + c_2^2 + c_3^2 + c_4^2 = c_1^2 + (1.618c_1)^2 + (1.618c_1)^2 + c_1^2 = 7.236c_1^2$, and $c_1 = 0.372$. Then $c_2 = 1.618c_1 = 0.602$, $c_3 = 1.618c_1 = 0.602$, and $c_4 = c_1 = 0.372$. The HMO corresponding to $x = -1.618$ is then $\varphi = 0.372f_1 + 0.602f_2 + 0.602f_3 + 0.372f_4$. The energy of this HMO is given by (15.114) as $e = \alpha + 1.618\,\beta$.

Substitution of each of the three remaining roots into (15.118) yields three more HMO's. One finds the following four normalized HMO's:

$$
\begin{aligned}
\varphi_1 &= 0.372f_1 + 0.602f_2 + 0.602f_3 + 0.372f_4 \\
\varphi_2 &= 0.602f_1 + 0.372f_2 - 0.372f_3 - 0.602f_4 \\
\varphi_3 &= 0.602f_1 - 0.372f_2 - 0.372f_3 + 0.602f_4 \\
\varphi_4 &= 0.372f_1 - 0.602f_2 + 0.602f_3 - 0.372f_4
\end{aligned}
\tag{15.119}
$$

The HMO energies are

$$e_1 = \alpha + 1.618\beta, \qquad e_2 = \alpha + 0.618\beta$$
$$e_3 = \alpha - 0.618\beta, \qquad e_4 = \alpha - 1.618\beta$$

(15.120)

The MO φ_1 in (15.119) has no nodes (other than the molecular plane) and leads to maximum charge buildup between the atoms; clearly this must be the lowest-energy π MO; its energy is $\alpha + 1.618\beta$ and, therefore, the bond integral β must be negative. [See also Eq. (13.58).] The HMO's and energy levels are sketched in Fig. 15.16. The pattern of nodes is the same as for the FE MO wave functions of butadiene; the number

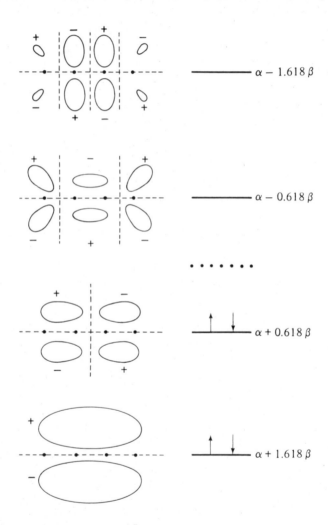

Figure 15.16 *HMO's for butadiene. Nodal planes for the π MO's are indicated by dashed lines. Note the symmetry of the bonding and antibonding energy levels about α (dotted line). The ground-state π-electron MO configuration is shown.*

of vertical nodal planes is zero for the ground MO and increases by one for each higher MO. From the figure it is clear that these MO's are orthogonal. We can approximate the energy of an electron in a carbon $2p\pi$ AO in the molecule by $\int f_i^* \hat{H}^{\text{eff}} f_i \, dv = \alpha$; an HMO is classified as bonding or antibonding according to whether its energy is less than or greater than α.

Conjugated polyenes   For the conjugated polyene (15.95) with n_C carbon atoms, the HMO secular equation involves a continuant similar to (15.113) but of order n_C; Eq. (8.56) with $b = c = 1$, $a = x$, and $n = n_C$ gives $x = 2\cos[j\pi/(n_C + 1)]$, where $j = 1, \ldots, n_C$. Since the values of x for $j = k$ and for $j = n_C + 1 - k$ are simply the negatives of each other, we can write

$$x = -2\cos\frac{j\pi}{n_C + 1} \tag{15.121}$$

$$e_j = \alpha + 2\beta\cos\frac{j\pi}{n_C + 1}, \qquad j = 1, 2, \ldots, n_C \tag{15.122}$$

Since β is negative, e_1 is the lowest π energy level. All the π-electron levels are nondegenerate. The HMO coefficients are (Problem 15.34)

$$c_{jr} = \left(\frac{2}{n_C + 1}\right)^{1/2} \sin\frac{jr\pi}{n_C + 1} \tag{15.123}$$

In the ground electronic state of (15.95), the highest occupied and lowest vacant π MO's have $j = \frac{1}{2}n_C$ and $(\frac{1}{2}n_C + 1)$, respectively; HMO theory predicts the longest wavelength band of the electronic absorption spectrum of a conjugated polyene to occur at

$$\frac{1}{\lambda} = -\frac{4\beta}{hc}\sin\frac{\pi}{2n_C + 2} \tag{15.124}$$

where we used

$$\cos a - \cos b = -2\sin\tfrac{1}{2}(a + b)\sin\tfrac{1}{2}(a - b) \tag{15.125}$$

The bond integral β is a semiempirical parameter and is adjusted to give the best fit to experimental data. If we use the observed butadiene absorption wavelength $\lambda = 2170$ Å, we find $|\beta|/hc = 37,300$ cm^{-1}, $|\beta| = 4.62$ eV. With this value of β, we then calculate wavelengths for the polyenes (15.95). The predicted values do show the correct trend of increase in λ with increase in n_C, but agreement with experiment is poor; predicted $1/\lambda$ values show an average absolute deviation of 9300 cm^{-1} for the first several members of the series (Problem 15.73).

An obvious defect of the Hückel approximation for conjugated polyenes is the use of a single value of β for each pair of adjacent carbon atoms. The bond lengths in these molecules alternate, and we expect β to be larger for doubly bonded carbons than for singly bonded carbons. The refinement of using two polyene bond integrals β_1 and β_2 for C—C and C=C bonds, respectively, was introduced by Lennard-Jones.[24] With two β's the mathematics becomes complicated, so we shall

---

[24]   J. E. Lennard-Jones, *Proc. Roy. Soc.*, **A158**, 280 (1937).

omit it. With

$$\beta_1 = -3.32 \text{ eV}, \qquad \beta_2 = -4.20 \text{ eV}$$

agreement with experiment is much improved over the single-β predictions. The average absolute deviation of predicted values is reduced to 3000 cm^{-1}.

Benzene   For benzene

the HMO secular equation is

$$\begin{vmatrix} x & 1 & 0 & 0 & 0 & 1 \\ 1 & x & 1 & 0 & 0 & 0 \\ 0 & 1 & x & 1 & 0 & 0 \\ 0 & 0 & 1 & x & 1 & 0 \\ 0 & 0 & 0 & 1 & x & 1 \\ 1 & 0 & 0 & 0 & 1 & x \end{vmatrix} = 0 \qquad (15.126)$$

where x is given by (15.114). Equation (8.53) gives

$$\prod_{k=1}^{6} (x + e^{2\pi i k/6} + e^{-2\pi i k/6}) = 0, \qquad i = \sqrt{-1}$$

where we used $\exp(2\pi i k \cdot 5/6) = \exp(-2\pi i k/6)$. We have

$$x = -2 \cos\left(\frac{2\pi k}{6}\right), \qquad k = 1, \ldots, 6 \qquad (15.127)$$

$$x = -1, +1, +2, +1, -1, -2$$

$$e_i = \alpha + 2\beta, \ \alpha + \beta, \ \alpha + \beta, \ \alpha - \beta, \ \alpha - \beta, \ \alpha - 2\beta \qquad (15.128)$$

There are two nondegenerate and two doubly degenerate Hückel levels. Figure 15.17 shows the HMO ground-state π-electron configuration.

The HMO coefficients can be found by solving the usual set of simultaneous equations, but it is simpler to use molecular symmetry. The $\hat{O}_{C_6}$ symmetry operator commutes with the π-electron Hamiltonian, so that we can choose each MO to be an eigenfunction of this 60° rotation. Since $(\hat{O}_{C_6})^6 = \hat{1}$, the eigenvalues of $\hat{O}_{C_6}$ are the six sixth roots of unity (Problem 7.13):

$$e^{2\pi i k/6}, \qquad k = 0, 1, \ldots, 5 \qquad (15.129)$$

(Since $\hat{O}_{C_6}$ has some complex eigenvalues, it certainly is not Hermitian; only operators representing physical quantities need be Hermitian, and $\hat{O}_{C_6}$ does not

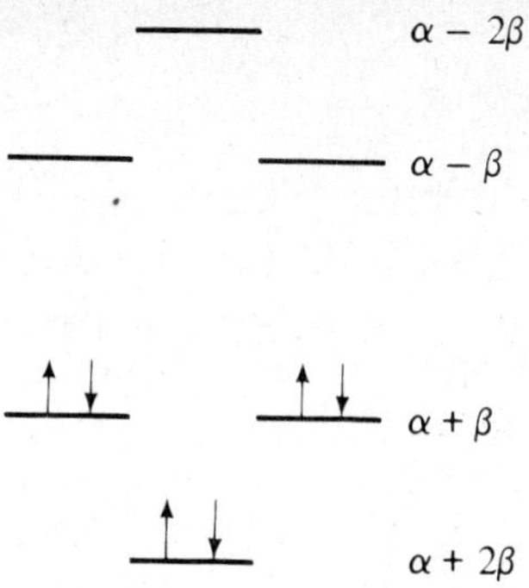

Figure 15.17 *Hückel MO energies for benzene*

correspond to any physical property of the molecule.) We have [Eq. (15.104)]

$$\hat{O}_{C_6}\varphi_j = e^{2\pi ik/6}\varphi_j$$

$$\sum_{r=1}^{6} c_{jr}\hat{O}_{C_6}f_r = \sum_{r=1}^{6} c_{jr}e^{2\pi ik/6}f_r$$

$$\sum_{r=1}^{6} c_{jr}f_{r-1} = \sum_{r=1}^{6} c_{jr}e^{2\pi ik/6}f_r$$

$$\sum_{r=1}^{6} c_{j,r+1}f_r = \sum_{r=1}^{6} c_{jr}e^{2\pi ik/6}f_r \qquad (15.130)$$

where $f_0 \equiv f_6$ and $c_{j7} \equiv c_{j1}$. Equating the coefficients of corresponding AO's in (15.130), we have

$$c_{j,r+1} = e^{2\pi ik/6}c_{jr} \qquad (15.131)$$

The normalization condition (15.112) is

$$\sum_{r=1}^{6} |c_{jr}|^2 = 1 \qquad (15.132)$$

We see from (15.131) that each coefficient in the jth MO has the same absolute value. Hence

$$|c_{jr}| = 1/\sqrt{6}, \qquad r = 1, 2, \ldots, 6 \qquad (15.133)$$

By setting $c_{j1} = 1/\sqrt{6}$ and using (15.131), we obtain the desired coefficients. To determine which coefficients go with which energy, we evaluate the variational integral:

$$\int \varphi_j^* \hat{H}^{\text{eff}}\varphi_j \, dv = \sum_{r=1}^{6}\sum_{s=1}^{6} c_{jr}^* c_{js} \int f_r^* \hat{H}^{\text{eff}} f_s \, dv$$

$$= \sum_{r=1}^{6} |c_{jr}|^2\alpha + \sum_{r=1}^{6} c_{jr}^* c_{j,r+1}\beta + \sum_{r=1}^{6} c_{jr}^* c_{j,r-1}\beta$$

$$= \alpha + e^{2\pi ik/6}\sum_{r=1}^{6} |c_{jr}|^2\beta + e^{-2\pi ik/6}\sum_{r=1}^{6} |c_{jr}|^2\beta$$

$$= \alpha + 2\beta\cos(2\pi k/6), \qquad k = 0, \ldots, 5 \qquad (15.134)$$

which agrees with (15.128). From (15.131), (15.133), and (15.134), the Hückel MO's and energies for benzene are

$$\varphi_1 = 6^{-1/2}(f_1 + f_2 + f_3 + f_4 + f_5 + f_6)$$

$$\varphi_2 = 6^{-1/2}(f_1 + e^{\pi i/3}f_2 + e^{2\pi i/3}f_3 - f_4 + e^{4\pi i/3}f_5 + e^{5\pi i/3}f_6)$$

$$\varphi_3 = 6^{-1/2}(f_1 + e^{-\pi i/3}f_2 + e^{-2\pi i/3}f_3 - f_4 + e^{-4\pi i/3}f_5 + e^{-5\pi i/3}f_6)$$

$$\varphi_4 = 6^{-1/2}(f_1 + e^{2\pi i/3}f_2 + e^{4\pi i/3}f_3 + f_4 + e^{2\pi i/3}f_5 + e^{4\pi i/3}f_6) \qquad (15.135)$$

$$\varphi_5 = 6^{-1/2}(f_1 + e^{-2\pi i/3}f_2 + e^{-4\pi i/3}f_3 + f_4 + e^{-2\pi i/3}f_5 + e^{-4\pi i/3}f_6)$$

$$\varphi_6 = 6^{-1/2}(f_1 - f_2 + f_3 - f_4 + f_5 - f_6)$$

$$e_1 = \alpha + 2\beta, \qquad e_2 = \alpha + \beta, \qquad e_3 = \alpha + \beta$$

$$e_4 = \alpha - \beta, \qquad e_5 = \alpha - \beta, \qquad e_6 = \alpha - 2\beta$$

The condition (15.131) determining the π-MO coefficients for benzene was derived solely from symmetry considerations, without use of the Hückel approximations. Thus the MO's $\varphi_1, \ldots, \varphi_6$ are (except for normalization constants) the correct minimal-basis-set SCF π-electron MO's for benzene. (The Hückel energies $e_1, \ldots, e_6$ are, however, not the true SCF orbital energies; the Hückel method ignores electron repulsions and takes the total π-electron energy as the sum of orbital energies. The SCF method takes electron repulsions into account in an average way, and the total SCF energy is not the sum of orbital energies.) A similar situation occurs for ethylene, where the minimal-basis-set π MO's (15.79) and (15.80) are determined solely by symmetry. (An extended-basis-set benzene calculation would mix in $3p\pi$, $3d\pi$, etc., carbon AO's, and the contributions of these AO's must be determined by an explicit SCF calculation.)

The MO's in (15.135) for the degenerate π levels are complex. Often, real forms of these MO's are used. The two MO's of each degenerate level are complex conjugates of each other; by adding and subtracting these MO's, we get the real forms:

$$\varphi_{2,\text{real}} = 2^{-1/2}(\varphi_2 + \varphi_3) \qquad \varphi_{3,\text{real}} = -2^{-1/2}i(\varphi_2 - \varphi_3)$$

$$\varphi_{2,\text{real}} = 12^{-1/2}(2f_1 + f_2 - f_3 - 2f_4 - f_5 + f_6)$$

$$\varphi_{3,\text{real}} = \tfrac{1}{2}(f_2 + f_3 - f_5 - f_6) \qquad (15.136)$$

$$\varphi_{4,\text{real}} = 12^{-1/2}(2f_1 - f_2 - f_3 + 2f_4 - f_5 - f_6)$$

$$\varphi_{5,\text{real}} = \tfrac{1}{2}(f_2 - f_3 + f_5 - f_6)$$

Figure 15.18 shows the real benzene π-electron MO's. Note the charge buildup between the nuclei for the bonding MO's.

The symmetry species of the benzene π MO's are (*Schonland*, p. 210)

MO	φ_1	φ_2	φ_3	φ_4	φ_5	φ_6
Symmetry Species	a_{2u}	e_{1g}	e_{1g}	e_{2u}	e_{2u}	b_{2g}

The ground-state π-electron configuration is $(1a_{2u})^2(1e_{1g})^4$.

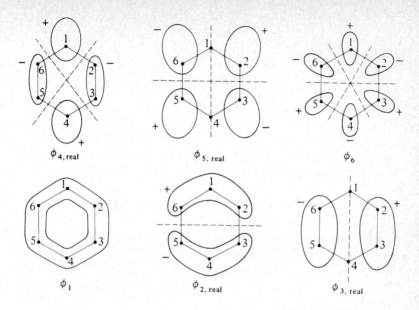

$\phi_{4,\,\text{real}}$ $\phi_{5,\,\text{real}}$ ϕ_6

ϕ_1 $\phi_{2,\,\text{real}}$ $\phi_{3,\,\text{real}}$

Figure 15.18 *Benzene π MO's (real form). A top view is shown; the π MO's change sign on reflection in the molecular plane (which is a nodal plane for them). Dashed lines indicate nodal planes perpendicular to the molecular plane.*

Monocyclic conjugated polyenes   For the monocyclic planar conjugated polyene C_nH_n, we can use the same treatment as for benzene; replacing 6 by n_C in (15.134), (15.131), and (15.133), we find the HMO energies and coefficients:

$$e_k = \alpha + 2\beta \cos \frac{2\pi k}{n_C}, \qquad k = 0, \ldots, n_C - 1 \tag{15.137}$$

$$c_{kr} = \frac{1}{\sqrt{n_C}} \exp\left[\frac{2\pi i(r-1)k}{n_C}\right], \qquad i = \sqrt{-1} \tag{15.138}$$

$$\varphi_k = \frac{1}{\sqrt{n_C}} \sum_{r=1}^{n_C} \exp\left[\frac{2\pi i(r-1)k}{n_C}\right] f_r \tag{15.139}$$

where $f_r = C_r 2p\pi$. Note that the index k in these equations does not correspond to the actual order of the MO's: The lowest MO has $k = 0$; next are the MO's with $k = 1$ and $k = n_C - 1$; next are the MO's with $k = 2$ and $k = n_C - 2$; and so on. A most amusing mnemonic device[25] is available for the HMO energies (15.137). One inscribes a regular polygon of n_C sides in a circle of radius $2|\beta|$, putting an apex at the bottom of the circle. If a vertical scale of energy is set up with energy α coinciding with the center of the circle, then each polygon vertex is located at an HMO energy (Fig. 15.19).

---

[25] A. A. Frost and B. Musulin, *J. Chem. Phys.*, **21**, 572 (1953).

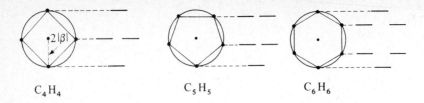

C_4H_4 C_5H_5 C_6H_6

Figure 15.19 *Mnemonic device for HMO energies of monocyclic planar hydrocarbons. The levels shown are for cyclobutadiene, the cyclopentadienyl radical, and benzene.*

The method gives the correct degeneracies and spacings of the Hückel levels of the ring hydrocarbon C_nH_n (Problem 15.37).

Consider the Hückel energy levels for the monocyclic polyene C_nH_n. The lowest subshell consists of a nondegenerate level and holds two electrons. Each of the remaining low-lying subshells consists of a doubly degenerate level and holds four electrons. (If n_C is even, the highest π energy level is nondegenerate, but this level is not occupied in the ground state.) To have a stable filled-subshell π-electron configuration, we see that the number of π electrons must satisfy

$$n_\pi = 4m + 2, \qquad m = 0, 1, 2, \ldots \qquad \textbf{(15.140)}$$

This is Hückel's famous $4m + 2$ rule, which ascribes extra stability to monocyclic conjugated systems that satisfy (15.140). With $4m + 1$ or $4m - 1$ π electrons, the compound is a free radical. With $4m$ π electrons, there are two electrons in a subshell that can hold four electrons, and Hund's rule predicts a triplet (diradical) ground state.

Benzene satisfies the $4m + 2$ rule. The cyclopentadienyl radical $\cdot C_5H_5$ is one electron short of satisfying (15.140); the cyclopentadienyl anion $C_5H_5^-$ satisfies the $4m + 2$ rule; the cation $C_5H_5^+$ is predicted to have a triplet ground state and be highly reactive. These predictions are borne out; $C_5H_5^-$ is found to be considerably more stable than either $C_5H_5^+$ or $\cdot C_5H_5$. Similarly, $C_7H_7^+$ should be more stable than $\cdot C_7H_7$ or $C_7H_7^-$, as is verified experimentally; e.g., the salt $C_7H_7^+Br^-$ is readily prepared.

Cyclobutadiene, C_4H_4, is predicted to have a triplet ground state. After decades of unsuccessful efforts to synthesize this compound, cyclobutadiene was finally prepared[26,27] in 1965. It turns out to be a highly reactive compound that dimerizes at temperatures above 35 K. In the dozen years following its synthesis, various experimental observations and theoretical calculations gave conflicting results about the nature of the C_4H_4 ground electronic state, but it is now generally agreed that the ground state is a singlet with a rectangular geometry, the bond lengths being close to ordinary single- and double-bond lengths [see the references cited in D. W. Whitman and B. K. Carpenter, *J. Am. Chem. Soc.*, **102**, 4272 (1980)]. The reason for

[26] L. Watts, J. D. Fitzpatrick, and R. Pettit, *J. Am. Chem. Soc.*, **87**, 3253 (1965).
[27] L. T. Scott and M. Jones, *Chem. Rev.*, **1972**, 181.

this violation of Hund's rule is discussed in H. Kollmar and V. Staemmler, *J. Am. Chem. Soc.*, **99**, 3583 (1977).

HMO theory predicts that planar cyclooctatetraene, C_8H_8, would have a triplet ground state. Experimentally, C_8H_8 has a nonplanar "tub" structure with alternating single and double bonds. The nonplanarity results from steric strain in the planar geometry due to deviation from the 120° bond angle at the sp^2-hybridized carbons; therefore C_8H_8 does not provide a good test of the $4m + 2$ rule. The dianion $C_8H_8^{2-}$ satisfies the $4m + 2$ rule. The salt $K_2C_8H_8$ has been prepared, and the evidence is that $C_8H_8^{2-}$ is planar. The extra stability arising from delocalization of the π electrons is sufficient to overcome the steric strain.

The monocyclic compounds C_nH_n are called *annulenes*. Benzene is [6]annulene. The annulenes with $n = 10, 12, 14,$ and 16 suffer steric strain and substantial repulsions between nonbonded hydrogens, thereby preventing planarity and a clear-cut test of the $4m + 2$ rule. One finds that [18]annulene is nearly planar with nearly equal bond lengths and shows chemically aromatic behavior, in agreement with the $4m + 2$ rule. The stabilization gained by π-electron delocalization becomes negligible as n goes to infinity in the aromatic annulenes $C_{4m+2}H_{4m+2}$; see the discussion of delocalization energy later in this section.

In the preceding discussion, we used the same HMO's for a neutral compound and for its related ions. The Hückel method takes no account of electron repulsions, and therefore the HMO's are unchanged when π electrons are added or removed.

The $4m + 2$ rule is sometimes applied to polycyclic systems; however, the pattern (15.137) of HMO levels holds only for a monocyclic system, and use of the $4m + 2$ rule for polycyclic systems is not justified.

The $4m + 2$ rule actually does not depend on the Hückel assumptions (15.107)–(15.110). The C_nH_n π MO's (15.139) were derived solely by symmetry considerations and are the correct SCF minimal-basis-set π MO's. With $k = 0$, we have an MO with all plus signs in front of the AO's; clearly this MO has a lower energy than any of the others. For the remaining MO's, the pair with $k = j$ and $k = n_C - j$ are complex conjugates of each other and must have the same energy. (Since $\hat{H}$ is Hermitian, we have $\int \varphi^* \hat{H} \varphi \, dv = \int (\varphi^*)^* \hat{H} \varphi^* \, dv$.) Thus the excited MO's occur in pairs (except that when n_C is even, the MO with $k = \frac{1}{2}n_C$ is nondegenerate; with alternating plus and minus signs in front of the AO's, this is the highest-energy MO). The pattern of a nondegenerate lowest π level followed by doubly degenerate π levels thus holds for the minimal-basis-set SCF MO's. Of course, the energies (15.137) are not the correct SCF orbital energies.

Naphthalene   Now consider the HMO treatment of naphthalene. For butadiene and benzene, we set up the secular equation without bothering with the intermediate step of constructing symmetry orbitals from the $2p\pi$ AO's; for these molecules the secular equation was easy enough to solve without the simplifications introduced by symmetry orbitals. For naphthalene the 10×10 secular determinant is difficult to deal with, and we first find symmetry orbitals. The point group of naphthalene (Fig. 15.20) is $\mathscr{D}_{2h}$. The possible symmetry species are (15.76). The $C2p\pi$ AO's all have eigenvalue -1 for reflection in the molecular (yz) plane. Each symmetry orbital will be some linear combination of AO's that are permuted among one another by the

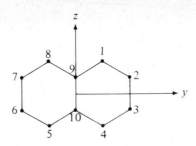

Figure 15.20 *Axes for naphthalene. The x axis is perpendicular to the molecular plane.*

symmetry reflections (recall ethylene). Hence to aid in finding the symmetry orbitals we examine the effects of the $\hat{\sigma}(xy)$ and $\hat{\sigma}(xz)$ operations on the π AO's. We find the π AO's to fall into three sets:

$$1, 4, 5, 8 \qquad 2, 3, 6, 7 \qquad 9, 10 \qquad (15.141)$$

where the members of each set are permuted among one another by the symmetry reflections. (Of course, these are also the chemically equivalent carbons.) Each symmetry orbital must be some linear combination of the AO's of a given set. Naphthalene and ethylene have the same point group, and the pattern of the hydrogen symmetry orbitals in (15.77) gives us the symmetry orbitals of the first two sets of AO's in (15.141). The symmetry orbitals and their readily verified symmetry species are then

$$b_{3u}: g_1 = \tfrac{1}{2}(f_1 + f_4 + f_5 + f_8) \qquad g_2 = \tfrac{1}{2}(f_2 + f_3 + f_6 + f_7)$$
$$g_3 = 2^{-1/2}(f_9 + f_{10})$$
$$a_u: g_4 = \tfrac{1}{2}(f_1 - f_4 + f_5 - f_8) \qquad g_5 = \tfrac{1}{2}(f_2 - f_3 + f_6 - f_7)$$
$$b_{2g}: g_6 = \tfrac{1}{2}(f_1 - f_4 - f_5 + f_8) \qquad g_7 = \tfrac{1}{2}(f_2 - f_3 - f_6 + f_7)$$
$$g_8 = 2^{-1/2}(f_9 - f_{10})$$
$$b_{1g}: g_9 = \tfrac{1}{2}(f_1 + f_4 - f_5 - f_8) \qquad g_{10} = \tfrac{1}{2}(f_2 + f_3 - f_6 - f_7) \qquad (15.142)$$

The constants $\tfrac{1}{2}$ and $1/\sqrt{2}$ normalize the symmetry orbitals, provided the approximation (15.110) is used.

   Instead of using the f_r's as basis functions, we set up the secular equation using the symmetry orbitals as basis functions:

$$\det\left[\langle g_p | \hat{H}^{\text{eff}} | g_q \rangle - \langle g_p | g_q \rangle e_k\right] = 0 \qquad (15.143)$$

The secular determinant in (15.143) is in block-diagonal form (Section 15.4), and we have two third-degree and two second-degree equations to solve. From (15.142) and (15.110), it follows that

$$\langle g_p | g_q \rangle = \delta_{pq} \qquad (15.144)$$

when g_p and g_q belong to the same symmetry species. Also

$$\langle g_1|\hat{H}^{\text{eff}}|g_1\rangle = \tfrac{1}{4}\langle f_1+f_4+f_5+f_8|\hat{H}^{\text{eff}}|f_1+f_4+f_5+f_8\rangle = \tfrac{1}{4}(\alpha+\alpha+\alpha+\alpha) = \alpha$$

$$\langle g_1|\hat{H}^{\text{eff}}|g_2\rangle = \tfrac{1}{4}\langle f_1+f_4+f_5+f_8|\hat{H}^{\text{eff}}|f_2+f_3+f_6+f_7\rangle = \tfrac{1}{4}(\beta+\beta+\beta+\beta) = \beta$$

Evaluating the remaining matrix elements, we find as the secular equation for the b_{3u} MO's

$$\begin{vmatrix} \alpha-e_k & \beta & \sqrt{2}\beta \\ \beta & \alpha+\beta-e_k & 0 \\ \sqrt{2}\beta & 0 & \alpha+\beta-e_k \end{vmatrix} = 0 \qquad (15.145)$$

$$\begin{vmatrix} x & 1 & \sqrt{2} \\ 1 & x+1 & 0 \\ \sqrt{2} & 0 & x+1 \end{vmatrix} = 0$$

$$(x+1)(x^2+x-3) = 0$$

$$x = -1, \qquad -\tfrac{1}{2}\pm\tfrac{1}{2}\sqrt{13} \qquad (15.146)$$

Solution of the three remaining secular equations is left as an exercise (Problem 15.43).

The HMO coefficients are found by solving the appropriate sets of simultaneous equations. The resulting HMO's and energies for naphthalene are given in Table 15.3.

Electronic transitions   The predicted wave number for a transition between the highest occupied and lowest vacant HMO's of a conjugated hydrocarbon is

$$\frac{1}{\lambda} = \frac{|\beta|}{hc}\Delta x \qquad (15.147)$$

Table 15.3 *The Hückel MO's of naphthalene*

$1b_{3u}$: $\varphi_1 = 0.301(f_1+f_4+f_5+f_8)$
 $\qquad\quad +0.231(f_2+f_3+f_6+f_7)+0.461(f_9+f_{10})$ $\qquad\qquad e_1 = \alpha+2.303\beta$

$1b_{1g}$: $\varphi_2 = 0.263(f_1+f_4-f_5-f_8)+0.425(f_2+f_3-f_6-f_7)$ $\qquad e_2 = \alpha+1.618\beta$

$1b_{2g}$: $\varphi_3 = 0.400(f_1-f_4-f_5+f_8)$
 $\qquad\quad +0.174(f_2-f_3-f_6+f_7)+0.347(f_9-f_{10})$ $\qquad\qquad e_3 = \alpha+1.303\beta$

$2b_{3u}$: $\varphi_4 = 0.408(f_2+f_3+f_6+f_7)-0.408(f_9+f_{10})$ $\qquad\qquad e_4 = \alpha+\beta$

$1a_u$: $\varphi_5 = 0.425(f_1-f_4+f_5-f_8)+0.263(f_2-f_3+f_6-f_7)$ $\qquad e_5 = \alpha+0.618\beta$

$2b_{1g}$: $\varphi_6 = 0.425(f_1+f_4-f_5-f_8)-0.263(f_2+f_3-f_6-f_7)$ $\qquad e_6 = \alpha-0.618\beta$

$2b_{2g}$: $\varphi_7 = -0.408(f_2-f_3-f_6+f_7)+0.408(f_9-f_{10})$ $\qquad\qquad e_7 = \alpha-\beta$

$3b_{3u}$: $\varphi_8 = 0.400(f_1+f_4+f_5+f_8)$
 $\qquad\quad -0.174(f_2+f_3+f_6+f_7)-0.347(f_9+f_{10})$ $\qquad\qquad e_8 = \alpha-1.303\beta$

$2a_u$: $\varphi_9 = 0.263(f_1-f_4+f_5-f_8)-0.425(f_2-f_3+f_6-f_7)$ $\qquad e_9 = \alpha-1.618\beta$

$3b_{2g}$: $\varphi_{10} = 0.301(f_1-f_4-f_5+f_8)$
 $\qquad\quad -0.231(f_2-f_3-f_6+f_7)-0.461(f_9-f_{10})$ $\qquad\qquad e_{10} = \alpha-2.303\beta$

---

where Δx is the difference in x values [Eq. (15.114)] for the two MO's. For naphthalene, $\Delta x = 1.236$, and the observed $1/\lambda$ is $34,700\ cm^{-1}$. Choosing β to fit the observed wavelength for naphthalene (because of the orbital degeneracy of its first excited term, benzene is atypical), we find

$$|\beta|/hc = 28,000\ cm^{-1}, \qquad |\beta| = 3.48\ eV \qquad (15.148)$$

Comparison of predicted and observed longest-wavelength absorptions for a large number of benzenoid hydrocarbons shows only fair agreement with experiment, with many deviations of a few thousand cm^{-1}.

Improved agreement can be obtained if we fit the frequencies to a straight line that does not pass through the origin; i.e, we use

$$\frac{1}{\lambda} = \frac{1}{hc}|\beta|\,\Delta x + a \qquad (15.149)$$

A least-squares fit of benzenoid hydrocarbon data gives[28]

$$a = 8200\ cm^{-1}, \qquad |\beta|/hc = 21,900\ cm^{-1}, \qquad |\beta| = 2.72\ eV \quad (15.150)$$

These constants give a good fit to the data; the standard deviation is $600\ cm^{-1}$. (Of course, with two parameters instead of one, the agreement is bound to be improved. The fact that a semiempirical theory with several adjustable parameters gives a good fit to experimental data cannot be taken as overwhelming proof of the validity of the theory.)

We can partially justify the constant term in (15.149) as follows. The Hückel method neglects electron repulsions and therefore does not distinguish between singlet and triplet terms. Hence we should compare (15.147) with the energy difference between the ground state and the average of the energies of the singlet and triplet terms of the configuration with one electron excited to the lowest vacant HMO. The experimental frequencies are for singlet-singlet transitions. If we assume that the singlet-triplet splitting of the first excited configuration is reasonably constant for aromatic hydrocarbons, then $a = 8200\ cm^{-1} = 1.0\ eV$ can be interpreted as one-half this singlet-triplet splitting. Experimental values[29] for this splitting are available for many aromatic hydrocarbons; typically, the value of half the singlet-triplet splitting is 0.7 to 0.8 eV, in reasonable agreement with $a = 1.0\ eV$.

Delocalization energy and aromaticity   There are several properties of conjugated hydrocarbons that can be defined using HMO theory. We begin with *delocalization energy*. The corresponding VB term is *resonance energy*. The energy of the occupied π HMO (15.79) of ethylene is

$$\int \frac{1}{\sqrt{2}}(f_1 + f_2)^* \hat{H}^{eff} \frac{1}{\sqrt{2}}(f_1 + f_2)\,dv = \alpha + \beta \qquad (15.151)$$

There are two π electrons in this MO, and the total Hückel π-electron energy for

---

[28] E. Heilbronner and J. N. Murrell, *J. Chem. Soc.*, **1962**, 2611.
[29] G. B. Porter and M. W. Windsor, *Proc. Roy. Soc.*, **A245**, 238 (1958).

ethylene is $2\alpha + 2\beta$. For butadiene the total Hückel π-electron energy is

$$2(\alpha + 1.618\,\beta) + 2(\alpha + 0.618\beta) = 4\alpha + 4.472\,\beta \qquad (15.152)$$

If butadiene had two isolated double bonds, its π-electron energy would be twice that of ethylene, namely $4\alpha + 4\beta$. The effect of delocalization is to change the butadiene π-electron energy by

$$4\alpha + 4.472\,\beta - (4\alpha + 4\beta) = 0.472\beta \qquad (15.153)$$

Since β is negative, butadiene is stabilized by π-electron delocalization, and $0.472\,|\beta|$ is its delocalization energy.

For benzene the π-electron energy is

$$2(\alpha + 2\beta) + 4(\alpha + \beta) = 6\alpha + 8\beta \qquad (15.154)$$

as compared to $6\alpha + 6\beta$ for the π-electron energy of three isolated double bonds. The delocalization energy of benzene is $2|\beta|$. An "experimental" delocalization energy for benzene can be calculated as follows. The heat of hydrogenation of cyclohexene to cyclohexane is 28.6 kcal/mole. If benzene had three isolated double bonds, one would expect its heat of hydrogenation to be 3×28.6 kcal/mole $= 85.8$ kcal/mole. The observed value is only 49.8 kcal/mole, indicating that benzene is more stable by 36 kcal/mole than it would be if its double bonds were isolated. (A similar delocalization energy is arrived at by adding up the bond energies of six C—H bonds, three C—C bonds, and three C═C bonds and comparing this to the heat of formation of benzene from its atoms; see Problem 15.41.) Setting $2|\beta| = 36$ kcal/mole, we get $|\beta| = 18$ kcal/mole $= 0.8$ eV/molecule. This is far less than the value 2.72 eV/molecule $= 63$ kcal/mole found by spectroscopic observations [Eq. (15.150)]. Part of the discrepancy is explainable as follows. In the hydrogenation of cyclohexene, the carbon–carbon double-bond length becomes a single-bond length; the figure of 85.8 kcal/mole applies to the hydrogenation of a hypothetical molecule with three isolated double bonds and three single bonds, with alternating bond lengths. We must therefore also consider the strain energy needed to compress three single bonds and stretch three double bonds to the benzene bond length. Even with this correction (Problem 15.40), the thermochemical value of β differs sharply from the spectroscopic value. The difference is to be attributed to the crudity of the HMO method. One finds that a different value of β is required for each different physical property that is being considered. Moreover, the optimum values of β differ for chain and ring conjugated hydrocarbons.

The just-discussed traditional method of calculating Hückel delocalization energies of conjugated hydrocarbons by comparing a molecule's HMO π-electron energy E_π with $n_d(2\alpha + 2\beta)$, where n_d is the number of carbon–carbon double bonds, has serious shortcomings. This method predicts a substantial delocalization energy for the linear polyenes (15.95), whereas experiment shows the delocalization stabilization in these molecules to be rather small. For example, comparison of twice the heat of hydrogenation of 1-butene with the heat of hydrogenation of 1,3-butadiene gives the 1,3-butadiene delocalization energy as only 4 kcal/mole. Moreover, this method predicts substantial delocalization stabilization for certain cyclic polyenes that experiment reveals to be unstable, with no aromatic character.

(A cyclic conjugated polyene is said to be *aromatic* when it shows substantially more stability than a hypothetical structure in which the double bonds do not interact with one another and when it undergoes substitution, rather than addition, when treated with electrophilic reagents like Br_2.)

To produce more reliable predictions of aromaticity, Hess and Schaad (following a suggestion of Dewar) calculated delocalization (resonance) energies of cyclic hydrocarbons by comparing the compounds' Hückel-theory E_π with a value calculated for a hypothetical acyclic conjugated polyene with the same number and kinds of bonds as in a localized structure of the cyclic hydrocarbon. [B. A. Hess and L. J. Schaad, *J. Am. Chem. Soc.*, **93**, 305, 2413 (1971); **94**, 3068 (1972); **95**, 3907 (1973); B. A. Hess, L. J. Schaad, and C. W. Holyoke, *Tetrahedron*, **28**, 3657, 5299 (1972); L. J. Schaad and B. A. Hess, *J. Chem. Educ.*, **51**, 640 (1974).]

These workers found that the Hückel π-electron energies of noncyclic conjugated polyenes could be accurately calculated as $E_\pi \approx \sum_b n_b E_{\pi,b}$, where n_b is the number of bonds of a given type, $E_{\pi,b}$ is an empirical parameter, and the sum goes over all the types of carbon–carbon bonds. A conjugated hydrocarbon has three types of carbon–carbon single bonds and five types of double bonds, the types differing in the numbers of H atoms bonded to the carbons. In calculations of the delocalization energy, the terms involving α always cancel, so Hess and Schaad measured energies relative to α. In the following discussion, α will be omitted.

The Hess–Schaad values of $E_{\pi,b}$ for the various bond types are

$CH_2{=}CH$	$CH{=}CH$	$CH_2{=}C$	$CH{=}C$	$C{=}C$
2.0000β	2.0699β	2.0000β	2.1083β	2.1716β

$CH{-}CH$	$CH{-}C$	$C{-}C$
0.4660β	0.4362β	0.4358β

For 1,3-butadiene, $CH_2{=}CH{-}CH{=}CH_2$, these values give an E_π of $2(2.0000\beta) + 0.4660\beta = 4.466\beta$, as compared to the accurate Hückel value 4.472β in (15.152). Thus this approach makes the delocalization energy of noncyclic conjugated polyenes essentially zero. The resonance energy of a cyclic polyene is then found as the difference between the cyclic polyene's Hückel π-electron energy E_π and the quantity $\sum_b n_b E_{\pi,b}$. The method thus gives the resonance stabilization of cyclic polyenes relative to noncyclic polyenes.

For example, for C_6H_6 the Hess–Schaad resonance energy is found by comparing benzene's HMO π-electron energy (with the α term omitted) of 8β [Eq. (15.154)] with $3(2.0699\beta) + 3(0.4660\beta) = 7.6077\beta$. The Hess–Schaad resonance energy of benzene is therefore $0.3923|\beta|$. More significant is the resonance energy per π electron (REPE), which is $0.3923|\beta|/6 = 0.065|\beta|$. For cyclobutadiene we compare an HMO π-electron energy of 4β [Fig. 15.19 or Eq. (15.137)] with the quantity $2(2.0699\beta) + 2(0.4660\beta) = 5.0718|\beta|$; this gives a resonance energy of $-1.0718|\beta|$ and a REPE of $-0.268|\beta|$.

A compound with a substantially positive REPE value (greater than, say, $0.01|\beta|$) is predicted to be *aromatic*. A compound with a near-zero REPE is *nonaromatic*. A compound with a substantially negative REPE is predicted to be *antiaromatic*, being less stable than if its double bonds were isolated from one another. (Some antiaromatic hydrocarbons are cyclobutadiene and fulvalene.)

For the annulenes, C_nH_n, Hess–Schaad REPE versus n values are

n	4	6	8	10	12	14	16	18	20	22
REPE/$|\beta|$	-0.268	0.065	-0.061	0.026	-0.024	0.016	-0.011	0.012	-0.005	0.010

As n increases, the REPE becomes negligible, and the aromatic $4m + 2$ compounds and antiaromatic $4m$ compounds become nonaromatic.

As originally formulated the Hess–Schaad method did not apply to conjugated ions or radicals, but by modifying their approach somewhat, Hess and Schaad were able to treat ions and radicals also. [B. A. Hess and L. J. Schaad, *Pure Appl. Chem.*, **52**, 1471 (1980).]

Comparison with experiment shows that the Hess–Schaad method is highly successful in predicting aromaticity.

> Another method of calculating resonance energies of cyclic compounds has been developed [I. Gutman, M. Milun, and N. Trinajstic, *J. Am. Chem. Soc.*, **99**, 1692 (1977); J. Aihara, *J. Am. Chem. Soc.*, **98**, 2750 (1976)]. This method gives a *topological resonance energy* (TRE), found as the difference between E_π and $E_{\pi, ac}$, where $E_{\pi, ac}$ is an acyclic Hückel π-electron energy found using graph theory. In the sense used here, a graph is a figure consisting of points and lines joining some or all of the points and has no connection with the graph of a function. (See A. Balaban, ed., *Chemical Applications of Graph Theory*, Academic Press, New York, 1976.) Unfortunately, the TRE method fails in a significant number of cases; see I. Gutman, *Theor. Chim. Acta*, **56**, 89 (1980).

π-electron charges and bond orders Another defined quantity is the *π-electron charge*. The probability density for an electron in the HMO (15.104) is

$$|\varphi_i|^2 = \sum_r \sum_s c_{ir}^* c_{is} f_r^* f_s \tag{15.155}$$

The normalization requirement (15.112) reads

$$\sum_r |c_{ir}|^2 = 1 \tag{15.156}$$

Since the HMO coefficients satisfy (15.156), it is natural to say that an electron in the MO φ_i has the probability $|c_{ir}|^2$ of being in the vicinity of the rth carbon atom. If $n_i (= 0, 1, \text{ or } 2)$ is the number of electrons in the MO φ_i, then the total π-electron charge q_r in the region of carbon-atom r is defined to be

$$q_r \equiv \sum_i n_i |c_{ir}|^2 \tag{15.157}$$

where the sum is over the π MO's. [q_r is often called the "π-electron density"; this name is misleading, since q_r is neither a charge density (which has dimensions of charge/volume) nor a probability density (which has dimensions of 1/volume); rather, q_r is a pure number that gives the approximate number of π electrons in the vicinity of carbon atom r.]

For butadiene

$$q_1 = 2|c_{11}|^2 + 2|c_{21}|^2 = 2(0.372)^2 + 2(0.602)^2 = 1.000$$

$$q_2 = 2|c_{12}|^2 + 2|c_{22}|^2 = 1.000 \tag{15.158}$$

(We shall see later in this section that butadiene is an example of an alternant hydrocarbon and that $q_r = 1$ for every carbon of a neutral alternant hydrocarbon.)

q_r values can sometimes be used to predict the position(s) at which substitution occurs in a conjugated cyclic hydrocarbon. An *electrophilic* ("electron-loving") substitution reaction occurs by attack by a positive ion (e.g., $NO_2{}^+$) or by the positive end of a polarized species. If we neglect variations in σ-electron distribution, we expect electrophilic substitution to occur preferentially at positions where the π-electron charge q_r is greatest. For azulene (Fig. 15.21) the HMO q_r is greatest at positions 1 and 3, and one finds that electrophilic substitution by Cl_2 occurs predominantly at these positions. In contrast, nucleophilic substitution (which involves attack by an electron-rich, positive-charge-seeking species) occurs predominantly at carbons 4 and 8 of azulene, which have the smallest q_r values. Although the q_r values correctly predict the positions of electrophilic and nucleophilic attack in azulene, they fail in fluoranthene. Here the q_r values lie in the order $8 > 2 > 7 > 3 > 1$, but electrophilic attack occurs in the order $3 > 8 > 7 > 1 > 2$, where the numbers are the carbon positions (see *Murrell and Harget*, pp. 77–78).

Because of failures like this and because the method cannot be used in the many systems where the q_r values are all 1, many other reactivity indices have been devised using Hückel theory; see *Murrell, Kettle, and Tedder*, Chapter 17; *Streitwieser*, Chapters 11–15; *Salem*, Chapter 6.

We saw in the population-analysis discussion of Section 15.5 that for a real MO φ_i expanded as the linear combination $\sum_r c_{ir} f_r$ of AO's f_r, the quantity $2c_{ir}c_{is}S_{rs}$ in Eq. (15.31) (where S_{rs} is the overlap integral) is a reasonable measure of the contribution of an electron in MO φ_i to the bonding overlap between AO's f_r and f_s. In the HMO method, overlap integrals are neglected, so we cannot use this population-analysis expression. The carbon–carbon bond distances in a conjugated compound are all reasonably similar, so we expect the $C2p\pi$-$C2p\pi$ overlap integral S_{rs} to have similar values for all pairs of bonded carbons; for nonbonded carbons, S_{rs} should be quite small. Since S_{rs} is approximately constant for bonded atoms, we can ignore the factor S_{rs} (and the constant factor 2) and take $c_{ir}c_{is}$ as the contribution of an electron in the real HMO φ_i to the π-electron bonding between bonded atoms r and s. If the MO φ_i is complex rather than real, one finds on integrating $|\varphi_i|^2$ to produce the equation corresponding to (15.31) that the quantity $\frac{1}{2}(c_{ir}^* c_{is} + c_{is}^* c_{ir})$ occurs instead of $c_{ir}c_{is}$. Coulson therefore defined the *π-electron* (or *mobile*) *bond order* p_{rs} for the bond between bonded atoms r and s as

$$p_{rs} \equiv \sum_i n_i \tfrac{1}{2}(c_{ir}^* c_{is} + c_{is}^* c_{ir}) \tag{15.159}$$

where the sum is over the π MO's. When the coefficients are all real, (15.159) reduces to $p_{rs} = \sum_i n_i c_{ir} c_{is}$. (That this definition is reasonable is indicated by the fact that it gives $p = 1$ for the π bond in ethylene.)

Addition of the σ electrons' single bond gives the *total bond order* P_{rs} as

$$P_{rs} = 1 + p_{rs} \qquad (15.160)$$

For butadiene we have $p_{12} = 2(0.372)(0.602) + 2(0.602)(0.372) = 0.894$ and $p_{23} = 2(0.602)(0.602) + 2(0.372)(-0.372) = 0.447$. The total bond orders are

$$CH_2 \overset{1.894}{\longrightarrow} CH \overset{1.447}{\longrightarrow} CH \overset{1.894}{\longrightarrow} CH_2 \qquad (15.161)$$

The central bond has some double-bond character, and the end bonds have some single-bond character. (In VB theory, this is explained by contributions from such resonance structures as $\bar{C}H_2$—CH=CH—$\overset{+}{C}H_2$.) The sum of the bond orders is 5.235, exceeding 5. [This is because the π-electron energy of butadiene exceeds that of two isolated double bonds—see Eq. (15.164).] For benzene each carbon–carbon bond order is found to be $5/3 = 1.667$.

Naturally, we expect a relation between the bond order and the bond length R_{rs} for carbon–carbon bonds. The simplest assumption is a linear relation:

$$R_{rs} = a + bP_{rs} \qquad (15.162)$$

where a and b are constants. The π-electron MO coefficients for ethylene and for benzene are determined completely by symmetry and are independent of the Hückel approximations. We therefore use the bond orders and bond lengths of ethylene (2; 1.335 Å) and benzene (5/3; 1.397 Å) to find a and b; we get

$$R_{rs} = [1.707 - 0.186P_{rs}] \text{ Å} = [1.521 - 0.186p_{rs}] \text{ Å} \qquad (15.163)$$

Equation (15.163) works reasonably well. Thus the predicted naphthalene 1–2, 2–3, 1–9, and 9–10 bond lengths (in Å), as compared to the experimental values (in parentheses), are 1.386 (1.361), 1.409 (1.421), 1.418 (1.425), and 1.425 (1.410), respectively. Note that the rings are not regular hexagons. In phenanthrene six of the nine predicted bond lengths are correct to 0.01 Å, but the errors in the other three are 0.02, 0.04, and 0.05 Å.

Ab initio, all-electron SCF MO calculations on naphthalene and azulene have been carried out [R. J. Buenker and S. D. Peyerimhoff, *Chem. Phys. Lett.*, **3**, 37 (1969); these monumental calculations required the evaluation of nearly 10^9 two-electron integrals]. The SCF MO bond orders for naphthalene correctly predict the observed order of increasing bond lengths (which the HMO bond orders do not quite do); however, the calculated SCF MO azulene bond orders bear little relation to the observed azulene bond lengths. Clearly other factors besides conjugation are important in determining bond lengths.[30]

A much-investigated question is whether the carbon–carbon bond lengths in very large conjugated linear and cyclic polyenes ($C_{2n}H_{2n+2}$ and $C_{4m+2}H_{4m+2}$ for m and n large) are equal or alternate in length. This question is still unsettled; see M. Kertesz et al., *J. Chem. Phys.*, **67**, 1180 (1977).

---

[30] For discussions on carbon–carbon bond lengths, see D. R. Lide, Jr., *Tetrahedron*, **17**, 125 (1962); B. P. Stoicheff, *Tetrahedron*, **17**, 135; L. S. Bartell, *Tetrahedron*, **17**, 177; E. B. Wilson, Jr., *Tetrahedron*, **17**, 191.

The Hückel π-electron energy E_π is related to the π bond orders p_{rs} and π-electron densities q_r; in fact (Problem 15.44),

$$E_\pi = \alpha \sum_r q_r + 2\beta \sum_{s-r} p_{rs} \tag{15.164}$$

where the first sum is over the carbon atoms and where the second is over the carbon–carbon bonds. For example, for butadiene

$$E_\pi = \alpha(1+1+1+1) + 2\beta(0.894 + 0.447 + 0.894) = 4\alpha + 4.47\beta$$

which agrees with (15.152).

For conjugated species with partly filled degenerate MO's, there is an ambiguity in the q_r and p_{rs} values if the real forms of the MO's are used. For example, for $C_6H_6{}^-$, if we put the unpaired electron in $\varphi_{4,\text{real}}$ of (15.136), then this electron contributes $\frac{1}{3}$ to q_1, $\frac{1}{12}$ to q_2, $\frac{1}{12}$ to q_3, $\frac{1}{3}$ to q_4, $\frac{1}{12}$ to q_5, and $\frac{1}{12}$ to q_6, but if we put the odd electron in $\varphi_{5,\text{real}}$, then this electron contributes 0 to q_1 and q_4 and $\frac{1}{4}$ to each of q_2, q_3, q_5, and q_6. This ambiguity can be avoided by using the complex MO's, which have the symmetry of the molecule. Putting the odd electron in either φ_4 or φ_5 of (15.135), we get a contribution of $\frac{1}{6}$ to each q. (Alternatively, we can average the contributions from $\varphi_{4,\text{real}}$ and $\varphi_{5,\text{real}}$ to give a $\frac{1}{6}$ contribution at each carbon.)

Free valence   Another defined quantity is Coulson's *free-valence index* F_r, for carbon atom r in a planar conjugated organic compound:

$$F_r \equiv 3 + \sqrt{3} - \sum_s{}' P_{rs} = \sqrt{3} - \sum_s{}' p_{rs} \tag{15.165}$$

where the sum is over all atoms (including hydrogens) bonded to carbon atom r. The quantity $3 + \sqrt{3} = 4.732$ is the value of $\sum' P_{rs}$ for the central carbon of the diradical $C(CH_2)_3$, whose central atom has the largest value for this sum of any trigonally bonded carbon. [The central carbon of the diradical $HC{=}C{=}CH$ has a larger value for this sum, namely $2 + 2^{3/2} = 4.828$; see Problem 15.47 and C. Finder, *J. Org. Chem.*, **32**, 1672 (1967).] For butadiene

$$\sum_s{}' P_{1s} = 1 + 1 + 1.89 = 3.89$$

$$F_1 = F_4 = 4.73 - 3.89 = 0.84 \tag{15.166}$$

$$F_2 = F_3 = 4.73 - 4.34 = 0.39$$

F_r is a measure of the unused bonding power of atom r, and serves as a rough indicator of the susceptibility of a given atom to attack by a neutral free radical; the end carbons in butadiene are attacked preferentially by free radicals. A study of the attack of the free radical $\cdot CCl_3$ on several aromatic hydrocarbons found that a plot of the log of the rate constant versus the free valence of the atom with the largest free valence gave a nearly straight line [E. C. Kooyman and E. Farenhorst, *Trans. Faraday Soc.*, **49**, 58 (1953)].

Alternant hydrocarbons   An *alternant hydrocarbon* is a planar conjugated hydrocarbon in which the carbon atoms can be divided into a starred set and an unstarred set, with starred carbons bonded only to unstarred carbons and vice versa (Fig. 15.21). All planar conjugated hydrocarbons are alternants except those containing a ring with an odd number of carbons. Alternants are divided into even-alternants (e.g., benzene) and odd-alternants (e.g., benzyl) according to whether n_C is even or odd.

The equations determining the HMO coefficients of the orbital φ_i with energy $e_i = \alpha - x_i\beta$ are

$$(\alpha - e_i)c_{ir} + \beta \sum_{s \to r} c_{is} = 0$$

$$x_i c_{ir} + \sum_{s \to r} c_{is} = 0, \qquad r = 1, \ldots, n_C \tag{15.167}$$

where the summation is over carbons bonded to carbon r. Let the hydrocarbon be an alternant. From the definition of an alternant, it follows that if r is a starred atom, then s in (15.167) is unstarred, and vice versa. We shall use the left superscripts $*$ and $\circ$ to denote coefficients of starred and unstarred carbons, respectively. For those atoms r that are unstarred, we rewrite (15.167) as

$$(-x_i)(-{}^\circ c_{ir}) + \sum_{s \to r} ({}^* c_{is}) = 0 \tag{15.168}$$

For r starred we multiply (15.167) by -1 to get

$$(-x_i)({}^* c_{ir}) + \sum_{s \to r} (-{}^\circ c_{is}) = 0 \tag{15.169}$$

If we let

$$x_j = -x_i, \qquad {}^* c_{jr} = {}^* c_{ir}, \qquad {}^\circ c_{jr} = -{}^\circ c_{ir}, \qquad r = 1, \ldots, n_C \tag{15.170}$$

we can combine (15.168) and (15.169) into the single equation

$$x_j c_{jr} + \sum_{s \to r} c_{js} = 0 \tag{15.171}$$

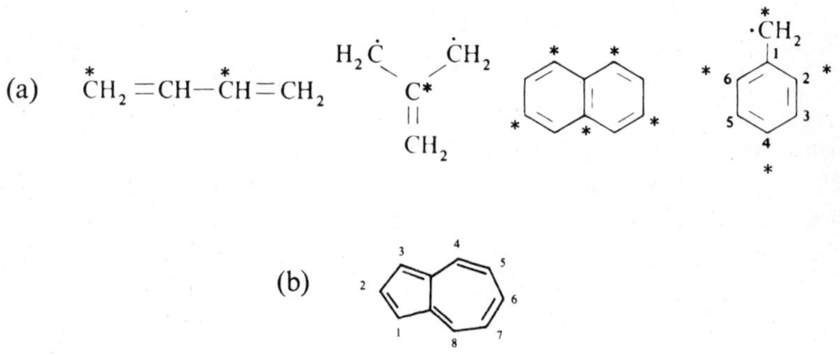

Figure 15.21 (a) *Some alternant hydrocarbons.* (b) *Azulene, a nonalternant hydrocarbon.*

Equation (15.171) has the same form as (15.167), and thus (15.171) shows that for each alternant HMO φ_i with x value x_i, there is an HMO φ_j with x value $-x_i$ and coefficients the same as φ_i except that the unstarred-atom coefficients are changed in sign. The HMO's of an alternant hydrocarbon are *paired*, with pairs having energies $\alpha + x\beta$ and $\alpha - x\beta$. The coefficients of an antibonding HMO of an alternant hydrocarbon can be obtained simply by changing the signs of the unstarred-atom coefficients in the paired bonding HMO. These results may be checked for butadiene, benzene, and naphthalene. For an odd-alternant hydrocarbon, the HMO with energy α is paired with itself, and (15.170) implies that the coefficients of the unstarred atoms are zero for this HMO.

The preceding derivation may bother the reader in that we found both x_j and the c_{jr}'s from the simultaneous equations (15.167) [corresponding to (15.105)], whereas the usual procedure is to find x_j from the determinantal secular equation (15.106) and then solve the simultaneous equations (15.105) for the coefficients. Actually the procedure of finding both x_j and the c_{jr}'s from (15.105) is perfectly valid. The only values of x_j that allow a nontrivial solution of (15.105) are those that satisfy the secular equation (15.106). (See Section 8.4.) Having found an x_j value that allows the nontrivial solution (15.170) for the c_{jr}'s, we can be sure it satisfies (15.106). Equation (15.106) is simply a *device* that can be dispensed with if we are able to solve (15.105) another way.

Benzyl (Fig. 15.21) is an odd-alternant with seven conjugated carbons and hence seven HMO's; from the above discussion, three MO's are bonding, three are antibonding, and one is nonbonding ($e_i = \alpha$). The bonding MO's are filled with six π electrons, and the unpaired electron is in the nonbonding MO. We can find the nonbonding MO coefficients without solving the secular equation. With $x_i = 0$, Eq. (15.167) becomes

$$\sum_{s \to r} c_{is} = 0, \qquad r = 1, \ldots, n_C \qquad \textbf{(15.172)}$$

where the sum is over carbons bonded to carbon r. The CH_2 carbon of benzyl is bonded to carbon number 1 only, and (15.172) gives $c_{i1} = 0$ for the nonbonding MO. Since C_2 is bonded to C_1 and C_3, we have $c_{i3} = 0$; similarly, $c_{i5} = 0$. The remaining four coefficients are readily found—see Problems 15.45 and 15.63.

For the ground state of a *neutral* alternant hydrocarbon, all the π-electron charges q_r are 1. (See *Pilar*, p. 600.)

Heteroatomic conjugated molecules   So far, we have applied the HMO method to hydrocarbons only. For planar conjugated molecules that involve π bonding to noncarbon atoms, the Coulomb and bond integrals for the heteroatoms must be modified from the carbon values. For heteroatoms X and Y, we write

$$\alpha_X = \alpha_C + h_X \beta_{CC}$$
$$\beta_{XY} = k_{XY} \beta_{CC} \qquad \textbf{(15.173)}$$

where h_X and k_{XY} are certain constants. The best values for these constants vary, depending on which molecular property is being considered, and the values used are based on a mixture of theory and guesswork. One set of values (*Streitwieser,* Chapter 5) is

Atom	—N=	—N̈—	=N̈—	—O—	=O	—F	—Cl	—Br
h_X	0.5	1.5	2	2	1	3	2	1.5

(15.174)

Bond	C—N̈—	aromatic CN	C=N̈—	N—O—	C—O—	C=O	C—F	C—Cl	C—Br
k_{XY}	0.8	1	1	0.7	0.8	1.0	0.7	0.4	0.3

These values should not be taken too seriously—e.g., different workers recommend values for $h_{=O}$ ranging from 0.15 to 2.

Consider pyridine and pyrrole as examples. Pyridine

pyridine   pyrrole

has six π electrons, one of which is contributed by nitrogen; pyrrole has six π electrons, two of which are contributed by nitrogen. If we use the values (15.174), the HMO secular determinant for pyridine is the same as (15.126) for benzene, except that the element x in the first row and first column is replaced by $x + 0.5$. For pyrrole the HMO secular equation is

$$
\begin{vmatrix}
x+1.5 & 0.8 & 0 & 0 & 0.8 \\
0.8 & x & 1 & 0 & 0 \\
0 & 1 & x & 1 & 0 \\
0 & 0 & 1 & x & 1 \\
0.8 & 0 & 0 & 1 & x
\end{vmatrix} = 0
$$

We might also use different values of β for the carbon–carbon single and double bonds in pyrrole; Streitwieser recommends

$$\beta_{C-C} = 0.9\,\beta_{CC}, \qquad \beta_{C=C} = 1.1\,\beta_{CC} \qquad \textbf{(15.175)}$$

where β_{CC} is for an aromatic carbon–carbon bond (e.g., benzene).

The HMO method can be applied to the ions (15.100); however, the FE MO method is simpler to use, and we omit the HMO calculation—see Problem 15.52. [The molecule (15.100) is not planar. However, the out-of-plane methyl protons are but a slight perturbation on an otherwise planar molecule, and the σ–π classification holds to a very good approximation.]

For molecules with heteroatoms, Eq. (15.164) becomes

$$E_\pi = \sum_{\text{atoms}} q_r \alpha_r + 2 \sum_{\text{bonds}} p_{rs} \beta_{rs} \tag{15.176}$$

Inclusion of overlap   Aside from using the one-electron Hamiltonian (15.102), perhaps the most serious approximation of the simple HMO method is that of taking all overlap integrals equal to zero. Wheland proposed using a nonzero value S for the overlap integral of carbons bonded together. Inclusion of overlap replaces each element β in the HMO secular equation (15.106) with $\beta - Se_i$. With overlap the benzene HMO secular equation becomes

$$\begin{vmatrix} w & 1 & 0 & 0 & 0 & 1 \\ 1 & w & 1 & 0 & 0 & 0 \\ 0 & 1 & w & 1 & 0 & 0 \\ 0 & 0 & 1 & w & 1 & 0 \\ 0 & 0 & 0 & 1 & w & 1 \\ 1 & 0 & 0 & 0 & 1 & w \end{vmatrix} = 0 \tag{15.177}$$

where

$$w \equiv \frac{\alpha - e_i}{\beta - Se_i} \tag{15.178}$$

$$e_i = \alpha - \frac{w(\beta - S\alpha)}{1 - Sw} = \alpha - \frac{w}{1 - Sw}\gamma \tag{15.179}$$

$$\gamma \equiv \beta - S\alpha \tag{15.180}$$

Since (15.177) is identical in form to (15.126), the values of w are the same as the values (15.127) for x, and no additional calculation is required. Provided we use a common value of S for any two bonded carbons in a given molecule, this situation holds for any planar conjugated hydrocarbon. To get the HMO energy levels with overlap included, we use Eq. (15.179), where w takes on the same values as found for x when we solve the HMO secular equation with overlap omitted. The HMO coefficients in the Wheland approximation differ only in the normalization constants from those obtained with zero overlap (Problem 15.57).

To estimate S, we use the following formula (Problem 15.24) for the overlap integral between two $2p\pi$ STO's with orbital exponent ζ:

$$S_{2p\pi-2p\pi} = e^{-\zeta R}\left[1 + \zeta R + \frac{2\zeta^2 R^2}{5} + \frac{\zeta^3 R^3}{15} \right] \tag{15.181}$$

where R is the internuclear separation in bohrs. Slater's rules give $\zeta = 1.625$ for a carbon $2p$ AO. (Since Slater's ζ is the effective nuclear charge divided by n, this value corresponds to $Z_{\text{eff}} = 3.25$.) We have

$R/\text{Å}$	1.20	1.335	1.397	1.53
S	0.338	0.272	0.246	0.195

$$\tag{15.182}$$

At the benzene bond distance 1.397 Å, the overlap integral is 0.25.

The Wheland orbital energies for benzene (with S taken as 0.25) are

$$\alpha + 1.33\gamma, \qquad \alpha + 0.80\gamma, \qquad \alpha + 0.80\gamma,$$
$$\alpha - 1.33\gamma, \qquad \alpha - 1.33\gamma, \qquad \alpha - 4.0\gamma$$

Inclusion of overlap destroys the symmetric spacing of the energy levels about α that holds for alternant hydrocarbons in the simple HMO theory.

Let w_{lv} and w_{ho} be the w values for the lowest vacant and highest occupied π MO's. The transition frequency between these two Wheland MO's is

$$\frac{hc}{\lambda} = \frac{\gamma w_{lv}}{1 - S w_{lv}} - \frac{\gamma w_{ho}}{1 - S w_{ho}}$$

$$\frac{1}{\lambda} = \frac{\gamma}{hc} \frac{(w_{lv} - w_{ho})}{[1 + S^2 w_{lv} w_{ho} - S(w_{lv} + w_{ho})]} \qquad (15.183)$$

For alternant hydrocarbons $w_{lv} + w_{ho} = 0$, and (15.183) becomes

$$\frac{1}{\lambda} = \frac{|\gamma|}{hc} \times \frac{\Delta w}{1 - \frac{1}{4}S^2 (\Delta w)^2} \qquad (15.184)$$

where the Δw values are identical to the Δx values calculated without overlap. Since most Δw values are close to 1, we have $\frac{1}{4}S^2 (\Delta w)^2 \approx 1/64 \ll 1$, and (15.184) is very nearly the same formula as (15.147), except that the proportionality constant is interpreted as $\beta - S\alpha$, rather than β. Also, since the overlap correction is roughly the same ($\approx 64/63$) for most aromatic hydrocarbons, it is largely taken care of by using an empirical value of γ. Inclusion of overlap would thus give only slight changes in predicted transition frequencies.

With inclusion of overlap, the definitions (15.157) and (15.159) for the π-electron charges and bond orders must be modified. With the modified definitions, it can be shown that inclusion of overlap gives the *same* charges and bond orders for hydrocarbons as found when overlap is neglected.[31] (This result does not hold for molecules with heteroatoms.)

Even though inclusion of overlap is not difficult, simple LCAO calculations on π-electron systems usually omit overlap. This gives no error in calculated hydrocarbon bond orders and charges and but little change in the predicted longest-wavelength transitions of alternant hydrocarbons. (However, inclusion of overlap does give better agreement between the spectroscopic and thermochemical values of γ than is found between the corresponding values of β when overlap is omitted; see Problem 15.50.)

Summary   Because of the simplicity of the method, carrying out HMO calculations became a favorite pastime of organic chemists, and the results of HMO calculations

---

[31] B. H. Chirgwin and C. A. Coulson, *Proc. Roy. Soc.*, **A201**, 196 (1950).

have been tabulated for hundreds of compounds.[32] The HMO method has been widely used to rationalize and predict the properties and reactivities of conjugated compounds. In view of the crudity of the HMO approximations, the fact that the method works as well as it does is surprising and not yet fully explained.

The development of semiempirical theories (Sections 15.12 and 15.14) more sophisticated than the HMO theory led some workers to argue that "Hückel theory has largely outlived its usefulness" (*Murrell and Harget*, p. v). However, these more sophisticated theories have their failings (as we shall see), and the success of the Hess–Schaad use of Hückel theory to predict aromaticity indicates that HMO theory is still useful, "especially on a qualitative level as a guide ... in planning and interpreting experiments" (N. Trinajstic in *Segal*, Part A, Chapter 1).

15.12 *THE PARISER–PARR–POPLE METHOD*

Although the Hückel theory can be used to predict the longest-wavelength bands of aromatic hydrocarbons, it would be hopeless to try to use HMO theory to predict the complete electronic spectrum of an aromatic hydrocarbon. For example, Hückel theory, which neglects interelectronic repulsions, gives no separation between singlet and triplet electronic terms arising from the same configuration. Experimentally, separations of one or two eV are observed between such terms. Several semiempirical π-electron theories have been developed to take electron repulsion into account and thereby improve on the Hückel method.

The most widely used of these theories is the *Pariser–Parr–Pople* (PPP) *method*. Here, the π-electron Hamiltonian (15.86) including electron repulsions is used, and the π-electron wave function is written as an antisymmetrized product of spin-orbitals. The π MO's are taken as LCAO's, with $2p\pi$ carbon orbitals used for hydrocarbons. The Roothaan equations are used to find SCF π MO's. However, instead of attempting a rigorous SCF calculation, the Pariser–Parr–Pople method makes certain approximations for some integrals and uses experimental data to evaluate other integrals. As in Hückel theory, overlap is neglected:

$$S_{rs} = \langle f_r(1)|f_s(1) \rangle = \delta_{rs} \tag{15.185}$$

where δ_{rs} is the Kronecker delta. The bond integral

$$\beta_{rs}^{\text{core}} = \langle f_r(1)|\hat{H}_{\text{core}}(1)|f_s(1) \rangle \tag{15.186}$$

is taken as an empirical parameter for two bonded atoms; for nonbonded atoms β_{rs}^{core} is assumed to vanish. The integrals

$$\alpha_r^{\text{core}} = \langle f_r|\hat{H}_{\text{core}}|f_r \rangle \tag{15.187}$$

---

[32] C. A. Coulson and A. Streitwieser, Jr., *Dictionary of π-Electron Calculations*, Freeman, San Francisco, 1965; A. Streitwieser, Jr., and J. I. Brauman, *Supplemental Tables of Molecular Orbital Calculations*, Pergamon Press, New York, 1965; Heilbronner and Straub, *Hückel Molecular Orbitals*, Springer-Verlag, New York, 1966.

are allowed to differ for different atoms and are evaluated with a certain approximate formula.

We also have electron repulsion integrals:

$$\int\int f_r^*(1)f_s(1)\frac{1}{r_{12}}f_t^*(2)f_u(2)\,dv_1\,dv_2 \equiv (rs|tu) \tag{15.188}$$

The widely used notation of (15.188) should not be misinterpreted as an overlap integral. Other notations, some mutually contradictory, are used for the electron repulsion integrals, so it is wise to always check an author's definition.

Consistent with (15.185), the Pariser–Parr–Pople method makes the assumption of *zero differential overlap* (ZDO):

$$f_r(1)f_s(1)\,dv_1 = 0 \qquad \text{for } r \neq s \tag{15.189}$$

From this approximation it follows that

$$(rs|tu) = \delta_{rs}\delta_{tu}(rr|tt) = \delta_{rs}\delta_{tu}\gamma_{rt} \tag{15.190}$$

where the notation of (15.188) is used and where γ_{rt} is defined as $(rr|tt)$. Thus the method ignores many (but not all) of the electron repulsion integrals, thereby greatly simplifying the calculation. In particular, all three- and four-center electron repulsion integrals are ignored. The integrals γ_{rt} are often treated as empirical parameters, rather than being evaluated theoretically.

The ZDO approximation is at first sight rather drastic. However, a partial theoretical justification for it can be given by reinterpreting the AO's used to express the MO's as orthogonalized AO's (rather than ordinary AO's). Each orbital in a set of orthogonalized AO's is a linear combination of ordinary AO's, the coefficients being chosen so that the members of the set are mutually orthogonal. There are many ways to choose the linear combinations to produce an orthogonal set. One approach is to make the orthogonalized AO's (OAO's) resemble the ordinary AO's as much as possible by minimizing the sum of the squares of the deviations of the OAO's from the ordinary AO's; one minimizes $\sum_i \int |\chi_{i,OAO} - \chi_i|^2\,dv$, where the sum goes over the set of AO's and where χ_i and $\chi_{i,OAO}$ are the ordinary and the orthogonalized AO's. This produces what are called *symmetrically orthogonalized* (or Löwdin) AO's (see *Pilar*, Section 17-7; *Parr*, pp. 51–52). One finds that in each symmetrically orthogonalized AO, the coefficient of one ordinary AO is substantially greater than the coefficients of the other ordinary AO's, so the symmetrically orthogonalized AO's are not drastically different from the ordinary AO's. One finds that with symmetrically orthogonalized AO's, the integrals not neglected in the ZDO approximation undergo only small changes in value as compared to their values with ordinary AO's; these changes in value can be partly allowed for by the fact that many integrals are taken as empirical parameters. Moreover, with symmetrically orthogonalized AO's, the electron repulsion integrals neglected in the ZDO approximation are generally found to be quite small (and all overlap integrals are, of course, zero). For details, see the references cited on p. 27 of *Murrell and Harget*.

With these approximations, the Roothaan equations are solved iteratively until self-consistency is attained. For alternant hydrocarbons the result $q_r = 1$ and the pairing properties of MO energies and coefficients hold in the PPP theory. Configuration interaction may be included to improve the results. The PPP treatment gives a good account of the electronic spectra of many, but not all, aromatic hydrocarbons.[33]

15.13 π-ELECTRON SYSTEMS

The literature on quantum-mechanical studies of π-electron systems is huge, mainly because of the relative ease with which semiempirical treatments of such molecules can be carried out. Part of the great interest results from the biological importance of conjugated molecules.

The five bases (adenine, cytosine, guanine, thymine, and uracil) that form the genetic code[34] in DNA and RNA are conjugated π-electron systems, as are many other biomolecules.

The carotenoid pigments contain systems of conjugated double bonds: β-carotene (the orange pigment in carrots) has eleven conjugated double bonds. The body cleaves β-carotene to produce vitamin A, which has five conjugated double bonds; vitamin A is converted by the body to 11-cis-retinal (retinene), which also has five conjugated double bonds; retinene combines with the protein opsin to form the photosensitive pigment rhodopsin found in the retinas of animals. When light strikes rhodopsin, it isomerizes 11-cis-retinal to all-trans-retinal, which does not "fit" the protein opsin and dissociates from it.

Most organic dyes owe their color to π-electron transitions. If there is only one strong absorption band in the visible region, the observed color of a compound is related to the wavelength of the electronic absorption maximum as follows:

$\lambda/\text{Å}$	4000	4250	4500	4900	5300	5900	6400	7300
Observed Color	greenish-yellow	yellow	orange	red	violet	blue	blue-green	green

Molecules related to porphin (which has a conjugated π-electron system) are of widespread biological occurrence; substituted porphins are called porphyrins. The protein hemoglobin contains four heme groups; each heme group is an iron complex of a porphyrin. Chlorophyll is a magnesium complex of a porphyrin; the presence of low-lying vacant MO's, which is characteristic of conjugated π-electron systems, allows chlorophyll to absorb visible light.

A. and B. Pullman have attributed the biological importance of conjugated molecules to two factors: the extra stability associated with electron delocalization and the functional advantages conferred by conjugated π-electron systems.

Löwdin suggested that naturally occurring mutations may be caused by

---

[33] For further discussion of the method, see *Parr*, Chapter III; *Pilar*, Chapter 19; *Murrell and Harget*, Chapter 2; *Offenhartz*, Chapter 11.

[34] J. Watson, *Molecular Biology of the Gene*, 3rd ed., Benjamin, New York, 1976; *The Double Helix*, Atheneum, New York, 1968.

tunneling of one or more protons of DNA hydrogen bonds from a base on one strand to the paired base on the other strand.

Clementi and co-workers did an ab initio all-electron SCF MO calculation on the guanine–cytosine $(C_5N_5H_5O–C_4N_3H_5O)$ base pair of DNA. [E. Clementi, J. Mehl, and W. von Niessen, *J. Chem. Phys.*, **54**, 508 (1971).] Computations were done for twenty-seven different geometries to determine the potential-energy function for motion of hydrogen atoms in the hydrogen bonds. For this 136-electron 29-nuclei system, the basis set used was 334 GTO's; 7×10^{10} two-electron repulsion integrals were evaluated, and the computation took eight days of computer time. Unfortunately, despite the enormous size of this computation, the basis set used was not large enough to approach the Hartree–Fock limit, and definitive conclusions about Löwdin's ideas could not be reached.

The relation between electronic structure and the carcinogenicity of certain aromatic hydrocarbons has been studied using π-electron theory, but definitive conclusions have not been reached.[35]

Significant biological applications of quantum chemistry are on the increase. A conference on quantum chemistry in the biomedical sciences [*Ann. N.Y. Acad. Sci.*, **367**, 1–552 (1981)] included papers on applications of quantum chemistry to drug design, enzyme–substrate and drug–receptor interactions, conformations of macro-molecules, and solvation of biological molecules.

For nonconjugated closed-subshell molecules (e.g., H_2O), we can view the electrons as residing either in the delocalized canonical MO's or in the energy-localized bond, inner-shell, and lone-pair orbitals. The latter picture allows one to accurately approximate the molecular binding energy as a sum of bond energies. For conjugated molecules (e.g., 1,3-butadiene, benzene), the molecular binding energy exceeds the sum of bond energies by the delocalization energy. We therefore expect that the energy-localized bonding MO's in conjugated systems are not as well localized as those in nonconjugated systems. Newton and Switkes calculated the energy-localized MO's for *s-trans*-butadiene and benzene, starting from all-electron minimal-basis-set SCF canonical MO's. [M. D. Newton and E. Switkes, *J. Chem. Phys.*, **54**, 3179 (1971).] In butadiene they found the localized double bonds to be the bent banana bonds (as Kaldor found for ethylene), rather than $\sigma–\pi$ bonds. They defined a percentage of delocalization d of an energy-localized bond MO that measures the degree to which orbitals centered on atoms other than the bonding pair of atoms contribute to the bond MO. For the butadiene energy-localized double-bond MO's, d is $8\frac{1}{2}$ percent, only slightly more than the 6 percent value for the ethylene localized double-bond MO's. Thus conjugation in butadiene leads to only a slight increase in delocalization of the localized double-bond orbitals.

For benzene the energy-localization procedure led to two equivalent solutions, in each of which single and double bonds alternate (as in the Kekulé structures), with each double bond composed of two bent banana bonds. (The total electron density in a localized MO wave function must have the same symmetry as the density calculated from canonical MO's; for a Kekulé-like localized benzene wave

---

[35] B. Pullman, *Int. J. Quantum Chem.*, **16**, 669 (1979).

function, the tails of the localized double-bond MO's are sufficiently large to maintain $\mathscr{D}_{6h}$ symmetry.) In benzene each energy-localized double-bond MO was found to have $d = 19$ percent, indicating large contributions from AO's of atoms other than the bonding pair. The high degree of delocalization is, of course, expected. Another significant point was that for benzene, the bent banana double bonds were only slightly preferred (in terms of energy localization) to σ–π double-bond MO's.

15.14 SEMIEMPIRICAL MO TREATMENTS OF NONPLANAR MOLECULES

Semiempirical MO theories fall into two categories: those using a Hamiltonian that is the sum of one-electron terms, and those using a Hamiltonian that includes two-electron repulsion terms, as well as one-electron terms. The Hückel method is a one-electron theory, whereas the Pariser–Parr–Pople method is a two-electron theory.

For nonplanar molecules, the simplifying approximation of treating the π electrons separately is not available, and all the valence electrons must be considered together.

Of course, all the semiempirical treatments of this section are applicable to planar (as well as nonplanar) compounds, and they represent improvements on semiempirical π-electron theories, since they explicitly allow for the effects of the σ valence electrons.

The extended Hückel method   The most important one-electron semiempirical MO method for nonplanar molecules is the *extended Hückel theory*; an early version was used by Wolfsberg and Helmholz in treating inorganic complex ions; the method was further developed and widely applied by Hoffmann.[36]

The extended Hückel (EH) method begins with the approximation of treating the valence electrons separately from the rest (Section 13.15). The valence-electron Hamiltonian is taken as the sum of one-electron Hamiltonians:

$$\hat{H}_{val} = \sum_i \hat{H}_{eff}(i) \tag{15.191}$$

where $\hat{H}_{eff}(i)$ is not specified explicitly. The MO's are approximated as linear combinations of the valence AO's f_j of the atoms:

$$\varphi_i = \sum_j c_{ij} f_j \tag{15.192}$$

In the simple Hückel theory of planar hydrocarbons, each π MO contains contributions from one $2p\pi$ AO on each carbon atom; in the extended Hückel treatment of nonplanar hydrocarbons, each valence MO contains contributions from

---

[36] R. Hoffmann, *J. Chem. Phys.*, **39**, 1397 (1963); **40**, 2745, 2474, 2480 (1964); *Tetrahedron*, **22**, 521, 539 (1966); M. Wolfsberg and L. Helmholz, *J. Chem. Phys.*, **20**, 837 (1952).

four AO's on each carbon atom (one $2s$ and three $2p$'s) and one $1s$ AO on each hydrogen atom. The AO's used are usually Slater-type orbitals with fixed orbital exponents. For the simplified Hamiltonian (15.191), the problem separates into several one-electron problems:

$$\hat{H}_{\text{eff}}(i)\varphi_i = e_i\varphi_i \qquad (15.193)$$

$$E_{\text{val}} = \sum_i e_i \qquad (15.194)$$

Application of the variation theorem to the linear trial function (15.192) gives as the secular equation and the equations for the MO coefficients

$$\det\,(H_{jk}^{\text{eff}} - e_i S_{jk}) = 0 \qquad (15.195)$$

$$\sum_k\,(H_{jk}^{\text{eff}} - e_i S_{jk})c_{jk} = 0, \qquad j = 1, 2, \ldots \qquad (15.196)$$

All this is similar to simple Hückel theory. However, the extended Hückel theory does not neglect overlap. Rather *all* overlap integrals are explicitly evaluated using the forms chosen for the AO's and the internuclear distances at which the calculation is being done. (Formulas for overlap integrals of STO's are available.[37]) Thus off-diagonal e_i's are present in the secular determinant. A computer is used to solve (15.195).

Since $\hat{H}_{\text{eff}}(i)$ is not specified, there is the problem of what to use for the integrals H_{jk}^{eff}. Generally, the integrals with $j = k$ are approximated by minus the appropriate atomic valence-state ionization potentials. (The justification is by analogy to Koopmans' theorem of SCF theory.) Valence-state ionization potentials differ from ordinary atomic ionization potentials; they will be defined in Section 15.17. For carbon and hydrogen atoms, the valence-state ionization-potential parametrization gives

$$\langle C2s|\hat{H}_{\text{eff}}|C2s\rangle = -20.8\text{ eV}, \qquad \langle C2p|\hat{H}_{\text{eff}}|C2p\rangle = -11.3\text{ eV}$$

$$\langle H1s|\hat{H}_{\text{eff}}|H1s\rangle = -13.6\text{ eV} \qquad (15.197)$$

For the off-diagonal matrix elements $H_{jk}^{\text{eff}}, j \neq k$, Wolfsberg and Helmholz wrote

$$H_{jk}^{\text{eff}} = K\frac{H_{jj}^{\text{eff}} + H_{kk}^{\text{eff}}}{2} S_{jk} \qquad (15.198)$$

where K is a numerical constant (values ranging from 1 to 3 have been used) and where the remaining quantities are evaluated as above. Since H_{jj}^{eff} and H_{kk}^{eff} are usually negative, (15.198) gives H_{jk}^{eff} as negative. The arithmetic and geometric means of two quantities of comparable magnitude do not differ greatly, and instead of (15.198), Ballhausen and Gray used

$$H_{jk}^{\text{eff}} = -K(H_{jj}^{\text{eff}}H_{kk}^{\text{eff}})^{1/2} S_{jk} \qquad (15.199)$$

---

[37] D. M. Bishop, M. Randič, and J. R. Morton, *J. Chem. Phys.*, **45**, 1880 (1966); footnotes 12–18 of this paper give references to available tabulations.

Equations (15.199) and (15.198) are the most widely used approximations, but many others have been proposed. In contrast to the simple Hückel theory, H_{jk}^{eff} is nonzero for all pairs of orbitals (unless S_{jk} vanishes for symmetry reasons).

Once the H_{jk}^{eff} and S_{jk} integrals have been evaluated, the secular equation is solved for the orbital energies and then the MO coefficients are found.

The total valence-electron energy is given by (15.194) as the sum of orbital energies. To predict molecular geometry using the EH method, one carries out a series of calculations over a range of bond distances and angles and looks for the nuclear configuration that minimizes (15.194). (This procedure is like that used in the Walsh-diagram method of Section 15.8, where one looks for the bond angle that minimizes the sum of orbital energies.) Note that (15.194) omits both electron–electron repulsions and nuclear–nuclear repulsions. It might be thought that such a theory would be useless for predicting molecular geometry, but this is not so.

Allen and Russell compared EH results with those of ab initio SCF calculations for twelve small molecules.[38] They used (15.199) with $K = 1.5$. They found that for compounds whose bonds are not highly polar, the extended-Hückel-theory graph of (15.194) versus bond angle and the graph of SCF MO total energy (including nuclear repulsion) versus bond angle have minima at essentially the same angles. Since the SCF MO method yields generally accurate bond angles, so does the extended Hückel method for compounds of low polarity. The cases where the extended Hückel method failed (H_2O, Li_2O, LiOH, FOH) all involve highly polar bonds.

Allen and Russell found the order of MO energies predicted by the EH method to be usually reliable for low-polarity compounds. Hoffmann's EH calculation of benzene predicted that the lowest π MO, the $1a_{2u}$ MO, would lie below some of the occupied σ MO's, namely, the $3e_{2g}$ and $1b_{2u}$ MO's. Near-Hartree–Fock calculations of benzene show the $1a_{2u}$ π MO to have higher energy than the $1b_{2u}$ MO but very slightly lower energy than the $3e_{2g}$ pair of σ MO's[39]; this MO order is confirmed experimentally by the photoelectron spectrum of benzene. Thus Hoffmann's prediction is partly confirmed.

For bond lengths Allen and Russell state that empirical formulas involving bond radii are more reliable than the extended Hückel method.

The extended Hückel MO coefficients are found by Allen and Russell to deviate significantly from those of ab initio wave functions. Thus the extended Hückel method does not give the accurate predictions of such properties as dipole moments and rotational barriers achieved by the SCF MO method.

The EH method has been widely applied to predict conformations of organic compounds, and it was once believed to be fairly reliable for such work. However, a careful EH study of benzene showed that the $\mathcal{D}_{6h}$ benzene geometry predicted by EH theory corresponds to a saddle point (rather than a true minimum) on the potential-energy surface for nuclear motion; in fact, EH theory erroneously predicts benzene to be unstable with respect to dissociation to three acetylenes. [See W. L.

---

[38] L. C. Allen and J. D. Russell, *J. Chem. Phys.*, **46**, 1029 (1967); S. D. Peyerimhoff, R. J. Buenker, and L. C. Allen, *J. Chem. Phys.*, **45**, 734 (1966).
[39] W. C. Ermler, R. S. Mulliken, and E. Clementi, *J. Am. Chem. Soc.*, **98**, 388 (1976); W. C. Ermler and C. W. Kern, *J. Chem. Phys.*, **58**, 3458 (1973).

Bloemer and B. L. Bruner, *Chem. Phys. Lett.*, **17**, 452 (1972).] Furthermore, Pullman found EH-predicted conformations of several pharmacologically active molecules to be in serious disagreement with the results of ab initio calculations and experimental results (B. Pullman in *Advances in Quantum Chemistry*, vol. 10, Academic Press, New York, 1977, pp. 251–328); Pullman concluded that these disagreements "practically disqualify [the EH method] as a valid tool for conformational studies."

Because the EH method gives poor predictions of such molecular properties as bond lengths, dipole moments, energies, and rotational barriers, Jug concluded that this method "is obsolete" [K. Jug, *Theor. Chim. Acta*, **54**, 263 (1980)]. However, this judgment is too harsh, since the EH method has been used by Hoffmann and others to provide valuable qualitative insights into chemical bonding [e.g., P. J. Hay, J. C. Thibeault, and R. Hoffmann, *J. Am. Chem. Soc.*, **97**, 4884 (1975)]. Gimarc noted that "the real value of the extended Hückel method is not in its quantitative results, which have never been impressive, but rather in the qualitative nature of the results and in the interpretations those results can provide" (B. M. Gimarc, *Molecular Structure and Bonding*, Academic Press, New York, 1979, p. 216).

The CNDO, INDO, and NDDO methods Several two-electron semiempirical MO theories for nonplanar molecules have been developed; these are generalizations of the Pariser–Parr–Pople theory. The *complete neglect of differential overlap* (CNDO) *method* was developed by Pople, Santry, and Segal in 1965; the *intermediate neglect of differential overlap* (INDO) *method* was developed by Pople, Beveridge, and Dobosh in 1967.[40] (If you learn enough abbreviations you can convince some people that you know quantum chemistry.) Both methods treat only the valence electrons explicitly. The valence-electron Hamiltonian has the same form as (15.86):

$$\hat{H}_{\text{val}} = \sum_{i=1}^{n} (-\tfrac{1}{2}\nabla_i^2 + V_i) + \sum_{i=1}^{n} \sum_{j>i} \frac{1}{r_{ij}} \qquad (15.200)$$

where there are n valence electrons and where V_i is the potential energy of valence electron i in the field of the nuclei and inner-shell electrons; the quantity in parentheses is $\hat{H}_{\text{core}}(i)$. Both methods are SCF methods that iteratively solve the Hartree–Fock–Roothaan equations using the Hamiltonian (15.200), subject to certain approximations for the integrals.

The CNDO method uses a minimal basis set of valence Slater AO's with fixed orbital exponents on each atom. When each MO is expressed as a linear combination of AO's, the molecular Coulomb and exchange integrals that occur during solution of the Roothaan equations become linear combinations of two-electron repulsion integrals $(rs|tu)$ over atomic orbitals [Eq. (15.188)]. The CNDO method makes the zero differential overlap approximation [Eq. (15.189)] for all pairs of AO's in two-electron integrals. Thus the only nonvanishing repulsion integrals in CNDO are of the form $(rr|tt)$; since only two AO's occur in these integrals, all three- and four-center

---

[40] J. A. Pople, D. P. Santry, and G. A. Segal, *J. Chem. Phys.*, **43**, S129 (1965); J. A. Pople and G. A. Segal, *J. Chem. Phys.*, **43**, S136; **44**, 3289 (1966); J. A. Pople, D. L. Beveridge, and P. A. Dobosh, *J. Chem. Phys.*, **47**, 2026 (1967).

integrals are neglected. In the PPP treatment of conjugated hydrocarbons, there is only one basis AO per atom—the $2p\pi$ AO. In the CNDO method, there are several basis valence AO's on every atom (except hydrogens), and the ZDO approximation leads to neglect of electron-repulsion integrals containing the product $f_r(1)f_s(1)$ where f_r and f_s are different AO's centered on the *same* atom. The CNDO method also neglects overlap integrals [Eq. (15.185)]. The core-Hamiltonian integrals (15.186) and (15.187) are approximated by using data such as atomic ionization potentials and electron affinities and by taking some of them as semiempirical parameters adjusted so that CNDO calculations will give a good overall fit with the results of minimal basis ab initio SCF calculations. The original version of CNDO is called CNDO/1; a version using an improved parametrization is called CNDO/2.

The INDO method is an improvement on CNDO; here, differential overlap between AO's on the same atom is not neglected in one-center electron-repulsion integrals, but is still neglected in two-center electron-repulsion integrals. Thus fewer two-electron integrals are neglected, as compared to CNDO; otherwise, the two methods are the same. The INDO method gives an improvement on CNDO results, especially where electron spin distribution is important (e.g., in calculating electron-spin-resonance spectra).

As to results, the CNDO and INDO methods give fairly good bond lengths and angles, somewhat erratic dipole moments, and poor dissociation energies. (For details on the CNDO and INDO methods, see *Pople and Beveridge*; G. Klopman and R. C. Evans in *Segal*, Part A, p. 29.)

Versions of CNDO and INDO parametrized to predict electronic spectra are called CNDO/S and INDO/S. These methods include some configuration interaction; although the ground state of a closed-shell molecule is generally well represented by a single-determinant wave function, one typically requires CI for accurate representation of excited states. For details of these methods, see R. L. Ellis and H. H. Jaffé in *Segal*, Part B, p. 49; J. Michl in *Segal*, Part B, p. 99.

The *neglect of diatomic differential overlap* (NDDO) method (suggested by Pople, Santry, and Segal in 1965) is an improvement on INDO in which differential overlap is neglected only between AO's centered on different atoms: $f_r(1)f_s(1)\,dv_1 = 0$ only when AO's r and s are on different atoms. The degree of neglect of differential overlap in NDDO is more justifiable than in CNDO or INDO. A few initial attempts at parametrizing the NDDO method gave results that were rather disappointing (see G. Klopman and R. C. Evans in *Segal*, Part A, Chapter 2), and the method was little used until 1977, when Dewar and Thiel modified it to give the MNDO method, discussed later in this section.

The PRDDO method   The *partial retention of diatomic differential overlap* (PRDDO) method was developed by Halgren and Lipscomb [T. A. Halgren and W. N. Lipscomb, *J. Chem. Phys.*, **58**, 1569 (1973)]. This is a semiempirical minimal-basis-set SCF MO method that includes all electrons but neglects some integrals and approximates others. The method uses symmetrically orthogonalized AO's (Section 15.12) as basis functions. Whereas the CNDO method neglects all electron repulsion integrals containing the product $f_r(1)f_s(1)$ or $f_r(2)f_s(2)$, where f_r and f_s are different AO's, the PRDDO method neglects only electron-repulsion integrals that contain the product $f_r(1)f_s(1)f_t(2)f_u(2)$, where f_r and f_s are different basis functions and f_t and f_u are different basis functions. (When $r = t$ and $s = u$, the integral is not neglected.) Fewer integrals are neglected than in NDDO (which in turn neglects fewer integrals than CNDO and INDO). In PRDDO (as in CNDO and INDO) some of the integrals are taken as parameters whose values are chosen so as to reproduce results of ab initio calculations on small molecules. The PRDDO method is much more accurate than CNDO and INDO in reproducing results of ab initio calculations and gives results pretty close to those of an ab initio STO-3G calculation in most cases [T. A. Halgren et al., *J. Am. Chem. Soc.*, **100**, 6595 (1978)].

The MINDO, MNDO, and SINDO1 methods   The aim of Pople and co-workers in the CNDO and INDO methods and of Halgren and Lipscomb in PRDDO was to reproduce as well as possible the results of minimal-basis-set SCF MO ab initio calculations with theories that require much less computational effort than an ab initio treatment. Since these methods use approximations in the SCF calculation, we can expect their results to be similar to but less accurate than minimal-basis ab initio SCF results. Thus these methods do pretty well on molecular geometry and do poorly on binding energies. Dewar and co-workers devised several semiempirical SCF MO theories which closely resemble the INDO theory. However, Dewar's aim was not to reproduce ab initio SCF wave functions, but to have a theory that would give molecular binding energies with chemical accuracy (i.e., to within 1 kcal/mole) and that would be applicable to large molecules without a lot of computation. It might seem unlikely that one could devise an SCF theory that involves approximations to the ab initio Hartree–Fock procedure but that succeeds in an area (binding energies) where the Hartree–Fock theory fails. However, by proper choice of the parameters in the semiempirical SCF theory, one can actually get better results than ab initio SCF calculations, because the choice of suitable parameters can compensate for the partial neglect of electron correlation in ab initio SCF theory.

In 1967 Dewar and Klopman put forth the PNDO (partial neglect of...) theory; in 1969, Baird and Dewar published the MINDO/1 (first version of modified INDO) theory.[41] The parameters in these theories were chosen so as to have calculated molecular heats of formation fit experimental data as well as possible. Although these theories worked well for heats of formation, they were quite poor for geometry. Therefore in 1970 Dewar and Haselbach proposed the MINDO/2 theory.[42] In MINDO/2 a careful choice of parameters allowed molecular geometries and heats of formation to be calculated rather accurately in most cases. Since MINDO/2 gave large errors in dipole moments and in bond lengths to hydrogen atoms, an improved version, MINDO/2', was published in 1972 [N. Bodor, M. J. S. Dewar, and D. H. Lo, *J. Am. Chem. Soc.*, **94**, 5303 (1972)].

Dewar's MINDO work culminated in MINDO/3, a substantially improved

[41] M. J. S. Dewar and G. Klopman, *J. Am. Chem. Soc.*, **89**, 3089 (1967); N. C. Baird and M. J. S. Dewar, *J. Chem. Phys.*, **50**, 1262 (1969); N. C. Baird, M. J. S. Dewar, and R. Sustmann, *J. Chem. Phys.*, **50**, 1275 (1969).

[42] M. J. S. Dewar and E. Haselbach, *J. Am. Chem. Soc.*, **92**, 590 (1970); N. Bodor, M. J. S. Dewar, A. Harget, and E. Haselbach, *J. Am. Chem. Soc.*, **92**, 3854.

version published in 1975 [R. C. Bingham, M. J. S. Dewar, and D. H. Lo, *J. Am. Chem. Soc.*, **97**, 1285, 1294, 1302, 1307 (1975); Dewar et al., *J. Am. Chem. Soc.*, **97**, 1311; Dewar, *Science*, **187**, 1037 (1975)]. The MINDO/3 method has been parametrized for compounds containing C, H, O, N, B, F, Cl, Si, P, and S. Calculations have been done on molecules as large as lysergic acid diethylamide, $C_{20}H_{25}N_3O$, a 49-atom molecule for which the MINDO/3 calculation required the optimization of 141 geometrical parameters. A MINDO calculation yields the gas-phase molecular heat of atomization; one then uses experimental values of heats of formation of C, H, O, etc., atoms from the elements in their standard states (graphite, H_2, O_2, etc.) at 25 °C to calculate the molecule's gas-phase heat of formation at 25 °C, $\Delta H^{\circ}_{f,298}$. Note that $\Delta H^{\circ}_{f,298}$ contains rotational and vibrational contributions (Section 13.1). Rather than being calculated explicitly, these contributions are allowed for by the choice of parameters used in the calculation. For a sample of a large number of compounds, the average absolute errors in MINDO/3-calculated properties are 11 kcal/mole in heats of formation, 0.02 Å in bond lengths, 5° in bond angles, 0.4 D in dipole moments, and 0.8 eV in ionization potentials [D. N. Nanda and K. Jug, *Theor. Chim. Acta*, **57**, 95 (1980)]. Large errors in heats of formation occur for small-ring compounds, compounds containing triple bonds, aromatic compounds, compact globular molecules, boron compounds, and molecules containing adjacent atoms with lone pairs. The errors in heats of formation are larger than Dewar was aiming for, but MINDO/3 is still a significant achievement.

The MINDO/3 method has been widely applied to calculate properties of ground-state organic molecules and to calculate potential-energy surfaces of chemical reactions (see Section 15.19). However, the usefulness of MINDO/3 came under attack by advocates of ab initio SCF calculations [J. A. Pople, *J. Am. Chem. Soc.*, **97**, 5306 (1975); W. J. Hehre, *J. Am. Chem. Soc.*, **97**, 5308 (1975); M. J. S. Dewar, *J. Am. Chem. Soc.*, **97**, 6591 (1975)].

Because MINDO/3 did not meet Dewar's original aims (and because it proved difficult to extend MINDO/3 to include metallic elements), Dewar and co-workers developed the MNDO (modified neglect of diatomic overlap) method, published in 1977 [M. J. S. Dewar and W. Thiel, *J. Am. Chem. Soc.*, **99**, 4899, 4907 (1977); Dewar and H. S. Rzepa, *J. Am. Chem. Soc.*, **100**, 58, 777, 784 (1978); Dewar and M. L. McKee, *J. Am. Chem. Soc.*, **99**, 5231 (1977)]. The MNDO method is based on the NDDO approximation discussed earlier in this section; as noted, the NDDO approximation is more justifiable than the INDO approximation used in MINDO/3. Like MINDO/3, MNDO is parametrized to reproduce gas-phase $\Delta H^{\circ}_{f,298}$ values. The MNDO method has been parametrized for compounds containing H, Be, B, C, N, O, F, Si, P, S, and Cl. For compounds containing only H, B, C, N, O, and F, the number of parameters in MNDO is 31, as compared to 61 in MINDO/3. (One criticism of MINDO/3 was the large number of adjustable parameters used.) MNDO gives substantially improved results as compared to MINDO/3; average absolute MNDO errors are 9 kcal/mole in heats of formation, 3° in bond angles, 0.025 Å in bond lengths, 0.35 D in dipole moments, and 0.5 eV in ionization potentials [D. N. Nanda and K. Jug, *Theor. Chim. Acta*, **57**, 95 (1980)]. The calculation time in MNDO is only slightly greater than that in MINDO/3.

In 1980 Jug and Nanda published the SINDO1 (symmetrically orthogonalized

INDO) method [D. N. Nanda and K. Jug, *Theor. Chim. Acta*, **57**, 95 (1980); K. Jug and D. N. Nanda, *Theor. Chim. Acta*, **57**, 107, 131 (1980)]. This semiempirical method is based on the INDO approximation and (like Dewar's methods) uses parameters designed to reproduce molecular binding energies and molecular geometries, rather than results of ab initio calculations. Rather than using ordinary AO's as basis functions, SINDO1 calculates certain integrals using symmetrically orthogonalized AO's (Section 15.12) and uses a pseudopotential (see later in this section) to eliminate the inner-shell electrons. SINDO1's accuracy is similar to that of MNDO, average errors being 8 kcal/mole in binding energies, $2\frac{1}{2}°$ in bond angles, 0.02 Å in bond lengths, 0.4 D in dipole moments, and 0.8 eV in ionization potentials.

The PCILO method   The PCILO (perturbative configuration interaction using localized orbitals) method was developed in 1969 by a group of French workers [see J. P. Malriu in *Segal*, Part A, Chapter 3 and references cited therein]. In the PCILO method, one chooses a reasonable set of localized bonding, lone-pair, and antibonding orbitals for the molecule and uses these orbitals to construct configuration functions for a CI treatment. Perturbation theory is used to calculate the energy. To simplify the calculations, one uses the CNDO approximations to evaluate integrals (sometimes the INDO approximations are used). The method gives an energy at least as accurate as that of the CNDO SCF method and is much faster than CNDO. The PCILO method has been extensively applied to calculate conformations of biological molecules.

The Xα method   In 1951 Slater proposed simplifying Hartree–Fock calculations of atomic wave functions by approximating the exchange term $- \sum_j \hat{K}_j(1)\varphi_i(1)$ in the Hartree–Fock equations (13.176) by $V_X\varphi_i(1)$, where Slater's exchange potential V_X is equal to $- 3(3\rho/8\pi)^{1/3}$, where ρ is the total electron probability density at a given point of space; later workers introduced a multiplicative parameter α to give the approximate exchange potential as $V_{X\alpha} = - 3\alpha(3\rho/8\pi)^{1/3}$. This approximation was used in calculations on atoms and solids.

Because molecules have neither the spherical symmetry of atoms nor the periodic symmetry of crystals, further development of the method was needed to apply it to molecules. This development was done by Johnson about 1970, following a suggestion of Slater's.

In the molecular (SCF) Xα method, one divides space up into several regions by imagining spheres around each atom and a sphere around the entire molecule. With the exchange potential replaced by $V_{X\alpha}$, one solves for the orbitals in each of the regions of space using numerical procedures (rather than expansion in a basis set); the most common procedure is called the *multiple scattering method*, or the *scattered wave* (sw) *method*. At the boundaries between the regions, one applies the conditions that the wave function be continuous and have continuous derivatives. (In the spherical region around each atom, one uses an α value appropriate to that atom; these are taken as the α values needed to make the Xα method give the correct Hartree–Fock energy in an Xα calculation on the isolated atom. The α values mainly lie between 0.78 and 0.69.)

The Xα method is not an ab initio method since the Xα expression for the total energy does not use the exact molecular Hamiltonian. The method uses no parameters evaluated from experimental data. However, it does use parameters (the α's) whose values are chosen to reproduce ab initio atomic Hartree–Fock results. In this sense it could be considered a semiempirical method, but it contains far less empiricism than the methods discussed earlier in this section.

The computer time required for an $X\alpha$ calculation is substantially less than for an ab initio SCF calculation.

The $X\alpha$ method has been applied to many transition-metal complex ions, where it gives impressively accurate values of orbital energies and ionization potentials. Chemisorption has been studied through $X\alpha$ calculations of the interaction between the chemisorbed molecule and a small number of metal atoms. An $X\alpha$ calculation of ethane gave an accurate rotational barrier of 2.9 kcal/mole, but further work on other molecules showed the method to give generally poor results for barriers [U. Wahlgren, *Chem. Phys. Lett.*, **20**, 246 (1973)]. $X\alpha$ calculations on benzene and other conjugated molecules found the calculated ionization potentials and orbital energies to be in good agreement with experiment (with a few exceptions) and found the accuracy of $X\alpha$ values of one-electron properties to lie between that of minimal-basis and double-zeta SCF values [D. A. Case, M. Cook, and M. Karplus, *J. Chem. Phys.*, **73**, 3294 (1980)].

The molecular $X\alpha$ method is still undergoing development, and definitive conclusions about its value are not yet available. References on the method include: J. W. D. Connolly in *Segal*, Part A, Chapter 4; K. H. Johnson et al. in *Computational Methods for Large Molecules and Localized States in Solids*, F. Herman et al., eds., Plenum, New York, 1973, p. 161; D. Case, *Ann. Rev. Phys. Chem.*, **33**, 151 (1982).

The DIM method   The diatomics-in-molecules (DIM) method was developed by Ellison in 1963 [F. O. Ellison, *J. Am. Chem. Soc.*, **85**, 3540 (1963)] and later given a more general formulation by Tully (J. C. Tully in *Segal*, Part A, Chapter 6). The DIM method expresses the molecular wave function in terms of wave functions for atomic and diatomic fragments and calculates the molecular electronic energy from energies of these fragments using a certain approximation; electron correlation is explicitly allowed for.

The method can be applied in an ab initio manner by performing ab initio CI calculations on the atomic and diatomic fragments. More commonly, the method is implemented semiempirically by using experimental atomic and diatomic energies. Alternatively, one can treat the atomic and diatomic fragment properties as parameters whose values are chosen so as to reproduce known properties of a few polyatomic systems as well as possible; one then uses these fragment parameters to treat other polyatomics. DIM has been used to calculate potential-energy surfaces (Section 15.19) for reactions of small molecules. The method is still under development, and it is too soon to assess its usefulness.

Pseudopotentials   Ab initio calculations on molecules containing atoms of intermediate and high atomic number are extremely time-consuming because of the large number of basis functions and resulting large number of integrals. To reduce the labor, one can attempt to eliminate the inner-shell electrons from the calculation.

The *pseudopotential* method uses an electronic Hamiltonian that has the form $\hat{T}_e + \hat{V}_{ee} + \hat{V}_{Ne} + \hat{V}_{e,\,\text{pseudo}}$, where $\hat{T}_e$ is the sum of kinetic-energy operators for the valence electrons only, $\hat{V}_{ee}$ consists of repulsions between the valence electrons, $\hat{V}_{Ne}$ consists of attractions between the valence electrons and the nuclei, and $\hat{V}_{e,\,\text{pseudo}}$ is a relatively simple term meant to allow for the influence of the inner-shell electrons on the valence electrons. (Of course, such an approach is an approximation.) Many different forms for $\hat{V}_{e,\,\text{pseudo}}$ have been tried. Whatever its form, $\hat{V}_{e,\text{pseudo}}$ contains parameters whose values are chosen either to give agreement with results of all-electron ab initio calculations or to give agreement with experimental quantities (e.g., ionization potentials). Having chosen $\hat{V}_{e,\text{pseudo}}$, one chooses a basis set for the valence electrons and uses an SCF MO approach to solve for valence-electron MO's and energies. A fair degree of success has been attained in pseudopotential calculations, and this approach is undergoing further development; see R. N. Dixon and I. L. Robertson in *Specialist Periodical Reports, Theoretical Chemistry*, vol. 3, The Chemical Society, London, 1978, Chapter 4.

15.15 *THE EMPIRICAL FORCE FIELD (MOLECULAR MECHANICS) METHOD*

The *empirical force field* (EFF) *method* (also called *molecular mechanics*) is quite different from the semiempirical methods described in the last section. The EFF method is not really a quantum-mechanical method at all, since it does not deal with an electronic Hamiltonian or wave function. Instead, the method starts out with a model of a molecule as composed of atoms held together by bonds; provision is also made for interactions between nonbonded atoms. The method was developed by Westheimer, Hendrickson, Wiberg, Allinger, Warshel, and others.

The molecular electronic energy U is written as $U = E_{\text{str}} + E_\theta + E_{\text{vdW}} + E_\omega$. The energy of bond stretching is taken as the following quadratic function of the displacement of each bond length l_i from its expected equilibrium length $l_{i,0}$: $E_{\text{str}} = \frac{1}{2}\sum_i k_{s,i}(l_i - l_{i,0})^2$, where $k_{s,i}$ is the force constant for stretching bond i. (Recall from Section 4.3 that for small displacements of the nuclei from their equilibrium positions U is approximately a quadratic function of $l_i - l_{i,0}$.) The energy of bond bending E_θ is taken as $E_\theta = \frac{1}{2}\sum_i k_{\theta,i}(\theta_i - \theta_{i,0})^2$, where $\theta_i, \theta_{i,0}$, and $k_{\theta,i}$ are the angle, the expected equilibrium angle, and the bending force constant for bond angle i. The expected lengths and angles $l_{i,0}$ and $\theta_{i,0}$ are taken from known equilibrium geometries of small unstrained molecules. (For example, carbon–carbon single-bond lengths are typically 1.53 Å.) The energy of van der Waals interactions, E_{vdW}, is written as the sum of interactions between pairs of nonbonded atoms; each pair interaction is taken as the sum of a long-range attraction and a short-range repulsion; the Lennard–Jones 6–12 potential $a/R^{12} - b/R^6$ (where a and b are constants and R is the interatomic distance) is often used. The term E_ω is present in molecules with internal rotation about single bonds; for ethane, which has a threefold rotational barrier, E_ω has the form $\frac{1}{2}V_0(1 - \cos 3\omega)$, where V_0 is the barrier height and ω is the angle of torsion about the carbon–carbon bond.

If the molecule has one or more small rings, additional terms are added to account for the resulting steric strain. If the molecule has polar groups, terms are added to represent the electrostatic interactions between these groups. For aromatic compounds one also incorporates information such as bond orders taken from quantum-mechanical π-electron calculations. If there is intramolecular hydrogen bonding, terms are included to allow for this.

The force constants and other parameters in U are chosen so as to give a good fit to the known geometries, energies, and vibrational spectra of small molecules.

After writing down the expression for U, one varies the molecular geometry so as to locate the structure (or structures) that minimizes U. This calculation is done simply and rapidly on a computer; molecules of one or two hundred atoms are readily handled. The method thus allows one to calculate the geometry of the various possible conformations of a molecule and also gives the energy differences between different conformations. Since the potential-energy function U for molecular vibration is generated, one can use the EFF method to calculate molecular vibrational frequencies.

Note that the zero level of energy is taken as corresponding to all bond lengths and bond angles having their customary values l_0 and θ_0, there being no nonbonded van der Waals interactions or internal-rotation interactions. In such a hypothetical state, one can well approximate the molecular binding energy as the sum of empirical bond energies. To estimate the heat of formation of a gas-phase compound from its atoms, one adds the EFF-calculated equilibrium-geometry energy U_{eq} (which is called the *steric energy*) to the sum of empirical bond energies for the bonds in the molecule.

To apply the EFF method, one needs sufficient data to choose values for the parameters. For this reason, most applications have been to hydrocarbons; molecules with isolated polar groups are also readily treated.

For hydrocarbons the EFF method gives very good results. For example, ΔH_f° values have typical errors of only 1 or 2 kcal/mole (as compared to an average absolute error of 6 kcal/mole for hydrocarbons in MNDO, the most accurate semiempirical quantum-mechanical method for ΔH_f° calculations); bond lengths are usually within 0.01 Å and bond angles within 2° of experimental values.

A few applications have been made to the calculation of potential-energy surfaces for chemical reactions, but the value of EFF here is not yet known.

For a review of EFF, see N. L. Allinger, *Adv. Phys. Org. Chem.*, **13**, 1 (1976).

An important application of the EFF method is to intramolecular motions in proteins. Karplus and co-workers constructed an empirical potential-energy function for a certain protein with fifty-eight amino acid residues. They then used this function to solve the classical-mechanical equations of motion for the protein's atoms, given kinetic energies corresponding to a temperature of 295 K. They found that certain portions of the protein molecule underwent large-amplitude fluidlike motions governed mainly by collisions between nonbonded atoms. This sort of motion was later confirmed by NMR and X-ray diffraction studies. [See *Physics Today*, Nov. 1979, p. 17; J. A. McCammon and M. Karplus, *Ann. Rev. Phys. Chem.*, **31**, 29 (1980).]

15.16 *COMPARISONS OF QUANTUM-MECHANICAL METHODS*

This section compares the accuracies of various semiempirical and ab initio methods in calculating molecular properties; the discussion is restricted to ground-state closed-subshell molecules of first- and second-row (H–Ne) atoms.

We first consider energy changes in chemical reactions. The previously discussed failure of ab initio SCF calculations to predict molecular dissociation energies might seem to prevent the use of SCF energies to calculate reaction energies. However, for certain types of reactions, we *can* use ab initio SCF energies to estimate the reaction energy. Recall that pretty good barriers to internal rotation can usually be obtained from minimal-basis-set SCF MO calculations, because of a near cancellation in correlation energies between different molecular conformations. As ethane goes from staggered to eclipsed, the number of chemical bonds of each type does not change. More generally, one might hope that a similar near cancellation of correlation energies might occur for an *isodesmic* chemical reaction; an isodesmic reaction (Greek *isos*, "equal"; *desm*, "bond") is one in which the number of bonds of each type does not change [W. J. Hehre et al., *J. Am. Chem. Soc.*, **92**, 4796 (1970)]. For example, the isodesmic reaction $CH_2\!\!=\!\!CHCH_2OH + CH_2\!\!=\!\!O \rightarrow CH_2\!\!=\!\!CHOH + CH_3CH\!\!=\!\!O$ has seven CH bonds, one CC double bond, one CC single bond, one CO double bond, one CO single bond, and one OH bond on each side.

A special kind of isodesmic reaction is a *bond-separation reaction*. Here, one starts with a molecule and converts it to products, each of which contains only one bond between nonhydrogen atoms. For example, starting with $CH_3\!\!-\!\!CH\!\!=\!\!C\!\!=\!\!O$, one would form the products $CH_3\!\!-\!\!CH_3$, $CH_2\!\!=\!\!CH_2$, and $CH_2\!\!=\!\!O$, in which the C—C, C=C, and C=O bonds are separated from one another. To balance the reaction, one adds an appropriate number of hydride molecules (e.g., CH_4, NH_3, H_2O) to the left side. The bond-separation reaction for CH_3CHCO is then $CH_3\!\!-\!\!CH\!\!=\!\!C\!\!=\!\!O + 2CH_4 \rightarrow C_2H_6 + C_2H_4 + CH_2O$. The bond-separation reaction for benzene is $C_6H_6 + 6CH_4 \rightarrow 3C_2H_6 + 3C_2H_4$. (If the energy of a molecule could be represented as the sum of bond energies that were invariant from molecule to molecule, then the energy change for any isodesmic reaction would be zero. The energy change for a bond-separation reaction measures the interactions between the bonds in the molecule.)

If a minimal-basis-set SCF MO calculation could predict the energy change for the bond-separation reaction of a large molecule, then one could use the known energies of the small product molecules like C_2H_6, etc., to get a good estimate for the energy of the large molecule from a minimal-basis-set calculation, without having to use the prohibitively expensive extended basis sets or configuration interaction.

A comparison of theoretically calculated with experimental bond-separation energies for fifteen compounds of C, H, O, and N (most containing three or four nonhydrogen atoms) has been done (M. C. Flanigan, A. Komornicki, and J. W. McIver in *Segal*, Part B, Chapter 1). For ab initio SCF calculations, the minimal STO-3G basis set (Section 15.3) gives results accurate to within 5 kcal/mole for most of the reactions but is subject to substantial errors (20 kcal/mole) for reactions involving small rings and is in error by 17 kcal/mole for the H_2NCHO bond-separation reaction; the STO-3G average absolute error is 7 kcal/mole for the fifteen reactions. The 4-31G basis set is accurate to within 5 kcal/mole except for small-ring compounds, where errors of 12 kcal/mole can occur. The 6-31G* basis set gives 5 kcal/mole accuracy for bond-separation reactions, including those involving small rings [M. D. Newton in *Schaefer, Applications of Electronic Structure Theory*, p. 234].

As to semiempirical methods, the INDO method gives generally poor results; the average absolute error for the fifteen reactions is 27 kcal/mole, and errors as high as 95 kcal/mole occurred. Another comparative study [T. A. Halgren et al., *J. Am. Chem. Soc.*, **100**, 6595 (1978)] concluded that INDO and CNDO/2 "contain serious deficiencies which frequently render them unsuitable for use in energy calculations." The MINDO/3 method is accurate to within 10 kcal/mole for two-thirds of the fifteen reactions but is subject to 30 kcal/mole errors for reactions involving small rings and is in error by 57 kcal/mole for the C_6H_6 bond separation; this reaction has $6CH_4$ on the left, and MINDO/3 gives a rather poor CH_4 energy. The average absolute MINDO/3 error for the fifteen reactions is $14\frac{1}{2}$ kcal/mole. The MNDO method is rather more accurate than MINDO/3 but is subject to 15–20 kcal/mole errors for small-ring compounds and compounds containing a CO double bond; the average absolute MNDO error for the fifteen reactions is 12 kcal/mole. SINDO1 gives a 14 kcal/mole average absolute error for the fifteen reactions.

An important question is the relative stabilities of isomers. The transformation between two isomers is not necessarily an isodesmic reaction (an example is cyclopropane → propene). A comparison (M. C. Flanigan et al. in *Segal*, Part B, Chapter 1) of isomerization energies of three- and four-carbon hydrocarbons found that for ab initio SCF calculations, the STO-3G basis set gives rather unreliable results (errors of 10 to 20 kcal/mole), the 4-31G basis set gives reliable results (2 or 3 kcal/mole error) except for small-ring compounds (where errors of 10 or 20 kcal/mole occur), and the 6-31G* basis set gives reliable results (2 or 3 kcal/mole error) for all isomers including small-ring compounds. As to semiempirical methods, INDO was completely unreliable with errors of as much as 50 to 100 kcal/mole, and the MINDO/3 method was rather unreliable with errors running 5 to 20 kcal/mole. MNDO is more reliable than MINDO/3, being able to predict isomerization energies typically to within 5 kcal/mole if small rings are not involved; however, MNDO occasionally gives large errors (for example, the error for $CH_3CN \rightarrow CH_3NC$ is 26 kcal/mole and that for $CH_3NO_2 \rightarrow CH_3ONO$ is 42 kcal/mole).

In summary, one can use ab initio SCF calculations to obtain energy changes for isodesmic reactions of molecules composed of first- and second-row atoms to an accuracy of 5 kcal/mole and to obtain isomerization energies accurate to 3 kcal/mole if a basis set of suitable size is used—4-31G when small rings are not involved and 6-31G* when small rings are involved. (Unfortunately, the minimal STO-3G basis set is generally not reliable here.) The INDO and CNDO/2 methods are useless for such energy calculations. The MINDO/3 method is of some value here, but its overall performance is disappointing in view of the fact that it was parametrized to reproduce experimental $\Delta H°$ values. The MNDO method is better than MINDO/3 but not as good as ab initio SCF calculations with a proper-size basis set. Of course, MNDO can handle much larger molecules than can extended-basis-set ab initio calculations.

(Except for MINDO and MNDO, which are parametrized to give gas-phase

ΔH_{298}° values, a quantum-mechanical calculation of an energy change applies to all the reactants and products at their minima in potential energy. If the quantum-mechanical result is corrected for the zero-point vibrational energies of the molecules, one obtains ΔU_0°, the gas-phase thermodynamic internal energy change at absolute zero. To get the gas-phase ΔH_{298}° value, one must allow for occupation of excited translational, rotational, and vibrational levels; this is readily done if good estimates for the molecular structure and vibrational frequencies are available.)

Now consider dipole moments. The accuracy of ab initio SCF dipole moments calculated with the STO-3G and 4-31G basis sets is fair; the 6-31G* basis set gives substantially more accurate results. One study [K. Jug and D. N. Nanda, *Theor. Chim. Acta*, **57**, 107, 131 (1980)] found an average absolute error of 0.6 D for STO-3G dipole moments; a second study (R. Ditchfield in D. R. Lide and M. Paul, eds., *Critical Evaluation of Chemical and Physical Structural Information*, Natl. Acad. Sci., Washington, D.C., 1974) found average absolute errors of 0.5 D for 4-31G and 0.24 D for 6-31G*. Although sometimes accurate, CNDO/2 and INDO dipole moments often show very substantial errors, being two or three times too large for boron hydrides [T. A. Halgren et al., *J. Am. Chem. Soc.*, **100**, 6595 (1978)] and grossly in error for a number of third-row compounds (M. S. Gordon et al., *J. Am. Chem. Soc.*, **100**, 2670). MNDO, MINDO/3, and SINDO1 are reasonably reliable for dipole moments, with average absolute errors of 0.35 D for MNDO, 0.4 D for MINDO/3, and 0.4 D for SINDO1 [D. N. Nanda and K. Jug, *Theor. Chim. Acta*, **57**, 95 (1980)].

Now consider molecular geometry. SCF calculations with the STO-3G basis set give quite good geometry predictions for compounds of first- and second-row elements; one sample of results (M. C. Flanigan et al. in *Segal*, Part B, Chapter 1) found an STO-3G average absolute error of 0.02 Å in bond lengths and 2.4° in bond angles. A second study [J. S. Binkley, J. A. Pople, and W. J. Hehre, *J. Am. Chem. Soc.*, **102**, 939 (1980)] found errors of 0.028 Å and 2.4°. [For third-row (Na–Ar) compounds, the STO-3G basis set is less adequate, with average absolute errors of 0.06 Å and 3°; addition of a set of d functions on each third-row atom to give the STO-3G* basis set reduces the average errors to 0.04 Å and $1\frac{1}{2}°$; J. B. Collins et al., *J. Chem. Phys.*, **64**, 5142 (1976).] The 4-31G basis set gives more accurate bond lengths than STO-3G but less accurate bond angles. The 6-31G* bond lengths and angles are more accurate than those of STO-3G and are pretty close to those of Hartree–Fock wave functions, but the 6-31G* basis set is too large for routine use. INDO gives only fair geometries; one study (M. C. Flanigan et al. in *Segal*) found average errors of 0.06 Å and 4° in bond lengths and angles. MINDO/3 is quite good for bond lengths but only fair for bond angles. MNDO and SINDO1 are quite good for both bond lengths and bond angles. A study [D. N. Nanda and K. Jug, *Theor. Chim. Acta*, **57**, 95 (1980)] found average absolute errors for first- and second-row compounds of 0.02 Å and 5.0° for MINDO/3, 0.025 Å and 2.9° for MNDO, and 0.023 Å and 2.6° for SINDO1.

Calculation of the equilibrium geometry of a ten-atom molecule requires optimization of as many as twenty-four geometrical parameters; if this is done by calculating the energy at many geometries, a huge amount of calculation is needed. Instead of calculating the energy at many geometries, a much more efficient procedure is to calculate the energy and the first derivatives of the energy with respect to the nuclear coordinates (the set of such derivatives constitutes the *energy gradient*) at a relatively small number of points and then use these data to find the equilibrium geometry. Because there are $3N - 6$ independent nuclear coordinates for an N-atom molecule, calculation of the gradient provides information equivalent to energy calculations at $3N - 6$ geometries but is much less costly in computer time. Calculation of the energy gradient from SCF wave functions is relatively straightforward [P. Pulay in *Schaefer, Applications of Electronic Structure Theory*, Chapter 4; P. Pulay et al., *J. Am. Chem. Soc.*, **101**, 2550 (1979)]. Methods to calculate the energy gradient from MCSCF wave functions and from CI wave functions have been developed [J. D. Goddard, N. C. Handy, and H. F. Schaefer, *J. Chem. Phys.*, **71**, 1525 (1979); B. R. Brooks et al., *J. Chem. Phys.*, **72**, 4652 (1980); R. Krishan, H. B. Schlegel, and J. A. Pople, *J. Chem. Phys.*, **72**, 4654]. For semiempirical MO methods based on neglect of differential overlap (e.g., MINDO, MNDO), calculation of the energy gradient is much faster than for the ab initio SCF method. The energy gradient can be used to determine structures of transition states in chemical reactions [J. W. McIver and A. Komornicki, *J. Am. Chem. Soc.*, **94**, 2625 (1972)].

Even with the use of the energy gradient, calculation of the equilibrium geometry of a medium-size molecule using the ab initio SCF MO method is a formidable task. Since the STO-3G method works well for geometries but not so well for energies of isomers and isodesmic reactions, a frequently used approach is to optimize the geometry of each compound using the STO-3G basis set and then use the larger 4-31G basis set to calculate the energy at the STO-3G equilibrium geometry.

Molecular (first) ionization potentials are usually estimated by taking minus the orbital energy of the highest-occupied MO (Koopmans' theorem—Section 15.4). Ab initio SCF (e.g., STO-3G) ionization potentials have typical errors of 1 or $1\frac{1}{2}$ eV; CNDO/2 and INDO ionization potentials are not reliable, having typical errors of 4 or 5 eV [T. A. Halgren et al., *J. Am. Chem. Soc.*, **100**, 6595 (1978)]. MNDO, MINDO/3, and SINDO1 give quite accurate ionization potentials; one study [D. N. Nanda and K. Jug, *Theor. Chim. Acta*, **57**, 95 (1980)] found mean absolute errors of 0.8 eV for MINDO/3, 0.5 eV for MNDO, and 0.8 eV for SINDO1.

Hydrogen bonding has been studied with semiempirical and ab initio methods. MINDO/3, MNDO, and SINDO1 are quite unable to properly describe hydrogen bonding [M. J. S. Dewar and G. P. Ford, *J. Am. Chem. Soc.*, **101**, 5558 (1979); S. Scheiner, *Theor. Chim. Acta*, **57**, 71 (1980); D. N. Nanda and K. Jug, *Theor. Chim. Acta*, **57**, 95 (1980)]. These methods give dimerization energies of only 0–1 kcal/mole for the hydrogen-bonded linear dimer $(H_2O)_2$, as compared to the experimental value 5.4 ± 0.7 kcal/mole [L. A. Curtiss et al., *J. Chem. Phys.*, **71**, 2703 (1979)], and they give *much* too long hydrogen bonds. CNDO/2 gives a quite incorrect description of the interaction of two H_2O molecules [M. J. S. Dewar and G. P. Ford, *J. Am. Chem. Soc.*, **101**, 5558 (1979)]. As to ab initio SCF methods, the H-bond energy in linear $(H_2O)_2$ is 6, 8, and $5\frac{1}{2}$ kcal/mole for STO-3G, 4-31G, and 6-31G*, respectively; near-Hartree–Fock wave functions give a 4 kcal/mole H-bond energy. Two ab initio CI calculations gave $5\frac{1}{2}$ and 6 kcal/mole [see S. Scheiner, *Theor. Chim. Acta*, **57**, 71 (1980)]. Thus the ab initio SCF and CI results for the H-bond energy are pretty good. The STO-3G and 4-31G oxygen–oxygen bond distances in linear $(H_2O)_2$ are too short by 0.2 Å, but 6-31G* and near-Hartree–Fock wave functions give a distance accurate to 0.02 Å. The angle describing the orientation of the plane of the hydrogen-accepting molecule with respect to the line of the hydrogen bond is hard to calculate accurately, and ab initio SCF and CI calculations show errors of about $20°$.

The equilibrium geometries and relative energies of the different conformations of a molecule are of considerable interest, especially for biological molecules. One finds that ab initio SCF calculations usually give reliable results for molecular conformations. For example, the dihedral angles in noncyclic compounds containing two, three, and four nonhydrogen atoms are generally predicted well by ab initio STO-3G and 4-31G calculations (J. A. Pople in *Schaefer, Applications of Electronic Structure Theory*, pp. 11–12, 16–17), and 4-31G calculations give good predictions of the relative stabilities of different conformers [L. Radom, W. J. Hehre, and J. A. Pople, *J. Am. Chem. Soc.*, **93**, 289 (1971)]. However, STO-3G calculations substantially underestimate the nonplanarity of cyclobutane and cyclopentane; improvement is obtained with the 4-31G basis set and further improvement with the 6-31G* set (Pople in *Schaefer, Applications of Electronic Structure Theory*, pp. 17–18).

Despite some early successes, it is now realized (as noted in Section 15.14) that the extended Hückel method is not reliable for conformational studies.

A number of researchers have concluded that the CNDO and INDO methods are not reliable for conformational work [A. Veillard, *Chem. Phys. Lett.*, **33**, 15 (1975); O. Gropen and H. M. Seip, *Chem. Phys. Lett.*, **11**, 445 (1971); L. L. Combs and M. Holloman, *J. Phys. Chem.*, **79**, 512 (1975); A. R. Gregory and M. N. Paddon-Row, *J. Am. Chem. Soc.*, **98**, 7521 (1976)].

The PCILO method is substantially more reliable than CNDO for conformational work [D. Perahia and A. Pullman, *Chem. Phys. Lett.*, **19**, 73 (1973)] and has given pretty good results in calculations of conformations of some biological molecules [B. Pullman, *Adv. Quantum Chem.*, **10**, 251 (1977)]. A study of PCILO versus ab initio STO-3G conformations of three amino acids [P. R. Lawrence and C. Thomson, *Theor. Chim. Acta*, **58**, 121 (1981)] found that all conformations corresponding to energy minima in the STO-3G calculation were also minima in the PCILO calculation; however, PCILO predicted minima for certain conformations that were not minima in the STO-3G calculation. The study concluded that "while PCILO does not reproduce exactly the results of *ab initio* calculations, a preliminary study of a molecule using PCILO may be very useful [as a starting point for more accurate calculations]."

Dihedral angles calculated by MNDO are pretty accurate in most cases, but there are some substantial errors and the number of dihedral angles calculated is not large enough to reach a definitive conclusion about the value of MNDO here. Dewar has noted that "ring conformations are not reproduced well by either MNDO or MINDO/3" [M. J. S. Dewar and W. Thiel, *J. Am. Chem. Soc.*, **99**, 4907 (1977)].

Semiempirical methods give very substantial savings in computation time over ab initio methods. The CNDO, INDO, MINDO/3, and MNDO methods all require roughly the same amount of computer time. For a single calculation on a medium-size molecule (something with on the order of forty minimal-basis AO's; e.g., B_6H_{10}), very approximate relative computer times are as follows: 1 for CNDO, etc., 6 for PRDDO, 100 for ab initio STO-3G, 600 for ab initio 4-31G, and 3000 for ab initio

6-31G* [T. A. Halgren et al., *J. Am. Chem. Soc.*, **100**, 6595 (1978); W. J. Hehre, *J. Am. Chem. Soc.*, **97**, 5308 (1978)]. For larger molecules the time advantage for the semiempirical methods becomes even more pronounced. The same is true if geometry optimization is done, because of the rapidity with which the energy gradient is calculated using semiempirical methods.

Because of the much shorter computation time, semiempirical methods are applicable to much larger molecules than ab initio methods. A 1980 discussion (S. Stinson, *Chem. Eng. News*, March 17, 1980, p. 34) stated that STO-3G calculations are limited to molecules with ten or fewer nonhydrogen atoms when done on a maxicomputer (Sec. 15.20) and to twenty or fewer nonhydrogens when done on a minicomputer. However, STO-3G calculations of equilibrium geometries are not feasible with molecules this large; for example, in a study of carbonium ions, the STO-3G geometries of the $C_8H_9^+$ isomers were only partially optimized by calculation (W. J. Hehre in *Schaefer, Applications of Electronic Structure Theory*, Chapter 7). When full geometry optimization is not possible, one uses some assumed bond lengths and angles, based on experimental data and calculations on similar smaller molecules.

In summary, one might say that although the extended Hückel, CNDO, and INDO methods may each be useful for certain restricted applications, the overall reliability of these methods for calculating molecular properties is not very good. The overall reliability of the ab initio SCF MO method is generally good for ground-state closed-shell molecules, provided one uses a basis set of suitable size (the size required depends on the property being calculated) and provided one restricts oneself to energy changes between isomers and energy changes in isodesmic reactions. Of course, one is quite restricted in the size of the molecules for which ab initio calculations can be done. The MNDO method is very substantially better than CNDO or INDO in reproducing molecular properties, but it does not seem to be as reliable as ab initio calculations.

Some workers are mistrustful of methods like MINDO and MNDO. A review article stated that although "very promising results" can be obtained for molecules similar to those used to parametrize the method, "there is no theoretical analysis to tell us the limits beyond which the method will give poor predictions." Moreover, "these methods may give us an incorrect or inadequate understanding of experimental phenomena, even though they may give good quantitative results" (B. J. Duke in *Theoretical Chemistry*, vol. 2, *Specialist Periodical Reports*, The Chemical Society, London, 1975, p. 185).

15.17 *VALENCE-BOND TREATMENT OF POLYATOMIC MOLECULES*

The valence-bond treatment of polyatomic molecules is closely tied to chemical ideas of structure.[43] One begins with the atoms that form the molecule and

---

[43] For a review of VB theory, see M. Simonetta, "Forty Years of Valence Bond Theory," in A. Rich and N. Davidson, eds., *Structural Chemistry and Molecular Biology*, Freeman, San Francisco, 1968.

pairs up the unpaired electrons to form chemical bonds. There are generally several ways of pairing up the electrons; each pairing scheme gives a VB *structure*. The molecular wave function is taken as a linear combination of Heitler–London-type functions (called *bond eigenfunctions*) for each structure, and the variation principle is applied. The VB wave function is said to be a *resonance hybrid* of the various structures; the weight of each resonance structure is usually taken as proportional to the square of its coefficient in the molecular wave function. Because the bond eigenfunctions are not mutually orthogonal, the electron probability density is not equal to the weighted sum of the probability densities of the various structures, and the weights of resonance structures are somewhat lacking in physical significance.

Consider water as an example. The ground-state configuration of oxygen is $1s^2 2s^2 2p^4$ with an unpaired electron in each of the AO's $2p_y$ and $2p_z$. We thus assume that these AO's along with the hydrogen $1s$ AO's will form electron-pair bonds. There are three possible ways of pairing these four AO's to get covalent structures:

$$\begin{array}{ccc}
\text{O}2p_y \quad \text{O}2p_z & \text{O}2p_y \text{\textemdash} \text{O}2p_z & \text{O}2p_y \quad \text{O}2p_z \\[4pt]
\Big| \quad\quad \Big| & & \times \\[4pt]
\text{H}_1 1s \quad \text{H}_2 1s & \text{H}_1 1s \text{\textemdash} \text{H}_2 1s & \text{H}_1 1s \quad \text{H}_2 1s \\[4pt]
A & B & C
\end{array} \qquad (15.201)$$

Let

$$s_1 \equiv \text{H}_1 1s, \qquad s_2 \equiv \text{H}_2 1s, \qquad p_y \equiv \text{O}2p_y, \qquad p_z \equiv \text{O}2p_z$$

The normalized VB function corresponding to structure A is

$$\Phi_A = N | \cdots \widehat{p_y s_1} \widehat{p_z s_2} | = N | \cdots p_y \overline{s_1} \, p_z \overline{s_2} | - N | \cdots \overline{p_y} s_1 \, p_z \overline{s_2} |$$
$$- N | \cdots p_y \overline{s_1} \, \overline{p_z} s_2 | + N | \cdots \overline{p_y} s_1 \, \overline{p_z} s_2 | \qquad (15.202)$$

where the signs are determined by the rule in Section 13.13; the dots stand for $\text{O}1s\overline{\text{O}1s}\,\text{O}2s\overline{\text{O}2s}\,\text{O}2p_x\overline{\text{O}2p_x}$. Similarly,

$$\Phi_B = N | \cdots \widehat{p_y p_z} \widehat{s_1 s_2} |, \qquad \Phi_C = N | \cdots \widehat{p_y s_2} \widehat{p_z s_1} | \qquad (15.203)$$

We then take as a trial variation function ψ, a linear combination of the bond eigenfunctions of structures A, B, and C. However, the functions Φ_A, Φ_B, and Φ_C are not linearly independent; we have (Problem 15.59)

$$\Phi_C = -(\Phi_A + \Phi_B) \qquad (15.204)$$

It is wasted effort to include all three structures; we shall drop structure C, taking $\psi = c_A \Phi_A + c_B \Phi_B$.

For a molecule with n valence AO's to be paired, where n is even, Rumer showed in 1932 that the following procedure gives the linearly independent covalent structures for singlet states: The n AO's are arranged in a ring and lines are drawn between pairs of AO's; those structures where no lines cross are linearly independent. These are the VB *canonical* covalent structures of the molecule. Any structure with lines crossing is linearly dependent on the canonical structures and is omitted from the VB wave function. The number of ways of drawing $\frac{1}{2}n$ noncrossing lines between n points on a circle is[44]

$$\frac{n!}{(\frac{1}{2}n)!\,(\frac{1}{2}n+1)!} \tag{15.205}$$

For water there are $4!/2!\,3! = 2$ canonical covalent structures, and these are A and B. (Actually, which structures are taken as the canonical ones is arbitrary, since the orbitals can be arranged in various ways on the ring.) To use Rumer's method when the number of orbitals to be paired is odd, one adds a "phantom" orbital, whose contribution is subtracted at the end of the calculation.

Rumer's procedure is easily justified. Let $\Phi(|\quad|)$, $\Phi(\boxminus)$, and $\Phi(\times)$ be three bond eigenfunctions that involve any number of AO's, but that differ only in the way they pair up a certain subset of four AO's—each of these functions corresponds to one of the three different ways of pairing these four AO's [see (15.201)]. By a slight extension of (15.204), it follows that

$$\Phi(\times) = -[\Phi(|\quad|) + \Phi(\boxminus)] \tag{15.206}$$

Any pairing scheme involving lines that cross can be shown by repeated application of (15.206) to be a linear combination of structures with no lines crossing. For example, consider the following pairing scheme for six orbitals:

$$\tag{15.207}$$

Application of (15.206) to the set of four AO's 1, 5, 3, 6 shows that (15.207) is a linear combination of

$$\text{and} \tag{15.208}$$

Use of (15.206) then shows each of the structures in (15.208) to be a linear combination of two structures with no lines crossing. Thus (15.207) has been reduced to a linear combination of canonical structures.

The structures corresponding to the pairing schemes A and B in (15.201) for water are

$$\tag{15.209}$$

$$A \qquad\qquad\qquad B$$

---

[44] For a proof of (15.205), see J. Barriol, *Elements of Quantum Mechanics with Chemical Applications*, Barnes and Noble, New York, 1971, pp. 281–282.

The separation between H_1 and H_2 is considerably greater than between O and each hydrogen; the small overlap between the hydrogen $1s$ AO's makes structure A far more significant than structure B. Since the $2p_y$ and $2p_z$ AO's are at 90° to each other, the VB structure A predicts a bond angle of 90°, since this allows maximum overlap between the bonding oxygen and hydrogen AO's. The observed angle of $104\frac{1}{2}°$ can be rationalized by considering electrostatic repulsions between the hydrogen atoms (ionic structures) and by allowing some mixing in (hybridization) of the $2s$ oxygen AO with the two bonding $2p$ AO's.

For H_2 the Heitler–London function was improved by inclusion of small contributions from ionic structures. Because of the considerable electronegativity difference between O and H, we expect ionic structures to be important in H_2O. [Recall, however, that the "covalent" structure A of (15.209) contains some ionic character—see Section 13.20.] Ionic structures for water include

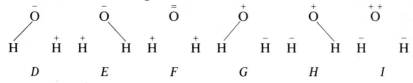

$$\qquad D \qquad\qquad E \qquad\qquad F \qquad\qquad G \qquad\qquad H \qquad\qquad I$$

Because oxygen is much more electronegative than hydrogen, the contributions of G, H, and I should be negligible. A reasonable variation function for water is

$$c_A \Phi_A + c_B \Phi_B + c_D \Phi_D + c_E \Phi_E + c_F \Phi_F \qquad (15.210)$$

The ionic functions are written by analogy to (13.190); for example,

$$\Phi_D = |\cdots \widehat{p_y s_1} p_z \overline{p_z}| = |\cdots p_y \overline{s_1} p_z \overline{p_z}| - |\cdots \overline{p_y} s_1 p_z \overline{p_z}| \qquad (15.211)$$

The linear variation function (15.210) leads to a secular equation

$$\det (H_{ij} - ES_{ij}) = 0 \qquad (15.212)$$

The AO's are expressed as, say, STO's, and the matrix elements and overlap integrals between the functions $\Phi_A, \ldots, \Phi_F$ are calculated. The lowest root of the secular equation gives an approximation to the ground-state energy (higher roots correspond to excited singlet states); evaluation of the corresponding coefficients gives the VB ground-state wave function. The calculation can be done semiempirically using experimental data to evaluate some of the integrals. Calculation of the matrix elements is involved, and approximations are often made; for example, overlap integrals between different AO's (but *not* between different structures) are neglected and exchange integrals involving interchange of the coordinates of more than two electrons are neglected. Systematic procedures for evaluating the matrix elements have been developed.[45]

One of the rare ab initio all-electron VB calculations on polyatomic molecules is Peterson and Pfeiffer's calculation on H_2O (Table 15.1). They used a Gaussian-lobe AO basis set and included ten covalent and thirty-nine ionic VB structures in their

[45] L. Pauling, *J. Chem. Phys.*, **1**, 280 (1933); M. Simonetta and V. Schomaker, *J. Chem. Phys.*, **19**, 649 (1951).

wave function. (The large number of structures arises because they considered structures arising from such oxygen configurations as $1s^2 2s2p^5$ and $1s^2 2p^6$, as well as $1s^2 2s^2 2p^4$.) The contribution of ionic structures was found to be 62 percent. They also performed calculations on OH and O, so as to calculate the first and second bond dissociation energies of H_2O; the calculated dissociation energies were not in good agreement with experiment.

Consider the VB treatment of methane. The carbon ground-state configuration $1s^2 2s^2 2p^2$ has two unpaired electrons and would seem to indicate a valence of 2. To get the well-known tetravalence of carbon, we assume that a $2s$ electron is promoted to the vacant $2p$ orbital, giving the configuration $1s^2 2s2p^3$. If we then assume one bond is formed with the $2s$ electron and three bonds are formed with the $2p$ electrons, the bonds are not all equivalent as they are known to be. Hence Pauling proposed that the L-shell s and p functions be linearly combined to form *hybridized* sp^3 orbitals, of the form

$$b_i(C2s) + d_i(C2p_x) + e_i(C2p_y) + f_i(C2p_z), \qquad i = 1, \ldots, 4 \qquad \textbf{(15.213)}$$

For maximum overlap we want each function to point to a vertex of a tetrahedron. From the discussion following Eq. (15.57), the constants d_i, e_i, f_i are proportional to the direction cosines in (15.68). Also, each orbital should have the same value of b_i, so that the four hybrid orbitals will be equivalent. Thus the orbitals (15.213) are of the form (15.69) with $a = 0$ and $2a_1, 1t_{2x}$, etc., replaced by $C2s, C2p_x$, etc. If we impose the requirement that the hybrid AO's be orthonormal, we have

$$b^2 + c^2 = 1, \qquad b^2 - \tfrac{1}{3}c^2 = 0 \qquad \textbf{(15.214)}$$

$$c = \tfrac{1}{2}\sqrt{3}, \qquad b = \tfrac{1}{2} \qquad \textbf{(15.215)}$$

The four equivalent sp^3 hybrid carbon AO's are then

$$\text{te}_1 = \tfrac{1}{2}[C2s + C2p_x + C2p_y + C2p_z]$$
$$\text{te}_2 = \tfrac{1}{2}[C2s + C2p_x - C2p_y - C2p_z]$$
$$\text{te}_3 = \tfrac{1}{2}[C2s - C2p_x + C2p_y - C2p_z] \qquad \textbf{(15.216)}$$
$$\text{te}_4 = \tfrac{1}{2}[C2s - C2p_x - C2p_y + C2p_z]$$

where "te" stands for tetrahedral. A typical sp^3 hybrid AO contour is shown in Fig. 15.22. (The three-dimensional shape is obtained by rotation about a horizontal axis through carbon.) Carbon sp^2 and sp hybrid AO's have similar shapes. [See I. Cohen and T. Bustard, *J. Chem. Educ.*, **43**, 187 (1966).]

There are many canonical covalent VB structures for methane, as well as ionic structures; however, chemical intuition tells us that the main contribution to the wave function is from the following covalent structure:

$$|C1s\overline{C1s}\,\widehat{s s_1}\,\text{te}_1\,\widehat{s_2}\,\text{te}_2\,\widehat{s_3}\,\text{te}_3\,\widehat{s_4}\,\text{te}_4| \qquad \textbf{(15.217)}$$

where $s_1 = H_1 1s$, etc. The structure (15.217) is a linear combination of $2^4 = 16$ determinants. [Taking the CH_4 VB wave function as the single covalent structure (15.217) or the H_2O VB wave function as the function corresponding to structure A in (15.209) gives what is called the *perfect pairing* approximation.]

Node

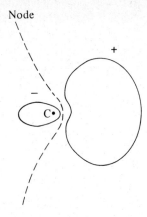

Figure 15.22 *Carbon sp³ hybrid orbital*

For ethylene the vb method uses sp^2-hybridized carbon AO's to form the σ bonds by overlap with the $1s$ hydrogen AO's; this leaves a p orbital on each carbon to form the π bond. The HCH bond angle is predicted to be 120°, in reasonable agreement with the observed value 116° or 117°. For acetylene each carbon is sp hybridized.

Now consider benzene. We use the π-electron approximation. The σ bonds are formed by sp^2 carbon hybrid AO's and $1s$ hydrogen orbitals; this leaves a p orbital on each carbon to form π bonds. There are $6!/3!4! = 5$ canonical covalent structures for pairing the π orbitals, and these are

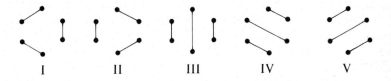

| I | II | III | IV | V |

Structures I and II are the **Kekulé** structures, and III, IV, and V are the Dewar structures. (The puckered compound bicyclo[2.2.0]hexa-2, 5-diene has been

synthesized and is called *Dewar benzene*. The vb Dewar "structures" of benzene are formal ways of pairing up electrons in AO's; each vb Dewar "structure" is based on a regular hexagonal arrangement of carbon atoms and is to be distinguished from the compound "Dewar benzene." Likewise, the structures I and II correspond to a regular hexagon of carbons, and differ from the hypothetical cyclohexatriene with alternating bond lengths. Each vb resonance "structure" is based on the same internuclear distances, but a different electron pairing scheme.)

The following types of singly polar ionic structures occur:

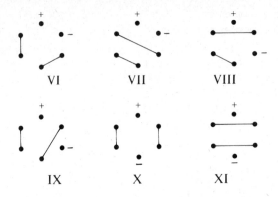

| | VI | VII | VIII |
| IX | X | XI |

There are twelve structures of the form VI (the plus sign can be put on each of six carbons, and the minus sign can be put on the preceding or following carbon), twelve of the form VII, twelve of the form VIII, twelve of the form IX, six of the form X, and six of the form XI. There are also contributions from doubly polar and triply polar structures. Of course, we include only linearly independent ionic structures; an ionic structure that involves lines crossing is a linear combination of two ionic structures with no lines crossing and is omitted. To simplify things, we include only the ionic structures that make the largest contributions; thus we omit doubly and triply polar structures; also the six singly polar structures like XI will be omitted, since they involve two "long" bonds between nonadjacent carbons.

Simonetta[46] did a semiempirical calculation of benzene, including the five canonical covalent structures and fifty-four singly polar ionic structures (VI–X). As for H_2 (Section 13.9), a vb calculation of polyatomic molecules involves Coulomb and exchange integrals. Simonetta took these integrals as empirical parameters whose values are adjusted to fit such experimental data as electronic spectra and thermodynamically determined resonance energies of π-electron systems. He found for the ground-state π-electron vb wave function of benzene

$$0.56(I + II) + 0.24(III + IV + V) + 0.11(VI + 11 \text{ others})$$

$$+ 0.05(VII + 11 \text{ others}) + 0.02(VIII + 11 \text{ others})$$

$$+ 0.02(IX + 11 \text{ others}) + 0.02(X + 5 \text{ others})$$

The Kekulé structures make the greatest contribution. Even though the coefficient of each ionic structure is small, the large number of ionic structures makes their contribution significant.

Recall that increasing the number of functions f_i in a linear variation function $\sum c_i f_i$ improves the variation function; i.e., lowers the value of the variational integral. If we were to consider structure I only, then the energy obtained would be considerably higher than when several vb structures are considered. The difference

---

[46] M. Simonetta, *J. Chim. Phys.*, **49**, 68 (1952).

between the energy for structure I and that found when all vb structures are considered is the *resonance energy* of benzene; the resonance energy is the vb analog of the mo delocalization energy. One says that benzene is "stabilized by resonance," but of course resonance is not a real phenomenon.

Such concepts as "configuration interaction," "resonance," "exchange," "hybridization," etc., are not real physical phenomena, but only artifacts of the approximations used in the calculations. Even the concept of orbitals is but an approximation; strictly speaking, orbitals do not exist. Around 1950 resonance theory came under attack by some Russian chemists as being "vicious," "perverse," "bourgeois," etc. The basis for the attack was the fact that vb resonance structures and the resonance "phenomenon" have no physical reality; hence resonance theory was deemed incompatible with Marxist–Leninist dialectical materialism.[47]

For cyclobutadiene, C_4H_4, a vb treatment including only the two canonical covalent π-electron structures gives considerable resonance energy and thus predicts stability for the molecule; however, when singly and doubly polar ionic structures are included, the vb method does predict lack of stability.[48]

Using the π-electron approximation for butadiene, we have $4!/2!3! = 2$ canonical covalent vb structures:

$$1: \quad CH_2{=}CH{-}CH{=}CH_2 \qquad 2: \quad \overline{CH_2{-}CH{=}CH{-}CH_2}$$

There are twelve singly polar ionic structures (Problem 15.64) of which two have a central double bond. Simonetta's fourteen-structure semiempirical calculation[49] gave the partial double-bond character of the central bond as due mainly to structure 2.

Naphthalene has $10!/5!6! = 42$ canonical covalent structures for the π electrons. There are three Kekulé structures (no "long" bonds), sixteen Dewar structures (one long bond), nineteen structures with two long bonds, and four structures with three long bonds (Fig. 15.23). The number of long bonds is called the

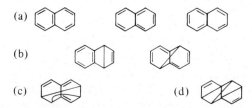

Figure 15.23 *Some vb π-electron structures for naphthalene. (a) The three Kekulé structures. (b) Two Dewar structures. (c) A doubly excited structure. (d) A triply excited structure.*

[47] For further discussion see I. M. Hunsberger, *J. Chem. Educ.*, **31**, 504 (1954). For translations of Russian attacks on resonance, see I. S. Bengelsdorf, *J. Chem. Educ.*, **29**, 2, 13 (1952).

[48] R. McWeeny, *Proc. Roy. Soc.*, **A227**, 288 (1955).

[49] M. Simonetta, *J. Chim. Phys.*, **49**, 68 (1952); see also M. Simonetta, E. Gianinetti, and I. Vandoni, *J. Chem. Phys.*, **48**, 1579 (1968)—this paper formulates vb theory for nonsinglet states.

degree of excitation of a vʙ structure; a Kekulé structure is unexcited, a Dewar structure singly excited, etc. (There is no connection between the degree of excitation of a vʙ structure and the excited electronic states of a molecule.) The greater the degree of excitation, the smaller the contribution of a structure. Each vʙ π-electron structure of naphthalene is a linear combination of $2^5 = 32$ determinants, and the calculation of matrix elements between the forty-two covalent structures and perhaps some ionic structures as well is involved. Obviously, the мо approach is computationally much simpler than the vʙ method for aromatic hydrocarbons. The naphthalene ionic structures are usually omitted in calculations, not because their contribution is negligible, but because their inclusion makes the calculation quite difficult. A semiempirical naphthalene calculation[50] that includes the forty-two covalent structures and no ionic structures showed the first Kekulé structure in Fig. 15.23a to be the most important one.

As in the мо method, one can define vʙ π-electron bond orders, free valences, etc.

Organic chemists have found the ideas of vʙ (resonance) theory valuable for rationalizing in a simple manner the facts of organic structure and reactions; no doubt the reader is familiar with such applications from organic-chemistry courses. However, quantitative application is much harder for the vʙ method than for the мо method. In most cases, the single-determinant Hartree–Fock мо wave function (or an accurate scf approximation to it) furnishes a good description of the molecular ground state; in contrast, the vʙ ground-state wave function frequently must be taken as a linear combination of many covalent and ionic structures. For quantitative calculations the мо method has overshadowed the vʙ method. The vʙ resonance approach is perhaps best suited to qualitative discussion of molecular structure.[51]

The vʙ method is often described as having the advantage of being based on familiar chemical ideas of bonding and structure. While it is true that the canonical мо's are not related to chemical bonds, the energy-localized мо's (which are becoming available for an increasing number of molecules) do furnish a convenient description in terms of chemical bonds.

Next we consider the concept of the *valence state* of an atom. By the valence state of an atom for a given molecular electronic state, we mean the state in which the atom exists in the molecule. Since individual atoms do not really exist in molecules, the valence-state concept is an approximate one. The vʙ approximation constructs molecular wave functions from wave functions of the individual atoms; we use the vʙ wave function to define the valence state of an atom as the wave function obtained on removing all other atoms to infinity, while keeping the form of the molecular wave function invariant. (This process is purely hypothetical, and the valence state is not in general a stationary atomic state.)

The simplest example is the H_2 ground state. The Heitler–London function is

$$N|1\widehat{s_a 1s_b}| = N[|1s_a \overline{1s_b}| - |\overline{1s_a} 1s_b|]$$

[50] T. H. Goodwin, *Theor. Chim. Acta*, **2**, 315 (1964).
[51] See L. Pauling, *The Nature of the Chemical Bond*, Third Edition, Cornell University Press, Ithaca, 1960; G. W. Wheland, *Resonance in Organic Chemistry*, Wiley, New York, 1955.

Removal of hydrogen atom b leaves (at large internuclear separation, the normal-ization constant for each Slater determinant becomes $1/\sqrt{2}$)

$$2^{-1/2}1s_a - 2^{-1/2}\overline{1s_a} \qquad (15.218)$$

We have a hydrogen atom with a $1s$ spatial function and a 50 percent probability of having each of spin α or spin β. The valence-state ionization potential of hydrogen is thus 13.6 eV, as in (15.197).

A less trivial example is water. The largest contribution to the VB wave function is from the bond eigenfunction (15.202). Removal of the hydrogen atoms from (15.202) gives

$$N[|\cdots 2p_y 2p_z| - |\cdots \overline{2p_y} 2p_z| - |\cdots 2p_y \overline{2p_z}| + |\cdots \overline{2p_y}\, \overline{2p_z}|] \qquad (15.219)$$

Each determinant in (15.219) belongs to the oxygen configuration $1s^2 2s^2 2p^4$. This configuration gives the terms ^3P, ^1D, and ^1S (Table 11.2). The first and last determinants are eigenfunctions of $\hat{S}_z$ with eigenvalues $+1\hbar$ and $-1\hbar$, respectively; hence these two determinants must correspond to states of the ^3P term (which has $S = 1$). Analysis (which we omit) of the other two determinants shows each of them to be an equal mixture (coefficients $1/\sqrt{2}$) of states belonging to the ^1D and ^3P terms. Thus the valence state is a mixture of states of the terms ^1D and ^3P and is definitely not a stationary state of the atom. The valence-state wave function ψ_{vs} is not an eigenfunction of the atomic Hamiltonian; we can, however, calculate an average energy:

$$\langle E \rangle_{vs} = \int \psi_{vs}^* \hat{H}\psi_{vs}\, d\tau \Big/ \int \psi_{vs}^* \psi_{vs}\, d\tau$$

From $\psi_{vs} = c_1 \psi\,(^3P) + c_2 \psi\,(^1D)$, we have

$$\langle E \rangle_{vs} = \frac{|c_1|^2 E(^3P) + |c_2|^2 E(^1D)}{|c_1|^2 + |c_2|^2}$$

From the above arguments, we have for c_1 and c_2 in the unnormalized valence-state wave function

$$|c_1|^2 = 1^2 + (2^{-1/2})^2 + (2^{-1/2})^2 + 1^2 = 3$$

$$|c_2|^2 = (2^{-1/2})^2 + (2^{-1/2})^2 = 1$$

The valence-state energy is then

$$\langle E \rangle_{vs} = \tfrac{3}{4}E(^3P) + \tfrac{1}{4}E(^1D) \qquad (15.220)$$

$\langle E \rangle_{vs}$ can then be calculated from Moore's tables of atomic energy levels (Section 11.6).

If $2s$ hybridization is included in the $2p$ bonding oxygen orbitals of water, the oxygen valence state is found[52] to be a linear combination involving terms of the configurations $1s^2 2s^2 2p^4$, $1s^2 2s2p^5$, and $1s^2 2p^6$. Hybridization gives a mixing of configurations in the valence state.

---

[52] M. Kotani, K. Ohno, and K. Kayama, "Quantum Mechanics of Electronic Structure," in S. Flugge, ed., *Encyclopedia of Physics/Handbuch der Physik*, Springer, Berlin, 1961, vol. 37, pp. 110–115.

The valence state of carbon in methane is of considerable importance. It is sometimes carelessly stated that the carbon sp^3 valence state corresponds to the ^5S term of the carbon atom $2s2p^3$ configuration. This is not correct. The $M_S = 2$ state of the $2s2p^3\ {}^5S_2$ level has one $2s$ and three $2p$ electrons, with all four of these electrons having spin α. It is true that we can use the procedure of Section 15.6 to form linear combinations of the $2s$ and $2p$ orbitals and thereby put each of the four outer electrons of the $^5S_2\ M_S = 2$ state into an sp^3 hybrid AO without changing the wave function. However, each such hybrid AO would still have spin α. On the other hand, when we remove the hydrogen AO's from the CH_4 VB wave function (15.217), we are left with a linear combination of sixteen determinants in which each sp^3 hybrid AO has spin α in eight determinants and spin β in eight determinants. Thus the carbon valence state differs from the $2s2p^3\ {}^5S_2\ M_S = 2$ state and in fact differs from the other states of this term. The valence state obtained on removal of the H's from (15.217) turns out to be a mixture of states of the ^5S, ^3D, and ^1D terms of the $2s2p^3$ atomic configuration. [When other CH_4 VB structures besides (15.217) are included, we also get contributions from terms of the $2p^4$ and $2s^2 2p^2$ configurations to the C valence state.[53]] The valence-state energy of carbon is well above the energy of the ground ^3P term of the $2s^2 2p^2$ configuration, but the energy gained by forming four bonds instead of two more than compensates for the energy needed to form the valence state.

Valence-state ionization potentials are used to estimate integrals in semi-empirical calculations, such as the extended Hückel and Pariser–Parr–Pople methods. The valence-state ionization potential for a $2p$ electron in an sp^3-hybridized carbon atom [Eq. (15.197)] is the energy difference between the valence state of sp^3-hybridized C and the valence state of sp^2-hybridized C^+. Tables of valence-state ionization potentials are available.[54] Valence-state ionization potentials and electron affinities are also used to calculate electronegativities from Mulliken's definition of half the sum of the valence-state ionization potential and electron affinity.

Valence-bond theory has been used to describe the electronic structure of transition-metal complex ions, with such concepts as $d^2 sp^3$ hybridization of the metal orbitals. However, the simple VB treatment of complex ions is not fully satisfactory and has been replaced by ligand-field theory, which is MO theory applied to species whose atoms have d (or f) electrons (see *Murrell, Kettle, and Tedder*, Chapter 13; *Offenhartz*, Chapter 9).

15.18 *THE GENERALIZED VALENCE-BOND METHOD*

The generalized valence-bond (GVB) method was developed about 1970 by Goddard and co-workers [W. J. Hunt, P. J. Hay, and W. A. Goddard, *J. Chem. Phys.*,

[53] H. H. Voge, *J. Chem. Phys.*, **4**, 581 (1936).
[54] J. Hinze and H. H. Jaffé, *J. Am. Chem. Soc.*, **84**, 540 (1962); G. Pilcher and H. A. Skinner, *J. Inorg. Nucl. Chem.*, **24**, 937 (1962); L. C. Cusachs, J. W. Reynolds, and D. Barnard, *J. Chem. Phys.*, **44**, 835 (1966).

57, 738 (1972); P. J. Hay, W. J. Hunt, and W. A. Goddard, *J. Am. Chem. Soc.*, **94**, 8293 (1972); W. A. Goddard, T. H. Dunning, W. J. Hunt, and P. J. Hay, *Acc. Chem. Res.*, **6**, 368 (1973)].

The Heitler–London VB wave function for ground-state H_2 is [Eq. (13.107)] $1s_a(1) 1s_b(2) + 1s_a(2) 1s_b(1)$ multiplied by a normalization constant and a spin function. The GVB ground-state H_2 wave function replaces this spatial function by $f(1)g(2) + f(2)g(1)$, where the functions f and g are found by minimization of the variational integral. To find f and g, one expands each of them in terms of a basis set of AO's and finds the expansion coefficients by iteratively solving one-electron equations that resemble the equations of the SCF method.

Clearly, the GVB method will give a lower energy than the simple VB wave function. The GVB method allows for the change in the AO's that occurs on molecule formation by solving variationally for f and g. In the VB method, this change is allowed for by adding to the wave function terms that correspond to ionic and other resonance structures. The GVB wave function is thus much simpler than a VB wave function with resonance structures and the calculations are simpler.

The GVB method gives a D_e of 4.12 eV for ground-state H_2, as compared to 3.20 eV for the Heitler–London VB function, 3.78 eV for the Heitler–London function with an optimized orbital exponent, and 4.75 eV for the experimental value. At very large internuclear distance R, the GVB functions f and g approach the atomic orbitals $1s_a$ and $1s_b$. Thus, like the VB ψ (but unlike the MO ψ), the GVB wave function shows the correct behavior on dissociation. At intermediate distances f is a linear combination of AO's which has its most important contribution from $1s_a$ but which has a significant contribution from $1s_b$ and lesser contributions from the other AO's (these contributions reflect the polarization of the $1s_a$ AO that occurs on molecule formation).

In the MO method, there is one orbital (an MO) for each electron pair. In the GVB method, there are two orbitals (f and g in the H_2 example) for each electron pair.

For CH_4 the VB wave function with resonance structures omitted is (15.217). For CH_4 the GVB wave function is

$$|i_a i_b \, b_{1a} b_{1b} \, b_{2a} b_{2b} \, b_{3a} b_{3b} \, b_{4a} b_{4b}|$$

The electrons are divided into pairs, and each pair is given two orbitals. In the VB method, the inner-shell orbitals i_a and i_b are each assumed to be $1s$ AO's on carbon, the bonding orbital b_{1a} is assumed to be an sp^3-hybridized carbon AO pointing toward H_1, and b_{1b} is assumed to be a $1s$ AO on hydrogen number 1. In the GVB method, no assumptions are made about the nature of the orbitals. One simply expands each of them in terms of the chosen basis set and solves the GVB equations until self-consistency is attained.

To simplify the calculation, the GVB method restricts orbitals in different pairs to be orthogonal to one another. (For example, b_{1a} and b_{2a} are assumed orthogonal, but b_{1a} and b_{1b} are not assumed orthogonal.) This should be a good assumption since Pauli repulsion between pairs keeps them well separated spatially; moreover, a few test calculations have shown the orthogonality requirement to lead to little error.

For CH_4 one finds that i_a and i_b are essentially $C1s$ AO's; b_{1a} is an orbital centered mainly on C with some contribution from H_1; b_{1a} points toward H_1 and has a carbon AO hybridization of $sp^{2.1}$ (using a minimal basis set). The hybridization differs from the sp^3 hybridization of the VB wave function as a result of the contribution of H_1 1s to b_{1a}. The orbital b_{1b} is found to be mainly an H_1 1s AO with some contribution of carbon AO's mixed in, thereby polarizing the orbital toward C.

Since inner-shell electrons are little changed on molecule formation, one can simplify the GVB calculation by assuming each of i_a and i_b to be a C1s AO, as is done in the VB wave function. This gives little loss in accuracy.

For C_2H_6 a GVB calculation with a minimal basis set gave a 3.1 kcal/mole rotational barrier, in good agreement with the 2.9 kcal/mole experimental value.

For C_2H_4 a GVB calculation gave a description of the double bond as composed of one σ and one π bond, in contrast to the energy-localized MO's (Section 15.6), which are two equivalent bent banana bonds.

For H_2O a GVB calculation produced an inner-shell pair on oxygen, two equivalent bonding pairs, and two equivalent lone pairs; with a double-zeta plus polarization on oxygen basis set, an energy of -76.11 hartrees was obtained, which is below the Hartree–Fock limit of -76.07 hartrees (Table 15.1).

The GVB method (like the VB method) gives a description in terms of localized inner-shell, bonding, and lone pairs, whereas one must carry out a time-consuming special procedure to find localized MO's from canonical SCF MO's.

The GVB method is most applicable to molecules for which a single VB covalent structure is a good approximation.

The GVB method has been used to develop qualitative descriptions of chemical bonding; see W. A. Goddard and L. B. Harding, *Ann. Rev. Phys. Chem.*, **29**, 363 (1978). GVB plus CI calculations have been used to study catalysis by transition metals; see A. K. Rappé and W. A. Goddard, *J. Am. Chem. Soc.*, **104**, 297, 448 (1982).

15.19 *CHEMICAL REACTIONS*

The course of a chemical reaction is determined by the potential-energy function for nuclear motion $U(q_\alpha)$ (Section 13.1), where q_α indicates the coordinates of the N nuclei of the reactant molecules. For two atoms coming together to form a diatomic molecule, U is a function of one variable, the internuclear distance R, and $U(R)$ is the usual potential-energy curve. For more interesting chemical reactions, U is a function of $3N - 6$ variables; we subtract 6 because the three translational and three rotational degrees of freedom of the system leave U unchanged. (For a two-atom system, there are only two rotational degrees of freedom.) U gives what is called the *potential-energy surface* for the reaction. If U depends on two variables, then a plot of $U(q_1, q_2)$ in three dimensions gives a surface in ordinary three-dimensional space. Because of the large number of variables, $U(q_\alpha)$ is a "surface" in an abstract "space" of $3N - 5$ dimensions. To find $U(q_\alpha)$, we must solve the electronic Schroedinger equation at a large number of nuclear configurations, a truly formidable task.

Once U is found, we look for the path of minimum potential energy on U connecting reactants and products. The point of maximum potential energy U on the minimum-energy path is called the *transition state*; this is a saddle point on the

U surface, since it is a maximum point on a minimum-energy path. The transition state is not a stable molecule, and the transition from reactants to products is a smooth one. (However, in certain theories of reaction rates, it is convenient to ascribe properties such as entropy, free energy, vibrational frequencies, etc., to the transition state.) The energy difference between the transition state and the reactants (omitting zero-point vibrational energies) is called the (*classical*) *barrier height* for the forward reaction. (For the reverse reaction, the surface U is the same as for the forward reaction.)

When the U surface is known, it is possible (at least in principle) to calculate the reaction rate constant k as a function of temperature. Such calculations are extremely difficult. One must allow for quantum-mechanical tunneling through the barrier; tunneling is important in reactions involving light species (e^-, H^+, H, H_2), including reactions that transfer such species between heavy molecules. Moreover, there is significant probability for molecules to traverse the reaction surface on paths that deviate somewhat from the minimum-energy path. Thus one must perform averaging over the various possible paths. (In other words, we must consider different approaches of the reactant molecules, calculate the probability of reaction occurring for each approach, and then suitably average these probabilities.) The rate constant thus depends not only on the barrier height, but on the whole shape of the reaction potential-energy surface. (Speaking thermodynamically, one considers the activation entropy as well as the activation energy.) For qualitative discussion we can use the fact that a large barrier height means a small rate constant, and a low barrier means a fast reaction.

By measuring the experimental rate constant k as a function of temperature, one can determine an experimental *activation energy* E_a, where $k = A \exp(-E_a/RT)$. The experimental quantity E_a *differs* from the barrier height on the U surface. [See I. Shavitt, *J. Chem. Phys.*, **49**, 4048 (1968); M. Menzinger and R. Wolfgang, *Angew. Chem. Int. Ed. Engl.*, **8**, 438 (1969); *Levine, Physical Chemistry*, Section 23.4.]

In the 1960s several workers applied orbital symmetry rules to predict the course of chemical reactions. The most fruitful applications have come from the *Woodward–Hoffmann rules*, which predict the preferred path and stereochemistry for many important classes of organic reactions. As an example of the application of these rules, we consider the cyclization of a substituted *s-cis*-butadiene to a substituted cyclobutene. There are two possible steric courses the reaction can take, described as *conrotatory* or *disrotatory*, depending on whether the terminal groups rotate in the same or opposite senses as the reaction proceeds. Note the difference in products in Fig. 15.24.

One finds that when the reaction is thermally induced, the butadiene cyclization process is conrotatory, but when the reaction is photochemically induced, the process is disrotatory. To explain these facts, Woodward and Hoffmann examined the symmetry of the highest occupied MO of the substituted *s-cis*-butadiene; this is the π MO φ_2 in Fig. 15.16. Figure 15.25 shows that for φ_2, a conrotatory motion of the π AO's on carbons 1 and 4 causes the positive lobe of $C_1 2p\pi$ to overlap the positive lobe of $C_4 2p\pi$; this overlap gives a bonding interaction of these AO's and leads to formation of the 1–4 σ bond of the cyclobutene. On the other hand, a disrotatory motion of these two AO's causes overlap of the positive lobe of

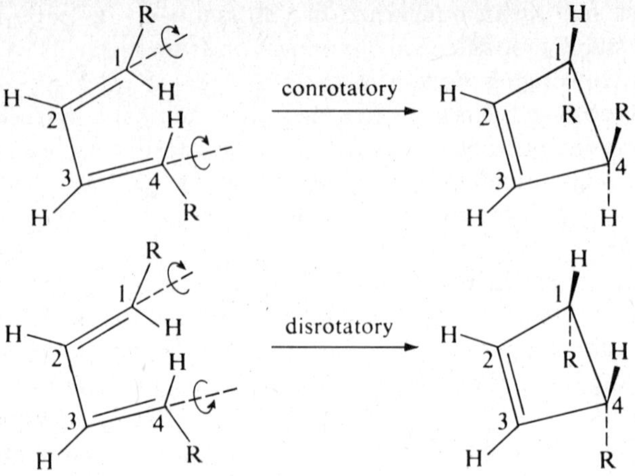

Figure 15.24 *Conrotatory and disrotatory cyclizations. All atoms lie in the same plane except those shown with dashed or heavy bonds.*

one π AO with the negative lobe of the other π AO; this gives an antibonding interaction. Thus we expect the conrotatory process to have a lower activation energy than the disrotatory and to be preferred over the disrotatory. As a further check, we note that conrotatory motion leads to an antibonding interaction between the carbon 1 and 4 π AO's in the lowest π MO φ_1; this leaves the π AO's of carbons 2 and 3 in φ_1 to form the 2–3 π bond in the cyclobutene. (Note that if the butadiene is less symmetrically substituted than in Fig. 15.24, clockwise and counterclockwise conrotatory motions lead to a mixture of two products.)

When the above reaction is induced photochemically, an absorbed photon excites an electron from φ_2 to the butadiene π MO φ_3 (Fig. 15.16). The highest

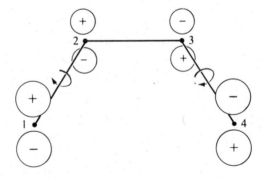

Figure 15.25 *Relative phases of AO's in the highest occupied butadiene MO*

occupied MO is now φ_3, in which the $2p\pi$ AO's on carbons 1 and 4 have the same phases (rather than opposite phases as in φ_2) and in which a disrotatory motion produces positive bonding overlap between these AO's. Thus we predict disrotatory ring closure, as observed. The same reasoning correctly predicts the courses of other polyene ring closures (Problem 15.71).

Provided the reactants and products do not differ greatly in energy, a high barrier for the forward reaction implies a high barrier for the reverse reaction (in which the reaction path is traversed in the opposite direction). Thus the above reasoning also applies in determining the course of the reverse, ring-opening, reactions.

For further details of the Woodward–Hoffmann rules and their application to a wide variety of organic reactions, see R. B. Woodward and R. Hoffmann, *Angew. Chem. Intern. Ed.*, **8**, 781 (1969); *The Conservation of Orbital Symmetry*, Academic Press, New York, 1970; R. E. Lehr and A. P. Marchand, *Orbital Symmetry*, Academic Press, New York, 1972.

Pearson has applied orbital symmetry concepts to inorganic reactions. As two reactant molecules approach, electrons begin to flow from the highest occupied molecular orbital (HOMO) of one to the lowest unoccupied MO (LUMO) of the other. These two MO's are called *frontier orbitals*. For a low activation energy, we require a positive overlap between these two MO's.

An example is the $H_2 + F_2 \rightarrow 2HF$ reaction. Consider a proposed mechanism in which the two molecules collide broadside to give a four-center transition state. The HOMO of H_2 is the $\sigma_g 1s$ MO; the LUMO of F_2 is the $\sigma_u^* 2p$ MO (Sections 13.5 and 13.6). Flow of electrons out of H_2 $\sigma_g 1s$ toward F_2 $\sigma_u^* 2p$ would lead to breaking of the H—H bond and formation of two H—F bonds. However, Fig. 15.26a shows that these two MO's do not have a positive overlap. Hence electron flow from H_2 to F_2 is forbidden by symmetry. Figure 15.26b shows the HOMO of F_2 and LUMO of H_2. Here there is a positive overlap, but flow of electrons out of the antibonding π_g^* MO would strengthen rather than weaken the F—F bond. We conclude that this bimolecular one-step mechanism has a high activation energy and is not favored. (The same reasoning applies to the famous $H_2 + I_2$ reaction.)

Further applications of orbital symmetry to inorganic reactions may be

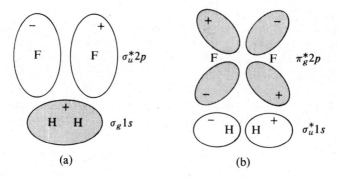

Figure 15.26 HOMO's and LUMO's for H_2 and F_2. Occupied MO's are shaded.

found in R. G. Pearson, *Chem. Eng. News*, Sept. 28, 1970, p. 66; *Acc. Chem. Res.*, **4**, 152 (1971); *J. Am. Chem. Soc.*, **94**, 8287 (1972); *Symmetry Rules for Chemical Reactions*, Wiley, New York, 1976.

We now consider ab initio calculation of reaction surfaces. As we have seen, the Hartree–Fock method does not generally describe correctly the process of molecular dissociation; hence (except in certain special cases) we must include CI in kinetics calculations. Together with the large number of variables, this makes calculation of $U(q_a)$ a very difficult problem. Nevertheless, about 1970 meaningful ab initio kinetics calculations started to become available.

The $F + H_2 \rightarrow FH + H$ reaction has been studied experimentally using molecular beams and laser spectroscopy. Bender and co-workers did ab initio CI calculations for this reaction using a large basis set. Preliminary calculations had shown that the reaction path is linear, with the F atom approaching along the H—H internuclear line. Bender et al. determined the reaction path by investigating over 150 linear geometries. The saddle point was found to occur early in the course of the reaction, with the H—H distance slightly lengthened from its calculated equilibrium value of 0.75 Å to 0.77 Å; the saddle-point F—H distance is 1.53 Å, compared to 0.93 Å in HF. Thus the transition state resembles H_2 much more than HF. The experimental activation energy is 1.7 kcal/mole. A Hartree–Fock calculation gave a barrier of 34 kcal; the large error here is, of course, expected. With a 338-configuration wave function, Bender et al. found a barrier height of 1.7 kcal, in agreement with the experimental E_a. (The perfect agreement may be fortuitous, since these two quantities can be expected to differ somewhat, as we noted earlier.) These workers conclude that their potential surface is probably quite realistic. Since the reaction path is linear, $U(q_a)$ is a function of two variables, which can be taken as the H—H and F—H distances, and U can be conveniently plotted in three dimensions. One can also make two-dimensional contour diagrams of U (as we did for orbitals). Bender and co-workers give some beautiful three-dimensional plots of U for this reaction. See C. F. Bender, P. K. Pearson, S. V. O'Neil, and H. F. Schaefer, *J. Chem. Phys.*, **56**, 4626 (1972); *Science*, **176**, 1412 (1972); *Schaefer, The Electronic Structure of Atoms and Molecules*, pp. 272–281.

Hsu, Buenker, and Peyerimhoff did an ab initio CI study of the interconversion of cyclobutene and *s-cis*-butadiene. Their wave functions contained up to 260 configurations, but because of the size of the molecule, their basis set was not as good as one would ideally want. The potential surface is twenty-four-dimensional, so obviously they had to restrict the geometry variations to be considered. They calculated points corresponding to conrotatory (and disrotatory) rotations of the CH_2 groups and variations in R_{14}, the 1–4 carbon–carbon distance. These are clearly the geometrical parameters that undergo the greatest change in the course of the reaction. In agreement with experiment and the Woodward–Hoffmann rules, they found the conrotatory path to be the favored one; the disrotatory path had a barrier 14 kcal/mole higher than the conrotatory. The minimum-energy path they found has an interesting stepwise process. For ring opening, first R_{14} increases to a value about midway between cyclobutene and butadiene; then the CH_2 groups abruptly rotate 90° in a conrotatory manner; finally, R_{14} increases to its value in butadiene. It was previously believed that the CH_2 groups rotated gradually and continuously as

R_{14} varied. [K. Hsu, R. J. Buenker, and S. D. Peyerimhoff, *J. Am. Chem. Soc.*, **93**, 2117 (1971).]

Perhaps the most famous reaction surface is that for $H + H_2 \rightarrow H_2 + H$. The H_3 surface has been calculated to an accuracy of $\frac{3}{4}$ kcal/mole at 267 points using CI wave functions [B. Liu, *J. Chem. Phys.*, **58**, 1925 (1973); P. Siegbahn and B. Liu, *J. Chem. Phys.*, **68**, 2457 (1978)], and an analytical potential function has been fitted to these points [D. G. Truhlar and C. J. Horowitz, *J. Chem. Phys.*, **68**, 2466 (1978)]. From this surface, rate constants in reasonably good agreement with experiment have been calculated [J. C. Sun et al., *J. Chem. Phys.*, **73**, 6095 (1980)].

Because of the enormous difficulties involved in ab initio reaction-surface calculations, it would be very desirable to have a semiempirical method that gave reliable reaction-surface results. MINDO/3 has been used to calculate potential-energy surfaces for many reactions. The MINDO/3-calculated barrier heights are in pretty good agreement with the experimentally observed activation energies in the large majority of cases [K. Jug, *Theor. Chim. Acta*, **54**, 263 (1980)]. However, the reliability of MINDO/3 for kinetics work has been questioned. For example, for the vinylcyclopropane rearrangement reaction, MINDO/3 calculations are "directly contrary to experimental fact" regarding the reaction path [G. D. Andrews and J. E. Baldwin, *J. Am. Chem. Soc.*, **98**, 6706 (1976)]; for the Diels–Alder reaction of ethylene and 1,3-butadiene, MINDO/3 gives results opposite to those of ab initio calculations [R. E. Townshend, *J. Am. Chem. Soc.*, **98**, 2190 (1976)]. It has been argued that even though MINDO/3 gives pretty good equilibrium geometries, it is subject to serious errors when used to calculate transition-state geometries [M. V. Basilevsky et al., *J. Am. Chem. Soc.*, **99**, 1369 (1977)]. Thus the reliability of conclusions reached by MINDO/3 kinetics calculations is unclear. (The MNDO method has not yet been applied to a substantial number of chemical reactions.)

Reaction mechanisms often involve more than one elementary step and hence involve reaction intermediates. (A reaction intermediate should not be confused with a transition state. A transition state is the maximum-energy point on the minimum-energy path of an elementary reaction. A reaction intermediate is a product in one elementary step and a reactant in a subsequent elementary step; a reaction intermediate lies at a minimum in potential energy for all nuclear displacements.) Reaction intermediates are frequently too short-lived to allow spectroscopic determination of their structures. Hence a very significant application of quantum chemistry has been the study of the geometries and relative energies of reaction intermediates. For example, ab initio STO-3G geometry calculations and 4-31G and 6-31G* energy calculations have been done for a wide variety of carbonium ions containing as many as eight carbons (W. J. Hehre in *Schaefer, Applications of Electronic Structure Theory*, Chapter 7).

15.20 COMPUTERS AND QUANTUM CHEMISTRY

An ab initio calculation of the wave function of a medium-size molecule may involve evaluation of 10^6 integrals and solution of secular equations of high degree; a computer is clearly indispensable.

Digital computers use electronic circuits to perform arithmetic and logical operations. The advantages of a computer are the great speed and the accuracy with which they do arithmetic. Any task to be performed by a computer (whether it be evaluation of an integral or translation from one language to. another) must be reduced to a problem in arithmetic or symbol manipulation. One must know how to solve a problem by hand calculation before it can be programmed for a computer.

Computers are classified as maxicomputers, minicomputers, and microcomputers. *Maxicomputers* (also called mainframe computers) are very large (room-size), expensive machines. *Microcomputers* are small, inexpensive machines based on a microprocessor, a single integrated circuit on a tiny chip of etched silicon, containing the equivalent of tens of thousands of transistors; the microprocessor functions as the central processing unit of the microcomputer, carrying out all the required calculations. Microcomputers for home and small-business use became popular in the late 1970s. *Minicomputers* are intermediate in size (typically as big as a home refrigerator) and cost. (The term *midicomputers* is sometimes used to describe those in the size range between mini and maxi.)

Until the late 1970s, quantum chemists relied on maxicomputers for their calculations, but with advances in speed and memory storage, it became feasible to use minicomputers for quantum-chemistry calculations. (See P. Lykos, ed., *Minicomputers and Large Scale Calculations*, ACS Symposium Series No. 57, American Chemical Society, Washington, D.C., 1977.) Although minicomputers are substantially slower than maxicomputers, a quantum chemist can buy a minicomputer and have it run full time for hours or days on a single calculation. CI calculations involving as many as 23,000 configuration functions have been done on minicomputers.

A stored program provides for optimal use of a digital computer. A *program* is a series of instructions fed into the computer. After the program has been stored, the data required for the calculation are fed in and the desired calculations are performed in accordance with the program's instructions. The computer can only "understand" instructions written in machine language, a language in which each step to be performed is explicitly specified according to a certain code. Writing a machine-language program is extremely tedious, and the probability of error is high; moreover, machine language is peculiar to the type of computer being used. To avoid the difficulty of dealing with machine language, experts have devised various higher-level languages; many of these are similar in appearance to ordinary algebra, and they enable complex programs to be written easily. Higher-level languages include FORTRAN and ALGOL, devised for scientific applications; COBOL, used for business applications; PL/1, which combines elements of FORTRAN and COBOL; BASIC, a rather simple language used for microcomputers; and Pascal, a language designed to produce efficient, logically structured programs. In the United States, FORTRAN (*formula translation*) is the language commonly used for computer applications in the sciences. A FORTRAN program is usable (with perhaps minor changes) on most computers. Statements in FORTRAN cannot be understood by the computer; hence, after being read in, a FORTRAN program is translated by the computer (using an auxiliary program supplied by the manufacturer) to machine language.

On many computers, calculations are done to seven or eight significant figures, although provision for "double precision" sixteen-figure calculations is usually available. A calculation such as the solution of a high-degree secular equation involves many arithmetic steps, and although each step is performed to eight-place accuracy, the final answers may be accurate to, say, only five places, because of accumulated round-off errors.

Despite their high speed and accuracy, computers do only what they are instructed to do and are not capable of exercising independent judgment to compensate for human errors in the program or input data. There is a tendency to consider a calculation that has been performed by a computer as having some special validity beyond what it would have if done by hand; however, every step of a computer calculation is performed in response to the directions given in the program, and the results can be no better or worse than the input data and program. A computer is no substitute for intelligent thought. (There are in the literature several quantum-chemistry computer calculations that are totally in error, because of errors in the programs used.) A favorite maxim of computer programmers is GIGO—garbage in, garbage out.

The reader interested in doing molecular calculations can (for a fee) obtain programs from the Quantum Chemistry Program Exchange, Chemistry Department, Indiana University, Bloomington, Indiana 47401. Available programs include the HMO, EH, PPP, CNDO, INDO, MINDO/3, MNDO, PCILO, molecular mechanics, and ab initio SCF MO methods, as well as programs in spectroscopy and kinetics.

The utility of computers for performing lengthy calculations is unquestioned. However, their capabilities in some other areas have sometimes been exaggerated. Despite considerable effort expended in devising programs for translating from one language to another, a committee appointed by the U.S. National Academy of Sciences reported in 1966 that machine translations of scientific texts were of such poor quality as to require extensive editing by humans; translation without the use of computers was found to be less expensive. The report stated, "There has been no [useful] machine translation of general scientific text . . . ." [Language and Machines, National Academy of Sciences, National Research Council, Washington, D.C., 1966]. Following this discouraging report, interest in machine translation diminished, but some work continued and some progress was made. A 1978 review [W. J. Hutchins, J. Doc., 34, 119 (1978)] stated that machine translation provides "reasonably acceptable translations within restricted subject fields" but that most workers believed that fully automatic high-quality translation is not a feasible goal. The article stated that "the quality of the translations of existing operational [machine translation] systems is not . . . good enough on the whole," since substantial costly postediting by humans is usually required, but the article also spoke of a "mood of quiet optimism" in machine-translation research. [See also New Scientist, 85, 836 (1980).]

In 1957 two experts on computers predicted that within ten years a computer program would be the chess champion of the world [H. A. Simon and A. Newell, Operations Research, 6, 1 (1958)]. In 1967 the best chess program was Mac Hack VI, which played a mediocre game and had a class C rating. Chess ratings are grandmaster (over 2500), master (2200–2500), expert (2000–2200), class A (1800–2000), class B (1600–1800), class C (1400–1600), class D (1200–1400), class E

(0–1200); the maximum rating ever attained by a human is 2800. In 1968 four computer scientists bet the Scottish chess master (and computer-science professor) David Levy £1000 that a computer would beat him in a chess match within ten years. Substantial progress in computer chess began in 1975, and by 1978 the best program, Chess 4.7, had achieved an expert rating of 2040. The match between Levy and Chess 4.7, played in 1978, was won by Levy with three wins, one draw, and one loss. (Levy later complained that one of the computer scientists had failed to pay off his share of the wager.) Belle, the winner of the 1980 world computer chess championship, had a rating of 2160 in 1982. Belle uses a "brute force" approach, examining every possible move of each player to as great a depth as is possible within the time limit set for moves. This is quite unlike how humans play chess; a good human chess player examines relatively few moves in each position and is guided by a long-range strategy. In 1980 the Fredkin foundation established a $100,000 prize (administered by Carnegie-Mellon University) for the first computer program to become the world chess champion. Although some might think it frivolous to devote much effort to computer chess, Hans Berliner (former world correspondence chess champion and now a researcher on artificial intelligence) has noted that computer chess programs are more competent at what they do than is any other example of artificial intelligence. For more on computer chess, see *Chess Skill in Man and Machine*, P. W. Frey, ed., Springer-Verlag, New York, 1977; M. M. Newborn, *Adv. Computers*, **18**, 59 (1979); H. J. Berliner, *Nature*, **274**, 745 (1978) and *Artif. Intell.*, **10**, 201 (1978); *Science*, **204**, 1396 (1979); *Scientific American*, Sept. 1979, p. 80; *Scientific American*, April 1981, p. 83.

Despite the so far relatively unimpressive results of those working on artificial intelligence (AI), Marvin Minsky of MIT stated[55] in 1970 that "in from three to eight years we will have a machine with the general intelligence of an average human being . . . . After all, the human brain is just a computer that happens to be made out of meat." The extent to which tasks performed by human intelligence can be performed by computer programs is a hotly debated question.[56] Objectivity in this area is hard to come by, since AI is related to the larger question: Are we machines?

A few people have deplored the introduction of computers into science and mathematics, claiming that creative thought is thereby stifled. While it is true that computers have sometimes been misused, they offer the only possibility for obtaining accurate molecular wave functions. Computers have also become important in processing the data of spectroscopy experiments.

---

[55] *Life*, Nov. 20, 1970, p. 58B. Minsky was a technical adviser to the film *2001*, whose most interesting character was a computer named HAL.

[56] See H. Dreyfus, *What Computers Can't Do*, Harper and Row, New York, 1972; I. Benson, *New Scientist*, **51**, 525 (1971); K. Gunderson, *Mentality and Machines*, Doubleday Anchor, New York, 1971; J. Weizenbaum, *Computer Power and Human Reason*, Freeman, San Francisco, 1976; M. Boden, *Artificial Intelligence and Natural Man*, Basic Books, New York, 1977; P. McCorduck, *Machines Who Think*, Freeman, San Francisco, 1979; N. Ringle, ed., *Philosophical Perspectives in Artificial Intelligence*, Humanities Press, Atlantic Highlands, N.J., 1979; D. R. Hofstadter, *Gödel, Escher, Bach*, Basic Books, New York, 1979; G. Kolata, *Science*, **217**, 1237 (1982); D. L. Waltz, *Scientific American*, Oct. 1982, p. 118.

15.21 *MATRICES*

A *matrix* is a rectangular array of numbers. Matrices furnish a very convenient way to formulate much of the theory of quantum mechanics, and matrix methods are the most practical way to solve the equations that occur in the SCF MO method. Although we have not used matrices in this book, some familiarity with them is needed to read much of the literature of quantum chemistry. Therefore this section gives a brief introduction to matrices.

The numbers that compose a matrix are called the *matrix elements*. Let the matrix **A** have m rows and n columns, and let a_{ij} ($i = 1, 2, \ldots, m$ and $j = 1, 2, \ldots, n$) denote the element in row i and column j. Then

$$\mathbf{A} = \begin{pmatrix} a_{11} & a_{12} & \cdots & a_{1n} \\ a_{21} & a_{22} & \cdots & a_{2n} \\ \cdot & \cdot & \cdots & \cdot \\ a_{m1} & a_{m2} & \cdots & a_{mn} \end{pmatrix}$$

A is said to be an m by n matrix. (Do not confuse **A** with a determinant. A matrix need not be square and is not equal to a single number.)

A *row matrix* (also called a *row vector*) is a matrix having only one row. A *column matrix* has only one column.

Two matrices **R** and **S** are *equal* if they have the same number of rows, the same number of columns, and have corresponding elements equal. If $\mathbf{R} = \mathbf{S}$, then $r_{ij} = s_{ij}$ for $i = 1, \ldots, m$ and $j = 1, \ldots, n$, where m and n are the dimensions of **R** and **S**. A matrix equation is thus equivalent to mn scalar equations.

The *sum* of two matrices **A** and **B** is defined as the matrix formed by adding corresponding elements of **A** and **B**; the sum is defined only if **A** and **B** have the same dimensions. If $\mathbf{C} = \mathbf{A} + \mathbf{B}$, then we have the mn scalar equations $c_{ij} = a_{ij} + b_{ij}$ for $i = 1, \ldots, m$ and $j = 1, \ldots, n$.

The product of the scalar k and the matrix **A** is defined as the matrix formed by multiplying every element of **A** by k. If $\mathbf{D} = k\mathbf{A}$, then $d_{ij} = ka_{ij}$.

If **A** is an m by n matrix and **B** is an n by p matrix, the *matrix product* $\mathbf{C} = \mathbf{AB}$ is defined to be the m by p matrix whose elements are calculated by the formula

$$c_{ij} = a_{i1}b_{1j} + a_{i2}b_{2j} + \cdots + a_{in}b_{nj} = \sum_{k=1}^{n} a_{ik}b_{kj} \qquad \textbf{(15.221)}$$

To calculate c_{ij}, we take row i of **A** (this row's elements are $a_{i1}, a_{i2}, \ldots, a_{in}$), multiply each element of this row by the corresponding element in column j of **B** (this column's elements are $b_{1j}, b_{2j}, \ldots, b_{nj}$), and add the n products. For example, suppose

$$\mathbf{A} = \begin{pmatrix} -1 & 3 & \frac{1}{2} \\ 0 & 4 & 1 \end{pmatrix} \quad \text{and} \quad \mathbf{B} = \begin{pmatrix} 1 & 0 & -2 \\ 2 & 5 & 6 \\ -8 & 3 & 10 \end{pmatrix}$$

The number of columns of **A** equals the number of rows of **B**, so the matrix product is defined. The product $\mathbf{C} = \mathbf{AB}$ is a 2 by 3 matrix. The element c_{21} is found from the second row of $\mathbf{A}$ and the first column of $\mathbf{B}$ as follows: $c_{21} = 0(1) + 4(2) + 1(-8) = 0$. Calculation of the remaining elements gives

$$\mathbf{C} = \begin{pmatrix} 1 & 16\frac{1}{2} & 25 \\ 0 & 23 & 34 \end{pmatrix}$$

Matrix multiplication is not commutative; the products $\mathbf{AB}$ and $\mathbf{BA}$ need not be equal. (In the preceding example, the product $\mathbf{BA}$ happens to be undefined.)

A matrix with equal numbers of rows and columns is a *square matrix*; the *order* of a square matrix equals the number of rows.

The elements $a_{11}, a_{22}, \ldots, a_{nn}$ of a square matrix of order n lie on its *principal diagonal*. A *diagonal matrix* is one having zero as the value of each element not on the principal diagonal.

A diagonal matrix whose diagonal elements are each equal to 1 is a *unit matrix*. The (i, j)th element of a unit matrix is the Kronecker delta δ_{ij}; $(\mathbf{I})_{ij} = \delta_{ij}$, where $\mathbf{I}$ is a unit matrix. Let $\mathbf{B}$ be a square matrix of the same order as $\mathbf{I}$. The (i, j)th element of the product $\mathbf{IB}$ is given by (15.221) as $(\mathbf{IB})_{ij} = \sum_k (\mathbf{I})_{ik} b_{kj} = \sum_k \delta_{ik} b_{kj} = b_{ij}$. Since the (i, j)th elements of $\mathbf{IB}$ and of $\mathbf{B}$ are equal for all i and j, we have $\mathbf{IB} = \mathbf{B}$. Similarly, one finds $\mathbf{BI} = \mathbf{B}$. Multiplication by a unit matrix has no effect.

The *determinant* of a square matrix is defined as the determinant whose elements equal those of the matrix.

The *complex conjugate* $\mathbf{A}^*$ of matrix $\mathbf{A}$ is formed by replacement of every element of $\mathbf{A}$ by its complex conjugate.

A *real matrix* is a matrix whose elements are all real numbers.

The *transpose* $\tilde{\mathbf{A}}$ of matrix $\mathbf{A}$ is formed by interchange of the rows and columns of $\mathbf{A}$, so that column 1 becomes row 1 and row 1 becomes column 1, etc. The relation between the elements of $\tilde{\mathbf{A}}$ and $\mathbf{A}$ is $\tilde{a}_{ij} = a_{ji}$.

A square matrix that equals its transpose is called a *symmetric matrix*. For a symmetric matrix, $\mathbf{A} = \tilde{\mathbf{A}}$, so $a_{ij} = \tilde{a}_{ij}$; however, $\tilde{a}_{ij} = a_{ji}$, so $a_{ij} = a_{ji}$ for a symmetric matrix. The matrix is symmetric about its principal diagonal.

The *Hermitian conjugate* (or *conjugate transpose*) $\mathbf{A}^\dagger$ of matrix $\mathbf{A}$ is equal to the complex conjugate of the transpose of $\mathbf{A}$; we have $\mathbf{A}^\dagger = \tilde{\mathbf{A}}^*$. The elements of $\mathbf{A}^\dagger$ are given by $a_{ij}^\dagger = a_{ji}^*$.

A square matrix that equals its Hermitian conjugate is called a *Hermitian matrix*. For a Hermitian matrix, $\mathbf{A} = \mathbf{A}^\dagger$ and $a_{ij} = a_{ij}^\dagger$; however, $a_{ij}^\dagger = a_{ji}^*$, so $a_{ij} = a_{ji}^*$ for a Hermitian matrix. [Note the similarity to Eqs. (7.15) and (7.18) for Hermitian operators.] A real Hermitian matrix is also a symmetric matrix.

If $\mathbf{A}$ is a square matrix and if $\mathbf{AG} = \mathbf{GA} = \mathbf{I}$, where $\mathbf{I}$ is a unit matrix, then $\mathbf{G}$ is said to be the *inverse* of matrix $\mathbf{A}$, and we write $\mathbf{G} = \mathbf{A}^{-1}$. (It turns out that $\mathbf{A}^{-1}$ exists only when the determinant of $\mathbf{A}$ is nonzero.)

An *orthogonal matrix* is a matrix whose inverse is equal to its transpose; if $\mathbf{A}^{-1} = \tilde{\mathbf{A}}$, then $\mathbf{A}$ is orthogonal.

A *unitary matrix* is a matrix whose inverse is equal to its Hermitian conjugate: $\mathbf{A}^{-1} = \mathbf{A}^\dagger$. (If $\mathbf{A}$ is real, then $\mathbf{A}^\dagger = \tilde{\mathbf{A}}$, and a real unitary matrix is an orthogonal matrix.)

After this barrage of definitions, we consider the application of matrices to the SCF MO method. The Roothaan equations (13.183) read

$$\sum_k F_{jk} c_{ki} = \sum_k S_{jk} c_{ki} \varepsilon_i \qquad \text{for } j = 1, 2, \ldots$$

The coefficients c_{ki} relate the MO's φ_i to the basis functions χ_k according to $\varphi_i = \sum_k c_{ki} \chi_k$, and F_{jk} and S_{jk} are defined by (13.184). Let $\mathbf{C}$ be the square matrix whose elements are the coefficients c_{ki}. Let $\mathbf{F}$ be the square matrix whose elements are $F_{jk} = \langle \chi_j | \hat{F} | \chi_k \rangle$. Let $\mathbf{S}$ be the square matrix whose elements are $S_{jk} = \langle \chi_j | \chi_k \rangle$. Let $\boldsymbol{\varepsilon}$ be the diagonal square matrix whose diagonal elements are the orbital energies ε_1, $\varepsilon_2, \ldots$, so that the elements of $\boldsymbol{\varepsilon}$ are $\varepsilon_{ij} = \delta_{ij} \varepsilon_i$, where δ_{ij} is the Kronecker delta. Use of (15.221) gives element ki of the matrix product $\mathbf{C}\boldsymbol{\varepsilon}$ as $(\mathbf{C}\boldsymbol{\varepsilon})_{ki} = \sum_m c_{km} \varepsilon_{mi} = \sum_m c_{km} \delta_{mi} \varepsilon_i = c_{ki} \varepsilon_i$. Hence the Roothaan equations read

$$\sum_k F_{jk} c_{ki} = \sum_k S_{jk} (\mathbf{C}\boldsymbol{\varepsilon})_{ki} \qquad (15.222)$$

From (15.221) the left side of (15.222) equals the (j, i)th element of $\mathbf{FC}$, and the right side equals the (j, i)th element of $\mathbf{S}(\mathbf{C}\boldsymbol{\varepsilon})$. Since the general element of $\mathbf{FC}$ equals the general element of $\mathbf{SC}\boldsymbol{\varepsilon}$, we conclude that these matrices are equal:

$$\mathbf{FC} = \mathbf{SC}\boldsymbol{\varepsilon} \qquad (15.223)$$

This is the matrix form of the Roothaan equations.

The set of AO's χ_i used to expand the MO's is not an orthonormal set. However, we can use the Schmidt (or some other) procedure to form linear combinations of the basis functions to give a new set of basis functions χ_i' that is an orthonormal set; $\chi_i' = \sum_k b_k \chi_k$ and $S_{ij}' = \langle \chi_i' | \chi_j' \rangle = \delta_{ij}$. With this orthonormal set, the overlap matrix is a unit matrix, and the Roothaan equations (15.223) have the simpler form

$$\mathbf{F}'\mathbf{C}' = \mathbf{C}'\boldsymbol{\varepsilon} \qquad (15.224)$$

where $F_{ij}' = \langle \chi_i' | \hat{F} | \chi_j' \rangle$ and where $\mathbf{C}'$ is the matrix of coefficients relating the MO's φ_i to the orthogonal basis functions χ_k': $\varphi_i = \sum_k c_{ki}' \chi_k'$. Because the Fock operator $\hat{F}$ is Hermitian, the Fock matrix $\mathbf{F}'$ is a Hermitian matrix. In the common case of real basis functions, F' is a real Hermitian matrix, i.e., $\mathbf{F}'$ is a symmetric matrix. Note the resemblance of (15.224) to the operator eigenvalue equation (3.16). One says that the numbers ε_i are the *eigenvalues* of the matrix $\mathbf{F}'$ and that each column of the matrix $\mathbf{C}'$ is an *eigenvector* of $\mathbf{F}'$.

We want the MO's φ_i to be orthonormal. Substitution of $\varphi_i = \sum_k c_{ki}' \chi_k'$ into $\langle \varphi_i | \varphi_j \rangle = \delta_{ij}$ and use of the orthonormality of the χ_k' functions gives $\delta_{ij} = \langle \sum_k c_{ki}' \chi_k' | \sum_m c_{mj}' \chi_m' \rangle = \sum_k \sum_m c_{ki}'^* c_{mj}' \delta_{km} = \sum_k c_{ki}'^* c_{kj}'$. But $c_{ki}'^* = c_{ik}'^\dagger$, so $\sum_k c_{ik}'^\dagger c_{kj}' = \delta_{ij}$. The elements $c_{ik}'^\dagger$, c_{kj}', and δ_{ij} are the matrix elements of the matrices $\mathbf{C}'^\dagger$, $\mathbf{C}'$, and the unit matrix $\mathbf{I}$; use of (15.221) then gives $\mathbf{C}'^\dagger \mathbf{C}' = \mathbf{I}$. Therefore the coefficient matrix $\mathbf{C}'$ is a unitary matrix. (A real unitary matrix is an orthogonal matrix, so for the common case of real coefficients, $\mathbf{C}'$ is orthogonal as well as unitary.)

Multiplication of (15.224) on the left by $\mathbf{C}'^{-1}$ gives $\mathbf{C}'^{-1} \mathbf{F}' \mathbf{C}' = \mathbf{C}'^{-1} \mathbf{C}' \boldsymbol{\varepsilon}$. Use of $\mathbf{C}'^{-1} \mathbf{C}' = \mathbf{I}$, $\mathbf{I}\boldsymbol{\varepsilon} = \boldsymbol{\varepsilon}$, and $\mathbf{C}'^{-1} = \mathbf{C}'^\dagger$ gives

$$\mathbf{C}'^\dagger \mathbf{F}' \mathbf{C}' = \boldsymbol{\varepsilon} \qquad (15.225)$$

We know the $\mathbf{F}'$ matrix; given $\mathbf{F}'$, our problem is to find the unitary matrix $\mathbf{C}'$ such that $\mathbf{C}'^\dagger \mathbf{F}' \mathbf{C}'$ is a diagonal matrix $\boldsymbol{\varepsilon}$. Once $\mathbf{C}'$ is found, then its elements c'_{ki} give us the MO's $\varphi_i = \Sigma_k c'_{ki} \chi'_k$, and calculation of $\mathbf{C}'^\dagger \mathbf{F}' \mathbf{C}'$ gives the $\boldsymbol{\varepsilon}$ matrix, whose diagonal elements are the orbital energies. Use of $\chi'_i = \Sigma_k b_k \chi_k$ then gives the MO's in terms of the original nonorthogonal basis functions χ_k.

There are two common procedures used to find the matrix $\mathbf{C}'$ that will reduce the real symmetric matrix $\mathbf{F}'$ to diagonal form: the Jacobi method and the Givens method. For details, see *Lowe*, Section 9-4 and Appendix 9.

15.22 *THE FUTURE OF QUANTUM CHEMISTRY*

In the 1950s there was a general belief that meaningful ab initio calculation of molecular properties for all except very small molecules was out of the question. Quantum-chemistry books written in this period contain such statements as "we cannot hope ever to make satisfactory *ab initio* calculations [for organic compounds]" and "It is wise to renounce at the outset any attempt at obtaining precise solutions of the Schroedinger equation for systems more complicated than the hydrogen molecule ion." In 1959 Mulliken and Roothaan identified the "bottleneck" holding up accurate quantum-mechanical calculations on polyatomic molecules as the difficulty in evaluating multicenter integrals.[57] This bottleneck has now been eliminated. The mood of many quantum chemists has changed to one of cautious optimism. At present, the principal limitations on the size of a molecule for which one can do an accurate ab initio SCF MO calculation are those imposed by the storage capacity and speed of the available electronic computers; as larger and faster computers are developed, it will become feasible to treat larger molecules.[58] (Recall the ab initio treatments of naphthalene, azulene, and guanine-cytosine mentioned earlier.)

Although many-center integrals are computed rapidly and accurately using GTO's, the large number of GTO basis functions needed to represent an AO accurately leads to huge numbers of electron-repulsion integrals. However, for large molecules many of these integrals (those involving AO's centered on nuclei far from one another) are extremely small and can be neglected without significant loss of accuracy. Although the total number of electron-repulsion integrals in an SCF calculation is proportional to m^4 (where m is the number of basis functions), Ahlrichs showed that for a large molecule the number of electron-repulsion integrals that are larger than some fixed threshold value is proportional to only m^2 [R. Ahlrichs, *Theor. Chim. Acta*, **33**, 157 (1974)]. Therefore, application of ab initio SCF calculations to large molecules is not out of the question. Several workers have implemented

---

[57] R. S. Mulliken and C. C. J. Roothaan, *Proc. Natl. Acad. Sci. U.S.*, **45**, 394 (1959).

[58] For references to calculations done subsequent to the publication of this book, consult the review articles on quantum chemistry in volumes of the *Annual Review of Physical Chemistry*. Useful compilations include: W. G. Richards et al., *A Bibliography of Ab Initio Molecular Wave Functions*, Oxford University Press, New York, 1971; *Supplement for 1970–73*, Oxford University Press, 1974; *Supplement for 1974–77*, Oxford University Press, 1978; *Supplement for 1978–80*, Oxford University Press, 1981.

computer programs that test electron-repulsion integrals to get their order of magnitude before they are calculated accurately; those smaller than a certain threshold value can be neglected, thereby permitting considerable savings in computation time [R. Ahlrichs, *Theor. Chim. Acta*, **33**, 157 (1974); E. Clementi et al., *J. Chem. Phys.*, **58**, 4699 (1973)].

The electronic wave function depends on $3n$ spatial and n spin coordinates. Since the Hamiltonian (15.12) contains only one- and two-electron spatial terms, the energy can be written in terms of integrals involving only six spatial coordinates (Problem 15.66). In a sense, the wave function of a many-electron molecule contains more information than is needed and is lacking in direct physical significance. This has prompted the search for functions that involve fewer variables than the wave function and that can be used to calculate the energy and other properties. The molecular energy can be expressed in terms of the first-order spinless density matrix (which is a function of the spatial coordinates of one electron) and the second-order spinless density matrix (which is a function of the spatial coordinates of two electrons). Unfortunately, no convenient principle (analogous to the variation principle used to calculate wave functions) has been developed that would allow direct calculation of these density matrices without first requiring calculation of the wave function.[59]

It has been proven that for the ground state of a molecule, the energy is uniquely determined by the electron density $\rho(x, y, z)$, a function of only three variables [P. Hohenberg and W. Kohn, *Phys. Rev.*, **136**, B864 (1964)].

Although a near-Hartree–Fock wave function allows accurate calculation of many equilibrium molecular properties (equilibrium geometry, dipole moment, rotational barrier), it must still deviate significantly from the true wave function. Computer programs to apply configuration interaction to medium-size molecules have been developed, and significant CI wave functions for polyatomic molecules are now appearing, at least for relatively small polyatomics. (Examples mentioned earlier include H_2O, CH_2, H_3.) Chemical kinetics is beginning to benefit from ab initio CI calculations of reaction surfaces (Section 15.19).

Application of configuration interaction to a polyatomic-molecule SCF MO wave function is computationally difficult because of the many configurations that must be included to get significant improvement in the wave function. Hence other techniques have been proposed to deal with electron correlation. The localized MO's are well separated from each other, thus minimizing repulsions between them; so, as noted earlier, correlation is greatest between two electrons in the same MO. Hurley, Lennard-Jones, and Pople therefore suggested the *separated-pair approximation*; here the molecular wave function is approximated not as an antisymmetrized product of orthogonal one-electron functions (called orbitals by Mulliken) but as an antisymmetrized product of orthogonal localized two-electron functions (called *geminals* by Shull); use of geminals takes into account much of the electron correlation omitted in an MO wave function. The geminal function for a given chemical bond (e.g., O—H) is expected to be approximately unchanged from molecule to molecule. [A GVB wave function (Section 15.18) is a special case of a

---

[59] For more on density matrices, see *Pilar*, Section 13-4.

geminal wave function, where each geminal is restricted to the form $f(1)g(2) + f(2)g(1)$.] Some applications of geminals have been made.[60] The separated-pair approximation allows for intraorbital correlation but not interorbital correlation. We noted in the Section 15.7 discussion of ethane that interorbital correlation is of substantial magnitude. Therefore the separated-pair approximation omits quite a substantial portion of the correlation energy and hence is not a really satisfactory quantitative approach to the correlation problem.

The separated-pair approximation can be considered to be a special case of a more general approach called *pair correlation theory*. (For a review of pair correlation theory, see W. Kutzelnigg in *Schaefer, Methods of Electronic Structure Theory*, Chapter 5.) A pair correlation theory substantially more accurate than the separated-pair approximation was developed by Sinanoglu and Nesbet in the 1960s. [Their work is related to the *many-body perturbation theory* (MBPT) developed by the physicists Brueckner and Goldstone to treat correlation in calculations of nuclear wave functions and extended to atomic and molecular wave functions by Kelly; for an MBPT calculation on H_2O, see J. H. Miller and H. P. Kelly, *Phys. Rev. A*, **4**, 480 (1971).] The Sinanoglu–Nesbet approach has been called the *independent electron-pair approximation* (IEPA). It assumes that atomic and molecular correlation energy can be well approximated as the sum of correlation energies over all pairs of electrons. (Thus for the Be $1s^2 2s^2$ configuration, the pair correlations are the $1s\alpha-1s\beta$ and $2s\alpha-2s\beta$ intraorbital correlations and the $1s\alpha-2s\alpha$, $1s\alpha-2s\beta$, $1s\beta-2s\alpha$, $1s\beta-2s\beta$ interorbital correlations.) One then attempts to calculate these pair correlation energies one at a time and then sums them to get the total correlation energy. Since interorbital correlation is included, IEPA is an improvement over the separated-pair approximation. Unfortunately, calculations on some small molecules (e.g., HF, CO, H_2O) show that the sum of pair correlation energies is not a truly accurate approximation to the correlation energy; one finds that the sum of pair correlation energies can be in error by ± 20 percent or more from the true correlation energy. (See W. Kutzelnigg in *Schaefer, Methods of Electronic Structure Theory*; *Schaefer, The Electronic Structure of Atoms and Molecules*, pp. 182–190, 323–325.) Of course, IEPA is a very substantial improvement over the Hartree–Fock energy.

An improvement over IEPA is the *coupled electron-pair approximation* (CEPA) of Meyer; CEPA produces a more accurate estimate of the correlation energy by allowing for the interactions (coupling) between the $n(n-1)/2$ pairs of the n-electron molecule, as well as the correlations within each of the $n(n-1)/2$ pairs. The CEPA method has been used in combination with pair natural orbitals (these are discussed later in this section) to give the PNO-CEPA method. Results of a PNO-CEPA calculation on H_2O are listed in Table 15.1.

An approach related to CEPA is Meyer's *self-consistent electron-pairs* (SCEP) *method* [W. Meyer, *J. Chem. Phys.*, **64**, 2901 (1976); C. E. Dykstra, H. F. Schaefer, and W. Meyer, *J. Chem. Phys.*, **65**, 2740 (1976)]. This approach allows for interactions between pairs, uses perturbation theory, and calculates a correlated

---

[60] *Pilar*, Section 17-8; *Parr*, Section 18; *Schaefer, The Electronic Structure of Atoms and Molecules*, pp. 190–197.

wave function directly from integrals over basis functions, without constructing the Hamiltonian matrix. It yields the equivalent of a CI calculation including all single and double excitations. The method is faster than conventional CI and is easily carried out on a minicomputer because it does not involve large Hamiltonian matrices. As originally formulated, SCEP applied only to closed-shell molecules, but it has since been generalized [C. E. Dykstra, *J. Chem. Phys.*, **67**, 4716 (1977); **72**, 2928 (1980)]. An application of SCEP to the isomers and transition states of N_2H_2 calculated wave functions equivalent to CI wave functions containing as many as 176,000 configuration functions [C. A. Parsons and C. E. Dykstra, *J. Chem. Phys.*, **71**, 3025 (1979)].

Because the conventional CI method suffers from slow convergence, several workers have developed CI methods designed to be more efficient than conventional CI. We now discuss some of these approaches.

An accurate CI calculation requires a very large number of configuration functions. A wave function with 10^4 configuration functions produces a secular determinant with 10^8 elements, which is difficult to work with. The standard matrix diagonalization techniques (Section 15.21), the Jacobi and Givens methods, cannot be used with such large matrices (special techniques have been developed to find the lowest few eigenvalues of a large matrix), nor can the complete matrix be stored in the fast part of computer storage. To avoid the problems of dealing with large matrices, Roos developed the *direct configuration interaction method* in 1972 (see B. O. Roos and P. E. M. Siegbahn, in *Schaefer, Methods of Electronic Structure Theory*, Chapter 7). The direct CI method avoids explicit calculation of the integrals H_{ij} between configuration functions and avoids the need to solve the secular equation (11.19); instead, the CI expansion coefficients in (11.18) and the energy are calculated directly from the one- and two-electron integrals over the basis functions. This allows the use of very large numbers of configuration functions to give very accurate results. For example, a direct CI calculation of the water dimer $(H_2O)_2$ included 56,268 configuration functions, a record at the time [G. H. F. Diercksen, W. P. Kraemer, and B. Roos, *Theor. Chim. Acta*, **36**, 249 (1975)]. The Liu–Siegbahn high-accuracy calculation of the H_3 potential-energy surface (Section 15.19) used direct CI. As initially developed, the direct CI method was not applicable to excited states, but the method has been generalized to include such states [N. C. Handy, J. D. Goddard, and H. F. Schaefer, *J. Chem. Phys.*, **71**, 426 (1979)].

The vector method (VM) of configuration interaction also calculates the CI wave function without construction and manipulation of the Hamiltonian matrix; see R. F. Hausman and C. F. Bender in *Schaefer, Methods of Electronic Structure Theory*, Chapter 8.

The use of approximate natural orbitals to speed up the convergence of a CI calculation was mentioned in Section 13.18. An approach called the *pair natural-orbital* or *pseudonatural-orbital* (PNO) *method* calculates approximate natural orbitals for separate pairs of electrons and then uses these in a CI calculation. These pseudonatural orbitals are not the true natural orbitals, since the latter must be calculated from the wave function of all electrons considered simultaneously, not just for separate pairs of electrons. A PNO-CI calculation by Meyer on the H_2O ground electronic state gave 83 percent of the correlation energy with only 237 configuration functions (Table 15.1). For more on PNO-CI, see W. Meyer in *Schaefer, Methods of Electronic Structure Theory*, Chapter 11.

A CI method called the *loop-driven graphical unitary-group approach* (GUGA) has been developed. GUGA allows efficient evaluation of the Hamiltonian matrix elements by identifying matrix elements that are equal and by establishing relationships between different matrix elements. GUGA has been found to be significantly faster than conventional CI [B. R. Brooks and H. F. Schaefer, *J. Chem. Phys.*, **70**, 5092 (1979); B. R. Brooks et al., *Physica Scripta*, **21**, 312 (1980)]. Moreover, this approach is more applicable to determination of the energy gradient (Section 15.16) than conventional CI. GUGA is also valuable in calculating multiconfiguration SCF wave functions (Section 13.18) containing very large numbers of configuration functions [B. R. Brooks et al., *J. Chem. Phys.*, **72**, 3837 (1980)]. A generalization of the direct CI method that is based on the GUGA approach has been published [P. E. M. Siegbahn, *J. Chem. Phys.*, **72**, 1647 (1980)].

A version of direct CI called the *alchemy CI method* (named after the computer program that implements it) uses special techniques for determining Hamiltonian matrix elements and is not subject to certain restrictions present in GUGA; see B. Liu and M. Yoshimine, *J. Chem. Phys.*, **74**, 612 (1981).

It is not yet clear which of the many pair-correlation methods and improved CI approaches will prove to be most useful, but the development of these methods is clearly a very promising sign.

The very substantial progress in quantum chemistry in recent years has made quantum-mechanical calculations a valuable tool to help decide a wide variety of questions of real chemical interest. Whereas years ago quantum-mechanical calculations on molecules were largely confined to journals read mainly by theoretical chemists, nowadays such calculations appear in large numbers in the *Journal of the American Chemical Society*, probably the most prestigious and widely read chemistry journal in the world. Quantum chemistry is being applied to such problems as the hydration of ions in solution, surface catalysis, the structures and energies of reaction intermediates, and the conformations of biological molecules. In many cases, theoretical calculations may not give definitive answers, but they are frequently good enough to allow for a very fruitful interaction of theory and experiment. Moreover, qualitative concepts like the Woodward–Hoffmann rules and Walsh diagrams have provided considerable insight into the course of chemical reactions and into chemical bonding.

In 1929 Dirac wrote, "The underlying physical laws necessary for the mathematical theory of . . . the whole of chemistry are thus completely known, and the difficulty is only that the exact application of these laws leads to equations much too complicated to be soluble." Application of high-speed digital computers to quantum chemistry has overcome to a significant degree the difficulties referred to by Dirac. Of course, just a small fraction of chemically important problems have been successfully treated by quantum mechanics, but future prospects are bright. To sum up, we quote Parr[61]: "Accurate descriptions of the electronic structure of molecules are upon us."

---

[61] *Parr*, p. 123.

PROBLEMS

15.1 The following argument has occasionally been used to "prove" the impossibility of obtaining accurate wave functions for many-electron systems: The electronic wave function of an n-electron system depends on $3n$ spatial coordinates (and n spin coordinates). To accurately tabulate the wave function, we must specify ψ for, say, 100 different values of each spatial coordinate; for a twenty-electron system, we need a table with $100^{60} = 10^{120}$ entries; this number is greater than the number of atoms in the observable universe (there are perhaps 10^{11} galaxies within 10^{10} light-years of us; a typical galaxy has 10^{11} stars; a typical star weighs 10^{33} g). Hence it is impossible to obtain accurate many-electron wave functions. Point out the fallacy in this argument.

15.2 Let $\hat{A} = n\delta(x_1 - x)\delta(y_1 - y)\delta(z_1 - z)$, where x_1, y_1, z_1 are the coordinates of electron 1, (x, y, z) is an arbitrary point in space, n is the number of electrons in the molecule, and δ is the Dirac delta function. If we evaluate $\langle \psi | \hat{A} | \psi \rangle$, what physically observable property is obtained?

15.3 Verify that (15.5) are the possible symmetry species for $\mathscr{C}_{2v}$.

15.4 Work out the possible symmetry species for $\mathscr{D}_2$.

15.5 For formaldehyde: (a) Work out the symmetry orbitals for a minimal-basis-set calculation; give the symmetry species of each symmetry orbital. (Choose the x axis perpendicular to the molecular plane.) (b) How many σ and how many π canonical MO's will result from a minimal-basis-set calculation? How many occupied σ and occupied π MO's are there for the ground state? (c) For each of the eight energy-localized MO's, state which AO's will make significant contributions. (d) What is the maximum-size secular determinant that occurs in finding the minimal-basis-set canonical MO's?

15.6 Work out the minimal-basis symmetry orbitals and their symmetry species for cis-1,2-difluoroethylene. (Choose the x axis perpendicular to the molecular plane.)

15.7 Sketch the $1a_1$, $3a_1$, $1b_1$, $4a_1$, and $2b_2$ MO's of water.

15.8 Give the point group of: (a) staggered C_2H_6; (b) eclipsed C_2H_6; (c) a C_2H_6 conformation that is neither eclipsed nor staggered; (d) s-trans-butadiene; (e) s-cis-butadiene. (f) Give the point group of C_2H_4 for $0°$, $30°$, and $90°$ twist angles of one CH_2 group relative to the other.

15.9 Frequently the x and z axes of ethylene are interchanged as compared to Fig. 15.10. What change does this cause in the MO labeling? [The symmetry-species designations (15.76) are retained.]

15.10 How many independent molecular wave functions correspond to: (a) a ^3E term? (b) a ^1E term?

15.11 Give the form of the normalization constants for the symmetry orbitals (15.18) and (15.19).

15.12 Suppose a ground-state calculation gives us some virtual orbitals for the molecule M. In which one of the following species would an excited electron occupy an MO that was well approximated by a virtual orbital of ground-state M? (a) M; (b) M^+; (c) M^-. Explain.

15.13 (a) Does having the coefficient of $C1s$ greater than 1.0 in the $1a_1$ methane MO in (15.56) violate the condition that the MO be normalized? Explain. (b) Use the results of Problem 13.28 to express the $2a_1$ methane MO of (15.56) using a nonorthogonal $2s$ STO. (c) Verify that the $2a_1$ MO found in part (b) is normalized. [The needed overlap integrals can be found by interpolating in the tables of R. S. Mulliken, C. A. Rieke, D. Orloff, and H. Orloff, *J. Chem. Phys.*, **17**, 1248 (1949).]

15.14 Let E_{HF} be the Hartree–Fock energy of a closed-subshell molecule; let $E^+_{k, HF, approx}$ be the approximate Hartree–Fock energy of the ion formed by removal of an electron from the kth MO of this molecule, this energy being calculated using the MO's of the un-ionized molecule. For both the molecule and the ion, the Hartree–Fock wave function is a single determinant. Use Eqs. (11.87)–(11.89) [where (11.89) is modified to allow for the presence of several nuclei] and Eq. (13.180) to show that $E_{HF} - E^+_{k,HF,approx} = \varepsilon_k$, where ε_k is the orbital energy of MO k.

15.15 Use the tables referred to in Problem 15.13c (or one of the other available overlap-integral tables—see footnote 37 in Chapter 15) to verify the values of the H_2O overlap integrals given in Section 15.5.

15.16 (a) Verify the values of the total overlap populations given for the H_2O MO's in Section 15.5. (b) Verify the total gross populations for the H_2O AO's given in Section 15.5.

15.17 Call the ethylene symmetry orbitals in (15.77) g_1 to g_{14}, in order. For each of the eight canonical ethylene MO's in (15.78), decide, as best you can, which symmetry orbitals will make major contributions; give the sign of each such symmetry orbital in the MO expression. (*Hint:* Decide how many inner-shell and how many bonding canonical MO's there are; combine the symmetry orbitals so as to build up charge density between the nuclei for the bonding MO's. *One further hint:* The third symmetry orbital in (15.77) makes no significant contribution to $3a_g$.) Sketch the canonical MO's. Assume an orthogonalized 2s orbital is used.

15.18 Predict the ground-state geometry of: (a) $SnBr_2$; (b) $HgBr_2$; (c) $TeCl_2$; (d) OF_2; (e) XeF_2; (f) H_2S; (g) I_3^-; (h) $HOCl$; (i) ClO_2^-; (j) GaI_3; (k) BrF_3; (l) PF_3; (m) BF_3; (n) H_3O^+; (o) ClO_3^-.

15.19 Predict the ground-state geometry of: (a) BrF_4^-; (b) SnH_4; (c) SeF_4; (d) XeF_4; (e) BH_4^-; (f) PCl_5; (g) $SbCl_5$; (h) BrF_5; (i) SF_6.

15.20 Give the point group of each of the molecules of Problem 15.19.

15.21 Predict the ground-state geometry of each of the following molecules, considering the electrons in a multiple bond as a single pair occupying a "large orbital": (a) ONF; (b) NO_2^+; (c) SO_2; (d) CO_2; (e) ClO_2^-; (f) NO_2^-; (g) F_2CO; (h) SO_3; (i) HNO_2; (j) XeO_3; (k) $SOBr_2$; (l) POF_3; (m) $FClO_3$; (n) $F_2IO_2^-$; (o) XeO_4; (p) $XeOF_4$; (q) SOF_4; (r) IOF_5.

15.22 Use the Walsh diagram of Fig. 15.14 to predict the ground-state geometry of: (a) BH_2^+; (b) CH_2^+; (c) AsH_2; (d) CH_2^-; (e) OH_2^+; (f) NH_2^-; (g) H_2F^+.

15.23 Use Fig. 15.14 to predict the geometry of the lowest excited state of BH_2.

15.24 Derive Eq. (15.181). Use confocal elliptic coordinates, but note the difference in choice of axes between a diatomic and a planar polyatomic molecule.

15.25 Is Eq. (15.79) an accurate representation of the Hartree–Fock $1b_{3u}$ MO of ethylene? Explain.

15.26 When an extended STO basis set is used to calculate a near Hartree–Fock ψ for H_2O, one includes $3d$ functions on oxygen. Use symmetry species arguments to decide which of the following $3d$ oxygen AO's must make no contribution to the occupied ground-state H_2O MO's: $3d_{z^2}$, $3d_{xz}$, $3d_{yz}$, $3d_{x^2-y^2}$, $3d_{xy}$.

15.27 Slater's rules for finding approximate orbital exponents of K-, L-, and M-shell Slater AO's are as follows. The orbital exponent ζ is taken as $(Z - s_{nl})/n$, where n is the principal quantum number and Z is the atomic number. The screening constant s_{nl} is calculated as follows: The AO's are divided into the following groups:

$$(1s) \qquad (2s, 2p) \qquad (3s, 3p) \qquad (3d)$$

To find s_{nl}, the following contributions are summed: (a) 0 from electrons in groups outside the one being considered; (b) 0.35 from each other electron in the group considered, except that 0.30 is used in the 1s group; (c) for an s or p orbital, 0.85 from each electron with n one less than the orbital considered and 1.00 from each electron still further in; for a d orbital, 1.00 for each electron inside the group.

Calculate the orbital exponents according to Slater's rules for the atoms H, He, C, N,

O, S, Ar. The optimum values of ζ to use when approximating an AO as a single STO have been calculated[62]; compare these values with the values found by Slater's rules.

15.28 The FE MO model usually takes the π electrons of a conjugated polyene as moving in a one-dimensional box. A better picture would be a long, thin, rectangular three-dimensional box. Answer the following questions for a three-dimensional-box model. (a) What restriction must be imposed on the quantum numbers of an orbital to give it the required π symmetry? (b) Show that this model still gives the expression (15.94) for the longest wavelength transition.

15.29 Calculate the longest wavelength electronic transition of:

(a) $CH_2{=}CH{-}CH{=}CH{-}\bar{C}H_2$;

(b) $CH_2{=}CH{-}CH{=}CH{-}\overset{+}{C}H_2$;

(c) β-carotene. For each species use either the FE MO model or the FE MO model with a sinusoidal potential added, whichever is more appropriate. Predict the color of each species.

15.30 Calculate the limiting value of the longest wavelength transition of the polyene (15.95) as $n_C \rightarrow \infty$, using: (a) the simple FE MO model; (b) the FE MO model with a sinusoidal potential added.

15.31 Express the real forms of the spherical harmonics as linear combinations of the complex functions Y_l^m. Three separate equations are required.

15.32 Apply the FE MO model to benzene by solving the Schroedinger equation for a particle confined to move on a circle. (There is only one variable, the angle φ.) Draw the pattern of π-electron FE MO energies and compare to the Hückel method. Pick a reasonable value for the radius and calculate the position of the longest wavelength band of benzene; compare with the experimental value 2038 Å.

15.33 Write down symmetry orbitals for a minimal-basis-set calculation of H_2.

15.34 (a) Verify that the HMO coefficients (15.123) satisfy the HMO set of simultaneous equations. [*Hint:* Use the identity $\sin a + \sin b = 2 \sin \frac{1}{2}(a+b) \cos \frac{1}{2}(a-b)$.] (b) Verify that the coefficients (15.123) give a normalized HMO. [To evaluate the needed sum, express the sine function as exponentials and then use the formula for the sum of a geometric series.]

15.35 Calculate the HMO total bond orders and the π-electron densities for the lowest excited state of butadiene.

15.36 Since only the topology of the carbon framework is of significance in the HMO method, the full symmetry of s-*trans*-butadiene need not be used to get the maximum simplification possible in the HMO method; instead, it is sufficient to use only the C_2 axis. (a) Write down the two possible symmetry species for the group $\mathscr{C}_2$. (b) Construct π-electron symmetry orbitals for butadiene, classifying them according to the symmetry species of $\mathscr{C}_2$. (c) Set up and solve the two Hückel secular equations for butadiene using the symmetry orbitals of (b) as basis functions.

15.37 Verify that the geometric construction of Fig. 15.19 gives the correct HMO energies of the cyclic polyene $C_n H_n$.

15.38 For the allyl radical $\cdot CH_2{-}CH{=}CH_2$, find: (a) the HMO's and energies; (b) the mobile bond orders; (c) the π-electron densities; (d) the free valences; (e) the delocalization energy.

15.39 Calculate the quantities (a) through (e) of Problem 15.38 for the allyl cation and anion, $[CH_2 CHCH_2]^+$ and $[CH_2 CHCH_2]^-$. Which ion is predicted to be more stable?

15.40 (a) Calculate the energy needed to compress three carbon–carbon single

---

[62] E. Clementi and D. L. Raimondi, *J. Chem. Phys.*, **38**, 2686 (1963); E. Clementi, D. L. Raimondi, and W. P. Reinhardt, *J. Chem. Phys.*, **47**, 1300 (1967). [For $n = 4$, Slater took ζ as $(Z - s_{nl})/3.7$; however, Slater's rules are generally unreliable for $n \geqslant 4$.]

bonds and stretch three carbon–carbon double bonds to the benzene bond length 1.397 Å. Assume a harmonic-oscillator potential-energy function for bond stretching and compression. Typical carbon–carbon single- and double-bond lengths are 1.53 and 1.335 Å; typical stretching force constants for carbon–carbon single and double bonds are 5 and 9.5 millidynes/Å. (b) Use the result of (a) to calculate an improved β value for benzene from the data following Eq. (15.154).

15.41 Typical bond energies (in kcal/mole) are

$$\begin{array}{ccc} C-H & C-C & C=C \\ 99 & 83 & 146 \end{array}$$

The heat of formation of benzene from six hydrogen and six carbon atoms is 1323 kcal/mole. Calculate the "experimental" delocalization energy of benzene using these data, first omitting the strain-energy correction, then including it. (Because bond energies vary by several kcal from compound to compound, this procedure is very rough.)

15.42 Estimate the carbon–carbon bond length in: (a) $C_5H_5^-$; (b) $C_7H_7^+$; (c) $C_8H_8^{2-}$.

15.43 (a) Set up and solve the a_u, b_{2g}, and b_{1g} HMO secular equations for naphthalene. (b) Find the coefficients of the lowest naphthalene HMO.

15.44 Derive Eq. (15.164). [*Hint:* Start with E_π as the sum of the orbital energies and use $e_i = \langle \varphi_i | \hat{H}^{\text{eff}} | \varphi_i \rangle$.]

15.45 Work out the coefficients of the nonbonding HMO of benzyl.

15.46 Sketch the HMO's of naphthalene; indicate the nodal planes by dashed lines.

15.47 (a) Calculate the quantities (a) through (e) of Problem 15.38 for the diradical trimethylenemethane $C(CH_2)_3$. (The degenerate MO's can be taken as either real or complex, similar to the benzene degenerate MO's.) (b) Do the same as in (a) for the propargyl diradical $HC \equiv C = CH$. Save time by using the results of Problem 15.38. Note that this linear molecule has two sets of spatially perpendicular π MO's.

15.48 Use Table 15.3 to calculate the naphthalene bond orders, free valences, delocalization energy, and predicted bond lengths.

15.49 The ionization potentials (in eV) of the first few polyacenes are

benzene	naphthalene	anthracene	tetracene
9.4	8.3	7.6	7.0

(a) Use these data in conjunction with the HMO method to calculate values of α and β; compare the value of β obtained with (15.150). (b) Predict the ionization potential of pentacene. (The HMO Δx values for HOMO's and LUMO's of these polyacenes are 2.0, 1.236, 0.828, 0.590, 0.439.)

15.50 (a) Compute the π-electron energy of ethylene when overlap is included in the HMO method. (b) Compute in terms of γ the delocalization energy of benzene when overlap is included in the HMO method. (c) Using the procedure of Problem 15.40, evaluate γ; compare with the spectroscopic value 2.72 eV [Eq. (15.150)].

15.51 Write down the simple HMO secular determinant for each of the following molecules: (a) phenol; (b) phenanthrene; (c) furan.

15.52 Find the HMO energies for the polymethine ion (15.100) with $k = 0$. Use averages of the values in (15.174); to simplify the work, use symmetry orbitals. What value of β is required to fit the observed lowest energy transition?

15.53 (a) For P_{rs} equal to 1 and to 3, compare the bond-length predictions of Eq. (15.163) with experimental carbon–carbon single- and triple-bond lengths. (b) Look up (in one of the tabulations mentioned in Section 15.11) the Hückel bond orders of azulene and compare the predicted bond lengths with the experimental values. [The experimental data can be found in R. J. Buenker and S. D. Peyerimhoff, *Chem. Phys. Lett.*, **3**, 37 (1969).]

15.54 (a) What is the HMO delocalization energy of cyclobutadiene? (b) Calculate the HMO total bond orders in cyclobutadiene.

15.55 Show that

$$\sum_r q_r = n_\pi$$

15.56 Calculate the Wheland π MO energies of butadiene; use $S = 0.25$.

15.57 Prove that the Wheland MO coefficients b_{jr} for conjugated hydrocarbons are related to the HMO coefficients by

$$b_{jr} = \frac{c_{jr}}{(1 - Sw_j)^{1/2}}$$

15.58 Apply the extended Hückel method to H_2. Use $K = 1.5$; use the expression from Chapter 13 of S as a function of R; plot the valence-electron energy as a function of R; compare the predicted value of R_e with the experimental value. (To avoid solving a secular equation, use symmetry orbitals.) Use 1.0 as the orbital exponent.

15.59 Verify Eq. (15.204).

15.60 (a) Set up the 8×8 extended Hückel secular determinant for methane at the equilibrium configuration; do not use symmetry orbitals. Take $K = 1.5$; use (15.198); use Slater's rules (Problem 15.27) for the orbital exponents; evaluate the overlap integrals from the reference of Problem 15.13. (b) Do the same as in (a), but now use symmetry orbitals.

15.61 The three equivalent sp^2 hybrid AO's of carbon point to the corners of an equilateral triangle. Derive expressions for them, assuming orthonormality.

15.62 The two equivalent sp hybrid carbon AO's make an angle of $180°$ with each other. Derive them.

15.63 Draw VB resonance structures of the benzyl radical with the odd electron on the CH_2 carbon, on an *ortho* carbon, on a *meta* carbon, and on a *para* carbon. Explain why the contribution of the structure with the odd electron on a *meta* carbon should be quite small. Is this consistent with the MO results (Problem 15.45)?

15.64 Draw the twelve butadiene singly polar ionic structures.

15.65 (a) Which bonds in naphthalene does the resonance method predict to be the shortest? (b) Which bond in azulene does the resonance method predict to be the longest? (Consider only Kekulé structures.)

15.66 Express the electronic energy of a molecule in terms of integrals involving no more than six spatial coordinates.

15.67 Consider the following statement[63]: "Semiempirical quantum chemistry violates the fundamental canon of a *scientific theory* to the effect that statements have to be expressed in a manner that allows for the possibility of disproof." Give arguments for or against this statement.

15.68 Let the AO's i and k be on the same atom in a molecule and let the AO s be on a different atom. Which of the following integrals are neglected in the CNDO method? Which are neglected in the INDO method? (a) $(ii|kk)$; (b) $(ik|ik)$; (c) $(ii|ss)$; (d) $(is|is)$; (e) $(ii|is)$.

15.69 Four of the six calculated pair correlation energies for the Be ground state are 0.00081, 0.00212, 0.04183, 0.04535 hartrees. Each of the two pair correlation energies not given has the same value as one of the four listed. (a) Which of these numbers is the $1s\alpha-2s\beta$ pair correlation energy? (b) Which of these values is the $1s\alpha-2s\alpha$ pair correlation energy? (c) What are the values of the two pair correlation energies not given?

15.70 Consider a closed-shell wave function with n electrons occupying $n/2$ MO's.

---

[63] H. Primas, *Int. J. Quantum Chem.*, **1**, 493 (1967).

(a) How many intraorbital pair correlations are there? (b) How many interorbital pair correlations are there? (c) What is the total number of pair correlations?

15.71 Consider the thermal reaction 1,3,5-hexatriene → 1,3-cyclohexadiene. (a) Use the symmetry of the polyene HOMO to predict whether the reaction path is conrotatory or disrotatory. (The HMO's need not be found explicitly; all that is needed is the signs of the AO's in the HOMO and these can be found by noting that the HMO nodal pattern is the same as the FE MO nodal pattern.) (b) Do the same as in (a) when the reaction occurs photochemically. (c) State the general rules for the cyclization of the polyene (15.95) with n_π π electrons.

15.72 Examine the frontier orbitals and decide whether each of the following elementary reactions should have a high or low activation energy for a four-center broadside-collision reaction path. (a) $H_2 + D_2 \rightarrow 2HD$; (b) $N_2 + O_2 \rightarrow 2NO$; (c) $F_2 + Br_2 \rightarrow 2FBr$; (d) $H_2 + C_2H_4 \rightarrow C_2H_6$.

15.73 For the polyenes (15.95), observed longest wavelength electronic absorption bands occur at the following $1/\lambda$ values: 61500, 46100, 37300, 32900, 29900, 27500, 25600, 24400, and 22400 cm^{-1} for $k = 0, 1, 2, 3, 4, 5, 6, 7$, and 9, respectively. (a) Compare these values to those predicted by the FE MO equation (15.97) and calculate the average absolute deviation. (b) Do the same using (15.99). (c) Do the same using the Hückel equation (15.124).

15.74 For the ions (15.100), observed longest wavelength electronic absorption bands occur at $1/\lambda$ values of 44600, 32000, 24000, 19300, 16000, 13600, and 11800 cm^{-1} for $k = 0, 1, 2, 3, 4, 5$, and 6, respectively. Compare these values with those calculated from the FE MO equation (15.101) and calculate the average absolute deviation.

15.75 For each of the following, calculate the Hess–Schaad resonance energy and the REPE, and predict whether the compound will be aromatic, nonaromatic, or antiaromatic: (a) naphthalene (see Table 15.3); (b) planar [8]annulene; (c) planar [18]annulene; (d) azulene (Fig. 15.21), for which the x values in (15.114) of the occupied HMO's are -2.3103, -1.6516, -1.3557, -0.8870, and -0.4773.

15.76 (a) Write the bond-separation reaction for cyclopropene. (b) Calculated 4-31G SCF MO energies for CH_3CHO, CH_4, C_2H_6, and H_2CO are -152.68631, -40.13938, -79.11562, and -113.69209 hartrees, respectively. The zero-point vibrational energies of CH_3CHO, CH_4, C_2H_6, and H_2CO are 0.05298, 0.04320, 0.07214, and 0.02567 hartrees, respectively. Calculate the value of ΔH_0° predicted by the 4-31G basis set for the gas-phase CH_3CHO bond-separation reaction and compare to the experimental value 9.5 kcal/mole.

15.77 Calculate the matrix products **CD** and **DC**, where

$$\mathbf{C} = \begin{pmatrix} 5 \\ 0 \\ -1 \end{pmatrix} \quad \text{and} \quad \mathbf{D} = (i \quad 2 \quad 1)$$

15.78 Find $\mathbf{A}^*$, $\tilde{\mathbf{A}}$, and $\mathbf{A}^\dagger$, where

$$\mathbf{A} = \begin{pmatrix} 1 & 2+i \\ -3 & i \end{pmatrix}$$

15.79 The *trace* $\text{tr}(\mathbf{A})$ of a square matrix $\mathbf{A}$ is the sum of the elements on the principal diagonal. Use (15.221) to prove that $\text{tr}(\mathbf{BC}) = \text{tr}(\mathbf{CB})$, where $\mathbf{B}$ and $\mathbf{C}$ are square matrices of order n.

APPENDIX

Table A.1 *Physical constants*

Constant and symbol[1]		SI value	Gaussian value
Speed of light in vacuum	c	2.9979246×10^8 m/s	2.9979246×10^{10} cm/s
Proton charge	e	1.60219×10^{-19} C	
	e'		4.80324×10^{-10} statC
Permittivity of vacuum	ε_0	8.85419×10^{-12} C^2/N-m^2	
Avogadro constant	N_0	6.0220×10^{23} mole^{-1}	6.0220×10^{23} mole^{-1}
Electron rest mass	m	9.1095×10^{-31} kg	9.1095×10^{-28} g
Proton rest mass	m_p	1.67265×10^{-27} kg	1.67265×10^{-24} g
Neutron rest mass	m_n	1.67495×10^{-27} kg	1.67495×10^{-24} g
Planck constant	h	6.6262×10^{-34} J s	6.6262×10^{-27} erg s
	$\hbar$	1.05459×10^{-34} J s	1.05459×10^{-27} erg s
Faraday constant	F	96485 C/mole	
Permeability of vacuum	μ_0	$4\pi \times 10^{-7}$ N C^{-2} s^2	
Bohr radius	a_0	5.29177×10^{-11} m	0.529177×10^{-8} cm
Bohr magneton	β_e	9.2741×10^{-24} J/T	
Nuclear magneton	β_N	5.05082×10^{-27} J/T	
Electron g value	g_e	2.00231931	2.00231931
Proton g value	g_p	5.58569	5.58569
Gas constant	R	8.314_4 J/mole-K	$8.314_4 \times 10^7$ erg/mole-K
Boltzmann constant	k	1.3807×10^{-23} J/K	1.3807×10^{-16} erg/K
Gravitational constant	G	6.67×10^{-11} m^3/kg-s^2	6.67×10^{-8} cm^3/g-s^2

Adapted from E. R. Cohen and B. N. Taylor, *J. Phys. Chem. Ref. Data*, **2**, 663 (1973).

---

[1] $\hbar = h/2\pi$, $\quad F = N_0 e$, $\quad e' = e/(4\pi\varepsilon_0)^{1/2}$, $\quad a_0 = \hbar^2/me'^2 = 4\pi\varepsilon_0\hbar^2/me^2$, $\quad \beta_e = e\hbar/2m$, $\beta_N = e\hbar/2m_p$, $\quad k = R/N_0$.

Table A.2 *Energy conversion factors*

		J	erg	eV	cm^{-1}	hartree	kcal/mole
1 J	=	1	10^7	6.2415×10^{18}	5.0340×10^{22}	2.2937×10^{17}	1.4393×10^{20}
1 erg	=	10^{-7}	1	6.2415×10^{11}	5.0340×10^{15}	2.2937×10^{10}	1.4393×10^{13}
1 eV	=	1.6022×10^{-19}	1.6022×10^{-12}	1	8065.5	3.6749×10^{-2}	23.060
1 cm^{-1}	=	1.9865×10^{-23}	1.9865×10^{-16}	$1.2398_5 \times 10^{-4}$	1	4.55634×10^{-6}	2.8591×10^{-3}
1 hartree	=	4.3598×10^{-18}	4.3598×10^{-11}	27.212	219474.6	1	627.51
1 kcal/mole	=	6.9478×10^{-21}	6.9478×10^{-14}	4.3364×10^{-2}	349.75	1.5936×10^{-3}	1

NOTE: The relationships involving cm^{-1} or kcal/mole are correspondences rather than actual equalities. The wave number $1/\lambda$ (commonly expressed in cm^{-1}) is often used as a measure of energy; a photon with energy E of 1.99×10^{-23} J has a wave number $1/\lambda = E/hc$ of $1 cm^{-1}$. If a mole of molecules has an energy of 1 kcal, then each molecule has an energy of 0.043 eV.

Table A.3 *Relations between Gaussian and SI quantities*

Quantity	Gaussian		SI
Electric charge	Q'	$=$	$Q/(4\pi\varepsilon_0)^{1/2}$
Electric field strength	$\mathbf{E}'$	$=$	$(4\pi\varepsilon_0)^{1/2}\mathbf{E}$
Electric displacement	$\mathbf{D}'$	$=$	$(4\pi/\varepsilon_0)^{1/2}\mathbf{D}$
Electric potential	ϕ'	$=$	$(4\pi\varepsilon_0)^{1/2}\phi$
Electric dipole moment	$\mathbf{d}'$	$=$	$\mathbf{d}/(4\pi\varepsilon_0)^{1/2}$
Electric polarizability	α'	$=$	$\alpha/4\pi\varepsilon_0$
Magnetic flux density	$\mathbf{B}'$	$=$	$(4\pi/\mu_0)^{1/2}\mathbf{B}$
Magnetic field strength	$\mathbf{H}'$	$=$	$(4\pi\mu_0)^{1/2}\mathbf{H}$
Magnetic dipole moment	$\boldsymbol{\mu}'$	$=$	$(\mu_0/4\pi)^{1/2}\boldsymbol{\mu}$
Magnetic vector potential	$\mathbf{A}'$	$=$	$(4\pi/\mu_0)^{1/2}\mathbf{A}$
Magnetization	$\mathbf{M}'$	$=$	$(\mu_0/4\pi)^{1/2}\mathbf{M}$

Table A.4 *Integrals*

$$\int x \sin bx\, dx = \frac{1}{b^2}\sin bx - \frac{x}{b}\cos bx \tag{A.1}$$

$$\int xe^{bx}\, dx = \frac{e^{bx}}{b^2}(bx-1) \tag{A.2}$$

$$\int x^2 e^{bx}\, dx = e^{bx}\left(\frac{x^2}{b} - \frac{2x}{b^2} + \frac{2}{b^3}\right) \tag{A.3}$$

$$\int_0^\infty x^n e^{-qx}\, dx = \frac{n!}{q^{n+1}}, \qquad n > -1,\, q > 0 \tag{A.4}$$

$$\int_0^\infty e^{-bx^2}\, dx = \frac{1}{2}\left(\frac{\pi}{b}\right)^{1/2} \tag{A.5}$$

$$\int_0^\infty x^{2n} e^{-bx^2}\, dx = \frac{1\cdot 3\cdots(2n-1)}{2^{n+1}}\left(\frac{\pi}{b^{2n+1}}\right)^{1/2}, \qquad n = 1, 2, 3,\ldots \tag{A.6}$$

$$\int_t^\infty z^n e^{-az}\, dz = \frac{n!}{a^{n+1}}\, e^{-at}\left(1 + at + \frac{a^2 t^2}{2!} + \cdots + \frac{a^n t^n}{n!}\right), \qquad n = 0, 1, 2,\ldots \tag{A.7}$$

BIBLIOGRAPHY

Anderson, J. M., *Introduction to Quantum Chemistry*, Benjamin, New York, 1969.

Anderson, J. M., *Mathematics for Quantum Chemistry*, Benjamin, New York, 1966.

Atkins, P. W., *Molecular Quantum Mechanics*, Oxford University Press, New York, 1970.

Bates, D. R., ed., *Quantum Theory*, 3 vols., Academic Press, New York, 1961.

Bethe, H. A., and R. W. Jackiw, *Intermediate Quantum Mechanics*, 2nd ed., Benjamin, New York, 1968.

Bethe, H. A., and E. E. Salpeter, *Quantum Mechanics of One- and Two-Electron Atoms*, Academic Press, New York, 1957.

Blokhintsev, D. I., *Principles of Quantum Mechanics*, Allyn and Bacon, Boston, 1964.

Cotton, F. A., *Chemical Applications of Group Theory*, 2nd ed., Wiley, New York, 1970.

Davidson, N., *Statistical Mechanics*, McGraw-Hill, New York, 1962.

Davis, J. C., Jr., *Advanced Physical Chemistry*, Ronald Press, New York, 1965.

Dewar, M. J. S., *Molecular Orbital Theory of Organic Chemistry*, McGraw-Hill, New York, 1969.

Dicke, R. H., and J. P. Wittke, *Introduction to Quantum Mechanics*, Addison-Wesley, Reading, Mass., 1960.

Dirac, P. A. M., *The Principles of Quantum Mechanics*, 4th ed., Oxford University Press, New York, 1958.

Eyring, H., J. Walter, and G. E. Kimball, *Quantum Chemistry*, Wiley, New York, 1944.

Fong, P., *Elementary Quantum Mechanics*, Addison-Wesley, Reading, Mass., 1962.

Gatz, C. R., *Introduction to Quantum Chemistry*, Merrill, Columbus, Ohio, 1971.

Goodisman, J., *Contemporary Quantum Chemistry*, Plenum, New York, 1977.

Halliday, D., and R. Resnick, *Physics*, 3rd ed., Wiley, New York, 1978.

Hameka, H. F., *Introduction to Quantum Theory*, Harper and Row, New York, 1967.

Hameka, H. F., *Quantum Theory of the Chemical Bond*, Hafner, New York, 1975.

Hanna, M. W., *Quantum Mechanics in Chemistry*, 3rd ed., Benjamin, Menlo Park, Cal., 1981.

Jaffé, H. H., and M. Orchin, *Symmetry in Chemistry*, Wiley, New York, 1965.

Jammer, M., *The Conceptual Development of Quantum Mechanics*, McGraw-Hill, New York, 1966.

Johnson, C. S., and L. G. Pedersen, *Problems and Solutions in Quantum Chemistry and Physics*, Addison-Wesley, Reading, Mass., 1974.

Karplus, M., and R. N. Porter, *Atoms and Molecules*, Benjamin, New York, 1970.

Kauzmann, W., *Quantum Chemistry*, Academic Press, New York, 1957.

Landau, L. D., and E. M. Lifshitz, *Quantum Mechanics: Non-Relativistic Theory*, 2nd ed., Pergamon, Elmsford, New York, 1965.

La Paglia, S. R., *Introductory Quantum Chemistry*, Harper and Row, New York, 1971.

Levine, I. N., *Molecular Spectroscopy*, Wiley-Interscience, New York, 1975.

Levine, I. N., *Physical Chemistry*, 2nd ed., McGraw-Hill, New York, 1983.

Lowe, J. P., *Quantum Chemistry*, Academic Press, New York, 1978.

Margenau, H., and G. M. Murphy, *The Mathematics of Physics and Chemistry*, 2nd ed., Van Nostrand, New York, 1956.

Mathews, J., and R. L. Walker, *Mathematical Methods of Physics*, 2nd ed., Benjamin, New York, 1970.

Merzbacher, E., *Quantum Mechanics*, 2nd ed., Wiley, New York, 1970.

Mulliken, R. S., and W. C. Ermler, *Diatomic Molecules*, Academic Press, New York, 1977.

Mulliken, R. S., and W. C. Ermler, *Polyatomic Molecules*, Academic Press, New York, 1981.

Murrell, J. N., and A. J. Harget, *Semi-empirical Self-Consistent-Field Molecular Orbital Theories of Molecules*, Wiley-Interscience, New York, 1971.

Murrell, J. N., S. F. A. Kettle, and J. M. Tedder, *Valence Theory*, 2nd ed., Wiley, New York, 1970.

Offenhartz, P. O'D., *Atomic and Molecular Orbital Theory*, McGraw-Hill, New York, 1970.

Park, D., *Introduction to the Quantum Theory*, 2nd ed., McGraw-Hill, New York, 1974.

Parr, R. G., *Quantum Theory of Molecular Electronic Structure*, Benjamin, New York, 1963.

Pauling, L., and E. B. Wilson, Jr., *Introduction to Quantum Mechanics*, McGraw-Hill, New York, 1935.

Pilar, F. L., *Elementary Quantum Chemistry*, McGraw-Hill, New York, 1968.

Pople, J. A., and D. L. Beveridge, *Approximate Molecular Orbital Theory*, McGraw-Hill, New York, 1970.

Salem, L., *The Molecular Orbital Theory of Conjugated Systems*, Benjamin, New York, 1966.

Schaefer, H. F., ed., *Applications of Electronic Structure Theory* (vol. 4 of *Modern Theoretical Chemistry*, W. Miller et al., eds.), Plenum, New York, 1977.

Schaefer, H. F., *The Electronic Structure of Atoms and Molecules*, Addison-Wesley, Reading, Mass., 1972.

Schaefer, H. F., ed., *Methods of Electronic Structure Theory* (vol. 3 of *Modern Theoretical Chemistry*, W. Miller et al., eds.), Plenum, New York, 1977.

Schiff, L., *Quantum Mechanics*, McGraw-Hill, New York, 1968.

Schonland, D. S., *Molecular Symmetry*, Van Nostrand, New York, 1965.

Segal, G. A., ed., *Semiempirical Methods of Electronic Structure Calculation, Parts A and B* (vols. 7 and 8 of *Modern Theoretical Chemistry*, W. Miller et al., eds.), Plenum, New York, 1977.

Slater, J. C., *Quantum Theory of Atomic Structure*, vols. I and II, McGraw-Hill, New York, 1960.

Slater, J. C., *Quantum Theory of Molecules and Solids*, vol. I, *Electronic Structure of Molecules*, McGraw-Hill, New York, 1963.

Sokolnikoff, I. S., and R. M. Redheffer, *Mathematics of Physics and Modern Engineering*, 2nd ed., McGraw-Hill, New York, 1966.

Streitwieser, A., Jr., *Molecular Orbital Theory for Organic Chemists*, Wiley, New York, 1961.

Taylor, A. E., and W. R. Mann, *Advanced Calculus*, 2nd ed., Wiley, New York, 1972.

Tinkham, M., *Group Theory and Quantum Mechanics*, McGraw-Hill, New York, 1964.

ANSWERS TO SELECTED PROBLEMS

1.1 4.0×10^4 dyn $= 0.40$ N.　　**1.2** (a) 3.32 A; (b) relativistic.　　**1.3** (a) g cm/sec^2;
(b) g cm^2/sec^2; (c) g$^{1/2}$ cm$^{3/2}$/sec; (d) kg m s^{-2}; (e) kg m^2 s^{-2}; (f) A s.　　**1.4** (a) 3.92 eV;
(b) 544 nm.　　**1.6** None; the function in (c) is not normalized.　　**1.7** 2.22×10^{-6} deg.
1.8 2.24, 0.0126.　　**1.9** 13/25.　　**1.10** 3.1 Å.　　**1.13** (a) 1, $\pi/2$; (b) $|a|$, $\pi/3$; (c) $5^{1/2}$, 296.6°.

2.1 $y = ae^x + be^{-2x}$.　　**2.2** (b) $y = ae^x + bxe^x$.　　**2.3** 3.0×10^{26}.　　**2.4** (b) 3;
(c) $\frac{1}{4}$.　　**2.5** (b) 1.1×10^2 A; (c) UV.　　**2.10** 3200 Å.　　**2.14** 2.　　**2.16** 4.02 eV, 13.6 eV.

3.3 (b) $\hat{A}$ and $\hat{B}$ linear and commute.　　**3.4** $\hat{1}$.　　**3.5** (a) linear; (b) nonlinear;
(c) linear; (d) nonlinear; (e) linear.　　**3.6** (a) complex conjugation; (b) *hint*: consider an
operator that is the product of operators.　　**3.9** $2a + (4ax + 2b)d/dx$.　　**3.11** (a) $i\hbar$;
(b) $2\hbar^2 \partial/\partial x$; (c) 0; (d) 0; (e) $(\hbar^2/m)\partial/\partial x$.　　**3.12** (a) Yes; (b) $1/p$; (c) $1/(p-a)$.　　**3.13** (a) Yes;
(b) $1 - 2x$.　　**3.15** (a) Yes, 1; (b) no; (c) yes, -1; (d) yes, -1; (e) yes, -1.　　**3.16** (a) $a/2$;
(b) $b/2$, $c/2$; (c) 0; (d) $(1 - 3/2n_x^2\pi^2)(a^2/3)$, no, yes.　　**3.17** (a) nondegenerate; (b) 6; (c) 4.
3.19 (a) cm$^{-1/2}$.　　**3.20** (a), (c), (d), (g).　　**3.21** (b), (c).　　**3.22** (a) No; (b) yes; (c) yes;
(d) yes; (e) no.　　**3.26** (a) $i\hbar^3 \partial^3/\partial y^3$.　　**3.30** (a) 17; (b) 6.　　**3.31** (b) 12.

4.1 $h\nu/4$, $h\nu/4$.　　**4.2** (a) Odd; (b) even; (c) odd; (d) neither.　　**4.5** $(v_x + \frac{1}{2})h\nu_x +$
$(v_y + \frac{1}{2})h\nu_y + (v_z + \frac{1}{2})h\nu_z$.　　**4.13** (b) $c_4 = -3c_0/8$;　　$c_5 = -3c_1/40$.　　**4.16** Use　　parity.
4.19 (c) Yes.　　**4.22** (a) 4.77 mdyn/Å.　　**4.24** (a) $\nu_e/c = 2989.96$ cm^{-1}, $\nu_e x_e/c = 51.99$ cm^{-1};
(b) 8346.0 cm^{-1}.

5.2 $\Delta x = (h/8\pi^2 m\nu)^{1/2}$, $\Delta p_x = (mh\nu/2)^{1/2}$.　　**5.5** $|\mathbf{A}| = 7$, $\mathbf{A} \cdot \mathbf{B} = 13$, $\mathbf{A} \times \mathbf{B} = -32\mathbf{i}$
$- 18\mathbf{j} + 10\mathbf{k}$.　　**5.8** $\nabla^2 f = 6$.　　**5.9** 3.　　**5.12** 35.3°, 65.9°, 90°, 114.1°, 144.7°.　　**5.13** 1, 0.707.

6.1 (a) 1.13 Å.　　**6.2** (a) 0.24; (c) 2.66a.　　**6.3** 14.　　**6.4** (a) $3a/2Z$; (b) a/Z;
(c) $5a/Z$.　　**6.5** At the origin (nucleus).　　**6.8** $-e'^2/a$, $e'^2/2a$, $-\frac{1}{2}$.　　**6.9** -6.8 eV.　　**6.14** Yes.
6.16 1/137.0.　　**6.19** s states.　　**6.20** -108.8 eV.　　**6.21** (a) Spheres centered at the
origin; (b) m.　　**6.23** (f) 1, 3, 6.　　**6.28** (a) All; (b) $\hat{H}$, $\hat{L}^2$; (c) all.

7.1 (a) F; (b) T; (c) F; (d) F; (e) F; (f) T; (g) F; (h) F.　　**7.4** id/dx, $4d^2/dx^2$.
7.8 $\pi^3/32 = 1 - 1/3^3 + 1/5^3 - 1/7^3 + \cdots$.　　**7.10** (c) Yes; no.　　**7.13** The n nth roots of 1.
7.14 (b) and (c).　　**7.17** (c) and (d).　　**7.21** (a) 0, 1, 0; (b) $\frac{1}{2}$, 0, $\frac{1}{2}$.　　**7.23** (a) 1; (b) 0; (c) 1;
(d) 0.　　**7.28** Possible outcomes: $n^2h^2/8ml^2$. Probabilities: 0 for n even, $960/n^6\pi^6$ for n odd.
7.29 $\pi^6/960$.

8.3 $3h^2/2\pi^2 ml^2$.　　**8.5** (a) $c = 8/9\pi$, 15 percent error; (b) $c = 8/729\pi$.　　**8.7** -84.
8.9 (a) $x = y = z = 0$.　　**8.10** $n!$　　**8.13** 1.3 percent, 6.4 percent.

9.1 $3b$, $15b$, where $b = dh^2/64\pi^4 v^2 m^2$.　　**9.2** $E_n^{(1)} = k/2 - (k/2n\pi)[\sin(3n\pi/2)$
$- \sin(n\pi/2)]$.　　**9.3** 1.2×10^{-8} eV.　　**9.5** -12.86 eV.　　**9.8** $1s3s$, two nondegenerate levels;
$1s3p$, two triply degenerate levels; $1s3d$, two fivefold degenerate levels.　　**9.9** $E^{(0)} =$
-77.5 eV; $E^{(1)} = 0$.　　**9.11** -27.2 eV.　　**9.12** First.　　**9.13** 0, 0, $\pm 3e\mathscr{E}a_0$; $2p_1$, $2p_{-1}$,
$2^{-1/2}(2s \mp 2p_0)$.　　**9.17** (a) True; (b) false.

10.1 Ground state: $1s(1)1s(2)1s(3)$.　　**10.3** (a) $2^{-1/2}(1 - \hat{P}_{12})$.　　**10.4** One of my
students gave the answer "To put the wave into the function."　　**10.6** $2J_{1s2s} + J_{1s1s}$.
10.8 54.7° (note that the component in the xy plane exceeds the z component).
10.9 (1) Neither; (2) antisymmetric; (3) symmetric.　　**10.11** (a) $\pm \hbar$; (b) $(\alpha + \beta)/2^{1/2}$,
$(\alpha - \beta)/2^{1/2}$; (c) 0.5 for $\hbar/2$, 0.5 for $-\hbar/2$.

11.1 (a) $2n^2$; (b) $4l + 2$; (c) 2; (d) 1.　　**11.2** $^2S_{1/2}$, $^1S_0$, $^2S_{1/2}$, $^1S_0$, $^2P_{1/2}$, $^3P_0$, $^4S_{3/2}$,
$^3P_2$, $^2P_{3/2}$, $^1S_0$.　　**11.3** ^1F, ^1G, ^1H, ^3F, ^3G, ^3H.　　**11.8** (a) 45°.　　**11.10** No.　　**11.11** $15\hbar^2/4$,
$\frac{3}{2}\hbar$.　　**11.13** 22.　　**11.16** Cobalt.　　**11.17** (a) 15; (b) 36.　　**11.18** 4.5×10^{-5} eV.
11.19 7.7×10^{-6} eV.　　**11.20** There are at least five such configurations.　　**11.22** B, N, F.

12.1 After solving this problem, consider the case where cigars (blunt at one end,
pointed at the other end) are used instead of coins.　　**12.2** (a) $2\sigma_v$, C_2; (b) $3\sigma_v$, C_3; (g) none.
12.3 (a) Yes; (b) no; (c) yes.　　**12.4** (b) Lies along the C_3 axis; (e) is zero; (g) no
information.　　**12.6** (a) The regular tetrahedron.　　**12.8** $\mathscr{C}_1$, $\mathscr{C}_s$, $\mathscr{C}_n$, $\mathscr{C}_{nv}$.　　**12.9** $\mathscr{C}_1$, $\mathscr{C}_n$, $\mathscr{D}_n$,
$\mathscr{T}$, $\mathscr{O}$, $\mathscr{I}$.　　**12.10** (b) 2, 6, 10, 14, ...; (c) 3, 5, 7, ...　　**12.11** (a) $\mathscr{T}_d$; (b) $\mathscr{C}_{3v}$; (c) $\mathscr{C}_{2v}$;
(d) $\mathscr{C}_{3v}$; (e) $\mathscr{C}_h$; (f) $\mathscr{C}_{4v}$; (g) $\mathscr{D}_{4h}$; (h) $\mathscr{C}_{3v}$.　　**12.13** (a) $\mathscr{D}_{6h}$; (b) $\mathscr{C}_{2v}$; (c) $\mathscr{C}_{2v}$; (d) $\mathscr{C}_{2v}$; (e) $\mathscr{D}_{2h}$.
12.15 (a) $\mathscr{C}_h$; (b) $\mathscr{C}_{4v}$; (c) $\mathscr{D}_{2d}$; (d) $\mathscr{C}_h$; (e) $\mathscr{C}_{3v}$; (f) $\mathscr{D}_{2h}$.　　**12.17** The answer is not $\mathscr{D}_2$.
12.18 (a) $\mathscr{C}_{4v}$; (b) $\mathscr{C}_{\infty v}$; (c) $\mathscr{D}_{4h}$; (d) $\mathscr{C}_{4v}$; (e) $\mathscr{D}_{\infty h}$; (j) the answer is not $\mathscr{C}_{2v}$.
12.19 (b) No, since its eigenvalues are not all real.　　**12.23** 10, 3, 14.

13.1 (a) 1836.1; (b) -1; (c) 2π; (d) -2; (e) 4.134×10^{16} (one atomic unit of time
$= ha_0/e'^2 = 2.419 \times 10^{-17}$ sec); (f) 137.04; (g) -0.49973; (h) 0.3934.　　**13.4** (a) Ellipsoids
of revolution; (c) half planes.　　**13.6** 15.425 eV.　　**13.10** $-(3k^3/\pi)^{1/2}e^{-kr}\cos\theta$.
13.12 $R^4/(1 - e^{-R})$.　　**13.15** (a) Li_2; (b) C_2; (c) O_2^+; (d) F_2^+. (Actually, the dissociation energy of Li_2^+ is greater than that of Li_2.)　　**13.16** (b) 3; (d) 2.　　**13.18** $j + k$
even. **13.26** VB.　　**13.32** (a) F; (b) F.　　**13.40** (a), (c), (f), (g).

14.1 (b) 0; (c) -2; (d) $\frac{3}{2}$.　　**14.2** State 1.　　**14.8** $\langle V \rangle = 3\frac{1}{3}$ eV.　　**14.10** 1.8, 4.1,
6.9, and 10 Å. General formula: $R_e[(Z_a Z_b)^{1/2} - Z_a]$.

15.5 (a) a_1: $H_1 1s + H_2 1s$, $C1s$, $C2s$, $C2p_z$, $O1s$, $O2s$, $O2p_z$. b_1: $C2p_x$, $O2p_x$. b_2: $C2p_y$,
$O2p_y$, $H_1 1s - H_2 1s$. (b) ten σ, two π; seven σ, one π. (d) 7×7.　　**15.6** There are nine a_1, nine
b_2, two b_1, and two a_2 orbitals.　　**15.8** (c) $\mathscr{D}_3$; (d) $\mathscr{C}_{2h}$; (e) $\mathscr{C}_{2v}$.　　**15.9** Interchanges the
subscripts 1 and 3 for the b symmetry species.　　**15.10** (b) 2.　　**15.12** (c).　　**15.13** (a) No.
15.18 (a) Bond angle somewhat less than 120°; (b) linear; (c) bond angle somewhat less
than $109\frac{1}{2}$°; (j) planar, bond angles 120°; (k) planar, T shaped, two bond angles somewhat less
than 90°.　　**15.19** (a) Square planar; (b) tetrahedral.　　**15.20** SeF_4—$\mathscr{C}_{2v}$; XeF_4—$\mathscr{D}_{4h}$.
15.21 (a) Bond angle near 120°; (b) linear; (g) FCF angle somewhat less than 120°.
15.22 (a) Linear; (b) bent; (c) bent.　　**15.23** Linear.　　**15.25** No.　　**15.26** $3d_{xy}$.

15.27 N: $1s$—6.7; $2s$—1.95; $2p$—1.95   **15.29** (a) 30,100 cm^{-1}, colorless; (b) 21,500 cm^{-1}, orange-red; (c) 22,000 cm^{-1}, orange.   **15.30** (a) 0 cm^{-1}; (b) 16,100 cm^{-1}.   **15.32** $E = J^2 h^2 / 2ml^2$, where $J = 0,\ \pm 1,\ \pm 2, \ldots$ and l is the circumference; 51,800 cm^{-1}.
15.38 (a) $\alpha + 2^{1/2}\beta$, α, $\alpha - 2^{1/2}\beta$; (b) 0.707, 0.707; (c) 1, 1, 1; (d) 1.025, 0.318, 1.025; (e) 0.828β.   **15.39** (a) Same as previous problem; (b) 0.707, 0.707; 0.707, 0.707.
15.40 (a) 27 kcal/mole;   (b) 1.4 eV.   **15.41** 42 kcal/mole,   69 kcal/mole.
15.42 (a) 1.40_1 Å.   **15.45** *Ortho* and *para* carbons, $7^{-1/2}$; CH_2 carbon, $2/7^{1/2}$; others, 0.
15.48 Free valences: 1, 0.452; 2, 0.404; 9, 0.104; delocalization energy, $3.683|\beta|$.
15.49 (a) $\alpha = -6$ eV; $\beta = -4$ eV.   **15.58** Predicted $R_e = 0$.   **15.61** $(\frac{1}{3})^{1/2} 2s + (\frac{2}{3})^{1/2} 2p_z$, $(\frac{1}{3})^{1/2} 2s - (\frac{1}{6})^{1/2} 2p_z \pm (\frac{1}{2})^{1/2} 2p_y$.   **15.62** $2^{-1/2}(2s \pm 2p_z)$.   **15.65** (a) 1–2, 7–8, 3–4, 5–6.
15.68 CNDO: (b), (d), (e); INDO: (d), (e).   **15.70** (a) $n/2$; (b) $\frac{1}{2}n^2 - n$; (c) $\frac{1}{2}n(n-1)$.
15.71 (a) Disrotatory; (b) conrotatory.   **15.72** (a) High.   **15.75** (a) $0.055|\beta|$, aromatic. (One averages the REPE values for the three structures of Fig. 15.23a.) (b) $-0.061|\beta|$, antiaromatic. (d) $0.023|\beta|$, aromatic.   **15.76** 12.3 kcal/mole.   **15.77** Both products are defined.